Natural Hazards

Earthquakes, Volcanoes, and Landslides

Natural Hazards

Earthquakes, Volcanoes, and Landslides

Edited by
Ramesh P. Singh
Darius Bartlett

CRC Press
Taylor & Francis Group
Boca Raton London New York

CRC Press is an imprint of the
Taylor & Francis Group, an **informa** business

CRC Press
Taylor & Francis Group
6000 Broken Sound Parkway NW, Suite 300
Boca Raton, FL 33487-2742

First issued in paperback 2020

ISBN-13: 978-0-367-57191-7 (pbk)
ISBN-13: 978-1-138-05443-1 (hbk)

Library of Congress Cataloging-in-Publication Data

Names: Singh, R. P. (Ramesh P.), author. | Bartlett, Darius J., 1955- author.
Title: Natural hazards : earthquakes, volcanoes, and landslides / Ramesh Singh and Darius Bartlett.
Description: Boca Raton, Florida : Taylor & Francis, 2018. | "A CRC title, part of the Taylor & Francis imprint, a member of the Taylor & Francis Group, the academic division of T&F Informa plc."
Identifiers: LCCN 2017038609 | ISBN 9781138054431 (hardback : acid-free paper)
Subjects: LCSH: Natural disasters.
Classification: LCC GB5014 .S566 2018 | DDC 363.34--dc23
LC record available at https://lccn.loc.gov/2017038609

Visit the Taylor & Francis Web site at
http://www.taylorandfrancis.com

and the CRC Press Web site at
http://www.crcpress.com

Ramesh P. Singh dedicates his two books to his late father, Professor Rama N. Singh, on the tenth anniversary of his death. I remember his love, affection and motivation for my efforts using satellite data to explore the Earth system and natural hazards.

Contents

Foreword

Around the globe, there continues to be horrific impacts of natural hazards. Although hazardous events arising from earthquakes, volcanoes and landslides are fewer in number than weather–climate events, their typical impacts are unfortunately larger. To reduce the impacts of hazards, we need to address the factors in risk – the hazard itself and the exposure and vulnerability of communities and other impacted areas. This book, *Natural Hazards: Earthquakes, Volcanoes, and Landslides*, is an excellent book of 20 chapters addressing the natural science issues of earthquakes, volcanoes and landslides. It is essential that there be improved understanding of the processes leading to these hazardous events in order to predict their occurrence. These predictions need to be both in the near term to alert populations and in the longer term so that actions can be taken to reduce the population's exposure and vulnerability, leading to the major reductions in impacts. To reduce the impacts on the societal structure and assets, we need projections of risks of these events so that societies can reduce their exposure and vulnerability. The book chapters present research information and analyses on, for example, satellite geophysical data and radar imaging and integrated approaches. The book also has chapters focusing on dust storms, coastal subsidence and remote sensing mapping. Other chapters deal with the roles of remote sensing related to landslides and volcanoes.

The Sendai Framework for Disaster Risk Reduction 2015–2030 has identified priority areas for Disaster Risk Reduction, which include understanding disaster risk and enhancing disaster preparedness for effective response, and to "Build Back Better" in recovery, rehabilitation and reconstruction. The science addressed in this book directly contributes to the knowledge that is needed as a basis for these actions. The international Integrated Research on Disaster Risk (IRDR) Programme, co-sponsored by the International Council for Science, is guided by research objectives, including the following: addressing the gaps in knowledge, methodologies and types of information that are preventing the effective application of science to averting disasters and reducing risk; and reducing risk and curbing losses through knowledge-based actions. The legacy of the IRDR Programme 'would be an enhanced capacity around the world to address hazards and make informed decisions on actions to reduce their impacts'. This book, edited by Ramesh Singh and Darius Bartlett, provides the information for moving ahead on the research program and directly contributes to the Sendai Framework and hence on reducing risk on planet Earth.

Professor Gordon McBean, C.M., O.Ont, PhD, FRSC, FAGU, FIUGG
President, International Council for Science
Institute for Catastrophic Loss Reduction
Professor Emeritus, Department of Geography, Western University, London, Canada

Preface

Natural hazards are associated with land, ocean and atmospheric processes, and their impacts on human societies. Over the years, the interactions between land, ocean, biosphere and atmosphere have increased, mainly due to population growth and anthropogenic activities, which have impacted the climate and weather conditions at local, regional and global scales. Due to population growth, the changes in land use and land cover and underground natural resources, in situ stress, pore pressure and surface albedo have led to various kinds of natural hazards associated with land (such as earthquakes, volcanoes, landslides, subsidence, desertification and droughts), oceans (cyclones, typhoons and hurricanes; harmful algal blooms; and tsunamis) and the atmosphere (lightning and dust storms). Natural hazards significantly impact human life and health on different spatiotemporal scales and also have socioeconomic bearings. In recent years, satellite data have been widely used by many developed and developing countries, in an effort to better understand and characterize the various underlying processes influencing natural hazards, and to carry out related impact assessments. Efforts have also been made to launch dedicated satellite missions for monitoring hazards and studying changes in land, ocean and atmospheric parameters. Satellite remote sensing, in general, is now routinely used to collect and analyze global and regional data for understanding Earth system processes, ranging from subsurficial features to upper atmospheric composition.

This book provides an overview of the physical processes associated with earthquakes, volcanoes, landslides, subsidence, desertification and dust storms. The Gujarat Earthquake of 26 January 2001 was deadly, killing more than 20,000 people and resulting in multimillion-dollar damages to properties and businesses. Widespread liquefaction, rupture and surface manifestations were observed, which are discussed in the chapters authored by Thakkar.

Susan Bartels and colleagues discuss medical complications associated with earthquakes. T.J. Majumdar explains satellite-retrieved geophysical data as possible precursors of earthquakes. Matt Pritchard and Sang-Ho Yun cover satellite radar imaging techniques for applications to natural hazards. Michel Parrot examines DEMETER satellite data and their observations to infer early warning signals associated with seismic and volcanic activity.

Luca Piroddi provides information on thermal infrared anomalies associated with earthquakes. Kalpna Gahalaut examines the role of fluids in triggering earthquakes that are common in intraplate earthquakes, which are believed to trigger earthquakes in shield and stable continental areas. Brijesh Bansal and Mithila Verma provide an overview of earthquake precursory studies in India.

Niki Evelpidou et al. discuss the geomorphic features associated with erosion, and Priyabrata Santra et al. provide details of the Thar Desert as a source for dust storms, its ground and satellite monitoring, and its impact on dust storms over the Indo-Gangetic plains and surrounding regions.

Andrea Taramelli et al. examine the causes, mapping and monitoring of coastal subsidence, and Mukesh Gupta explains the use of InSAR technology in subsidence studies. Landslides associated with earthquakes in Pakistan are presented by Shah Khan et al. The cause, distribution and monitoring of landslides in Jamaica are covered by Servel Miller et al., and landslides in Malaysia are covered by Omar Althuwaynee and Biswajeet Pradhan. Michel Jaboyedoff et al. discuss the use of LIDAR for mapping and monitoring landslides.

Zhong Lu and Daniel Dzurisin present information on the use of radar monitoring of volcanic activities, and satellite remote sensing and monitoring of active volcanoes are examined by Nicola Pergola and his group. Matthew Blackett discusses the basic principles of thermal remote sensing and its application in many kinds of natural hazards associated with land (earthquakes, volcanoes, forest fires, landslides, heat waves and storms).

Ramesh P. Singh, PhD
Professor
Earth System Science and Remote Sensing
Schmid College of Science and Technology
Chapman University
Orange, California
E-mail: rsingh@chapman.edu

Darius Bartlett
Research Associate and Retired Lecturer
Department of Geography
University College Cork
Cork, Ireland
E-mail: d.bartlett@ucc.ie

Acknowledgements

The editors acknowledge their appreciation of and thank each contributor for writing chapters and revising them many times, taking care to address comments made by the referees and editors. Further, the editors are grateful to the following referees for their help in review process:

Jan Altman (Czech Republic)
B.R. Arora (India)
Namrata Batra (United States)
Partha Bhattacharjee (United States)
Matthew Blackett (United Kingdom)
Laxmansingh S. Chamyal (India)
Masahiro Chigira (Japan)
Simona Colombelli (Italy)
Leah Courtland (United States)
Gilda Currenti (Italy)
Minakshi Devi (India)
Nicolas D'Oreye (Luxembourg)
Paula Dunbar (United States)
Qi Feng (China)
A.M. Foyle (United States)
Carolina Garcia (Italy)
Ritesh Gautam (United States)
Adrian Grozavu (Romania)
Mukesh Gupta (Canada and Spain)
M. Hayakawa (Japan)
Kenneth Hewitt (Canada)
James Hower (United States)
Fuqiong Huang (China)
S.K. Jain (India)
Vikrant Jain (India)
Ezatollah Karami (Iran)
Anabel Alejandra Lamaro (Argentina)
A.M. Martinez-Granna (Spain)
Rachel G. Mauk (United States)
Yoav Me-Bar (Israel)
N.C. Mishra (United States)
U.S. Panu (Canada)
Gerassimos Papadopoulos (Greece)
Guido Pasquariello (Italy)
T.A. Przylibski (Poland)
B.K. Rastogi (India)
Rasik Ravindran (India)
E. Sansosti (Italy)
Sudipta Sarkar (United States)
Devendra Singh (India)
T.N. Singh (India)

Ajay Srivastav (India)
Hans von Suchodoletz (Germany)
C. Wauthier (Belgium)

The editors are grateful to all the authors who have made substantial efforts in contributing their studies. We would also like to thank Irma Shagla-Britton and her team at CRC Press/Taylor & Francis for encouraging us to take on the project, and for guidance and support at all stages, from conception to final production.

Ramesh Singh acknowledges the help of his wife, Alka Singh, for sacrificing her time to support him during the preparation of this book.

Darius Bartlett would like to acknowledge, with thanks, the friendship, encouragement and support of his former colleagues in the Department of Geography at University College Cork, both prior to his retirement and subsequently, and of course, as always, his wife, Mary-Anne.

We also thank Susan Lee and Molly Pohlig, T&F Encyclopedia Program, Taylor & Francis, for their help with completing the review process.

Editors

Ramesh P. Singh has been a professor in the School of Life and Environmental Sciences, Chapman University, Orange, California, since 2009. He graduated from Banaras Hindu University, Varanasi, India. He was a postdoctoral and Alberta Oil Sands Technology and Research Authority (AOSTRA) fellow with the Department of Physics, University of Alberta, Canada (1981–1986). He joined the Department of Civil Engineering, Indian Institute of Technology, Kanpur, India, as a faculty member in 1986 and remained there until 2007.

Dr. Singh joined George Mason University as a distinguished visiting professor from 2003 to 2005 and later as a full professor from 2007 to 2009. He then moved to Chapman University, where he has been since 2009.

Dr. Singh has published more than 200 research papers, and supervised several MTech and PhD students. He was chief editor of the *Indian Journal of Remote Sensing* from 1999 to 2007; chief editor of *Geomatics, Natural Hazards and Risk*; and associate editor of the *International Journal of Remote Sensing*, the latter two of which are published by Taylor & Francis. He has edited books and several special issues of journals.

Dr. Singh is a recipient of the Alexander von Humboldt Fellowship and a JSPS Fellowship. He is the recipient of Indian National Remote Sensing, Indian National Mineral and Hari Om Ashram Prerit awards, as well as a fellow of the Indian Remote Sensing Society and the Indian Geophysical Union. He is a member of the editorial board of Aerosol Air Quality Research.

Dr. Singh has been a member of the International Union of Geodesy and Geophysics Commission on GeoRisk since 2004 and has served as vice president. He is currently (2017–2018) president of the American Geophysical Union Natural Hazards Focus Group. He is also a member of the International Union of Geodesy and Geophysics–Electromagnetic Studies of Earthquakes and Volcanoes Bureau.

Darius Bartlett is an earth scientist and geomorphologist by training, with particular specialism in geographic information systems and the application of geoinformation technologies to coastal zone science. Now retired from active teaching, he is currently researching sites of possible historical tsunami impacts on the coast of Ireland. He is the co-editor of *Marine and Coastal Geographical Information Systems* (with Dawn Wright, 2000), *GIS for Coastal Zone Management* (with Jennifer Smith, 2005) and *Geoinformatics for Marine and Coastal Management* (with Louis Celliers, 2017), all published with CRC Press/Taylor & Francis.

Contributors

Antonio Abellán
Risk Analysis Group
Institut des Sciences de la Terre (ISTE)
Faculty of Geosciences and Environment
University of Lausanne
Lausanne, Switzerland

and

Scott Polar Research Institute
Geography Department
University of Cambridge
Cambridge, United Kingdom

Omar F. Althuwaynee
Department of Town and Regional Planning
University of Johannesburg
Johannesburg, South Africa

Brijesh K. Bansal
Ministry of Earth Sciences
New Delhi, India

Susan A. Bartels
Department of Emergency Medicine
Beth Israel Deaconess Medical Center
Boston, Massachusetts
and
Harvard Humanitarian Initiative
Cambridge, Massachusetts

Scarlet Benson
Department of Emergency Medicine
Beth Israel Deaconess Medical Center
Boston, Massachusetts

Matthew Blackett
Coventry University
Coventry, United Kingdom

Lyndon Brown
Earthquake Unit
University of the West Indies
Mona, Jamaica

Dario Carrea
Risk Analysis Group
Institut des Sciences de la Terre (ISTE)
Faculty of Geosciences and Environment
University of Lausanne
Lausanne, Switzerland

Loreta Cornacchia
NIOZ Royal Netherlands Institute for Sea
    Research
Department of Estuarine and Delta Systems
and
Utrecht University
Yerseke, The Netherlands

Marc-Henri Derron
Risk Analysis Group
Institut des Sciences de la Terre (ISTE)
Faculty of Geosciences and Environment
University of Lausanne
Lausanne, Switzerland

Daniel Dzurisin
U.S. Geological Survey
Vancouver, Washington

Laura Ebbeling
Department of Emergency Medicine
Beth Israel Deaconess Medical Center
Boston, Massachusetts

Niki Evelpidou
Faculty of Geology and Geoenvironment
National and Kapodistrian University of
    Athens
Panepistimiopolis, Athens, Greece

Kalpna Gahalaut
CSIR–National Geophysical Research Institute
Hyderabad, India

Mukesh Gupta
Centre for Earth Observation Science
University of Manitoba
Winnipeg, Manitoba, Canada

and

Institut de Ciències del Mar—CSIC
Barcelona Expert Center on Remote Sensing
    (BEC)
Barcelona, Spain

Norman Harris
Mines and Geology Division
Kingston, Jamaica

Michel Jaboyedoff
Risk Analysis Group
Institut des Sciences de la Terre (ISTE)
Faculty of Geosciences and Environment
University of Lausanne
Lausanne, Switzerland

Ulrich Kamp
Department of Natural Sciences
University of Michigan–Dearborn
Dearborn, Michigan

Isidoros Kampolis
Faculty of Geology and Geoenvironment
National and Kapodistrian University of
    Athens
Panepistimiopolis, Athens, Greece

Anna Karkani
Faculty of Geology and Geoenvironment
National and Kapodistrian University of
    Athens
Panepistimiopolis, Athens, Greece

Muhammad Asif Khan
University of Peshawar
Peshawar, Pakistan

Shah F. Khan
National Centre of Excellence in Geology
University of Peshawar
Peshawar, Pakistan

Suresh Kumar
Division of Integrated Land Use Management
    and Farming Systems
ICAR—Central Arid Zone Research Institute
Jodhpur, Rajasthan, India

Zhong Lu
Southern Methodist University
Dallas, Texas

T.J. Majumdar
Space Applications Centre (ISRO)
Ahmedabad, India

Ciro Manzo
CNR Italian National Council of Research
Institute for Atmospheric Pollution Research
Rome, Italy

Francesco Marchese
National Research Council
Institute of Methodologies for Environmental
    Analysis
Tito Scalo, Italy

Battista Matasci
Risk Analysis Group
Institut des Sciences de la Terre (ISTE)
Faculty of Geosciences and Environment
University of Lausanne
Lausanne, Switzerland

Clément Michoud
Risk Analysis Group
Institut des Sciences de la Terre (ISTE)
Faculty of Geosciences and Environment
University of Lausanne
Lausanne, Switzerland

Servel Miller
Department of Geography and Development
    Studies
University of Chester
Chester, United Kingdom

Lewis A. Owen
Department of Geology
University of Cincinnati
Cincinnati, Ohio

Michel Parrot
LPC2E/CNRS and University of Orléans
Orléans, France

Nicola Pergola
National Research Council
Institute of Methodologies for Environmental
  Analysis
Tito Scalo, Italy

Luca Piroddi
Department of Civil Engineering,
  Environmental Engineering and
  Architecture
University of Cagliari
Cagliari, Italy

Biswajeet Pradhan
School of Systems, Management and
  Leadership
Faculty of Engineering and Information
  Technology
University of Technology Sydney
Ultimo, New South Wales, Australia

Matt E. Pritchard
Department of Earth and Atmospheric Sciences
Cornell University
Ithaca, New York

Dionne Richards
Mines and Geology Division
Kingston, Jamaica

M.M. Roy
ICAR—Central Arid Zone Research Institute
Jodhpur, Rajasthan, India

Eugenio Sansosti
National Research Council
Istituto per il Rilevamento Elettromagnetico
  dell'Ambiente
Napoli, Italy

Priyabrata Santra
Division of Agricultural Engineering for Arid
  Production Systems
ICAR—Central Arid Zone Research Institute
Jodhpur, Rajasthan, India

Anestoria Shalkowski
Department of Geography and Geology
University of the West Indies
Mona, Jamaica

Andrea Taramelli
IUSS—University Institute for Advanced Study
  of Pavia
Pavia, Italy
and
ISPRA Institute for Environmental Protection
  and Research
Rome, Italy

M.G. Thakkar
Department of Earth and Environmental
  Science
KSKV Kachchh University
Bhuj–Kachchh, Gujarat, India

Emiliana Valentini
ISPRA Institute for Environmental Protection
  and Research
Rome, Italy

Michael J. VanRooyen
Department of Emergency Medicine
Brigham and Women's Hospital
Boston, Massachusetts
and
Harvard Humanitarian Initiative
Cambridge, Massachusetts
and
Department of Global Health and Population
Harvard School of Public Health
Boston, Massachusetts

Mithila Verma
Ministry of Earth Sciences
New Delhi, India

Sang-Ho Yun
Jet Propulsion Laboratory
California Institute of Technology
Pasadena, California

1 Gujarat Earthquake
Ground Deformation

M.G. Thakkar

CONTENTS

1.1   INTRODUCTION

The 2001 Bhuj–Gujarat Earthquake was one of the largest events in the stable continental regions (SCRs) of the world. Comparable events are the New Madrid earthquakes of 1811–1812, the Charleston Earthquake of 1886, and the 1819 Kachchh (Allah Bund) Earthquake (Figure 1.1). The 2001 Bhuj–Gujarat Earthquake (Mw 7.7) is an important event for the comparative study of two big earthquakes in the same tectonic province. Kachchh has a long history of earthquakes and has experienced two major events and tens of smaller events in the past 200 years. In this chapter, various types of surface deformations observed during the 2001 Gujarat Earthquake are discussed. Each feature, with its preserved stratigraphy in the subsurface strata, is especially significant for palaeoseismic studies in comparable tectonic environments elsewhere in the world. On-fault, off-fault,

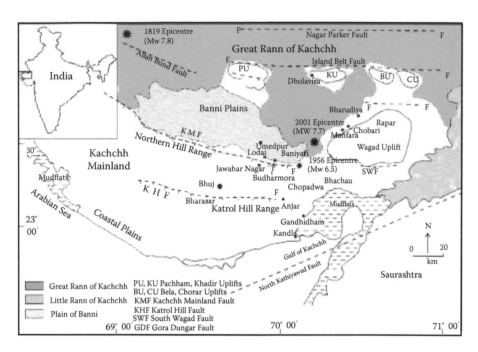

FIGURE 1.1   Map showing study locations of 2001 Gujarat Earthquake ground deformation sites around the meizoseismal area with the major physiographic units and basin-bounding faults. Note the locations of major earthquakes in the past 200 years in the Kachchh Basin, with their magnitudes.

co-seismic, and post-seismic deformation features, like sand blows, lateral spreads, pressure ridges, mole tracks, dextral thrusts, monoclinal humps, extensional cracks, secondary faults, ground collapse, rockfalls, displacement in the Quaternary apron, soft sediment deformation, shattered bedrock ridges and sand dykes, are some of the observed features associated with the earthquake, which are important for intraplate compressional tectonic environments in SCRs. The activeness of the Kachchh intraplate basin is confirmed by the neotectonic studies of the last two decades. Palaeoseismic investigations prove historic movements along major seismogenic faults, like the Kachchh Mainland Fault (KMF), South Wagad Fault (SWF), Island Belt Fault (IBF) and Katrol Hill Fault (KHF). The movements along various faults played an important role in shaping the landscape in the Quaternary period. The study of the tectonic nature of the 2001 Gujarat Earthquake provides an incomparable example for the entire basin evolution of Kachchh, especially using the nature and genesis of its ground deformation features. The secondary surface ruptures in the 2001 Gujarat Earthquake were found to be reactivated at the same sites, so palaeoseismic studies provide past geological information.

The Kachchh rift evolved during the breakup of Gondwanaland, in the Mesozoic period, followed by the northward drift of the Indian plate (Biswas 1987). This breakup was controlled by movements along a series of normal faults, resulting in multiple horsts and grabens, which gave rise to the basin and range-type morphology to this region. The Rann of Kachchh occupies grabens, formed as part of this rifting (Biswas 1987) (Figure 1.1). A reversal in the stress regime, from extension to compression, was probably initiated 40 Ma ago, subsequent to the collision of India with Asia (Biswas 1987; Talwani and Gangopadhyay 2001; Biswas and Khattri 2002). Early stages of contractile deformation were probably aided by a stretched and thermally weak lithosphere (Turcotte and Schubert 2002).

Exploratory drilling in the Kachchh Basin suggests that a 3 km thick succession of Mesozoic sediments was deposited in a sheltered gulf overlying the Precambrian granitic basement, an extension of the peninsular continental crust (Biswas 1987). The Tertiary and Quaternary sediments are characterized by intervals of non-deposition and erosional unconformities. These major surface features, like the KMF, IBF, Kathiawad Fault and Wagad Fault, have been active during the Holocene (Bodin and Horton 2004). For example, the ground-penetrating radar (GPR) profiles across one such structure (KHF) imply a reverse fault (Patidar et al. 2006). The Kachchh region has experienced large earthquakes in the historic past, mostly restricted to the western part of the Great Rann (Rajendran and Rajendran 2001). Palaeoseismologic investigations in the meizoseismal area of the 1819 Kachchh–Allah Bund Earthquake have indicated the occurrence of one earlier earthquake close to this source between 800 and 1000 years BP (Rajendran and Rajendran 2001). These investigations also indicated the general absence of large earthquakes in the eastern parts of the Great Rann, at least during the last 4000 years (Rajendran and Rajendran 2003a,b). Palaeoseismic studies based on secondary deformation structures of the 2001 Gujarat Earthquake in the eastern Kachchh suggest similar earthquake struck ~4000 years BP (Rajendran et al. 2008). The archaeological evidence suggests that an earthquake caused minor damage to the Dholavira (Indus valley civilization site in Kachchh) (Figure 1.1) settlement in 4150–4450 years BP (Joshi and Bisht 1994). Oldham (1926), in a classic paper based on the observed earthquake effects, stated that the eastern part of the Rann is more vulnerable to earthquakes originating from the 2001 source zone, rather than the 1819 event (Joshi and Bisht 1994). An attempt has been made to correlate the intensity of liquefaction (in terms of liquefaction-induced sediment thickness) to the epicentre location, distance from active faults, size of the earthquake and availability of liquefiable sediments for the 2001 Bhuj–Gujarat Earthquake (Thakkar and Goyal 2004). Here, efforts have been made to discuss observed ground deformation features in the epicentral zone; further, their genetic relations with the established regional structures are discussed. Some features are incomparable and inimitable of active intraplate regions undergoing regional compression.

1.2   2001 GUJARAT EARTHQUAKE: GROUND DEFORMATION STUDY

The epicentre of the 2001 Gujarat Earthquake is located north of Bhachau (23.40° N, 70.28° E) (Kayal et al., 2002; Antoliok and Dreger 2003; Bodin and Horton 2004; Mandal et al. 2004; Singh et al. 2004) (Figure 1.1). The focal mechanism suggests that it is close to the well-exposed KMF (Figure 1.1), but its thrust-type focal mechanism immediately ruled out any causal association with this normal fault. The tight cluster of aftershocks is confined to a relatively small area (60–40 km^2), with their focal depths ranging from 10 to 45 km and defining a 45°–50° south dipping westerly striking plane (Bodin and Horton 2004; Mandal et al. 2004). The surface projection of the fault plane emerged 20–25 km north of the KMF at the intersection of 'Banni' grasslands and the low-lying salt playa known as the 'Rann' (Gahalaut and Burgmann 2004; Mandal et al. 2004). The 2001 earthquake, despite its large magnitude, did not generate any pronounced primary surface rupture, nor did it develop any escarpment, as in the case of the 1819 earthquake in the same tectonic province. However, the strata in the hanging wall showed signs of intense deformation by way of flexures, folds, extensional cracks, uplifts, ground slumping, ground fissures and tear faults, in addition to widespread liquefaction and lateral spreads (Rajendran et al. 2001; Tuttle et al. 2002; Schweig et al. 2003; Rastogi 2004). McCalpin and Thakkar (2003) prefer to call at least some of these co-seismic features indicators of primary faulting. However, they retain ambiguity in interpretation by invoking the possibility that such features could also be formed by 'shaking-triggered-slip' on pre-existing structures, not necessarily the primary fault. The largest surface deformation associated with this earthquake is the 16 km long east-west zone of cracks and lateral spreads between Budharmora and Chobari (Figure 1.2), which most of the workers consider a surficial landslide and lateral spreads (Tuttle et al. 2002; Rastogi 2004). The 8 km long Manfara Fault and the 2 km long linear mounds and small thrust scarp at Bhrudiya are the other deformational features associated with this earthquake (McCalpin and Thakkar 2003). The strong ground shaking from the seismogenic rupture would cause the slip

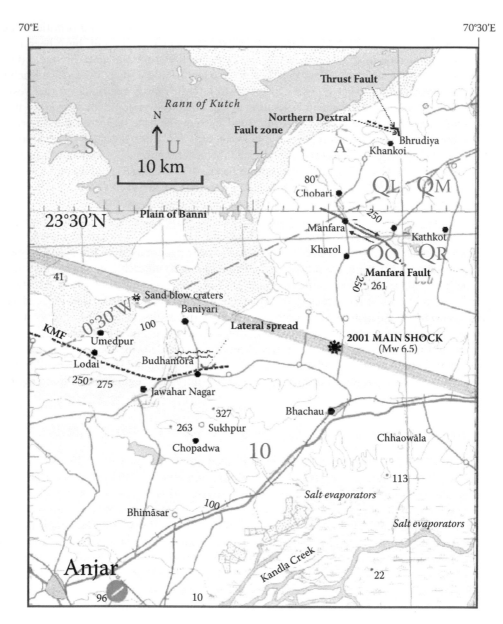

70°E   70°30′E

FIGURE 1.2 Detailed map of the meizoseismal zone of the 2001 Gujarat Earthquake showing the locations of the villages and types of deformations. Note that the dextral secondary Manfara Fault and northern bedrock deformation zones are insignificant in terms of size but are important proxies for the palaeoseismic studies.

and the surface ruptures to die out with depth. In other words, the features would remain rootless, showing no connection with the seismogenic fault plane (Rajendran et al. 2008). The co-seismic effects reported from Budharmora indicate high vertical acceleration at the site of the lateral spread, where people were reportedly pushed up into the air and thrown 3 m to the northwest (McCalpin and Thakkar 2003), adding credence to the high vertical acceleration theory. Rastogi (2004) compared earthquake intensity with ground deformation, showing isoseismals II to X+ drawn on the Modified Mercalli Intensity (MMI) scale. Isoseismal VII is elongated towards southeast-northwest, covering soft soil areas up to Surat in India and Hyderabad (Sindh) in Pakistan. The other isoseismals trend northeast-southwest or north-south. The major and minor axes of the isoseismal areas of intensity X+, X, IX, VIII, VII, VI, V, IV, III and II were 40 and 20 km, 100 and 43 km, 180 and 125 km, 300 and 215 km, 660 and 450 km, 1150 and 800 km, 1500 and 1100 km, 1800 and 1800 km, 2700 and 2500 km, and 3500 and 2700 km, respectively (Rastogi 2004). The maximum damage due to the Bhuj Earthquake is of intensity X+ on the Medvedev–Sponheuer–Karnik (MSK) or MMI scale. It covers an area of 40 × 20 km that trends north-northeast to south-southwest. At Bhachau, even very small structures have collapsed. At one place along a ground crack in Bhachau, the railway line was also slightly bent (Rastogi 2004). Excavations along the key field sites in the zone of intense ground deformation suggest the formation of near-surface features that do not relate to a deeper fault plane. Slip along the bedding planes above the tip of a blind thrust (Roering et al. 1997) and varying slip distribution as the blind thrust propagates through extreme lateral heterogeneities (Burgmann et al. 1994) are other possible explanations for the origin of secondary features. At Bhrudiya, a secondary strike-slip fault near Manfara, a lateral spread at Budharmora, a shattered ridge at Chopadva, rockfalls at Kas Hills and sand blows in the Great Rann of Kachchh (Figures 1.1 and 1.2) were produced during the 2001 earthquake. Among the 2001 deformation features, Bhrudiya bedrock ruptures are still visible on the surface, somewhat retaining the originally described characteristics.

1.3 PRIMARY DEFORMATION STRUCTURES

The earthquake deformation features related to the primary fault or causative fault exposed on the surface are termed primary structures (McCalpin 1996). However, when the fault hits the surface, it may change its angle, geometry and surface morphology, making the interpretation more difficult. Fault scarps, fissures and folds along the trace of a fault are typical examples of primary, on-fault, instantaneous (geomorphic) evidence.

1.3.1 GROUND FISSURES

Ground fissures have been reported around Dhrang, Lodai, Bhachau, Dudhai, Jawaharnagar, Rapar, Bhrudiya, Anjar, Ratnal and Gandhidham in the intensity IX to X+ area and at Mandvi in the intensity VIII area (Rastogi 2004). Cracks have also occurred out of the Kachchh region at Deesa, Suigam (Banaskantha District), and Viramgam and Nal Sarovar (Ahmedabad District) in the intensity VII area (Karanth et al. 2001). Most of the ground fissures with no slip or lateral spread are observed near the KMF (Rastogi 2001, 2004; Singh et al. 2001a,b). The co-seismic fissures are mostly observed on the hanging wall while shaking. Thick alluvial deposit north of the KMF is an ideal terrain for deep ground fissures, lateral spreads and pressure ridges. Hundreds of metres long and a few centimetres to a half metre wide, fissures trending east-west to northeast-southwest are observed at Budharmora and Baniyari villages (Figures 1.2 and 1.3).

1.3.2 MONOCLINAL HUMPS, MOLE TRACKS AND TENT STRUCTURES

The co-seismic features, like monoclinal humps in the Quaternary sediments, 2 km north of Bhrudiya village, were found by Rajendran et al. (2001) (Figure 1.4a). Further, a detailed investigation was carried out by McCalpin and Thakkar (2003) that explained the mechanism of the total deformation structure with the observed co-seismic features that are the indicators of primary faulting. McCalpin and Thakkar (2003) have further reported some of the characteristic features of compressional tectonic environments, such as mole tracks, pressure ridges, tent structures, bedrock back thrust and scarps, to the north of Bhrudiya village, where the 2001 rupture plane falls on the surface if projected up by mathematical means (Figure 1.4b,c). Displacements in the Quaternary channel deposits at the site have confirmed low-acute-angle thrust that dips south (McCalpin and Thakkar 2003). A zone about 80 m long and 10 m wide in an agricultural field near Bhrudiya is

FIGURE 1.3   Ground fissures developed in the thick alluvium, showing an en échelon pattern to the north of Baniyari village.

elevated by about 80 cm and shows a gentle northeasterly dip (Figure 1.4a). Crops in the arched portion of the field have dried due to the sudden change in the moisture content. Several ground cracks oriented in a northwest direction have developed on the upwarped portion. This structure is reported as a monoclinal fold formed as a result of compression and deformation of the surface layers (Rajendran et al. 2001).

1.3.3   2001 THRUST EXPOSURE IN A STREAM BANK (FAULTED QUATERNARY APRON)

About 2 km north of Bharudiya village, a bedrock ridge is exposed amidst the alluvial terrain. Nearly 275 m west of the bedrock ridge, the 2001 thrust scarp crosses a 1–1.5 m deep stream channel and the fault plane is exposed in the western stream bank. The fault cuts Quaternary fluvial

(a)

FIGURE 1.4   (a) Monoclinal humps in the ploughed farm field north of Bhrudiya village.   (*Continued*)

(b)

(c)

FIGURE 1.4 (CONTINUED)   (b) Mole tract and back thrust faults 2 km north of Bhrudiya village. (c) Tent structure – atypical of the compressional tectonic environment seen to the north of Bhrudiya village, probably a primary surface deformation feature.

deposits and creates a 22 to 25 cm high mole track on the surface with tent structures. The thrust dips 13° south and displaces a small channel margin with a net slip of 22 cm (Figure 1.5a,b). This fortuitous exposure permitted the author to observe how tent structures were created by low-angle thrusting and how their dimensions are related geometrically to fault slip (McCalpin and Thakkar 2003). From the stream bank exposure, the sinuous thrust scarp continues west, characterized by mole tracks and hanging wall collapse scarps 15–30 cm high. About 120 m west of the stream bank, the scarp continues up and over a north trending, 2.3 m high earth dam and forms a degraded scarp at its crest 20 cm high. The fault trace migrates 3.25 m northward as it climbs the 2.3 m high embankment, resulting in an apparent dip of about 36° south. This angle is considerably

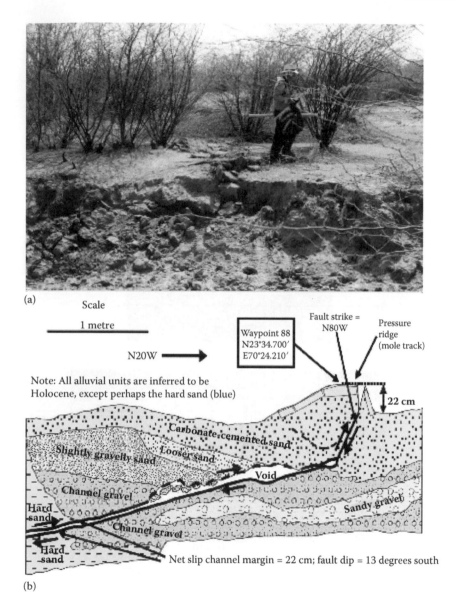

FIGURE 1.5   (a) Photograph of a stream bank north of Bhrudiya village. The low-angle rupture plane of 2001 passes across this stream and has displaced the Quaternary alluvium. Jim McCalpin is seen in the picture as a scale. The photo was taken in May 2001. (b) Log of the stream bank (in (a)) showing the low-angle thrust fault displace gravelly sediments, which is manifested as a popping-up structure on the surface.

steeper than that observed in the stream bank. From the dam, the thrust scarp continues west as a sinuous hanging wall collapse scarp or mole track. Individual thrust scarp lobes have a wavelength of 5–10 m and an amplitude of 2–3 m in plan view. In this area, thrust scarps reach their greatest height (27–35 cm, at 23°34.701′ N, 70°24.073′ E). Farther west, the thrust scarp or mole track dies out and is replaced by a N 20° W trending zone of small and widely spaced dextral cracks. These dextral cracks average 1–2 cm of net slip and are about 50 m long, after which they step left and form a small north-facing thrust scarp or mole track of similar length. Northwest about 23°34.812′ N, 70°24.019′ E, even the dextral cracks vanish, and no surface faulting was traced (McCalpin and Thakkar 2003).

1.3.4 DISPLACEMENT ESTIMATES

Since the vertical exposures of the surface faulting were rare, it was difficult to measure the displacement on the 2001 thrust fault. The first estimates assumed that a fault plane dipping south at 15°–35° had created a simple thrust scarp 15–35 cm high, which implied a net slip vector ranging from 26 cm (35° fault dip and 15 cm high scarp) to 135 cm (15° fault dip and 35 cm high scarp) (McCalpin and Thakkar 2003) (Figure 1.5b). This wide discrepancy shows the difficulty in making an accurate net slip estimate solely from scarp height. However, the stream bank exposure shows that the low-angle thrust fault did not continue propagating at 13° once it hit the base of the hardpan layer (carbonate-cemented sand in Figure 1.5b), but instead steepened to subvertical, probably following a subvertical crack created during earthquake shaking of the hardpan. This refraction of the fault plane created head-on compression of the broken hardpan layer, resulting in upward arching of hardpan slabs and the resultant tent structure (McCalpin and Thakkar 2003). Based on the stream bank exposure, McCalpin and Thakkar (2003) proposed a simple geometric model that relates horizontal tectonic shortening to the surface dimensions of the tent structure (Figure 1.6). They also observed that the ground surface in the tent structure is uplifted and tilted but not horizontally compressed, and in elastic deformation, it appears to be limited to thickening of the hardpan layer in the tent structure. Therefore, the amount of horizontal shortening (D) can be related to the tent structure as follows:

$$D = 2\left(L - (L \cos \alpha)\right) \tag{1.1}$$

where D = horizontal shortening, L = length of one limb of the tent structure, and α = dip angle of the limb. The height of the tent structure is given by

$$Z = L \sin \alpha \tag{1.2}$$

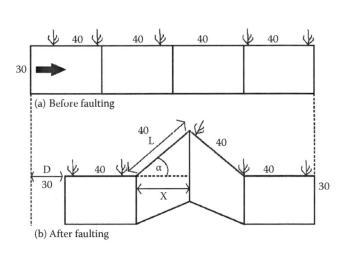

FIGURE 1.6 Schematic diagrams showing the geometry of an idealized tent structure. (a) The hardpan layer is 30 cm thick and is arbitrarily divided into four 40 cm long sections. (b) The hardpan layer is subjected to horizontal shortening (D), which creates a tent structure with height (Z), half length (L) and limb dip angle (α). After shortening, field observations show that the length of each 40 cm long section remains constant at ground level. Thus, the tented sections must thicken as shown, in order to conserve volume (this also minimizes the space problem below the centre of the tent structure). The horizontal distance X beneath each half of the tent structure is now less than 40 cm, by half the amount of horizontal shortening (D/2). In this example, dimensions are taken from the stream-cut locality on the thrust fault northwest of Bharodia village (Figure 1.5), where W = 40 cm and Z = 25 cm, resulting in a computed slab dip angle (α) of 39° and computed horizontal shortening of 18 cm. Equations in the text describe relationships between D, W, X, Z and α. (From McCalpin, J. P., and Thakkar, M. G., *Ann. Geophys.*, 46, 937–956, 2003.)

where Z = height of the tent structure. For example, 5 m west of the stream bank exposure (where net slip was 22 cm on a 13° dipping fault) is a symmetrical tent structure with Z = 25 cm and L = 40 cm. According to Equation 1.2, the dip of the limbs is 39°. Entering that value into Equation 1.1, we can compute a horizontal shortening of 18 cm. This compares with the 21 cm of horizontal shortening observed in the stream bank exposure. The value computed from the tent structure is much smaller than the net slip required for an unrefracted 13° dipping fault to create a scarp 25 cm high. In contrast, the computed 18 cm of horizontal shortening is only 15% smaller than the value measured in a vertical cut, which suggests that a small amount of inelastic shortening did occur in the hardpan layer in the tent structure. Farther east, the tent structure shown in Figure 1.5a is symmetrical, with a right (north) limb of L = 90 cm and α = 25°. By applying Equation 1.1, a horizontal shortening of 17 cm can be predicted (McCalpin and Thakkar 2003).

1.3.5   Summary of Primary Surface Faulting

The thrust scarp described previously is ~830 m long, which clearly resulted from tectonic faulting, because it involves bedrock, crosses natural and man-made topography (earth fill dam structure) indiscriminately and is not associated with any gravitational failures. It is similar to that of primary surface ruptures during other thrust earthquakes (Philip et al. 1992; McCalpin 1996; Karanth et al. 2001 Rastogi 2001). However, McCalpin and Thakkar (2003) do not say that this thrust scarp is the surface expression of the 26 January 2001 main shock plane but give several possible origins: (a) the surface expression of the 26 January main shock plane (b) represents a shaking-triggered slip on a pre-existing fault updip from the 26 January main shock plane (i.e. induced displacement); (c) is the surface expression of an aftershock plane or (d) represents shaking-triggered slip on a pre-existing fault updip from an aftershock plane (i.e. induced displacement). Based on their limited reconnaissance, they favour origin (a), but cannot rule out origins (b) through (d) without further detailed study.

1.4   SECONDARY DEFORMATION STRUCTURES

1.4.1   Off-Fault Co-Seismic Deformation

An intriguing aspect about the Bhuj Earthquake is that it did not produce surface rupture indicative of reverse motion. However, as the strata in the hanging wall deformed, several extensional as well as compressive features were produced. These include flexures, folds, extensional cracks and tear faults, categorized as secondary off-fault and on-fault co-seismic deformation features commonly associated with thrust-faulting earthquakes (McCalpin 1996). These features were well developed in the regions north of Bhachau, presumably part of the hanging wall block. A few of them are discussed in the following sections.

1.4.2   Secondary Deformation Structures by Seismic Shaking

Several post-earthquake reconnaissance teams observed northwest trending dextral surface faults east of the main shock epicentre. Their orientation (northwest) and sense of slip (right lateral) do not match the focal mechanism of the 26 January main shock (northward thrusting on an east-west plane), but they were apparently formed wholly or partly during the main shock. Seeber et al. (2001) and Wesnousky et al. (2001) concluded that the longest of these faults, the 8 km long Manfara Fault (described in the next section), was a secondary tear fault that bounded the eastern margin of the north-thrusted block. It is a 5 km long N 25° W trending rupture with a right lateral slip of 15–35 cm reported by Wesnousky et al. (2001) that extends southward from Manfara and cuts the Kharoi–Rapar road 5 km from Kharoi (see Figure 1.14). Wesnousky et al. (2001) described the northwest-striking right lateral strike-slip fault as showing a maximum offset of

32 cm and reflecting a pure shear or tear in the hanging wall, and further noticed that there is a consistent northward contraction in the bedrock structure and the focal mechanism of the event. Therefore, two zones of dextral faulting are described. The first zone (the northern dextral fault zone) was discovered and mapped by McCalpin and Thakkar (2001), 2 km north of Bhrudiya village, and the second is the Manfara Fault described by Seeber et al. (2001) and later studied in detail by Rajendran et al. (2008).

1.4.3   Landslides: Lateral Spreads, Pressure Ridges and Sand Dykes at Budharmora

Most of the ground failure during the 2001 Gujarat Earthquake was related to lateral spreads. These are produced by liquefaction, generally develop on very gentle slopes (most commonly between 1° and 3°) and move downslope towards a free face. Flows may consist of completely liquefied soils or blocks of intact material, riding on layers of liquefied soil, with the whole process causing deformation of the soil layers, often leading to ground failure. At many locations in the epicentral area, lateral spreads had given rise to a series of ground cracks showing steplike displacements and compressional features at the toe of the mass movement (Figure 1.7a,b). At Budharmora (23.34° N, 70.19° E), a topographic profile was surveyed across a disturbed zone about 140 m wide and 400 m long, with a gentle (~1°) northerly slope. The morphology of the lateral spread includes a toe ridge, 1 m wide extensional cracks and back-rotated soil blocks, combined with ejection of sand at some location in the southern side. While the extensional features at the head of the lateral spread included 1.3 m horizontal displacement of an irrigation pipe, an extension of the same pipe close to the toe appeared offset by 0.8 m, both horizontally and vertically (Rajendran et al. 2001) (Figure 1.7b). McCalpin and Thakkar (2003), while admitting that this feature is a lateral spread, suggested an overall structural control of the adjacent KMF on this feature (Figure 1.1). A verbatim report says that water sloshed out of the reservoir and spilled out of its northwest corner. Farm fields here were cracked at several places, and extensional cracks were developed with an east-west orientation, while 150–200 m north of the extensional cracks, there was a 600 m long zone of east-west ompressional pressure ridges (Figure 1.7c). Lateral spreads and downslips have been prominently seen at the Rudramata reservoir basin and in the coastal region near Kandla and Navlakhi (Figure 1.1). At Chopadva, lateral spreading is observed along an east-west-running road, where the north side ground has subsided by about a metre (Rastogi 2004). However, the large ground accelerations at Budharmora may also indicate that the lateral spread overlies a fault (probably the KMF) that moved during the 2001 main shock, as well as in prehistoric times, as explained in the trench descriptions.

The excavation work at the Budharmora lateral spread site did not reveal any fault plane that might have been activated during the 2001 earthquake, although a failure plane at the gravelly bed was detected; this was interpreted as an older surface on which a previous lateral spread had occurred, prior to 2001. McCalpin and Thakkar (2003) also imply an earlier event of lateral spread, as an alternative to their hypothesis of a previous faulting episode at this site. However, the trenching studies at Bhudharmora revealed significant palaeoseismological information that is explained below. A French reconnaissance team dug two backhoe trenches across the toe thrust of the lateral spread prior to May 2001, perhaps because shortly after the earthquake, this ridge was thought to represent a primary tectonic surface rupture of the KMF. McCalpin and Thakkar (2003) went a little deeper and relogged it and suggested palaeolateral spread at the site, but they did not provide a date for it. Rajendran et al. (2008) excavated two more trenches east and west of the French trench to see if fault geometry supported a tectonic or landslide origin.

1.4.3.1   Trench 1

The log of the trench (Figure 1.8) shows that the fault planes beneath the compressional ridge flatten about 1.5 m beneath the ridge crest, and continue south at a dip of about 10°, parallel to bedding in the Holocene sandy alluvium. Rajendran et al. (2008) further evoked that in reality, the 2001 fault

(a)

(b)

(c)

FIGURE 1.7   (a) Lateral spread type of ground deformation on the upper apron of Quaternary alluvium at Budharmora north of the KMF. (b) An irrigation pipe passing from the farm field, a metre below the surface, is displaced by ~80 cm horizontally and 30 cm laterally. (c) Significant pressure ridges near the lateral spreads sites (~8 km in length east-west) at Budharmora north of the KMF.

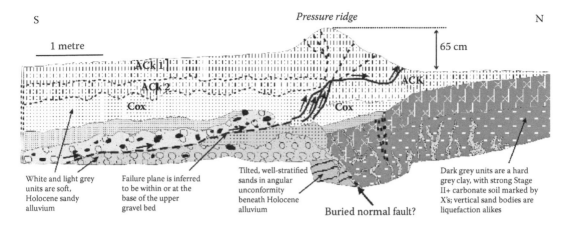

FIGURE 1.8 Log of trench 1 excavated across the lateral spread and compressional ridge at Budharmora showing an earlier generation of sand dykes and an older lateral spread that has disturbed the settlement horizon. (From McCalpin, J. P., and Thakkar, M. G., *Ann. Geophys.*, 46, 937–956, 2003.)

plane was extremely difficult to see once it paralleled bedding, but they inferred that the failure plane lay at the base of a gravelly sand unit in the trench wall. Because the fault does not displace obvious stratigraphic markers and deformation beneath the pressure ridge was partly ductile, they could not measure net displacement. A relatively weak soil profile is developed on the sandy alluvium, consisting of ACk1 and ACk2 horizons (weakly organic horizons with Stage I calcium carbonate, after the classification of Gile et al. [1996]) and a Cox horizon (oxidized parent material) (Figure 1.8). This degree of soil development suggests a mid-Holocene age for the alluvial deposit. The surprising feature in the trench is the strong soil profile at its northern end, which is bent down to the south and then truncated by an older, north dipping fault that was not reactivated in 2001 (Figure 1.8). This soil is developed on a hard grey clay parent material that is also truncated by the older fault. The lower half of the clay is riddled with subvertical, rootlike bodies of sand that look like dykes of liquefied sand injected into the clay. However, these sand dykes are also overprinted by soil calcium carbonate, so they were formed in the Middle Holocene or earlier, not in the 2001 earthquake. The combination of an older fault and older liquefaction features indicates that prehistoric deformation occurred here, but it is unclear exactly when that deformation occurred or whether it was associated with a lateral spread (as occurred in 2001) or with tectonic surface faulting of the KMF. It appears that the 2001 lateral spread failure plane was forced to the surface at this location by the reinforcing effect of this hard clay deposit (Rajendran et al. 2008).

1.4.3.2 Trench 2

Rajendran et al. (2008) opened another trench (trench 2) 50 m east of the earlier one at Budharmora to see the lateral variations in style of deformation and whether any previous deformation was available. The log of this 3 m deep trench (Figure 1.9) revealed a relatively simple stratigraphy consisting of three sedimentary units: (1) a bottom layer of blackish clay with a calcretized top; (2) a middle layer of light brown gritty sand, interspersed with indurated unburned sandstone slabs (10–20 cm long, 10 cm wide); and (3) a top layer of brownish yellow silty sand (Figure 1.9).

1.4.4 SAND DYKES AND THEIR CHRONOLOGY AT BUDHARMORA

What stands out in the second trench wall section are the closely spaced and parallel placed vertical dykes of sand intrusions at various levels (Figure 1.9). The subvertical dykes contain medium-sized yellowish sand, different from the surrounding sediments and they tend to widen at the base.

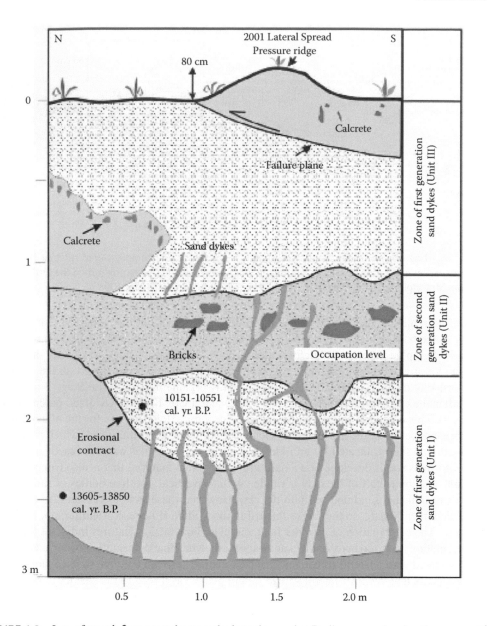

FIGURE 1.9 Log of trench 2 excavated across the lateral spread at Budharmora, showing three generations of sand dykes. The second event has disturbed an occupation level. (From Rajendran, C. P. et al., *J. Geophys. Res.*, 113, 1–17, 2008.)

Their characteristics, such as wide distribution in the trench, strong parallel alignment, widening base, vertical orientation and upward termination at the unconformities, and their morphological similarities to the 2001 sand dykes in the same site suggest that they may be related to earlier episodes of lateral spreading, rather than root moulding or animal burrowing (Obermeier 1996). Furthermore, the blackish host sediments are impregnated with the pervasive venting of yellowish sand. Three generations of feeder dykes were noticed in this trench; the oldest of the intrusions breaks the lowest erosional contact (Figure 1.9). However, these dykes do not cut through the gritty sand above, which is probably a subsequent deposition. A second set of feeder dykes has cut through all three layers, reaching about 1 m below the present ground surface. The third and latest generation of sand dykes (of 2001) can be seen within the yellowish sand, which forms the top layer in the stratigraphic column (Figure 1.9). The earlier generation of sand dykes abruptly terminate within the black clay, most of them reaching its calcretized top (Rajendran et al. 2008). This suggests earlier lateral spreading and liquefaction events, which migrated towards the present location of the 2001 compressional bulge. This layer has been fluidized, and flow structures are preserved at the bottom part. Judging from the randomly distributed (rotated) sandstone slabs, which probably belong to the occupation level, the age bracket for this region is 3850–4700 years BP (Joshi and Bisht 1994). As observed in trench 1, the fluidization has disturbed the occupation level. Based on the degree of soil development, McCalpin and Thakkar (2003) suggest a mid-Holocene age for the alluvial deposit. They further suggest that since the sand dykes are overprinted by soil carbonate (6750–7180 calibrated years BP), they must have been formed in the mid-Holocene or earlier. The maximum age of the earlier liquefaction event recorded in this trench is 4700 years BP because the event is younger than the occupation level (Rajendran et al. 2008).

1.4.5 Sand Blows in the Rann

Sand blows were the most common liquefaction features observed in the meizoseismal area of the 2001 Gujarat Earthquake. Sand blows are constructional cones of mostly sand vented with water to the ground surface through ground cracks. They range in size from tens of centimetres to tens of metres in length and up to tens of centimetres in thickness. Sand blows are dominant in the plain of Banni, Great Rann of Kachchh and their boundary zone. Umedpar is one such site where sand blows are reported to be ~60 m in length, 12 m in width and 14 cm in thickness (Rajendran et al. 2002) (Figure 1.10). Small-sized sand blows are reported 50 km north-northwest of the epicentre in the Great Rann of Kachchh near Khadir Island (KU in Figure 1.1). Fifty kilometres south of the epicentre, many small to moderate-sized sand blows formed in the mudflats along the northern and southern margin of the Gulf of Kachchh. Tuttle et al. (2002) documented the largest sand blow near Kandla; it is about 15 m long, 10 m wide, and 20 cm thick (Figure 1.1).

1.4.6 Ground Fissures and Extensional Cracks

Immediately after the earthquake, Rajendran et al. (2001) observed a series of ground cracks at the location south of Manfara village. They are oriented N 60° to 90° E, and a few of them cross the road to Manfara, forming a bulge in the tar layer of the road. Some of the larger cracks extend to a distance of about 1 km and show vertical offsets of 10–15 cm, with the northern and southern

FIGURE 1.10 Largest liquefaction crater and sand blow with an exploratory shallow pit (for the thickness of liquefaction sediments) at Umedpar.

sets showing motions in the opposite sense. On the largest and most prominent ground crack that extends through the road, the south side is up by 10 cm, whereas a parallel crack located 100 m to the north in the same direction shows movement of about 15 cm on the northern block. The block within these two cracks is cut by several parallel fractures, with little or no vertical displacement reported by Rajendran et al. (2001). These cracks may be part of a graben-type extensional feature formed on the crest of the folded surface, as exemplified by some large earthquakes resulting from reverse faulting.

1.4.7 GROUND SUBSIDENCE: COLLAPSE OF GROUND

Subsidence of the ground due to the subsurface liquefaction events is the most common ground deformation in the area north of the KMF or on the hanging wall of the 2001 rupture plane, where alluvial fan sediments dominate and have wedging relations with widespread clay formations of the Banni Plains (Figures 1.1 and 1.2). At many places on this alluvial apron, subsidence is observed wherein a few metres of alluvial soil collapse because the subsurface sand layers lose their capacity due to liquefaction. This is the major cause of building collapse in urban areas like Bhuj, Anjar and Gandhidham. A large collapse structure was reported at Baniyari village north of Bhachau during the field documentation immediately after the earthquake carried out by the author of this chapter (Figure 1.11).

1.4.8 SOFT SEDIMENT DEFORMATION

The Rann of Kachchh is an ideal ground for earthquake-induced liquefaction features due to its thick pile of clay, sand and silt-dominated soft sediments. Hence, this geomorphic unit preserves seismite-like features, such as convolutions of finer sediments, folds, flame structures and sand

FIGURE 1.11   View of co-seismic ground subsidence due to subsurface liquefaction at village Baniyari, north of the KMF. A very thin sand layer on the surface (1 mm) is reported here due to the eruption of saturated sand from a shallow depth.

(a)

(b)

FIGURE 1.12 (a) Convolution type of seismically induced soft sediment deformation structure to the north of Umedpar. (b) 'Flame structure'-type seismites exposed on shallow pit walls at Chobari village.

dykes. Excavations at large sand blow craters near Umedpar are made for their source parameters, and characterization reveals seismite-type deformations in clayey sediments of the Banni plains (Figure 1.12).

1.4.9 SAND BLOW CRATERS

Sand blow craters are characterized by constructional cones of vented sand. They have large central craters formed by the removal of surface soil and sediments. Craters are not as frequent as sand blows in the meizoseismal area of the Gujarat Earthquake. The largest documented sand blow crater is located near Umedpar, about 45 km west-northwest of the epicentre (Figures 1.2 and 1.10). There were four craters at the site, but the central couplet crater was reactivated several times during aftershocks subsequent to the main shock, and was observed and documented by Thakkar et al. (2012) as displaying geometry with non-reactivated craters. A dry crater is also reported at the site, apart from sand blow craters, where the capping soils were blown about 26 m to the west, indicating gas activities during the 2001 earthquake (Rajendran et al. 2001; Singh et al. 2001b).

1.4.10 ROCKFALLS AND SHATTERED HILLS

Rockfalls are the most common co-seismic and post-seismic and on-fault and off-fault features, especially in the orogenic and active block-faulted tectonic regions. Kachchh, being the active intraplate region, has bedrock topography and tectonically controlled linear hill ranges; the precipitous scarps are common geomorphic features. These escarpments are subjected to shatter and produce voluminous rockfalls at Kas Hill escarpments 20 km west-southwest of the epicentre and 1–2 km from the KMF (Figures 1.1, 1.2 and 1.13a). Rocks 3.05 × 2.43 × 1.52 m in size have also been reported just after the earthquake (Bendick et al. 2001). The verbatim report of the local villagers says that for three nights following the earthquake, on 26 January, they heard dreary sounds of rockfalls from the straight scarps behind the village, indicating a continuous process of falling rocks. The shattering of hills on the upthrusting hanging wall is a common on-fault and off-fault instantaneous feature observed in many earthquakes (McCalpin 1996). The 2001 Bhuj–Gujarat Earthquake also exhibited similar shattering at Chopadva village, south of Bhachau (Figure 1.13b). The shattered hill is surrounded by numbers of small hillocks and low elevated rocky mounds that remained unshattered. It is entirely composed of backed boulders and fault breccias, indicating the old faulted and sheared zone. Interestingly, this zone is a few kilometres away from the KMF zone and trending

FIGURE 1.13 (a) 'Rockfall' type of co-seismic and post-seismic off-fault deformations reported at the Kas Hill scarps, seen as light-coloured patches on the steep colluvial slopes below the scarps. (b) 'Shattered hill' is an exclusive deformation feature on the hanging wall. A shattered hill seen near Chopadva village is an example of the 2001 surface deformation.

nearly subparallel to the KMF. For those interested in further study of the past shattering events and old faulting in this area, detailed work on the buried rock and its dating, to prepare an entire model of the 2001 deformation zone, especially between the SWF and KMF (Figure 1.1), is suggested.

1.4.11 SECONDARY FAULTS

The secondary surface faulting does not match the orientation of the main shock focal plane, so it presumably represents some type of tear faulting above the eastern margin of the thrusted block. However, when measuring the dimensions of the 2001 surface faulting, it would probably include both the primary and secondary faulting, because in databases such as those of Wells and Coppersmith (1994), no distinction is made between the two. Dimensional parameters of surface faulting are based on detailed site mapping carried out immediately after the earthquake (Seeber et al. 2001; Rajendran et al. 2008).

1.4.11.1 Manfara Fault

Close to Manfara village, at 23.46° N, 70.38° E, the Manfara Fault ruptured during the earthquake and was first reported by Seeber et al. (2001). The fault strikes northwest and runs for 8 km, with a maximum 32 cm of dextral displacement and 19 cm of vertical displacement, typically east side up (Figure 1.14). Therefore, the fault was interpreted as a 'tear fault' in the hanging wall of the master thrust (Rajendran et al. 2001). The surface trace is composed of metre-scale linear segments of different strikes that give the rupture a zigzag appearance. This shape may be inherited from pre-existing fractures in the bedrock that underlies the thin alluvium along the fault trace. The part of the rupture discussed here is characterized by a series of right-stepping en échelon fractures showing right lateral movement, with a maximum offset of ~16 cm and an extensional net slip of ~10 cm (Rajendran et al. 2001). The east-west-flowing river

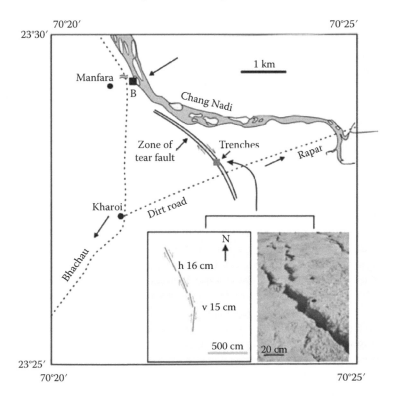

FIGURE 1.14 Location of the Manfara fault and the site of the trenches. Left Inset: Map view of the tear fault. Right Inset: View of the tear fault showing the right lateral offset with an oblique component.

(Chang Nadi) (Figure 1.14), takes a northwesterly swing north of this rupture zone, near the Manfara village (Figure 1.2). The rupture zone mapped follows the same trend as the river for about 1 km, possibly suggesting a long-standing structural feature.

1.4.11.1.1 *Trenching at Manfara Fault*

A study of the aftershocks of the 2001 earthquake by Mandal et al. (2004) suggested at least a half dozen focal mechanism solutions showing strike-slip movement along near-vertical planes, near Manfara. With their focal depths of >10 km, these earthquakes appear to be associated with deeper sources, and the tear faults may be considered manifestations of movement at depth. If indeed these faults were pre-existing features, and if they had slipped in the past, we would expect to see evidence for previous movements, this being the rationale for the trenching investigations across the fault (Rajendran et al. 2008). Further, the right-stepping, right lateral faults create transtension in the step-over zones, which are accompanied by normal faulting. In the following sections, various types of features observed in the three trenches excavated across and along the Manfara Fault, which may have bearing on the previous activity at the 2001 source zone, are discussed (Rajendran et al. 2008).

1.4.11.1.1.1 *Trench 1*   Trenching investigations were conducted in the deformed zone at Manfara, in the plough fields across the road (Figure 1.15). The shallow stratigraphy in trench 1 (Figure 1.15) consists of a semi-indurated unit of sand (the optically stimulated luminescence [OSL] date from an adjacent trench indicates that it could be mid-Holocene [Rajendran et al. 2001]), generally uniform and devoid of structures, overlain by brown sandy soil, possibly defining an erosional contact between them. The trench walls expose vertical fissures, as well as a network of sand intrusions, that abruptly truncate at the top of the lower sand unit, defined by an erosional unconformity (Figure 1.15b). These fissure fills differ from the host rock, as they contain very fine-grained pure brownish sand, with the dykes extending downwards to deeper levels into the source sand reservoir, a characteristic that argues against root moulding and animal burrowing. However, it is possible that the pre-existing planes of weakness in the host sediment could have offered paths of least resistance for the upward propagation and emplacement of sand. An earthquake origin is ascribed to such sand-filled fissures elsewhere, and they are termed 'incipient craters' (Obermeier 1996). They are considered to represent the initial phase of development of craters, but are too weak to break the overlying material. The fact that these sand-filled fissures end abruptly at the erosional contact suggests that they belong to an earlier event (Rajendran et al. 2001).

1.4.11.1.1.2 *Trench 2*   Rajendran et al. (2008) excavated the second trench 6 m south of trench 1 (Figure 1.15c), revealing a bowl-shaped sand-filled crater associated with a truncated sheet of sand layer and a vertical sand dyke at a depth of 2 m (Figure 1.15c). The crater contained a fining upward sequence of brownish sand. The ~15 cm wide crater contains a few small clasts at the margins. However, the presence of clasts within the crater (C) is minimal, probably because of the lack of thick clay cap. An 'out-of-sequence' silty sand (S in Figure 1.15c) was also observed, with internal layering of fine sand and silt at the upper margin of the crater. It is likely that this layer and the crater are part of a single sand blow, although the disturbed nature of the stratigraphy precludes such an interpretation. The adjacent vertical dyke (Figure 1.15b,c) contains alternating layers of coarse and fine-grained sand; the lower part contains a few larger-sized clasts. Sharp margins with characteristic stratification indicate that this is an independent feature. The strong stratification suggests gravitational settling during different pulses of venting. Another interesting feature paralleling the vertical dyke is a zone (Figure 1.15c) containing clay-rich patches, which also coincides with the fault zone. The overall characteristics of these features include the presence of clasts, fining upward sequences and internal layering. Interestingly, the clasts within the vertical dyke tend to be parallel to the sidewalls, indicating that they are derived from the sidewalls and transported up the dyke. The graded fill sequence within the dyke suggests a short-duration churning type of upwelling from the vent – a process generally associated with earthquake-related liquefaction (Obermeier 1994, 1996). Further,

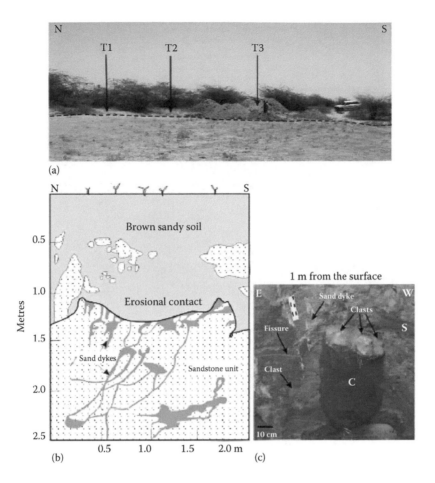

FIGURE 1.15 (a) View of the Manfara Fault; general trend shown by the dotted line and the locations of the trenches 1–3. (b) Log of trench 1 exposing the pre-2001 generation of sand dykes. (c) Photograph of the lower part (1 m from the surface) of the section from trench 2. C, crater; S, sandy layer.

the brownish sand fill of these features must have been emplaced from the underlying sequence, as no such sand is associated with the surrounding sedimentary sequence. The presence of clasts within the dyke argues against the role of long-term artesian springs in generating such features (Obermeier 1994). Rotated or rolled clasts or rubble of the host sediment embedded within the top part of the crater (C) suggests that it was derived from the free face and embedded at the top of the sand blow. It is possible that the fissures at the base of the scarp face were generated first as a part of a lateral spreading episode, as these features tend to be characteristically parallel to each other in the vertical section. This could happen if liquefaction occurs within the underlying sandy layer and the surface layer moves laterally down the slope, breaking into blocks bounded by fissures, and the liquefied sand from the underlying layer is emplaced. The stratigraphic position of the aforementioned features suggests that an earlier earthquake produced those (Rajendran et al. 2008). Liquefaction and the venting of sand must have been facilitated by more favourable ground conditions, including a shallower water table and wetter environment. The palaeoenvironmental studies of this region point to the possibility that such conditions prevailed until about 4000–5000 years ago (Gaur and Vora 1999).

1.4.11.1.1.3 Trench 3: Fault Stratigraphy and Age Determinations Rajendran et al. (2008) opened a 3 m deep trench (trench 3) (Figure 1.16) for evidence of earlier faulting events, such as upward terminations of fractures, vertical offsets of marker beds, colluvial wedges and fissure

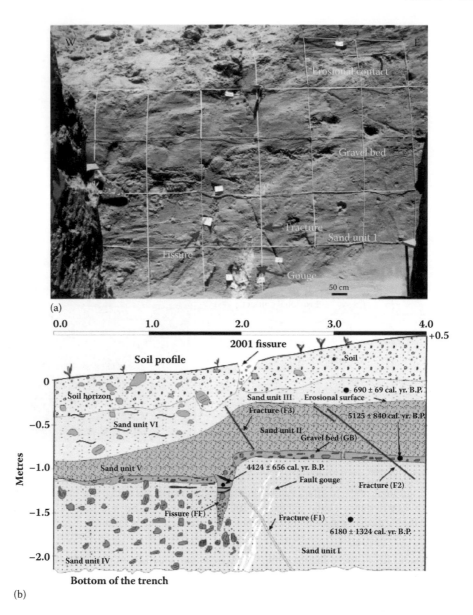

FIGURE 1.16 (a) Photograph of the western wall of trench 3, excavated across the Manfara Fault, showing the fault gouge at the base of the trench, an old fissure fill and fractures. (b) Log of the western wall of trench 3, across the Manfara Fault. The upthrown side corresponds to the footwall block, and the downthrown side is the hanging wall.

fillings (Figures 1.15 and 1.16), which are characteristic of regions of repeated earthquake activity (Sieh 1978; McCalpin 1996). Stratigraphically, the section mainly shows two units of semi-indurated sand (unit I of medium to fine sand and unit II of fine sand), giving OSL ages of 6180 ± 1324 years and 5125 ± 840 years, respectively, as shown in Figures 1.15 and 1.16, with a break in deposition defined by a thin layer of gravel. This gravel bed is representative of mid-Holocene palaeosurface, sandwiched between the two units of sandy sequences. The scattered, thin layer of gravel also characterizes the present-day ground surface, and thus it appears that such surfaces are quite typical of this region. Unit II above the gravel layer is a later deposition; the top layer of silt (aeolian) occurs above an erosional contact, quite well defined on the eastern part of the trench. Clear evidence of vertical displacement of one unit is documented, as shown by the shift in the lower gravel layer (GB), with a maximum offset of ~20 cm (Figure 1.16). The OSL date of the sample taken from the gravel bed is 5125 ± 840 years, and the top part of sand unit II gave an OSL date of 690 ± 69 years (Rajendran et al. 2001, 2008). Development of a fissure (FF), with a vertical long axis and with fills of clasts and the presence of pebbles derived from the gravel layer above, close to the zone of vertical offset, suggests an earthquake-induced origin for both these features. Clearly, such fissures with fill derived from the top could have formed only if the surface was exposed. The gravels and pebbles are part of the hard iron crust that usually develops in association with weathering, similar to what marks the present-day ground surface. These fissures were formed before sand unit II was deposited, as there are no pieces of sand unit II in the fissure. The OSL date of the material from the fissure fill is estimated to be 4424 ± 656 years (Figures 1.15 and 1.16) (Rajendran et al. 2001). The 2001 earthquake also caused a few similar near-surface fissures, filled with loose material derived from the top, as well as deposition of debris (unit VI) on the footwall, as a result of shaking. This event also left a large amount of debris on the western footwall block. Although there are rubbles at the base of the offset on the side of the hanging wall, it is difficult to identify clear boundaries of a colluvial wedge. The sedimentary sequences (units IV, V and VI) (Figures 1.15 and 1.16) on the hanging wall side are generally composed of loose sediments with embedded rubble and clasts, in contrast to the layered medium to fine-grained sandy facies on the footwall side (Figures 1.15 and 1.16).

Rajendran et al. (2008) noticed a diffused and disturbed strand of whitish material, across the fracture plane, FP-1, a feature that has been observed in many of the trenches excavated across the Manfara Fault. These are crushed grains of quartz and rock flour of siliceous material, generally indicative of frictional wear, and it is a highly localized zone corresponding with the linear north-south trend of the Manfara Fault (Figure 1.16). The x-ray diffraction analyses indicate that these patches of finer whitish rock flour are entirely composed of siliceous material without any clayey minerals. This zone terminates at the gravel layer. The ubiquitous presence of this gouge zone in this and other trenches, almost trending in the fault direction (north-south), suggests recurrent activity of crushing, shearing and consolidation during previous faulting episodes. The fracture FP-1 does not seem to displace any units, and this fracture is terminated at the displaced layer of gravels. The second fracture (FP-2) terminates at the top layer of fine-grained sand. The breakage in this layer from a pre-2001 earthquake is evident from the well-preserved broken edge on the hanging wall. However, the downthrown layer has been highly modified, possibly due to the severity of shaking and mixing with the in situ material, disturbing the stratigraphic boundaries. Thus, the exact amount of slip on this plane could not be estimated. The 2001 earthquake caused a vertical displacement of only 5–10 cm on the surface. The slip observed at the gravel bed (GB in Figure 1.16) is 20 cm, and it is quite possible that the slip here is cumulative, and a product of the previous event and the 2001 earthquake. Thus, the OSL date of 4424 ± 656 years is assumed to be the minimum age of the penultimate event. If the fissure and the fill had formed immediately after the faulting, then this age could be considered contemporaneous (Rajendran et al. 2008).

A thin layer of debris formed from talus, derived from the top of the footwall block, was observed in the central part of the trench. The loose matrix with large clasts and poor stratification is characteristic of shaking derived debris, rather than a product of normal deposition (as observed in unit VI in Figure 1.16). From its composition, the lower part of the deposit (unit IV) also appears to have been deposited by an older shaking event. The whole zone is quite disturbed, and it is difficult to interpret the sequence of events and the order of deposition. Differentiating the earthquake-derived colluvial wedges from normal talus (wash facies) at the free face of the low displacement fault is not easy, as has been discussed by many (Machette et al. 1992; Nelson 1992; McCalpin et al. 1993). However, the context of these deposits and their spatial association with a vertical fissure and its coincidence with the 2001 rupturing are indicative of earthquake shaking and ground failure. Some of the known criteria were applied to identify the reactivation of this fault, which include the presence of a gouge zone, crack fill, upward termination of fractures and differential displacement at the fault plane (McCalpin 1996). These lines of evidence, along with the age data, converge to suggest that an earthquake affected this area ~4000 years ago. Stratigraphically, it is evident that this event occurs just below an undisturbed facies, indicating a quiet interval before the occurrence of the 2001 event, and therefore the ~4000-year event is considered the penultimate event in this region. Because of the lack of clear-cut stratigraphic evidence in the lower part of the trench, zeroing out a still older event is difficult, except a partially preserved section of colluvial facies mixed with clasts (unit IV), underneath the gravel bed (GB) on the hanging wall block (Figure 1.16).

1.5   CONCLUSION

The co-seismic, post-seismic, on-fault and off-fault ground deformation features produced during the 2001 Bhuj–Gujarat Earthquake are exclusive of active intraplate seismic zones, and hence can be used as ideal features for compressive intraplate tectonic environments. The 75 km long south dipping rupture plane is not exposed on the surface, barring a 2 km stretch of low-angle thrust scarps found north of Bhrudiya village. The secondary faults, sand blow craters, liquefactions, landslide-type lateral spreads and pressure ridge type of deformations on the surface were documented just after the earthquake by the author with collaborators. Further detailed study on evidence of stratigraphical proxies, with the chronology of the past earthquakes comparing with that of the 2001 Gujarat Earthquake, has been carried out in subsequent years to find out the palaeoseismic characterization of the specific faults and Kachchh as a whole. Further, to understand the nature of past and present seismicity in an SCR such as Kachchh, Quaternary tectonic features are the best indicators. Thrusting in Kachchh is mostly along the pre-Tertiary faults (such as the KMF and KHF) developed in an older extensional regime.

The following are the major conclusions based on the present study.

The 2001 thrust scarp and the mole tracks mapped north of Bhrudiya village are parallel to the major plane of rupture of the main shock, and are linked to two zones of dextral faulting (the northern dextral fault zone and the Manfara Fault).

If these dextral fault zones define the boundary of the north-thrusted block, it follows that the 2001 seismogenic fault, which underlies the thrusted block, must project to the surface somewhat north of the dextral faults. This constraint appears to rule out the KMF as the seismogenic fault, as does the fact that the 2001 epicentre lies slightly north of the surface trace of the KMF.

The dextral slip on these faults rules out the SWF as a plane of rupture. Instead, the 2001 main shock plane must have been a nearly blind thrust parallel to but 30 km north of the KMF. McCalpin and Thakkar (2003) say 'nearly blind thrust' because the 0.8 km long thrust scarp was created, and may represent the only location where primary displacement propagated all the way to the surface.

However, based on the reconnaissance data collected by McCalpin and Thakkar (2003) in May 2001, the possibility that the Mw 7.7 event was completely blind, and that the mapped thrust scarp was secondary (passive) thrusting induced on a pre-existing fault by strong, channelled ground motion extending updip from the seismogenic plane, cannot be ruled out.

Further, based on the detailed trenching excavation at the secondary faults at Manfara and the significant ground deformation zone of Budharmora, where trenches were opened for the palaeolateral spread and landslides, it has been established that even secondary surface ruptures can be useful for studying the palaeoseismic characteristics of any region.

The ground deformation features produced due to the 2001 Gujarat Earthquake will be helpful in studying the similar co-seismic ground deformation features of the past earthquakes in an intraplate region like Kachchh. The secondary faulting, lateral spreads and sand blows generated in the deformation zone offered possibilities to search for similar features formed in the past. Features such as the dextural tear faults formed near Manfara and the lateral spread at Budharmora are useful for understanding the previous activity in the epicentral zone of the 2001 Bhuj Earthquake.

The 2001 Gujarat Earthquake ground deformation features are considered the most distinctive, as the primary fault remains blind, while most of the secondary features are useful in palaeoseismic investigations and become the modern analogue of palaeoearthquakes.

The excavations of a 2001 lateral spread in close vicinity of its epicentre exposed an older feature that seems to have disturbed the cultural horizons. The chronological constraints of these events indicate two previous events, which occurred around 4000 and 9000 years ago.

Palaeoseismic studies at the secondary Manfara Fault carried out using three trenches where older sand blows, ground deformation, colluvial wedges and their stratigraphic displacements are observed, which in turn were dated by various techniques and two such events in the past, have been identified ~4 and ~9 ka.

ACKNOWLEDGEMENTS

The author is grateful to the Department of Science and Technology, New Delhi (Project Nos. DST/23/(257) SU/2001, SR/S-4/ES-21/Kachchh Window/P3/2005 and DST/SR/S4/ES-TG/02-2008), for providing financial assistance to work in the field and laboratory. He is also thankful to C. P. Rajendran, Kusala Rajendran and L. S. Chamyal for their valuable guidance throughout the work, since the 2001 Bhuj–Gujarat Earthquake and subsequent events and episodes of field and classroom discussions. The author also acknowledges M. P. Tuttle, Memphis, Tennessee, and J. P. McCalpin, GEO-HAZ Consulting Inc., Crestone, Colorado, for their edifying field teachings. Professor Ramesh P. Singh of Chapman University, Orange, California, was a great instigator for submitting this work.

REFERENCES

Antoliok, M., Dreger, D. S. (2003). Rupture process of the 26 January 2001 Mw 7.6 Bhuj, India, earthquake from teleseismic broadband data. *Bull. Seismol. Soc. Am.*, 93, 1235–1248.

Bendick, R., Bilham, R., Fielding, E., Gaur, V. K., Hough, S. E., Kier, G., Kulkarni, M. N., Martin, S., Mueller, K., Mukul, M. (2001). The 26 January 2001 'Republic Day' earthquake, India. *Seismol. Res. Lett.*, 72, 328–335.

Biswas, S. K. (1987). Regional tectonic framework, structure and evolution of the western marginal basins of India. *Tectonophysics*, 135, 307–327.

Biswas, S. K., Khattri, K. N. (2002). A geological study of earthquakes in Kutch, Gujarat, India. *J. Geol. Soc. India*, 60, 131–142.

Bodin, P., Horton, S. (2004). Source parameters and tectonic implications of the aftershocks of the Mw 7.6 Bhuj earthquake of January 26, 2001. *Bull. Seismol. Soc. Am.*, 94, 818–827.

Burgmann, R. D., Pollard, D., Martel, S. J. (1994). Slip distributions of faults: Effects of stress gradients, inelastic deformation, heterogeneous host-rock stiffness, and fault interaction. *J. Struct. Geol.*, 16, 1670–1675.

Gahalaut, V. K., Burgmann, R. (2004). Constraints on the source parameters of the 26 January 2001 Bhuj, India, earthquake from satellite images. *Bull. Seismol. Soc. Am.*, 94, 2407–2413.

Gaur, A. S., Vora, K. H. (1999). Ancient shorelines of Gujarat, India, during the Indus civilization (Late Mid-Holocene): A study based on archaeological evidence. *Curr. Sci.*, 77, 180–185.

Gile, L. H., Peterson, F. F., Grossman, R. B. (1996). Morphological and genetic sequences of carbonate accumulation in desert soils. *Soil Sci.*, 101, 347–360.

Joshi, V. P., Bisht, R. S. (1994). *India and the Indus Civilization*. New Delhi: National Museum Institute, pp. 22–31.

Karanth, R. V., Sohoni, P. S., Mathew, G., Khadkikar, A. S. (2001). Geological observations of 26th January 2001 Bhuj earthquake. *J. Geol. Soc. India*, 58(3), 193–202.

Kayal, J. R., Zhao, D., Mishra, O. P., De, R., Singh, O. P. (2002). The 2001 Bhuj earthquake: Tomographic evidence for fluids at the hypocenter and its implications for rupture nucleation. *Geophys. Res. Lett.*, 29(24), 2152.

Machette, M. N., Personius, S. F., Nelson, A. R. (1992). Paleoseismology of the Wasatch fault zone: A summary of recent investigations, interpretations, and conclusions. In *Assessment of Regional Earthquake Hazards and Risk along the Wasatch Front, Utah*, ed. P. L. Gori, W. W. Hays. U.S. Geological Survey Professional Paper 1500-A. Reston, VA: U.S. Geological Survey, pp. A1–A71.

Mandal, P., Rastogi, B. K., Satyanarayana, H. V. S., Kausalya, M., Vijayraghvan, R., Satyamurty, C., Raju, I. P., Sarma, A. N. S., Kumar, N. (2004). Characterization of the causative fault system for the 2001 Bhuj earthquake of Mw 7.7. *Tectonophysics*, 378, 105–121.

McCalpin, J. P., ed. (1996). *Paleoseismology.* 2nd ed. New York: Academic Press, p. 588.

McCalpin, J. P., Thakkar, M. G. (2003). 2001 Bhuj-Kachchh earthquake: Surface faulting and its relation with neotectonics and regional structures, Gujarat, Western India. *Ann. Geophys.*, 46, 937–956.

McCalpin, J. P., Zuchiewicz, W., Jones, L. C. (1993). Sedimentology of fault-scarp-derived colluvium from the 1983 Borah Peak rupture, central Idaho. *J. Sediment. Petrol.*, 63, 120–130.

Nelson, A. R. (1992). Lithofacies analysis of colluvial sediments – An aid in interpreting the recent history of Quaternary normal faults in the Basin and Range province, western United States. *J. Sediment. Petrol.*, 62, 607–621.

Obermeier, S. F. (1994). Using liquefaction induced features for paleoseismic analysis. U.S. Geological Survey Open File Report 94-663. Reston, VA: U.S. Geological Survey, pp. 1–58.

Obermeier, S. F. (1996). Use of liquefaction-induced features for paleoseismic analysis – An overview of how seismic liquefaction features can be distinguished from other features and how their regional distribution and properties of source sediment can be used to infer the location and strength of Holocene paleo-earthquakes. *Eng. Geol.*, 44, 1–76.

Oldham, R. D. (1926). The Cutch (Kachh) earthquake of the 16th June, 1819 with a revision of the great earthquake of the 12th June, 1897. *India Geol. Surv. Mem.*, 46, 71–147.

Patidar, A. K., Maurya, D. M., Thakkar, M. G., Chamyal, L. S. (2006). Fluvial geomorphology and neotectonic activity based on field and GPR data, Katrol hill range, Kachchh, western India. *Quat. Int.*, 159, 74–92.

Philip, H., Rogozhin, E., Cisternas, A., Bousquet, J. C., Borisoy, B., Karkhanian, A. (1992). The American earthquake of 1988 December 7; faulting and folding, neotectonics and paleoseismicity. *Geophys. J. Int.*, 110, 141–158.

Rajendran, C. P., Rajendran, K. (2001). Characteristics of deformation and past seismicity associated with the 1819 Kutch earthquake, northwestern India. *Bull. Seismol. Soc. Am.*, 91(3), 407–426.

Rajendran, C. P., Rajendran, K. (2003a). The surface deformation and earthquake history associated with the 1819 Kachchh earthquake. *Geol. Soc. India Mem.*, 54, 87–142.

Rajendran, C. P., Rajendran, K. (2003b). Studying earthquake recurrence in the Kachchh region, India. *Eos Trans. AGU*, 84(48), 529, 533, 536.

Rajendran, C. P., Rajendran, K., Thakkar, M. G., Goyal, B. (2008). Assessing the previous activity at the source zone of the 2001 Bhuj earthquake based on the near-source and distant paleoseismological indicators. *J. Geophys. Res.*, 113, 1–17.

Rajendran, K., Rajendran, C. P., Thakkar, M. G., Gartia, R. (2002). The sand blows from the 2001 Bhuj earthquake reveal clues on past seismicity. *Curr. Sci.*, 83(5), 603–610.

Rajendran, K., Rajendran, C. P., Thakkar, M. G., Tuttle, M. P. (2001). The 2001 Kachchh (Bhuj) earthquake: Coseismic surface features and significance. *Curr. Sci.*, 80, 1397–1405.

Rastogi, B. K. (2001). Ground deformation study of Mw 7.7 Bhuj earthquake of 2001. *Episodes*, 24(3), 160–165.

Rastogi, B. K. (2004). Damage due to the Mw 7.7 Kutch, India earthquake of 2001. *Tectonophysics*, 390, 85–103.

Roering, J. R., Cooke, M. L., Pollard, D. D. (1997). Why blind thrust faults do not propagate to the earth's surface: Numerical modeling of coseismic deformation associated with thrust-related anticlines. *J. Geophys. Res.*, 102(B6), 11901–11912.

Schweig, E., Gomberg, J., Petersen, M., Ellis, M., Bodin, P., Mayrose, L., Rastogi, B. K. (2003). The M 7.7 Bhuj earthquake: Global lessons for earthquake hazard in intra-plate regions. *Geol. Soc. India*, 61, 277–282.

Seeber, N., Ragona, D., Rockwell, T., Babu, S., Briggs, R., Wesnousky, S. G. (2001). Field observations bearing on the genesis of the January 26, 2001 Republic Day earthquake of India resulting from a field survey of the epicentral region. Web document, 28 February 2001 version. http://neotectonics.seismo .unr.edu/Bhuj/Report.html.

Sieh, K. E. (1978). Prehistoric large earthquakes produced by slip on the San Andreas Fault at Pallet Creek, California. *J. Geophys. Res.*, 83, 3907–3939.

Singh, R. P., Bhoi, S., Sahoo, A. K., Raj, U., Ravindran, S. (2001a). Surface manifestations after the Gujarat earthquake. *Curr. Sci.*, 81, 164–166.

Singh, R. P., Sahoo, A. K., Bhoi, S., Kumar, M. G., Bhuiyan, C. S. (2001b). Ground deformation of the Gujarat earthquake of 26 January 2001. *J. Geol. Soc. India*, 58, 209–214.

Singh, S. K., Pacheco, J. F., Bansal, B. K., Perez-Campos, X., Dattatrayam, R. S., Suresh, G. (2004). A source study of the Bhuj, India, earthquake of 26 January 2001 (Mw 7.6). *Bull. Seismol. Soc. Am.*, 94, 1195–1206.

Talwani, P., Gangopadhyay, A. (2001). Tectonic framework of the Kachchh earthquake of 26 January 2001. *Seismol. Res. Lett.*, 72, 336–345.

Thakkar, M. G., Goyal, B. (2004). On relation between magnitude and liquefaction dimension at the epicentral zone of 2001 Bhuj earthquake. *Curr. Sci.*, 87(6), 811–817.

Thakkar, M. G., Goyal, B., Maurya, D. M., Chamyal, L. S. (2012). Internal geometry of reactivated and non-reactivated sand blow craters related to 2001 Bhuj earthquake, India: A modern analogue for interpreting paleosandblow craters. *Geol. Soc. India*, 79, 367–375.

Turcotte, D. L., Schubert, G. (2002). *Geodynamics*. New York: Cambridge University Press.

Tuttle, M. P., Hengesh, J., Tucker, K. B. (2002). Observations and comparisons of liquefaction features and related effects induced by the Bhuj earthquake. *Earthquake Spectra*, 18(Suppl. A), 79–100.

Wells, D. L., Coppersmith, K. J. (1994). Empirical relationships among magnitude, rupture length, rupture area, and surface displacement. *Bull. Seismol. Soc. Am.*, 84, 974–1002.

Wesnousky, S. G., Seeber, L., Rockwell, T. K., Thakur, V., Briggs, R., Kumar, S., Ragona, D. (2001). Eight days in Bhuj: Field report bearing on surface rupture and genesis of the January 26, 2001 republic day earthquake of India. *Seismol. Res. Lett.*, 72, 514–524.

2 Gujarat Earthquake
Liquefaction

M.G. Thakkar

CONTENTS

2.1 INTRODUCTION

The evolution of the Kachchh Basin is related to the breakup of Gondwanaland in the Late Triassic and Early Jurassic and the subsequent spreading history of the eastern Indian Ocean (Biswas 1982, 1987). The main tectonic event took place during the Late Cretaceous, when the drift motion was at its acme, with an average rate exceeding 15 cm/year (Powell 1979). The basin was formed by subsidence of a block between Nagarparkar Hills and the southwest extension of the Aravalli Range (Biswas 1982, 1987). Kachchh graben became marine basin during the Middle Jurassic period (Biswas 1981). In the Late Cretaceous, uplift of the Jurassic sediments took place in the Kachchh Basin. At the end of this period, extensive subaerial eruption of trappean lava took place through a number of volcanic craters in the Cambay graben, Saurashtra and Kachchh (Biswas and Despande 1973). The present profile of the continental margins of India evolved during the Early Tertiary, when India collided with Asia and stabilized its present position. Reverse movement of the blocks along steep faults (normal faults during rifting) took place in the Kachchh Basin. The uplifts came into existence by upthrust of the footwall blocks; during the drift stage, the horizontal stress increased and wrench-related structures developed (Biswas 2005). This tectonic movement has continued episodically from the Early Tertiary until the present day, as evident from the neotectonic movements (Thakkar et al. 1999, 2001a; Maurya et al. 2003). The Banni and Gulf of Kachchh half grabens were the main depositional domains separated by the Kachchh Mainland Fault (KMF), which acted as the principal intra-rift fault along the rift axis (Figure 2.1). The latter subsided most along the master fault accommodating thicker sediment fill. Evidently, the basin evolved in two stages: (1) an extensional rift stage, when Mesozoic sedimentation took place in the subsiding basin, and (2) a compressive inversion stage,

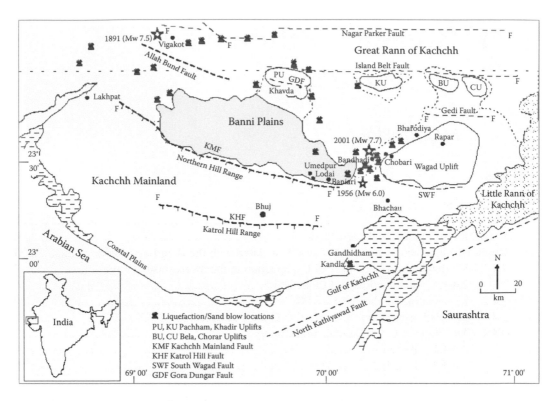

FIGURE 2.1 Location map of the study area showing major physiographic features and basin-bounding faults. The locations of three major earthquakes in the basin and their liquefaction sites are shown.

when the present structural style evolved. Due to near-vertical fault planes, strike-slip movements dominated the inversion stage. The same set of faults was first reactivated as normal faults during the rift phase, and later as strike-slip faults during the inversion stage (Biswas 2005). The Great Rann of Kachchh (Hindi *ran*, 'wilderness') formed on the half graben between Nagar Parkar Fault (KPF) in the north and Island Belt Fault (IBF) in the south during the Quaternary period (Figure 2.1). The Rann of Kachchh was once an extension of the Arabian Sea, and then later a navigable lake, but today, due to the deposition of silt as well as tectonic uplift of the landscape, it is a vast swampland (Merh and Patel 1988). During the dry season, the Rann becomes a sandy salt basin, prone to dust storms, while in the rainy season, it is inundated with floodwaters and the whole Kachchh becomes an isolated island. The geochemical characteristics of the Rann sediments and sandy islets within it suggest that the Indus River was a major contributor of sediment into the western Great Rann, which was largely routed through Kori Creek (Figure 2.1). After around 2.2 ka and before 1.4 ka, a combination of climate and tectonic activity probably led to the withdrawal of the intertidal environment from the major part of the western Great Rann (Tyagi et al. 2012).

2.2 LIQUEFACTION

Liquefaction is a phenomenon wherein a mass of soil loses a large percentage of its shear resistance, when subjected to monotonic, cyclic or shock loading, and flows in a manner resembling a liquid until the shear stresses acting on the mass are as low as the reduced shear resistance (Sladen et al. 1985). Liquefaction is most commonly observed in shallow, loose, saturated deposits of cohesionless soils subjected to strong ground motions in large-magnitude earthquakes. Unsaturated soils are not subjected to liquefaction because the volume compression does not generate excess pore pressure. Since liquefaction is associated with the tendency for soil grains to rearrange when sheared, anything that impedes the movement of soil grains will increase the liquefaction resistance of a soil deposit. Particle cementation, soil fabric and ageing – all related to the geologic formation of a deposit – are important factors that can hinder particle rearrangement (Seed 1979; Seed et al. 1983). Soils deposited prior to the Holocene epoch (>10 ka old) are usually not prone to liquefaction (Youd and Perkins 1978). The Ranns are recent continental basins where Pleistocene and Holocene deposits are dominant. The basin contains very soft materials, known to be the best grounds for the preservation of soft sediment deformations caused by the historical earthquakes. Such kinds of basins are more prone to liquefaction during strong to major earthquakes due to their shallow groundwater and clay capping. Therefore, the Rann areas are the best place for palaeoseismic studies.

2.3 KACHCHH EARTHQUAKES AND LIQUEFACTION

The Bhuj Earthquake (Mw 7.7) of 26 January 2001 is one of the largest seismic events among the intraplate earthquakes in the post-instrumental era (Rajendran et al. 2001, 2002). The energy was released on a 90 km long east-west trending and 55° dipping fault plane (Rajendran et al. 2001). This earthquake generated negligible traces of the primary surface deformation; however, secondary deformations, like liquefaction, lateral spreads, sand blows and rockfalls, are common in meizoseismal areas (Rajendran et al. 2001). One noteworthy aspect was the large liquefaction field covering nearly a 10,000 km^2 area in Kachchh. The Ranns provided the best ground for liquefaction during the earlier earthquakes (1819 Kachchh–Allah Bund earthquake and AD 893 earthquake) (Rajendran et al. 2001; Rajendran and Rajendran 2001, 2002). A linear track of Rann sediments, 85 km long and 15 km wide, was uplifted during the 1819 earthquake and blocked the water from the northern river system to debouch into the Arabian Sea, and hence is known as 'Allah Bund' – 'a dam of God'. It also generated large-scale deformation and flexuring in the form of extensive uplift and subsidence of Rann sediments (Oldham 1926; Rajendran et al. 1998). Palaeoseismic studies near the epicentre of the 1819 earthquake zone at the Indo-Pak border in the north of the Kachchh Basin revealed massive liquefaction layers that eventually cut by the 2001 earthquake liquefaction pipes; even older liquefactions, ~893 AD, have also been reported (Rajendran et al. 2001).

2.4 OBSERVED LIQUEFACTION AND SAND BLOWS ASSOCIATED WITH THE BHUJ EARTHQUAKE

Earthquake-induced liquefaction is a process by which saturated, granular sediment temporarily loses its strength due to earthquake ground shaking (Ngangom et al. 2012). A large amount of seismic deformation is reported on the plain of Banni along the KMF zone and in the Great Rann of Kachchh along the Allah Bund Fault (Rajendran and Rajendran 2001, 2002) (Figure 2.1). Shallow trenching at various sites in the Rann revealed a comparative picture of the dimension of liquefaction at the 2001 epicentral zone. Massive and widespread liquefaction has been mapped using satellite and ground studies in the Rann of Kachchh (Figure 2.2) (Rajendran and Rajendran 2001, 2002; Singh et al. 2001a,b; Tuttle et al. 2001, 2002; Thakkar and Goyal 2004). Palaeoseismological studies in the 2001 epicentre zone using pre- and post-earthquake satellite imagery show that the eastern part of the Great Rann was flooded with water, and sediments were ejected that vented through hundreds of sand blows and large craters ubiquitously spread in the Rann (Rajendran and Rajendran 2001, 2002; Thakkar and Goyal 2004). This earthquake produced massive liquefaction far from the epicentre in the western Allah Bund area, where sand blows have been found to rework the 1819 sand vent and its predecessor. Large craters at the Indo-Pak border disclosed 1819 and 893 AD liquefaction layers below the 2001 sand cover (Rajendran and Rajendran 2001). Liquefaction is observed along the Indo-Pak border, especially between the Rann and the Nagar Parkar hill range. Many buried channels within the dry and salty Rann surface have reappeared as violet and blue rivulets amid the salt playas; nevertheless, they lack a definite slope of the terrain, and the ejected water remained for some time (Figure 2.2). Largely affected and approachable places of liquefaction

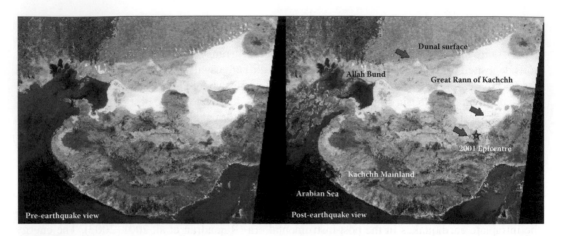

FIGURE 2.2   Multiangle imaging spectroradiometer (MISR) pre- (15 January 2001) and post- (31 January 2001) earthquake images. False colour images are 275 km wide and 218 km high. The red dot shows the epicentre of the 2001 Bhuj Earthquake of Mw 7.7. The liquefaction effects in the post-earthquake imagery are seen by the purple patches and infant streams within the white salt encrustations in the eastern Kachchh and the geographic boundary between the Rann and northern dune surface. (From https://www.jpl.nasa.gov/spaceimages/details.php?id=PIA03403.)

within the eastern Rann of Kachchh include the Chobari, Amarsar, Bandhadi and Bharodiya villages and the area of the Rann between the mainland of Kachchh and the islands to the north (Figure 2.1). The creek area to the south of Kandla and the eastern Kachchh coastal segment also experienced shallow sea liquefactions in the creek and coastal zones (Tuttle et al. 2002).

2.4.1   SHALLOW SUBSURFACE STUDY OF LIQUEFACTION IN THE MEIZOSEISMAL AREA

2.4.1.1   Shallow Pits and Liquefaction

Sites for shallow pits and deep trenches in the soft sediments are identified based on minor observations in pre- and post-earthquake satellite imagery, while the subsurface characterizations of liquefaction craters have been identified using ground-penetrating radar (GPR) techniques (Maurya et al. 2006). Liquefied sand layers, after thorough physical examinations, are dated by ^{14}C, thermoluminescence (TL) or optically stimulated luminescence (OSL), and accelerator mass spectrometry (AMS) techniques to premise the chronology of the past seismic events (Rajendran et al. 2008). Large liquefaction craters of the 2001 Bhuj Earthquake have been found to be reactivated during the subsequent aftershocks, while some craters were not reactivated, and their geometries are characterized (Thakkar et al. 2012). Massive surface deformation and liquefactions are reported in the villages located on the edge of the geographical boundary between mainland Kachchh and the plain of Banni, where a basin-bounding fault (KMF) happens to pass. Baniari, Lodai and Umedpar are situated on the KMF and have a long history of devastation due to earthquakes. The name of some villages represents the history of earthquakes in the region. In the local language, *Lodai* means the 'place which constantly shakes'. The eastern edge of the Wagad upland is also subjected to massive liquefaction, where a major subsurface fault passes northeast-southwest and holds several villages that were devastated in the 2001 earthquake. Baniari is located on the edge of the Rann of Kachchh, where compositionally varied sands were ejected during the 2001 earthquake. Coarse yellow sand, 5–8 cm thick, blown out with water from N 75° E trending cracks, has been observed. An elongated crater of 2.5 m long, located 50 m southwest of the en échelon cracks, ejected yellowish, gritty and coarse fluvial sand of 25–27 cm thickness with a lateral dimension of 20 × 6 m (Thakkar and Goyal 2004). Enormous liquefaction is observed 1 km northwest of this crater in the Rann of Kachchh, where a shallow pit shows a sand vent with 35 cm of liquefaction sand, which includes 5–6 cm of brown sand on the top, which is underlain by 4–6 cm of thick, fine, whitish and buff sand with clay, while the bottom is composed of fine sand with brown silt (Figure 2.3a). A large liquefaction field and sand blows were documented northeast of Amarsar village located near the epicentre (Figure 2.2). In Chobari village, half of the total 400 bore wells failed due to the ground shaking

(a)

(b)

FIGURE 2.3 (a) A shallow pit exposes a ~20 cm thick liquefaction vent in the Great Rann of Kachchh near Baniari village. Fine-to-coarse sand with silt is observed in sand blows. (b) The linings of a dug well at Baniari village are collapsed by almost 3 m. Subsurface liquefaction of the sand layers is responsible for such events. The well linings subside 2 m from the ground level.   (*Continued*)

(c)

FIGURE 2.3 (CONTINUED)   (c) A large sand blow crater, 5 m diameter and 1.75 m deep, was documented near Baniari village. A very thin layer of yellow, coarse sands was blown from the crater, along with water (noted by the locals). Large craters on the surface are indicative of massive subsurface liquefaction.

and liquefaction. A 12.2 ft deep dug well had been filled with liquefaction sand and turned into a flat ground (Rajendran et al. 2001). In another well, the linings have collapsed and subside 2.0 m from the ground level due to the subsurface liquefaction (Figure 2.3b). A large sand blow crater, 5 m diameter and 1.75 m depth, was documented near Baniari (Figure 2.3c). Large liquefaction grounds, craters and minor sand blows are reported from the Rann near Chobari village; the shallow pits were opened up to document the liquefaction pipes and their relations with the host layers (Figures 2.1 and 2.4a–c) (Thakkar and Goyal 2004).

2.4.2   LIQUEFACTION STUDIES USING GPR

Efforts were made to carry out GPR studies using the SIR-20 system manufactured by Geophysical Survey Systems Inc. to understand the source zone of the sand ejected from the vents of large craters. The locations of the GPR transects are selected along trench 1 to trench 3 (T1, T2 and T3) for verification of the data and accuracy of the sampling (Figure 2.5). Several sample transects were taken and matched with the sediments exposed in the trenches, as the GPR usage for the liquefaction studies in Kachchh Basin has not been reported and needs to be matched with exposed sediments to get accurate subsurface data. The GPR data revealed three centrally emplaced sand vents, one in the oblong western crater and two in the circular crater (Maurya et al. 2006). The vents are of roughly similar widths and continue vertically downward in the profile. The continuation of vents through the profile suggests the presence of a liquefied source horizon deeper than 6.5 m. Therefore, the data suggest that the large craters in the plain of Banni related to the 2001 Bhuj–Gujarat Earthquake formed due to the liquefaction of sediments located at >6.5 m depth, which is also evident by the absence of any horizon of comparable lithology in the host sediments. The internal geometry of the sand blow craters delineated in the present study appears to validate the model of a typical sand blow by Obermeier (1996). The GPR studies show that the fine-grained clay-rich sediments of the Banni Plains acted as a 'finer-grained cap' above the liquefiable sand horizon at >6.5 m depth. The finer-grained cap suffered only initial liquefaction, as evidenced by the well-preserved stratification of the sand-rich layers within the sand blow craters (see Figure 2.7b).

(a)

(b)

FIGURE 2.4 (a) Large liquefaction grounds with craters are reported from the Rann near Chobari village; one such crater is almost 10 m deep. The liquefaction grounds at this site were perilous for walking. (b) Yellow to brown-coloured sandy sediments are traversed by grey-coloured carbonaceous and coarse to fine liquefaction sand vents in the Great Rann of Kachchh near Chobari village. (*Continued*)

2.4.3 MORPHOLOGY OF LARGE LIQUEFACTION CRATERS AT UMEDPAR

Two of the four liquefaction craters are unusually large and amalgamated; they are named craters 1 and 2. The third one is comparatively smaller, and the fourth one is a dry crater, which mainly released gas (Figure 2.5). The amalgamated large craters were reactivated several times during strong aftershocks in February and March 2001, while crater 3 did not reactivate (Thakkar and Goyal 2004). The excessive rains modified the surface geometry of the couplet craters. Craters 1–3 show visible vents that ejected massive sand and clay on the surface. Clays generally do not liquefy, but it has been found that they occur at the mouth of craters 2 and 3 and do not match the host stratigraphy that is constructed out of the other trench in the area (Thakkar et al. 2012).

(c)

FIGURE 2.4 (CONTINUED)   (c) A thin liquefaction vent of black to grey-coloured fine to coarse sand cuts across the yellow to brown-coloured fluvial sand in the Great Rann of Kachchh near Chobari. The source sand was not found in the shallow pits of ~1 m deep.

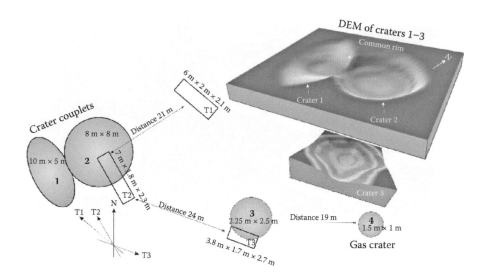

FIGURE 2.5   Sketch map of the site of investigation near Umedpar showing the locations of four craters and the trenches (T1 to T2) across them. The DEMs and dimensions of all craters and the trenches of three craters are mentioned. The distance between both reactivated craters (1 and 2) and non-reactivated crater. (From Thakkar, M. G. et al., *J. Geol. Soc. India*, 79, 367–375, 2012.)

The lateral spread is, however, restricted to topographic lows; the seismic dykes and vents formed due to hydraulic fracturing are generally independent of proximity to a topographic low or slope (Obermeier 1996). However, the topography around Umedpar craters is flat and featureless; therefore, the surface geology, permeability, thickness and spatial relations of sedimentary deposits appear to have influenced the location and mode of ground failure (McCalpin 1996). The shallow water table and proximity to a major basinal fault (KMF) control the geologic environment. GPR profiles across the craters confirm the downward extension of liquefaction vents to more than 6.5 m in crater 1 (Maurya et al. 2006). The two trenches, across craters 2 and 3, and the third one, about 21 m northeast of crater 2, revealed the nature of the host and emplaced sediments.

2.4.4 TRENCHING AT LARGE LIQUEFACTION CRATERS

Massive liquefaction is observed 6.5 km northeast of Umedpar village in the Great Rann of Kachchh (Figure 2.1). The digital elevation model (DEM) of the area with the sketch map shows four large liquefaction craters near Umedpar during the 2001 earthquake and explicitly provides the details regarding their size and relative distances (Figure 2.5). The size of the liquefaction craters exceeds 3 m in diameter, while the lateral dimension of the sand blows is as large as 70–80 m (Table 2.1). The sediments observed in the trench are 18–20 cm of varying composition, which reveals different episodes of liquefaction, respectively, during the main shock and subsequent strong aftershocks. Near the biggest crater, gas ejection is also reported, where its cap was blown 15–16 m away from the crater and the blowing direction of the cap was opposite that of the epicentre (Rajendran et al. 2001). Two craters (1 and 2) have been reactivated and ejected sands, but the third one has not been found to be affected by subsequent aftershocks and earthquakes (Figure 2.5). Therefore, an attempt has been made to characterize the internal geometry of these craters by opening trenches to 2.5–3.0 m in depth. The trench studies have been found useful in understanding the two-dimensional geometry of these craters and characterizing the reactivated and non-reactivated craters, which are considered a modern analogue for palaeoseismic studies.

2.4.4.1 Trench 1

To understand the stratigraphy of the host sediments, a trench measuring 6 × 2 × 2.1 m was opened 21 m northeast of crater 1. The measured log of the western wall of the trench provides the stratigraphy of the host sediments (Figure 2.6a). The trench revealed three major units of clay with variable proportions of sand. The upper unit of 40 cm is composed of silty clay. The middle unit of 90 cm is composed of finely laminated clay. The lower unit of 80 cm is composed of alternate bands of clay and sand. The thin layers of clay and sand are found to be convoluted and traversed by thin vents at the trench wall, showing earthquake deformations (Figure 2.6b). The surface is covered by a thin veneer of coarse sand ejected from the nearby sand blow and spread a considerable distance.

2.4.4.2 Trench 2

A 7 × 1.8 × 2.3 m trench was opened from the centre of crater 2 to beyond its eastern margin (Figure 2.7a). The western wall of the trench indicates that the sediments inside the crater are entirely different than those of trench 1, extensively deformed and remixed because of massive liquefaction during the 2001 Bhuj Earthquake and subsequent strong aftershocks (Figure 2.7b). The ejected and intruding sediments are characterized by the big clasts of sands and clays of the host, as well as of surface sand, liquefied during the 2001 earthquake (Figure 2.7b). The clay and sand intrusions remixed with chunks of sandy and clayey sediments make the geometry of the crater complex due to repeated venting. The dyke is composed of fine to coarse sand with a minor proportion of silt and clay. Numerous large clasts of clay and sand are encased in well-stratified, medium to coarse-grained sands in the basal part of the crater. Clasts of the cap material occur in the central part of the crater (Figure 2.7b). The cross-cutting relationship between the clasts of sands and clays within the dyke represents more than one episode of venting. Wherever a large amount of sand is vented to the surface, downwarping of cap material towards the dyke is observed (McCalpin 1996). In trench 2, host layers of laminated sand and silt are downwarped (Figure 2.7b). The other side of the trench wall clearly shows the downwarping event in more than one layer. The middle unit of the host sediments of crater 2 shows vertical minuscule vents, miniature folding and convolutions in finely laminated silt, clay and fine sand layers (Thakkar and Goyal 2004).

2.4.4.3 Trench 3

Crater 3 is comparatively small, through which very thick and immobile sticky clay came out. It oozed only water for some days after the main shock. A 3.8 × 1.7 × 2.7 m east-west trending

TABLE 2.1

Measurements of Liquefaction Features Produced during the 26 January 2001 Earthquake in Kachchh

Name of the Trench	North Latitude	East Longitude	Type of Deformation Structure	Arial Distance from the Epicentre (in km)	Direction from the Epicentre	Maximum Thickness of Ejected Sand (in cm)	Geomorphic Setting	Original Sediments
Northwest of Amarsar (A-1)	23°25′26.5″	70°15′28.8″	Sand blow	6.25	N 64° W	16–18	Boundary of Rann and Banni Plains	Clay and fluvial sand
Northwest of Amarsar (A-2)	23°26′10.1″	70°15′27.1″	Sand blow	7.25	N 58° W	7–8	Rann	Hard, swelling clay + desiccation cracks
Northwest of Baniari (B-1)	23°24′24.8″	70°09′52.1″	Sand blow and ground cracks	16.1	N 83.5° W	6–7	Banni Plains	Very hard clay (alluvium)
Northwest of Baniari (B-2)	23°24′21.2″	70°09′51.4″	Liquefaction crater and sand blow	16.1	N 84° W	28–30	Banni Plains	Very hard clay (alluvium)
Northwest of Baniari (B-3)	23°24′43.5″	70°09′39.6″	Liquefaction ground	16.25	N 85° W	32–35	Rann	Silty clay of Rann
North of Chobari (C-1)	23°34′07.8″	70°20′44.1″	Liquefaction ground	18.6	N 9° E	6–8	Rann	Silt and clay
North of Chobari (C-2)	23°33′55.8″	70°20′57.9″	Small sand blow	17.75	N 10° E	3–4	Boundary of Rann and Banni Plains	Hard clay and sand
North of Chobari (C-3)	23°35′03.4″	70°20′53.8″	Liquefaction ground	20.4	N 8° E	8–12	Rann	Silty clay
Northeast of Umedpur (U-1)	23°29′17.5″	69°54′54.6″	Liquefaction ground and big craters		N 64° W	18–20	Rann	Silty clay
Bhrudiya	23°57′308″	70°41′320″	Flexture	–	–	–	Banni Plains	–
Vighakot 1[a]	24°27′483″	69°33′316″	Sand blow craters	–	–	–	Great Rann	–
Vighakot 2[a]	24°27′166″	69°34′250″	Sand blow craters	–	–	–	Great Rann	–

Source: Thakkar, M. G., and Goyal, B., *Curr. Sci.*, 87 (6),811–817, 2004.

[a] Vighakot is located at Indo-Pak international border at the fringe of the Great Rann of Kachchh and sand dunes of Thar Desert; it is also an ancient revenue collection centre on the Bank of Nara channel, a distributory of Indus River. The 2001 earthquake liquefaction craters are found at this location.

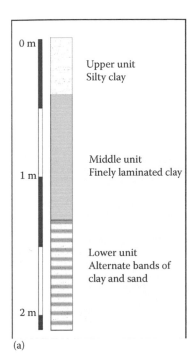

(a)   (b)

FIGURE 2.6   (a) Litholog of trench 1 exhibiting typical host sediments at the Banni–Rann interface area, where all four craters are found. (b) View of a small section of a wall of trench 1 showing thinly bedded white sands with clay lamina. They show seismite-type convolution and a small network of veins, indicating that intensive liquefaction occurred in several sand layers.

trench was opened across this crater. The host sediments of the southern wall of trench 3 are mainly non-stratified fine sand. The source zone of ejected sands and clay capping was not encountered, but it revealed a peculiar funnel-shaped geometry of the entire crater and vent (Figure 2.8a,b). The upper part of the trench exhibits 1 m thick, hard clay in a funnel-shaped opening and was ejected first and later formed a cap during the low-energy phase. The vent shows a fining upward sequence (Figure 2.8c). Most of the liquefied sediments on the surface are fine to coarse yellow sands, similar to those of the vent. The vent exposed on the southern wall is 1.5 m wide, gradually decreasing to 20 cm at the bottom, forming the shape of a funnel. The vent has sharp contact with the host strata, which is very fine, uniformly composed, non-stratified silty sand. The lower part of the vent below the sticky clay is dominated by vertical stratification with bands of varied sands (Figure 2.8d). Four distinct vertical layers of sand are noticeable on the southern wall of the trench. Over it is a 10 cm thick, very fine brown sand layer in a half-lobe shape, pinching towards the centre of the vent. Two subsidiary vents of 1–1.5 mm thickness originate from this layer (Figure 2.8a,b). Overlying it is a 30 cm brown, silty clay layer studded with clasts of dark brown sand. The sand clasts are about 10 cm in diameter and are derived from host sediments.

The cone-shaped dyke extends to the north as a long irregular fissure with two lenses exposed in the northern wall (Figure 2.9a,b). The lenses are 10 and 25 cm thick and are located at heights of 70 and 130 cm from the bottom of the trench. The upper lens contains very coarse and gritty, whitish to brown sand in its lower part, while a uniform deposit of clay and fine sand with an irregular grey-coloured upper boundary is observed in the middle part (Figure 2.9c). The lower lens is divisible into four sand layers, which vary in colour and grain size. The long patches of clay remnants within it indicate initial liquefaction of the clay.

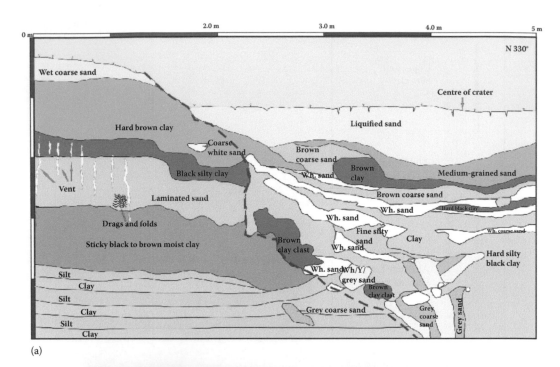

(a)

(b)

FIGURE 2.7   (a) Log of the western wall of trench 2 across the main crater (reactivated). It specifies the middle portion of the crater where the sediments within the bowl-shaped crater are deformed and remixed and show collapsed blocks of the cap material. The hard silty clay is confined to the core, surrounded by a variety of ejected sands. Coarse whitish sand and grey sand form big clasts within the host sediments, which are mainly silt and clay, showing drags, microfolds and microvents in laminated sediments. (b) Photograph of the northern wall of trench 2 showing collapsed blocks of hard cap sediments. The blocks also contain coarse to fine sands that were ejected during the earlier episodes of liquefaction. Note that the complex mixture of liquefaction sands and clays with minor convoluted (shown by the arrow) beds indicates reactivation of the vent many times during several aftershocks. (From Thakkar, M. G. et al. *J. Geol. Soc. India*, 79, 367–375, 2012.)

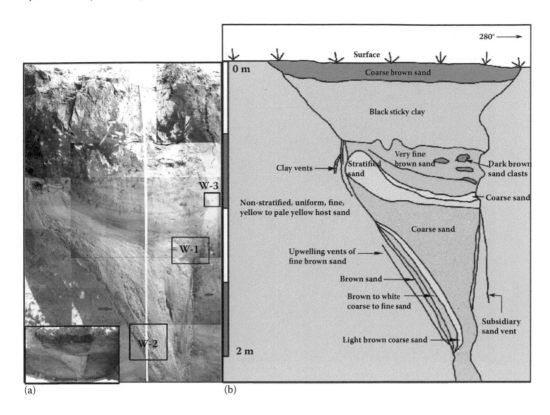

FIGURE 2.8 (a and b) Mosaic picture of the southern wall of trench 3 across the non-reactivated crater 3. The inset picture portrays the entire vent in a cone shape, while the enlarged part illustrates details of the ejected sand and clay. (From Thakkar, M. G. et al., *J. Geol. Soc. India*, 79, 367–375, 2012.)   (*Continued*)

2.5 DISCUSSION

The liquefaction associated with the 26 January 2001 Bhuj–Gujarat intraplate earthquake generated many patterns and dimensions in the soft sediments of Great Rann, the plain of Banni and the coastal plains of the Kachchh Basin. These features are considered ideal evidence of an intraplate earthquake in the post-instrumental era. Further, instrumental records in this period are correlated with the well-documented geographical and stratigraphical changes that occurred in the meizoseismal area. Pre- and post-earthquake satellite image-based studies help to locate the overall sites of earthquake-induced liquefactions where massive liquefaction and liquefaction-related craters have been found and deep trenches are opened up. Attempts have been made to understand the sand sources of the liquefaction craters and vents at Umedpar using GPR; however, only up to 6.5 m depth has been encountered due to the higher salinity and availability of antenna frequency (Maurya et al. 2006). Therefore, the source depth has not been decisively found at the site near Umedpar. In the shallow pit study at this site, the varied nature of the sediments and structures has been identified. It was also observed that the same craters have been repeatedly reactivated during the strong aftershocks for 3 years, hence allowing us to anticipate the characteristic geometry of the craters if the sections are taken. Three trenches have been opened up for the study of the two-dimensional crater geometry of the non-reactivated and activated craters. The liquefied sediments suggest heterogeneous sources with a discrepancy in depth. The couplet crater oozed water and sediments even during the strong aftershocks in February and March 2001; conversely, the other two craters did not. Trenches opened across craters 2 and 3 show variations in sediments, from highly impermeable sticky mud to very coarse sand. The detailed trench studies indicate that trench 2 (crater 2) shows the complex geometry of the reactivated crater, while trench 3 (crater 3)

(c)

(d)

FIGURE 2.8 (CONTINUED) Note that a window (W-1) in this figure is enlarged in (d) for specific observation of the fining upward sequence in the liquefaction deposits. Window 2 is enlarged in (c), which distinguishes host and vent sediments. Note the thin vent running down but parallel to the vent wall. Clay and sand clasts within the vent are indicated by green and yellow arrows, respectively. Millimetre-sized vents are shown by red arrows on both sides of the funnel-shaped vent.

(a)

FIGURE 2.9 (a) Mosaic picture of the northern wall of crater 3 showing a very thin vent and two sand lenses with a fining upward sequence in each of them. The cap of the vent is made up of black, high-density clay. Note that a thin vein runs upward but vanishes at about halfway up the trench. Also note that the host sediments in this trench are uniform, non-stratified and fine, unlike the typical host stratigraphy of the site.   (*Continued*)

revealed the nature and geometry of a non-reactivated crater. The collapsed blocks of the cap, as well as surface liquefaction layers, found within the core of crater 2 suggest the effects of subsequent strong aftershocks and monsoons. The presence of clay and silty sand at about 1–1.5 m depth suggests that initially clay and silty sand were liquefied. The liquefied clay in the core of crater 2 and the sand blow deposits on the surface suggest liquefaction during the initial moments due to intense pore water pressure. The wider dyke in crater 2 is interpreted to have been formed by highly mobile sediments, which flowed into fissures between the blocks of the silty clay caps. Subsequently, the fissures opened up as the blocks shifted laterally in response to back-and-forth shaking. Further, the sharp upturn of the host clay and silt layers in trench 2 suggests repeated liquefaction in the crater. In host sediments of trench 2, the lamination and convolutions, flexures, microfolds and small vertical vents not reaching the surface indicate that they were not formed in a single episode. The clay layers in the host stratigraphy in trench 2 became instrumental in acquiring pore pressure in less time, creating an environment for repeated liquefaction. The absence of collapsed blocks, as well as the presence of distinct homogenous zones and intact conical crater geometry, in trench 3 is an example of non-reactivated crater geometry. The presence of clay clasts, graded sand deposits and vertical laminations near the walls of the vents suggests continuous outpouring of liquefied sediments from the same vent. Periods of quiescence provided time for sorting of the liquefied sands, forming gradation and fining upward sequences. The well-sorted

(b)

(c)

FIGURE 2.9 (CONTINUED)   (b) Close view of the sand lens found near the top of a thin vent on the northern wall of trench 3. Note the ideal fining upward sequence in the lens, having finer sands and clay on the top and coarser at the bottom. (c) Close-up of a lower lens along the thin vent of trench 3. The separation of sand particles is quite distinctive. The fining upward sequence is also visible within the lens. (From Thakkar, M. G. et al., *J. Geol. Soc. India*, 79, 367–375, 2012.)

and fining upward sequence and the uniform host sediment suggest the non-reactivation of the crater. These craters are located about 41 km away from the epicentre of the 2001 Bhuj Earthquake and about 8 km to the north of the KMF. The proximity to the KMF and the availability of shallow groundwater and a liquefiable sand source played a major role in the size and dimensions of the liquefactions (Thakkar and Goyal 2004). The strong reaction of seismic shocks, lithological variations in host sediments and widespread liquefaction with various patterns in the reactivated and non-reactivated crater geometries can be used as a modern analogue for understanding the palaeoseismicity of the tectonically active intraplate areas.

Three trenches north of Chobari village reveal extensive sand blows in the Rann near Chobari. These sites are close to the assumed surface projection of the seismogenic fault plane of the 2001 Bhuj Earthquake. Chobari is located on a major geomorphic lineament that separates the Rann of Kachchh and the Wagad Uplift. Possible palaeoliquefaction sand suggests that the site shows great potential for comparative study of various earthquakes in Kachchh. Unusual liquefaction craters are reported at the Indo-Pak border, northeast of Vigakot, which is 160 km northwest of the epicentre. At the site of the 1819 Kachchh earthquake, fluvial sand and estuarine deposits of the old mouth of the Nara and Hakra channels of the Indus River became the source of liquefaction. Thus, if the source sand is abundant and the area is close to the active fault system, huge liquefaction may occur even if it is far from the epicentre (Rajendran et al. 2001). It has also been concluded that for the liquefaction features of the 2001 Bhuj Earthquake, the proximity to the active fault system and the availability of shallow groundwater and a liquefiable sand source have played a major role in the size and dimension of the liquefaction.

2.6   CONCLUSION

The main conclusion drawn, based on the observed liquefaction in the Bhuj (Gujarat) Earthquake, is that the present liquefaction features and their characteristics can be useful in assessing the magnitudes and epicentre location of older events.

Shallow trenching around Umedpur, Chobari, Baniyari and Amarsar villages reveals a comparative picture of the dimension of liquefaction at the epicentral zone. The thickness of the sand blow deposits was found to decrease with an increase in epicentral distance in some cases, but this relation does not hold in some other sites in the epicentral area. It is not always true in major active fault zones because huge liquefaction is reported on the plain of Banni along the KMF zone and in the Great Rann of Kachchh along the Allah Bund Fault.

The characteristics of the reactivated and non-reactivated craters in the active intraplate seismic zones depend on the presence or absence of collapsed clasts of host sediments, cap material and liquefied sediments in the vent. The upturns in the cap material at different junctures within the vent indicate more than one episode of seismic blows.

The seismic deformation features, like lamination and convolutions, flexures, microfolds and small vertical vents not reaching the surface, are characteristic of a reactivated crater.

Non-reactivated craters reflect the tranquil nature of deposition, like fining sequence and laminations within the vent, as well as uninterrupted host and crater sediments.

Liquefaction patterns in an intraplate earthquake area with many active faults depend largely on terrain characteristics, proximity to the seismogenic faults – irrelevant to the fault that is responsible for the present earthquake, season and depth of the groundwater table, intensity of the earthquake and amount of liquefiable material.

As far as the liquefaction-related hazard is concerned, in an intraplate earthquake zone with a number of seismogenic faults, the areas with thin alluvial cover and enormously thick inland playas must be considered most vulnerable for strategic planning.

ACKNOWLEDGEMENTS

The author is grateful to the Department of Science and Technology, New Delhi (Project Nos. DST/23/(257) SU/2001, SR/S-4/ES-21/Kachchh Window/P3/2005 and DST/SR/S4/ES-TG/02-2008), for providing financial assistance to work in the field and laboratory. He is also thankful to C. P. Rajendran, Kusala Rajendran and L. S. Chamyal for their valuable guidance all throughout the work, since the 2001 Bhuj–Gujarat Earthquake and subsequent events and episodes of field and classroom discussions. The author also acknowledges M. P. Tuttle, Memphis, Tennessee, and

J. P. McCalpin, GEO-HAZ Consulting Inc., Crestone, Colorado, for their edifying field teachings. Lastly, the formatting and figures drawn by Bhanu Goyal and Mamata Ngangom cannot be forgotten.

REFERENCES

Biswas, S. K. (1981). Basin framework, palaeoenvironment and depositional history of the Mesozoic sediments of Kachchh, Western India. *Q. J. Geol. Min. Met. Soc. India*, 53, 56–85.

Biswas, S. K. (1982). Rift basins in western margin of India and their hydrocarbon prospects with special reference to Kutch basin. *Am. Assoc. Petr. Geol. Bull.*, 66, 1467–1513.

Biswas, S. K. (1987). Regional tectonic framework, structure and evolution of the western marginal basins of India. *Tectonophysics*, 135, 307–327.

Biswas, S. K. (2005). A review of structure and tectonics of Kutch basin, Western India, with special reference to earthquake. *Curr. Sci.*, 88(10), 1592–1600.

Biswas, S. K., Despande, S. V. (1973). Mode of eruption of Deccan traps lavas with special reference to Kutch. *J. Geol. Soc. India*, 14, 134–141.

Maurya, D. M., Goyal, B., Patidar, A., Mulchandani, N., Thakkar, M. G., Chamyal, L. S. (2006). GPR imaging of two large sand blow craters related to 2001 Bhuj earthquake, Western India. *J. Appl. Geophys.*, 60, 142–152.

Maurya, D. M., Thakkar, M. G., Chamyal, L. S. (2003). Implications of transverse fault system on tectonic evolution of Mainland Kachchh, Western India. *Curr. Sci.*, 85(5), 661–667.

McCalpin, J. P., ed. (1996). *Paleoseismology*. 2nd ed. New York: Academic Press, p. 588.

Merh, S. S., Patel, P. P. (1988). Quaternary geology and geomorphology of the Rann of Kutch. In *Proceedings of the National Seminar on Recent Quaternary Studies in India*, eds. M. P. Patel, N. D. Desai M. S. Univ. Baroda, pp. 377–391.

Ngangom, M., Thakkar, M. G., Bhushan, R., Juyal, N. (2012). Continental-marine interaction in the vicinity of the Nara River during the last 1400 years from the western Great Rann of Kachchh, Gujarat. *Curr. Sci.*, 103(11), 1339–1342.

Obermeier, S. F. (1996). Use of liquefaction-induced features for paleoseismic analysis—An overview of how seismic liquefaction features can be distinguished from other features and how their regional distribution and properties of source sediment can be used to infer the location and strength of Holocene paleoearthquakes. *Eng. Geol.*, 44, 1–76.

Oldham, R. D. (1926). The Cutch (Kachh) earthquake of the 16th June, 1819 with a revision of the great earthquake of the 12th June, 1897. *India Geol. Surv. Mem.*, 46, 71–147.

Powell, C. A. (1979). Speculative tectonic history of Pakistan and surroundings; some constraints from the Indian Ocean. *Geodynamics of Pakistan: Pakistan Geological Survey*, 5–24.

Rajendran, C. P., Rajendran, K. (2001). Characteristics of deformation and past seismicity associated with the 1819 Kutch earthquake, northwestern India. *Bull. Seismol. Soc. Am.*, 91(3), 407–426.

Rajendran, C. P., Rajendran, K. (2002). Historical constraints on previous seismic activity and morphologic changes near the source zone of the 1819 Rann of Kachchh earthquake: Further light on the penultimate event. *Seismol. Res. Lett.*, 73(4), 470–479.

Rajendran, C. P., Rajendran, K., John, B. (1998). Surface deformation related to the 1819 Kachchh earthquake: Evidence for recurrent activity. *Curr. Sci.*, 75(6), 623–626.

Rajendran, C. P., Rajendran, K., Thakkar, M. G., Goyal, B. (2008). Assessing the previous activity at the source zone of the 2001 Bhuj earthquake based on the near-source and distant paleoseismological indicators. *J. Geophys. Res.*, 113, 1–17.

Rajendran, K., Rajendran, C. P., Thakkar, M. G., Gartia, R. (2002). The sand blows from the 2001 Bhuj earthquake reveal clues on past seismicity. *Curr. Sci.*, 83(5), 603–610.

Rajendran, K., Rajendran, C. P., Thakkar, M. G., Tuttle, M. P. (2001). The 2001 Kachchh (Bhuj) earthquake: Coseismic surface features and significance. *Curr. Sci.*, 80, 1397–1405.

Seed, H. B. (1979). Soil liquefaction and cyclic mobility evaluation for level ground during earthquakes. *J. Geotech. Eng. Div.*, 105(2), 201–255.

Seed, H. B., Idriss, I. M., Arango, I. (1983). Evaluation of liquefaction potential using field performance data. *J. Geotech. Eng.*, 109, 458–482.

Singh, R. P., Bhoi, S., Sahoo, A. K., Raj, U., Ravindran, S. (2001a). Surface manifestations after the Gujarat earthquake. *Curr. Sci.*, 81, 164–166.

Singh, R. P., Sahoo, A. K., Bhoi, S., Kumar, M. G., Bhuiyan, C. S. (2001b). Ground deformation of the Gujarat earthquake of 26 January 2001. *J. Geol. Soc. India*, 58, 209–214.

Sladen, J. A., D'Hollander, R. D., Krahn, J. (1985). The liquefaction of sands, a collapse surface approach. *Can. Geotech. J.*, 22, 564–578.

Thakkar, M. G., Goyal, B. (2004). On relation between magnitude and liquefaction dimension at the epicentral zone of 2001 Bhuj earthquake. *Curr. Sci.*, 87(6), 811–817.

Thakkar, M. G., Goyal, B., Maurya, D. M., Chamyal, L. S. (2012). Internal geometry of reactivated and non-reactivated sandblow craters related to 2001 Bhuj earthquake, India: A modern analogue for interpreting paleosandblow craters. *J. Geol. Soc. India*, 79, 367–375.

Thakkar, M. G., Maurya, D. M., Raj, R., Chamyal, L. S. (2001a). Morphotectonic analysis of Khari River Basin of Mainland Kachchh: Evidence for neotectonic activity along transverse fault. *Bulletin of the Indian Geological Association of Structure and Tectonics of Indian Plate, Chandigarh*, 205–220.

Thakkar, M. G., Maurya, D. M., Raj, R., Chamyal, L. S. (2001b). Quaternary tectonic history and terrain evolution of the area around Bhuj, Mainland Kachchh, Western India. *J. Geol. Soc. India*, 53(5), 601–610.

Tuttle, M. P., Hengesh, J., Tucker, K. B., Lettis, W., Deaton, S. L., Frost, J. D. (2002). Observations and comparisons of liquefaction features and related effects induced by the Bhuj earthquake. *Earthquake Spectra*, 18(Suppl. A), 79–100.

Tuttle, M. P., Johnston, A., Rajendran, C., Rajendran, K., Thakkar, M. (2001). Liquefaction features induced by the Republic Day earthquake, India, and comparison with features related to the 1811–1812 New Madrid, Missouri, earthquake. *American Geophysical Union, Transactions*, EOS. 82(100).

Tyagi, A. K., Shukla, A. D., Bhushan, R., Thakkar, P. S., Thakkar, M. G., Juyal, N. (2012). Mid-Holocene sedimentation and landscape evolution in the western Great Rann of Kachchh, India. *Geomorphology*, 151–152, 89–98.

Youd, T. L., Perkins, D. M. (1978). Mapping liquefaction-induced ground failure potential. *J. Geotech. Eng. Div.*, 104(4), 433–446.

3 Earthquakes and Medical Complications

Scarlet Benson, Laura Ebbeling, Michael J. VanRooyen and Susan A. Bartels

CONTENTS

3.1 INTRODUCTION

Earthquakes frequently impact populous urban areas with poor structural standards, resulting in high death rates and mass casualties with numerous traumatic injuries. These injuries and medical complications extend across multiple organ systems, including musculoskeletal, renal and neurologic, as well as mental health, and can be classified as acute (in the first post-earthquake week), subacute (in the first post-earthquake month) or long term (in the first post-earthquake year). Provision of primary care for chronic health problems must also be continued post-earthquake, which can be challenging if the physical environment and healthcare infrastructure have been interrupted. Furthermore, rescuers and healthcare providers can suffer from both earthquake-related injuries and mental health disorders as a result of their rescuing efforts.

Major earthquakes often occur in densely populated areas of the world and cause many traumatic injuries in a short period of time without advance warning, making them some of the most catastrophic natural disasters. More than a million earthquakes occur worldwide each year, averaging approximately two earthquakes every minute (Naghii 2005). In the past decade alone, disasters have caused more than 780,000 deaths, with earthquakes accounting for nearly 60% of all disaster mortalities (https://www.unisdr.org/archive/12470) (Table 3.1). In addition to lives lost, nearly 2 billion people are estimated to have been directly affected by earthquakes in the past decade. It is likely that earthquakes will continue to be a major threat since 8 out of the 10 most populous cities in the world are located on earthquake fault lines (http://www.unisdr.org/files/20108_mediabook .pdf) (Figure 3.1). The extent of injury caused by an individual earthquake will depend on the magnitude of the earthquake; the epicentre's proximity to large, urban centres; the degree of earthquake preparedness; and the extent to which mitigation measures have been implemented (Naghii 2005).

TABLE 3.1
Earthquakes with >1000 Deaths Since the Year 2000

Location	Date	Magnitude	Estimated Death Toll
India (Gujarat)	2001 January 26	7.6	20,085
Afghanistan (Hindu Kush)	2002 March 25	6.1	1,000
Algeria	2003 May 21	6.8	2,266
Iran (Bam)	2003 December 26	6.6	31,000
Sumatra	2004 December 26	9.1	227,898
Indonesia (Sumatra)	2005 March 28	8.6	1,313
Pakistan	2005 October 8	7.6	86,000
Indonesia	2006 May 26	6.3	5,749
China (Sichuan)	2008 May 12	7.9	87,587
Indonesia (Sumatra)	2009 September 30	7.5	1,117
Haiti	2010 January 12	7.0	316,000
China (Qinghai)	2010 April 13	6.9	2,698
Japan	2011 March 11	9.0	20,352

Source: Adapted from Bartels, S. A., and VanRooyen, M. J., *Lancet*, *379*(9817), 748–757, 2012.

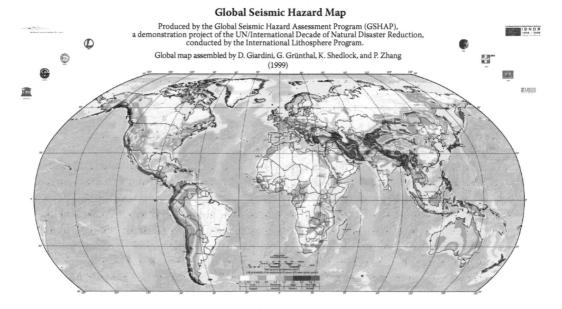

FIGURE 3.1   World map highlighting degree of seismic hazards. (From the Global Seismic Hazard Assessment Program, http://static.seismo.ethz.ch/GSHAP/global/.)

The traumatic injuries and medical complications of large earthquakes are compounded by significant damage to hospitals and healthcare facilities and by disruption of healthcare infrastructure. Massive earthquakes may result in casualty rates which range from 1% to 8% of the at-risk population (Wood and Cowan 1991). The reported ratios of death to injury vary depending on the earthquake and depending on how *injury* is defined. However, across multiple studies the ratio of death to injury appears to approximate 1:3 (Alexander 1996). Following an earthquake, there are 'immediate deaths', which are fatalities resulting from catastrophic injuries at the time of the earthquake (Schultz et al. 1996). Severe brain and spinal cord injuries are a common cause of such immediate deaths. Other earthquake victims die within the first few hours of the quake, from injuries such as subdural hematomas, liver or splenic lacerations or pelvic fractures. Mortality within this group can be significantly reduced with prompt medical and surgical intervention. The patients with the greatest potential for survival, however, are those who die more slowly over days to weeks from crush injuries, exsanguination and multisystem organ failure (Schultz et al. 1996).

Experience has suggested that injured victims usually seek emergency medical care in the first 3–5 after the earthquake (Naghii 2005), and most patients who require hospitalization are admitted in the first 6 days (Schultz et al. 1996). Thus, when there are large numbers of earthquake casualties, the greatest demand for healthcare providers is within the first week. However, medical equipment and supplies, as well as healthcare responders, are needed beyond the first week to reinstate the pre-hospital healthcare system, to treat delayed complications and less acute injuries and also to resume management of pre-existing chronic illnesses. Although primary care medical complaints increase in frequency in the days after an earthquake, care of acutely injured victims will usually predominate for several weeks following the disaster (L. Zhang et al. 2011; Tan et al. 2012).

There is significant overlap for the time periods in which various earthquake-related medical complications occur. However, for the purposes of this chapter, the medical complications of earthquakes have been divided into the three time periods in which those particular complications are more likely to be brought to medical care: (1) within the first week, (2) within the first month and (3) within the first year. Earthquake-induced complications of chronic diseases and medical complications from earthquakes will also be discussed (Figure 3.2).

3.2 TRAUMATIC INJURIES AND MEDICAL COMPLICATIONS IN THE FIRST WEEK POST-EARTHQUAKE

3.2.1 MUSCULOSKELETAL INJURIES

Among earthquake victims, the most common musculoskeletal injuries are lacerations (65%), fractures (22%) and soft tissue contusions or sprains (6%) (Mulvey et al. 2008). Multiple bone fractures and comminuted fractures commonly result from earthquake crush injuries (T. W. Chen et al. 2009), and the lower extremities are more commonly affected. Open fractures and fractures complicated by vascular injuries, nerve injuries, compartment syndrome and wound infections are frequent (Liu et al. 2010). Pelvic fractures are also relatively common in the post-earthquake setting (Tahmasebi et al. 2005).

Earthquakes are estimated to cause crush injuries in 3%–20% of victims (Gonzalez 2005), and the lower extremity is again most commonly affected (Briggs 2006; Ardagh et al. 2012). In crush injuries, compartment syndrome is a frequent concern and there is a high incidence of sepsis, disseminated intravascular coagulation and adult respiratory distress syndrome (Hosseini et al. 2009). Within the first 5 days of an earthquake-related crush injury, hypovolaemia and hyperkalaemia are the major causes of death, and multiorgan failure from sepsis is the most common cause of delayed mortality (Oda et al. 1997).

Fasciotomy for crush injury victims is an area of considerable controversy. Proponents of fasciotomy argue that crush-injured limbs should be actively decompressed in an attempt to improve circulation and reverse muscle necrosis (Duman et al. 2003; Quan et al. 2009). Other experts advise against fasciotomy for crush injuries, largely due to the risk of infection (Better 1989, 1990; Better and Stein 1990; Michaelson 1992; Finkelstein et al. 1996). In some case series, fasciotomies have

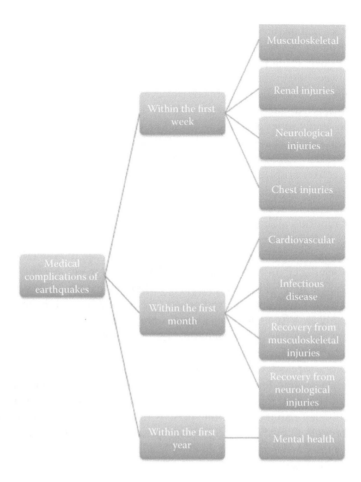

FIGURE 3.2   Medical complications of earthquakes organized according to the post-earthquake time period in which patients usually seek medical care.

been complicated by wound infections in up to 81% of patients (Gunal et al. 2004). Furthermore, fasciotomies are a significant risk factor for sepsis, which in turn is a significant risk factor for death (Sever et al. 2002a). Others believe that fasciotomy is necessary only under certain circumstances. Michaelson (1992), for instance, believes that fasciotomy should only be performed for crush injuries that are believed to be at high risk for becoming gangrenous. Other experts argue for the measurement of compartment pressures, using a pressure in excess of 30–50 mmHg as an indication for fasciotomy (Finkelstein et al. 1996). While logical, this recommendation may not be feasible in a mass casualty situation since measuring pressures in all injured compartments can be time consuming and impractical (Duman et al. 2003). Amputation of limbs in crush syndrome patients is likewise controversial. Necrotic tissue can release large amounts of myoglobin, potassium and thromboplastin, rendering retention of a non-salvageable limb potentially life threatening (Malinoski et al. 2004). For this reason and also because necrotic tissue can serve as a nidus for infection, some experts recommend early amputation in those cases where amputation is inevitable (Malinoski et al. 2004). Other experts disagree, however, claiming that amputation should play a limited role in the management of crush syndrome since it often does not improve outcomes (Gonzalez 2005) and has been associated with death (Sever et al. 2002a). A 2003 consensus statement concluded that there is no evidence to support prophylactic amputation to prevent crush syndrome, and that even severely crushed limbs can recover to gain full function (Porter and Greaves 2003). Porter and Greaves therefore believe that amputation should be reserved for those infrequent situations where a limb barely remains attached or where a patient's life is in danger due to limb entrapment. In reality, the correct approach to amputation will likely depend on the specific setting and local context.

3.2.2 RENAL INJURIES

When pressure is released from a crush-injured body part, calcium, sodium and water diffuse into the necrotized muscle and there is release of potassium, lactic acid, myoglobin and creatine phosphokinase (Demirkiran et al. 2003). The resultant intratubular obstruction by myoglobin and uric acid, in combination with impaired kidney perfusion, can result in acute kidney injury (Sever et al. 2006). Ensuing acute renal failure, with its concomitant hyperkalaemia, acidosis and hypovolaemic shock, carries an extremely high mortality. For this reason, crush injury is one of the most frequent causes of mortality after earthquakes (Vanholder et al. 2000). Given the prevalence of crush injuries following earthquakes and their associated high mortality, experts recommend the use of urine dipsticks to check for myoglobin during triage (Yoshimura et al. 1996; Sever et al. 2002a). Reported rates of renal failure following earthquakes vary from 0.5% to almost 25% of casualties (Najafi et al. 2009).

According to the International Society of Nephrology's Renal Disaster Relief Task Force (RDRTF) (Vanholder et al. 2008), early and vigorous administration of intravenous fluids is of critical importance in preventing and treating renal failure among crush injury victims (Better 1990, 1993; Noji 1992; Gunal ct al. 2004; Yokota 2005). All such patients should be aggressively fluid resuscitated even if their initial vital signs are normal (Jagoginski et al. 2010). Because release of pressure on a crushed extremity can cause acute hypovolaemia and reperfusion metabolic derangements (Briggs 2006), some experts recommend that rehydration be initiated before the patient is extricated from the rubble (Noji 1992; Gonzalez 2005). Pre-extrication intravenous fluids may be particularly important for crush injury victims who have been trapped for more than 4 hours and for those with abnormal neurological or vascular exams. If pre-treatment is not feasible, some recommend tourniquet application to the crushed limb for up to 2 hours, until intravenous fluids can be initiated (Briggs 2006; Jagoginski et al. 2010). However, other experts believe that there is no credible evidence to support tourniquet use in the prevention of reperfusion injury, and it is generally accepted that tourniquets should never be applied for longer than 2 hours (Porter and Greaves 2003).

Alkalinization, with the addition of 50 mEq of sodium bicarbonate to every second or third litre of hypotonic saline to maintain a urinary pH of at least 6.5 (Porter and Greaves 2003; Sever et al. 2006), is recommended to prevent deposition of myoglobin and uric acid within the renal tubules. If urine output can be maintained above 20 mL/hour, the RDRTF also recommends adding 50 mL of 20% mannitol to each litre of intravenous fluid (1–2 g/kg of body weight per day to a total of 120–200 g) (Porter and Greaves 2003; Sever et al. 2006). Mannitol serves as an osmotic diuretic and a free radical scavenger (Jagoginski et al. 2010) and is thought to protect injured muscle and reduce the leakage of nephrotoxins such as myoglobin (Better 1999). Urine output should ideally be maintained at >300 mL/hour, a goal which may require administration of up to 12 L of fluid per day (Sever et al. 2006). This protocol should be continued until myoglobinuria has resolved both clinically and biochemically, which is usually day 3 (Sever et al. 2006). Some experts recommend administering acetazolamide 250 mg as an adjunct if the blood pH is >7.45 (Michaelson 1992).

Electrolyte disturbances are common in crush-injured patients, and hyperkalaemia, hypocalcaemia and oliguria can quickly precipitate cardiac arrhythmias and cardiac arrest (Jagoginski et al. 2010). Consequently, potassium-containing intravenous solutions should not be administered and potassium levels should ideally be checked three or four times daily for the first few days (Sever et al. 2006). If such frequent potassium checks are not feasible, EKGs can be used as a screening tool to identify hyperkalaemia-associated changes (Yoshimura et al. 1996). Patients who require prolonged transfer to medical care and those who must wait for available resources in order to receive dialysis should be treated empirically with potassium binding resins such as sodium polystyrene sulphonate (Sever et al. 2006; Jagoginski et al. 2010).

Creatine kinase concentrations above 5000 U/L are statistically the best markers for the development of acute renal failure (Huang et al. 2002; Jagoginski et al. 2010). It is estimated that approximately half of all patients with crush syndrome develop acute renal failure, and approximately half of all those with acute renal failure require dialysis (Briggs 2006). Since crush injuries are so common after severe earthquakes, dialysis services can become overwhelmed. If and when it becomes necessary to prioritize who will get dialysis, highest priority should be given to those with oliguric renal failure, and hyperkalaemic patients should be medically managed until they can receive dialysis (Amundson et al. 2010).

For those who require renal replacement therapy, there are several dialysis options. Peritoneal dialysis does not require electricity or tap water but is difficult to use in patients with abdominal or chest injuries, requires relatively large quantities of sterile dialysate and may cause peritonitis when used under non-hygienic field conditions (Sever et al. 2006). With continuous renal replacement therapy, only one patient can be treated per machine, and experienced personnel, electricity and large volumes of substitute fluid are required (Sever et al. 2006). Some experts advocate for continuous arteriovenous haemofiltration (CAVH) in mass casualty situations located in rural areas because it is simple and does not require electricity, pumps or delivery systems (Better 1993). Intermittent haemodialysis offers the advantage of being able to treat several patients each day with the same dialysis machine and is also more efficient at removing potassium (Sever et al. 2006). The disadvantages of intermittent haemodialysis include the need for technical support, electricity and water supplies (Sever et al. 2006). Twice and even three times daily dialysis may be needed, and some patients may require more than one dialysis modality (Sever et al. 2002b). The average duration of dialysis for crush injury patients is 13–18 days (Sever et al. 2006).

3.2.3 Neurological Injuries

Patients with neurological injuries also present in the first week after an earthquake, and spinal trauma accounts for many of the earthquake-induced neurological injuries. Thoracolumbar spinal injuries are most common (T. Li et al. 2009; Karamouzian et al. 2010), followed by isolated lumbar spine injuries (Rathore et al. 2007; T. Li et al. 2009). Multilevel injuries are also frequently observed (22%–29.5%) (R. Chen et al. 2009; T. Li et al. 2009). Vigilance is essential in evaluating patients with spinal trauma, not only because the spine can be affected at different levels but also because many of these patients have concurrent non-spinal injuries that can be overlooked if a thorough exam is not conducted (R. Chen et al. 2009).

Spine fractures and fracture dislocations are approximately equal in frequency (Rathore et al. 2007). Of earthquake-induced spine fractures, burst fractures are the most common (49%–55%) (R. Chen et al. 2009), followed by compression fractures (approximately 33%) (T. Li et al. 2009) and then wedge fractures (13%) (Karamouzian et al. 2010). More than half of the patients with earthquake-related spine injuries (53.8%) have some form of neurologic deficit, and thoracic injuries were responsible for the majority of these (R. Chen et al. 2009). Lumbar spine fractures, in contrast, rarely resulted in neurologic deficits. Patients with earthquake-induced spinal injuries might have better outcomes than patients with similar injuries from other types of trauma, and this difference may be explained by differences in velocity of the traumatic injury (Karamouzian et al. 2010).

The other common earthquake-related neurological injury resulting in early disability is traumatic head injury. After the 1999 Taiwan earthquake (20 September, 7.7 magnitude with 2500 casualties) (https://earthquake.usgs.gov/earthquakes/eventpage/usp0009eq0#region-info), 30% of victims died from head injuries (Liang et al. 2001) and head injuries were the second-most frequent type of trauma (following lower limb injuries) after China's 2008 earthquake (L. Wang et al. 2009). With regards to severity, the majority of head-injured patients have mild (55%) (Bhatti et al. 2008) or mild to moderate injuries (85%) (Jia et al. 2010). Within this group, scalp lacerations are common, ranging from 43% (L. Wang et al. 2009) to 65% (Jia et al. 2010), and concussions are described in about 4% (Jia et al. 2010). Skull fractures are identified in 8% (L. Wang et al. 2009) to 28% (Jia et al.

2010) of head-injured patients, 11% have basal skull fractures (Jia et al. 2010) and 10% have intra-cranial bleeding (Jia et al. 2010). Because many of these head injuries are minor in severity, surgical intervention is often unnecessary and the overall prognosis for head injuries is usually good. One Chinese group described a 2% mortality rate for their head-injured patients and reported that 82% of their head injury patients made a good neurological recovery (Jia et al. 2010).

Nevertheless, severe head injuries do occur as a result of earthquakes. In one series, 15% of all earthquake-induced head injuries were intracranial (L. Wang et al. 2009). Of these intracranial injuries, 1% was subarachnoid haemorrhage, 2% were primary brainstem injuries and 3% were diffuse axonal injuries (Jia et al. 2010). Approximately 34% of patients with severe earthquake-related head injuries require a major surgery, and burr hole evacuation has been described as an essential life-saving procedure in situations where early evacuation to a trauma centre is not feasible (Bhatti et al. 2008).

3.2.4 CHEST INJURIES

Chest injuries vary in incidence and severity after major earthquakes. Following the 1995 Japan earthquake, approximately 13% of patients referred to tertiary care had chest injuries (Yoshimura et al. 1996). Many of these chest injuries (47/63) were mild to moderate in severity, with lacerations and contusions being the most common. Patients with mild to moderately severe chest injuries were often discharged from the emergency room after receiving initial treatment (Yoshimura et al. 1996). On the other hand, earthquakes also cause more serious chest injuries, and 8 of the 63 chest-injured patients in Japan had severe chest trauma resulting in death. Furthermore, in Turkey 7%–10% of victims had thorax and lung injuries (Ozdogan et al. 2001). Pneumothoraces accounted for 52.4% and haemothoraces accounted for 19%. Another 19% of patients developed crush syndrome and acute respiratory distress syndrome (ARDS) (Ozdogan et al. 2001). ARDS is reported to be a serious complication of chest injuries (Gonzalez 2005), and following the Marmara Earthquake, it was a significant predictor of death (odds ratio [OR] = 4.53, $p < 0.0001$) (Sever et al. 2002a). Approximately 21% of patients with chest injuries develop respiratory failure, requiring mechanical ventilation (Hu et al. 2010). Risk factors for respiratory failure include flail chest, pulmonary contusion and crush syndrome (Hu et al. 2010).

Computed tomography (CT) scans provide more details on the nature of chest injuries, and in one Chinese series, 67% of chest-injured patients had at least one rib fracture and the mean number of rib fractures per patient was 11 (Dong et al. 2010). Forty-five patients had a flail chest, 12 had sternal fractures and 48 patients had either scapular or clavicular fractures. One hundred and seventeen patients had pulmonary parenchymal injuries, with approximately 95% being contusions and the remainder being lacerations (Dong et al. 2010). Of the 146 patients with pleural injuries, 63% were haemothorax, 32% were haemopneumothorax and 5% were pneumothorax. Additional chest injuries included pneumomediastinum (5.6%), as well as one patient with aortic dissection and one patient with a diaphragmatic hernia (Dong et al. 2010).

3.3 MEDICAL COMPLICATIONS

The medical complications of earthquakes include cardiovascular events and infections. Additionally, medical complications arising from the musculoskeletal and neurological injuries can be long-standing and may have a significant impact on outcomes and quality of life.

3.3.1 FIRST MONTH POST-EARTHQUAKE

3.3.1.1 Cardiovascular System

The effects of earthquakes on the cardiovascular system have been studied in a number of different settings with varying results (Bhattacharyya and Steptoe 2007). The most commonly reported cardiovascular events following earthquakes are myocardial infarctions, cerebral vascular accidents

(CVAs), cardiac arrhythmias, cardiomyopathies and elevations in blood pressure. Following some earthquakes, the increased incidence of cardiovascular events was limited to only a few days (Gold et al. 2007), while in other post-earthquake settings, the increased number of cardiovascular events lasted for at least 6 weeks (Tsai et al. 2004). These differences in duration of the earthquake-induced cardiovascular events may be related to variation in the study populations, variation in the extent of earthquake damage and variation in the study population's proximity to the epicentre (Kario et al. 2003). When considering earthquake-induced adverse cardiac events, the presence or absence of pre-earthquake illnesses and the severity of these illnesses may better determine outcome than the severity of post-earthquake psychological stress (Vieweg et al. 2011).

Following the California earthquake in 1994 (17 January, 6.7 magnitude with 60 casualties) (https://earthquake.usgs.gov/learn/today/index.php?month=1&day=17&submit=View+Date), there was a 35% increase in the incidence of acute myocardial infarctions in the week following the earthquake compared with the week prior (Kloner 2006). Following the 1999 Taiwan earthquake, the rate of hospitalization due to acute myocardial infarction increased significantly in the 6-week period following the earthquake compared with the same 6-week period the previous year (Tsai et al. 2004). In contrast, following the 1989 California earthquake, there was no detectable increase in the number of acute coronary syndromes (Brown 1999).

The incidence of cardiac arrhythmias has also been reported to increase following severe earthquakes. After the 2008 earthquake in China, the event rate for haemodynamically unstable ventricular arrhythmias among inpatients increased to 67 events/10,000 person-days from a baseline of 7 and 14 events/10,000 person-days during each of two control periods (Zhang et al. 2009).

Earthquakes have also been found to be associated with an increased incidence of cardiomyopathies. In the month following the 2004 earthquake in Japan (23 October, 6.6 magnitude with 40 casualties) (https://walrus.wr.usgs.gov/geotech/Niigata/index.html), Takotsubo cardiomyopathy (a stress-induced cardiomyopathy or so-called 'broken-heart syndrome') was diagnosed in 16 patients (1 man and 15 women, with mean age of 71.5 years) (Sato et al. 2006). Sixty-nine percent of patients with cardiomyopathy developed symptoms on the day of the earthquake. The incidence of Takotsubo cardiomyopathy 1 month after the earthquake was approximately 24-fold higher near the epicentre than it had been before the earthquake (Sato et al. 2006).

Furthermore, earthquakes have been known to adversely affect blood pressure in patients with hypertension. For two weeks after the 1995 Japan earthquake, during the period when aftershocks were common, mean systolic blood pressure was 14 ± 16 mm Hg greater and mean diastolic blood pressure was 6 ± 10 mmHg greater than baseline readings in patients wearing ambulatory blood pressure monitors (Kario et al. 2001). These blood pressure elevations typically resolved by 3–5 weeks after the earthquake except in patients with microalbuminuria, for whom the increases persisted for at least 2 months (Kario et al. 2001). Increased cardiovascular morbidity and mortality following earthquakes is likely multifactorial in nature. Increased physical and psychological stress with an associated sympathetic nervous system activation, interruption of prescribed cardiovascular drug regiments and disruption of pre-hospital care and emergency medical services are all likely contributing factors (Hata 2009).

Mortality from cerebrovascular accidents is also reportedly higher following major earthquakes. The number of stroke deaths increased significantly among elderly subjects in the first 3 months after the 1995 Japan earthquake (58 stroke deaths after the earthquake vs. 31 in the same time period of the previous year) (Kario and Ohashi 1998).

3.3.1.2 Infectious Diseases

Earthquakes can be associated with infectious diseases and outbreaks of respiratory and waterborne illnesses for several reasons, including disruption of water and sanitation systems (Floret et al. 2006; Watson et al. 2007) and large-scale population displacement, leading to the formation of overcrowded camps and temporary shelters (Watson et al. 2007; Kouadio et al. 2012). Furthermore, seismic events often disrupt public services, such as electricity, potable water and emergency medical services, which may also contribute to disease outbreaks (Watson et al. 2007).

Despite the disruption of public health services, epidemics after earthquakes remain uncommon. In a 2006 review of the literature, Floret and others (2006) concluded that the only disease outbreak that could be clearly attributed to an earthquake was a coccidiomycosis outbreak following the 1994 Northridge earthquake in California. Other experts agree that epidemics are rare following earthquakes (de Ville de Goyet 2000; Noji 2005; VanRooyen and Leaning 2005), and studies have shown that natural disasters do not introduce diseases that were not already present in the affected area prior to the disaster (Chan et al. 2003; Kouadio et al. 2012).

However, not all experts agree that there is minimal risk of epidemics following earthquakes. Karmakar et al. (2008) reported an outbreak of rotavirus gastroenteritis following the 2005 Indian earthquake, and Korten (2009) described outbreaks of acute watery diarrhoea, acute respiratory infections, hepatitis E and measles after the 2005 Pakistan earthquake. Baseline surveillance data are often not available to definitively conclude that infectious disease cases such as these truly constitute outbreaks.

The presence of cadavers following natural disasters is frequently perceived as a potential source of infection and epidemics. The fear of epidemics has sometimes led to the immediate disposal of bodies through mass burial without proper identification of the deceased and without the opportunity for family members to have the closure of viewing their loved one's remains (Gionis et al. 2007). However, a 2004 review concluded that cadavers resulting from natural disasters are unlikely to have 'epidemic-causing' infections, and that the public risk of infection from dead bodies is negligible (Morgan 2004). Morgan further concluded that only people who are in close physical contact with cadavers, such as rescue workers and those responsible for disposing of remains, are at risk of disease transmission from those cadavers. Other experts have similarly concluded that there is no evidence that corpses pose an infection risk (Kang et al. 2008; Kouadio et al. 2012). Experts with knowledge and experience in disaster settings should educate the public and the media about the risks or lack thereof of epidemics after earthquakes (Kirkis 2006).

Soft tissue infections are common among earthquake victims and are associated with a high mortality, particularly if the wound infection leads to sepsis (Keven et al. 2003). It is important to consider the microbiology of such infections, and a 2010 review reported that patients admitted after natural disasters often have polymicrobial infections with atypical bacteria, as well as fungi, and that multidrug resistance is common (Ambrosioni et al. 2010). Gram-negative bacteria are reported more commonly than Gram-positive bacteria in post-earthquake wounds. Prevalent post-earthquake pathogens include *Pseudomonas aeruginosa* (H. Y. Wang et al. 2009; Ran et al. 2010), *Enterobacter cloacae* (Xie et al. 2009; Ran et al. 2010), *Acinetobacter baumannii* (H. Y. Wang et al. 2009; Ran et al. 2010) and *Escherichia coli* (H. Y. Wang et al. 2009; Xie et al. 2009). In addition to Gram-negative bacteria, *Staphylococcus* spp. are also commonly identified, and many infections are thought to be nosocomial in origin (Keven et al. 2003). Post-disaster fungal infections have been described as particularly difficult to treat (Ambrosioni et al. 2010). It is important to initiate mass vaccination campaigns for tetanus, prioritizing patients with wounds and then the general population (Chan et al. 2003; Yasin et al. 2009).

Kang et al. (2008) identified the following risk factors for outbreak following an earthquake: safety of food and water, sanitation and population density. Generally, proper water and sanitation management, combined with enhanced disease surveillance, is sufficient to prevent and control potential epidemics in the post-earthquake setting (Chan et al. 2003).

3.3.1.3 Recovery from Musculoskeletal Injuries

The morbidity associated with musculoskeletal injuries can be significant and can result in serious disability if appropriate rehabilitation resources are not available. Since musculoskeletal injuries are the most commonly sustained earthquake injuries, the recovery and long-term prognosis for these patients should be taken into account when planning longer-term rescue and recovery efforts. For instance, the functional recovery of earthquake survivors with tibia fractures is positively associated with rehabilitation intervention and negatively correlated with immobilization duration

(Xiao et al. 2011). Despite the recognized importance of rehabilitation services in the recovery phase, multiple studies have demonstrated a serious deficit in rehabilitation resources for injured patients, including physical and occupational therapy (Iezzoni and Ronan 2010). After the Haiti 2010 earthquake (12 January, 7.0 magnitude with 222,570 casualties) (https://earthquake.usgs.gov /learn/today/index.php?month=1&day=12), 12% of earthquake-injured patients required transfer to another medical or rehabilitation facility upon discharge from the hospital (Centers for Disease Control and Prevention 2011), and after the China 2008 earthquake (12 May, 7.9 magnitude with 69,195 casualties) (https://earthquake.usgs.gov/learn/today/index.php?month=5&day=12), 72% of patients with fractures were restricted in their daily activities due to their injuries (J. L. Zhang et al. 2011). Traumatic amputations are frequent following major earthquakes, and the shortage of prosthetic limbs and a lack of skilled personnel to fit prosthetic limbs are major impediments to effective rehabilitation (Mallick et al. 2010). In underdeveloped countries where resources are often already strained, there are few providers properly trained to accommodate the increased demand for specialized rehabilitative services (Landry 2010; Landry et al. 2010a,b). It is therefore important not only to mobilize physical and occupational therapists post-earthquake, but also to incorporate rehabilitation services into longer-term disaster planning.

3.3.1.4   Recovery from Neurological Injuries

Patients with neurological injuries as a result of an earthquake often develop complications and have longer-term healthcare needs extending beyond the immediate treatment period. Following the 2005 earthquake in Pakistan, 20% of patients with spinal injuries developed pressure ulcers, 15% developed bowel complications, 2% developed deep vein thrombosis and urinary tract infections were seen in almost all patients (Tauqir et al. 2007). These patients frequently require rehabilitation services, which are often lacking in the post-earthquake recovery period. Experts argue that rehabilitation services are an important aspect of disaster planning (Tauqir et al. 2007), and propose that dedicated rehabilitation centres, staffed by appropriately trained personnel, are essential in preparing patients to return to their communities, particularly since many of these patients will return to devastated communities in which it can be extremely difficult to adapt to a new physical disability (Rathore et al. 2008).

Unfortunately, no instruments exist to specifically assess rehabilitation needs for patients injured in a natural disaster. After the 2010 earthquake in Haiti, nearly all patients who had suffered a traumatic brain or spinal cord injury demonstrated problems with mobility (Rauch et al. 2011). Additionally, wheelchair accessibility and health professional availability were all major barriers to patients' performance in the post-earthquake setting (Rauch et al. 2011). While many spinal cord injury survivors report improvement in physical independence, mobility and quality of life over time, only 15% return to work, suggesting that cognitive and emotional functioning are an important aspect of rehabilitation for these patients (Hu et al. 2012). Furthermore, 80% of earthquake survivors with head trauma report symptoms of depression when interviewed 18 months following the disaster (Xu et al. 2011a). Thus, a multidisciplinary approach with physiatrists, physical therapists, occupational therapists, nurses, psychologists, social workers and urban planners is needed to optimize management for earthquake-injured patients requiring rehabilitation.

3.3.2   Medical Complications in the First Year Post-Earthquake

3.3.2.1   Mental Health

The mental health complications of earthquakes are significant, and the potential for long-term morbidity is greater for mental health than it is for many other medical consequences of earthquakes. Despite this, mental health services remain a low priority in many post-earthquake settings. Mental health issues are often underestimated, and the prevalence is often derived from convenience samples at health clinics and medical centres because many people with mental health problems do not seek care (Bourque et al. 2002).

Different phases of disaster and recovery are associated with different mental health problems. The first reactions to an earthquake often consist of emotional numbness, a loss of the sense of reality and an abnormal sense of time (Shinfuku 2002). Several days after the quake, these reactions are followed by anxiety and fear of aftershocks. Within a week of the earthquake, poor sleep and other somatic signs begin (Kokai et al. 2004). A few weeks later, depressive symptoms start and can persist for months to years. A year after the earthquake, the dominant problems can become social – including lack of support and consumption of alcohol (Kokai et al. 2004). This 'social dysfunction' is particularly common among older earthquake survivors, and in some cases, coping abilities are still remarkably impaired 2 years after the earthquake (Toyabe et al. 2007). To reduce the impact of disaster on survivors' mental health, experts advocate for prompt return to daily routine and re-establishing important family and friend support networks (Thomas 2006). This can be difficult, however, especially if the disaster has destroyed community infrastructure, disrupted communication or caused displacement.

Researchers have found that the degree of concern about safety and health is inversely correlated with the level of devastation caused by the earthquake in that particular area (S. Li et al. 2009). For instance, individuals living in extremely devastated areas report less concern than those geographically further from the epicentre. It is recommended that this phenomenon, referred to as the 'psychological typhoon eye', be taken into consideration when planning mental health interventions (Li et al. 2010). Cultural differences may also contribute to the psychological impact of an earthquake. For instance, following the Wenchuan Earthquake in 2008, Tibetans experienced fewer acute stress reaction symptoms, like avoidance and numbness, than did the Han ethnic group. This discrepancy between ethnic groups was thought to be related to cultural values, religious beliefs and living conditions before the earthquake (Xu et al. 2011b).

Depression is common following earthquakes, with reported frequencies varying from 6% in bereaved quake survivors in Taiwan (Kuo et al. 2003) to greater than 72% in young adults in Turkey (Vehid et al. 2006). Females are reportedly more likely to be affected by post-earthquake depression (Chan et al. 2006; Chadda et al. 2007; Z. Zhang et al. 2011). Krug et al. (1998) reported that of all natural disasters, earthquakes most significantly increase the rate of suicide. After the Marmara Earthquake in Turkey (17 August 1999, 7.6 magnitude with 17,118 casualties) (https://earthquake.usgs.gov/learn/today/index.php?month=8&day=17), the overall rate of suicidal ideation was found to be 17% and suicidal thoughts were reportedly 1.76 times higher among students who were injured and among students whose relatives had serious earthquake-related injuries (Vehid et al. 2006). Suicidal ideation was higher among students who had family members die in the disaster and among students who experienced extensive destruction to their homes and property (Vehid et al. 2006). In contrast to depression, for which rates tend to be higher in females, suicidal thoughts are more common among male earthquake survivors (Vehid et al. 2006).

Earthquakes are also commonly associated with post-traumatic stress disorder (PTSD). However, in general, prevalence rates of PTSD are lower for victims of natural disaster, such as earthquakes, than they are for victims of conflict and technological disasters (Neria et al. 2007). Frequencies of PTSD range from 3% (Chadda et al. 2007) to 91% (Hagh-Shenas et al. 2006) in various post-earthquake settings. A multitude of risk factors have been identified for the development of PTSD after a major earthquake, including being physically injured and experiencing the death of a close family member (Hsu et al. 2002; Kun et al. 2009a), female sex, significant loss and strong initial sense of fear (Z. Zhang et al. 2011), as well as low household income, being an ethnic minority and living in a shelter (Kun et al. 2009a,b). In a longitudinal study in China, PTSD rates were lower in villages with greater initial exposure to the earthquake and with a higher level of post-earthquake support (Wang et al. 2000). Other studies have similarly found that the severity of PTSD is negatively correlated with perceived support both within and outside the family (Y. F. Yang et al. 2010). These findings suggested that prompt and effective intervention and assistance, even if not focused on mental health, could mitigate the probability of PTSD occurrence (Wang et al. 2000). The natural course of PTSD following earthquakes is largely unknown (Neria et al. 2007).

However, some experts recommend reassuring victims of catastrophic disasters that PTSD symptoms are common in the short term, but that the symptoms can be expected to dissipate with time (North and Smith 1990).

Psychological distress and complicated grief reactions have also been reported following major earthquakes. A reported 58% of survivors suffer from psychological distress (three times higher than that of the general population) (Montazeri et al. 2005), and 76% of earthquake survivors are found to have complicated grief (grief that lingers and becomes debilitating to the patient) (Ghaffari-Nejad et al. 2007). Risk factors for complicated grief include female sex, lower education levels, being present in a city at the time of the quake, observing the burial of corpses, destruction of the residential home, residential concerns after the disaster and loss of at least one first-degree relative (Ghaffari-Nejad et al. 2007). Furthermore, children who stay in alternative living arrangements for longer periods of time following an earthquake are found to have higher rates of behavioural difficulties and emotional problems (Situ et al. 2009). Experts therefore suggest that rebuilding homes and addressing residential problems could aid psychological recovery following earthquakes (Ghaffari-Nejad et al. 2007). This suggestion was also supported by Yasamy et al. (2005), who reported a 55-fold greater chance of psychological recovery in adults who received psychosocial interventions than in those who did not receive interventions 7 months post-earthquake.

Mental health has a significant impact on the physical health and well-being of survivors long after the earthquake. Studies suggest that earthquake-related loss and depression are independent predictors of event-free survival in heart failure patients (Huang et al. 2011). After the Christchurch earthquake, women with non-cardiac chest pain scored higher on anxiety and neuroticism scales, while those with stress cardiomyopathy had mild coronary artery disease and scored the highest on both scales (Zarifeh et al. 2012). Some experts suggest that earthquake-induced mental health problems resolve with time, and that the quality of life reported by survivors returns to being positive (Priebe et al. 2011). However, for many earthquake survivors, there is still a marked decreased in self-reported quality of life when compared with the norm (Wen et al. 2012). These findings reinforce the need for awareness regarding the impact of mental health on the quality of life of survivors.

3.4 MANAGEMENT OF PRE-EXISTING DISEASES IN THE POST-EARTHQUAKE PERIOD

Earthquake-related traumatic injuries are not the only healthcare priority following earthquakes since patients with pre-existing diseases and chronic medical problems must also continue to receive care. After the 2008 Wenchuan Earthquake, 77% of patients presenting to a triage clinic had chronic medical problems, and lack of medication for chronic conditions was a frequent presenting complaint (Chan and Kim 2011). A similar finding was reported in Japan, where 60% of people seeking emergency aid were elderly patients with chronic medical problems, requiring primary care rather than management of acute earthquake-related injuries and illnesses (Kohsaka et al. 2012). This is particularly problematic in unofficial settlement camps for displaced persons, as these populations generally have less resources and access to medical care, and have been shown to self-report worse health outcomes (Chan and Kim 2011). Since primary care providers often offer mental health services for everyday patients, the increased risk for PTSD, addiction and other mental health disorders after an earthquake generates an even greater demand for primary healthcare resources (Sullivan and Wong 2011).

3.4.1 CHRONIC KIDNEY DISEASE

In post-earthquake settings, careful consideration must also be given to chronic haemodialysis patients. These patients are at particular risk following earthquakes because dialysis is almost always performed in hospitals or dialysis units, which can be heavily damaged by the earthquake and because dialysis is a complex procedure requiring trained staff, water and electricity

(Sever et al. 2004). Furthermore, dialysis supplies (lines, dialyzers and concentrates) may be lost in the quake. In some earthquake settings, it is necessary to use a dialyzer reuse programme, whereby designated filters are cleaned between treatments and reused for each patient (Amundson et al. 2010). As an additional challenge, dialysis personnel and their families are often also affected by the disaster, making it difficult for dialysis staff to report to work (Sever et al. 2004). These restrictions, coupled with the new increased demand for dialysis services as a result of earthquake-induced crush injuries, can quickly overwhelm the available resources. If responders must choose to dialyze acute versus chronic renal failure patients, some experts recommend that priority be given to the acute patients, whose kidneys almost always recover and who are therefore candidates for a healthy life (assuming these acute patients will otherwise survive their traumatic injuries) (Sever et al. 2004). At least some chronic haemodialysis patients can be dialyzed once weekly, allowing more patients to be treated each week. Although this practice carries the risk of complications, many patients do well if they are able to restrict their fluid and dietary intake (Sever et al. 2004).

Chronic haemodialysis patients require instruction on what to do in the event of an earthquake. For instance, patients should be trained to restrict fluid and diet intake and not to panic should they miss one or two dialysis sessions following the disaster (Sever et al. 2009). End-stage renal failure patients should also store potassium exchange resins, such as sodium polystyrene sulphonate, at home as a prophylactic measure against hyperkalaemia should their dialysis be delayed or reduced in frequency (Sever et al. 2004). Patients on peritoneal dialysis should be cautioned about non-hygienic conditions and about the need to have a surplus of dialysis solutions on hand (Sever et al. 2009).

3.4.2 Non-Infectious Respiratory Disease

Respiratory complications following major earthquakes include exacerbation of chronic illnesses, such as asthma. After the Japan earthquake in 2000, 11% of patients had an asthma exacerbation in the first month (Tomita et al. 2005). The authors reported that post-earthquake asthma attacks were more likely to occur within the first week following the disaster and suggested that the ratio of forced expiratory volume in 1 second (FEV1%) to forced vital capacity be used as a predictor of airflow restriction following earthquakes (Tomita et al. 2005).

3.5 MEDICAL COMPLICATIONS FOR EARTHQUAKE RESPONDERS

Earthquake responders often work in difficult environments due to collapsed structures, falling debris, exposure to hazardous materials, lack of water and electricity, poor sanitation and extreme weather. These austere conditions are frequently made even more challenging by interruptions in healthcare infrastructure, communication networks and transportation systems (Koyama et al. 2011). Furthermore, the psychological stress of dealing with severe injuries, significant mortality and widespread devastation, particularly over long periods of time, can greatly impact responders' health and well-being.

Within the first 2 weeks of an earthquake, the most common medical complaints among responders are respiratory problems, skin problems, neurologic problems, intestinal infections and emotional distress (O'Ryan 2010; W. Q. Zhang et al. 2011). Natural disasters in countries where there is limited clean drinking water, food, shelter and healthcare contribute to the development of infectious diseases (Migl and Powell 2010). People who are in close physical contact with cadavers, such as rescue workers and those responsible for disposing of remains, may be at risk of infections such as hepatitis B and C, HIV, enteric pathogens and *Mycobacterium tuberculosis* (Morgan 2004). The potential for the transmission of cadaver infections to responders can be mitigated by universal precautions, by use of disposable gloves and body bags, by practicing good hygiene and by vaccination of all responders against hepatitis B and tuberculosis (Morgan 2004).

There is significant controversy in the literature regarding the risks of psychological complications for disaster responders. In a comparison of mental health problems between soldiers, peacekeepers and relief workers, relief work was not found to be a risk factor for PTSD or substance abuse (Connorton et al. 2011). In a comparison of volunteer rescuers and professional responders, volunteers tended to have more mental health complaints, and contributing factors were thought to include a personal relationship with victims, severity of event exposure, anxiety sensitivity and lack of social support (Thormar et al. 2010). When rescue workers were surveyed pre- and 3 months post-earthquake, symptoms of depression, anxiety, PTSD, substance abuse and mental health service utilization were not significantly different before or after the earthquake (van der Velden et al. 2012).

Other studies, however, report that earthquake responders experience significant psychological stress, often higher than the norm (Ehring et al. 2011; W. Q. Zhang et al. 2011). Some medical professionals who fill the role of rescuer during disaster response report feeling overwhelmed and underprepared (Y. N. Yang et al. 2010). Rescue personnel in the hospitals and front lines are under significant stress. Concerns about job safety due to physical assault is the most frequently reported fear by disaster responders, and 40% of respondents report dissociation when their patients die after failed resuscitation attempts (Ogłodek and Araszkiewicz 2011). PTSD rates among responders range from 6.5% (Wang et al. 2011) to as high as 42.6% (Ehring et al. 2011). Depression and anxiety rates are estimated to be approximately 20% among responders (Ehring et al. 2011). Risk factors for PTSD in rescue workers include severity of earthquake experience (Ehring et al. 2011; Soffer et al. 2011; Wang et al. 2011), as well as past traumatic history, low social support and female sex (Ehring et al. 2011). Other social factors that contribute to the development of PTSD symptoms include single-child family, having been raised apart from family, no psychological counselling and a history of alcohol use (Wang et al. 2011).

3.6   DONATED MEDICAL SUPPLIES

Following a major earthquake, it is necessary to regulate the entry of medical goods into the affected region. The 2007 earthquake in Peru (15 August, 8.0 magnitude with 512 casualties) (https://earthquake.usgs.gov/learn/today/index.php?month=8&day=15) highlights this need for regulation. Thousands of tons of unsorted and unlabelled medical donations arrived in Peru, often inappropriate for treating the predominant diseases and injuries. The volume of donation, the lack of utility of some of the supplies and the poorly organized manner in which the donations arrived made it difficult to provide patient care and to access proper medical equipment (Bambaren 2010).

3.7   CONCLUSION

Earthquakes are one of the most potentially devastating natural disasters and result in significant annual morbidity and mortality worldwide. The medical complications of earthquakes extend across multiple organ systems, including musculoskeletal, renal and neurologic, as well as mental health. Earthquake-related medical complications can be acute (in the first post-earthquake week), subacute (in the first post-earthquake month) or long term (in the first post-earthquake year). Appropriate care for these medical complications requires experienced medical personnel and ample medical supplies. Provision of primary care for chronic health problems must also be continued post-earthquake, which can be challenging if the physical environment and healthcare infrastructure have been interrupted. Furthermore, earthquake victims are not the only individuals directly affected. Rescuers and healthcare providers can suffer from both earthquake-related injuries and mental health disorders as a result of their rescuing efforts, and their health is often overlooked in the aftermath of an earthquake. A multidisciplinary response is critical and should include sufficient numbers of rescuers, nurses, physicians, specialists, rehabilitation services, mental health providers and requisite ancillary staff to provide the appropriate care.

REFERENCES

Alexander, D. (1996). The health effects of earthquakes in the mid-1990s. *Disasters*, *20*(3), 231–247.

Ambrosioni, J., Lew, D., and Uckay, I. (2010). Infectious diseases and infection control after natural disasters. *Int J Infect Dis*, *14*(Suppl. 1), e16.

Amundson, D., Dadekian, G., Etienne, M., Gleeson, T., Hicks, T., Killian, D., and Miller, E. J. (2010). Practicing internal medicine onboard the USNS COMFORT in the aftermath of the Haitian earthquake. *Ann Intern Med*, *152*(11), 733–737. http://annals.org/aim/fullarticle/745808/practicing-internal-medicine-onboard-usns-comfort-aftermath-haitian-earthquake.

Ardagh, M. W., Richardson, S. K., Robinson, V., Than, M., Gee, P., Henderson, S., and Deely, J. M. (2012). The initial health-system response to the earthquake in Christchurch, New Zealand, in February, 2011. *Lancet*, *379*(9831), 2109–2115. Retrieved from http://www.embase.com/search/results?subaction=viewrecord&from=export&id=L364903606.

Bambaren, C. (2010). Legal issues of humanitarian assistance after the 2007 earthquake in Pisco, Peru. *Prehosp Disaster Med 25*(3), 203–206. Retrieved from http://www.embase.com/search/results?subaction=viewrecord&from=export&id=L359616787.

Bartels, S. A., and VanRooyen, M. J. (2012). Medical complications associated with earthquakes. *Lancet*, *379*(9817), 748–757. http://doi.org/10.1016/S0140-6736(11)60887-8.

Better, O. S. (1989). Traumatic rhabdomyolysis ('crush syndrome') – Updated 1989. *Isr J Med Sci*, *25*(2), 69–72. Retrieved from http://www.ncbi.nlm.nih.gov/entrez/query.fcgi?cmd=Retrieve&db=PubMed&dopt=Citation&list_uids=2649451.

Better, O. S. (1990). The crush syndrome revisited (1940–1990). *Nephron*, *55*(2), 97–103. Retrieved from http://www.ncbi.nlm.nih.gov/entrez/query.fcgi?cmd=Retrieve&db=PubMed&dopt=Citation&list_uids=2194135.

Better, O. S. (1993). Acute renal failure in casualties of mass disasters. *Kidney Int Suppl*, *41*, S235–S236.

Better, O. S. (1999). Rescue and salvage of casualties suffering from the crush syndrome after mass disasters. *Mil Med*, *164*(5), 366–369.

Better, O. S., and Stein, J. (1990). Early management of shock and prophylaxis of acute renal failure in traumatic rhabdomyolysis. *N Engl J Med*, *322*, 825–829.

Bhattacharyya, M. R., and Steptoe, A. (2007). Emotional triggers of acute coronary syndromes: Strength of evidence, biological processes, and clinical implications. *Prog Cardiovasc Dis*, *49*(5), 353–365. http://doi.org/10.1016/j.pcad.2006.11.002.

Bhatti, S. H., Ahmed, I., Qureshi, N. A., Akram, M., and Khan, J. (2008). Head trauma due to earthquake October, 2005 – Experience of 300 cases at the Combined Military Hospital Rawalpindi. *J Coll Physicians Surg Pak*, *18*(1), 22–26. https://www.jcpsp.pk/archive/2008/Jan2008/06.pdf.

Bourque, L. B., Siegel, J. M., and Shoaf, K. I. (2002). Psychological distress following urban earthquakes in California. *Prehosp Disaster Med*, *17*(2), 81–90. Retrieved from http://www.ncbi.nlm.nih.gov/pubmed/12500731.

Briggs, S. M. (2006). Earthquakes. *Surg Clin North Am*, *86*(3), 537–544. http://doi.org/10.1016/j.suc.2006.02.003.

Brown, D. L. (1999). Disparate effects of the 1989 Loma Prieta and 1994 Northridge earthquakes on hospital admissions for acute myocardial infarction: Importance of superimposition of triggers. *Am Heart J*, *137*(5), 830–836. http://www.ahjonline.com/article/S0002-8703(99)70406-0/abstract.

Centers for Disease Control and Prevention. (2011). Post-earthquake injuries treated at a field hospital – Haiti, 2010. *MMWR Morb Mortal Wkly Rep*, January 7. Retrieved from https://www.cdc.gov/mmwr/pdf/wk/mm5951.pdf.

Chadda, R. K., Malhotra, A., Kaw, N., Singh, J., and Sethi, H. (2007). Mental health problems following the 2005 earthquake in Kashmir: Findings of community-run clinics. *Prehosp Disaster Med*, *22*(6), 541–545; discussion 546. Retrieved from http://www.ncbi.nlm.nih.gov/entrez/query.fcgi?cmd=Retrieve&db=PubMed&dopt=Citation&list_uids=18709944.

Chan, C. C., Lin, Y. P., Chen, H. H., Chang, T. Y., Cheng, T. J., and Chen, L. S. (2003). A population-based study on the immediate and prolonged effects of the 1999 Taiwan earthquake on mortality. *Ann Epidemiol*, *13*(7), 502–508. http://doi.org/10.1016/S1047-2797(03)00040-1. Retrieved from http://ezp-prod1.hul.harvard.edu/login?url=http://search.ebscohost.com/login.aspx?direct=true&db=c8h&AN=2004018806&site=ehost-live&scope=site.

Chan, E. Y., and Kim, J. (2011). Chronic health needs immediately after natural disasters in middle-income countries: The case of the 2008 Sichuan, China earthquake. *Eur J Emerg Med*, *18*(2), 111–114. http://doi.org/10.1097/MEJ.0b013e32833dba19.

Chan, Y. F., Alagappan, K., Gandhi, A., Donovan, C., Tewari, M., and Zaets, S. B. (2006). Disaster management following the Chi-Chi earthquake in Taiwan. *Prehosp Disaster Med*, *21*(3), 196–202.

Chen, R., Song, Y., Kong, Q., Zhou, C., and Liu, L. (2009). Analysis of 78 patients with spinal injuries in the 2008 Sichuan, China, earthquake. *Orthopedics*, *32*(5), 322. https://www.healio.com/orthopedics/spine/journals/ortho/2009-5-32-5/%7B28061f45-cca5-458a-bf1a-3c894e728794%7D/analysis-of-78-patients-with-spinal-injuries-in-the-2008-sichuan-china-earthquake.

Chen, T., Yang, Z., Wang, Q., Dong, Z., Yu, J., Zhuang, Z., et al. (2009). Crush extremity fractures associated with the 2008 Sichuan earthquake: Anatomic sites, numbers and statuses evaluated with digital radiography and multidetector computed tomography. *Skeletal Radiol*, *38*(11), 1089–1097. http://doi.org/10.1007/s00256-009-0743-5.

Connorton, E., Perry, M. J., Hemenway, D., Miller, M., and Professor, A. (2011). Occupational trauma and mental illness: Combat, peacekeeping or relief work and the NCS-R. *J Occup Environ Med*, *53*(12), 1360–1363. http://doi.org/10.1097/JOM.0b013e318234e2ec.

Demirkiran, O., Dikmen, Y., Utku, T., and Urkmez, S. (2003). Crush syndrome patients after the Marmara earthquake. *Emerg Med J*, *20*(3), 247–250. Retrieved from https://www-ncbi-nlm-nih-gov.ezp-prod1.hul.harvard.edu/pmc/articles/PMC1726105/pdf/v020p00247.pdf.

de Ville de Goyet, C. (2000). Stop propagating disaster myths. *Lancet*, *356*(9231), 762–764. Retrieved from http://www.ncbi.nlm.nih.gov/pubmed/11085709.

Dong, Z. H., Yang, Z. G., Chen, T. W., Feng, Y. C., Chu, Z. G., Yu, J. Q., and Wang, Q. L. (2010). Crush thoracic trauma in the massive Sichuan earthquake: Evaluation with multidetector CT of 215 cases. *Radiology*, *254*(1), 285–291. http://doi.org/10.1148/radiol.09090685.

Duman, H., Kulahci, Y., and Sengezer, M. (2003). Fasciotomy in crush injury resulting from prolonged pressure in an earthquake in Turkey. *Emerg Med J*, *20*(3), 251–252. Retrieved from https://www-ncbi-nlm-nih-gov.ezp-prod1.hul.harvard.edu/pmc/articles/PMC1726110/pdf/v020p00251.pdf.

Ehring, T., Razik, S., and Emmelkamp, P. M. (2011). Prevalence and predictors of posttraumatic stress disorder, anxiety, depression, and burnout in Pakistani earthquake recovery workers. *Psychiatry Res*, *185*(1–2), 161–166. http://doi.org/10.1016/j.psychres.2009.10.018.

Finkelstein, J. A., Hunter, G. A., and Hu, R. W. (1996). Lower limb compartment syndrome: Course after delayed fasciotomy. *J Trauma*, *40*(3), 342–344. Retrieved from http://www.ncbi.nlm.nih.gov/entrez/query.fcgi?cmd=Retrieve&db=PubMed&dopt=Citation&list_uids=8601846.

Floret, N., Viel, J. F., Mauny, F., Hoen, B., and Piarroux, R. (2006). Negligible risk for epidemics after geophysical disasters. *Emerg Infect Dis*, *12*(4), 543–548. Retrieved from http://www.cdc.gov/ncidod/EID/vol12no04/pdfs/05-1569.pdf.

Ghaffari-Nejad, A., Ahmadi-Mousavi, M., Gandomkar, M., and Reihani-Kermani, H. (2007). The prevalence of complicated grief among Bam earthquake survivors in Iran. *Arch Iran Med*, *10*(4), 525–528. http://citeseerx.ist.psu.edu/viewdoc/download?doi=10.1.1.518.6379&rep=rep1&type=pdf.

Gionis, T. A., Wecht, C. H., Marshall Jr., L. W., and Hagigi, F. A. (2007). Dead bodies, disasters, and the myths about them: Is public health law misinformed? *Am J Disaster Med*, *2*(4), 173–188. Retrieved from http://www.ncbi.nlm.nih.gov/entrez/query.fcgi?cmd=Retrieve&db=PubMed&dopt=Citation&list_uids=18488831.

Gold, L. S., Kane, L. B., Sotoodehnia, N., and Rea, T. (2007). Disaster events and the risk of sudden cardiac death: A Washington State investigation. *Prehosp Disaster Med*, *22*(4), 313–317. Retrieved from http://www.ncbi.nlm.nih.gov/pubmed/18019098.

Gonzalez, D. (2005). Crush syndrome. *Crit Care Med*, *33*(1), S34–S41. http://doi.org/10.1097/01.CCM.0000151065.13564.6F.

Gunal, A. I., Celiker, H., Dogukan, A., Ozalp, G., Kirciman, E., Simsekli, H., and Sever, M. S. (2004). Early and vigorous fluid resuscitation prevents acute renal failure in the crush victims of catastrophic earthquakes. *J Am Soc Nephrol*, *15*, 1862–1867. http://doi.org/10.1097/01.ASN.0000129336.09976.73.

Hagh-Shenas, H., Goodarzi, M. A., Farajpoor, M., and Zamyad, A. (2006). Post-traumatic stress disorder among survivors of Bam earthquake 40 days after the event. *East Mediterr Health J*, *12*, S118–S125. Retrieved from http://ezp-prod1.hul.harvard.edu/login?url=http://search.ebscohost.com/login.aspx?direct=true&db=c8h&AN=2009378111&site=ehost-live&scope=site.

Hata, S. (2009). Cardiovascular disease caused by earthquake-induced stress: Psychological stress and cardiovascular disease. *Circ J*, *73*(7), 1195–1196. https://www.jstage.jst.go.jp/article/circj/73/7/73_CJ-09-0305/_pdf.

Hosseini, M., Safari, S., and Saghafinia, M. (2009). Wide spectrum of traumatic rhabdomyolysis in earthquake victims. *Acta Med Iran*, *47*(6), 459–464.

Hsu, C.-C., Chong, M.-Y., Yang, P., and Yen, C.-F. (2002). Posttraumatic stress disorder among adolescent earthquake victims in Taiwan. *J Am Acad Child Adolesc Psychiatry*, *41*(7), 875–881. http://doi.org/10.1097/00004583-200207000-00022.

Hu, X., Zhang, X., Gosney, J. E., Reinhardt, J. D., Chen, S., Jin, H., and Li, J. (2012). Analysis of functional status, quality of life and community integration in earthquake survivors with spinal cord injury at hospital discharge and one-year follow-up in the community. *J Rehabil Med*, *44*(3), 200–205. http://doi.org/10.2340/16501977-0944.

Hu, Y., Tang, Y., Yuan, Y., Xie, T.-P., and Zhao, Y.-F. (2010). Trauma evaluation of patients with chest injury in the 2008 earthquake of Wenchuan, Sechuan, China. *World J Surg*, *34*(4), 728–732. http://doi.org/10.1007/s00268-010-0427-2.

Huang, K., Deng, X., He, D., Huang, D., Wu, Q., Wen, S., and Chen, M. (2011). Prognostic implication of earthquake-related loss and depressive symptoms in patients with heart failure following the 2008 earthquake in Sichuan. *Clin Cardiol*, *34*(12), 755–760. Retrieved from http://www.embase.com/search/results?subaction=viewrecord&from=export&id=L363050095.

Huang, K. C., Lee, T. S., Lin, Y. M., and Shu, K. H. (2002). Clinical features and outcome of crush syndrome caused by the Chi-Chi earthquake. *J Formos Med Assoc*, *101*(4), 249–256.

Iezzoni, L. I., and Ronan, L. J. (2010). Disability legacy of the Haitian earthquake. *Ann Intern Med*, *152*(12), 812–814. Retrieved from http://www.embase.com/search/results?subaction=viewrecord&from=export&id=L359231375.

Jagoginski, N., Weerasinghe, C., Porter, K., and Jagodzinski, N. A. (2010). Crush injuries and crush syndrome – A review. Part 1: The systemic injury. *Trauma*, *12*, 69–88.

Jia, L., Li, G.-P., You, C., Li, H., Huang, S.-Q., Yang, C.-H. et al. (2010). The epidemiology and clinical management of craniocerebral injury caused by the Sichuan earthquake. *Neurol India*, *58*(1), 85–89. http://doi.org/10.4103/0028-3886.60406.

Kang, D. Y., Hong, Q., Li, Y. P., Hu, D., and Wei, M. L. (2008). Management of dead bodies and epidemic control after earthquake – An evidence-based approach. *Chin J Evid Based Med*, *8*(8), 575–580. Retrieved from http://www.cjebm.org.cn/OA/pdfdow.aspx?Sid=080802.

Karamouzian, S., Saeed, A., Ashraf-Ganjouei, K., Ebrahiminejad, A., Dehghani, M. R., and Asadi, A. R. (2010). The neurological outcome of spinal cord injured victims of the Bam earthquake, Kerman, Iran. *Arch Iran Med*, *13*(4), 351–354. http://citeseerx.ist.psu.edu/viewdoc/download?doi=10.1.1.915.1858&rep=rep1&type=pdf.

Kario, K., Matsuo, T., Shimada, K., and Pickering, T. G. (2001). Factors associated with the occurrence and magnitude of earthquake-induced increases in blood pressure. *Am J Med*, *111*(5), 379–384. http://www.amjmed.com/article/S0002-9343(01)00832-4/abstract.

Kario, K., McEwen, B. S., and Pickering, T. G. (2003). Disasters and the heart: A review of the effects of earthquake-induced stress on cardiovascular disease. *Hypertens Res*, *26*(5), 355–367. Retrieved from http://www.ncbi.nlm.nih.gov/pubmed/12887126.

Kario, K., and Ohashi, T. (1998). After a major earthquake, stroke death occurs more frequently than coronary heart disease death in very old subjects. *J Am Geriatr Soc*, *46*(4), 538.

Karmakar, S., Rathore, A. S., Kadri, S. M., Dutt, S., Khare, S., and Lal, S. (2008). Post-earthquake outbreak of rotavirus gastroenteritis in Kashmir (India): An epidemiological analysis. *Public Health*, *122*(10), 981–989.

Keven, K., Ates, K., Sever, M. S., Yenicesu, M., Canbakan, B., Arinsoy, T., and Erek, E. (2003). Infectious complications after mass disasters: The Marmara earthquake experience. *Scand J Infect Dis*, *35*(2), 110–113. http://doi.org/10.1080/0036554021000027013.

Kirkis, E. J. (2006). A myth too tough to die: The dead of disasters cause epidemics of disease. *Am J Infect Control*, *34*(6), 331–334.

Kloner, R. A. (2006). Natural and unnatural triggers of myocardial infarction. *Prog Cardiovasc Dis*, *48*(4), 285–300. http://doi.org/10.1016/j.pcad.2005.07.001.

Kohsaka, S., Endo, Y., Ueda, I., Namiki, J., and Fukuda, K. (2012). Necessity for primary care immediately after the March 11 tsunami and earthquake in Japan. *Arch Intern Med*, *172*(3), 290–291. Retrieved from http://www.embase.com/search/results?subaction=viewrecord&from=export&id=L364317548.

Kokai, M., Fujii, S., Shinfuku, N., and Edwards, G. (2004). Natural disaster and mental health in Asia. *Psychiatry Clin Neurosci*, *58*(2), 110–116. http://onlinelibrary.wiley.com/doi/10.1111/j.1440-1819.2003.01203.x/epdf.

Korten, V. (2009). Earthquakes. *Clin Microbiol Infect*, 15(S14).

Kouadio, I. K., Aljunid, S., Kamigaki, T., Hammad, K., and Oshitani, H. (2012). Expert review of anti-infective therapy infectious diseases following natural disasters: Prevention and control measures. *Expert Rev Anti Infect Ther*, *10*(1), 95–104. http://doi.org/10.1586/ERI.11.155.

Koyama, A., Fuse, A., Hagiwara, J., Matsumoto, G., Shiraishi, S., Masuno, T., and Yokota, H. (2011). Medical relief activities, medical resourcing, and inpatient evacuation conducted by Nippon Medical School due to the Fukushima Daiichi Nuclear Power Plant accident following the Great East Japan Earthquake 2011. *J Nippon Med Sch*, *78*(6), 393–396. Retrieved from http://www.embase.com/search/results?subaction=viewrecord&from=export&id=L363112614.

Krug, E. G., Kresnow, M., Peddicord, J. P., Dahlberg, L. L., Powell, K. E., Crosby, A. E., and Marcie-Jo, K. (1998). Suicide after natural disasters. *N Engl J Med*, *338*(6), 373–378. http://doi.org/10.1056/NEJM199802053380607.

Kun, P., Chen, X., Han, S., Gong, X., Chen, M., Zhang, W., and Yao, L. (2009a). Prevalence of post-traumatic stress disorder in Sichuan Province, China after the 2008 Wenchuan earthquake. *Public Health*, *123*(11), 703–707. http://doi.org/10.1016/j.puhe.2009.09.017.

Kun, P., Han, S., Chen, X., and Yao, L. (2009b). Prevalence and risk factors for posttraumatic stress disorder: A cross-sectional study among survivors of the Wenchuan 2008 earthquake in China. *Depress Anxiety*, *26*(12), 1134–1140. http://doi.org/10.1002/da.20612.

Kuo, C.-J. J., Tang, H.-S. S., Tsay, C.-J. J., Lin, S.-K. K., Hu, W.-H. H., and Chen, C.-C. C. (2003). Prevalence of psychiatric disorders among bereaved survivors of a disastrous earthquake in Taiwan. *Psychiatr Serv*, *54*(2), 249–251. Retrieved from http://www.ncbi.nlm.nih.gov/pubmed/12556609.

Landry, M. D. (2010). Physical therapists in post-earthquake Haiti: Seeking a balance between humanitarian service and research. *Phys Ther*, *90*(7), 974–976. http://doi.org/10.2522/ptj.2010.90.7.974.

Landry, M. D., O'Connell, C., Tardif, G., and Burns, A. (2010a). Post-earthquake Haiti: The critical role for rehabilitation services following a humanitarian crisis. *Disabil Rehabil*, *32*(19), 1616–1618. http://doi.org/10.3109/09638288.2010.500345.

Landry, M. D., Quigley, A., Nakhle, A., and Nixon, S. A. (2010b). Implications of a gap between demand and supply for rehabilitation in post-earthquake Haiti. *Physiother Res Int*, *15*(3), 123–125. http://doi.org/10.1002/pri.488.

Li, S., Rao, L.-L., Bai, X.-W., Zheng, R., Ren, X.-P., Li, J.-Z., and Zhang, K. (2010). Progression of the "psychological typhoon eye" and variations since the Wenchuan earthquake. *PLoS One*, *5*(3), e9727. http://doi.org/10.1371/journal.pone.0009727.

Li, S., Rao, L. L., Ren, X. P., Bai, X. W., Zheng, R., Li, J. Z., and Liu, H. (2009). Psychological typhoon eye in the 2008 Wenchuan earthquake. *PLoS One*, *4*(3), e4964. http://doi.org/10.1371/journal.pone.0004964.

Li, T., Zhou, C., Liu, L., Gong, Q., Zeng, J., Liu, H., and Song, Y. (2009). Analysis of spinal injuries in Wenchuan earthquake [in Chinese]. *Zhongguo Xiu Fu Chong Jian Wai Ke Za Zhi*, *23*(4), 415–418. Retrieved from http://www.ncbi.nlm.nih.gov/entrez/query.fcgi?cmd=Retrieve&db=PubMed&dopt=Citation&list_uids=19431977.

Liang, N. J., Shih, Y. T., Shih, F. Y., Wu, H. M., Wang, H. J., Shi, S. F., and Wang, B. B. (2001). Disaster epidemiology and medical response in the Chi-Chi earthquake in Taiwan. *Ann Emerg Med*, *38*(5), 549–555. http://doi.org/10.1067/mem.2001.118999.

Liu, L., Tang, X., Pei, F. X., Tu, C. Q., Song, Y. M., Huang, F. G. et al. (2010). Treatment for 332 cases of lower leg fracture in "5.12" Wenchuan earthquake. *Chin J Traumatol*, *13*(1), 10–14.

Malinoski, D., Slater, M., and Mullins, R. (2004). Crush injury and rhabdomyolysis. *Crit Care Clin*, *20*, 171–192.

Mallick, M., Aurakzai, J. K., Bile, K. M., and Ahmed, N. (2010). Large-scale physical disabilities and their management in the aftermath of the 2005 earthquake in Pakistan. *East Mediterr Health J*, *16 Suppl*, S98–S105.

Michaelson, M. (1992). Crush injury and crush syndrome. *World J Surg*, *16*(5), 899–903. Retrieved from http://www.ncbi.nlm.nih.gov/pubmed/1462627.

Migl, K. S., and Powell, R. M. (2010). Physical and environmental considerations for first responders. *Crit Care Nurs Clin North Am*, *22*(4), 445–454. http://doi.org/10.1016/j.ccell.2010.10.002.

Montazeri, A., Baradaran, H., Omidvari, S., Azin, S. A., Ebadi, M., Garmaroudi, G., and Shariati, M. (2005). Psychological distress among Bam earthquake survivors in Iran: A population-based study. *BMC Public Health*, *5*, 4. http://doi.org/10.1186/1471-2458-5-4.

Morgan, O. (2004). Infectious disease risks from dead bodies following natural disasters. *Rev Panam Salud Publica*, *15*(5), 307–312.

Mulvey, J. M., Awan, S. U., Qadri, A. A., and Maqsood, M. A. (2008). Profile of injuries arising from the 2005 Kashmir earthquake: The first 72 h. *Injury*, *39*(5), 554–560. http://doi.org/10.1016/j.injury.2007.07.025.

Naghii, M. R. (2005). Public health impact and medical consequences of earthquakes. *Pan Am J Pub Health*, *18*(3), 216–221.

Najafi, I., Safari, S., Sharifi, A., Sanadgol, H., Hosseini, M., Rashid-Farokhi, F., and Boroumand, B. (2009). Practical strategies to reduce morbidity and mortality of natural catastrophes: A retrospective study based on Bam earthquake experience. *Arch Iran Med*, *12*(4), 347–352. http://www.ams.ac.ir/AIM /NEWPUB/09/12/4/005.pdf.

Neria, Y., Nandi, A., and Galea, S. (2007). Post-traumatic stress disorder following disasters: A systematic review. *Psychol Med*, *38*(4), 467–480. http://doi.org/10.1017/S0033291707001353.

Noji, E. K. (1992). Acute renal failure in natural disasters. *Ren Fail*, *14*(3), 245–249.

Noji, E. K. (2005). Public health issues in disasters. *Crit Care Med*, *33*(Suppl.), S29–S33. http://doi .org/10.1097/01.CCM.0000151064.98207.9C.

North, C. S., and Smith, E. M. (1990). Post-traumatic stress disorder in disaster survivors. *Compr Ther*, *16*(12), 3–9.

Oda, J., Tanaka, H., Yoshioka, T., Iwai, A., Yamamura, H., Ishikawa, K., and Sugimoto, H. (1997). Analysis of 372 patients with Crush syndrome caused by the Hanshin-Awaji earthquake. *J Trauma*, *42*(3), 470–475; discussion 475–476. Retrieved from http://www.ncbi.nlm.nih.gov/pubmed/9095115.

Ogłodek, E., and Araszkiewicz, A. (2011). Chronic and post-traumatic stress in the profession of medical rescuer worker. *Pol Merkur Lekarski*, *31*(182), 97–99. Retrieved from http://www.ncbi.nlm.nih.gov /pubmed/21936345.

O'Ryan, G. M. (2010). Reflections and feelings from a team of volunteers from the Ministry of Health and the Chilean Medical Association, after the recent earthquake [in Spanish]. *Rev Med Chil*, *138*(3), 270–273. http://www.scielo.cl/pdf/rmc/v138n3/art02.pdf.

Ozdogan, S., Hocaoglu, A., Caglayan, B., Imamoglu, O. U., and Aydin, D. (2001). Thorax and lung injuries arising from the two earthquakes in Turkey in 1999. *Chest*, *120*(4), 1163–1166.

Porter, K., and Greaves, I. (2003). Crush injury and crush syndrome: A consensus statement. *Emerg Nurse*, *11*(6), 26–30. Retrieved from http://ezp-prod1.hul.harvard.edu/login?url=http://search.ebscohost.com /login.aspx?direct=true&db=c8h&AN=2004090666&site=ehost-live&scope=site.

Priebe, S., Marchi, F., Bini, L., Flego, M., Costa, A., and Galeazzi, G. (2011). Mental disorders, psychological symptoms and quality of life 8 years after an earthquake: Findings from a community sample in Italy. *Soc Psychiatry Psychiatr Epidemiol*, *46*(7), 615–621. http://doi.org/10.1007/s00127-010-0227-x.

Quan, Y., Pan, X., Deng, S., Lu, S., Tao, S., Zhou, J., and Kang, X. (2009). Features of crush injury in Wenchuan earthquake and the corresponding operational methods [in Chinese]. *Zhongguo Xiu Fu Chong Jian Wai Ke Za Zhi*, *23*(5), 549–551.

Ran, Y.-C., Ao, X.-X., Liu, L., Fu, Y.-L., Tuo, H., and Xu, F. (2010). Microbiological study of pathogenic bacteria isolated from paediatric wound infections following the 2008 Wenchuan earthquake. *Scand J Infect Dis*, *42*(5), 347–350. http://doi.org/10.3109/00365540903510682.

Rathore, F. A., Farooq, F., Muzammil, S., New, P. W., Ahmad, N., and Haig, A. J. (2008). Spinal cord injury management and rehabilitation: Highlights and shortcomings from the 2005 earthquake in Pakistan. *Arch Phys Med Rehabil*, *89*(3), 579–585. http://doi.org/10.1016/j.apmr.2007.09.027.

Rathore, M. F. A., Rashid, P., Butt, A. W., Malik, A. A., Gill, Z. A., and Haig, A. J. (2007). Epidemiology of spinal cord injuries in the 2005 Pakistan earthquake. *Spinal Cord*, *45*(10), 658–663. http://doi .org/10.1038/sj.sc.3102023.

Rauch, A., Baumberger, M., Moise, F. G., von Elm, E., and Reinhardt, J. D. (2011). Rehabilitation needs assessment in persons with spinal cord injury following the 2010 earthquake in Haiti: A pilot study using an ICF-based tool. *J Rehabil Med*, *43*(11), 969–975. http://doi.org/10.2340/16501977-0896.

Sato, M., Fujita, S., Saito, A., Ikeda, Y., Kitazawa, H., Takahashi, M., and Aizawa, Y. (2006). Increased incidence of transient left ventricular apical ballooning (so-called "Takotsubo" cardiomyopathy) after the mid-Niigata Prefecture earthquake. *Circ J*, *70*(8), 947–953. https://www.jstage.jst.go.jp /article/circj/70/8/70_8_947/_pdf.

Schultz, C. H., Koenig, K. L., and Noji, E. K. (1996). A medical disaster response to reduce immediate mortality after an earthquake. *N Engl J Med*, *334*(7), 438–444. http://doi.org/10.1056/NEJM 199602153340706.

Sever, M. S., Erek, E., Vanholder, R., Akoglu, E., Yavuz, M., Ergin, H., and Lameire, N. (2002a). Clinical findings in the renal victims of a catastrophic disaster: The Marmara earthquake. *Nephrol Dial Transplant*, *17*(11), 1942–1949. Retrieved from http://www.ncbi.nlm.nih.gov/pubmed/12401851.

Sever, M. S., Erek, E., Vanholder, R., Kalkan, A., Guney, N., Usta, N., and Lameire, N. (2004). Features of chronic hemodialysis practice after the Marmara earthquake. *J Am Soc Nephrol*, *15*(4), 1071–1076. Retrieved from http://www.ncbi.nlm.nih.gov/entrez/query.fcgi?cmd=Retrieve&db=PubMed&dopt=Cit ation&list_uids=15034111.

Sever, M. S., Erek, E., Vanholder, R., Koc, M., Yavuz, M., Ergin, H., and Lameire, N. (2002b). Treatment modalities and outcome of the renal victims of the Marmara earthquake. *Nephron*, 92(1), 64–71. https://www.karger.com/Article/Pdf/64487.

Sever, M. S., Lameire, N., and Vanholder, R. (2009). Renal disaster relief: From theory to practice. *Nephrol Dial Transplant*, 24(6), 1730–1735. http://doi.org/10.1093/ndt/gfp094.

Sever, M. S., Vanholder, R., and Lameire, N. (2006). Management of crush-related injuries after disasters. *N Engl J Med*, 354, 1052–1063. Retrieved from http://www.nejm.org.ezp-prod1.hul.harvard.edu/doi /pdf/10.1056/NEJMra054329.

Shinfuku, N. (2002). Disaster mental health: Lessons learned from the Hanshin Awaji earthquake. *World Psychiatry*, 1(3), 158–159. Retrieved from http://www.ncbi.nlm.nih.gov/entrez/query.fcgi?cmd=Retriev e&db=PubMed&dopt=Citation&list_uids=16946841.

Situ, M. J., Zhang, Y., Zou, K., Gao, X., Fang, H., Jing, L. S., and Huang, Y. (2009). The mental health of children and adolescents from earthquake affected areas [in Chinese]. *Sichuan Da Xue Xue Bao Yi Xue Ban*, 40(4), 712–715, 723. Retrieved from http://www.ncbi.nlm.nih.gov/entrez/query.fcgi?cmd=Retrieve &db=PubMed&dopt=Citation&list_uids=19764580.

Soffer, Y., Jacob Wolf, J., and Ben-Ezra, M. (2011). Correlations between psychosocial factors and psychological trauma symptoms among rescue personnel. *Prehosp Disaster Med*, 26(3), 166–169. http://doi .org/10.1017/S1049023X11006224.

Sullivan, S., and Wong, S. (2011). An enhanced primary health care role following psychological trauma: The Christchurch earthquakes. *J Prim Health Care*, 3(3), 248–251.

Tahmasebi, M. N., Kiani, K., Mazlouman, S. J., Taheri, A., Kamrani, R. S., Panjavi, B., and Harandi, B. A. (2005). Musculoskeletal injuries associated with earthquake. A report of injuries of Iran's December 26, 2003 Bam earthquake casualties managed in tertiary referral centers. *Injury*, 36(1), 27–32. http://doi .org/10.1016/j.injury.2004.06.021.

Tan, C. M., Lee, V. J., Chang, G. H., Ang, H. X., and Seet, B. (2012). Medical response to the 2009 Sumatra earthquake: Health needs in the post-disaster period. *Singapore Med J*, 53(2), 99–103. Retrieved from http://www.embase.com/search/results?subaction=viewrecord&from=export&id=L364281044.

Tauqir, S. F., Mirza, S., Gul, S., Ghaffar, H., and Zafar, A. (2007). Complications in patients with spinal cord injuries sustained in an earthquake in northern Pakistan. *J Spinal Cord Med*, 30(4), 373–377.

Thomas, C. R. (2006). Psychiatric sequelae of disasters. *J Burn Care Res*, 27(5), 600–605. http://doi.org /10.1097/01.BCR.0000235463.20759.A0.

Thormar, S. B., Gersons, B. P. R., Juen, B., Marschang, A., Djakababa, M. N., and Olff, M. (2010). The mental health impact of volunteering in a disaster setting: A review. *J Nerv Ment Dis*, 198, 529– 538. Retrieved from https://ovidsp-tx-ovid-com.ezp-prod1.hul.harvard.edu/sp-3.25.0a/ovidweb.cgi? WebLinkFrameset=1&S=EIJNFPDFJGDDGKJDNCGKAGOBPGHCAA00&returnUrl=ovidweb. cgi%3F%26Full%2BText%3DL%257cS.sh.22.23%257c0%257c00005053-201008000-00001%26S%3 DEIJNFPDFJGDDGKJDNCGKAGOBPGH.

Tomita, K., Hasegawa, Y., Watanabe, M., Sano, H., Hitsuda, Y., and Shimizu, E. (2005). The Totton-Ken Seibu earthquake and exacerbation of asthma in adults. *J Med Invest*, 52(1–2), 80–84. Retrieved from http://www.ncbi.nlm.nih.gov/entrez/query.fcgi?cmd=Retrieve&db=PubMed&dopt=Citation&list _uids=15751277.

Toyabe, S., Shioiri, T., Kobayashi, K., Kuwabara, H., Koizumi, M., Endo, T., and Akazawa, K. (2007). Factor structure of the General Health Questionnaire (GHQ-12) in subjects who had suffered from the 2004 Niigata-Chuetsu earthquake in Japan: A community-based study. *BMC Public Health*, 7, 175. http://doi .org/10.1186/1471-2458-7-175.

Tsai, C.-H. H., Lung, F.-W. W., and Wang, S.-Y. Y. (2004). The 1999 Ji-Ji (Taiwan) earthquake as a trigger for acute myocardial infarction. *Psychosomatics*, 45(6), 477–482. http://doi.org/10.1176/appi .psy.45.6.477.

van der Velden, P. G., van Loon, P., Benight, C. C., and Eckhardt, T. (2012). Mental health problems among search and rescue workers deployed in the Haiti earthquake 2010: A pre-post comparison. *Psychiatry Res*, 198(1), 100–105. Retrieved from http://www.embase.com/search/results?subaction=viewrecord&f rom=export&id=L51938687.

Vanholder, R., Sever, M. S., Erek, E., and Lameire, N. (2000). Acute renal failure related to the crush syndrome: Towards an era of seismo-nephrology? *Nephrol Dial Transplant*, 15(10), 1517–1521. Retrieved from http://www.ncbi.nlm.nih.gov/entrez/query.fcgi?cmd=Retrieve&db=PubMed&dopt=Citation&list _uids=11007816.

Vanholder, R., Sever, M. S., and Lameire, N. (2008). The role of the Renal Disaster Relief Task Force. *Nat Clin Pract Nephrol*, 4(7), 347. http://doi.org/10.1038/ncpneph0862.

VanRooyen, M., and Leaning, J. (2005). After the tsunami – Facing the public health challenges. *N Engl J Med*, *352*(5), 435–438. http://doi.org/10.1056/NEJMp058013.

Vehid, H. E., Alyanak, B., and Eksi, A. (2006). Suicide ideation after the 1999 earthquake in Marmara, Turkey. *Tohoku J Exp Med*, *208*(1), 19–24. https://pdfs.semanticscholar.org/f4ed/2201b97f23e38de374 2e08c6642bb08c951f.pdf.

Vieweg, W. V., Hasnain, M., Mezuk, B., Levy, J. R., Lesnefsky, E. J., and Pandurangi, A. K. (2011). Depression, stress, and heart disease in earthquakes and Takotsubo cardiomyopathy. *Am J Med*, *124*(10), 900–907. http://doi.org/10.1016/j.amjmed.2011.04.009.

Wang, H., Jin, H., Nunnink, S. E., Guo, W., Sun, J., Shi, J., Baker, D. G. (2011). Identification of post traumatic stress disorder and risk factors in military first responders 6 months after Wen Chuan earthquake in China. *J Affect Disord*, *130*(1–2), 213–219. http://doi.org/10.1016/j.jad.2010.09.026.

Wang, H. Y., Duan, X., Chen, Y., and Li, J. (2009). Microbiologic study on the pathogens isolated from wound culture among orthopaedic patients after Wenchuan earthquake [in Chinese]. *Zhongguo Gu Shang*, *22*(12), 910–912.

Wang, L., Lei, D., He, L., Liu, Y., Long, Y., Cao, J., and Zhao, Y. (2009). The association between roofing material and head injuries during the 2008 Wenchuan earthquake in China. *Ann Emerg Med*, *54*(3), e10–e15. http://doi.org/10.1016/j.annemergmed.2009.03.028.

Wang, X., Gao, L., Shinfuku, N., Zhang, H., Zhao, C., and Shen, Y. (2000). Longitudinal study of earthquake-related PTSD in a randomly selected community sample in north China. *Am J Psychiatry*, *157*(8), 1260–1266. Retrieved from http://www.ncbi.nlm.nih.gov/entrez/query.fcgi?cmd=Retrieve&db=PubMed&dop t=Citation&list_uids=10910788.

Watson, J. T., Gayer, M., and Connolly, M. A. (2007). Epidemics after natural disasters. *Emerg Infect Dis*, *13*(1), 1–5. Retrieved from http://www.pubmedcentral.nih.gov/articlerender.fcgi?artid=2725828&tool=p mcentrez&rendertype=abstract.

Wen, J., Shi, Y.-K., Li, Y.-P., Yuan, P., and Wang, F. (2012). Quality of life, physical diseases, and psychological impairment among survivors 3 years after Wenchuan earthquake: A population based survey. *PLoS One*, *7*(8), e43081. http://doi.org/10.1371/journal.pone.0043081.

Wood, D. P., and Cowan, M. L. (1991). Crisis intervention following disasters: Are we doing enough? (A second look). *Am J Emerg Med*, *9*(6), 598–602. Retrieved from http://www.ncbi.nlm.nih.gov/entrez/query .fcgi?cmd=Retrieve&db=PubMed&dopt=Citation&list_uids=1930406.

Xiao, M., Li, J., Zhang, X., and Zhao, Z. (2011). Factors affecting functional outcome of Sichuan-earthquake survivors with tibial shaft fractures: A follow-up study. *J Rehabil Med*, *43*(6), 515–520. http://doi .org/10.2340/16501977-0813.

Xie, Y., Chen, Z. X., Tao, C. M., Kang, M., Chen, H. L., Fan, H., and Wang, L. L. (2009). Etiology of infections in the wounded victims of Wenchuan earthquake [in Chinese]. *Zhonghua Yi Xue Za Zhi*, *89*(6), 366–370.

Xu, J., You, C., Zhou, L., Wu, B., Li, X., Li, Z., and Yuan, Y. (2011a). Long-term results of patients with head injuries treated in different hospitals after the Wenchuan, China, earthquake. *World Neurosurg*, *75*(3–4), 390–396. http://doi.org/10.1016/j.wneu.2011.02.006.

Xu, J., Zhang, Y., Chan, J., Li, N., Yang, Y., and Li, J. (2011b). A comparison of the acute stress reactions between the Han and Tibetan ethnic groups in responding to devastating earthquakes. *Int J Psychiatry Med*, *42*(2), 167–180.

Yang, Y. F., Ye, Y. L., Li, T., Liu, X. X., and Yuan, P. (2010). Mental health status among middle school students in Wenchuan earthquake region. *Zhonghua Yu Fang Yi Xue Za Zhi*, *44*(2), 134–139. Retrieved from http://www.embase.com/search/results?subaction=viewrecord&from=export&id=L358965766.

Yang, Y. N., Xiao, L. D., Cheng, H. Y., Zhu, J. C., and Arbon, P. (2010). Chinese nurses' experience in the Wenchuan earthquake relief. *Int Nurs Rev*, *57*(2), 217–223. Retrieved from http://www.embase.com /search/results?subaction=viewrecord&from=export&id=L359142534.

Yasamy, M., Farajpur, M., Gudarzi, S. S., Aminesmaeeli, M., Bahramnezhad, A., Mottaghipour, Y. et al. (2005). Second phase of psychological intervention in Bam. Report to UNICEF. Ministry of Health and Medical Education, Deputy for Health, Bureau for Psychosocial and School Health, Mental Health Office.

Yasin, M. A., Malik, S. A., Nasreen, G., and Safdar, C. A. (2009). Experience with mass casualties in a sub-continent earthquake. *Ulus Travma Acil Cerrahi Derg*, *15*(5), 487–492.

Yokota, J. (2005). Crush syndrome in disaster. *Japan Med Assoc J*, *48*(7), 341–352.

Yoshimura, N., Nakayama, S., Nakagiri, K., Azami, T., Ataka, K., and Ishii, N. (1996). Profile of chest injuries arising from the 1995 southern Hyogo Prefecture earthquake. *Chest*, *110*(3), 759–761. http://doi .org/10.1378/chest.110.3.759.

Zarifeh, J. A., Mulder, R. T., Kerr, A. J., Chan, C. W., and Bridgman, P. G. (2012). Psychology of earthquake-induced stress cardiomyopathy, myocardial infarction and non-cardiac chest pain. *Intern Med J*, *42*(4), 369–373. http://doi.org/10.1111/j.1445-5994.2012.02743.x.

Zhang, J. L., He, H. C., Lin, H. D., Luo, Q. L., Xia, L., Li, S. S., and He, C. Q. (2011). Motor function and activities of daily living capacity of patients with fractures sustained during the Wenchuan earthquake. *Chin Med J (Engl)*, *124*(10), 1504–1507.

Zhang, L., Liu, Y., Liu, X., and Zhang, Y. (2011). Rescue efforts management and characteristics of casualties of the Wenchuan earthquake in China. *Emerg Med J*, *28*(7), 618–622. http://doi.org/10.1136/emj.2009.087296.

Zhang, W. Q., Liu, C., Sun, T. S., Zhao, J., Han, J. Q., Yang, Y. H. et al. (2011). Physical and mental health status of soldiers responding to the 2008 Wenchuan earthquake. *Aust N Z J Public Health*, *35*(3), 207–211. http://doi.org/10.1111/j.1753-6405.2011.00680.x.

Zhang, X. Q., Chen, M., Yang, Q., Yan, S. D., Huang, D. J., and Huang, D. J. (2009). Effect of the Wenchuan earthquake in China on hemodynamically unstable ventricular tachyarrhythmia in hospitalized patients. *Am J Cardiol*, *103*(7), 994–997. http://doi.org/10.1016/j.amjcard.2008.12.009.

Zhang, Z., Shi, Z., Wang, L., and Liu, M. (2011). One year later: Mental health problems among survivors in hard-hit areas of the Wenchuan earthquake. *Public Health*, *125*(5), 293–300. http://doi.org/10.1016/j.puhe.2010.12.008.

4 Utilization of Satellite Geophysical Data as Precursors for Earthquake Monitoring

T.J. Majumdar

CONTENTS

4.1   INTRODUCTION

An earthquake is the trembling or shaking movements of the Earth's surface. Earthquakes usually begin with slight tremors, rapidly take the form of one or more violent shocks and end in vibrations of gradually diminishing force called aftershocks. Most earthquakes originate due to various reasons, which fall into two major categories (tectonic and non-tectonic). Most of the geophysical processes, such as earthquakes, landslides and glacier flows, are associated with crustal movements at the surface of the Earth. Measurement of such movements provides vital clues about the mechanism of the events, like earthquakes and landslides. Conventionally, ground-based geodetic techniques are used for the detection and measurement of small crustal movements, but these techniques have their own limitations, such as sparse sampling network and higher cost involved. Remote sensing from the space platforms provides an alternative to carry out such studies. It is possible to detect small crustal movements, including swelling and buckling of fault zones, residual displacement from seismic events and slowly progressing landslides, using space-based microwave remote sensing techniques (Tronin 1996; Majumdar and Massonnet 2002; Chatterjee et al. 2007b). The current tectonic activity in Asia is attributed to the underthrusting of the Indian plate beneath the Eurasian plate; stresses are being generated and released through earthquakes of various magnitudes (USGS website 1). At present, appreciable movements are taking place all along the Himalayas. Geodetic measurements carried out by the Survey of India show that both horizontal and vertical movement are taking place, practically all along the thrusts of the Himalayas. Earthquakes occurring in and around central India are driven by the interaction of collisional stresses associated with flexure of the Indian plate. The imposition of this stress field on the northward-moving Indian plate appears fundamental to explain the current distribution of intraplate earthquakes and their mechanisms, including Bhuj, Latur and Koyna Earthquakes. The overall flexural stress distribution provides a physical basis for earthquake hazards mapping and monitoring (Gaur 2003).

The devastating earthquake that struck the Indian Ocean near Sumatra (Indonesia) on 26 December 2004 was extremely powerful and immensely damaged a number of countries, including Indonesia, Sri Lanka and India. The devastating earthquake has also accelerated the Earth's rotation (Hopkin 2004). TOPEX/POSEIDON and Jason-1 satellite altimeters passed over the Bay of Bengal 2 h after the massive earthquake and captured the only measurements of the corresponding tsunami height in deep waters. The satellites recorded the first two wavefronts caused by the main earthquake, spaced 500–800 km apart. These fronts reached a maximum height of 50 cm in the open ocean (McKee 2005). At most subduction zones, seismicity involves very large earthquakes characterized by a thrust-faulting mechanism, expressing the overriding of the subducting oceanic plate by the other plate (Okal et al. 1980). Variations in the gravity field can be utilized as a predictor of seismic behaviour (Song and Simon 2003).

The Indian Ocean has experienced, along with three main phases of seafloor spreading, two major plate reorganizations from the late Jurassic to the present (Bhattacharya and Chaubey 2001; Ramana et al. 2001). No earthquake precursor signal exists which can provide an early warning about an impending earthquake. The method of development and study of earthquake patterns is extremely complicated, and hence it is difficult to come to meaningful conclusive information. Changes in surface temperature and gravity in the case of land and changes in gravity in the case of ocean can very well work as precursors for earthquake monitoring. With this background, four case studies with thermal and gravity precursors are discussed below, two over land and two over ocean.

4.2 SURFACE TEMPERATURE ANOMALIES OVER GUJARAT, INDIA, AND THEIR POSSIBLE CORRELATION WITH EARTHQUAKE OCCURRENCES

4.2.1 THERMAL PRECURSOR

A few days before the occurrence of an earthquake, the average temperature of the area keeps increasing. Weather report bulletins refer to temperatures above or below average by a few degrees. It is seen that where the area is heading for an earthquake, the average temperature increases (Qiang et al. 1997; Saraf and Choudhury 2005). At times, on the day of the earthquake it is about 5°C above the average normal temperature. Moderate Resolution Imaging Spectroradiometer (MODIS) thermal infrared (IR) data over Bhuj, Gujarat, have been studied, and sudden increases in surface temperatures were observed during the deadly Bhuj Earthquake of 26 January 2001 (National Aeronautics and Space Administration [NASA] website).

4.2.2 SURFACE TEMPERATURE DATA FROM THERMAL INFRARED

An IR remote sensing imagery records the surface radiance emitted from the Earth. If instrument calibration and atmospheric and topographic corrections are made, the resulting product represents the surface temperature (Sabins 1997; Majumdar and Mohanty 1998). The surface temperature is the most variable heat source, which is transferring heat across the Earth atmosphere interface. Utilization of thermal IR data for geothermal zoning for regional and petroleum geology on the basis of satellite IR thermal survey and for the studies of seismoactive regions has been discussed in detail elsewhere (Verma et al. 1968; Kahle 1977; Gupta 1981; Gorny et al. 1996; Tronin 1996; Nasipuri et al. 2006).

The main objectives include (1) the generation and study of the surface temperature patterns and anomalies over Gujarat using National Oceanic and Atmospheric Administration (NOAA) Advanced Very High Resolution Radiometer (AVHRR) daytime data for 6 days during the occurrences of tremors, and (2) utilization of the same for demarcation of change in surface temperature as a precursor to impending earthquakes. The area chosen for this study is shown in Figure 4.1.

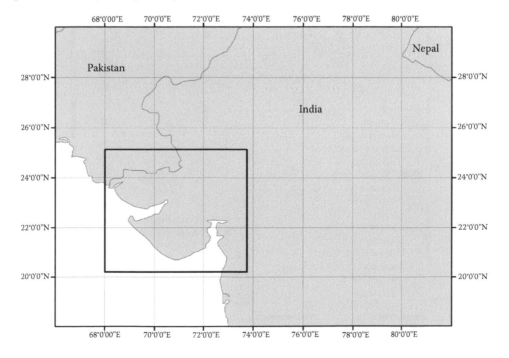

FIGURE 4.1   Location of the study area for thermal precursor studies.

4.2.3 GEOLOGY OF THE STUDY AREA

Geologically, Gujarat represents a wide spectrum of rock types of different ages (Figure 4.2). The geology of Gujarat consists of three distinct physiographic units (Merh 1995): (1) mainland Gujarat, (2) Saurashtra Peninsula and (3) the Kachchh region. The geology of the Kachchh region is discussed.

4.2.3.1 Kachchh Region

It is an important site of Mesozoic and Cenozoic sedimentation. Tectonically, Kachchh is situated at the area where the northwestern margin of the Indian continental shield meets the geosynclinal belt of the Sindh–Baluchistan. The Kachchh mainland is bounded by the following major regional faults which are responsible for its existing configuration, including (1) the Kachchh Mainland Fault striking east-southeast to west-northwest to east-west, forming the northern limit of the mainland; (2) the offshore West Coast Fault striking northwest-southeast and marking the western limit; and (3) the Gulf of Kachchh Fault and the Little Rann of Kachchh fault system bounding the southern and eastern limits of the Kachchh mainland, respectively (Biswas 1982).

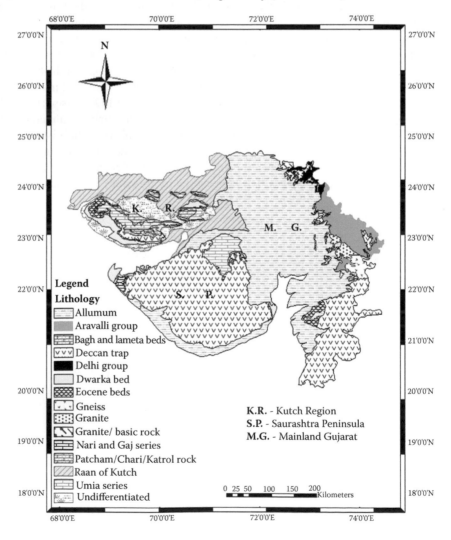

FIGURE 4.2 Geological map over Gujarat. (After Merh, S. S., The geology of Gujarat, Geological Society of India, Bangalore, 1995.)

It is now more or less established that in Kachchh, seismicity has been a manifestation of continued tectonism, mostly along pre-existing planes of weaknesses (faults, major joints, etc.). Figure 4.3 shows possible active faults recorded in the region of Kachchh (Karanth 2000). *Kachchh falls under seismically active zone V* of the Indian subcontinent. With the increasing socioeconomic development of the Kachchh region, it is essential to reliably estimate the local earthquake risk. Kachchh has produced two great earthquakes in recent history, the 1819 and 2001 earthquakes. Figure 4.4 shows the spatio-temporal distribution of earthquakes in the Kachchh region. The earthquake epicentres lie along the faults (Karanth 2000; Karanth and Gadhvi 2007).

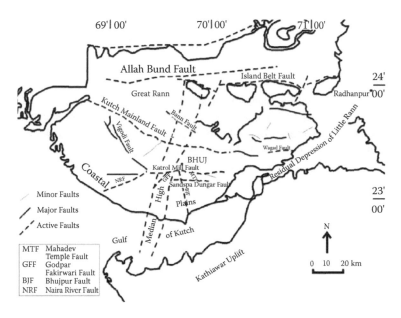

FIGURE 4.3 Map showing active faults in the Kachchh region. (After Karanth, R. V., Neotectonic and palaeoseismic studies in Kutchch and its adjoining areas, unpublished project report submitted to the Department of Science and Technology, New Delhi, 2000.)

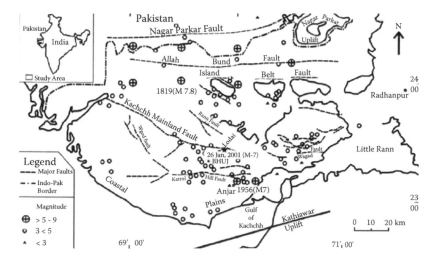

FIGURE 4.4 Spatio-temporal distribution of earthquakes in the Kachchh region. (After Karanth, R. V., Neotectonic and palaeoseismic studies in Kutchch and its adjoining areas, unpublished project report submitted to the Department of Science and Technology, New Delhi, 2000.)

TABLE 4.1

Occurrences of Tremors in Gujarat during 2006 (Up to April 2006)

Date	Magnitude
3 February 2006	4.4
7 March 2006	5.2
6 April 2006	5.5
6 April 2006	4.7
10 April 2006	4.8

4.2.4 Image Data Source

The NOAA series of polar orbiting meteorological satellites' onboard sensor, the AVHRR, generates data with a spatial resolution of 1.1 km at the nadir and a radiometric resolution of 10 bits. The Landsat Ground Station Operations Working Group (LGSOWG) data contain raw AVHRR spectral data as well as calibration coefficients, solar zenith angles, Earth location and other auxiliary data (NOAA 1990). ENVI 4.2 software (RSI 2001) has been used to extract images from the LGSOWG (BIL) format.

4.2.4.1 Earthquake Occurrences in Gujarat during January–April 2006

A list of mild tremors occurring in the Gujarat region with sources near Bhuj during January–April 2006 is given in Table 4.1. A set of NOAA 17 and 18 were acquired to analyze the earthquakes listed in Table 4.1.

4.3 METHODOLOGY

4.3.1 Geometric Correction of NOAA AVHRR Data

Extraction of the area of interest from NOAA AVHRR data and georeferencing the extracted image with eight ground control points using a second-order polynomial was carried out.

4.3.2 Calibration of AVHRR Data

AVHRR thermal data (channels 4 and 5) were converted to temperatures using a lookup table (LUT).

4.3.3 Retrieval of Brightness Temperature

The retrieval of brightness temperature from the thermal bands of channels 4 and 5 of NOAA AVHRR images was done by using inverse Planck's equation (Saraf et al. 1995):

$$T = c_2 \Big/ \left\{ \lambda \ln \left[1 + \left(\varepsilon \lambda c_1 \lambda^{-5} / E(\lambda) \right) \right] \right\}$$

(4.1)

where T is the temperature (K) for the energy value $E(\lambda)$, c_1 and c_2 are constants, $E(\lambda)$ is the radiance energy value in the thermal IR band and ε is the emissivity.

4.3.4 Retrieval of Land Surface Temperature

Terrestrial surface temperature measurements by sensors are attenuated by the Earth's atmosphere by decreasing the observed brightness temperature (due to water vapour) as received by the sensor. In a non-cloudy region of interest, the major perturbation factor is due to the existence of atmospheric water vapour and different techniques exist for its evaluation (Price 1984; Majumdar and Bhattacharya 1988; Prata et al. 1995). Among the existing techniques, the split-window channel method is the most popularly used one in calculating the surface temperature over oceans, as the case is simplified there due to the homogeneity of a pixel (Prabhakara et al. 1974). However, the situation becomes complex over land for the following reasons: (1) the land surface temperature is generally not homogeneous within a pixel, as in the case of the ocean; (2) the difference between land surface temperature and air temperature near the surface is larger for land surfaces than for the sea; and (3) land surface emissivities may be quite different from unity and spectrally variable, which is not the case for the sea (Cooper and Asrar 1988; Vidal 1991; Coll et al. 1994; Majumdar and Mohanty 1998).

4.3.5 ATMOSPHERIC ATTENUATION CORRECTION FOR RETRIEVING SURFACE TEMPERATURE

AVHRR channels 4 and 5 lie in the thermal region of the electromagnetic spectrum, where attenuation is relatively small; however, significant error is introduced in AVHRR-sensed surface temperature. Several split-window atmospheric attenuation correction models have been developed by a number of investigators (McClain et al. 1983; Price 1984; Singh 1984; Cooper and Asrar 1988; Majumdar and Mohanty 1998). A number of models have been used, of which the model proposed by Singh (1984) is considered one of the best in tropical environments:

$$Tc = 1.699\,(\text{ch}.4) - 0.699\,(\text{ch}.5) - 0.240 \qquad (4.2)$$

where Tc is the atmosphere-corrected surface temperature, ch.4 is the calibrated surface temperature from channel 4 and ch.5 is the calibrated surface temperature from channel 5.

All temperatures in the above equations are in degrees kelvin.

4.4 RESULTS

Figure 4.5 shows the surface temperature changes over Gujarat during tremors on 6 April 2006 using NOAA data after calibration and atmospheric correction using the model proposed by Singh (1984). Similarly, Figure 4.6 shows the surface temperature changes over Gujarat during tremors on 10 April 2006. Validation of the surface temperature patterns has been performed utilizing India Meteorological Department (IMD) data over meteorological stations including Bhuj, Rajkot and Ahmedabad. A comparison obtained on 6 April 2006 of daytime NOAA AVHRR surface temperature data with in situ surface temperatures was found to be satisfactory (IMD Weather Bulletin 2006; Majumdar and Rao 2009). A sharp increase in surface temperatures (~6°C–7°C) was observed during 3–6 April 2006, which was observed to decrease on 7 April (Figure 4.5). During 9–10 April, there was again a rise in surface temperatures (~6°C), which sharply decreased on 11 April 2006 (Figure 4.6). However, the surface temperature imagery of 7–11 April 2006 was contaminated with scattered clouds. Since the earthquakes occurred on 6 and 10 April 2006 (Table 4.1), surface temperature anomalies could be utilized as a precursor for such catastrophic events (Majumdar and Rao 2009). However, the surface temperature may suddenly increase due to other meteorological phenomena, so the increase in temperature cannot be considered an earthquake precursor in each case.

4.5 DISCUSSIONS

NOAA AVHRR daytime thermal IR data were calibrated and converted to surface temperature using a suitable model for atmospheric correction with a split-window channel computation technique. In the present condition, the model proposed by Singh (1984) was considered the best for the comparison of surface temperature data obtained from the IMD over a few meteorological stations

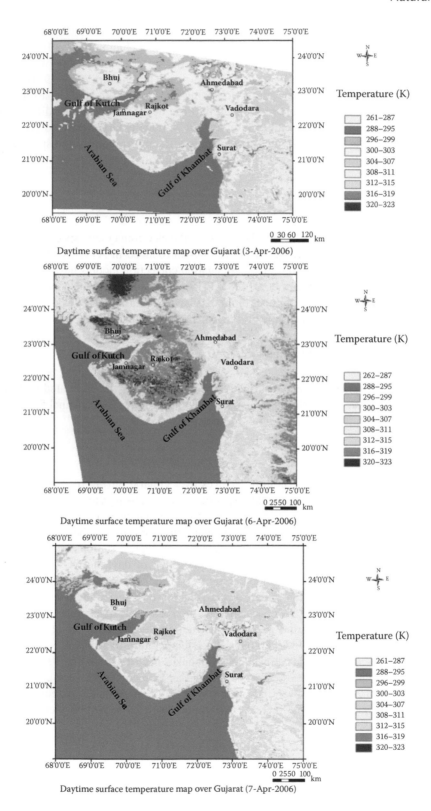

FIGURE 4.5   Surface temperature changes over Gujarat during tremors on 6 April 2006 using NOAA thermal IR data.

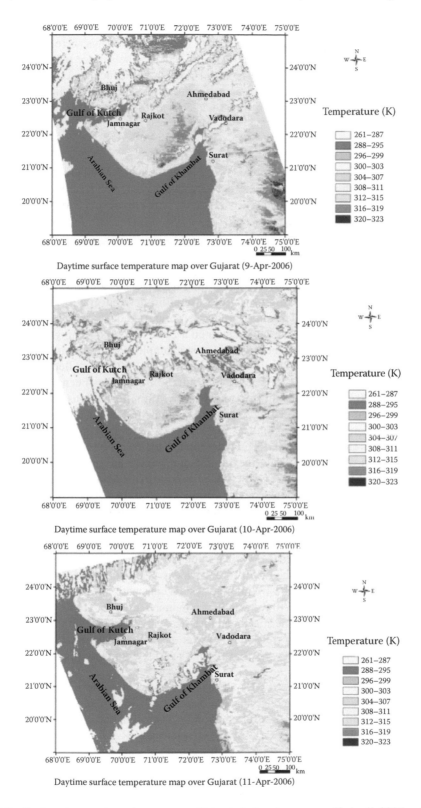

FIGURE 4.6 Surface temperature changes over Gujarat during tremors on 10 April 2006 using NOAA thermal IR data.

in Gujarat. After comparison of the surface temperature data for different dates before and after the mild tremors occurred near Bhuj in April 2006, it was observed that there was indeed a rise in the surface temperature (~6°C–7°C) around April 6 and 10. The surface temperatures decreased after the mild tremors occurred on April 6 and 10. However, it is difficult to conclude whether every seismic disturbance is preceded with surface temperature anomalies as a precursor, or whether each rise in surface temperature anomaly will be associated with the earthquakes or tremors.

4.6 GRAVITY PRECURSOR OVER LAND

The major objectives were (1) to generate and study the satellite-derived gravity over the Kachchh Earthquake region with reference to the severe earthquakes that occurred on 26 January 2001 in Bhuj and on 21 July 1956 in Anjar, and (2) to utilize the same for demarcation of change in gravity as a precursor to impending earthquakes.

4.6.1 Data Sources and the Study Area of Interest

The study area and the generalized geological map over Gujarat are shown in Figures 4.1 and 4.2, respectively. The details of high-resolution satellite-derived gravity data have been discussed by Hwang et al. (2002). They carried out very detailed data assimilation for calculation of the deflection of the vertical and then generated a 2 × 2 min (4 × 4 km) grid over oceans and geoidal gravity using a high-resolution geoid, for example, Earth Gravitational Model 1996 (EGM96), over land (Majumdar et al. 2001; Hwang et al. 2002). Detailed quality evaluation of marine gravity (Hwang et al. 2002) over the Indian offshore has been discussed elsewhere (Chatterjee et al. 2007a,b). Root mean square errors (RMSEs) for various satellite-derived profiles with ship-borne gravity over the Arabian Sea and the Bay of Bengal have been found to be within ±3–6 mGal, which is quite satisfactory (Chatterjee et al. 2007b). The National Geophysical Research Institute (NGRI), Hyderabad, has taken several in situ gravimeter surveys over Kachchh and other places in Gujarat and generated an in situ gravity map over this region (NGRI Map Series 1978). The NGRI land gravity data grid size is approximately 0.5° × 0.5°. Recently, Gravity Recovery and Climate Experiment (GRACE) gravity data (grid size ~0.1° × 0.1°) have been generated over land which are utilized over the Kachchh region (GRACE website 1). Figure 4.7a–c shows various gravity images over the Kachchh region after superimposition with the tectonic map.

GRACE (website 2) was launched in 2002, carrying the two satellites into orbit. The twin GRACE satellites – Tom and Jerry – orbit the Earth 16 times a day at an altitude of 311 miles. Separated by 137 miles (220 km), a precise microwave ranging system constantly measures the distance between them to within the equivalent of 1/10 the width of a human hair. That ability permits scientists on the ground to monitor changes in the speed and distance, which indicate differences in the mass of the Earth's surface below and the corresponding variations in its gravitational pull. GRACE obtains a gravity field map by looking at how the Earth's mass varies from place to place on the surface as the twin satellites pass over.

In the present study, we have used GRACE data processed at a grid interval of 0.1° × 0.1° for 2 years over the study area. This model is a combination of GRACE and the Laser Geodynamics Satellite (LAGEOS) mission plus 0.5° × 0.5° gravimetry and altimetry surface data and is complete to 360° and order in terms of spherical harmonic coefficients (GRACE website 1). Combined high-resolution gravity models are essential for the static gravity potential, and its gradients are needed in the medium- and short-wavelength spectrum. GRACE data analysis is given in detail (http://icgem .gfz-potsdam.de/home).

The Lacoste Romberg Gravimeter (LRG) is used by the NGRI, Hyderabad, India, for the generation of land gravity maps. Only land gravity data have been utilized to interpolate over nearby ocean regions as well, and no ship-borne data corrections were required (Figure 4.7c).

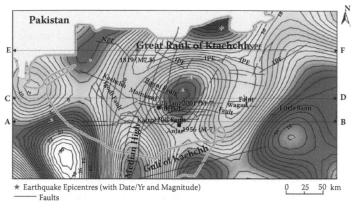

(a)

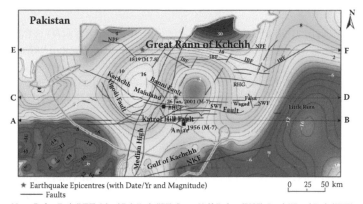

(b)

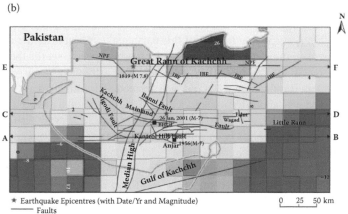

(c)

FIGURE 4.7 (a) GRACE gravity image of the Kachchh region after superimposition with the tectonic map. (After Biswas, S. K., *Curr. Sci.*, 88, 1592–1600, 2005.) (b) Hwang gravity image of the Kachchh region after superimposition with the tectonic map. (c) NGRI gravity image of the Kachchh region after superimposition with the tectonic map.

4.7   METHODOLOGY

A geoid is an equipotential surface over the Earth, and details of geoid undulation over the Indian land mass have been discussed by Moritz (1980) and Majumdar et al. (2001). EGM96 geoid data have been extensively used for geoid surface calculation over land. Theoretical geoids are converted to gravity using the following technique.

4.7.1   Gravity Anomaly Modelling Using Geoid

The fast Fourier transform (FFT) approach uses geoids for the computation of free-air gravity anomalies, which is based on a flat-Earth approximation and is derived from two fundamental equations, namely, Brun's equation and the equation of physical geodesy (Moritz 1980). The relation between gravity anomaly and geoid undulation, derived from the above two fundamental equations, is given by Chapman (1979):

$$F(\Delta g) = g_o |k| F(N) \tag{4.3}$$

where $F(\Delta g)$ is the Fourier transform of the free-air gravity anomaly, $F(N)$ is the Fourier transform of geoid undulation, g_o is the normal gravity and $|k|$ is the one-dimensional wave number associated with wavelength λ.

GRACE gravity data over Kachchh have been generated and utilized for further comparison with other gravity data as obtained by Hwang et al. (2002) and NGRI-derived in situ gravity data (NGRI Map Series 1978).

4.8   RESULTS AND DISCUSSION

High-resolution gravity images have been generated over a part of the Bhuj region and superimposed over a tectonic map (after Biswas 2005) with the epicentres of earlier earthquakes, including the earthquake of 26 January 2001 in Bhuj (epicentre: 23.5° N, 69.8° E/M_w 7.0) (Figure 4.7a–c) (Karanth 2000). The Anjar earthquake occurred on 21 July 1956. Considerable damage occurred at Anjar and a number of villages in the central mainland of Kachchh. Based on reports, the maximum Modified Mercalli (MM) Intensity was IX and the magnitude was 7. The epicentral location was 23.2° N, 70.0° E (Karanth 2000). A surprising feature of this earthquake is that after the initial rapid rapture, subsequent slip of the plate interface occurred with decreasing speed towards the north (Bilham and Gaur 2000; Bilham 2005). On 16 June 1819, the Kachchh Earthquake occurred at the Island Belt Fault (IBF) and the Allah Bund, or Dam of God, was formed. The bund had a steep slope, rose like a wall above the plain and was about 16 miles in length. This earthquake affected several places. The maximum MM Intensity was XI, and its epicentral location was 24.2° N, 69.2° E (Karanth 2000).

On land, the gravity anomaly has been reported to be as high as –37 mGal near the epicentral region near the IBF. The high-magnitude Gujarat Earthquakes (Bhuj, Anjar and the IBF) occurred due to the collision of the Indo-Australian and Eurasian plates at the rate of 2 cm per year, thus leaving pent-up pressure which was released instantly. All these earthquakes occurred in the Kutchch rift, an example of a failed rift that is now affected by stresses and not by pull. Fault plane solutions and other tectonic studies related to the Bhuj Earthquake of 26 January 2001 and other earthquakes have been discussed by Li et al. (2002).

High-resolution satellite gravity data have been utilized to infer subsurface geological structures in the area of devastating earthquakes that struck the Kachchh region in Gujarat in 1819, 1956 and 2001. Three latitudinal gravity profiles have been generated in the Bhuj, Anjar and IBF regions across the epicentre (23.5° N, 69.8° E/M_w 7.0 in 2001; 23.2° N, 70° E/M_w 7.0 in 1956; 24.2° N, 69.2° E/M_w 7.8 in 1819) regions (Figure 4.8a–c). The arrow marks in Figure 4.8a,b show the land–sea crossings, with C marked below. A substantial difference in gravity anomaly patterns as high as 37 mGal is

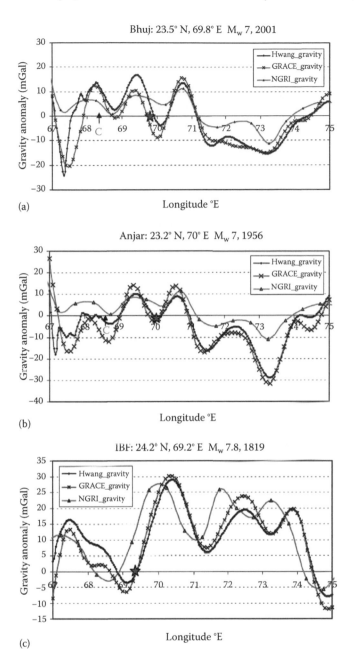

FIGURE 4.8 Latitudinal gravity profiles across epicentre regions: (a) Bhuj, (b) Anjar and (c) the IBF. The arrow marks in parts (a) and (b) show the land–sea crossings, with C marked below.

observed near the epicentral regions (Figure 4.8c). These gravitational differences might have been caused due to plate tectonic processes, whereas minor and major gravitational differences in other areas may be due to the changes in densities of different lithospheric zones and sedimentary layers (Mahapatra 1987). This change of 37 mGal is not caused due to the earthquake, but there is already an existing difference of 37 mGal near the epicentre which has the potential to cause such big earthquakes, and hence we call it a 'gravity precursor' (Majumdar and Bhattacharyya 2010). Hence, our study further supports the idea that drastic changes in gravity and/or topography anomalies can be considered precursors for the occurrences of large earthquakes in a seismically active zone.

Generated data grid sizes are similar for Huang and GRACE gravity data, whereas the gravity map prepared by the NGRI was comparatively poorer (0.5° × 0.5°). The first two could properly detect the changes and differences that exist near the epicentral regions (Figure 4.8a–c); however, NGRI gravity failed at places due to its poorer resolution.

Song and Simon (2003) considered variations in the gravity field as a predictor of seismic behaviour. They found that within subduction zones, areas where the attraction due to gravity is relatively high are less likely to experience large earthquakes than areas where the gravitational force is relatively low. Similar results were reported in the Sumatra region by Chatterjee et al. (2007b).

Second, temporal variations of gravity may also give further input. NGRI in situ data are as old as 1978, Hwang land gravity data are based on the EGM96 geoid and GRACE observations are from 2003 onwards, so the temporal variations can be observed. Some change observed between Hwang and GRACE gravity over Bhuj may be due to the devastating earthquake of 2001. In Anjar and the IBF, all data are collected after the earthquakes, and as such, Hwang and GRACE gravity are more or less similar, whereas in situ gravity does not pinpoint many changes (Figure 4.8a–c).

Gravity anomaly patterns over the Bhuj region, particularly the epicentres of earlier major earthquakes, have been generated and studied. It was interesting to note that substantial differences in gravity anomaly could be observed existing over these epicentral regions using GRACE, Hwang and NGRI gravity data, which has the potential to cause such big earthquakes, and hence we call them 'gravity precursors'. Temporal variations of the satellite-derived gravity and their probable relations with major earthquakes that have already occurred in this region have also been studied.

4.9 GRAVITY PRECURSORS OVER OCEAN

Satellite altimetry has recently emerged as an efficient alternative to expensive and hazardous ship-borne gravity surveys (Stewart 1985). The averaged sea surface height (SSH) as obtained from a satellite altimeter is a good approximation to the classical geoid, which contains information regarding mass distribution in the entire Earth. The underlying concept of the satellite gravity method is that SSH measured by satellite altimeter, when corrected for dynamic variability due to tides, waves, eddies, and so forth, corresponds to mass distribution below the Earth (Majumdar et al. 1998; Majumdar and Bhattacharyya 2004).

Haxby et al. (1983) generated digital images from the combined oceanic and continental datasets and specified their usages in tectonic studies. Sandwell and Smith (1997) applied gradient methods for the analysis of gravity and geoid information. Rapp (1983) developed a method for the prediction of the gravity anomaly using spherical harmonic coefficients up to 30° and order and above. Majumdar et al. (1998) developed a brief methodology for offshore structure delineation using altimeter data. An atlas for very high-resolution satellite geoid and gravity maps over the Indian offshore using Geosat GM (Geodetic Mission), TOPEX/POSEIDON, ERS-1 and Seasat altimeter data was published (Majumdar and Bhattacharyya 2004). In the present study, in addition to the development of methods and applications for satellite-derived geoid and gravity, the gravity signatures over the epicentre of recent severe earthquake (26 December 2004) and aftershock regions near Sumatra have been studied with very high-resolution satellite gravity. The main objectives were to generate and study very high-resolution satellite gravity over the Sumatran Earthquake region with reference to the recent severe earthquake on 26 December 2004 for gravity precursor-related studies.

4.9.1 DATA SOURCES AND STUDY AREA

Geosat GM, ERS-1/2, TOPEX/POSEIDON and Seasat altimeter data have been used to generate high-resolution satellite gravity over the study area comprising a part of the Bay of Bengal and the northeastern Indian ocean in general and the Andaman and Sunda Trench systems in particular (latitude 0–22° N, longitude 87–100° E) (Hwang et al. 2002). Because of being very high density in nature (off-track resolution ~3.33 km), sea surface parameters, as well as gravity, derived from these data are more accurate. Details of data processing and adjustment types for very high-resolution data have been discussed by Hwang et al. (2002). With regard to spatial resolution, Geosat GM (in the high-resolution dataset) showed the highest (~3.5 km), followed by ERS-1/2 (~35 km), Seasat (~100 km) and TOPEX (~250 km), whereas TOPEX has more accurate information than the others. Hwang et al. (2002) carried out very detailed data assimilation using these datasets and Levitus topography for calculation of the deflection of the vertical and then generated a 2 × 2 min (4 × 4 km) grid.

4.9.2 METHODOLOGY

The altimeter is a nadir-viewing instrument which transmits short-duration radar pulses (frequency ~13.0 GHz) with known power in a pencil beam towards the Earth's surface and then measures the reflected energy. The time delay (i.e. the two-way travel time of the pulse), when coupled with knowledge of the velocity of propagation through the ionosphere and wet troposphere, can be converted into a highly accurate measurement of the altitude of the satellite, and therefore a measurement of the sea surface topography (after due corrections for various parameters and assuming that the orbit ephemeris is accurately determined) (Stewart 1985).

The SSH with respect to the reference ellipsoid is computed for instrumental bias and atmospheric propagation delays (Majumdar et al. 1998).

The SSH observed by the altimeter is only an instantaneous sea surface topography, and deviations of SSH from the geoid are due to various dynamic variabilities, for example, ocean tide, solid tide, electromagnetic bias and inverse barometric pressure effect. Correction of the SSH for dynamic variability yields the mean sea surface height (MSH). By taking averages of repeat observations over time periods, it is possible to minimize the dynamic part of the sea surface topography in MSH.

4.9.2.1 Crossover Analysis

The near-repeat observations taken at the points of intersection of ascending and descending tracks are known as crossover points. The time-invariant geoidal part of the SSH at the crossover point should be constant. Any discrepancy at the crossover point, termed the crossover error, is mainly due to the orbital error (Tai 1988). The crossover errors were estimated and removed using a second-degree polynomial.

4.9.2.2 Deeper Earth Effects in Geoid

Contributions due to density variations below the lithosphere are manifested in the form of very long-wavelength components (>4000 km). The deeper earth effect was removed by spherical harmonic modelling of the geopotential field (Rapp 1983). Residual geoid was obtained after removing long-wavelength components, caused mainly due to the deeper earth effects, from classical geoid, thus containing information due to bathymetry as well as lithospheric anomalies. Details of the methodology for obtaining geoid and gravity from altimeter-derived SSH have been discussed elsewhere (Majumdar et al. 1998; Hwang et al. 2002; Majumdar and Bhattacharyya 2004). Gravity anomaly modelling using geoid was already discussed in Section 4.7.1.

4.9.3 RESULTS AND DISCUSSION

4.9.3.1 Study of Oceanic Processes over the Sumatran Earthquake Region

Geotectonic and gravimetric geoids over the Bay of Bengal have been discussed in detail by several authors (Kahle and Talwani 1973; Curray et al. 1982; Lundgren and Nordin 1988; Mukhopadhayay and Krishna 1991; Subramanium and Verma 1992; Gopala Rao et al. 1997; Ramana et al. 2001). The average sediment thickness in the northeastern Bay of Bengal is 18–22 km (Curray et al. 1982; Gopala Rao et al. 1997). In such high-sediment-thickness areas, the residual geoid anomaly-related studies are more effective than studies of the free-air gravity anomaly for mapping the basement undulations because residual geoid is a more regional phenomenon than the localized free-air gravity anomaly. The high geoidal undulations along the trench wall and the close proximity of the aseismic 90° E ridge in the northern region suggest that the existence of the 90° E ridge is a hindrance to the overall subduction processes (Cloetingh and Wortel 1985; Majumdar et al. 2001; Rajesh and Majumdar 2004). The presence of this ridge up to 10° N was reported earlier by Ramana et al. (2001), who carried out marine magnetic studies of the region.

4.9.3.2 Study of Gravity Signatures

A very high-resolution three-dimensional gravity image has been generated over a part of the eastern Indian offshore containing the 90° E ridge, Andaman subduction zone, Sumatra, and so forth (Figure 4.9) (Chatterjee et al. 2007b). A satellite-derived free-air gravity image shows the prominent gravity signatures that are associated with the Sunda Trench (distinct gravity low), Ninetyeast Ridge (subdued gravity high), continental shelf slope of the eastern margin of India (steep gravity low) and some isolated structural highs (gravity lows marked between eastern margin of India and the Ninetyeast Ridge) (Figures 4.9 and 4.10).

This high-resolution satellite gravity image has been superimposed over the earthquake occurrence tectonic map of 26 December 2004 near Sumatra (Figure 4.10) (Chatterjee et al. 2007b). A surprising feature of this earthquake is that after the initial rapid rapture, subsequent slip of the plate interface occurred with decreasing speed towards the north (Bilham and Gaur 2000; Bilham 2005). Four latitudinal (2°, 3°, 4° and 8.6° N) gravity profiles have been generated across the epicentre (3° N/M_w 9.0) and aftershock (4° N/M_w 5.8 and 8.6° N/M_w 6.0 and 6.6) regions and south of the epicentre (2° N) (Figure 4.9). The specified aftershocks occurred within 24 h of the major earthquake. A drastic change in gravity anomaly patterns (as high as 130 mGal) near the epicentre and the aftershock regions was observed (Figure 4.9), which slowly diminished farther away from the epicentre. It has been observed that ridges are close to isostatic equilibrium, but sinks are characterized by topographic trenches, for example, Sunda Trench, which shows the largest negative gravity anomalies (Rajesh and Majumdar 2004). This gravitational difference might have caused the plate instability near the subduction zone, leading to the severe earthquake, whereas minor and major gravitational differences in other areas may be due to the changes in densities of different lithospheric zones and sedimentary layers (Mahapatra 1987). Details of fault plane solutions in the case of the Sumatra 26 December 2004 devastating earthquake have been discussed by Scalera (2005). The conjugate centroid–moment–tensor (CMT) fault plane solution (strike = 129, dip = 83, slip = 87) was found to be the most likely real solution, as it fulfils the need for vertical displacement.

Song and Simon (2003) considered variations in the gravity field to be a predictor of seismic behaviour. They discovered that within subduction zones, areas where the attraction due to gravity is relatively high are less likely to experience large earthquakes than areas where the gravitational force is relatively low. They also examined existing data from satellite-derived observations of the gravity field in subduction zones. Comparing variations in gravity along the trenches with earthquake data from two different catalogues going back 100 years, they found that within a given subduction zone, areas with negative gravity and topography anomalies correlated with an increased large earthquake activity. We generated ETOPO2 bathymetry datasets over the region and found that the bathymetry anomaly varies, within the zone of seismic occurrence, from −800 to −1000 m (the threshold is −750 m as given by Song and Simon [2003]). Hence, our study further supports the utilization of gravity as a predictor for earthquake occurrences.

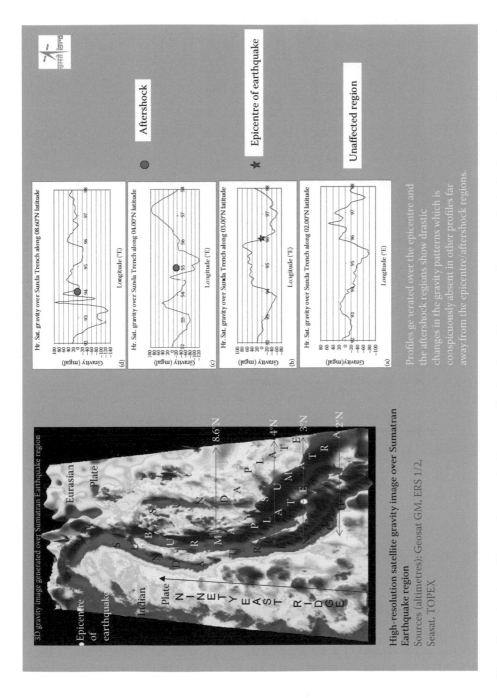

FIGURE 4.9 Three-dimensional satellite gravity over the Sumatran Earthquake and its surrounding areas. Latitudinal gravity profiles across epicentre, aftershock regions and south of the epicentre are also shown.

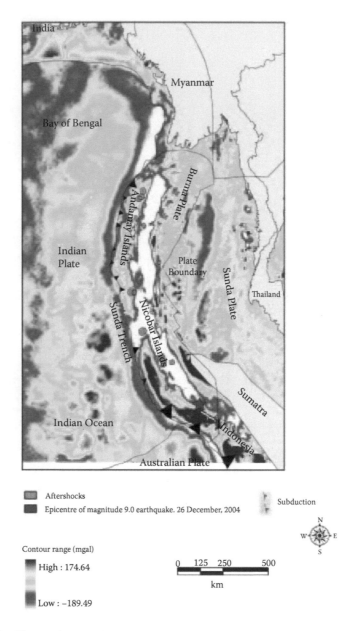

FIGURE 4.10   Satellite gravity superimposed over the earthquake occurrence tectonic map.

4.10   STUDY OF ANDAMAN SWARM USING GRACE GEOID ANOMALIES

Earthquake swarms are sequences of earthquakes closely clustered in space and time, in which no single earthquake dominates in size (Scholz 2002; Mogi 2003). Swarms originate in the crust from ambient stress generated by volcanic or tectonic activity (Mukhopadhayay and Dasgupta 2008). A large number of earthquakes have occurred beneath the Andaman Sea, east of the Nicobar Islands, starting 26 January 2005. The swarm activities were centred near 8° N and occurred mainly during 26–31 January 2005 following the great earthquake of 26 December 2004. GRACE gravity changes during December 2004 to January 2005 up to March 2005 show drastic changes in

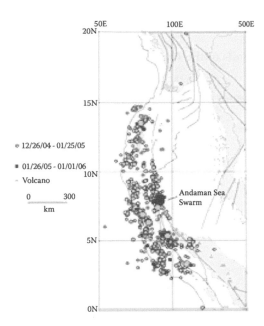

FIGURE 4.11   Andaman swarm region. (From USGS [U.S. Geological Survey] website 2, http://www.usgs
.gov/.)

this region, which has been correlated with the swarm activities. Related tectonic studies over the
Sumatra–Andaman arc region and the fault solutions have also indicated strong correlation at a
particular depth.

We have considered the study region bounded by latitude 7°–10° N and longitude 90°–96° E,
where strong swarm activities occurred during January 2005 (Figure 4.11) (USGS website 2).
GRACE time-variant geoid data for 2 years have been used at a grid interval of 0.01° × 0.01° for
the study area. This model is a combination of GRACE and LAGEOS missions plus 0.5° × 0.5°
gravimetry and altimetry surface data and is complete to 360° and order in terms of spherical harmonic coefficients. High-resolution combination gravity models are essential for the static gravity
potential, and its gradients are needed in the medium- and short-wavelength spectrum.

4.10.1   RESULTS AND DISCUSSION

Statistical details of the Andaman Sea 2005 swarm activity are given in Mukhopadhayay et al.
(2010) (Table 4.2). A total of 651 tremors between 26 and 31 January were recorded at crustal to
subcrustal depths with magnitudes between 3.9 and 5.2 in Richter scale. Different 10-day datasets
(from GRACE) were used for this study between 26 October 2004 and 3 May 2005. Figure 4.12a,b
shows different time-varying GRACE geoid data over the study area at two different ranges.
Corresponding geoid plots along the latitudinal profile at 8° N, for all the available 10-day datasets,
are shown in Figure 4.13. From Figures 4.12 and 4.13, it is observed that a drastic change has
occurred in the GRACE geoidal values during the 10-day geoid image covering 25 November
to 4 December 2004 and 24 January to 2 February 2005, before and after the deadly Sumatran
Earthquake on 26 December 2004, in terms of negative geoidal changes near 92°–98° E. A sudden surge in temporal geoid (along the 8° N profile) prior to the Great Sumatran Earthquake and a
temporal geoid low during and after the swarm event took place during 27–31 January 2005 (Figure
4.13) (Majumdar et al. 2012).

TABLE 4.2

Statistical Details of the Andaman Sea 2005 Swarm Activity

(a) 26–31 January 2005

Date	Number of Events/Day	mb (range)	Depth in km (range)
26	41	3.9–5.6	13–49
27	234	3.9–5.9	16–38
28	243	4.0–5.7	13–43
29	89	3.9–5.6	16–48
30	25	4.1–5.6	11–30
31	19	4.2–5.2	15–30

(b) February–August 2005

Month	Number of Events/Day	mb (range)	Depth in km (range)
February	43	4.2–5.9	11–56
March	9	4.1–5.2	30–37
April	5	4.2–5.0	3–30
May	3	4.0–4.5	30
June	18	4.2–5.3	20–66
July	8	4.3–4.9	16–46
August	4	4.2–5.5	19–46

Source: After Mukhopadhayay, B., and Dasgupta, S., *Acta Geophys.*, 56(4), 1000–1014, 2008.

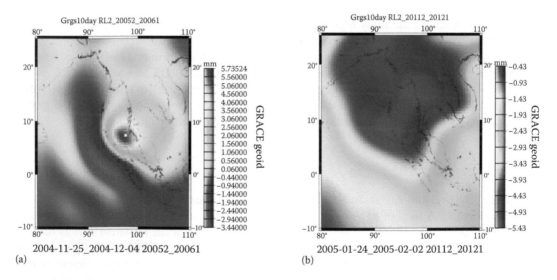

(a)   (b)

FIGURE 4.12   GRACE geoid imageries over the study area at two different dates: (a) 25 November 2004 to 4 December 2004 and (b) 24 January 2005 to 2 February 2005.

GRACE time-varying geoid data have been used to study the changes in the lithosphere in the Sumatra–Andaman region due to the deadly earthquake on 26 December 2004 and subsequent swarm activities in 27–31 January 2005. Drastic changes in geoid have been observed during both periods. Corresponding CMT solutions and *b*-value changes are correlated with the geoidal changes during these periods.

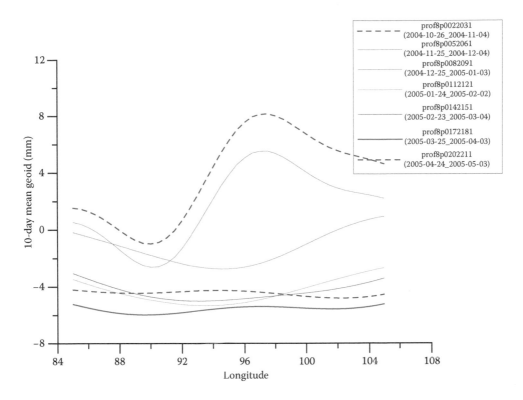

FIGURE 4.13   Profile plots of GRACE geoid data along 8° N latitude at different dates.

4.11   OVERALL CONCLUSION

A study of the thermal precursors of earthquakes over the Gujarat region was carried out, and surface temperature patterns were generated over Gujarat for 6 days during the occurrences of tremors; the same was utilized for demarcation of change in surface temperature as a precursor to impending earthquakes. Sharp changes in the surface temperature patterns (~6°C–7°C) were observed in the Bhuj region before the occurrences of these mild tremors during 6–10 April 2006. A sharp decrease of surface temperature was also observed after the occurrences of such tremors.

- For studies of gravity precursors over land, GRACE data have been generated and compared with Huang gravity and NGRI gravity data over the three vulnerable regions in Kachchh, namely, Bhuj, Anjar and along the IBF. Sharp changes in gravity patterns were observed for a few epicentres in the Kachchh region.
- For studies of gravity precursors over the ocean, a three-dimensional high-resolution gravity image was generated and superimposed over the tectonic map over the Ninetyeast Ridge and the Sumatran Earthquake region. Profiles generated over the epicentre and the aftershock regions show a drastic change (as high as 130 mGal) in the gravity patterns, which is conspicuously absent in other profiles far away from the epicentre and aftershock regions.
- GRACE time-varying geoid data have been used to study the changes in the lithosphere in the Sumatra–Andaman region due to the severe earthquake on 26 December 2004 and the subsequent swarm activities during 27–31 January 2005. Drastic changes in geoid have been observed during the periods of pre-occurrence and swarm occurrence phases. Corresponding CMT solutions and b-value changes are being correlated with the geoidal changes during these periods.

ACKNOWLEDGEMENTS

The author is thankful to A. S. Kiran Kumar, Chairman, Indian Space Research Organisation, for his keen interest in this study. He is also thankful to Dr. R. Bhattacharyya, Senior Geologist, Oil India Limited, Guwahati; S. Chatterjee, Western Geco, Navi Mumbai; B. Chakradhara Rao, Reliance Petroleum, Guwahati; Dr. P. Rajasekhar, Space Applications Centre, Ahmedabad; and Dr. M. Radhakrishna and G. S. Rao, Indian Institute of Technology, Mumbai, for their help in various ways. Thanks are also due to the Council for Scientific and Industrial Research, New Delhi, for the Emeritus Scientist Fellowship since January 2011.

REFERENCES

Bhattacharya, G. C., and Chaubey, A. K. (2001). Western Indian Ocean – A glimpse of the tectonic scenario. In *The Indian Ocean—A Perspective*, ed. Sen Gupta, R., Desa, E. New Delhi: Oxford and IBH Publishers, 691–729.

Bilham, R. (2005). A flying start, then a slow slip. *Science*, 308 (5725), 1126–1127.

Bilham, R., and Gaur, V. K. (2000). Geodetic contributions to the study of seismotectonics in India. *Current Science*, 79, 1259–1269.

Biswas, S. K. (1982). Rift basins in western margin of India and their hydrocarbon prospects with special reference to Kutchch Basin. *American Association for Petroleum Geologists Bulletin*, 66, 1497–1513.

Biswas, S. K. (2005). A review of structure and tectonics of Kutchch basin, western India, with special reference to earthquakes. *Current Science*, 88, 1592–1600.

Chapman, M. E. (1979). Techniques for interpretation of geoid anomalies. *Journal of Geophysical Research*, 84, 3793–3801.

Chatterjee, S., Bhattacharyya, R., Laju, M., Krishna, K. S., and Majumdar, T. J. (2007a). Validation of ERS-1 and high resolution satellite gravity with *in-situ* ship-borne gravity over the Indian offshore regions – Accuracies and implication to subsurface modelling. *Marine Geodesy*, 30, 197–216.

Chatterjee, S., Bhattacharyya, R., and Majumdar, T. J. (2007b). Generation and study of high resolution satellite gravity over the Sumatran earthquake region. *International Journal of Remote Sensing*, 28, 2915–2925.

Cloetingh, S., and Wortel, W. (1985). Regional stress field of the Indian plate. *Geophysical Research Letters*, 12, 77–80.

Coll, C., Caselles, V., Sobrino, J. A., and Valor, E. (1994). On the atmospheric dependence of the split-window equation for land surface temperatures. *International Journal of Remote Sensing*, 15, 105–122.

Cooper, D. I., and Asrar, G. (1988). Evaluating atmospheric correction models for retrieving surface temperature from AVHRR over a tallgrass prairie. *Remote Sensing of Environment*, 27, 93–102.

Curray, J. R., Emmel, F. J., Moore, D. G., and Raitt, R. W. (1982). Structure, tectonics and geological history of the north eastern Indian Ocean. In *The Ocean Basins and Margins*, ed. Nairn A. E. M., Stehli, F. G. Vol. 6. New York: Plenum Press, 399–450.

Gaur, V. K. (2003). The 2001 Bhuj earthquake revisited – Two years later. In *Proceedings of the International Workshop on Earth System Processes Related to Gujarat Earthquake Using Space Technology*, Kanpur, India, 27–29 January, 35–41.

Gopala Rao, D., Krishna, K. S., and Sar, D. (1997). Crustal evolution and sedimentation history of the Bay of Bengal since the Cretaceous. *Journal of Geophysical Research*, 102, 17747–17768.

Gorny, V. I., Kritsuk, S. G., Latipov, I. Sh., and Tronin, A. A. (1996). Terrestrial heat flux measuring and geothermal zoning for regional and petroleum geology on the base of satellite IR-thermal survey (STS). In *Proceedings of the Eleventh Thematic Conference on Geologic Remote Sensing (ERIM 1996)*, Las Vegas, NV, 27–29 February, vol. 1, 594–605.

GRACE website 1. http://icgem.gfz-potsdam.de/home (accessed June 2009).

GRACE website 2. http://www2.csr.utexas.edu/grace/ (accessed October 2017).

Gupta, M. L. (1981). Surface heat flow and igneous intrusion in Cambay basin, India. *Journal of Volcanology & Geothermal Research*, 10, 279–292.

Haxby, F., Karner, G. D., La Brecque, J. L., and Weissel, J. K. (1983). Digital images of combined oceanic and continental data sets and their uses in tectonic studies. *EOS Transactions, American Geophysical Union*, 64, 995–1004.

Hopkin, M. (2004). Sumatran quake sped up Earth's rotation. *Nature*, 30 December 2004. http://www.nature.com/news/2004/041229/full/041229-6.html (accessed February 2005).

Hwang, C., Hsu, H. Y., and Jang, R. J. (2002). Global mean surface and marine gravity anomaly from multi-satellite altimetry: Application of deflection-geoid and inverse Vening Meinesz formulae. *Journal of Geodesy*, 76, 407–418.

IMD Weather Bulletin. (2006). 3–11 April, New Delhi.

Kahle, A. B. (1977). A simple thermal model of the earth's surface for geologic mapping by remote sensing. *Journal of Geophysical Research*, 82, 1673–1680.

Kahle, H. G., and Talwani, M. (1973). Gravimetric Indian ocean geoid. *Geophysics*, 39, 167–187.

Karanth, R. V. (2000). Neotectonic and palaeoseismic studies in Kutchch and its adjoining areas. Unpublished project report submitted to the Department of Science and Technology, New Delhi.

Karanth, R. V., and Gadhvi, M. S. (2007). Structural intricacies: Emergent thrusts and blind thrusts of central Kutchch, western India. *Current Science*, 93, 1271–1280.

Li, Q., Liu, M., and Yang, Y. (2002). The 01/26/2001 Bhuj, India, earthquake: Intraplate or Interplate? In *Plate Boundary Zones*, ed. Stein, S., Freymuller, G. AGU Geophysical Monograph. Washington, DC: American Geophysical Union, 255–264.

Lundgren, B., and Nordin, P. (1988). Satellite altimetry – A new prospecting tool. In *Proceedings of the 6th Thematic Conference on Remote Sensing for Exploration Geology*, Houston, Texas, 16–19 May, 565–575.

Mahapatra, G. B. (1987). *A Text Book of Geology* [with special reference to India]. New Delhi: CBS Publishers.

Majumdar, T. J., and Bhattacharya, B. B. (1988). Derivation of surface temperatures on land after correction due to atmospheric water vapor – A case study with INSAT VHRR data. *Remote Sensing of Environment*, 26, 185–191.

Majumdar, T. J., and Bhattacharyya, R. (2004). An atlas of very high resolution satellite geoid/gravity over the Indian offshore. SAC Technical Note No. SAC/RESIPA/MWRG/ESHD/TR-21/2004. Ahmedabad: Space Applications Centre.

Majumdar, T. J., and Bhattacharyya, R. (2010). Generation and study of satellite gravity over Gujarat, India and their possible correlation with earthquake occurrences. *Geocarto International*, 25 (4), 269–280.

Majumdar, T. J., and Massonnet, D. (2002). D-InSAR applications for monitoring of geological hazards with special reference to Latur earthquake, 1993. *Current Science*, 83, 502–508.

Majumdar, T. J., and Mohanty, K. K. (1998). Derivation of land surface temperatures from MOS-1 VTIR data using split-window channel computation technique. *International Journal of Remote Sensing*, 19, 287–294.

Majumdar, T. J., Mohanty, K. K., Mishra, D. C., and Arora, K. (2001). Gravity image generation over the Indian subcontinent using NGRI/EGM96 and ERS-1 altimeter data. *Current Science*, 80, 542–554.

Majumdar, T. J., Mohanty, K. K., and Srivastava, A. K. (1998). On the utilization of ERS-1 altimeter data for offshore oil exploration. *International Journal of Remote Sensing*, 19, 1953–1968.

Majumdar, T. J., Rajsekhar, P., Radhakrishna, M., and Rao, G. S. (2012). Utilization of GRACE geoid anomaly over the Andaman-Sumatra convergence zone for study of swarm activities and assessment of their impact in the lithospheric zone. In *Proceedings of the National Seminar on Frontiers of Earth Science Research*, 5–6 May, Gulbarga, India, 11–15.

Majumdar, T. J., and Rao, B. (2009). Surface temperature anomalies over Gujarat, India and their possible correlation with earthquake occurrences. *NNRMS Bulletin*, 33, 48–57.

McClain, E. P., Pichel, W. G., Walton, C. C., Ahmed, Z., and Sutton, J. (1983). Multichannel improvements to satellite-derived global sea-surface temperatures. *Advances in Space Researches*, 2, 43–47.

McKee, M. (2005). Radar satellites capture tsunami wave height. *New Scientist*, 6 January. http://www.newscientist.com/article.ns?id=dn6854 (accessed 23 January 2007).

Merh, S. S. (1995). The geology of Gujarat. Bangalore: Geological Society of India.

Mogi, K. (2003). Some discussions on aftershocks, foreshocks and earthquake swarms – The fracture of semi-infinite body caused by inner stress origin and its relation to earthquake phenomena. *Bulletin of the Earthquake Research Institute, University of Tokyo*, 41, 615–658.

Moritz, H. (1980). *Advanced Physical Geodesy*. Kent, UK: Abacus Press.

Mukhopadhayay, M., and Krishna, M. R. (1991). Gravity field and deep structure of the Bengal fan and its surrounding continental margins, north east Indian ocean. *Tectonophysics*, 86, 365–386.

Mukhopadhayay, B., Acharyya, A., Mukhopadhayay, M., and Dasgupta, S. (2010). Relationship between earthquake swarm, rifting history, magmatism and pore pressure diffusion – An example from south Andaman sea, India. *Journal of the Geological Society of India*, 76 (2), 164–170.

Mukhopadhayay, B., and Dasgupta, S. (2008). Swarms in Andaman sea, India – A seismotectonic analysis. *Acta Geophysica*, 56 (4), 1000–1014.

NASA website. Satellite-derived images: 26 January 2001 Bhuj earthquake, Gujarat, India. http://www.nasa.gov/centers/ames/news/releases/2002/02images/quakes/earthquakes.html (accessed January 2007).

Nasipuri, P., Majumdar, T. J., and Mitra, D. S. (2006). Study of high resolution thermal inertia over western India oil fields using ASTER data. *Acta Astronautica*, 58, 270–278.

NGRI Map Series. (1978). Gravity map series of India (1:5 million). Hyderabad: National Geophysical Research Institute.

NOAA (National Oceanic and Atmospheric Administration). (1990). NOAA Polar Orbiter Data User's Guide. Washington, DC: U.S. Department of Commerce, NOAA, NESDI, NCDC, and the Satellite Data Service Division.

Okal, E. A., Talandier, J., Sverdrup, K. A., and Jordan, T. H. (1980). Seismicity and tectonic stress in the south-central Pacific. *Journal of Geophysical Research*, 85, 6479–6495.

Prabhakara, C., Dalu, G., and Kunde, V. G. (1974). Estimation of sea surface temperature from remote sensing in the 11 and 13 μm window regions. *Journal of Geophysical Research*, 79, 5039–5044.

Prata, A. J., Caselles, V., Coll, C., Sobrino, J. A., and Ottle, C. (1995). Thermal remote sensing of land surface temperature from satellites: Current status and future prospects. *Remote Sensing Reviews*, 12, 173–234.

Price, J. C. (1984). Land surface temperature measurements from the split window channels of the NOAA 7 Advanced Very High Resolution Radiometer. *Journal of Geophysical Research*, 89, 7231–7237.

Qiang, Z., Xu, X., and Dian, C. (1997). Thermal infrared anomaly precursor of impending earthquakes. *Pure and Applied Geophysics*, 149, 159–171.

Rajesh, S., and Majumdar, T. J. (2004). Generation of 3-D geoidal surface of the Bay of Bengal lithosphere and its tectonic implications. *International Journal of Remote Sensing*, 25, 2897–2902.

Ramana, M. V., Krishna, K. S., Ramprasad, T., Desa, M., Subrahmanyam, V., and Sarma, K. V. L. N. S. (2001). Structure and tectonic evolution of the northeastern Indian Ocean. In *The Indian Ocean—A Perspective*, ed. Sen Gupta, R., Desa, E. New Delhi: Oxford and IBH Publishers, 731–816.

Rapp, R. H. (1983). The determination of geoid undulations and gravity anomalies from Seasat altimeter data. *Journal of Geophysical Research*, 88, 1552–1562.

RSI (Research Systems Inc.). (2001). *ENVI User's Guide*. Boulder, CO: RSI.

Sabins, F. F., Jr. (1997). *Remote Sensing: Principles and Interpretation*. New York: W. H. Freeman & Co.

Sandwell, D. T., and Smith, W. H. F. (1997). Marine gravity anomaly from Geosat and ERS-1 satellite altimetry. *Journal of Geophysical Research*, 102, 10039–10054.

Saraf, A. K., and Choudhury, S. (2005). NOAA AVHRR detects thermal anomaly associated with the 26 January, 2001 Bhuj earthquake, Gujarat, India. *International Journal of Remote Sensing*, 26, 1065–1073. doi: 10.1080/01431160310001642368.

Saraf, A. K., Prakash, A., Sengupta, S., and Gupta, R. P. (1995). Landsat-TM data for estimating ground temperature and depth of subsurface coal fire in the Jharia coalfield, India. *International Journal of Remote Sensing*, 16, 2111–2124.

Scalera, G. (2005). The geodynamic meaning of the great Sumatran earthquake: Inferences from short time windows. *New Concepts in Global Tectonics Newsletter*, no. 35, 8–23.

Scholz, C. H. (2002). *The Mechanics of Earthquakes and Faulting*. 2nd ed. Cambridge: Cambridge University Press.

Singh, S. M. (1984). Removal of atmospheric effects on a pixel by pixel basis from thermal infrared data from instruments on satellites – The Advanced Very High Resolution Radiometer (AVHRR). *International Journal of Remote Sensing*, 5, 161–183.

Song, T. A., and Simon, M. (2003). Large trench-parallel gravity variations predict seismogenic behaviour in subduction zones. *Science*, 301, 630–633.

Stewart, R. H. (1985). *Methods of Satellite Oceanography*. Oakland: University of California Press.

Subramanium, C., and Verma, R. K. (1992). Geotectonics of the Bay of Bengal. *Indian Journal of Petroleum Geology*, 1, 161–180.

Tai, C. K. (1988). Geosat crossover analysis in the tropical Pacific: Constrained sinusoidal crossover adjustment. *Journal of Geophysical Research*, 93(C9), 10621–10629.

Tronin, A. A. (1996). Satellite thermal survey – A new tool for the studies of seismoactive regions. *International Journal of Remote Sensing*, 17, 1439–1455.

USGS (U.S. Geological Survey) website 1. Understanding plate motion. http://pubs.usgs.gov/publications/text/understanding.html (accessed June 2008).

USGS (U.S. Geological Survey) website 2. http://www.usgs.gov/ (accessed June 2011).

Verma, R. K., Gupta, M. L., Hamza, V. M., Rao, G. V., and Rao, R. U. M. (1968). Heat flow and crustal structure near Cambay, Gujarat, India. *Bulletin NGRI*, 6, 153–165.

Vidal, A. (1991). Atmospheric and emissivity correction of land surface temperature measured from satellite using ground measurements or satellite data. *International Journal of Remote Sensing*, 12, 2449–2460.

5 Satellite Radar Imaging and Its Application to Natural Hazards

Matt E. Pritchard and Sang-Ho Yun

CONTENTS

5.1 INTRODUCTION

Satellites provide critical global views for natural hazard characterization in advance of a crisis, as well as immediate damage assessment during or after a natural disaster (Voigt et al. 2007). While ground- and aircraft-based observations are also essential tools for hazard assessment and disaster response, satellites provide advantages to observe an area spanning hundreds of kilometres that may have had significant damage to infrastructure.

Radar images can be compared in multiple ways to measure ground changes before and during natural hazard events, like earthquakes, volcanic eruptions, floods, wildfires and landslides. Radar has unique capabilities compared with other remote sensing techniques; radar can 'see' through clouds and at night. More than a dozen civilian radar satellites are now available; the use of radar images has increased. New satellites with multisensors are required to monitor all potential hazards associated with land, ocean and atmosphere. We review five different change measurements that can be made by radar:

1. Amplitude
2. Phase (interferometry)
3. Along-track displacements and subpixel movements
4. Topography
5. Coherence

We discuss how each type of change detected can be applied to natural hazards or responding to natural disasters. A combination of different measurements is often necessary to avoid ambiguities in measurements (e.g. using both amplitude and coherence change to find the outline of flood zones under windy conditions, or using phase change and topography change for volcanic unrest).

We focus here on civilian satellite imaging radar systems that send electromagnetic pulses to the ground, record the reflections and then transmit the data back to Earth, where it can be formed into images. A commonly used method of combining many pulses together to make detailed images is called Synthetic Aperture Radar (SAR) (Figure 5.1). SAR is complementary to other types of remote sensing imagery, and in some cases it is superior because

- The satellite sends the electromagnetic signal, and images can be made day or night
- The radar signal penetrates through clouds
- Images taken at different times can be compared to forecast impending hazards or provide detailed maps of areas affected by disaster

In this chapter, we provide an overview of the natural hazard applications of imaging radar that complements recent technical reviews (Simons and Rosen 2015) and updates summaries of the hazard applications of SAR (Tralli et al. 2005; Lu et al. 2010a,b). We first briefly review five types of measurements made by radar that can be applied to the study of natural hazards (earthquake, volcano, flood, wildfire, cryosphere, landslide and coastal hazards) (Table 5.1), explain how the data can be used in rapid damage assessment during and in the immediate hours after a disaster, and then provide practical advice about the different types of radar data characteristics from the current and future satellites. We concentrate on applications on land or near land.

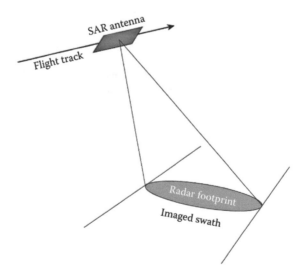

FIGURE 5.1 Typical side-looking imaging geometry of a SAR sensor. Unlike nadir-looking optical sensors, which resolve objects based on spatial (angular) separation of them, radar sensors differentiate objects based on the arrival time of backscattered echo.

TABLE 5.1

Summary of How the Five Different Radar Measurements Discussed in This Chapter Could Be Applied to the Study of Different Natural Hazards

	Amplitude	Phase	Pixel Tracking	DEMs	Coherence
Floods	X	X		X	X
Deformation before earthquakes and eruptions		X			
Wildfires	X				X
Deformation before river ice breakup		X			
Sea ice			X		
Landslides		X	X	X	X
Volcano eruptions and deposits (ash, lava and lahars)	X	X		X	X
Glacier change		X	X	X	
Building damage	X				X
Earthquake fault slip		X	X		X
Liquefaction					X
Water defence structures		X			X
Permafrost change		X			
Deformation before sinkhole collapse		X			
Subsidence – Coastal or human caused		X			

5.2 WHAT DOES RADAR MEASURE THAT IS USEFUL FOR NATURAL HAZARDS?

Change detection on the ground surface plays an essential role for natural hazard studies in both prediction and after-the-fact assessment. There are at least five different ways that SAR can be used to detect these changes. All these methods require at least two radar images acquired at different times (usually days to weeks apart) in order to determine the change between them. The power of imaging radar is that the change detection can usually occur on a pixel-by-pixel basis across the image, providing high spatial resolution maps (1–100 m/pixel) of changes over large areas spanning tens to hundreds of kilometres. In addition, for most applications, the calculation of change detection is done automatically, with little human intervention, and so can be applied over large regions to thousands of images or rapidly after a disaster. In the next sections, we provide overviews of the methods of change detection and some of their applications.

5.3 CHANGE IN THE AMPLITUDE OF THE RADAR SIGNAL

Each pixel in the SAR image records the amplitude of the radar signal returned to the satellite after being scattered by the ground. The amplitude signal will change over time if any of the following change: roughness, slope or dielectric constant (which depends on water content, among other things). Some ways that amplitude change can be used for natural hazards include the following:

- Detecting natural or man-made oil slicks on water (Jones et al. 2011; Leifer et al. 2012). The amplitude is lower on a smooth oil slick than on the rougher water surrounding the slick.

- Mapping the extent of flooding from rivers, tsunamis and storm surges. The surface rough-ness changes due to inundation – rough land surfaces are covered with a smooth water surface – causing flooded areas to often appear darker in a post-flood amplitude image compared with a pre-flood image (Figure 5.2). The existence of waves in the flood zone can complicate the interpretation, but the use of additional radar measurements like phase change (Section 5.4) or coherence change (Section 5.5) can isolate the wind effect (Strozzi et al. 2000). Flooding also tends to be accompanied by clouds, and thus can benefit from SAR observations. However, flood mapping is one of the most time-sensitive applications in natural disaster response, because the floodwaters drain away on a timescale of hours to days – often too short a time for SAR images to be acquired. The ephemeral character of the events also makes it challenging to find ground truth surveys or independent observa-tions for validation of the SAR flood maps.
- Mapping the extent of flooding in areas with vegetation (Alsdorf et al. 2000). The ampli-tude in vegetated flooded areas is higher than that in dry conditions because the radar signal undergoes a 'double bounce' off the water and the vegetation to return with a larger amplitude to the satellite.
- Radar amplitude change can detect building devastation (Brunner et al. 2010). Buildings are often good radar signal reflectors, especially when their facades are parallel to satel-lites' flight direction.
- Radar energy is double or triple bounced on the ground and building facades and reflected back to satellites. Significant damage to building structures can reduce the amount of such energy reflection, reducing the radar amplitude. On the other hand, however, massive debris from damaged buildings can increase the roughness of the buildings, increasing the scattered energy (Arciniegas et al. 2007). Thus, care needs to be taken when interpreting amplitude change for building damage assessment, and other measurements (like coher-ence change) may be necessary to robustly map the damage.
- Areas that are recently burned in a wildfire can have different radar amplitude than pre-burned areas (Figure 5.3), although whether the amplitude increases or decreases from the fire can depend on the angle of the radar beam and the characteristics of the burned area (Lu et al. 2010b). An advantage of using SAR is that it can 'see' through the smoke.

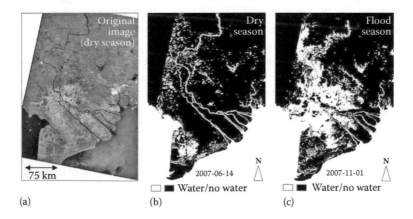

(a)   (b)   (c)

FIGURE 5.2 Mapping flood extent in the Mekong delta of south Vietnam using Envisat ScanSAR data. (a) Original radar image from 14 June 2007. (b) Interpretation of left image in terms of water and dry land. (c) Interpretation of radar image from 1 November 2007 at the end of the rainy season. (From Kuenzer, C. et al., *Remote Sens.*, 5(2), 687–715, 2013.)

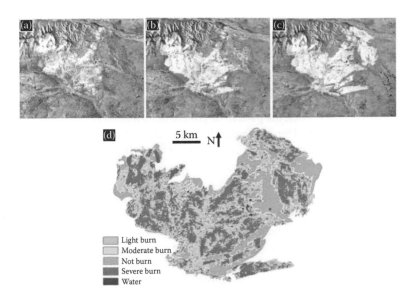

FIGURE 5.3 Radar amplitude images taken during a fire in the Yukon River Basin, Alaska, on sequential dates: (a) 17 August 2003, (b) 10 September 2003 and (c) 4 October 2003. (d) Once the fire was over, the amplitude and coherence change was used to determine a fire severity map. (All from Lu, Z. et al., *Int. J. Image Data Fusion*, 1(3), 217–242, 2010.)

- Fresh volcanic deposits – Lava, ash and mudflows (lahars) – can cause a change in amplitude. In some cases, the deposit absorbs more radar energy than before the deposition occurs (the amplitude decreases as for fine-grained lahars at Mt. Pinatubo, Philippines) (Chorowicz et al. 1997) or the deposit could scatter more radar energy than before – becoming radar bright like most lava flows (Wadge et al. 2012).

5.4 CHANGE IN THE DISTANCE FROM SATELLITE TO GROUND: INTERFEROMETRY

In addition to the amplitude of the returned radar signal, each pixel in a radar image records the returned phase of the radar signal with a value in radians between 0 and 2π. The phase values in an individual SAR image appear as white noise, because the pixel-to-pixel values of phase vary significantly depending on the scattering elements in a pixel. But when the phase difference between two images is calculated, systematic trends in the phase change are visible, creating an interferogram (Figure 5.4). This technique is generically called interferometric synthetic aperture radar (InSAR), and its power lies in the ability to measure phase changes of a fraction of a radar wavelength – a few millimetres to centimetres of distance change between the satellite and the ground (including a combination of both vertical and horizontal motion). While a single interferogram can provide precise observations of surface displacements on a pixel-by-pixel basis that can be valuable for natural hazards studies (e.g. providing information on ground motion during an earthquake), additional value is obtained by combining several interferograms spanning different time intervals together to calculate how deformation is changing with time. There are numerous reviews of the applications and limits of InSAR (Massonnet and Feigl 1998; Bürgmann et al. 2000; Rosen et al. 2000; Simons and Rosen 2015), including time-series methods (Ferretti et al. 2001; Berardino et al. 2002; Hooper et al. 2012), so we provide only a few illustrative examples, focusing on hazards:

- InSAR can measure slow deformation near faults, indicating that strain is building up that will probably be released in a future earthquake (Figure 5.5). Although the time of a future earthquake cannot be determined, the rate of deformation and how quickly the deformation decays with distance from the fault indicate how fast the fault is accumulating strain and how much of the fault might move – useful information for forecasting the earthquake magnitude (Fialko 2006).
- When an earthquake occurs, the pattern of ground deformation reveals which portions of a fault slipped (Figure 5.6). The detailed estimates of the location of a fault slip from InSAR are useful for several reasons, especially in the first few hours after the earthquake – the earthquake fault may not have been previously recognized (Talebian et al. 2004), seismic catalogues may place the earthquake in the wrong location (Lohman and Simons 2005), or depending on the details of the fault slip, the earthquake may make additional earthquakes on nearby faults more or less likely in the near term (Stramondo et al. 2011).
- Volcano uplift can indicate that magma is accumulating and an eruption may be possible within months to years (Chaussard and Amelung 2012; Lundgren et al. 2013). Not all volcanoes deform before eruption, and some volcanoes deform without eruption, but the vast majority of volcanoes that are not deforming are also not erupting (Biggs et al. 2014). Surface deformation before, during and after volcanic eruption has been observed with

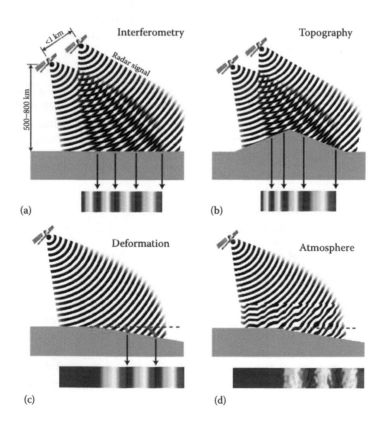

FIGURE 5.4   SAR interferometry. (a) Interference of two radar echoes from two sensors (or one sensor at different times) produces colour fringes. (b) Topography distorts the fringe pattern. (c) Ground deformation between two SAR acquisitions produces colour fringes. (d) Atmosphere (ionosphere and troposphere) distorts radio wavefronts. Satellites are flying into this page. (From Yun, S. H., A mechanical model of the large-deformation 2005 Sierra Negra volcanic eruption derived from InSAR measurements, PhD thesis, Stanford University, 147 pp, 2007.)

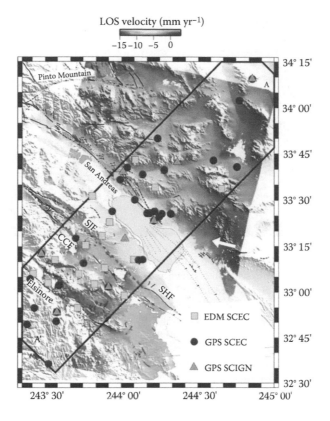

LOS velocity (mm yr⁻¹)

-15 -10 -5 0

FIGURE 5.5 Ground velocity in the direction of the satellite line of sight (LOS) between the years 1992 and 2000 along the southern San Andreas Fault showing the interseismic strain accumulation that will be released in an earthquake in the future. The location of ground-based global positioning system (GPS) and electronic distance measurements (EDMs) from the Southern California Earthquake Center (SCEC) and Southern California Integrated Geodetic Network (SCIGN) are also shown. Fault locations are shown as solid and dashed black lines are labelled. SJF, San Jacinto Fault; CCF, Coyote Creek Fault; SHF, Superstition Hills Fault. (From Fialko, Y., *Nature*, 441(7096), 968–971, 2006.)

InSAR measurements (Figure 5.7). Once a volcanic eruption begins, InSAR deformation observations could indicate where the magma is coming from, if the magma chamber is refilling and possibly the length of the eruption (Lundgren et al. 2013).

- Sinkholes are formed either naturally in karst regions where carbonate rock is dissolved into groundwater or due to human activities, such as mining. Many sinkholes occur rapidly over a small region, so it is difficult to capture precursory deformation using remote sensing techniques. In some cases, however, there may be slow deformation before sinkholes collapse catastrophically, indicating where a future collapse is possible (Castañeda et al. 2009; Paine et al. 2012; Jones and Blom 2014).

- Slow-moving landslides or mass movements pose numerous hazards and can be mapped using advanced InSAR techniques called persistent scatterers (also known as PSInSAR) (Hilley et al. 2004; Colesanti and Wasowski 2006; Delacourt et al. 2007) or small baseline subset (Calò et al. 2014), as in Figure 5.8. To date, satellite observations have been too infrequent to detect precursors before large catastrophic landslides.

- When levees (or other water defence structures) subside, there is a high risk of catastrophic flooding. Such subsidence was observed by PSInSAR before the Hurricane Katrina floods in New Orleans (Dixon et al. 2006). InSAR also detected the motion of embankments

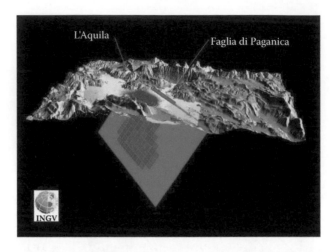

FIGURE 5.6   Ground displacement from the 6 April 2009 L'Aquila, Italy, earthquake (magnitude 6.3) observed by InSAR from the COSMOSkyMed satellite shown in colour on the Earth's surface, with red showing the maximum subsidence of 25 cm. Below the surface in blue is the fault plane inferred to have caused the earthquake (bottom of the fault is 12 km), with purple colours showing the amount of fault slip (maximum 90 cm at 4 km depth) determined by inversion of the surface displacements. The fault that caused the earthquake (Faglia di Paganica) is shown as a blue line at the Earth's surface and is about 20 km long, and the city of L'Aquila is also labelled. (From Salvi, S. et al., Measurement and modeling of co-seismic deformation during the L'Aquila Earthquake, preliminary results, Istituto Nazionale di Geofisica e Vulcanoligia, Rome, 2009, http://portale.ingv .it/primo-piano-1/news-archive/2009-news/april-6-earthquake/sar-prelimina ry-results/view?set_language=en.)

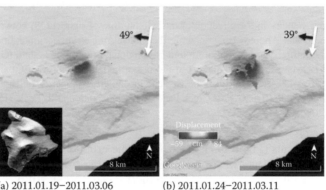

(a) 2011.01.19–2011.03.06   (b) 2011.01.24–2011.03.11

FIGURE 5.7   Ground deformation maps of the eastern part of Kilauea Volcano, Hawaii, observed by the PALSAR sensor mounted on the ALOS satellite. The varying colour represents the amount of ground surface displacement in the radar line of sight. The deformation was caused by magma intrusion, followed by fissure eruption on 5–9 March 2011. The white and black arrows indicate the satellite heading and look directions, respectively, and the approximate incidence angle, from vertical, in the centre of the image. Panel (a) includes only the first day of eruption, and panel (b) spans the entire eruption period, where part of the deformation signal was obscured due to the lava flow and localized large deformation. (Adapted from Lundgren, P. et al., *J. Geophys. Res. Solid Earth*, 118(1018), 897–914, 2013.)

before they failed catastrophically in Hungary, creating the worst environmental disaster in that country's history (Grenerczy and Wegmüller 2011).

• Subsidence along coastlines, often combined with sea level rise, is of special interest because it leads to increased flooding. InSAR has been used to detect and map coastal subsidence in several areas (Finnegan et al. 2008; Wang et al. 2012).

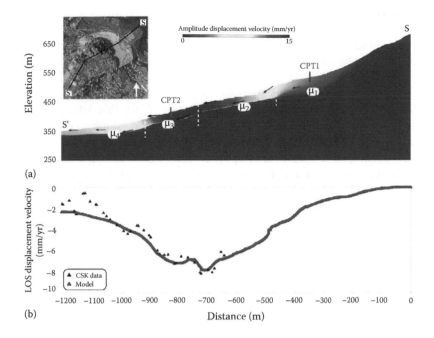

(a)

(b)

FIGURE 5.8   (a) Two-dimensional displacement field obtained from numerical modelling along the longitudinal cross section S-S′ of the Ivancich landslide in Assisi, Italy. μ1, μ2, μ3 and μ4 are the dynamic viscosities for the four shear band sectors. CPT1 and CPT2 indicate direction changes of the profile trace. The inset shows the modelled longitudinal cross section S-S′, superimposed on the full-resolution, ground deformation velocity map obtained processing COSMO-SkyMed data. (b) Comparison between the modelled velocity profile (red triangles) and the COSMO-SkyMed measurements (black triangles) along the longitudinal cross section of the landslide. (From Calò, F. et al., *Remote Sens. Environ.*, 142(0), 69–82, 2014.)

- Changes to permafrost causes coastal erosion, mass movements in mountainous areas (Kääb et al. 2005) and damage to human infrastructure. InSAR is capable of measuring long-term changes in the permafrost active layer thickness and its spatial variations – essential measurements for the state of health of the permafrost (Liu et al. 2012).
- The breakup of ice in rivers or along coastlines poses several hazards, and deformation that precedes the breakup can be detected by InSAR (Smith 2002; Vincent et al. 2004). River ice breakup can be hazardous for navigation (Vincent et al. 2004) and cause flooding (Smith 2002). Coastal ice also presents problems for navigation (both travelling over the ice and ships avoiding the ice) and increased storm damage when it is absent (Eicken et al. 2011).
- Glacier motions and/or melting can cause hazards to populations living near glaciers, as well as contribute to global sea level rise (Kääb et al. 2005). Glacier advance can threaten infrastructure, while glacier-controlled dams (usually below the surface) can fail catastrophically, causing glacial lake outburst floods (GLOFs). Glacier motions can be tracked with InSAR when satellites make repeat overflights within a few days or weeks (Goldstein et al. 1993; Joughin 2002). Subglacial lakes have been detected and monitored by InSAR (Capps et al. 2010), but observations are unlikely to be frequent enough to provide immediate warning for GLOFs.
- During flooding, in addition to mapping changes in radar amplitude that can be a proxy for flood extent (see Section 5.3 on amplitude change), the phase change measurement is related to changes in height–important for assessing flood stage (Alsdorf et al. 2000; Smith 2002). However, the height measurement is not absolute–only the change in height between measurements is possible, unless additional ground or space observations are

available (Alsdorf et al. 2001). Phase change measurements are only possible in flooded areas with vegetation where the radar signal undergoes a double bounce that also increases the radar amplitude in these areas. But as few as one or two emergent tree trunks within a radar pixel might be sufficient for the double-bounce effect (Alsdorf et al. 2001).

Various types of human activities can cause surface deformation (sometimes including catastrophic collapse) that can have an impact on infrastructure. While this deformation could be considered human-made instead of a natural disaster, it is not always simple to draw the line between the two – for example, are changes in permafrost due to anthropogenic global warming (human-made) or natural hazards? Examples of anthropogenic deformation that has been detected by InSAR and PSInSAR include mining (Lu and Wicks 2010; Ismaya and Donovon 2012), oil and gas production (Fielding et al. 1998), groundwater pumping (Amelung et al. 1999), geothermal production (Carnec and Fabriol 1999) and carbon sequestration (Vasco et al. 2010).

5.5   ALONG-TRACK DISPLACEMENT AND SUBPIXEL MOVEMENTS

Sometimes the ground motion from an earthquake or landslide, for example, is so large that it can be detected with a simple visual inspection of images before and after the event. More sophisticated analysis can be applied to hundreds or thousands of pixels to even detect motions that are a fraction of a pixel (Brown 1992), and this has been demonstrated for both SAR and optical images (Scambos et al. 1992; Michel et al. 1999) – we will call these subpixel displacement measurements pixel tracking. Thus, depending on the pixel size, ground motion as small as 1–100 cm could be detected. While these measurements are not as precise as InSAR, they have several advantages, including that they are often successful in areas of large change where InSAR fails (e.g. on glaciers that have significant melting or snowfall), and they are more sensitive to certain horizontal motions (particularly along the track of the satellite orbit).

Pixel tracking has been used to study earthquakes (Michel et al. 1999; Simons et al. 2002), volcanic eruptions (e.g. Yun et al. 2007), landslides (Raucoules et al. 2013) and glaciers (Joughin 2002), and are a critical tool for monitoring sea ice (Kwok et al. 1990). Time series of the pixel tracking (or pixel offsets) can be made to monitor temporal change of large ground deformation (Casu et al. 2011). Pixel tracking produces both across-track and along-track displacement measurements. This means that from a pair of SAR data, one can generate a two-dimensional ground displacement map – radar line of sight and along-track direction.

Another way of measuring along-track displacement is called multiple aperture interferometry (MAI) or along-track beam splitting. This analysis creates two intreferograms – one forward looking and the other backward looking. Differencing the two interferograms produces the along-track displacements (Bechor and Zebker 2006; Jung et al. 2009). The quality of MAI results depends on the quality of the regular InSAR – MAI produced poor results where InSAR quality is poor. The technique is not as sensitive as InSAR, but because of the differencing, the noise from the atmosphere is lessened with MAI. However, in the presence of ionospheric noise, MAI can be severely affected. This means that MAI can be used to correct for ionospheric noise in InSAR observations (Jung et al. 2013; Liu et al. 2013). Modern SAR missions tend to have higher resolution than their predecessors, and both pixel tracking and MAI benefit from high-resolution SAR.

5.6   CHANGES IN TOPOGRAPHY: DIGITAL ELEVATION MODELS

When comparing phase change in two radar images, if the satellites are not in the exact same location, part of the phase change is caused by topography on the ground (aka the parallax effect). This phase change from topography is usually removed with topographic maps (called digital elevation models [DEMs]). In fact, the InSAR sensitivity to topography can be used to create DEMs, particularly if the data acquisitions are taken close in time so that there is little chance for the ground to deform or for the atmosphere to change between satellite overflights. The best InSAR DEMs are created when two antennas are used to measure topography at the exact same time, as was done by the U.S. space shuttle in 2000 (the Shuttle Radar Topography Mission [SRTM]) (Farr et al. 2007) or between 2010 and 2014 by the German TanDEM-X twin satellites (Moreira et al. 2004). Useful DEMs can also be created when the overflights are separated by short time periods, especially when several images can be combined to reduce errors caused by changes in the atmosphere (Rufino et al. 1998; Lu et al. 2010a,b). For example, during the tandem phase of the European ERS-1 and ERS-2 satellites in 1995–1996, when there was 1 day between overflights. Improved DEMs can also be created in PSInSAR and other methods that use multiple interferograms (Lu et al. 2010b) and can be combined with DEMs made from LIDAR and stereo-optical images.

Several types of natural hazards require high-quality topographic maps (landslides, floods and wildfires), and InSAR is one method for mapping the change in topography related to the natural disaster. InSAR-created DEMs have been used to measure erosion from flooding (Smith 2002), emplacement of lava flows (Lu et al. 2010a,b), mining (Bhattacharya et al. 2012), landslides and glaciers (Kääb et al. 2005). In addition, DEMs are needed to correct for geometric distortions and convert imagery (optical, radar or other) to geographic coordinates. In other words, if a disaster causes a significant change in topography, a new DEM is necessary to properly interpret imagery.

5.7 COHERENCE CHANGE

We can quantify how similar the amplitudes and phases are between pairs of SAR images by calculating the coherence (Zebker and Villasenor 1992). Coherence has a value between zero (no coherence) and 1 (identical signal in the two images). Bare rock with no vegetation will maintain high coherence, while an area that just had a large landslide or was covered by a fresh lava flow will have low coherence. Coherence decrease, or decorrelation, was observed, for example, along the pathway of lava flows on the flank of the Kilauea volcano (Zebker et al. 1996) and along the surface rupture of an earthquake, where the ground was most severely disturbed (Simons et al. 2002). Other examples of the utility of coherence change are listed below:

- Large earthquakes often cause building damage, which is a major component of economic loss. Thus, a map of building damage can be very useful for early assessment of loss and fatality. The potential of interferometric coherence change has been tested with the 1995 Kobe earthquake in Japan (Yonezawa and Takeuchi 2001) and the 2003 Bam earthquake in Iran (Talebian et al. 2004; Fielding et al. 2005; Hoffmann 2007) by comparing two pairs of SAR images – the pair including the earthquake and the pair before the earthquake. This approach requires that the imaging conditions of the two pairs, specifically the separation of the satellites in space and time, be similar to each other. This requirement is later mitigated by equalizing the coherence statistics of the two pairs (Yun et al. 2011a). In principle, buildings damaged by windstorms (hurricanes or tornadoes) or other disasters could be measured by coherence change as well. While there are other ways to assess building damage (using satellite optical images), they usually underestimate the damage (Lemoine et al. 2013), and so additional methods, like coherence-based damage mapping, may improve damage assessment.
- Volcanic eruptions can cause ground surface change with lava flows and ash deposits. The resurfacing with fresh lava completely changes the radar scattering properties of the surface and often appears as a distinct low-coherence band along the path of lava flow (Zebker et al. 1996; Dieterich et al. 2012). Ash fall damage can spread and affect wider areas than lava. Depending on the depth of the ash deposit, the sensitivity of the radar signal to detect such change varies with radar wavelength (Yun et al. 2011b). Even without eruption, volcanoes can produce mudflows (lahars) that are deadly and damaging to infrastructure (Kerle and Oppenheimer 2002). Because these flows cause a loss of radar coherence, they can

be mapped rapidly after the event to assess their impact. In some cases, the flows can be detected by amplitude change (like after the 1991 Mt. Pinatubo, Philippines, eruption), but not always, so coherence change might be the best way to detect them. However, if images are not acquired frequently, coherence can be lost in vegetated areas, making it impossible to map the lahar (Kerle and Oppenheimer 2002).

- Unlike slow-moving landslides, the ground deformation due to catastrophic landslides cannot be imaged with InSAR observations due to the complete loss of coherence. Because of this disturbance, however, such events can be detected in a coherence change map. However, it is often challenging to produce such a map of mountain terrain because of (1) steep slopes, which cause the landslides, which also cause distortions and shadows in SAR images, and (2) dense vegetation that causes decorrelation even when there are no landslides.

- Earthquakes sometimes induce liquefaction, a phenomenon where sediments lose stiffness and behave like a liquid due to applied sudden stress change (Green et al. 2011; Cox et al. 2013). During earthquakes, liquefied soil, or silt, often oozes out of the ground, replacing the dry land with a wet silty surface. Such change in the surface material, along with the moisture content, was mapped with coherence change (Yun et al. 2011a), as in Figure 5.9.

- Both natural and man-made levees can be breached suddenly by weather events, earthquakes or human activities, or gradually by subsurface weakening processes due to hydraulic pressure. The breakage of levees can lead to substantial inundation of land, and sometimes also poses a significant threat to water resources for drinking and irrigation, if the flooding in river deltas reverses the water flow and shifts the saltwater boundary inland (Shwartz 2006). Levees usually maintain high coherence along their crowns (Hanssen and van Leijen 2008), so if coherence is lost, it could indicate damage to the structure.

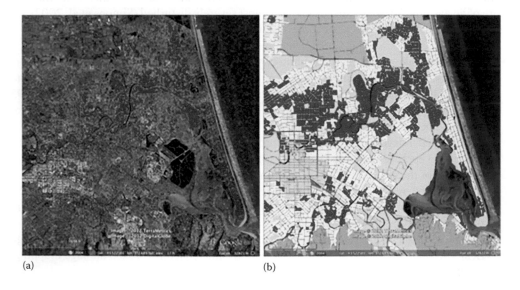

(a)   (b)

FIGURE 5.9   (a) Damage proxy map of Christchurch area, New Zealand, derived from ALOS PALSAR data acquired on 10 October 2010, 10 January 2011 and 25 February 2011. The red pixels represent change due to the February 2011 Christchurch earthquake. (b) Damage zone map released by the New Zealand government, where red polygons indicate significant damage on buildings or land. This ground truth map was produced 8 months after the earthquake, whereas the damage proxy map (a) was produced from remotely sensed radar data acquired 3 days after the earthquake. (From Yun, S. et al., Damage proxy map of M6.3 Christchurch Earthquake using InSAR coherence [abstract], presented at Fringe 2011 Workshop: Advances in the Science and Applications of SAR Interferometry from ESA and 3rd Party Missions, Frascati, Italy, 19–23 September, 2011, 2011a.)

However, the width of levees is often small, so high-resolution SAR imagery is usually required to monitor changes in levees (Jones et al. 2012).

- In addition to the changes in radar amplitude mentioned in Section 5.1, wildfires cause changes in coherence because of changes in water content, the number and type of vegetation and other factors (Lu et al. 2010b). By combining the radar amplitude and coherence changes, the severity of the fire burn can be assessed (Figure 5.3).

5.8 PRACTICAL CONSIDERATIONS: WHAT DATA TYPE WILL BE OF MOST UTILITY?

When considering whether an imaging radar will be useful for a specific study, several characteristics of the radar need to be considered:

- The wavelength of the radar – Existing systems (Table 5.2) have wavelengths that span 3–24 cm (corresponding to frequencies of 1–10 GHz), but most satellite systems fall into three categories: X-band (about 3 cm), C-band (about 5.6 cm) and L-band (about 23.5 cm). Each band has different advantages, in part related to the fact that the radar energy is scattered off the ground and back to the satellite by scatterers that are about the same size as the radar wavelength. Thus, short-wavelength X-band radar signals interact with the leaves at the tops of trees, while the longer L-band radar signals scatter off the larger parts of the trees – the trunks and branches. Because the leaves are more likely to change their orientation between SAR acquisitions than the branches, the coherence of the radar signal is higher in vegetated areas at L-band than X-band over the same time interval. Thus, to maintain coherence between observations, observations at short radar wavelengths must be more frequent than those at longer radar wavelengths. On the other hand, shorter radar wavelengths are less affected than longer radar wavelengths by distortions caused by changes in the ionosphere that occur in polar areas or near the magnetic equator. While longer-wavelength radars should be less sensitive to ground deformation than short-wavelength radars (i.e. L-band should be less sensitive than C-band), this effect is smaller than expected and may be overcome by the higher coherence of longer-wavelength systems (Sandwell et al. 2008).
- Spatial resolution of the radar image – This property can vary between about 1 and 100 m even for a given satellite (Table 5.1), depending on the type of observation mode for the radar. One of the lowest-resolution modes is called ScanSAR, and it can be often combined with the intermediate-mode Stripmap data. The highest-resolution mode is typically called Spotlight, and it cannot be combined with the other modes. The Spotlight, Stripmap and ScanSAR modes are often subdivided even further into high- and low-resolution modes. For all modes, the resolution trades off with swath width – the lower spatial resolution corresponds to wider coverage, and only a small area can be imaged using the highest spatial resolution. For some applications, the highest-resolution modes are necessary for change detection – for example, individual buildings, small landslides or domes inside volcano craters – while the lower-resolution modes are better to see large areas affected by great earthquakes or floods. The general trend in recent years has been that newer systems have higher resolution – for example, a new mode on TerraSAR-X called Staring Spotlight has a spatial resolution of 0.25 m.
- Polarization. Most radar signals used for satellite remote sensing are linearly polarized – either horizontally or vertically. When these signals encounter objects on the ground, electric currents are induced in the objects. Depending on the dominant orientation of the object, the scattered radio wave back to the satellite may contain vertically polarized energy even when the original transmitted signal was purely horizontal. The polarization in a radar image can affect the coherence, particularly in vegetated areas (Alsdorf et al. 2000).

As polarization changes the amplitude and phase, only images with the same polarization should be compared for change detection.

- Repeat time between measurements. For the specific techniques and applications described here, a strong constraint is that the radar images should be taken from the same location in space by the same radar system. Thus, observations can only be made as frequently as the satellite repeats the same flight path over a given area (called the repeat time) – usually several weeks to more than a month (Table 5.2). It is increasingly common for constellations of satellites with the same radar system (like COSMO-SkyMed or Sentinel) to follow each other in orbit so that repeat observations can be made within a few days to about a week. But even without constellations, observations are more frequent than the repeat time would suggest for several reasons:
  - A given spot is imaged by the satellite as it travels in its orbit from north to south (called descending orbits) and south to north (ascending orbits).

TABLE 5.2
List of Civilian SAR Satellites

Mission	Agency	Life	Repeat Cycle (days)	Band	Wavelength (cm)	Resolution (m)
ERS1	ESA	July 1991–March 2000	3, 168, 35	C	5.66	20
JERS1	JAXA	March 1992–October 1998	44	L	23.5	20
ERS2	ESA	April 1995–September 2011	35	C	5.66	20
Radarsat1	CSA	November 1995–March 2013	24	C	5.66	10–100
Envisat	ESA	March 2002–April 2012	35, 30	C	5.63	20–100
ALOS1	JAXA	January 2006–April 2011	46	L	23.6	10–100
TerraSARX (TSX), TanDEMX	DLR	June 2007–present, June 2010–present	11	X	3.1	1–16
Radarsat2	CSA	December 2007–present	24	C	5.4	3–100
COSMOSkyMed	ASI	June 2007– present, December 2007–present, October 2008–present, November 2010–present	1, 3, 4, 8	X	3.1	1–100
KOMPSAT5	KARI	August 2013–present	28	X	3.2	1–20
PAZ	INTA/His deSAT	2017 (planned)	11 (5.5 days in constellation with TSX)	X	3.1	1–16
Sentinel 1a, 1b	ESA	April 2014–present, April 2016–present	12, 6	C	5.66	5–40
ALOS2	JAXA	May 2014–present	14 (selected tracks)	L	23.6	1–100
SAOCOM 1a, 1b	CONAE	2017, 2018 (planned)	16, 8	L	23.5	7–100
NISAR	ISRO/ NASA	2021 (planned)	12	L, S	23.8, 9.3	3–12

Note: ASI, Agenzia Spaziale Italiana (Italy); CONAE, Comisión Nacional de Actividades Espaciales (Argentina); CSA, Canadian Space Agency; DLR, Deutsches Zentrum für Luft- und Raumfahrt e.V. (Germany); ESA, European Space Agency; INTA, Instituto Nacional de Técnica Aeroespacial (Spain); ISRO, Indian Space Research Organization; JAXA, Japanese Aerospace Exploration Agency; KARI, Korea Aerospace Research Institute (South Korea).

- The orbital tracks start to converge near the poles so the tracks increasingly overlap at higher latitudes – thus, the frequency of imaging a certain patch of ground is latitude dependent.
- On many satellites, the radar antenna is steerable so that it can be focused on an area of interest – particularly over isolated ocean islands like Hawaii at the expense of imaging the open ocean.
- On many satellites, the swath width of the imaging radar overlaps with other swaths.

Through the combination of multiple satellite tracks (ascending and descending orbits, partly overlapping orbits and steerable beams), more dense time series of observations can be created, albeit with some added complexity of interpretation.

5.8.1   DATA AVAILABILITY

By 2018, there should be at least 14 functioning civilian radar satellites, from eight different space programmes, with data prices ranging from no cost to thousands of U.S. dollars per image. Each satellite has different mission objectives, ranging from a focus on specific geographic regions, to targeted commercial applications, to global monitoring. Thus, the amount of data available for a given area on the Earth varies a lot – in some areas, data are collected during every overpass, while in others, data may never have been acquired by certain satellites. But the good news is that with so many satellite systems, after a large disaster it should be possible to acquire an image by at least one of the satellites within 24 hours to assist with emergency response. After a disaster, the International Charter can be invoked to facilitate satellite data collection and access (www.disasterscharter.org). Furthermore, the Global Earth Observation System of Systems (GEOSS) seeks to coordinate and improve international satellite observations to prepare and respond to disasters (Lautenbacher 2006). This coordination of efforts is critical considering that all satellites have limited duty cycles and cannot collect data over all global areas that have natural hazards.

5.8.2   DATA LATENCY

Latency is one of the most critical parameters for disaster response, as the utility of images after the event diminishes with time. Data acquisition latency is the time interval between the disaster event and data acquisition by a radar sensor. This latency is completely controlled by satellite orbits and space agencies' data acquisition plan. Data discovery latency is the time interval between data acquisition and a data user's awareness of the existence of such data. This may be affected by the space agencies' cataloguing efficiency. Data access latency is defined as the time interval between such awareness and the moment of gaining access to the data. This may involve obtaining credentials for such access and file transfer speed. Data processing latency is the time interval between data access and the production of a high-level decision support product for response. This latency depends on the processing algorithm and automation level of the data system.

In response to superstorm Sandy, for example, a damage proxy map of New York City was produced by the Advanced Rapid Imaging and Analysis (ARIA) team at the Jet Propulsion Laboratory and California Institute of Technology. The COSMO-SkyMed constellation imaged the affected area in Stripmap mode 5 days after the landfall of Sandy on the East Coast of the United States. It took 3 days for the information about the data to appear in the online catalogue. The ARIA team contacted the Italian Space Agency and gained access to the data a day later.

The processing took 2 days until the product was delivered to responding agencies. Thus, data acquisition, discovery, access and processing latencies added up to 11 days, with every step implemented manually. This demonstrates the importance of coordinated effort to establish an end-to-end system that automatically handles all intermediate processes. Especially important are background observations in critical areas by all satellites, so that a pre-event image is available after a disaster.

Once the radar data are acquired, they must be processed and useful data products created. While there is an impression that radar data are difficult to interpret and process (Voigt et al. 2007), there are now several commonly used open-source and commercial software packages to process the data (ISCE, GMTSAR, ROI_PAC, Gamma, NEST, DORIS, SARScape and DIAPASON) and commercial companies that are generating data products. As the number of SAR satellites increases, the resources available to aid in interpretation of the data will continue to grow.

5.9 CONCLUSION

We have highlighted the variety of uses of imaging radar, from anticipating earthquakes, ice breakup and volcanic eruptions, to mapping the extent of floods in progress, to assessing building damage from earthquakes and windstorms (summarized in Table 5.1). While interpreting radar images can be non-unique – for example, damaged buildings or lahars can both increase and decrease radar signal amplitudes – because radar images can be analyzed in several ways, including phase and coherence change, some of the non-uniqueness can be removed at the expense of additional time and analysis. Thus, there is a critical need to find effective, efficient algorithms that take full advantage of the myriad types of imaging radar observables to maximize the full potential of the international constellation of more than a dozen radar satellites to reduce the impact of natural hazards. Yet, even with the large international constellation of satellites, they are still limited in the amount of data that can be collected – SAR data files are large and downlink capabilities are insufficient to download data over all land areas where natural hazards exist. International coordination and additional satellite missions are necessary to routinely monitor all areas of potential natural disasters.

ACKNOWLEDGEMENTS

We thank Paul Lundgren for providing the data used in Figure 5.6. M. P. was partly supported by the NASA Science Mission Directorate (Grant Number NNX12AO13G). S.-H.Y. was partly supported by the NASA Applied Sciences Program (Grant Number NNN13D788T).

REFERENCES

Alsdorf, D., Birkett, C., Dunne, T., Melack, J., and Hess, L. (2001). Water level changes in a large Amazon lake measured with spaceborne radar interferometry and altimetry. *Geophysical Research Letters*, *28*(14), 2671–2674.

Alsdorf, D. E., Melack, J. M., Dunne, T., Mertes, L. A., Hess, L. L., and Smith, L. C. (2000). Interferometric radar measurements of water level changes on the Amazon flood plain. *Nature*, *404*(6774), 174–177.

Amelung, F., Galloway, D. L., Bell, J. W., Zebker, H. A., and Laczniak, R. J. (1999). Sensing the ups and downs of Las Vegas: InSAR reveals structural control of land subsidence and aquifer-system deformation. *Geology*, *27*(6), 483–486.

Arciniegas, G. A., Bijker, W., Kerle, N., and Tolpekin, V. A. (2007). Coherence- and amplitude-based analysis of seismogenic damage in Bam, Iran, using Envisat ASAR data. *IEEE Transactions on Geoscience and Remote Sensing*, *45*(6), 1571–1581.

Bechor, N., and Zebker, H. (2006). Measuring two-dimensional movements using a single InSAR pair. *Geophysical Research Letters*, *33*, L16311. doi: 10.1029/2006GL026883.

Berardino, P., Fornaro, G., Lanari, R., and Sansosti, E. (2002). A new algorithm for surface deformation monitoring based on small baseline differential SAR interferograms. *IEEE Transactions on Geoscience and Remote Sensing*, *40*(11), 2375–2383.

Bhattacharya, A., Arora, M. K., and Sharma, M. L. (2012). Usefulness of synthetic aperture radar (SAR) interferometry for digital elevation model (DEM) generation and estimation of land surface displacement in Jharia coal field area. *Geocarto International*, *27*(1), 57–77.

Biggs, J., Ebmeier, S. K., Aspinall, W. P., Lu, Z., Pritchard, M. E., Sparks, R. S. J., and Mather, T. A. (2014). Global link between deformation and volcanic eruption quantified by satellite imagery. *Nature Communications*, 5, 3471. doi: 10.1038/ncomms4471.

Brown, L. G. (1992). A survey of image registration techniques. *ACM Computing Surveys*, 24(4), 325–376.

Brunner, D., Lemoine, G., and Bruzzone, L. (2010). Earthquake damage assessment of buildings using VHR optical and SAR imagery. *IEEE Transactions on Geoscience and Remote Sensing*, 48(5), 2403–2420.

Bürgmann, R., Rosen, P. A., and Fielding, E. J. (2000). Synthetic aperture radar interferometry to measure Earth's surface topography and its deformation. *Annual Review of Earth and Planetary Sciences*, 28(1), 169–209.

Calò, F., Ardizzone, F., Castaldo, R., Lollino, P., Tizzani, P., Guzzetti, F., Lanari, R., Angeli, M.-G., Pontoni, F., and Manunta, M. (2014). Enhanced landslide investigations through advanced DInSAR techniques: The Ivancich case study, Assisi, Italy. *Remote Sensing of Environment*, 142(0), 69–82.

Capps, D. M., Rabus, B., Clague, J. J., and Shugar, D. H. (2010). Identification and characterization of alpine subglacial lakes using interferometric synthetic aperture radar (InSAR): Brady Glacier, Alaska, USA. *Journal of Glaciology*, 56(199), 861–870.

Carnec, C., and Fabriol, H. (1999). Monitoring and modeling land subsidence at the Cerro Prieto geothermal field, Baja California, Mexico, using SAR interferometry. *Geophysical Research Letters*, 26(9), 1211–1214.

Castañeda, C., Gutiérrez, F., Manunta, M., and Galve, J. P. (2009). DInSAR measurements of ground deformation by sinkholes, mining subsidence, and landslides, Ebro River, Spain. *Earth Surface Processes and Landforms*, 34(11), 1562–1574.

Casu, F., Manconi, A., Pepe, A., and Lanari, R. (2011). Deformation time-series generation in areas characterized by large displacement dynamics: The SAR amplitude pixel-offset SBAS technique. *IEEE Transactions on Geoscience and Remote Sensing*, 49(7), 2752–2763.

Chaussard, E., and Amelung, F. (2012). Precursory inflation of shallow magma reservoirs at west Sunda volcanoes detected by InSAR. *Geophysical Research Letters*, 39(21), L21311. doi: 10.1029/2012GL053817.

Chorowicz, J., Lopez, E., Garcia, F., Parrot, J. F., Rudant, J. P., and Vinluan, R. (1997). Keys to analyze active lahars from Pinatubo on SAR ERS imagery. *Remote Sensing of Environment*, 62(1), 20–29.

Colesanti, C., and Wasowski, J. (2006). Investigating landslides with space-borne synthetic aperture radar (SAR) interferometry. *Engineering Geology*, 88(3–4), 173–199.

Cox, B. R., Boulanger, R. W., and Tokimatsu, K. (2013). Liquefaction at strong motion stations and in Urayasu city during the 2011 TohokuOki earthquake. *Earthquake Spectra*, 29(Suppl. 1), S55–S80.

Delacourt, C., Allemand, P., Berthier, E., Raucoules, D., Casson, B., Grandjean, P., Pambrun, C., and Varel, E. (2007). Remote-sensing techniques for analysing landslide kinematics: A review. *Bulletin de la Societe Geologique de France*, 178(2), 89–100.

Dietterich, H. R., Poland, M. P., Schmidt, D. A., Cashman, K. V., Sherrod, D. R., and Espinosa, A. T. (2012). Tracking lava flow emplacement on the east rift zone of Kīlauea, Hawai'i, with synthetic aperture radar coherence. *Geochemistry, Geophysics, Geo*systems, 13(5), Q05001. doi: 10.1029/2011GC004016.

Dixon, T. H., Amelung, F., Ferretti, A., Novali, F., Rocca, F., Dokka, R., Sella, G., Kim, S.-W., Wdowinski, S., and Whitman, D. (2006). Space geodesy: Subsidence and flooding in New Orleans. *Nature*, 441(7093), 587–588.

Eicken, H., Jones, J., Meyer, F., Mahoney, A., Druckenmiller, M. L., Rohith, M. V., and Kambhamettu, C. (2011). Environmental security in Arctic ice-covered seas: From strategy to tactics of hazard identification and emergency response. *Marine Technology Society Journal*, 45(3), 37–48.

Farr, T. G., Rosen, P. A., Caro, E., Crippen, R., Duren, R., Hensley, S., Kobrick, M. et al. (2007). The Shuttle Radar Topography Mission. *Reviews of Geophysics*, 45(2), RG2004. doi: 10.1029/2005RG000183.

Ferretti, A., Prati, C., and Rocca, F. (2001). Permanent scatterers in SAR interferometry. *IEEE Transactions on Geoscience and Remote Sensing*, 39(1), 8–20.

Fialko, Y. (2006). Interseismic strain accumulation and the earthquake potential on the southern San Andreas Fault system. *Nature*, 441(7096), 968–971.

Fielding, E. J., Blom, R. G., and Goldstein, R. M. (1998). Rapid subsidence over oil fields measured by SAR interferometry. *Geophysical Research Letters*, 25(17), 3215–3218.

Fielding, E. J., Talebian, M., Rosen, P. A., Nazari, H., Jackson, J. A., Ghorashi, M., and Walker, R. (2005). Surface ruptures and building damage of the 2003 Bam, Iran, earthquake mapped by satellite synthetic aperture radar interferometric correlation. *Journal of Geophysical Research: Solid Earth*, 110(B3), B03302. doi: 10.1029/2004JB003299.

Finnegan, N. J., Pritchard, M. E., Lohman, R. B., and Lundgren, P. R. (2008). Constraints on surface deformation in the Seattle, WA, urban corridor from satellite radar interferometry time-series analysis. *Geophysical Journal International*, 174(1), 29–41.

Goldstein, R. M., Engelhardt, H., Kamb, B., and Frolich, R. M. (1993). Satellite radar interferometry for monitoring ice sheet motion: Application to an Antarctic ice stream. *Science, 262*, 1525–1530.

Green, R. A., Wood, C., Cox, B., Cubrinovski, M., Wotherspoon, L., Bradley, B., Algie, T., Allen, J., Bradshaw, A., and Rix, G. (2011). Use of DCP and SASW tests to evaluate liquefaction potential: Predictions vs. observations during the recent New Zealand earthquakes. *Seismological Research Letters, 82*(6), 927–938.

Grenerczy, G., and Wegmüller, U. (2011). Persistent scatterer interferometry analysis of the embankment failure of a red mud reservoir using ENVISAT ASAR data. *Natural Hazards, 59*(2), 1047–1053.

Hanssen, R. F., and van Leijen, F. J. (2008). Monitoring water defense structures using radar interferometry. In *IEEE Radar Conference 2008 (RADAR '08)*, Rome, Italy, 26–30 May, 2008, pp. 1–4.

Hilley, G. E., Bürgmann, R., Ferretti, A., Novali, F., and Rocca, F. (2004). Dynamics of slow-moving landslides from permanent scatterer analysis. *Science, 304*(5679), 1952–1955.

Hoffmann, J. (2007). Mapping damage during the Bam (Iran) earthquake using interferometric coherence. *International Journal of Remote Sensing, 28*(6), 1199–1216.

Hooper, A., Bekaert, D., Spaans, K., and Arıkan, M. (2012). Recent advances in SAR interferometry time series analysis for measuring crustal deformation. *Tectonophysics, 514–517*(0), 1–13.

Ismaya, F., and Donovan, J. (2012). Applications of DInSAR for measuring mine-induced subsidence and constraining ground deformation model. In *GeoCongress 2012: State of the Art and Practice in Geotechnical Engineering*, Oakland, CA, 25–29 March, 2012, pp. 3001–3010.

Jones, C. E., Bawden, G., and Deverel, S. (2012). Study of movement and seepage along levees using DINSAR and the airborne UAVSAR instrument. In *SPIE Remote Sensing*. Bellingham, WA: International Society for Optics and Photonics, edited by C. Notarnicola, S. Paloscia, and N. Pierdicca, p. 85360E.

Jones, C. E., and Blom, R. G. (2014). Bayou Corne, Louisiana, sinkhole: Precursory deformation measured by radar interferometry. *Geology, 42*(2), 111–114.

Jones, C. E., Minchew, B., Holt, B., and Hensley, S. (2011). Studies of the Deepwater Horizon oil spill with the UAVSAR radar. In *Monitoring and Modeling the Deepwater Horizon Oil Spill: A Record-Breaking Enterprise*, ed. Liu, Y., Macfadyen, A., Ji, Z.-G., Weisberg, R. H. AGU Geophysical Monograph Series 195. Washington, DC: American Geophysical Union, pp. 33–50.

Joughin, I. (2002). Icesheet velocity mapping: A combined interferometric and speckle-tracking approach. *Annals of Glaciology, 34*(1), 195–201.

Jung, H., Lee, D., Lu, Z., and Won, J. (2013). Ionospheric correction of SAR interferograms by multiple-aperture interferometry. *IEEE Transactions on Geoscience and Remote Sensing, 51*(5), 3191–3199.

Jung, H., Won, J., and Kim, S. (2009). An improvement of the performance of multiple-aperture SAR interferometry (MAI). *IEEE Transactions on Geoscience and Remote Sensing, 47*(8), 2859–2869.

Kääb, A., Huggel, C., Fischer, L., Guex, S., Paul, F., Roer, I., Salzmann, N. et al. (2005). Remote sensing of glacier- and permafrost-related hazards in high mountains: An overview. *Natural Hazards and Earth System Science, 5*(4), 527–554.

Kerle, N., and Oppenheimer, C. (2002). Satellite remote sensing as a tool in lahar disaster management. *Disasters, 26*(2), 140–160.

Kuenzer, C., Guo, H., Huth, J., Leinenkugel, P., Li, X., and Dech, S. (2013). Flood mapping and flood dynamics of the Mekong Delta: ENVISATASARWSM based time series analyses. *Remote Sensing, 5*(2), 687–715.

Kwok, R., Curlander, J. C., McConnell, R., and Pang, S. S. (1990). An ice-motion tracking system at the Alaska SAR facility. *IEEE Journal of Oceanic Engineering, 15*(1), 44–54.

Lautenbacher, C. C. (2006). The Global Earth Observation System of Systems: Science serving society. *Space Policy, 22*(1), 8–11.

Leifer, I., Lehr, W. J., Simecek-Beatty, D., Bradley, E., Clark, R., Dennison, P., Hu, Y. et al. (2012). State of the art satellite and airborne marine oil spill remote sensing: Application to the BP Deepwater Horizon oil spill. *Remote Sensing of Environment, 124*(0), 185–209.

Lemoine, G., Corbane, C., Louvrier, C., and Kauffmann, M. (2013). Intercomparison and validation of building damage assessments based on post-Haiti 2010 earthquake imagery using multisource reference data. *Natural Hazards and Earth System Sciences Discussions, 1*(2), 1445–1486.

Liu, L., Schaefer, K., Zhang T., and Wahr, J. (2012). Estimating 1992–2000 average active layer thickness on the Alaskan North Slope from remotely sensed surface subsidence. *Journal of Geophysical Research: Earth Surface, 117*, F01005. doi: 10.1029/2011JF002041.

Liu, Z., Jung, H., and Lu, Z. (2013). Joint correction of ionosphere noise and orbital error in L-band SAR interferometry on interseismic deformation in Southern California. *IEEE Transactions on Geoscience and Remote Sensing, 52*(6), 3421–3427.

Lohman, R. B., and Simons, M. (2005). Locations of selected small earthquakes in the Zagros Mountains. *Geochemistry, Geophysics, Geosystems, 6*, Q03001. doi: 10.1029/2004GC000849.

Lu, Z., Dzurisin, D., Jung, H.-S., Zhang, J., and Zhang, Y. (2010a). Radar image and data fusion for natural hazards characterisation. *International Journal of Image and Data Fusion*, *1*(3), 217–242.

Lu, Z., and Wicks Jr., C. (2010). Characterizing 6 August 2007 Crandall Canyon mine collapse from ALOS PALSAR InSAR. *Geomatics, Natural Hazards and Risk*, *1*(1), 85–93.

Lu, Z., Zhang, J., Zhang, Y., and Dzurisin, D. (2010b). Monitoring and characterizing natural hazards with satellite InSAR imagery. *Annals of GIS*, *16*(1), 55–66.

Lundgren, P., Poland, M., Miklius, A., Orr, T., Yun, S., Fielding, E., Liu, Z., Tanaka, A., Szeliga, W., Hensley, S., and Owen, S. (2013). Evolution of dike opening during the March 2011 Kamoamoa fissure eruption, Kilauea volcano, Hawaii. *Journal of Geophysical Research: Solid Earth*, *118*(1018), 897–914. doi: 10.1002/jgrb.50108.

Massonnet, D., and Feigl, K. L. (1998). Radar interferometry and its application to changes in the earth's surface. *Reviews of Geophysics*, *36*, 441–500.

Michel, R., Avouac, J. P., and Taboury, J. (1999). Measuring ground displacements from SAR amplitude images: Application to the Landers earthquake. *Geophysical Research Letters*, *26*(7), 875–878.

Moreira, A., Krieger, G., Hajnsek, I., Hounam, D., Werner, M., Riegger, S., and Settelmeyer, E. (2004). TanDEM-X: A TerraSAR-X add-on satellite for single-pass SAR interferometry. In *2004 IEEE International Proceedings of the Geoscience and Remote Sensing Symposium (IGARSS '04)*, Anchorage, AK, 20–24 September, 2004, vol. 2, pp. 1000–1003.

Paine, J. G., Buckley, S. M., Collins, E. W., and Wilson, C. R. (2012). Assessing collapse risk in evaporite sinkhole-prone areas using microgravimetry and radar interferometry. *Journal of Environmental Engineering and Engineering Geophysics*, *17*, 75–87.

Raucoules, D., de Michele, M., Malet, J. P., and Ulrich, P. (2013). Time-variable 3D ground displacements from high-resolution synthetic aperture radar (SAR). Application to La Valette landslide (South French Alps). *Remote Sensing of Environment*, *139*, 198–204. http://dx.doi.org/10.1016/j.rse.2013.08.006.

Rosen, P. A., Hensley, S., Joughin, I. R., Li, F. K., Madsen, S. N., Rodriguez, E., and Goldstein, R. M. (2000). Synthetic aperture radar interferometry. *Proceedings of the IEEE*, *88*(3), 333–382.

Rufino, G., Moccia, A., and Esposito, S. (1998). DEM generation by means of ERS tandem data. *IEEE Transactions on Geoscience and Remote Sensing*, *36*(6), 1905–1912.

Salvi, S., Salvi, S., Stramondo, S., Hunstad, I., Chini, M., Atzori, S., Moro, M. et al. (2009). Measurement and modeling of coseismic deformation during the L'Aquila Earthquake, preliminary results. Rome: Istituto Nazionale di Geofisica e Vulcanoligia. http://portale.ingv.it/primo-piano-1/new-archive/2009 -news/april-6-earthquake/sar-preliminary-results/view?set_language=en (accessed 29 September 2014).

Sandwell, D. T., Myer, D., Mellors, R., Shimada, M., Brooks, B., and Foster, J. (2008). Accuracy and resolution of ALOS interferometry: Vector deformation maps of the Father's Day intrusion at Kilauea. *IEEE Transactions on Geoscience and Remote Sensing*, *46*(11), 3524–3534.

Scambos, T. A., Dutkiewicz, M. J., Wilson, J. C., and Bindschadler, R. A. (1992). Application of image cross-correlation to the measurement of glacier velocity using satellite image data. *Remote Sensing of Environment*, *42*(3), 177–186.

Shwartz, M. (2006). Experts fear impacts of quake on Delta. Stanford report, May 17.

Simons, M., Fialko, Y., and Rivera, L. (2002). Coseismic deformation from the 1999 Mw 7.1 Hector Mine, California, earthquake as inferred from InSAR and GPS observations. *Bulletin of the Seismological Society of America*, *92*(4), 1390–1402.

Simons, M., and Rosen, P. A. (2015). Interferometric synthetic aperture radar geodesy. In *Treatise on Geophysics*, ed. Shubert, G. 2nd ed., vol. 3. Amsterdam: Elsevier, pp. 339–385.

Smith, L. C. (2002). Emerging applications of interferometric synthetic aperture radar (InSAR) in geomorphology and hydrology. *Annals of the Association of American Geographers*, *92*(3), 385–398.

Stramondo, S., Kyriakopoulos, C., Bignami, C., Chini, M., Melini, D., Moro, M., Picchiani, M., Saroli, M., and Boschi, E. (2011). Did the September 2010 (Darfield) earthquake trigger the February 2011 (Christchurch) event? *Scientific Reports*, *1*, 98. doi: 10.1038/srep00098.

Strozzi, T., Dammert, P. B., Wegmuller, U., Martinez, J. M., Askne, J. I., Beaudoin, A., and Hallikainen, N. T. (2000). Landuse mapping with ERS SAR interferometry. *IEEE Transactions on Geoscience and Remote Sensing*, *38*(2), 766–775.

Talebian, M., Fielding, E. J., Funning, G. J., Ghorashi, M., Jackson, J., Nazari, H., Parsons, B., Priestley, K., Rosen, P. A., Walker, R., and Wright, T. J. (2004). The 2003 Bam (Iran) earthquake: Rupture of a blind strike-slip fault. *Geophysical Research Letters*, *31*(11), L11611. doi: 10.1029/2004GL020058.

Tralli, D. M., Blom, R. G., Zlotnicki, V., Donnellan, A., and Evans, D. L. (2005). Satellite remote sensing of earthquake, volcano, flood, landslide and coastal inundation hazards. *ISPRS Journal of Photogrammetry and Remote Sensing*, *59*(4), 185–198.

Vasco, D. W., Rucci, A., Ferretti, A., Novali, F., Bissell, R. C., Ringrose, P. S., Mathieson, A. S., and Wright, I. W. (2010). Satellite-based measurements of surface deformation reveal fluid flow associated with the geological storage of carbon dioxide. *Geophysical Research Letters*, *37*(3), L03303. doi: 10.1029/2009GL041544.

Vincent, F., Raucoules, D., Degroeve, T., Edwards, G., and Abolfazl Mostafavi, M. (2004). Detection of river/sea ice deformation using satellite interferometry: Limits and potential. *International Journal of Remote Sensing*, *25*(18), 3555–3571.

Voigt, S., Kemper, T., Riedlinger, T., Kiefl, R., Scholte, K., and Mehl, H. (2007). Satellite image analysis for disaster and crisismanagement support. *IEEE Transactions on Geoscience and Remote Sensing*, *45*(6), 1520–1528.

Wadge, G., Saunders, S., and Itikarai, I. (2012). Pulsatory andesite lava flow at Bagana volcano. *Geochemistry, Geophysics, Geosystems*, *13*(11), Q11011. doi: 10.1029/2012GC004336.

Wang, H., Wright, T. J., Yu, Y., Lin, H., Jiang, L., Li, C., and Qiu, G. (2012). InSAR reveals coastal subsidence in the Pearl River Delta, China. *Geophysical Journal International*, *191*(3), 1119–1128.

Yonezawa, C., and Takeuchi, S. (2001). Decorrelation of SAR data by urban damages caused by the 1995 Hyogoken-Nanbu earthquake. *International Journal of Remote Sensing*, *22*(8), 1585–1600.

Yun, S., Agram, P., Fielding, E., Simons, M., Webb, F., Tanaka, A., Lundgren, P., Owen, S., Rosen, P. A., and Hensley, S. (2011b). Damage proxy map from InSAR coherence applied to February 2011 M6.3 Christchurch earthquake, 2011 M9.0 Tohoku-oki earthquake, and 2011 Kirishima volcano eruption [abstract NH31A1533]. Presented at AGU Fall Meeting, San Francisco, CA, 5–9 December, 2011.

Yun, S., Fielding, E., Simons, M., Rosen, P. A., Owen, S. E., and Webb, F. (2011a). Damage proxy map of M6.3 Christchurch earthquake using InSAR coherence [abstract]. Presented at Fringe 2011 Workshop: Advances in the Science and Applications of SAR Interferometry from ESA and 3rd Party Missions, Frascati, Italy, 19–23 September, 2011.

Yun, S., Zebker, H. A., Segall, P., Hooper, A., and Poland, M. (2007). Interferogram formation in the presence of complex and large deformation. *Geophysical Research Letters*, *34*, L12305. doi: 10.1029/2007GL02974.

Zebker, H. A., Rosen, P. A., Hensley, S., and Mouginis-Mark, P. J. (1996). Analysis of active lava flows on Kilauea volcano, Hawaii, using SIRC radar correlation measurements. *Geology*, *24*(6), 495–498.

Zebker, H. A., and Villasenor, J. (1992). Decorrelation in interferometric radar echoes. *IEEE Transactions on Geoscience and Remote Sensing*, *30*(5), 950–959.

6 DEMETER Satellite and Detection of Earthquake Signals

Michel Parrot

CONTENTS

6.1 INTRODUCTION

Since the 1980s, many satellites have been used to find changes of ionospheric parameters in relation to seismic activity. These satellites were not dedicated to this scientific objective, and the observations were limited and not conclusive. Many projects have been undertaken in order to have a dedicated mission in the 1990s. It became a reality when the Centre National d'Etudes Spatiales (CNES) issued in 1999 an announcement of opportunity for a series of low-altitude microsatellites named Myriades. The project Detection of Electro-Magnetic Emissions Transmitted from Earthquake Regions (DEMETER) was selected to be the payload of the first microsatellite of this series. The launch was in June 2004, and the mission came to an end in December 2010.

The aim of this chapter is to present an overview of the results provided by DEMETER up to now. In Section 6.2, we briefly describe the objectives of the mission, the satellite and its experiments. Examples of ionospheric changes prior to earthquakes (EQs) are included in Section 6.3. In contrast to ground experiments, satellite experiments cover most seismic zones of the Earth, and statistical studies become meaningful because of the much larger number of recorded events.

Only a statistical study with many events will show the general behaviour of such perturbations and will help to define a signature of perturbations prior to EQs. The statistical analysis with various parameters is given in Section 6.4. In Section 6.5, the problem of EQ prediction is discussed, especially with the magnitude 8.8 Chile EQ, and conclusions are given in Section 6.6.

6.2   DEMETER MISSION

The main scientific objectives of the DEMETER experiment were to study the disturbances of the ionosphere due to the seismoelectromagnetic effects, and due to anthropogenic activities (power line harmonic radiation, very low-frequency (VLF) transmitters and high-frequency (HF) broadcasting stations). The seismoelectromagnetic effects are the electric and magnetic perturbations caused by natural geophysical activity, such as EQs and volcanic eruptions. It includes electromagnetic (EM) emissions in a large frequency range, perturbations of ionospheric layers, anomalies on the records of VLF transmitter signals, particle precipitation and night airglow observations. Such phenomena are of great interest, because they start a few hours before the shock and can be considered short-term precursors. EM emissions in the ultra-low-frequency (ULF), extremely low-frequency (ELF) or VLF range that are related to seismic or volcanic activity have been known about for a long time, but their generation mechanisms are not well understood.

Two types of emissions are considered:

1. Precursory emissions occur between a few days and a few hours before EQs, in a large frequency range from one hundredth of a hertz up to several megahertz.
2. Emissions observed after the shock are generally attributed to the propagation of acoustic-gravity waves (AGWs). However, all hypotheses concerning the generation mechanism of precursor emissions are also valid after the shock, when the Earth's crust returns to an equilibrium state.

The emissions can propagate up to the ionosphere, and observations made with low-altitude satellites have shown increases of ULF, ELF and VLF waves above seismic regions. The first ionospheric observations of EM emissions onboard satellites have been presented in Gokhberg et al. (1982) and Larkina et al. (1983). Larkina et al. (1983) reported case studies when the INTERKOSMOS-19 satellite passed over epicentres of EQs. The following observations were made by Larkina et al. (1983):

• An increase of the VLF wave intensity between a few tens of minutes and some hours before and after an EQ
• A longitude spreading of the zone where the waves are amplified, which can be related to the longitudinal drift of the ionospheric plasma
• A latitude focalization

A perturbation in the ULF range (~8 Hz), during an EQ of magnitude 5.2 (24 August 1981, 15:46:27 Universal Time [UT], latitude = 51.5°, longitude = 181.7°), has been described by Chmyrev et al. (1989, 1997). The observations were made with the INTERCOSMOS-BULGARIA 1300 satellite at an altitude of 860 km, and at the same invariant latitude as the EQ. Another increase of the low-frequency (<450 Hz) emissions during numerous aftershocks of the Spitak EQ in Armenia has been observed with COSMOS-1809 (Serebryakova et al. 1992).

Much evidence of electron density perturbations in the ionosphere after strong EQs has been reported (e.g. during the great Alaskan EQ in 1964). Ionospheric perturbations above the seismic zone have also been observed a few days before (e.g. see Kon et al. 2011). They are better detected during the night when the ionosphere is calm. An increase and a decrease of the critical frequencies are observed in different regions of the ionosphere before EQs. The ionospheric density variations can be induced by a change of the current in the global electric circuit between the bottom of the ionosphere and the Earth's surface where electric charges associated with the stressed rocks can appear (Kuo et al. 2011). Additional information provided by GPS, measurement such as total electron content (TEC) data, can be used (e.g. see Garcia et al. 2005; Liu et al. 2009; Heki 2011).

Wave emissions and electron density perturbations can be linked through various mechanisms in the ionosphere, and the same hypotheses of generation mechanism of precursor are valid for the two perturbations. These hypotheses are mainly related to wave production by compression of rocks, diffusion of water in the epicentral area and redistribution of electric charges at the surface of the Earth, and then in the Earth's atmospheric system.

A number of generation mechanisms have been suggested for the explanation of these precursors, in particular electrokinetic effects (Dobrovolsky et al. 1989; Fenoglio et al. 1995), depolarization currents (Varotsos and Alexopoulos 1986) and movement of charged dislocations (Slifkin 1993; Vallianatos and Tzanis 1998; Tzanis et al. 2000). The details of seismoelectromagnetic mechanisms, processes, hypotheses and modelling have been reported by many authors (Parrot and Johnston 1989, 1993; Hayakawa and Ogawa 1992; Molchanov 1993; Hayakawa 1996, 1997, 2002; Hayakawa et al. 2004; Freund et al. 2006; Pulinets 2009, 2012; Freund 2011; Pulinets and Ouzounov 2011; De Santis et al. 2015). In recent years, special attention has been drawn to gas release prior to EQs (Tributsch 1978; Toutain and Baubron 1998; Omori et al. 2007; Pulinets 2007; Harrison et al. 2010; Baragiola et al. 2011), atmospheric heating which can be revealed with infrared experiments onboard satellites (Tronin 2006; Ouzounov et al. 2007, 2011) and the electric field (Kelley et al. 2017). DEMETER was the first satellite with a payload especially dedicated to study perturbations induced by the seismic activities (Parrot 2006).

6.2.1 Scientific Payload

The payload of the DEMETER microsatellite allowed measuring the waves and also some important plasma parameters (ion composition, electron density and temperature, and energetic particles). The scientific payload was composed of several sensors: three electric and three magnetic sensors (six components of the EM field to investigate from direct current [DC] up to 3.5 MHz), a Langmuir probe, an ion spectrometer and an energetic particle analyzer. The experiment capabilities are given in Table 6.1. Details about the experiments can be found in Berthelier et al. (2006a,b), Lebreton et al. (2006), Parrot et al. (2006a) and Sauvaud et al. (2006). There were two modes of operation: (1) a survey mode to record low-bit-rate data and (2) a burst mode to record high-bit-rate data above

TABLE 6.1
Experiment Capabilities

Frequency range, B	10 Hz to 17 kHz
Frequency range, E	DC to 3.5 MHz
Sensibility B	$2 \cdot 10^{-5}$ nT Hz$^{-1/2}$ at 1 kHz
Sensibility E	0.2 μV Hz$^{-1/2}$ at 500 kHz
Particles: Electrons	30 keV to 10 MeV
Particles: Ions	90 keV to 300 MeV
Ionic density	$5 \cdot 10^2$ to $5 \cdot 10^6$ ions/cm^3
Ionic temperature	1000–5000 K
Ionic composition	H$^+$, He$^+$, O$^+$
Electron density	10^2 to $5 \cdot 10^6$ cm^{-3}
Electron temperature	500–3000 K

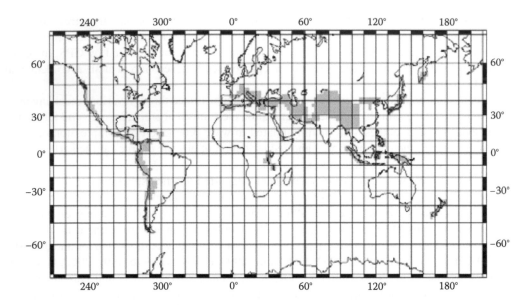

FIGURE 6.1   Map of the Earth where locations of the burst mode are indicated in grey. (From Pascal Bernard, Institut de Physique du Globe de Paris.)

seismic regions (Figure 6.1). In the survey mode, the telemetry was of the order of 950 Mb/day, and in burst mode, it was larger than 1 Gb/orbit.

For the wave experiment, the following data were recorded:

- During the burst mode
  - Waveforms of three electric components up to 15 Hz
  - Waveforms of the six components of the EM field up to 1 kHz
  - Waveforms of two components (1B + 1E) up to 17 kHz
  - Spectrum of one electric component up to 3.5 MHz
  - Waveform of one electric component up to 3.5 MHz (snapshots)
- During the survey mode
  - Waveforms of three electric components up to 15 Hz
  - Spectra of two components (1B + 1E) up to 17 kHz
  - Spectrum of one electric component up to 3.5 MHz
  - Results of a neural network to detect whistlers and sferics (Elie et al. 1999)

A specific tool to process the six components of the EM field during the burst mode has been provided by Santolík et al. (2006). For the other experiments, the difference between the burst and the survey modes only concerns the time resolution of the data.

The secondary mission of DEMETER was technological (Cussac et al. 2006). The purpose was to perform in-flight validation of

- An advanced system of payload telemetry incorporating a solid-state mass memory of 8 Gbits and an X-band transmitter at 16.8 Mbits/s
- A system of autonomous orbit control relying on a GPS receiver and an integrated navigator
- A pyro device firing by laser
- Thermal protection performance

6.2.2 Constraints

6.2.2.1 EMC Constraints

The high level of EM sensibility required by the payload brought specific constraints on DEMETER (Cussac et al. 2006):

- Sensors mounted at the end of the boom, bringing high inertia and low natural modes of the satellite. Attitude and orbit control system (AOCS) laws had to be optimized.
- Equipotentiality of the external surface of the satellite. Thermal protection, like multilayer insulation (MLI), was covered with specific conductive black painting. The solar generator was protected by indium tin oxide (ITO), cover glasses were interconnected and any current loop was compensated in order to reduce any generation of magnetic field.
- Magnetic shielding of equipment. The reaction wheels were placed in double shells made of μ metal. The electronic box of the star tracker was wrapped in a single sheet of μ metal. Finally, local shielding of the OBC (OnBoard Computer) was developed.
- Magneto-torquer activation limited to orbital period free from scientific measurements (terrestrial high invariant latitudes of >65° and <−65°).
- Solar generator rotation was stopped when the scientific instruments performed measurement, which was at an invariant latitude of <65°.

6.2.2.2 Constraints on the Mission

DEMETER recorded data all around the Earth and normally all the time. But there are some restrictions. Data were not recorded in the auroral zones (invariant latitude of >65°) because the seismic activity is low (except in Alaska), and the large level of the natural noise prevents observations of the signal due to seismic activity. This zone was reserved for the attitude control of the satellite because magneto-torquer activation perturbed magnetic measurements.

Due to various problems onboard the satellite, which was sometimes hit by heavy particles or disturbed by the full moon, the system was in safe mode (no data recording) for a few days from time to time. For example, this was the case just before the Sumatra EQ in December 2004. Nevertheless, the availability of the system was better than 90% during the mission.

6.2.3 Data Processing

The data were stored in a large onboard memory which was downloaded when the satellite was above Toulouse (two times per day). For the first time, the DEMETER telemetry was transmitted to a scientific mission centre (SMC) located at Orléans, where the data were processed. It included a check of the good health for the experiments, the production of quick looks, the transformation in physical units and archiving. For the second time, experimenters and guest investigators had access to the facility of a dedicated web server in order to download or display online selected data. Multiparameter plots were allowed. Details about the SMC are given by Lagoutte et al. (2006). The seismic data were received from the Institut de Physique du Globe de Paris (IPGP) and merged with the orbitography data in order to check the positions of the EQ epicentres relative to the satellite orbit. The SMC was also in relation to ground-based experiments. In Europe, special attention has been paid to Greece and dedicated ground stations have been installed in the Corinth Gulf by IPGP and close to La Réunion volcano by the Observatoire de Physique du Globe de Clermont-Ferrand (OPGC) (Zlotnicki et al. 2006). The satellite data have given an overview of the ionospheric parameters above the regions where these ground-based measurements were performed. In addition, maps of TEC were provided during the mission (Lognonné et al. 2006). The DEMETER data were stored in a centre dedicated to space missions (https://cdpp-archive.cnes.fr/).

6.2.4 ORBIT

DEMETER was a low-altitude satellite (710 km) launched in June 2004 onto a polar and circular orbit. The altitude of the satellite was decreased to 660 km in December 2005. The satellite science mission came to an end in December 2010, providing us with about 6.5 years of data in total. The orbit of DEMETER was nearly sun synchronous, and the upgoing half orbits correspond to night-time (22.30 LT), whereas the downgoing half orbits correspond to daytime (10.30 LT). Nearly sun synchronous means that every day the satellite does not return exactly above the same point, but above the same area (it could be at more than 1000 km from the point it flew over in the days before).

In order to optimize the filling up of the onboard memory, a specific software was implemented on ground to take into account the positions of the telemetry station and of the burst modes along the orbits. In the case of overflow before the dump, the duration of the burst modes has been accordingly reduced. This software prepared the telecommands for the experiments. The amount of telecommands was on the order of 600 octets per 3 days.

6.3 EXAMPLES OF PARTICULAR EVENTS

Many papers dealing with DEMETER data have been published showing particular events, that is, perturbations of ionospheric parameters in relation to EQs (Parrot et al. 2006b; Sarkar et al. 2007; Ouyang et al. 2008, 2011; Zhu et al. 2008; Bhattacharya et al. 2009; Zhang et al. 2009b,c, 2010b; Akhoondzadeh et al. 2010a,b; An et al. 2010a; Bankov et al. 2010; He et al. 2010; Sarkar and Gwal 2010; Singh et al. 2010; Zeren et al. 2010, 2012; Jiao et al. 2011; Li et al. 2011; Liu et al. 2011a; Onishi et al. 2011a; Priyadarshi et al. 2011; Sarkar et al. 2011, 2012; Stangl et al. 2011; Karia et al. 2012; Hobara et al. 2013; Ryu et al. 2014). Efforts have been made to analyze the DEMETER data of a few major EQs: the magnitude 8.8 Chile EQ (Huang et al. 2010; Liu et al. 2011b,c; Píša et al. 2011; Zhang et al. 2011; Z. Zhang et al. 2012) and the Wenchuan EQ (He et al. 2009; Zeng et al. 2009; Zhang et al. 2009a,d,e, 2010c; An et al. 2010b, 2011a; Blecki et al. 2010; Hau-Kun et al. 2010; Li et al. 2010; Onishi et al. 2011b; Wu et al. 2011; Zhu and Wang 2011; Yan et al. 2012; Zhang et al. 2012a). Zlotnicki et al. (2010) carried out analysis of DEMETER data during August 2004–December 2007 above volcanoes; the analysis shows that the electric and magnetic anomalies, as well as plasma density perturbations, may be observed when the projection of the orbit at the ground surface is less than 900 km from the volcanoes. They occur between 30 days prior to the eruption and 15 days after. Most of these perturbations have been observed during nighttime.

An ULF electric field is often associated with density fluctuations, and ULF and ELF electric field perturbations in the ionosphere have been widely observed by the satellites. A method to automatically distinguish this kind of disturbances has been developed with DEMETER data (Zhang et al. 2010a, 2011). It is based on the spectrum intensity and damping exponent with frequency in EM signals. This method is applied to DEMETER data processing around Chile EQs with a magnitude larger than 6.0. It was found that two-thirds of EQs have shown obvious ULF and ELF electric field perturbations in this region. The temporal and spatial distributions of electron density and temperature were compared with those of the electric field, which proved the existence of irregularities above the epicentre area.

Other studies in the ULF range have been carried out by Bhattacharya et al. (2007a,b) and by Athanasiou et al. (2011) related to the Haiti EQ. Electrostatic fluctuations, also called plasma turbulence, have been examined by Blecki et al. (2011, 2012) and Zhang et al. (2012b), which were often found to be associated with density variations. Several novel analysis methods for DEMETER data have been proposed; details are given in several papers (Ma et al. 2010; An et al. 2011b; Saradjian and Akhoondzadeh 2011; Zaourar et al. 2011). A number of papers have been published dealing with the change of signal intensity emitted by the ground-based VLF transmitters and received onboard DEMETER (Molchanov et al. 2006; Muto et al. 2008; Boudjada et al. 2008, 2010,

2012; Rozhnoi et al. 2007, 2008, 2010, 2012; He et al. 2009; Solovieva et al. 2009; Slominska et al. 2009). With the particle experiment onboard DEMETER, particle precipitation from the radiation belts has been studied (Li et al. 2010; Galper et al. 2011; Sidiropoulos et al. 2011; Anagnostopoulos et al. 2012) where similar data from another experiment are used.

Another example of ionospheric perturbation recorded by DEMETER is shown in Figure 6.2. It corresponds to an EQ occurring on 1 April 2007 at 20:39:58 UT with a magnitude equal to 8.1 and a depth equal to 24 km. Its position was −8.47° S, 157.04° E. From the top to the bottom, the panels show the electron density provided by the experiment ISL (Instrument Sonde de Langmuir), the ion density provided by the experiment IAP (Instrument Analyseur de Plasma), and the EQ occurrences along the satellite orbit. The bottom panel indicates the satellite's closest approach of past and future EQ epicentres that are within 2000 km of the DEMETER orbit. The Y-axis represents the distances D between the epicentres and the satellite, from 750 km up to 2000 km. The symbols are filled squares for post-seismic events and filled triangles for pre-seismic events. The scale on the right represents the time interval between the EQs and the DEMETER orbit, with a graduation from >30 days up to a 0- to 6-hour interval. The empty symbols have similar significance, except that they are related to the conjugate points of the epicentres (the distance D is then the distance between the conjugate points of the epicentres and the satellite). The symbol sizes correspond to EQs of magnitude 5–6, 6–7 and >7. From 12:21:15 to 12:22:20 UT, the cluster of triangles indicates the closest approach to the epicentres of this EQ and the many aftershocks. It can be observed that there is an increase with fluctuations of the electron density at this closest approach.

It was said that such a detection method of short-term precursors may not be valid because no co-seismic perturbation is observed in the ionospheric data, whereas it is obvious that it must occur (see Section 6.2.1). It is not easy to detect pre-seismic or co-seismic perturbations with a satellite. On one hand, the co-seismic perturbation occurs in the atmosphere and it takes time to reach the lower

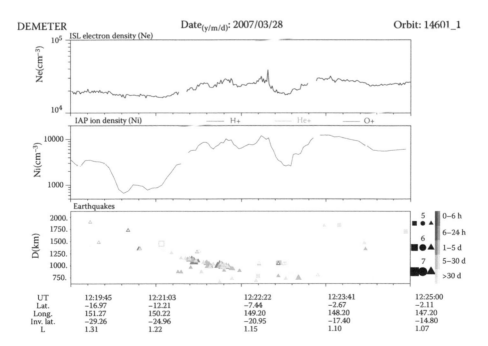

FIGURE 6.2 Density variations recorded 4 days before a magnitude 8.1 EQ. The top panel is related to the electron density, and the middle panel to the density of the ion O⁺. The bottom panel gives information about the EQ epicentres close to the satellite orbit (see the text for explanation). The parameters below the plots (the UT, the geographic latitude and longitude, the invariant latitude and the McIlwain parameter L) indicate that the observation takes place during nighttime along the rupture zone of the EQs.

ionosphere. On the other hand, the satellite must be at the right place at the right time to observe such perturbation quickly before it vanishes. However, during the lifetime of DEMETER, it was possible to find such events when the satellite was above the epicentre a few tens of minutes after the EQ. An example is shown in Figure 6.3. It corresponds to variations of electron and ion densities recorded on 1 May 2006 when the satellite is over an EQ epicentre 34 minutes after the shock. The EQ parameters are 02:49:42 UT, latitude −27.4°, longitude 289.36°, magnitude 5.0 and depth 30 km. It is shown that the perturbation is not located just above the epicentre but at the magnetically conjugate point of the epicentre in the same hemisphere (Marchand and Berthelier 2008).

Many perturbations have been observed in the DEMETER data which are close in time and space to EQs. However, one has to keep in mind that all measured parameters also display variations in the absence of seismic activity since the midlatitude and equatorial ionosphere is affected by a number of other sources of perturbations and primarily by solar activity (Onishi et al. 2011b). A statistical study with many events will show the general behaviour of such ionospheric perturbations and will help us to define a signature of ionospheric perturbations prior to EQs. This was possible with the DEMETER data because the lifetime of the mission was more than 6 years.

On the other hand, with satellite data we do not find ionospheric perturbations for all EQs. This could perhaps be due to the crust composition and soil configuration. But we also do not expect to have continuous ionospheric perturbations, and with a single satellite we are 'above' (here the term *above* means at a distance less than 1500 km) a given future epicentre only during 3 minutes per day (nighttime half-orbit). In such a case, possible perturbations can be missed. This is the main drawback of a satellite experiment in comparison with a ground-based experiment. For these two above-mentioned reasons, we prefer to search for a possible influence of the seismic activity on the ionosphere with a statistical analysis.

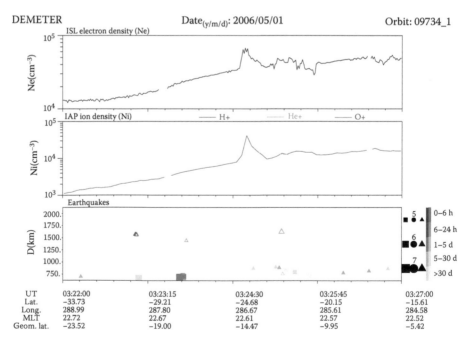

FIGURE 6.3 Density variations recorded approximately half an hour after an EQ. The top (middle) panel indicates the electron (O+ ion) density variation, whereas the bottom panel is similar to the one in Figure 6.2. In this last panel, the square symbol at ~03:23:50 UT indicates the closest approach to the EQ epicentre. It is observed that the variation is not right above the epicentre until later. This corresponds to the magnetically conjugate point of the epicentre at the altitude of the satellite.

6.4 STATISTICAL ANALYSIS

6.4.1 STATISTIC WITH THE ELECTRIC FIELD IN THE VLF RANGE

No specific event can be linked to the seismic activity in the VLF EM field. In fact, the spectrograms (frequency–time plots) are mainly dominated by whistlers which are associated with the thunderstorm activities. Nevertheless, a statistical analysis has been undertaken with the VLF DEMETER data of the first 2.5 years of the mission (Němec et al. 2008, 2009). They have shown that during the night, a small but statistically significant decrease of wave intensity is observed in the vicinity of EQs shortly (0–4 hours) before the time of the main shock. More recently, the method used by Němec et al. (2008) has been applied to the complete dataset (~6.5 years) acquired by DEMETER (Píša et al. 2012). The EQ data have been considered from the U.S. Geological Survey (USGS) catalogue (earthquake.usgs.gov). A total of about 9000 EQs with a magnitude larger than or equal to 5.0 and a depth lower than or equal to 40 km occurring all over the world have been considered. They used the same superposed epoch method developed and described by Němec et al. (2008, 2009). In the first step of the data processing, it was necessary to describe the distribution of the intensity of EM waves observed by DEMETER using all available data. This long-term distribution of intensities was calculated separately for each combination of parameters used to characterize the state of the ionosphere (Zhu 2010; Lu et al. 2011). The selected parameters and the number of bins used for each parameter were the same as those used by Němec et al. (2008), namely, frequency (16), geomagnetic latitude (66), geomagnetic longitude (36), magnetic local time (2), geomagnetic activity expressed by the K_p index (3) and season of the year (2). They constructed a multidimensional array with the number of dimensions equal to the number of the parameters. In each bin of this array, they stored a histogram of the wave intensities observed during the appropriate ionospheric conditions. The histogram is used to construct the experimental cumulative distribution function (CDF) of the wave intensity for each of the bins.

In the second step of the data processing, the data related to EQs (i.e. the data acquired close to the EQ epicentres in both space and time) were considered. These data were evaluated using the CDFs obtained in the first step of the data processing, and it was determined whether they were different from the expected distributions seismically unperturbed. In order to do so, they organized the data related to EQs in a grid as a function of the frequency, the time relative to the main shock and the distance from the epicentre. The values, which would be used more than once (typically the main shock and aftershocks), have been excluded from the analysis to avoid mixing of pre- and postseismic activity. Otherwise, one single measurement could be used more than once and it would be impossible to say to which EQ the observed data belong. It would cause bigger uncertainty in the statistic. For each measurement, the values of the corresponding CDF were evaluated. Afterward, they calculated the mean value of the CDFs in each bin of the grid ('probabilistic intensity'; 0.5 is subtracted in order to get the mean value equal to 0). The resulting mean values were then normalized to have the mean value equal to 0 and the standard deviation equal to 1 ('normalized probabilistic intensity'). A lower estimate of the standard deviation used for normalization was obtained by taking into account the uniform distribution of probabilistic intensity values and by assuming that all these values were independent. Since this was not exactly the case, an additional factor expressing what relative fraction of the data measured during the same half orbit could be considered as independent was introduced (Němec et al. 2008, 2009).

Figure 6.4 shows the frequency–time dependence of the normalized probabilistic intensity obtained for distances lower than 440 km from the epicentre. One can see that the main observed feature is a decrease of the normalized probabilistic intensity at the frequency of about 1.7 kHz shortly (0–4 hours) before the time of the main shocks. This is a confirmation of the result discussed by Němec et al. (2008), but with a much more important dataset. Němec et al. (2008) observed a decrease in frequency at about 1.7 kHz that corresponds approximately to the cut-off frequency of the first transverse magnetic (TM) mode (EM wave lacks magnetic field component in the direction of propagation) of the

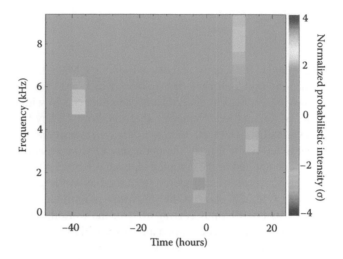

FIGURE 6.4 Frequency–time dependence of the normalized probabilistic intensity (see text) obtained from the nighttime electric field data measured within 440 km of the epicentre of the EQs with magnitudes larger than or equal to 5.0 and depth less than or equal to 40 km. (Adapted from Píša, D. et al., *Ann. Geophys.*, 55 (1), 157–163, 2012.)

Earth–ionosphere waveguide during the nighttime. An increase of this cut-off frequency would therefore necessarily lead to a decrease of the power spectral density of electric field fluctuations observed by DEMETER in the appropriate frequency range (Zhang et al. 2011). Such an increase in the cut-off frequency would correspond to a decrease of the height of the ionosphere. These results could therefore indicate that the height of the ionosphere is statistically lower above the epicentres of imminent EQs, which is likely consistent with the results obtained by Hayakawa et al. (2010) using subionospheric VLF and LF propagation. As the EM waves which are propagating in the Earth–ionosphere waveguide are mainly whistlers, this means that it is not a change of their intensities but a disturbance of their propagations above the epicentres of future EQs. Opposite to the results of past satellites, we have not observed a production (or intensification) of waves above the EQ epicentres.

6.4.2 Statistic with the Electron Density

A classical superposed epoch method with the electron density recorded by DEMETER during nighttime has been used by He et al. (2011a,b). They compared the data just before the EQs with data recorded well before the EQs at the same location. The Earth's surface was divided into cells with 2° resolution in both latitude and longitude. They have defined a square area centred on each EQ and containing 121 cells in all, which was considered the research zone. In order to explore the pre-EQ phenomena, only the data before the EQ occurrences were used. In each cell, the data with $K_p < 2+$ and the time interval 31–75 days before the EQs were used to construct the background, and the data with $K_p < 3+$ and the time interval 1–30 days before the EQs were analyzed for comparison. A low value of the index of magnetic activity K_p was chosen in order to avoid natural perturbations. For each EQ and each cell, a quantity R is calculated. It is the mean values of the data during background (–75 to –31 days) minus the mean values of the data during seismoactive times (–30 to –1 days) and normalized by the standard deviation of the background (He et al. 2011b). But the last step of the data processing concerns the final statistics with all seismic events. Ř, the mean value of R, is calculated for all EQs in each cell. The result is shown in Figure 6.5, where the data have been smoothed over the cells. It is observed that Ř has a maximum close to the epicentres of the EQs. This means that we statistically observe an increase of the electron density close to the epicentres before the EQs. It has been shown in He et al. (2011b) that (1) this effect disappears if a

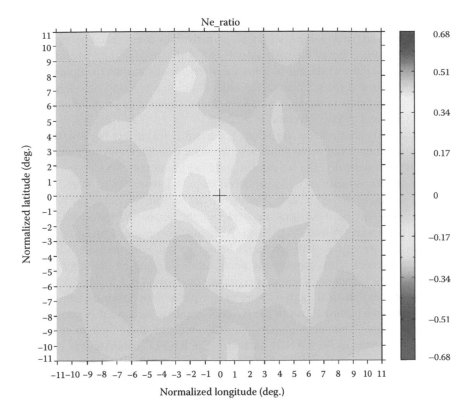

FIGURE 6.5   Results of the statistic with the electron density. The centre of the figure corresponds to the epicentres of the EQs. They are 5690 EQs of magnitude >5 and depth <60 km in the statistic. (Adapted from He, Y. et al. *Nat. Hazards Earth Syst. Sci.*, 11, 2173–2180, 2011.)

random position of events is used instead of the real position of epicentres, (2) this effect increases with the magnitude of the EQs and (3) this effect decreases if the depth of the EQs increases.

6.4.3   Statistic with the Ion Density

An automatic software to detect density fluctuations has been developed (Parrot 2011, 2012). The inputs are the EQ list and the DEMETER data during nighttime. The EQs have been selected with a magnitude larger than 4.8, and we have associated for each of them a parameter related to the position of the epicentres: below the sea or inland and close to a coast. During the considered period (August 2004–October 2009), there are 17,366 EQs. The software searches for the data of the orbits which are close to the epicentres (less than 1500 km) between 0 and 15 days before each EQ. To detect a variation, the software considers the DEMETER ion density data during 3 minutes around the closest approach to the future EQ epicentre. The data are smoothed and a variation (crest or trough) is searched using the change of the derivative sign.

   In order to evaluate the results of the automatic detection software working with the EQ list, we considered two other lists with random data. First, we took the list of EQs but randomly changed their latitudes and longitudes (keeping the same time). The resulting database after the software was applied is known as RAND1. Second, we considered the same list of EQs but only shifted their longitudes 25° to the west (keeping the same latitude and the same time). After the software was applied, a third database was generated, known as RAND2. This shift of the longitude is due to the fact that, on one hand, most of the EQs are concentrated around the equator, and on the other hand, it is known that during nighttime, the natural occurrence of ionospheric perturbations is also more concentrated around the equator. The RAND2 database fits more with the reality concerning the positions of the EQs and the possible ionospheric perturbations due to usual geophysical activities.

In order to reduce the effect of the solar activity, we eliminated the data when the K_p index was larger than 3+. We also did not take into account the aftershock data when the time of the aftershock was too close to the time of the main shock, in order to not mix pre- and post-seismic effects. It is known that at the exact time of the EQs, you have the propagation of an AGW which can perturb the ionosphere (Garcia et al. 2005; Iyemori et al. 2005).

This automatic software to detect variations is certainly not perfect because the shapes of the ionospheric perturbations could be very different. We have examples with very sharp peaks or very smooth peaks in the ion density. But we ran the software in the same way on the three datasets to search for ionospheric perturbations and produced three databases (EQs, RAND1 and RAND2) where the main parameter is the percentage of the variations with respect to the background.

Different analyses have been performed considering the number of perturbations and the intensity of these perturbations (Parrot 2011, 2012). A distinction has been made regarding the positions of the EQs (all epicentres, with epicentres below the sea, with inland epicentres or close to a coast). In order to check if the ionospheric perturbations depend on the magnitude, the results have been displayed as a function of several magnitude intervals (4.8, 5.0; 5.0, 5.5; 5.5, 6.0; and 6.0, 9.0). The results of this statistical analysis were in agreement with generally admitted behaviours. As expected, there were perturbations in the RAND1 database because it is known that there are ionospheric variations without seismic activity, and as expected, there were more perturbations in the RAND2 database than in the RAND1 database. The RAND2 database was used as reference. The statistics have shown that there are more perturbations in the EQ database than in the RAND2 database. It was observed in the EQ database that the highest perturbation value over the 15 days for one EQ increases with its magnitude, as expected. Using a statistical comparison with RAND2, it was shown that these results are more significant for powerful ($M > 6$) and inland EQs. It was also shown that, as expected, the perturbations are not so important for deep EQs (>40 km).

6.4.4 Discussions about the Statistics

A few examples of ionospheric perturbations prior to seismic activity have been discussed in this chapter. These perturbations occur in close vicinity of the EQ epicentres and a few hours or a few days before the shock. But many other phenomena can perturb the ionosphere, and as people may have doubts about the relation between the ionosphere and the Earth's crust, despite the various mechanisms (as discussed in Section 6.2), several statistical analyses have been performed with a huge number of events. The ionospheric perturbations are more important close to the epicentres of future inland EQs than prior to events at random positions. This indicates that the intensity of the perturbations is more important when the magnitude of the EQs increases. It is very clear that the ionosphere is influenced by the seismic activity prior to EQs, but this does not mean that it is possible to predict EQs.

6.5 IS IT POSSIBLE TO PREDICT EQs?

6.5.1 The Magnitude 8.8 Chile EQ

It is not possible to predict EQs up to now because we must find answers to the three following questions: When? Where? Which magnitude? But using the example of a powerful event, we examine in this section both the possibilities and the problems. A magnitude 8.8 EQ occurred in Chile on 27 February 2010 with an epicentre located at 35.85° S, 72.72° W. The data recorded before this event are fully described in Píša et al. (2011).

Figure 6.6 shows the electron density recorded on 10 February 2010, which indicates an abnormal variation in a specific area. This is known because background data have been recorded several

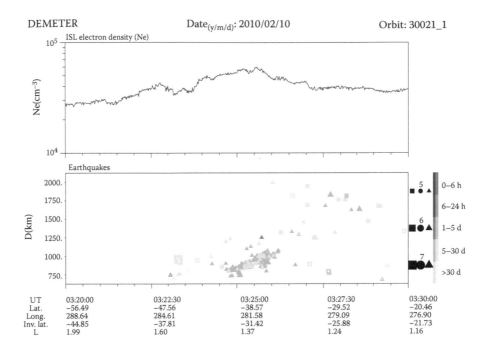

FIGURE 6.6 Density variation observed 17 days before the Chile EQ. The top panel shows the electron density, and the bottom panel the location of the main shock and the aftershocks along the orbit (cluster of triangles between 03:24:30 and 03:26:00 UT). One observes a density variation when the satellite is close to the epicentre.

years before at the same location and during the same season (Píša et al. 2011), and we can compare the current data with the past data. The satellite returns above the same area on 16 February 2010 (Figure 6.7), and again, similar variation of the electron density is observed. Similar variations were recorded on 18 February 2010 (Figure 6.8). We can plot on a map the orbits where the anomalies are observed. We indicate on each orbit the position of the highest amplitude of the perturbation because it must correspond to the closest approach of a possible event, shown in Figure 6.9 with stars. It is possible to draw a line passing through the three stars and to assume that, if an event must occur, it will be along this portion of line limited on the left by the 10 February orbit and on the right by the 18 February orbit. It could not occur on the right because otherwise nothing will be observed on the 10 February orbit, and similarly, it could not be on the left because nothing will be observed on the 18 February orbit. We have pinpointed a zone of occurrence which is *a posteriori* good (the future EQ epicentre is indicated by an asterisk in Figure 6.9), but the length of this line is about 1000 km. With our knowledge of this seismic area, it is possible to reduce this length. It is known that the area along Chile's coast is one of the most active seismic zones around the world. The Nazca Plate is going under the South American tectonic plate, and there is a known fault that we can consider as the location of a future EQ. Taking into consideration the dimension of this fault allows us to claim that an EQ will occur somewhere along a 250 km line. This is rather imprecise for a prediction, but moreover, we have no indication of the exact time and the exact magnitude. As time continues, all we can say is that the magnitude must be very important because the ionospheric perturbations start to appear a long time before. In fact, when an ionospheric perturbation occurs, we do not know if it may correspond to an EQ of small magnitude which will appear in a few hours or to an EQ of large magnitude which will appear in a few days. As an example, it can be seen in the same Chile area that an EQ with a small magnitude (10 February 2006; 17:51:54 UT; 32.52° S, 288.61° E; $M = 5.1$) can also be associated with a non-negligible ionospheric perturbation a few hours before the shock (Figure 6.10).

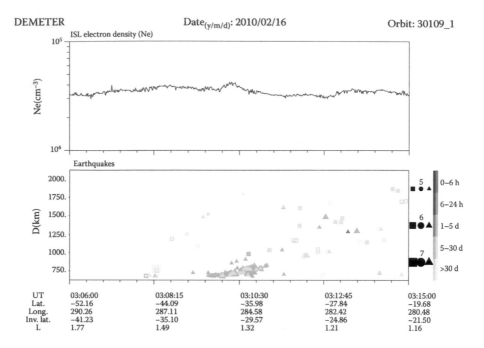

FIGURE 6.7 Density variation observed 11 days before the Chile EQ. The top panel shows the electron density, and the bottom panel the location of the main shock and the aftershocks along the orbit (cluster of triangles between 03:09:36 and 03:11:20 UT). One observes a density variation when the satellite is close to the epicentre.

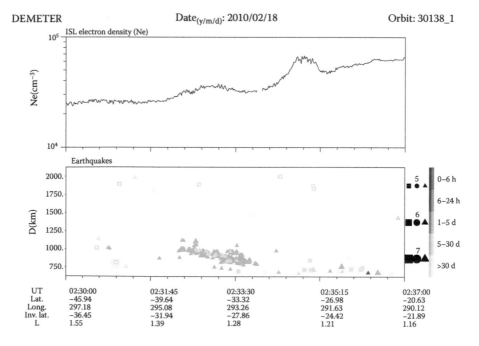

FIGURE 6.8 Density variation observed 9 days before the Chile EQ. The top panel shows the electron density, and the bottom panel the location of the main shock and the aftershocks along the orbit (cluster of triangles between 02:32:15 and 02:33:55 UT). One observes a density variation when the satellite is close to the epicentre, but another variation is also seen around 02:36:00 UT which corresponds to the conjugate location of the epicentre at the satellite altitude in the same hemisphere.

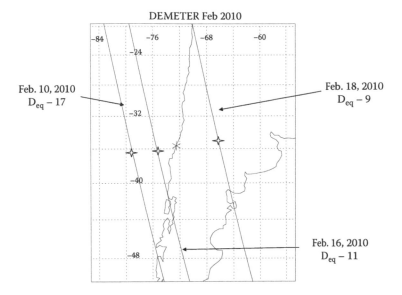

FIGURE 6.9 Ground projections of three DEMETER orbits above Chile. Days are indicated together with the delay to the magnitude 8.8 EQ. On these orbit traces the stars indicate the location where the ionospheric perturbations are highest. The asterisk indicates the epicentre of the future EQ.

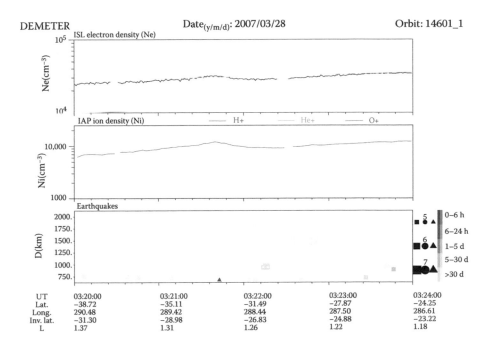

FIGURE 6.10 Density variations observed about 15 hours before an EQ of small magnitude (see text). The electron (ion O$^+$) density is shown in the top (second) panel. The bottom panel indicates the location of the main shock along the orbit (red triangle at ~03:21:42 UT).

6.5.2   A Prediction Attempt

In a first step, an automatic search for ionospheric density perturbations similar to the one explained in Section 6.4.3 has been done, but now using the complete DEMETER dataset (6.5 years) and disregarding the EQ occurrence. The perturbations occurring during high geomagnetic activity and those not above seismic areas have been eliminated. The outputs of this first step are the amplitude, time and location of the perturbations. In a second step, a list of EQs is used to automatically check if a given perturbation could be attributed to a given EQ. The parameters are

- The EQ magnitude (M) and the depth (d)
- The maximum distance between the EQ epicentre and the perturbation location (D)
- The maximum time between the perturbation occurrence and the EQ (T)

The outputs are the number of good detections (one or several perturbations could correspond to one EQ), the number of false alarms (one perturbation but no EQ) and the number of bad detections (no perturbation but one EQ). A preliminary result (Li and Parrot 2012) for $T < 7$ days, $D < 500$ km, $d < 20$ km and $M = 4.8$–5 gives a rate of good detections of 12.9% (a rate of bad detections of 87.1%) and a rate of false alarms equal to 96.6%. These false alarms are due to natural variations of the ionosphere. The number of wrong detections is also very high because, as it was said before, we do not expect to have continuous ionospheric perturbations and the satellite is 'above' a seismic area only a few minutes per day. This rate will decrease if data from several satellites could be involved. The only good point is that the number of EQs which are detected increases with the magnitude, as expected (Li and Parrot 2012). A good detection does not mean that we are able to predict the EQ.

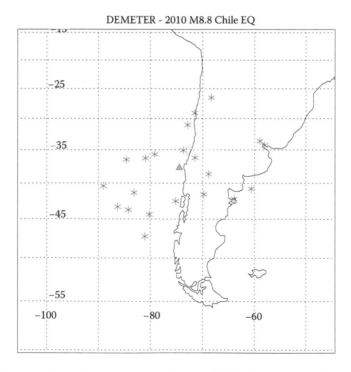

FIGURE 6.11   Attempt to determine the epicentre of a powerful EQ (the same as in Figure 6.9). The red stars are the positions of the automatically detected anomalies during 15 days, the blue triangle is the point which is at the minimum distance from all the red stars, and the green symbol is the real position of the epicentre. This means that if a cluster of perturbations appear in the data in a given area, it is possible to approximately determine the epicentre. (From Li, M., and Parrot, M., *J. Geophys. Res.*, 118, 3731–3739, 2013.)

Uncertainty about the position is large, as we have seen before. But, if we have several perturbations for one EQ, there is the possibility to use triangulation (Figure 6.11). However, uncertainties about the time and the magnitude are also large. At this time, some work is being done to try to increase the number of good detections and decrease the number of bad detections and false alarms.

6.6 CONCLUSION

A number of studies have shown ionospheric perturbations prior to seismic activities. These perturbations were observed in close vicinity of the EQ epicentres and a few hours or a few days before the EQ events. Ionospheric perturbations can also occur by many other mechanisms, so it becomes difficult to associate observed perturbations with an EQ event. A statistical analysis must be performed with a large number of EQ events. With the help of statistical analysis of DEMETER data, it has been shown that perturbations exist in the ionosphere which can be attributed to the seismic activities. Depending on the checked parameters, these perturbations occur between a few hours and a few days before EQs. Detailed studies are needed with the DEMETER data in particular to study the ionospheric behaviours in the conjugate region of the EQs because waves can propagate along the magnetic field lines up to the opposite hemisphere.

At this time, the three SWARM satellites (a European Space Agency [ESA] mission launched at the end of 2013) provide a unique opportunity to study the ionospheric perturbations in relation to the seismic activity. It is intended to study more carefully the EQs which produce perturbations in order to try to extract some common characteristics. The use of the three SWARM satellites, which will return more often above a given seismic area than a single satellite, will be of great help (De Santis et al. 2017). This attempt to predict EQs will of course generate false alarms and wrong detections, but it is expected that the number of true detections will be larger, at least for powerful EQs. Nevertheless, it will remain difficult to predict the exact position, time and magnitude of the EQs. Concerning the position, it will be useful to consider the fault system, which is already known at the Earth's surface. This will improve the accuracy on the epicentre location. Concerning the time and magnitude, it will be difficult, faced with an unusual ionospheric perturbation (i.e. not related to solar or magnetic activities), to say if it is due to a big EQ occurring in a few days or to a small EQ occurring in a few hours.

There are many other satellite projects in different countries especially dedicated to study this problem, and the Chinese satellite CSES is at a very advanced stage, with an expected launch in February 2018 (Shen et al. 2011). Their aim is to have several satellites in orbit by the year 2020.

ACKNOWLEDGEMENTS

This work was supported by the Centre National d'Etudes Spatiales in France. It is based on observations with the experiments embarked on with DEMETER. The author thanks J. J. Berthelier, J. P. Lebreton and J. A. Sauvaud, the principal investigators of these instruments, for the use of their data. The research leading to these results has also received funding from the European Community's Seventh Framework Programme (FP7/2007–2013) under Grant Agreement Number 262005.

REFERENCES

Akhoondzadeh, M., Parrot, M., and Saradjian, M. R. (2010a). Electron and ion density variations before strong earthquakes (M > 6.0) using DEMETER and GPS data. *Nat. Hazards Earth Syst. Sci.*, 10, 7–18.

Akhoondzadeh, M., Parrot, M., and Saradjian, M. R. (2010b). Investigation of VLF and HF waves showing seismo ionospheric anomalies induced by the 29 September 2009 Samoa earthquake (Mw = 8.1). *Nat. Hazards Earth Syst. Sci.*, 10, 1061–1067. doi: 10.5194/nhess-10-1061-2010.

An, Z., Du, X., Fan, Y., Liu, J., Tan, D., Chen, J., and Xie, T. (2011a). A study of the electric field before the Wenchuan 8.0 earthquake of 2008 using both space-based and ground-based observational data. *Chin. J. Geophys.*, 54 (6), 818–827.

An, Z., Du, X., Tan, D., and Chen, J. (2010a). Study of Wudu-Wenxian M5.0 earthquake in Gansu by using the DEMETER satellite data. *Plateau Earthquake Res.*, 22 (2), 63–68 (in Chinese).

An, Z., Fan, Y., Liu, J., Tan, D., Chen, J., Zheng, G., and Xie, T. (2010b). Analysis on ion temperature variation detected by DEMETER before 2008 Wenchuan Ms8.0 earthquake. *Acta Seismol. Sin.*, 32 (6), 754–759 (in Chinese with English abstract).

An, Z., Tan, D., Chen, J., Fan, Y., Liu, J., and Xie, T. (2011b). Discussion on the analysis method of magnetic field waveform data recorded by the DEMETER satellite. *South China J. Seismol.*, 1, 55–65 (in Chinese with English abstract).

Anagnostopoulos, G. C., Vassiliadis, E., and Pulinets, S. (2012). Characteristics of flux-time profiles, temporal evolution, and spatial distribution of radiation-belt electron precipitation bursts in the upper ionosphere before great and giant earthquakes. *Ann. Geophys.*, 55 (1), 21–36. doi: 10.4401/ag-5365.

Athanasiou, M. A., Anagnostopoulos, G. C., Iliopoulos, A. C., Pavlos, G. P., and David, C. N. (2011). Enhanced ULF radiation observed by DEMETER two months around the strong 2010 Haiti earthquake. *Nat. Hazards Earth Syst. Sci.*, 11, 1091–1098. doi: 10.5194/nhess-11-1091-2011.

Bankov, L. G., Parrot, M., Heelis, R. A., Berthelier, J.-J., Marinov, P. G., and Vassileva, A. K. (2010). DEMETER and DMSP satellite observations of the disturbed H+/O+ ratio caused by Earth's seismic activity in the Sumatra area during December 2004. *Adv. Space Res.*, 46 (4), 419–430.

Baragiola, R. A., Dukes, C. A., and Hedges, D. (2011). Ozone generation by rock fracture: Earthquake early warning? *Appl. Phys. Lett.*, 99, 204101. doi: 10.1063/1.3660763.

Berthelier, J. J., Godefroy, M., Leblanc, F., Malingre, M., Menvielle, M., Lagoutte, D., Brochot, J. Y. et al. (2006a). ICE, the electric field experiment on DEMETER. *Planet. Space Sci.*, 54 (5), 456–471.

Berthelier, J. J., Godefroy, M., Leblanc, F., Séran, E., Peschard, D., Gilbert, P., and Artru, J. (2006b). IAP, the thermal plasma analyzer on DEMETER. *Planet. Space Sci.*, 54 (5), 487–501.

Bhattacharya S., Sarkar, S., Gwal, A. K., and Parrot, M. (2007a). Observations of ULF/ELF anomalies detected by DEMETER satellite prior to earthquakes. *Indian J. Radio Space Phys.*, 36, 103–113.

Bhattacharya, S., Sarkar, S., Gwal, A. K., and Parrot, M. (2007b). Satellite and ground-based ULF/ELF emissions observed before Gujarat earthquake in March 2006. *Curr. Sci.*, 93 (1), 41–46.

Bhattacharya, S., Sarkar, S., Gwal, A. K., and Parrot, M. (2009). Electric and magnetic field perturbations recorded by DEMETER satellite before seismic events of the 17th July 2006 M 7.7 earthquake in Indonesia. *J. Asian Earth Sci.*, 34, 634–644.

Blecki, J., Kosciesza, M., Parrot, M., Savin, S., and Wronowski, R. (2012). Extremely low frequency plasma turbulence recorded by the DEMETER satellite in the ionosphere over the Abruzzi region prior to the April 6, 2009, L'Aquila earthquake. *Ann. Geophys.*, 55 (1), 37–47. doi: 10.4401/ag-5356.

Blecki, J., Parrot, M., and Wronowski, R. (2010). Studies of the electromagnetic field variations in ELF frequency range registered by DEMETER over the Sichuan region prior to the 12 May 2008 earthquake. *Int. J. Remote Sens.*, 31 (13), 3615–3629.

Blecki, J., Parrot, M., and Wronowski, R. (2011). Plasma turbulence in the ionosphere prior to earthquakes, some remarks on the DEMETER registrations. *J. Asian Earth Sci.*, 41 (4–5), 450–458. doi: 10.1016/j.jseaes.2010.05.016.

Boudjada, M. Y., Schwingenschuh, K., Al-Haddad, E., Parrot, M., Galopeau, P. H. M., Besser, B., Stangl, G., and Voller, W. (2012). Effects of solar and geomagnetic activities on the sub-ionospheric very low frequency transmitter signals received by the DEMETER micro-satellite. *Ann. Geophys.*, 55 (1), 49–55. doi: 10.4401/ag-5463.

Boudjada, M. Y., Schwingenschuh, K., Biernat, H. K., Berthelier, J. J., Blecki, J., Parrot, M., Stachel, M., Aydogar, O., Stangl, G., and Weingrill, E. (2008). Similar behaviors of natural ELF/VLF ionospheric emissions and transmitter signals over seismic Adriatic regions. *Nat. Hazards Earth Syst. Sci.*, 8, 1229–1236.

Boudjada, M. Y., Schwingenschuh, K., Döller, R., Rohznoi, A., Parrot, M., Biagi, P. F., Galopeau, P. H. M. et al. (2010). Decrease of VLF transmitter signal and chorus-whistler waves before L'Aquila earthquake occurrence. *Nat. Hazards Earth Syst. Sci.*, 10, 1487–1494. doi: 10.5194/nhess-10-1487-2010.

Chmyrev, V. M., Isaev, N. V., Bilichenko, S. V., and Stanev, G. (1989). Observation by space-borne detectors of electric fields and hydromagnetic waves in the ionosphere over an earthquake centre. *Phys. Earth Planet. Inter.*, 57, 110–114.

Chmyrev, V. M., Isaev, N. V., Serebryakova, O. N., Sorokin, V. M., and Sobolev, Ya. P. (1997). Small-scale plasma inhomogeneities and correlated ELF emissions in the ionosphere over an earthquake region. *J. Atmos. Terr. Phys.*, 59, 967–974.

Cussac, T., Clair, M. A., Ultré-Guerard, P., Buisson, F., Lassalle-Balier, G., Ledu, M., Elisabelar, C., Passot, X., and Rey, N. (2006). The DEMETER microsatellite and ground segment. *Planet. Space Sci.*, 54 (5), 413–427.

De Santis, A., Balasis, G., Pavón-Carrasco, F. J., Cianchini, G., and Mandea, M. (2017). Potential earthquake precursory pattern from space: The 2015 Nepal event as seen by magnetic Swarm satellites. *Earth Planet. Sci. Lett.*, 461, 119–126.

De Santis, A., De Franceschi, G., Spogli, L., Perrone, L., Alfonsi, L., Qamili, E., Cianchini, G. et al. (2015). Geospace perturbations induced by the earth: The state of the art and future trends. *Phys. Chem. Earth A/B/C*, 85–86, 17–33. doi: 10.1016/j.pce.2015.05.004.

Dobrovolsky, I. P., Gershenzon, N. I., and Gokhberg, M. B. (1989). Theory of electrokinetic effects occurring at the final stage in the preparation of a tectonic earthquake. *Phys. Earth Planet. Inter.*, 57, 144–156.

Elie, F., Hayakawa, M., Parrot, M., Pinçon, J. L., and Lefeuvre, F. (1999). Neural network system for the analysis of transient phenomena onboard the DEMETER micro-satellite. *J. IEICE*, E82-A (8), 1575–1581.

Fenoglio, M. A., Johnston, M. J. S., and Byerlee, J. D. (1995). Magnetic and electric fields associated with changes in high pore pressure in fault zones: Application to the Loma Prieta ULF emissions. *J. Geophys. Res.*, 100, 12951–12958.

Freund, F. (2011). Pre-earthquake signals: Underlying physical processes. *J. Asian Earth Sci.*, 41, 383–400.

Freund, F. T., Takeuchi, A., and Lau, B. W. S. (2006). Electric currents streaming out of stressed igneous rocks – A step towards understanding pre-earthquake low frequency EM emissions. *Phys. Chem. Earth A/B/C*, 31 (4–9), 389–396.

Galper, A. M., Koldashov, S. V., and Ulitin A. A. (2011). Local perturbations of the earth's radiation belt during the seismic event development in Japan on March 11, 2011. *Bull. Lebedev Phys. Inst.*, 38 (7), 209–214. doi: 10.3103/S1068335611070050.

Garcia, R., Crespon, F., Ducic, V., and Lognonné, P. (2005). 3D ionospheric tomography of post-seismic perturbations produced by the Denali earthquake from GPS data. *Geophys. J. Int.*, 163 (3), 1049–1064. doi: 10.1111/j.1365-246X.2005.02775.

Gokhberg, M. B., Pilipenko, V. A., and Pokhotelov, O. A. (1982). Satellite observation of the electromagnetic radiation above the epicentral region of an incipient earthquake. *Dokl. AN SSSR*, 268, 56–58.

Harrison, R. G., Aplin, K. L., and Rycroft, M. J. (2010). Atmospheric electricity coupling between earthquake regions and the ionosphere. *J. Atmos. Sol. Terrestr. Phys.*, 72, 376–381.

Hau-Kun, J., Ho, Y.-Y., Kakinami, Y., Liu, J.-Y., Oyama, K.-I., Parrot, M., Hattori, K., Nishihashi, M., and Zhang, D. (2010). Seismo-ionospheric anomalies of the GPS-TEC appear before the 12 May 2008 magnitude 8.0 Wenchuan earthquake. *Int. J. Remote Sens.*, 31 (13), 3579–3587.

Hayakawa, M., ed. (1996). Seismo-electromagnetic phenomena. *J. Atmos. Electr.* (Special Issue), 16 (3).

Hayakawa, M. (1997). Electromagnetic precursors of earthquakes: Review of recent activities. In *Review of Radio Science, 1993–1996*. Oxford: Oxford University Press, 807–818.

Hayakawa, M., ed. (2002). Seismo-electromagnetics. *J. Atmos. Electr.* (Special Issue), 22 (3).

Hayakawa, M., Kasahara, Y., Nakamura, T., Muto, F., Horie, T., Maekawa, S., Hobara, Y., Rozhnoi, A. A., Solovieva, M., and Molchanov, O. A. (2010). A statistical study on the correlation between lower ionospheric perturbations as seen by subionospheric VLF/LF propagation and earthquakes. *J. Geophys. Res.*, 115, A09305. doi: 10.1029/2009JA015143.

Hayakawa, M., Molchanov, O. A., Biagi, P. F., and Vallianatos, F., eds. (2004). Seismo electromagnetics and related phenomena. *Phys. Chem. Earth A/B/C* (Special Issue), 29 (4–9).

Hayakawa, M., and Ogawa, T., eds. (1992). Atmospheric electricity phenomena associated with earthquakes and volcanic eruptions. *Res. Lett. Atmos. Electr.* (Special Issue), 12 (3).

He, Y., Yang, D., Chen, H., Qian, J., Zhu, R., and Parrot, M. (2009). SNR changes of VLF radio signals detected onboard the DEMETER satellite and their possible relationship to the Wenchuan earthquake. *Sci. China D*, 52 (6), 754–763.

He, Y., Yang, D., Qian, J., and Parrot, M. (2011a). Anomaly of the ionospheric electron density close to earthquakes: Case studies of Pu'er and Wenchuan earthquakes. *Earthq. Sci.*, 24 (6), 549–555. doi: 10.1007/s11589-011-0816-0.

He, Y., Yang, D., Qian, J., and Parrot, M. (2011b). Response of the ionospheric electron density to different types of seismic events. *Nat. Hazards Earth Syst. Sci.*, 11, 2173–2180. doi: 10.5194/nhess-11-2173-2011.

He, Y., Yang, D., Zhu, R., Qian, J., and Parrot, M. (2010). Variations of electron density and temperature in ionosphere based on the DEMETER ISL data. *Earthq. Sci.*, 23 (4), 349–355. doi: 10.1007/s11589-010-0732-8.

Heki, K. (2011). Ionospheric electron enhancement preceding the 2011 Tohoku-Oki earthquake. *Geophys. Res. Lett.*, 38, L17312. doi: 10.1029/2011GL047908.

Hobara, Y., Nakamura, R., Suzuki, M., Hayakawa, M., and Parrot, M. (2013). Ionospheric perturbations observed by the low altitude satellite DEMETER and possible relation with seismicity. *J. Atmos. Electr.*, 33 (1), 21–29. doi: 10.1541/jae.33.21.

Huang, J. P., Liu, J., Ouyang, X. Y., and Li, W. J. (2010). Analysis to the energetic particles around the M8.8 Chile earthquake. *Seismol. Geol.*, 32 (3), 417–423 (in Chinese with English abstract).

Iyemori, T., Nose, M., Han, D., Gao, Y., Hashizume, M., Choosakul, N., Shinagawa, H. et al. (2005). Geomagnetic pulsations caused by the Sumatra earthquake on December 26, 2004. *Geophys. Res. Lett.*, 32, L20807. doi: 10.1029/2005GL024083.

Jiao, Q., Yan, R., and Zhang, J. (2011). Ionospheric disturbances observed on DEMETER satellite prior to the 2010 Ms7.1 Haiti earthquake. *Earthquake*, 31 (2), 68–78 (in Chinese with English abstract).

Karia, S., Sarkar, S., and Pathak, K. (2012). Analysis of GPS-based TEC and electron density by the DEMETER satellite before the Sumatra earthquake on 30 September 2009. *Int. J. Remote Sens.*, 33 (16), 5119–5134.

Kelley, M. C., Swartz, W. E., and Heki, K. (2017). Apparent ionospheric total electron content variations prior to major earthquakes due to electric fields created by tectonic stresses. *J. Geophys. Res. Space Phys.*, 122, 6689–6695. doi: 10.1002/2016JA023601.

Kon, S., Nishihashi, M., and Hattori, K. (2011). Ionospheric anomalies possibly associated with M ≥ 6.0 earthquakes in the Japan area during 1998–2010: Case studies and statistical study. *J. Asian Earth Sci.*, 41, 410–420. doi: 10.1016/j.jseaes.2010.10.005.

Kuo, C. L., Huba, J. D., Joyce, G., and Lee, L. C. (2011). Ionosphere plasma bubbles and density variations induced by pre-earthquake rock currents and associated surface charges. *J. Geophys. Res.*, 116, A10317. doi: 10.1029/2011JA016628.

Lagoutte, D., Brochot, J. Y., de Carvalho, D., Elie, F., Harivelo, F., Hobara, Y., Madrias, L. et al. (2006). The DEMETER science mission centre. *Planet. Space Sci.*, 54 (5), 428–440.

Larkina, V. I., Nalivayko, A. V., Gershenzon, N. I., Gokhberg, M. B., Liperovskiy, V. A., and Shalimov, S. L. (1983). Observations of VLF emissions related with seismic activity, on the Interkosmos-19 satellite. *Geomag. Aeron.*, 23, 684–687.

Lebreton, J. P., Stverak, S., Travnicek, P., Maksimovic, M., Klinge, D., Merikallio, S., Lagoutte, D., Poirier, B., Kozacek, Z., and Salaquarda, M. (2006). The ISL Langmuir probe experiment and its data processing onboard DEMETER: Scientific objectives, description and first results. *Planet. Space Sci.*, 54 (5), 472–486.

Li, L., Yang, J., Cao J., Lu, L., Wu, Y., and Yang, D. (2011). Statistical backgrounds of topside-ionospheric electron density and temperature and their variations during geomagnetic activity. *Chin. J. Geophys.*, 54 (10), 2437–2444 (in Chinese with English abstract). doi: 10.3969/j.issn.0001-5733.2011.10.001.

Li, M., and Parrot, M. (2012). "Real time analysis" of the ion density measured by the satellite DEMETER in relation with the seismic activity. *Nat. Hazards Earth Syst. Sci.*, 12, 2957–2963. doi: 10.5194/nhess -12-2957-2012.

Li, M., and Parrot, M. (2013). First attempt to use an ionospheric parameter for earthquake prediction. *J. Geophys. Res.*, 118, 3731–3739. doi: 10.1002/jgra.50313.

Li, X., Ma, Y., Wang, H., Lu, H., Zhang, X., Wang, P., Shi, F. et al. (2010). Observation of particle on space electro-magnetic satellite during Wenchuan earthquake. *Chin. J. Geophys.*, 53 (10), 2337–2344 (in Chinese with English abstract). doi: 10.3969/j.issn.0001-5733.2010.10.007.

Liu, J., Du, X., Zlotnicki, J., Fan, Y., An, Z., Xie, T., Zheng, G., Tan, D., and Chen, J. (2011a). The changes of the ground and ionosphere electric/magnetic fields before several great earthquakes. *Chin. J. Geophys.*, 54 (11), 2885–2897. doi: 10.3969/j.issn.0001-5733.

Liu, J., Ouyang, X., and Li, W. (2011b). Analysis of the energetic particles around the Chile earthquake of M8.8. *Earthq. Res. China*, 25 (2), 166–172 (in Chinese with English abstract).

Liu, J., Wan, W., Huang, J., Zhang, X., Zhao, S., Ouyang, X., and Zeren, Z. (2011c). Electron density perturbation before Chile M8.8 earthquake. *Chin. J. Geophys.*, 54 (11), 2717–2725 (in Chinese with English abstract). doi: 10.3969/j.issn.0001-5733.2011.11.001.

Liu, J. Y., Chen, Y. I., Chen, C. H., Liu, C. Y., Chen, C. Y., Nishihashi, M., Li, J. Z., Xia, Y. Q., Oyama, K. I., Hattori, K., and Lin, C. H. (2009). Seismo-ionospheric GPS total electron content anomalies observed before the 12 May 2008 Mw7.9 Wenchuan earthquake. *J. Geophys. Res.*, 114, A04320. doi: 10.1029/2008JA013698.

Lognonné, P., Artru, J., Garcia, R., Crespon, F., Ducic, V., Jeansou, E., Occhipinti, G., Helbert, E., and Moreaux, G. (2006). Ground based GPS tomography of ionospheric post-seismic signal during Demeter: The SPECTRE project. *Planet. Space Sci.*, 54 (5), 528–540.

Lu, L., Yang, J., Cao, J., Zhang, X., and Chen, H. (2011). Observational characteristics of ionospheric magnetic VLF wave in the solar minimum year. *Chin. J. Geophys.*, 54 (6), 1403–1420 (in Chinese with English abstract). doi: 10.3969/j.issn.0001-5733.2011.06.001.

Ma, L., Xu, F., Wang, X., and Tang, L. (2010). Earthquake prediction based on Levenberg-Marquardt algorithm constrained back-propagation neural network using DEMETER data. In *Knowledge, Science, Engineering and Management*, ed. Bi, Y., Williams, M. A. Lecture Notes in Computer Science 6291. Berlin: Springer, 591–596. doi: 10.1007/978-3-642-15280-1_57.

Marchand, R., and Berthelier, J. J. (2008). Simple model for post seismic ionospheric disturbances above an earthquake epicentre and along connecting magnetic field lines. *Nat. Hazards Earth Syst. Sci.*, 8, 1341–1347.

Molchanov, O. (1993). Wave and plasma phenomena inside the ionosphere and magnetosphere associated with earthquakes. In *Review of Radio Science 1990–1992*, ed. Stone, W. R. Oxford: Oxford University Press, 591–600.

Molchanov, O., Rozhnoi, A., Solovieva, M., Akentieva, O., Berthelier, J. J., Parrot, M., Lefeuvre, F., Biagi, P. F., Castellana, L., and Hayakawa, M. (2006). Global diagnostics of the ionospheric perturbations related to the seismic activity using the VLF radio signals collected on the DEMETER satellite. *Nat. Hazards Earth Syst. Sci.*, 6, 745–753.

Muto, F., Yoshida, M., Horie, T., Hayakawa, M., Parrot, M., and Molchanov, O. (2008). Detection of ionospheric perturbations associated with Japanese earthquakes on the basis of reception of LF transmitter signals on the satellite DEMETER. *Nat. Hazards Earth Syst. Sci.*, 8, 135–141.

Němec, F., Santolík, O., and Parrot, M. (2009). Decrease of intensity of ELF/VLF waves observed in the upper ionosphere close to earthquakes: A statistical study. *J. Geophys. Res.*, 114, A04303. doi: 10.1029/2008JA013972.

Němec, F., Santolík, O., Parrot, M., and Berthelier, J. J. (2008). Spacecraft observations of electromagnetic perturbations connected with seismic activity. *Geophys. Res. Lett.*, 35, L05109. doi: 10.1029/2007GL032517.

Omori, Y., Yasuoka, Y., Nagahama, H., Kawada, Y., Ishikawa, T., Tokonami, S., and Shinogi, M. (2007). Anomalous radon emanation linked to preseismic electromagnetic phenomena. *Nat. Hazards Earth Syst. Sci.*, 7, 629–635.

Onishi, T., Berthelier, J. J., and Kamogawa, M. (2011b). Critical analysis of the electrostatic turbulence enhancements observed by DEMETER over the Sichuan region during the earthquake preparation. *Nat. Hazards Earth Syst. Sci.*, 11, 561–570.

Onishi, T., Parrot, M., and Berthelier, J. J. (2011a). The DEMETER mission, recent investigations on ionospheric effects associated with man-made activities and seismic phenomena. *C. R. Physique*, 12 (2), 160–170. doi: 10.1016/j.crhy.2010.11.009.

Ouyang, X., Zhang, X., Shen, X., Huang, J., Liu, J., Zeren, Z., and Zhao S. (2011). Disturbance of O+ density before major earthquake detected by DEMETER satellite. *Chin. J. Space Sci.*, 31 (5), 607–617 (in Chinese with English abstract).

Ouyang, X., Zhang, X., Shen, X., Liu, J., Qian, J., Cai, J., and Zhao, S. (2008). Ionospheric Ne disturbances before 2007 Puer, Yunnan, China, earthquake. *Acta Seismol. Sin.*, 21 (4), 425–437. doi: 10.1007/s11589-008-0425-8.

Ouzounov, D., Liu, D., Chunli, K., Cervone, G., Kafatos, M., and Taylor P. (2007). Outgoing long wave radiation variability from IR satellite data prior to major earthquakes. *Tectonophysics*, 431 (1–4), 211–220.

Ouzounov, D., Pulinets, S., Romanov, A., Romanov, A., Tsybulya, K., Davidenko, D., Kafatos, M., and Taylor, P. (2011). Atmosphere-ionosphere response to the M9 Tohoku earthquake revealed by multi-instrument space-borne and ground observations: Preliminary results. *Earthq. Sci.*, 24, 557–564.

Parrot, M., ed. (2006). First results of the DEMETER micro-satellite. *Planet. Space Sci.* (Special Issue), 54 (5).

Parrot, M. (2011). Statistical analysis of the ion density measured by the satellite DEMETER in relation with the seismic activity. *Earthq. Sci.*, 24 (6), 513–521. doi: 10.1007/s11589-011-0813-3.

Parrot, M. (2012). Statistical analysis of automatically detected ion density variations recorded by DEMETER and their relation to seismic activity. *Ann. Geophys.*, 55 (1), 149–155. doi: 10.4401/5270.

Parrot, M., Benoist, D., Berthelier, J. J., Blecki, J., Chapuis, Y., Colin, F., Elie, F. et al. (2006a). The magnetic field experiment IMSC and its data processing onboard DEMETER: Scientific objectives, description and first results. *Planet. Space Sci.*, 54 (5), 441–455.

Parrot, M., Berthelier, J. J., Lebreton, J. P., Sauvaud, J. A., Santolík, O., and Blecki J. (2006b). Examples of unusual ionospheric observations made by the DEMETER satellite over seismic regions. *Phys. Chem. Earth A/B/C*, 31, 486–495. doi: 10.1016/j.pce.2006.02.011.

Parrot, M., and Johnston, M., eds. (1989). Seismoelectromagnetic effects. *Phys. Earth Planet. Int.* (Special Issue), 57 (1–2).

Parrot, M., and Johnston, M., eds. (1993). Seismoelectromagnetic effects. *Phys. Earth Planet. Int.* (Special Issue), 77 (1–2).

Píša, D., Němec, F., Parrot, M., and Santolík, O. (2012). Attenuation of electromagnetic waves at the frequency ~1.7 kHz in the upper ionosphere observed by the DEMETER satellite in the vicinity of earthquakes. *Ann. Geophys.*, 55 (1), 157–163. doi: 10.4401/ag-5276.

Píša, D., Parrot, M., and Santolík, O. (2011). Ionospheric density variations recorded before the 2010 Mw 8.8 earthquake in Chile. *J. Geophys. Res.*, 116, A08309. doi: 10.1029/2011JA016611.

Priyadarshi, S., Kumar, S., and Singh, A. K. (2011). Ionospheric perturbations associated with two recent major earthquakes (M>5.0). *Phys. Scr.*, 84, 045901. doi: 10.1088/0031-8949/84/04/045901.

Pulinets, S. (2012). Low-latitude atmosphere-ionosphere effects initiated by strong earthquakes preparation process. *Int. J. Geophys.*, 2012, 131842. doi: 10.1155/2012/131842.

Pulinets, S. A. (2007). Natural radioactivity, earthquakes, and the ionosphere. *EOS Trans. AGU*, 88 (20), 217.

Pulinets, S. A. (2009). Physical mechanism of the vertical electric field generation over active tectonic faults. *Adv. Space Res.*, 44 (6), 767–773. doi: 10.1016/j.asr.2009.04.038.

Pulinets, S. A., and Ouzounov, D. (2011). Lithosphere-atmosphere-ionosphere coupling (LAIC) model – An unified concept for earthquake precursors validation. *J. Asian Earth Sci.*, 41, 371–382. doi: 10.1016/j.jseaes.2010.03.005.

Rozhnoi, A., Molchanov, O., Solovieva, M., Gladyshev, V., Akentieva, O., Berthelier, J. J., Parrot, M., Lefeuvre, F., Hayakawa, M., Castellana, L., and Biagi, P. F. (2007). Possible seismo-ionosphere perturbations revealed by VLF signals collected on ground and on a satellite. *Nat. Hazards Earth Syst. Sci.*, 7, 617–624.

Rozhnoi, A., Solovieva, M., Molchanov, O., Akentieva, O., Berthelier, J. J., Parrot, M., Biagi, P. F., and Hayakawa, M. (2008). Statistical correlation of spectral broadening in VLF transmitter signal and low-frequency ionospheric turbulence from observation on DEMETER satellite. *Nat. Hazards Earth Syst. Sci.*, 8, 1105–1111.

Rozhnoi, A., Solovieva, M., Molchanov, O., Biagi, P.-F., Hayakawa, M., Schwingenschuh, K., Boudjada, M., and Parrot, M. (2010). Variations of VLF/LF signals observed on the ground and satellite during a seismic activity in Japan region in May–June 2008. *Nat. Hazards Earth Syst. Sci.*, 10, 529–534.

Rozhnoi, A., Solovieva, M., Parrot, M., Hayakawa, M., Biagi, P. F., and Schwingenschuh, K. (2012). Ionospheric turbulence from ground-based and satellite VLF/LF transmitter signal observations for the Simushir earthquake (November 15, 2006). *Ann. Geophys.*, 55 (1), 187–192. doi: 10.4401/ag-5190.

Ryu, K., Lee, E., Chae, J. S., Parrot, M., and Pulinets, S. (2014). Seismo-ionospheric coupling appearing as equatorial electron density enhancements observed via DEMETER electron density measurements. *J. Geophys. Res. Space Phys.*, 119, 8524–8542. doi: 10.1002/2014JA020284.

Santolík, O., Němec, F., Parrot, M., Lagoutte, D., and Madrias, L. (2006). Analysis methods for multi-component wave measurements on board the DEMETER spacecraft. *Planet. Space Sci.*, 54 (5), 512–527.

Saradjian, M. R., and Akhoondzadeh, M. (2011). Prediction of the date, magnitude and affected area of impending strong earthquakes using integration of multi precursors earthquake parameters. *Nat. Hazards Earth Syst. Sci.*, 11, 1109–1119.

Sarkar, S., Choudhary, S., Sonakia, A., Vishwakarma, A., and Gwal, A. K. (2012). Ionospheric anomalies associated with the Haiti earthquake of 12 January 2010 observed by DEMETER satellite. *Nat. Hazards Earth Syst. Sci.*, 12, 671–678. doi: 10.5194/nhess-12-671-2012.

Sarkar, S., and Gwal, A. K. (2010). Satellite monitoring of anomalous effects in the ionosphere related to the great Wenchuan earthquake of May 12, 2008. *Nat. Hazards*, 55 (2), 321–332. doi: 10.1007/s11069-010-9530-9.

Sarkar, S., Gwal, A. K., and Parrot, M. (2007). Ionospheric variations observed by the DEMETER satellite in the mid-latitude region during strong earthquakes. *J. Atmos. Sol. Terrestr. Phys.*, 69, 1524–1540.

Sarkar, S., Tiwari, S., and Gwal, A. K. (2011). Electron density anomalies associated with M ≥ 5.9 earthquakes in Indonesia during 2005 observed by DEMETER. *J. Atmos. Sol. Terrestr. Phys.*, 73 (16), 2289–2299. doi: 10.1016/j.jastp.2011.06.004.

Sauvaud, J. A., Moreau, T., Maggiolo, R., Treilhou, J. P., Jacquey, C., Cros, A., Coutelier, J., Rouzaud, J., Penou, E., and Gangloff, M. (2006). High energy electron detection onboard DEMETER: The IDP spectrometer, description and first results on the inner belt. *Planet. Space Sci.*, 54 (5), 502–511.

Serebryakova, O. N., Bilichenko, S. V., Chmyrev, V. M., Parrot, M., Rauch, J. L., Lefeuvre, F., and Pokhotelov, O. A. (1992). Electromagnetic ELF radiation from earthquakes regions as observed by low-altitude satellites. *Geophys. Res. Lett.*, 19 (2), 91–94.

Shen, X., Zhang, X., Wang, L., Chen, H., Wu, Y., Yuan, S., Shen, J., Zhao, S., Qian, J., and Ding, J. (2011). The earthquake-related disturbances in ionosphere and project of the first China seismo-electromagnetic satellite. *Earthq. Sci.*, 24 (6), 639–650. doi: 10.1007/s11589-011-0824-0.

Sidiropoulos, N. F., Anagnostopoulos, G., and Rigas, V. (2011). Comparative study on earthquake and ground based transmitter induced radiation belt electron precipitation at middle latitudes. *Nat. Hazards Earth Syst. Sci.*, 11, 1901–1913. doi: 10.5194/nhess-11-1901-2011.

Singh, V., Chauhan, V., Singh, O. P., and Singh, B. (2010). Ionospheric effect of earthquakes as determined from ground based TEC measurement and satellite data. *Indian J. Radio Space Phys.*, 39, 63–70.

Slifkin, L. (1993). Seismic electric signals from displacement of charged dislocations. *Tectonophysics*, 224 (1–3), 149–152.

Slominska, E., Blecki, J., Parrot, M., and Slominski, J. (2009). Satellite study of VLF ground-based transmitter signals during seismic activity in Honshu Island. *Phys. Chem. Earth A/B/C*, 34 (6–7), 464–473. doi: 10.1016/j.pce.2008.06.016.

Solovieva, M. S., Rozhnoi, A. A., and Molchanov, O. A. (2009). Variations in the parameters of VLF signals on the DEMETER satellite during the periods of seismic activity. *Geomag. Aeron.*, 49 (4), 532–541.

Stangl, G., Boudjada, M. Y., Biagi, P. F., Krauss, S., Maier, A., Schwingenschuh, K., Al-Haddad, E., Parrot, M., and Voller, W. (2011). Investigation of TEC and VLF space measurements associated to L'Aquila (Italy) earthquakes. *Nat. Hazards Earth Syst. Sci.*, 11, 1019–1024. doi: 10.5194/nhess-11-1019-2011.

Toutain, J.-P., and Baubron, J.-C. (1998). Gas geochemistry and seismotectonics: A review. *Tectonophysics*, 304, 1–27.

Tributsch, H. (1978). Do aerosol anomalies precede earthquakes? *Nature*, 276, 606–608. doi: 10.1038/276606a0.

Tronin, A. A. (2006). Remote sensing and earthquakes: A review. *Phys. Chem. Earth A/B/C*, 31, 138–142.

Tzanis, A., Vallianatos, F., and Makropoulos, K. (2000). Seismic and electrical precursors to the 17-1-1983, M7 Kefallinia earthquake, Greece: Signatures of a SOC system. *Phys. Chem. Earth A/B/C*, 25 (3), 281–287.

Vallianatos, F., and Tzanis, A. (1998). Electric current generation associated with the deformation rate of a solid: Preseismic and coseismic signals. *Phys. Chem. Earth A/B/C*, 23 (9–10), 933–939.

Varotsos, P., and Alexopoulos, K. (1986). Stimulated current emission in the earth: Piezo stimulated currents and related geophysical aspects. In *Thermodynamics of Point Defects and Their Relation with Bulk Properties*, ed. Amelinckx, S., Gevers, R., Nihoul, J. Amsterdam: North-Holland, 136–142.

Wu, A., Zhang, Y., Zhou, Y., Zhang, X., and Li, G. (2011). On the spatial-temporal characteristics of ionospheric parameters before Wenchuan earthquake with the MPI method. *Chin. J. Geophys.*, 54 (10), 2445–2457 (in Chinese with English abstract). doi: 10.3969/j.issn.0001-5733.2011.10.002.

Yan, X., Shan, X., Cao, J., Tang, J., and Wang, F. (2012). Seismoionospheric anomalies observed before Wenchuan earthquake using GPS and DEMETER data. *Seismol. Geol.*, 34 (1), 160–171 (in Chinese with English abstract). doi: 10.3969/j.issn.0253-4967.2012.01.015.

Zaourar, N., Mebarki, R., Hamoudi, M., and Parrot, M. (2011). La dynamique fractale des perturbations séismo-ionosphériques enregistrées par le micro-satellite DEMETER. *Télédétection*, 10 (2–3), 77–90 (in French with English abstract).

Zeng, Z., Zhang, B., Fang, G., Wang, D., and Yin, H. (2009). The analysis of ionospheric variations before Wenchuan earthquake with DEMETER data. *Chin. J. Geophys.*, 52 (1), 11–19 (in Chinese with English abstract).

Zeren, Z., Shen, X., Zhang, X., Cao, J., Huang, J., Ouyang, X., Liu, J., and Bingqing, L. (2012). Possible ionospheric electromagnetic perturbations induced by the Ms7.1 Yushu earthquake. *Earth Moon Planets*, 108, 231–241. doi: 10.1007/s11038-012-9393-z.

Zeren, Z., Zhang, X., Liu, J., Ouyang, X. Y., Xiong, P., and Shen, X. (2010). Ionospheric disturbances associated with strong earthquakes—Results from Langmuir probe onboard DEMETER satellite. *Seismol. Geol.*, 32 (3), 424–433 (in Chinese with English abstract).

Zhang, X., Battiston, R., Shen, X., Zeren, Z., Ouyang, X., Qian, J., Liu, J., Huang, J., and Miao, Y. (2010a). Automatic collecting technique of low frequency electromagnetic signals and its application in earthquake study. In *Knowledge, Science, Engineering and Management*, ed. Bi, Y., and Williams, M. A. Lecture Notes in Computer Science 6291. Berlin: Springer, 366–377. doi: 10.1007/978-3-642-15280-1_34.

Zhang, X., Chen, H., Liu, J., Shen, X., Miao, Y., Du, X., and Qian, J. (2012a). Ground-based and satellite DC-ULF electric field anomalies around Wenchuan M8.0 earthquake. *Adv. Space Res.*, 50 (1), 85–95.

Zhang, X., Ding, J., Shen, X., Wang, M., Liu, J., Yu, S., Wang, Y., and Ouyang, X. (2009a). Electromagnetic perturbations before Wenchuan M8 earthquake and stereo electromagnetic observation system. *Chin. J. Radio Sci.*, 24 (1), 1–8 (in Chinese with English abstract).

Zhang, X., Liu, J., Shen, X., Parrot, M., Qian, J., Ouyang, X., Zhao, S., and Huang, J. (2010b). Ionospheric perturbations associated with the M8.6 Sumatra earthquake on 28 March 2005. *Chin. J. Geophys.*, 53 (3), 567–575 (in Chinese with English abstract).

Zhang, X., Qian, J., Ouyang, X., Cai, J., Liu, J., Shen, X., and Zhao, S. (2009b). Ionospheric electro-magnetic disturbances prior to Yutian 7.2 earthquake in Xinjiang. *Chin. J. Space Sci.*, 29 (2), 213–221 (in Chinese with English abstract).

Zhang, X., Qian, J., Ouyang, X., Shen, X., Cai, J., and Zhao, S. (2009c). Ionospheric electromagnetic perturbations observed on DEMETER satellite before Chile M7.9 earthquake. *Earthq. Sci.*, 22, 251–255.

Zhang, X., Qian, J., Wang, Y., Zhao, X., Ouyang, X., and Zhao, S. (2008). Spectral features of geomagnetic low point displacement and its mechanism. *Acta Seismol. Sin.*, 21 (5), 474–484. doi: 10.1007/s11589 -008-0474-z.

Zhang, X., Shen, X., Liu, J., Ouyang, X., Qian, J., and Zhao, S. (2009d). Analysis of ionospheric plasma perturbations before Wenchuan earthquake. *Nat. Hazards Earth Syst. Sci.*, 9, 1259–1266.

Zhang, X., Shen, X., Liu, J., Ouyang, X., Qian, J., and Zhao, S. (2010c). Ionospheric perturbations of electron density before the Wenchuan earthquake. *Int. J. Remote Sens.*, 31 (13), 3559–3569.

Zhang, X., Shen, X., Ouyang, X., Cai, J., Huang, J., Liu, J., and Zhao, S. (2009e). Ionosphere VLF electric field anomalies before Wenchuan M8 earthquake. *Chin. J. Radio Sci.*, 24 (6), 1024–1032 (in Chinese with English abstract).

Zhang, X., Shen, X., Parrot, M., Zeren, Z., Ouyang, X., Liu, J., Qian, J., Zhao, S., and Miao, Y. (2012b). Phenomena of electrostatic perturbations before strong earthquakes (2005–2010) observed on DEMETER. *Nat. Hazards Earth Syst. Sci.*, 12, 75–83. doi: 10.5194/nhess-12-75-2012.

Zhang, X., Zeren, Z., Parrot, M., Battiston, R., Qian, J., and Shen, X. (2011). ULF/ELF ionospheric electric field and plasma perturbations related to Chile earthquakes. *Adv. Space Res.*, 47 (6), 991–1000. doi: 10.1016/j.asr.2010.11.001.

Zhang, Z., Li, X., Wu, S., Ma, Y., Shen, X., Chen, H., Wang, P., You, X., and Yuan, Y. (2012). DEMETER satellite observations of energetic particle prior to Chile earthquake. *Chin. J. Geophys.*, 55 (5), 1581–1590 (in Chinese with English abstract). doi: 10.6038/j.issn.0001-5733.2012.05.016.

Zhu, R., Yang, D., Jing, F., Yang, J., and Ouyang, X. (2008). Ionospheric perturbations before Pu'er earthquake observed on DEMETER. *Acta Seismol. Sin.*, 21 (1), 77–81. doi: 10.1007/s11589-008-0077-8.

Zhu, T. (2010). A preliminary study on characteristics of average power spectrum density of LF/MF electric field observed by DEMETER satellite. *Acta Seismol. Sin.*, 32 (4), 476–489 (in Chinese with English abstract).

Zhu, T., and Wang, L. (2011). LF electric field anomalies related to Wenchuan earthquake observed by DEMETER satellite. *Chin. J. Geophys.*, 54 (3), 717–727 (in Chinese with English abstract). doi: 10.3969/j.issn.0001-5733.2011.03.011.

Zlotnicki, J., Le Mouël, J. L., Kanwar, R., Yvetot, P., Vargemezis, G., Menny, P., and Fauquet, F. (2006). Ground-based electromagnetic studies combined with remote sensing based on Demeter mission: A way to monitor active faults and volcanoes. *Planet. Space Sci.*, 54 (5), 541–557.

Zlotnicki, J., Li, F., and Parrot, M. (2010). Signals recorded by DEMETER satellite over active volcanoes during the period 2004 August–2007 December. *Geophys. J. Int.*, 183 (3), 1332–1347.

7 TIR Anomaly as Earthquake Precursor

Luca Piroddi

CONTENTS

7.1 INTRODUCTION

Many earthquake studies have been performed using remote sensing technologies, both active and passive (Tronin 2010). Among them, thermal studies were done using data derived from multi-spectral sensors onboard meteorological satellites, especially geostationary ones. Various satellite thermal studies are based on thermal infrared (TIR) observations (Gorny et al. 1988; Tramutoli et al. 2001, 2005; Tronin et al. 2002, 2004; Ouzounov and Freund 2004; Ouzounov et al. 2006, 2007; Genzano et al. 2007, 2009; Saraf et al. 2008, 2012; Bleier et al. 2009; Wei et al. 2009; Piroddi and Ranieri 2012; Wu et al. 2012; Qin et al. 2013), in single-band brightness temperature or even in land surface temperature (LST). Despite these promising results, the use of thermal parameters in seismology is still under debate. Authors have pointed out the unreliability of thermal phenomena alone (Tronin et al. 2002; Tronin 2010), because of the different mechanisms of earthquake generation, the different magnitude and depth values, the surface conditions, the geological frameworks, the weather and human activities. Other authors only partially found thermal phenomena before an earthquake (one data analysis method between the three performed ones) but attribute the thermal anomaly to a lack of raw data before processing (Blackett et al. 2011). A report on operational earthquake forecasting, compiled by the International Commission on Earthquake Forecasting for Civil Protection and submitted to the Department of Civil Protection of Italy (Jordan et al. 2011), states that there is not yet enough reliability in thermal precursor studies to perform earthquake predictions. Jordan et al.'s (2011) report was mainly based on the measurements of thermal properties of faults in Southern California from remote sensing data for 7 years. Eneva et al. (2008) concluded that there exist apparent LST precursory thermal anomalies associated with warming, but they are not unique to the periods preceding earthquakes, so such thermal anomalies cannot be used for deterministic prediction of earthquakes. The observed systems are inherently complex for both the physics of earthquake preparation and the external parameters influencing possible precursory thermal phenomena, so that the approach to the problem must move from a deterministic to a probabilistic point of view (Tronin et al. 2002; Grandori and Guagenti 2009; Piroddi and Ranieri 2012).

Satellite thermal studies are often affected by poor visibility due to cloud cover and thermal variability due to weather effects. Statistical algorithms for the mitigation of weather and spatial effects are known (Tramutoli et al. 2015), which show better results by using data from geostationary sensors (even if polar satellite applications still exist). These methods, also called robust satellite techniques (RSTs), are still heavily influenced by cloud cover rate for both visible area availability and the presence of artefacts with limited availability of cloud-free pixels at the scenes (Genzano et al. 2009). The thermal observations are generally made using a single sample per day, usually at night.

Occasionally, thermal observations are carried out using more night samples to highlight the temperature dynamics (Ouzounov et al. 2006; Bleier et al. 2009; Piroddi and Ranieri 2012). The evolution of this new approach has recently allowed scientists to obtain significantly better resolution of the observed thermal phenomena, leading to the possibility of studying relations between thermal patterns and ground spatial properties (Piroddi and Ranieri 2012; Piroddi et al. 2014).

7.2 ORIGIN OF THERMAL ANOMALIES

The mechanism of generation of the thermal phenomena proposed as earthquake precursors and their physical relationship with the impending earthquake are not yet universally accepted. Many causes of thermal mechanisms have been proposed to explain the origin of thermal anomalies:

- Frictional heat release and stress–temperature coupling in the volumes of rocks around the hypocentre and transmission through conduction and convection up to the ground surface (Wu et al. 2006a,b).
- Release of gases and vapours (CO_2, CO, O_3, CH_4, H_2S, SO_2, HCl, H_2, H_2O, etc.) from the fractured zones with the generation of surface thermal effects as a consequence of a ground-level atmospheric temperature increase related to changes of state in vapours, to chemical reactions with the first layers of the atmosphere and in large part to greenhouse effects (Qiang et al. 1999; Tramutoli et al. 2001; Tronin et al. 2004; Cervone et al. 2006; Wu et al. 2006b; Genzano et al. 2007; Tronin 2010; Bonfanti et al. 2012).
- Release of radioactive gas (Rn) and its effect on the ground-level atmospheric temperature as a consequence of air ionization and subsequent water vapour condensation (Pulinets et al. 2006; Pulinets and Ouzounov 2011).
- Breaking the boundary layer of hydrothermal reservoirs that come in contact with shallower groundwater (Tronin et al. 2004; Saraf et al. 2008).
- Electrical and electromagnetic emissions (O'Keefe and Thiel 1995; Freund 2003; Carpinteri et al. 2010) by the rocks under stress of brittle fracture, and effects of the interaction of these fields radiated from the hypocentre volumes when reaching the lithosphere–atmosphere interface (Freund 2003, 2011; Ouzounov and Freund 2004; Freund et al. 2006; Rozhnoi et al. 2009).

Among these, the most reliable and general theories concern the release of gases and vapours (CO_2, CH_4, H_2, H_2O and Rn) into the atmosphere, and to electrical and electromagnetic emissions.

7.2.1 GAS AND VAPOURS

The release of gas and vapours refers to the model of dilatancy–diffusion, which links release rates to changes of porosity in rocks, as they are connected to the various phases of the elastic deformation and microfracturing before the final rupture (Scholz et al. 1973). This model requires the crossing of five phases of the earthquake generation to complete the seismic cycle:

- Increased elastic stress, as the initial condition.
- The elastic deformation causes dilation and the consequent increase in the volume of rocks. The opening of fractures develops minor seismic activity, and reduces the ratio V_P/V_S. In this stage, abnormal gas emissions begin.

- The infiltration of water in the fractures increases the fluid pressure and reduces the resistance of the rock, facilitating breaking, and again the relationship V_P/V_S grows. In this stage, we first have the maximum gaseous emission and then a subsequent gradual decrease.
- Breakage occurs; there is a reduction of stress in rocks and of the fluid pressure confined within. The earthquake happens.
- Aftershocks and resettlement of the system.

The amount of gas released in the dilatancy model is suspected to be responsible for the increase of temperature in the lower atmospheric layer, mainly by means of the greenhouse effect, radioactive ionization and subsequent water vapour condensation.

The anomalous degassing activity has been reported prior to the strong earthquakes by means of geochemical monitoring in groundwater, springs and soils, even with the appearance of anomalous dissolved salt concentrations. Sometimes, different observations of the same parameter do not present a clear law, mainly due to local different geological settings (Yechieli and Bein 2002; Grant et al. 2011; Inan et al. 2012).

7.2.2 Seismoelectromagnetic and Seismoelectric Effects

Seismoelectromagnetic effects refer to electromagnetic signals generated by fault failure processes in the Earth's crust. These may occur slowly (when associated with plate tectonic loading, slow earthquakes, post-seismic slip, etc.) or rapidly preceding, during and following earthquakes. The electromagnetic emissions at infrared wavelengths have been measured on dry laboratory specimen subjects to quasi-rupture or rupture strains (Freund 2003; Wu et al. 2006a,b), while transients in magnetic and electric fields have also been observed at radio wave frequencies (Freund 2003; Carpinteri et al. 2010).

Electromagnetic emissions at various frequencies (microhertz to megahertz) are reported for earthquake regions and laboratory specimens, but their transmission through the crust from the hypocentre to the ground level does not yet have a clear explanation, and noise suppression must be done carefully (Johnston 2002). Several different physical processes related to crustal failure can contribute to the generation of seismoelectromagnetic effects and the loading and rupture of water-saturated crustal rocks during earthquakes, together with fluid and gas movement, stress redistribution and changes in material properties. The primary mechanisms for the generation of electric and magnetic fields with crustal deformation and earthquake-related fault failure include piezomagnetism, stress and conductivity, electrokinetic effects, charge generation processes, charge dispersion, magneto-hydrodynamic effects and thermal remagnetization and demagnetization effects (Johnston 2002). The model that explains electromagnetic effects for most of the crust rock compositions (igneous, highly metamorphic and partly sedimentary) refers to the presence of charge gaps (positive hole pairs) in the crystal lattices of the rocks which are activated by high stress, bullet impacts or high temperature values (Freund 2003, 2011; Freund et al. 2006; Takeuchi et al. 2006). The surface charge carriers' recombination and partially subsequent air ionization leads to an increase of infrared emission and an apparent temperature that is even bigger than the kinematic temperature of the external surface of specimens (Freund et al. 2007).

7.2.3 Extension of Thermal Anomalies

Currently, neither a general theoretical nor empirical relation has been specifically found between the spatial and temporal features of thermal anomalies and the main characteristics of incoming earthquakes (magnitude and depth). Furthermore, it is generally accepted that thermal anomalies, like any other earthquake precursors, appear inside a radius, exponentially related to the magnitude, from the epicentre, which has been recently estimated on the basis of the surface deformation (Dobrovolsky et al. 1979):

$$R = 10^{0.43M} \qquad (7.1)$$

where M is the magnitude and R is the radius (in km) at which a deformation of 10^{-8} is observed for a homogeneous isotropic inclusion and for the position of the hypocentre on the surface.

Regarding the spatial extension of precursory phenomena and their relations with the earthquake generation, a classification has been proposed which distinguishes between physical precursors and tectonic precursors (Ishibashi 1988): the first are the precursors inherently related to the source region of the main shock, while the second are the precursors that appear immediately outside this region and are supposed to be related to the strain redistribution due to plate motions or resultant block movements.

For operational forecasting and prediction purposes, the area needs to be confined to a small multiple of fault rupture dimensions, like has been proposed for long-term precursory seismic activity variations and forecasting regions where earthquakes are likely to occur (Holliday et al. 2005).

7.3 ADVANCES IN TIR PRECURSOR STUDIES

Recent advances recognize simultaneous protocols for thermal multiparametric analyses, geostationary high temporal resolution data processing and linkage between an observed thermal anomaly pattern and its possible origin.

7.3.1 DEVIATION–TIME–SPACE–THERMAL CRITERIA

Multiple precursors observation is becoming a shared value in earthquake multiphysics studies. By the integration of various parameters, it is possible to have better statistical performances in the prediction of both enhanced and reduced hazards, which can be evaluated by their combined warnings. Furthermore, multiple and simultaneous observations are the basis with which to understand physics related to earthquakes, and the debate on coupling aspects is open. In order to enhance hazard estimation reliability, a combination method for thermal data measured from the Global Earth Observation System of Systems (GEOSS) has been proposed (Wu et al. 2012; Qin et al. 2013). Based on the detailed analysis using GEOSS data, three normalized indices were proposed to integrate the most important features of used thermal parameters: normalized deviation index (NDI), normalized synchronism index (NSI) and normalized adjacency index (NAI), which are calculated to quantify the deviation–time–space–thermal (DTS-T) criteria (Figure 7.1).

The NDI measures the deviation from the mean value of a measurement normalized to the signal standard deviation. NSI measures the fact that different parameters' time series have temporal synchronism in their anomaly appearance and NAI measures the spatial coherence of different signal anomalies.

To evaluate the reliability of the recognized thermal anomalies, a reliability index (RI) has been defined considering the three normalized indices:

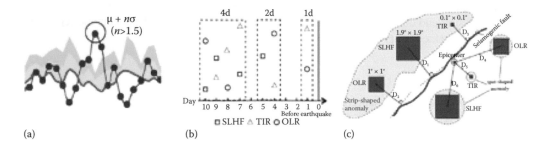

FIGURE 7.1   Deviation–time–space criteria for earthquake anomaly recognition with multiple parameters: (a) Deviation: notable enough; (b) Time quasi-synchronism; (c) Space: geo-adjacency. (From Qin et al. 2013.)

$$RI = \frac{(NDI + NSI + NAI)}{3} \times 100\% \qquad (7.2)$$

Six thermal parameters were considered for earthquake thermal anomaly analysis (Qin et al. 2013): outgoing long-wave radiation (OLR), TIR, surface latent heat flux (SLHF), skin temperature (ST), surface air temperature (SAT) and diurnal temperature range (DTR) (Wu et al. 2012; Qin et al. 2013). With non-earthquake years (the years without an $M > 5.5$ earthquake) providing background reference, the daily anomalous changes of each pixel were calculated by subtracting the non-earthquake day-average from the current day-value (Qin et al. 2013). Detailed analysis has been conducted on maps and single pixel time series for the M_S 7.1 Yushu Earthquake of 14 April 2010, which occurred 44 km northwest of Yushu county (33.2° N, 96.6° E), Qinghai, China, with a focal depth of 14 km. Time series were plotted for the potential epicentre pixel or pixel group, together with day-average, and $\mu + 1.5\sigma$ to help visualize a change in trend and to pick out anomaly candidates (Figure 7.2) (Qin et al. 2013).

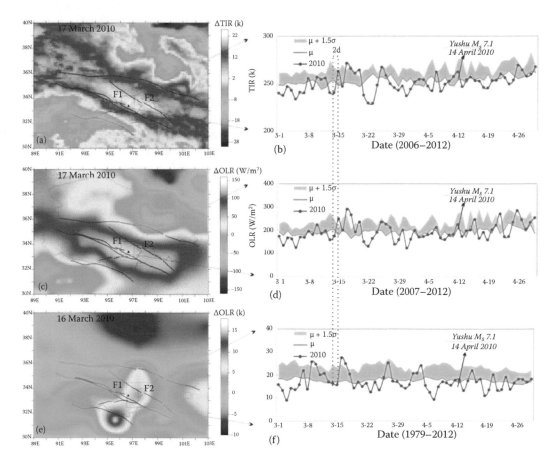

FIGURE 7.2 Multiple thermal parameters recognition for earthquake anomaly: (a) Spatial distribution of differential thermal infrared radiation (ΔTIR) on 17 March 2010. (b) Time series of TIR in the epicentre pixel group of the Yushu Earthquake. (c) Spatial distribution of differential outgoing long-wave radiation (ΔOLR) on 17 March 2010. (d) Time series of OLR in the epicentre pixel group of the Yushu Earthquake. (e) Spatial distribution of differential diurnal temperature range (ΔDTR) on 16 March 2010. (f) Time series of DTR in the epicentre pixel of the Yushu Earthquake. The black triangles indicate the epicentre. (From Qin, K. et al., *Remote Sensing*, 5(10), 5143–5151, 2013.)

7.3.2 Night Thermal Gradient

Recently, a new processing algorithm has been developed in relation to thermal precursors of earthquakes (Piroddi 2011; Piroddi and Ranieri 2012). Based on the high temporal resolution of geostationary satellite sensors, this new method has defined very narrow anomalous thermal patterns around the epicentre area of the L'Aquila Earthquake, 2009 April 6 (M_L 5.9, M_W 6.3, depth of about 8.3 km): the method involves the processing of temporal data distributed throughout the night (41 samples, every 15 minutes from 1800 to 0400 Universal Time Coordinated [UTC], in the case of L'Aquila) for the extraction of characteristic parameters of the temporal thermal behaviour of individual pixels. The thermal index proposed by the researchers is the night thermal gradient (NTG), which originated from the linear regression of the thermal data of each pixel (Figure 7.3 shows a flow chart of the algorithm).

In order to increase the signal-to-noise ratio and furthermore to reduce negative effects due to cloud cover, a moving average of the data in the correlation phase has been first applied by means of a stack procedure with a processing window of some days (9 in the case of L'Aquila). The representation of the NTG parameter is able to highlight areas where there is a warming night ($dT/dt > 0$), a clearly abnormal phenomenon compared with the behaviour of heat release in normal conditions at night (Piroddi 2011; Piroddi and Ranieri 2012). The time series of the anomalous thermal pattern is shown in Figure 7.4, and a zoom of the highest anomaly map is shown in Figure 7.5.

Figure 7.4 shows the dynamic of NTG maps from 29 March to 5 April, which is the night between 5 and 6 April when the earthquake occurred. They clearly show an anomaly (positive NTG values, red colours) gradually appearing in the middle of the map and reaching a maximum 2–3 days before the main shock; after this, the anomalies gradually disappear.

Figure 7.5 shows the map of 2 April (night from 2 to 3 April) when the maximum extension and intensity of the NTG anomaly was observed: the most important feature of this map is the relatively small dimension of the anomaly, which is also greatly focused around the incoming earthquake epicentre.

7.3.3 Spatial Features Linked to Thermal Anomalies

Starting from NTG analyses, a recent study has highlighted correlations of observed thermal anomalies of the L'Aquila Earthquake with geological, tectonic and topographic features of the area

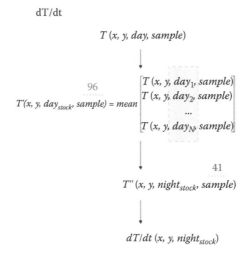

FIGURE 7.3 NTG flux diagram of processing steps. (From Piroddi, L., and Ranieri, G., *IEEE J. Sel. Top. Appl. Earth Obs. Remote Sen.*, 5(1), 307–312, 2012. Copyright © IEEE.)

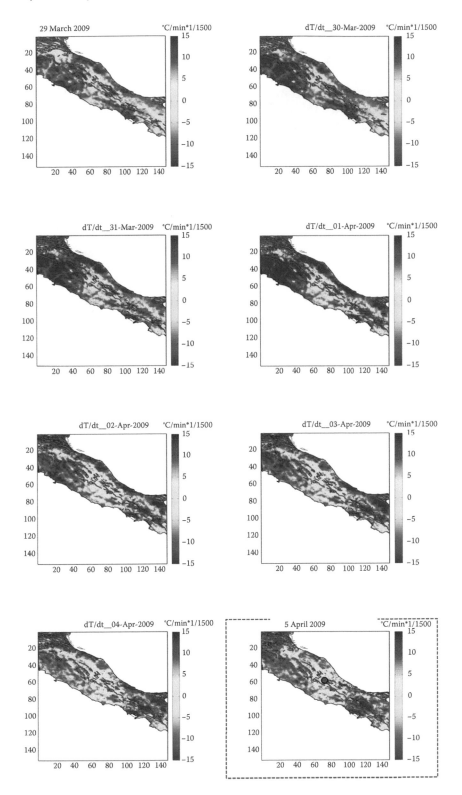

FIGURE 7.4 Time series of NTG maps eight nights before the L'Aquila Earthquake (night between 5 and 6 April 2009), starting from the night between 29 and 30 March 2009. (From Piroddi, L., and Ranieri, G., *IEEE J. Sel. Top. Appl. Earth Obs. Remote Sen.*, 5(1), 307–312, 2012. Copyright © IEEE.)

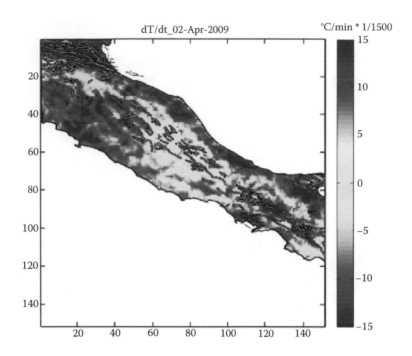

FIGURE 7.5   NTG map of the night between 2 and 3 April 2009 with the maximum amplitude and extension of anomalous patterns. (From Piroddi, L., and Ranieri, G., *IEEE J. Sel. Top. Appl. Earth Obs. Remote Sen.*, 5(1), 307–312, 2012. Copyright © IEEE.)

(Piroddi et al. 2014). One relevant result has been the fact that instead of a generally recognized local normal style of seismogenetic faults – and of the failure of the Paganica fault which caused the earthquake – thermal anomalies are localized in correspondence to the two major thrusts in the area, which are not considered seismogenetic and border homogeneous lithological areas (Figure 7.6).

A pixel-by-pixel time-series comparison has shown that anomalous thermal patterns do not originate from the epicentre but directly in the pixels of maximum anomaly. Another important result has been that fault density was not a parameter influencing the thermal anomaly distribution; however, the effect of topographic features has been heavily important (Figure 7.7). In fact, local and absolute maxima of NTG values (red curve) in Figure 7.7 clearly correspond to greater terrain heights surrounding the epicentre (Piroddi et al. 2014).

These results imply that, at least for the thermal anomalies observed in the case of the 2009 L'Aquila Earthquake with the NTG index, a big part of their origin could be due to electronic dormant charge activation and migration; volumes affected by this charge activation are expected to be wider than seismic nucleation volumes because observed thermal anomalies have been spatially stable and centred only around the epicentre.

The same thermal anomalies were recently analyzed by an independent group (Wu et al. 2016), confirming the validity of anomalous thermal patterns and the information layering approach for the study of possible origins of thermal phenomena. The authors noted that NTG anomalies before the L'Aquila 2009 main shock are also in good agreement with broadleaved deciduous forest distribution (Figure 7.8), and that forests are a preferential route for degassing activity, especially CO_2. Mainly based on these considerations and seismologic observations, they stated that there is an important role for CO_2 degassing.

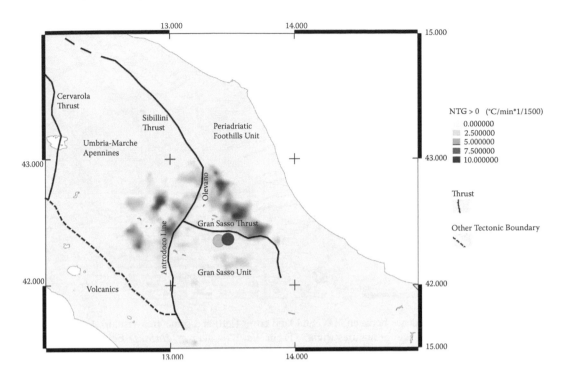

FIGURE 7.6 Superimposition of tectonic structures and NTG anomalies. (From Piroddi, L. et al., *Geophys. J. Int.*, 197(3), 1532–1536, 2014.)

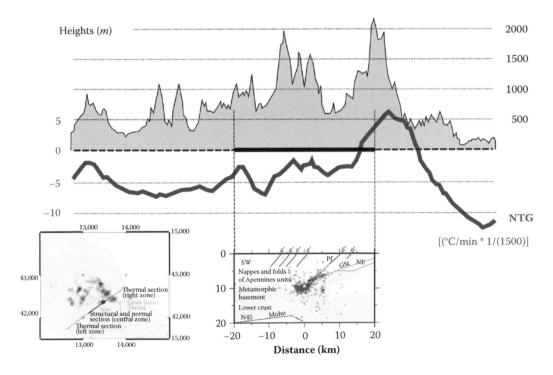

FIGURE 7.7 Correspondence between NTG and height profiles. (From Piroddi, L. et al., *Geophys. J. Int.*, 197(3), 1532–1536, 2014.)

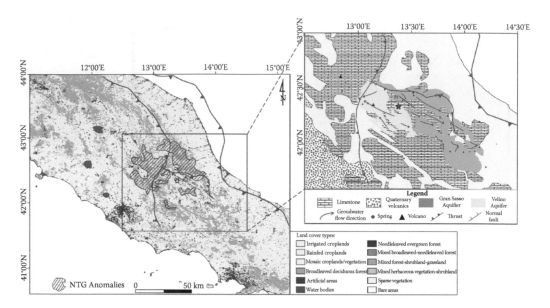

FIGURE 7.8 Correspondence between NTG and land cover (left) and same area visualization with identification of geological and aquifer features (right). (From Wu, L. et al., *Nat. Hazards Earth Syst. Sci.*, 16(8), 1859–1880, 2016.)

7.4 CONCLUSION

The satellite TIR monitoring of the Earth's surface has shown great potential in research applications in seismic-prone areas. Many independent observations, sometimes with well-documented seismic origin, exist even if some doubts have been raised. TIR anomalies can signal increasing stress status but sometimes are under the threshold of environmental noise and are not clearly distinguishable. Current state-of-the-art literature does not allow a deterministic approach to TIR-based prediction; however, it is possible to use TIR precursors in multiparametric real-time seismic hazard evaluation. The origin and nature of TIR precursors is still under debate and probably multiple sources should be considered.

REFERENCES

Blackett, M., Wooster, M. J., and Malamud, B. D. (2011). Exploring land surface temperature earthquake precursors: A focus on the Gujarat (India) earthquake of 2001. *Geophysical Research Letters*, 38(15), L15303.

Bleier, T., Dunson, C., Maniscalco, M., Bryant, N., Bambery, R., and Freund, F. (2009). Investigation of ULF magnetic pulsations, air conductivity changes, and infrared signatures associated with the 30 October Alum Rock M5.4 earthquake. *Natural Hazards and Earth System Sciences*, 9(2), 585–603.

Bonfanti, P., Genzano, N., Heinicke, J., Italiano, F., Martinelli, G., Pergola, N., Telesca, L., and Tramutoli, V. (2012). Evidence of CO2-gas emission variations in the central Apennines (Italy) during the L'Aquila seismic sequence (March–April 2009). *Bollettino di Geofisica Teorica ed Applicata*, 53(1), 147–168.

Carpinteri, A., Cardone, F., and Lacidogna, G. (2010). Energy emissions from failure phenomena: Mechanical, electromagnetic, nuclear. *Experimental Mechanics*, 50(8), 1235–1243.

Cervone, G., Maekawa, S., Singh, R. P., Hayakawa, M., Kafatos, M., and Shvets, A. (2006). Surface latent heat flux and nighttime LF anomalies prior to the Mw = 8.3 Tokachi-Oki earthquake. *Natural Hazards and Earth System Sciences*, 6(1), 109–114.

Dobrovolsky, I. P., Zubkov, S. I., and Miachkin, V. I. (1979). Estimation of the size of earthquake preparation zones. *Pure and Applied Geophysics*, 117(5), 1025–1044.

Eneva, M., Adams, D., Wechsler, N., Ben-Zion, Y., and Dor, O. (2008). Thermal properties of faults in Southern California from remote sensing data. SAIC No. NNH05CC13C. Washington, DC: NASA, March, 1–70.

Freund, F. (2003). Rocks that crackle and sparkle and glow – Strange pre-earthquake phenomena. *Journal of Scientific Exploration*, 17(1), 37–71.

Freund, F. T. (2011). Pre-earthquake signals: Underlying physical processes. *Journal of Asian Earth Sciences*, 41(4), 383–400.

Freund, F. T., Takeuchi, A., and Lau, B. W. S. (2006). Electric currents streaming out of stressed igneous rocks – A step towards understanding pre-earthquake low frequency EM emissions. *Physics and Chemistry of the Earth, Parts A/B/C*, 31(4–9), 389–396.

Freund, F. T., Takeuchi, A., Lau, B. W. S., Al-Manaseer, A., Fu, C. C., Bryant, N. A., and Ouzounov, D. (2007). Stimulated infrared emission from rocks: Assessing a stress indicator. *eEarth*, 2(1), 7–16.

Genzano, N., Aliano, C., Corrado, R., Filizzola, C., Lisi, M., Mazzeo, G., Paciello, R., Pergola, N., and Tramutoli, V. (2009). RST analysis of MSG-SEVIRI TIR radiances at the time of the Abruzzo 6 April 2009 earthquake. *Natural Hazards and Earth System Sciences*, 9(6), 2073–2084.

Genzano, N., Aliano, C., Filizzola, C., Pergola, N., and Tramutoli, V. (2007). A robust satellite technique for monitoring seismically active areas: The case of Bhuj–Gujarat earthquake. *Tectonophysics*, 431(1–4), 197–210.

Gorny, V. I., Salman, A. G., Tronin, A. A., and Shilin, B. B. (1988). The earth's outgoing IR radiation as an indicator of seismic activity. *Proceedings of the USSR Academy of Sciences*, 301, 67–69.

Grandori, G., and Guagenti, E. (2009). Prevedere I terremoti: La lezione dell'Abruzzo. *Ingegneria Sismica*, 26(3), 56–61.

Grant, R. A., Halliday, T., Balderer, W. P., Leuenberger, F., Newcomer, M., Cyr, G., and Freund, F. T. (2011). Ground water chemistry changes before major earthquakes and possible effects on animals. *International Journal of Environmental Research and Public Health*, 8(6), 1936–1956.

Holliday, J. R., Nanjo, K. Z., Tiampo, K. F., Rundle, J. B., and Turcotte, D. L. (2005). Earthquake forecasting and its verification. *Nonlinear Processes in Geophysics*, 12, 965–977.

Inan, S., Balderer, W. P., Leuenberger-West, F., Yakan, H., Özvan, A., and Freund, F. T. (2012). Springwater chemical anomalies prior to the Mw = 7.2 Van earthquake (Turkey). *Geochemical Journal*, 46(1), e11–e16.

Ishibashi, K. (1988). Two categories of earthquake precursors, physical and tectonic, and their roles in intermediate-term earthquake prediction. *Pure and Applied Geophysics*, 126(2–4), 687–700.

Johnston, M. J. S. (2002). Electromagnetic fields generated by earthquakes. *International Handbook of Earthquake and Engineering Seismology*. Vol. 81A. San Diego: Academic Press, 621–635.

Jordan, T., Chen, Y., Gasparini, P., Madariaga, R., Main, I., Marzocchi, W., Papadopoulos, G., Sobolev, G., Yamaoka, K., and Zschau, J. (2011). Operational earthquake forecasting. State of knowledge and guidelines for utilization. *Annals of Geophysics*, 54(4), 315–391.

O'Keefe, S. G., and Thiel, D. V. (1995). A mechanism for the production of electromagnetic radiation during fracture of brittle materials. *Physics of the Earth and Planetary Interiors*, 89(1), 127–135.

Ouzounov, D., Bryant, N., Logan, T., Pulinets, S., and Taylor, P. (2006). Satellite thermal IR phenomena associated with some of the major earthquakes in 1999–2003. *Physics and Chemistry of the Earth*, 31(4), 154–163.

Ouzounov, D., and Freund, F. (2004). Mid-infrared emission prior to strong earthquakes analyzed by remote sensing data. *Advances in Space Research*, 33(3), 268–273.

Ouzounov, D., Liu, D., Kang, C., Cervone, G., Kafatos, M., and Taylor, P. (2007). Outgoing long wave radiation variability from IR satellite data prior to major earthquakes. *Tectonophysics*, 431(1–4), 211–220.

Piroddi, L. (2011). Sistemi di telerilevamento termico per il monitoraggio e la prevenzione dei rischi naturali: Il caso sismico. PhD thesis, University of Cagliari. http://veprints.unica.it/550/.

Piroddi, L., and Ranieri, G. (2012). Night thermal gradient: A new potential tool for earthquake precursors studies. An application to the seismic area of L'Aquila (central Italy). *IEEE Journal of Selected Topics in Applied Earth Observations and Remote Sensing*, 5(1), 307–312.

Piroddi, L., Ranieri, G., Freund, F., and Trogu, A. (2014). Geology, tectonics and topography underlined by L'Aquila earthquake TIR precursors. *Geophysical Journal International*, 197(3), 1532–1536.

Pulinets, S., and Ouzounov, D. (2011). Lithosphere–atmosphere–ionosphere coupling (LAIC) model – An unified concept for earthquake precursors validation. *Journal of Asian Earth Sciences*, 41(4–5), 371–382.

Pulinets, S. A., Ouzounov, D., Karelin, A. V., Boyarchuk, K. A., and Pokhmelnykh, L. A. (2006). The physical nature of thermal anomalies observed before strong earthquakes. *Physics and Chemistry of the Earth, Parts A/B/C*, 31(4–9), 143–153.

Qiang, Z., Dian, C., Li, L., Xu, M., Ge, F., Liu, T., Zhao, Y., and Guo, M. (1999). Satellitic thermal infrared brightness temperature anomaly image-short-term and impending earthquake precursors. *Science in China Series D: Earth Sciences*, 42(3), 313–324.

Qin, K., Wu, L., Zheng, S., and Liu, S. (2013). A deviation-time-space-thermal (DTS-T) method for Global Earth Observation System of Systems (GEOSS)-based earthquake anomaly recognition: Criterions and quantify indices. *Remote Sensing*, 5(10), 5143–5151.

Rozhnoi, A., Solovieva, M., Molchanov, O., Schwingenschuh, K., Boudjada, M., Biagi, P. F., Maggipinto, T., Castellana, L., Ermini, A., and Hayakawa, M. (2009). Anomalies in VLF radio signals prior to the Abruzzo earthquake (M=6.3) on 6 April 2009. *Natural Hazards and Earth System Sciences*, 9(5), 1727–1732.

Saraf, A. K., Rawat, V., Banerjee, P., Choudhury, S., Panda, S. K., Dasgupta, S., and Das, J. D. (2008). Satellite detection of earthquake thermal infrared precursors in Iran. *Natural Hazards*, 47(1), 119–135.

Saraf, A. K., Vineeta Rawat, J. D., Mohammed, Z., and Kanika, S. (2012). Satellite detection of thermal precursors of Yamnotri, Ravar and Dalbandin earthquakes. *Natural Hazards*, 61(2), 861–872.

Scholz, C. H., Syke, L. R., and Aggarwal, Y. P. (1973). Earthquake prediction: A physical basis. *Science*, 181(4102), 803–809.

Takeuchi, A., Lau, B. W. S., and Freund, F. T. (2006). Current and surface potential induced by stress-activated positive holes in igneous rocks. *Physics and Chemistry of the Earth, Parts A/B/C*, 31(4–9), 240–247.

Tramutoli, V., Corrado, R., Filizzola, C., Genzano, N., Lisi, M., and Pergola, N. (2015). From visual comparison to robust satellite techniques: 30 years of thermal infrared satellite data analyses for the study of earthquake preparation phases. *Bollettino di Geofisica Teorica ed Applicata*, 56(2), 167–202.

Tramutoli, V., Cuomo, V., Filizzola, C., Pergola, N., and Pietrapertosa, C. (2005). Assessing the potential of thermal infrared satellite survey for monitoring seismically active areas: The case of Kocaeli (Izmit) earthquake, August 17, 1999. *Remote Sensing of Environment*, 96, 409–426.

Tramutoli, V., Di Bello, G., and Pergola, N. (2001). Robust satellite techniques for remote sensing of seismically active areas. *Annals of Geophysics*, 44(2), 295–312.

Tronin, A. A. (2010). Satellite remote sensing in seismology. A review. *Remote Sensing*, 2(1), 124–150.

Tronin, A. A., Biagi, P. F., Molchanov, O. A., Khatkevich, Y. M., and Gordeev, E. I. (2004). Temperature variations related to earthquakes from observation at the ground stations and by satellites in Kamchatka area. *Physics and Chemistry of the Earth*, 29, 501–506.

Tronin, A. A., Hayakawa, M., and Molchanov, O. A. (2002). Thermal IR satellite data application for earthquake research in Japan and China. *Journal of Geodynamics*, 33, 519–534.

Wei, L., Guo, J., Liu, J., Lu, Z., Li, H., and Cai, H. (2009). Satellite thermal infrared earthquake precursor to the Wenchuan MS 8.0 earthquake in Sichuan, China, and its analysis on geo-dynamics. *Acta Geologica Sinica*, 83(4), 767–775.

Wu, L., Qin, K., and Liu, S. (2012). GEOSS-based thermal parameters analysis for earthquake anomaly recognition. *Proceedings of the IEEE*, 100(10), 2891–2907.

Wu, L., Zheng, S., Santis, A. D., Qin, K., Mauro, R. D., Liu, S., and Rainone, M. L. (2016). Geosphere coupling and hydrothermal anomalies before the 2009 Mw 6.3 L'Aquila earthquake in Italy. *Natural Hazards and Earth System Sciences*, 16(8), 1859–1880.

Wu, L. X., Liu, S. J., and Wu, Y. H. (2006a). Precursors for rock fracturing and failure – Part I: IRR image abnormalities. *International Journal of Rock Mechanics and Mining Sciences*, 43(3), 473–482.

Wu, L. X., Liu, S. J., and Wu, Y. H. (2006b). Precursors for rock fracturing and failure – Part II: IRR T-curve abnormalities. *International Journal of Rock Mechanics and Mining Sciences*, 43(3), 483–493.

Yechieli, Y., and Bein, A. (2002). Response of groundwater systems in the Dead Sea Rift Valley to the Nuweiba Earthquake: Changes in head, water chemistry, and near-surface effects. *Journal of Geophysical Research: Solid Earth*, 107(B12), 2332.

8 Stress Change and Earthquake Triggering by Reservoirs
Role of Fluids

Kalpna Gahalaut

CONTENTS

8.1　INTRODUCTION

For decades, scientists have been investigating how human activity, like fluid injection or extraction, deep underground mining, quarrying and impoundment of reservoir behind dams, can influence or stimulate seismicity in a region. The topic gained more attention in the early 1960s, when hundreds of quakes were recorded in Colorado after fluid had been injected into a disposal well near the Rocky Mountain Arsenal. In this category of stimulated earthquakes, reservoir-triggered earthquakes are the most common. Ambient and reservoir stresses and pore pressure jointly control the stability of faults near a reservoir. The stresses and pore pressure induced by a reservoir are miniscule in comparison with the ambient crustal stresses in the hypocentral regions of earthquakes. When the ambient stresses approach critical levels for failure of an upper crustal fault, the reservoir-induced stresses and pore pressure may exert influence by assisting or opposing the attainment of criticality. Mathematical developments in porous elastic theory and concepts of Coulomb stress have become important tools to understand reservoir-triggered earthquakes.

A rock mechanics-based explanation for the Denver earthquakes of the early 1960s was a remarkable success for Terzaghi's concept of apparent normal stress or the role of pore water pressure in modifying normal stresses in water-filled porous rocks. Healy et al. (1968) have shown how precisely the pumping of water into the Rocky Mountain Arsenal well led to earthquakes on nearby subsurface faults. The Denver earthquakes come under the category of artificially stimulated earthquakes. Some examples of artificially stimulated earthquakes are fluid injection or extraction for various purposes, like waste disposal; solution mining; geothermal power generation and secondary oil recovery; deep underground mining; removal of a large volume of rocks during quarrying; and impoundment of large reservoirs behind dams. These kinds of stimulated earthquakes are termed induced or triggered earthquakes. According to McGarr and Simpson (1997), if the causative activity can account for either most of the stress change or most of the energy required to produce the earthquake, it is termed induced, and if causative activity accounts for only a small fraction of stress change or energy associated with the earthquake, it is termed triggered. In the case of triggered earthquakes, tectonic loading plays the primary role. When the modification of the stress state by an external stimulant, causing change in stress and/or pore pressure, is sufficient to cause failure, induced or triggered earthquakes occur. This can occur due to either an increase in the stress driving the fault or a decrease in the strength of the fault. In the case of fluid injection or extraction, seismicity results from a decrease in effective normal stress caused by increased formation pressure. This type of seismicity comes under the category of induced seismicity, as the increases in pressure that stimulate earthquake activity are a few megapascals, of the order of stress drops observed for shallow earthquakes. An increase in shear stress due to the creation of cavities in deep mining is the cause for mining-induced seismicity. The change in shear stress associated with this kind of seismicity is large, about 20–50 MPa, and hence it is categorized as induced seismicity. Reduction in normal stress due to removal of overburden in quarrying leads to a small stress change. Hence, seismicity caused by quarrying is categorized as triggered seismicity. There is evidence linking Earth tremors and water reservoir operation for more than 80 dams (Figure 8.1) (Gupta 1992). The loads imposed on the Earth's surface by the deepest reservoirs are not more than a few megapascals. Even at shallow hypocentral depths, the expected stress changes would be of the order of 0.1 MPa, significantly smaller than the stress drop for typical earthquakes. This comes under the category of triggered seismicity. The mechanism of reservoir-triggered seismicity (RTS) is more complex than

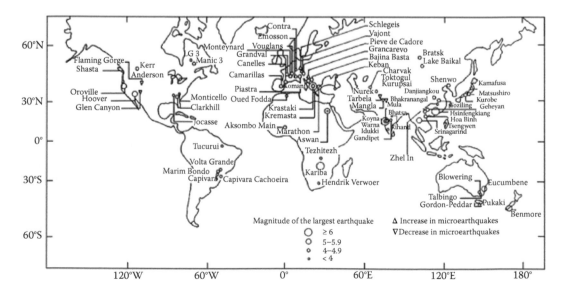

FIGURE 8.1 Worldwide distribution of RTS (modified after Gupta 2002). Three cases of RTS at Koyna–Warna, Aswan and Rihand are shown by red circles.

other types of triggered or induced seismicity. RTS can involve all the factors; that is, a reservoir impoundment can change shear stress, normal stress and pore pressure. Some investigators have also termed this hydroseismicity, as water plays an important role in the occurrence of such earthquakes (Costain et al. 1987; Costain and Bollinger 2010). This category also includes earthquakes triggered by rainfall and stream flow.

Among all these, the reservoir-triggered earthquakes are the most damaging due to their relatively larger magnitude. In this chapter, we discuss factors influencing reservoir-triggered earthquakes and their mechanism and quantification. Further, we present a few well-known cases of RTS with detailed numerical simulations to understand their occurrences.

8.2 RTS: A BRIEF HISTORICAL ACCOUNT

Gupta (1992) prepared a list of more than 80 instances of worldwide RTS (Figure 8.1). Here we discuss a few of them briefly. Due to their historical importance, Mead, Hsinfengkiang, Kariba and Kremasta reservoirs have been discussed repeatedly. The case of Nurek reservoir in the Tadjik Republic of the former USSR is of special interest because the reservoir is located in a region of active thrust-fault tectonics.

The first recognized case of RTS is that associated with the construction of the Hoover Dam, which impounds Lake Mead, across the Colorado River in California. Filling of Lake Mead began in 1935, and the region was apparently free of earthquakes at least prior to that time. However, historical records of earthquakes are very sketchy because the region was mostly uninhabited until the city of Boulder was founded when the construction of the dam was started (Gupta 1992). Still, going by the evidence available through the newspaper from Las Vegas, Nevada, there was a significant increase in felt earthquakes after impoundment started (Gupta 1992). Earthquakes were felt in the area for the first time in September 1936, when the reservoir level reached its maximum for that year. The main shock (magnitude 5.0) was recorded in 1939 when the reservoir had reached 80% capacity (Steinbrugge 1972). Most of the epicentres located by various investigators are within 25 km of Lake Mead. Carder (1945) was apparently the first to associate the earthquake activity with the filling of Lake Mead. There was a positive correlation between seasonal lake loads and seismic activity for a number of years initially. But subsequently, the correlation was negative and even zero (Gupta 1992).

Impoundment of the Hsinfengkiang reservoir in China started in October 1959. The occurrence of earthquakes in the vicinity of the reservoir started soon after. An earthquake of magnitude (M) 6.1 occurred in the reservoir area on 19 March 1962 (Gupta 1992). Although China lies in a seismically active part of the Earth, no destructive earthquake had been reported from the Hsinfengkiang region. Meade's (1991) contention is that the increase in seismicity in late 1963 was not due to the reservoir. In 1964, the water level rose and seismicity increased. In 1968, the water level rose abruptly, but there was no increase in seismicity. All earthquakes after 1965 seem unrelated to the reservoir since no significant earthquake occurred in the surrounding area (Meade 1991).

Kariba Lake on the Zambezi River in Central Africa began to fill up in 1958. The evidence is incomplete, but not many earthquakes occurred prior to the impoundment, although the occurrence of one earthquake near the upper end of the reservoir in 1956 was recorded (Gupta 1992). Earthquakes were first recorded near the lake in 1961 (Steinbrugge 1972), and seismic activity continued to increase and a strong earthquake sequence began on 14 August 1963, a few days after the reservoir water level reached to its maximum. An earthquake of magnitude 6.2 occurred in September 1963. All epicentres of the main seismic activity are concentrated within a 40 × 40 km area near the deepest part of the reservoir. Meade (1991) observed that earthquake activities in 1963 occurred during filling, but later events seem unrelated to the reservoir.

In the surroundings of the man-made Lake Kremasta in Greece was well known for its seismicity, but no earthquake had occurred since 1700; the filling of the lake started in July 1965. Earthquake shocks were felt in August 1965, and the rate of water loading increased rapidly from

November 1965 to January 1966. Seismic activity also increased rapidly in this period, culminating in an earthquake of magnitude 6.3 in February 1966, when the lake had almost reached to its maximum elevation (Steinbrugge 1972). However, Meade's (1991) objection was that the largest earthquake occurred outside the lake with a focal depth of 20 km. This earthquake fits into the normal seismicity pattern of central Greece (Meade 1991).

The filling of the Nurek reservoir was initiated in 1968, and the first sustained filling occurred in 1972 when the water depth reached up to 100 m for the first time. A second major filling episode occurred in 1976 when the water level reached 200 m. The final level of 300 m was reached in 1981. Both of the filling episodes in 1972 and 1976 were accompanied by a 5- to 10-fold increase in seismicity over the background level. The largest earthquakes had magnitudes of 4.6 and 4.1 in the first and second major phases of filling, respectively (Keith et al. 1982).

8.3   FACTORS INFLUENCING RTS

Simpson (1976) prepared the following fairly comprehensive list of factors controlling the incidence of reservoir-triggered earthquakes.

1. Pre-existing stress
   * Orientation: Tectonic regime (normal, strike slip and thrust)
   * Magnitude of stress: How close to failure
   * Rate of strain accumulation
2. Geological and hydrological conditions
   * Faults: Location with respect to reservoir
     - Orientation with respect to pre-existing and induced stresses
     - Failure conditions and permeability
   * Hydromechanical properties of rock
     - Lithology
     - Fracture strength
     - Frictional characteristics
     - Porosity
     - Permeability
   * Hydrological condition: Water table (seasonal variations)
     - Accessibility of reservoir water to groundwater system and faults
     - Storage capacity of groundwater system
3. The reservoir
   * Size: Depth (pressure)
     - Volume (mass)
     - Shape (spatial variations and concentration of stress)
   * Temporal changes in water level: Loading and offloading rate

8.3.1   Pre-Existing or In Situ Stress

Since the stress changes induced by man-made reservoirs are small in comparison with the ambient stresses in the Earth's crust, it is inevitable to conclude that reservoir-triggered earthquakes can only occur where the in situ stresses are already so high and the differences in principal stresses so large that pre-existing faults in the region are close to failure. The incremental stresses induced by the reservoir loads should be such that resolved normal and shear stress increments on the faults would promote failure. In addition, Simpson (1976) recalled the earlier work of Lomnitz (1974) and Castle et al. (1975) regarding the fact that the rate of strain accumulation is another consideration. It was argued that stress changes from a reservoir may be obvious in areas of moderate rates of strain accumulation compared with the areas of high rates. Simpson (1976) considered fault plane solutions for major earthquakes near reservoirs. All the fault plane mechanisms were strike slip or normal type, and they were consistent with the stress field deduced from other nearby earthquakes or aligned with geological features. Subsequent evidence had emerged, suggesting that thrust and reverse-type earthquakes also occur near reservoirs (Keith et al. 1982; Roeloffs 1988; Meade 1991).

8.3.2 GEOLOGICAL AND HYDROLOGICAL CONDITIONS

Simpson (1976) pointed out that there are little or no data available on the conditions that exist at hypocentral depths near man-made water reservoirs. We know little about the permeability of the fault zone and state of stress, temperature and pore pressure at depth. Monticello reservoir is unique in that a fairly extensive set of seismogenic and rock mechanics observations have been carried out (Zoback and Hickman 1982). Zoback and Hickman (1982) have included in situ measurements in wells of stresses, pore pressure, permeability and distribution of faults, fractures and joints in hypocentral zones near the reservoir. Talwani and Acree (1985) and Talwani et al. (2007) introduced the concept of seismogenic permeability, which is defined as an intrinsic property of fractures where pore pressure diffusion is associated with induced or triggered seismicity. After a very detailed study, they inferred that the permeability of fractures where fluid pressure flow triggers seismicity lies in a narrow range of more than two orders of magnitude, from 5×10^{-16} to 5×10^{-14} m^2, which corresponds to the hydraulic diffusivity in the range of 0.1–10 m^2/s.

8.3.3 RESERVOIRS

In view of the various cases all over the world, Gupta et al. (1972) inferred that in addition to the tectonic setting and geological conditions, among the factors affecting the earthquake frequency near the reservoir are (1) the rate of increase of the water level, (2) the maximum load attained and (3) the period for which higher levels are retained. Simpson (1976) suggested that earthquakes occur soon after impoundment in those cases where reservoirs have depths greater than 100 m, although such conditions are neither a necessary nor sufficient condition. There are many reservoirs deeper than 100 m without any seismic activity. Also, seismicity seems to have occurred near many reservoirs shallower than 100 m. Even the size of the major earthquake does not increase with the height of the dam or volume of the reservoir (Simpson 1976). With the exception of the seismicity at Oroville reservoir, most earthquakes near reservoirs have occurred directly beneath them and most activity was within 25 km of the reservoirs (Simpson 1976). In an altogether different vein, Simpson et al. (1988) presented evidence that the temporal variation of earthquakes following the filling of a large reservoir may show one of two main trends. At some reservoirs, earthquakes begin almost immediately following the first filling of the reservoir; in some cases, not much seismicity was observed until a number of seasonal filling cycles. Simpson et al. (1988) classified the earthquakes at Nurek, Kariba, Manic-3 and Monticello as examples of rapid response and the earthquakes at Koyna, Oroville and Aswan as examples of delayed response. Simpson et al. (1988) did find that both types of response may have also occurred at some of these sites. Simpson et al. (1988) observed that sites showing a rapid response tend to have seismicity that is shallow (<10 km), of low magnitude and was concentrated directly beneath or near the edges of the reservoir. In contrast, sites showing a delayed response have seismicity that was of larger magnitude, deeper (>10 km) and sometimes extending to distances of 10 km or more from the reservoir.

8.4 MECHANISM OF RESERVOIR-TRIGGERED EARTHQUAKES

8.4.1 RESERVOIR WATER LOAD AND PORE PRESSURE

The role of reservoir load in the occurrence of reservoir-triggered earthquakes may also have been considered earlier, but the first definite quantitative analysis was undertaken by Gough and Gough

(1970) to explain the earthquakes around Lake Kariba. Bock (1980) elaborated on the computational scheme and applied it to discuss reservoir earthquakes near a reservoir site in Austria. Bell and Nur (1978) and Roeloffs (1988) also considered the role of reservoir load. It was found that reservoir load induced stresses at relatively small seismogenic depths, and they only perturb the ambient stress fields. The role of load-induced stresses on subsurface faults could be stabilizing or destabilizing depending on the stress environment at seismogenic depths. The Boussinesq theory (Jaeger and Cook 1969) is at the core of analyses due to reservoir loads. Kuo (1969) has provided a matrix theory-based analysis for a finite load on a layered elastic half space. Developments during the 1960s and early 1970s regarding reservoir-triggered earthquakes and the seismicity associated with the Rocky Arsenal made it plausible that theoretical simulation related to RTS should also take into account the effect of water in pore spaces of rock, which results in a change in pore pressure. Since pore pressure leads to a decrease of compressive normal stresses, it promotes destabilization through the reduction of frictional stresses on all faults. Snow (1972) considered the combined influence of reservoir load and reservoir-induced pore pressure in the incidence of reservoir-triggered earthquakes. He estimated stresses and pore pressures at the subsurface by assuming a reservoir of infinite horizontal extent in the incidence of reservoir-triggered earthquakes. He thus argued that reservoir loading would destabilize normal and strike-slip faults but stabilize thrust faults. The possibility was entertained that pore water from the reservoir may take time to diffuse downward into rocks. Thus, it was postulated that load-induced stress changes would occur instantly throughout the affected rock, while pore pressures would develop with time in consonance with such rock properties as hydraulic diffusivity.

8.4.2 PorouS Elastic Solid and Theoretical Analyses

Biot (1941, 1955, 1956a,b), in a series of seminal papers dealing with stresses and pore pressure in fluid-saturated porous elastic media, explored the basis of the well-known theory of consolidation of soil mechanics. It is assumed that while the pores of the solid medium are compressible, the solid material itself and the pervading fluid are not. Rice and Cleary (1976) generalized the theory of Biot (1941, 1955, 1956a,b) by considering that fluids may be compressible. They also noted that the compressibility of a porous rock may arise from two distinct causes, namely, changes in pore volume under applied loads and compressibility of the rock substance itself. Thus, in the more exact coupled theory for porous elastic media, pore pressure changes have two components, namely, instantaneous changes due to elastic compression or undrained response of porous elastic media and delayed time-dependent changes due to the diffusion of water or drained response of porous elastic media. These two are called pore pressure due to compression and diffusion, respectively. This distinction is for convenience of reference only, because diffusion is also involved subsequently in pore pressure due to compression. This distinction is justified because pore pressure due to compression, as the name implies, arises due to compression but subsequently diffuses so as to reach an equilibrium situation.

A saturated porous homogeneous and isotropic elastic media was an idealized representation of wet rocks under and near a reservoir. But initial theoretical developments by Biot (1941) and Rice and Cleary (1976) made it a natural choice in the simulations of RTS. Bell and Nur (1978) and Roeloffs (1988) carried out fairly elaborate mathematical analyses of stress and pore pressure changes assuming reservoir loads which are finite in one horizontal direction and infinite in the perpendicular horizontal direction (two-dimensional [2D] simulations). They solved boundary value and initial value problems for partial differential equations derived by Rice and Cleary (1976). Bell and Nur (1978) used numerical procedures because they considered to some extent the lateral variation in medium properties. Roeloffs (1988) applied transform methods assuming sinusoidally varying reservoir loads. Roeloffs (1988) has identified three classes of mathematical problems of porous elastic media. The classification is at the level of governing partial differential equations. The classes identified are coupled, decoupled and uncoupled problems. The coupled and uncoupled problems represent extreme cases of no simplification and most significant simplifications of the partial differential equations. In physical terms, a coupled problem is one in which the elastic stresses and pore pressure influence each other appropriately. An uncoupled problem is one in which the elastic stresses and pore pressure are independent of each other. A decoupled problem is one in which we assume that elastic stresses influence pore pressure but not vice versa. The level of mathematical difficulty in solutions increases from uncoupled to coupled problems.

Bell and Nur (1978) and Roeloffs (1988) have shown that for finite reservoirs, stabilization and destabilization can occur on all types of faults for appropriate spatial relationships between reservoirs and faults. Also, stabilization and destabilization may be time dependent in general conditions. Kalpna and Chander (2000) have developed a Green's function-based algorithm for simulation of stresses and pore pressure in a homogeneous porous elastic half-space three-dimensional (3D) model of reservoirs. Their algorithm can be employed for actual time-varying water level changes of reservoir load.

Subsequently, attempts have been made to move forward from analyses with homogeneous media and consider the influence of spatial variations in rock properties (Simpson and Narasimhan 1990). The starting point of their work is the hypothesis that the occurrence of discrete earthquakes after a noticeable elapse of time following reservoir filling can be related reasonably to the time taken by water to move physically from the reservoir to the site of potential failure in a plane of weakness. But the occurrence of earthquake swarms after a relatively short time following reservoir filling requires recourse to medium heterogeneities which result in a combination of localized pore pressure increase due to undrained response and diffusive migration of water over short distances to the site of incipient failure. The spatial variations of interest here are with respect to the Skempton coefficient B and large stress increases at locales of stress concentration where weak and competent rocks coexist. In this connection, Gahalaut and Gupta (2008) developed an algorithm for simulation of pore pressure due to a 3D surface water reservoir, on a porous elastic half space with a heterogeneity of different permeability and hydraulic diffusivity using the integral equation technique.

Classification of RTS based on its temporal variation can be directly explained through its mechanism. Seismicity which results from the reservoir water load and/or sudden change in water load, that is, undrained response of pore pressure, occurs immediately after the initial impoundment of reservoir. This type of seismicity is classified as rapid seismicity. Seismicity which occurs mainly due to diffusion of pore pressure starts after some time of initial impoundment and is classified as delayed seismicity (Simpson et al. 1988; Rajendran and Talwani 1992). Talwani (1997) coined another term in this respect: protracted seismicity. In his classification, the seismicity which occurs after the initial impoundment and dies out afterwards is classified as initial seismicity, and seismicity which continues for several years after the initial filling is termed protracted seismicity. Gupta (2002) merged the above two classifications and gave a three-step classification that is rapid, delayed and continuing seismicity. Considering the recent development in understanding the intraplate seismicity, here we wish to add that aftershock and seismicity decay in the case of intraplate earthquakes is slower than that in the plate boundary regions (Stein and Liu 2009). Thus, the seismicity and aftershock occurrence continues for a longer duration in the case of intraplate earthquakes. In the presence of a reservoir in the intraplate region, where the initial seismicity or main shock was triggered by reservoir impoundment, the ongoing continuing seismicity and aftershocks are probably modulated by the reservoir effects.

8.5 CONCEPT OF COULOMB STRESS AND FAULT STABILITY IN RTS

To quantify the effect of stress and pore pressure changes due to the reservoir impoundment, in the presence of tectonic stresses, and their influence on earthquake causative fault planes, the Coulomb–Mohr frictional failure criterion of earthquake occurrence has widely been adopted in

RTS, according to which the change in the Coulomb stress, ΔS (Scholz 1990; King et al. 1994; Hardebeck et al. 1998), can be defined as

$$\Delta S = \Delta\tau - \mu(\Delta\sigma - \Delta P) \qquad (8.1)$$

where $\Delta\sigma$ and $\Delta\tau$ are the changes in normal and resolved shear stress due to the reservoir on the considered fault plane. ΔP is the change in pore pressure due to the reservoir operations, and μ is the coefficient of friction. $\Delta\tau$ is resolved in the slip direction derived from the earthquake focal mechanism. Positive $\Delta\tau$ and negative $\Delta\sigma$ promote failure. Accordingly, failure is encouraged, referred to as destabilization, if ΔS is positive and vice versa. The role of pore pressure is always to encourage failure by decreasing the normal stress. If an earthquake occurs in the vicinity of a reservoir, then to assess its role in triggering the earthquake, ΔS is calculated due to the reservoir impoundment on the fault plane of the earthquake at its hypocentre. If at the time of the earthquake ΔS is positive, this should have a destabilizing effect at the hypocentre that suggests that the reservoir has a positive role in the occurrence of the earthquake and vice versa. It is assumed that the tectonic stresses and background pore pressure did not change between the considered short time period of the reservoir impoundment and the earthquake occurrence, and changes in stresses and pore pressure in that period occurred due to the reservoir impoundment only.

Chander and Kalpna (1997) and Chander (1999) have used the concept of fault stability, ΔS, to categorize the triggered earthquakes near reservoirs. At any given instant of time, stresses due to many natural causes may act on the fault at the hypocentre of an impending tectonic earthquake. They include, at the very least, the stresses due to the weight of rock in and around the hypocentre and a group of stresses that are collectively called stresses of plate tectonic origin. Also, pore pressure may arise if fluids exist in the pore spaces of rocks around the hypocentre. While stresses due to the weight of the rock may be considered constant on the timescales of interest here, namely, decades to millennia, stresses of plate tectonic origin may vary and in general accumulate with time. These stresses and pore pressure due to natural causes are often referred to collectively as ambient stresses and ambient pore pressure, respectively. A natural tectonic earthquake will occur if and when, due to accumulation of ambient stresses and ambient pore pressure at the hypocentre, the ambient motive shear stress acting on the fault exceeds its frictional strength. Similarly, the condition for a tectonic earthquake near a reservoir is that the cumulative motive shear stress acting on the fault at the hypocentre should exceed its cumulative frictional strength. Cumulative motive shear stress is the result of ambient and reservoir-induced motive shear stresses, and the cumulative frictional strength of the fault is the frictional strength under the combined influence of ambient and reservoir-induced normal stresses and pore pressure. If the ambient stresses are stabilizing for the causative fault, then reservoir stresses will be destabilizing to the same degree for the occurrence of the earthquake. In this case, the role of the former is passive and that of the latter is active in the occurrence of the earthquake. This is a natural earthquake whose occurrence has been hastened by the construction of the reservoir. When ambient stresses are negative for the causative fault and reservoir stresses are negative, then the role of the reservoir is to delay the occurrence of the natural earthquake.

8.5.1 MAGNITUDES OF RESERVOIR-TRIGGERED EARTHQUAKES AND TRIGGERING THRESHOLD

The Coulomb–Mohr criterion has been adopted as the criterion for slip on faults and the occurrence of earthquakes. The failure criterion remains the same for earthquakes of all magnitudes. Apparently, the magnitude of the earthquake was determined by other factors, such as rupture area and stress drop, rather than the failure criterion. Thus, we cannot predict the magnitude of the reservoir-triggered earthquakes from stability analysis. The situation is the same for earthquakes under the ambient stresses away from reservoirs.

One of the important questions is the threshold for earthquake triggering. Stress change caused by the tides is considered to be very small, and even that stress change has been found to be capable of triggering earthquakes. Stein (2004) found that tidal stress change was of the order of 0.01 bar and was capable of earthquake triggering. However, Ziv and Rubin (2000) found no lower threshold for stress triggering. They found that stress change even less than that by the tides can trigger earthquakes.

8.6 APPLICATION OF THE CONCEPT OF THE POROUS ELASTIC THEORY AND COULOMB STRESS IN RTS: DETAILED CASE HISTORIES

There have been several cases in which reservoir operation has been considered to trigger earthquakes in its vicinity. However, quantitative analysis with respect to all such cases was very limited. Here three cases are discussed that show triggering of earthquakes due to impoundment of reservoirs.

8.6.1 RTS DUE TO RIHAND RESERVOIR, CENTRAL INDIA

The Govind Ballav Pant reservoir is located on the Rihand River, a tributary of Son River, India (Figure 8.2). The 92 m high Rihand Dam was built in 1962, and the reservoir is the second largest reservoir in India. On the basis of correlation between the times of high water level in the reservoir

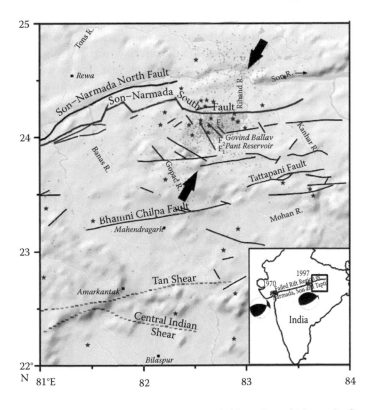

FIGURE 8.2 Broad tectonic features of part of the failed rift region of Narmada–Son and Tapti. Stars denote earthquakes reported in the ISC and IMD catalogues during 1984–2004 ($M \sim 3$), while small circles denote epicentres by DMG during January 1997–December 1999 ($M \leq 3$). Bold arrows show the direction of maximum compression due to plate movement (Gowd et al. 1992). Dark brown colour faults, marked as F_1, F_2 and F_3, and the Son–Narmada South fault are the neotectonic faults (GSI 2000). The 1970 Broach and 1997 Jabalpur earthquakes are shown in the inset.

and the occurrence of the maximum number of earthquakes, and the simulation of the effect of the reservoir on the nearby earthquakes' causative faults, Gahalaut et al. (2007) first time carried out analysis and pointed out that this was the case of RTS.

The reservoir is located in a failed rift zone, which is known as the Narmada–Son–Tapti failed rift zone. The east-northeast-west-southwest trending failed rift zone transects the Indian peninsular shield area into northern and southern blocks (Figure 8.2). It is seismically the most active region of the peninsular India. Many prominent faults have been mapped in the region, but the faults bounding the failed rift zone to the south are suggested to be active and have witnessed reactivation until the Quaternary period. Most of the earthquake activity is also suggested to be associated with the southern faults (Rao et al. 2002). The reservoir lies close to and south of the Son–Narmada South fault. Numerous northwest-southeast and northeast-southwest trending small neotectonic faults have also been mapped in the vicinity of the reservoir (GSI 2000); however, the sense of motion on these faults is not known (Figure 8.2).

The Department of Mines and Geology, Kathmandu, Nepal (DMG), which operates a 17-station network across Nepal, about 400 km north of the reservoir, has reported numerous small magnitude ($M \leq 3$) earthquakes from the study region. These earthquakes cluster close to the reservoir (Figure 8.2). The data from January 1997 to December 1999 only are available and have been used by Gahalaut et al. (2007). The epicentres of these earthquakes as reported by the India Meteorological Department (IMD), International Seismological Centre (ISC) and DMG, coincide within 10–15 km in a tight cluster around the reservoir, although their focal depths are unreliable. Temporal variation of the earthquakes reported by the DMG that occurred within 30 km of the reservoir and the maximum water levels in the reservoir are shown in Figure 8.3. The frequency of earthquake occurrence was the lowest during June–July in each year, and afterwards an increase in earthquake frequency was observed. The water level in the reservoir was at a minimum during late May, and afterwards it increased with the onset of monsoon each year. The cross-correlation coefficient between the two time series is 0.71 at an average lag of about a month between reservoir water level

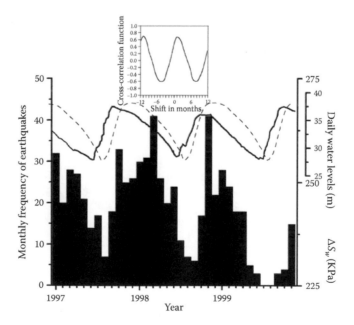

FIGURE 8.3   Monthly frequency of earthquakes reported by DMG ($M \leq 3$) that occurred within 30 km of the reservoir, the reservoir water levels and the cross-correlation function between the two time series. The correlation is the maximum (0.71) at a time lag of about 1 month. The graph with the dashed line shows the temporal variation of change in stress, ΔS, at a point located at a depth of 8 km under the reservoir and the prominent earthquake cluster.

and frequency of earthquake occurrence. Qualitative correlation, which also includes IMD and ISC catalogues, indicates that the majority of the earthquakes occurred in the periods after the high water level was attained in the reservoir. Thus, the good correlation between the water level and earthquakes (Figure 8.3) suggests that the frequency of earthquake occurrence is influenced by the annual changes in the reservoir water levels (Gahalaut et al. 2007). Gahalaut et al. (2007) calculated stress changes due to reservoir water load and pore pressure. As the focal mechanisms of the earthquakes that occurred close to the reservoir were not available, Gahalaut et al. (2007) first considered that these earthquakes accompanied similar motion as was involved during the 1997 Jabalpur and 1970 Broach earthquakes, which involved predominantly reverse motion on approximately south dipping steep planes (Singh et al. 1999; Gahalaut et al. 2004b), which is consistent with the north-east to north-northeast directed compressional regime of central and peninsular India (Gowd et al. 1992). The region of increased ΔS in both cases lies southeast of the reservoir, whereas the region of earthquake occurrence lies west of the reservoir. Thus, the two did not correlate in any significant way. Computations of ΔS even on the north dipping planes did not produce favourable results. Thus, they suggested that the reservoir impoundment does not destabilize the faults of the 1997 Jabalpur or 1970 Broach earthquake type. Even the east-west trending south dipping Son–Narmada South fault, the largest fault in the region, did not produce destabilization in the region of earthquake occurrence; destabilization was observed to the north of the fault. This was expected, as a reservoir located in the hanging wall will stabilize a thrust fault, rather than destabilize it (Roeloffs 1988); all the geologically mapped faults in the vicinity of the reservoir were considered for the analysis. Throughout the analysis, the sense of motion on geologically mapped faults was chosen in such a way so as to be consistent with north-northeast-south-southwest compression due to plate tectonic forces. This is also a condition in the analysis related to the RTS, as earthquakes will be triggered only on those faults which are critically stressed for failure under the ambient stresses (Simpson 1986; McGarr and Simpson 1997; Talwani 1997; Gupta 2002). Gahalaut et al. (1997) found that west-northwest-east-southeast oriented geologically mapped neotectonic faults (GSI 2000), shown as F_1, F_2 and F_3 in Figure 8.2, are destabilized by the reservoir load in the region of prominent earthquake cluster reported by the DMG and ISC (Figure 8.4). The computations of ΔS were shown at 8 km depth in Figure 8.4a, where the effect of the reservoir load is most pronounced in a vertical cross section (Figure 8.4b). They also performed the grid search to identify the strike, along with the dip and rake, of the fault and found that the above estimated planes are the only fault planes on which the effect of the reservoir is favourable in the region of intense earthquake activity. Gahalaut et al. (2007) computed the temporal variation of ΔS (Figure 8.3) at a point that lies at 8 km depth under the reservoir and where the increased ΔS is the maximum (Figure 8.4a). It can be seen that earthquakes do not appear to have occurred during the low water stand, even though the effect of water load was to destabilize the fault during that period (Figure 8.3). Thus, the analysis implies that the faults in the region are critically stressed for failure under a compressive regime, and a small increase in stress change due to reservoir operation, by about only 25 kPa, corresponding to an annual increase of water load by 10–15 m (Figure 8.3), triggers earthquakes on these faults. The presence of critically stressed faults is also supported by the observation that this region is the most seismically active region of peninsular India. Thus, the analysis of Gahalaut et al. (2007) pertaining to the quantitative effect of reservoir operation on the seismogenic faults suggests that the reservoir operations destabilize the nearby neotectonic faults, a case of triggered seismicity due to the reservoir loading.

8.6.2 RTS due to Aswan Reservoir, Egypt

The Aswan region in Egypt is located in the northeastern part of Africa. The reservoir of the Aswan Dam (second largest reservoir in the world) extends up to 500 km towards the south and covers an area of about 6000 km^2 along the Nile River (Figure 8.5). The reservoir filling started in 1964,

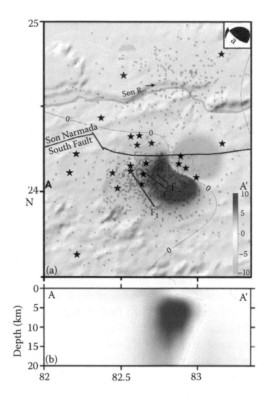

FIGURE 8.4 (a) Stress changes ΔS (in KPa) at 8 km depth on the south-southwest dipping planes (ϕ 120°, δ 70° and λ 120°, corresponding to plane a of the derived fault plane solution, shown at the top right) are shown on a LANDSAT image. Geologically mapped neotectonic faults having similar orientation (F_1, F_2 and F_3) are also shown. Earthquakes reported by DMG ($M \leq 3$) are shown by small circles, and those by ISC and IMD ($M \sim 3$) with stars. (b) ΔS in a vertical section along A-A'. Red and green denote an increase and decrease in ΔS, respectively, in this and all other subsequent figures. An increase in ΔS favours destabilization and vice versa.

and construction of the 111 m high dam was completed in 1968. The continuing seismicity for about 30 years near a large western embayment of Lake Nasser, about 50 km southwest of the Aswan Dam, has led to an obvious debate about its relation with the reservoir impoundment. Most of the Aswan seismicity is concentrated west of the reservoir in the Wadi Kalabsha area, a large embayment covered by the water of Lake Nasser, which is about 50 km southwest of the Aswan Dam (Figure 8.5). The largest event in the region occurred on 14 November 1981 ($M = 5.3$), 20 km beneath the Wadi Kalabsha embayment. During 1920–1981, in the ISC catalogue no earthquake within 200 km of the Aswan Dam was observed (Kebeasy et al. 1987). In October 1975, two seismic stations, located 60 and 200 km from the Kalabsha embayment, were installed (Kebeasy et al. 1987). During 1975–1976, these stations operated for only 160 days and only one earthquake was recorded from the Kalabsha area. Operation of these stations again started in mid-1980, and one or both stations were in operation for 217 days from August 1980 to August 1981. During this period, 20 earthquakes of magnitude 2.8–3.6 were recorded from the Kalabsha area (Kebeasy et al. 1987). Before the 1981 main shock, three foreshocks, two on 9 November 1981 ($M = 3.6$ and 4.2) and one on 11 November 1981 ($M = 4.5$), were recorded from the Kalabsha and Aswan region by the World-Wide Standardized Seismograph Network (WWSSN) station at Helwan, near Cairo (690 km from the epicentral area at Kalabsha). Two successive aftershocks were recorded with focal depths of 18 and 22 km on 2 January 1982 (Hassoup 2002). Since June 1982, a continuous recording of earthquakes by the Aswan seismological centre has been done by a network which consists of 13 field stations. Many spatio-temporal studies of earthquakes have been carried out since the installation of this

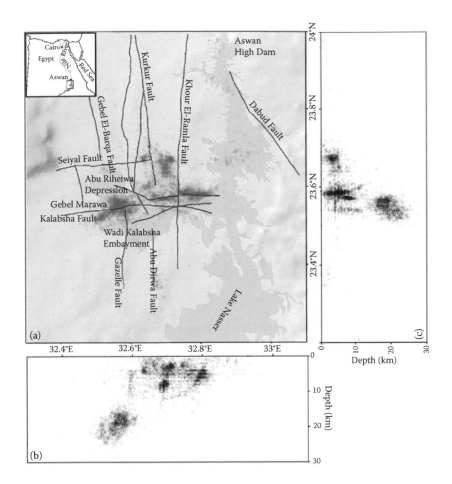

FIGURE 8.5 (a) Aswan seismicity and the reservoir on a topographic map with identified faults. The inset at the top of the figure shows the location of the reservoir. Vertical depth section of seismicity (b) along longitude and (c) along latitude.

network. According to Simpson et al. (1990) and Kebeasy and Gharib (1991), except the main shock and most of the immediate aftershocks, which occurred beneath Gebel Marawa at depths between 15 and 30 km, all other earthquakes occurred between 0 and 12 km (Figure 8.5). Deep seismicity followed the typical aftershock pattern and died down with time, while the shallow seismicity is still continuing (Hassib et al. 2010).

Gahalaut and Hassoup (2012) simulated the stresses and pore pressure due to the Aswan reservoir impoundment and quantified its influence on the nearby seismogenic faults. They calculated changes in stress and pore pressure due to the reservoir impoundment and inferred Coulomb stress (ΔS) on the fault planes responsible for the majority of the seismicity of the region (Figure 8.5). Figure 8.6 shows ΔS (with and without pore pressure) on the fault plane of the 14 November 1981 event, at 20 km depth. This earthquake and its aftershocks are associated with the Kalabsha Fault. Figure 8.7a shows ΔS again on the east-west trending fault plane at 8 km depth to see the effect of reservoir impoundment associated with the seismicity on the east segment of the Kalabsha Fault and other parallel east-west faults. Figure 8.7b shows ΔS (ΔS is calculated at the time epoch of maximum magnitude earthquake in the respective region) on a north-south trending fault at 5 km depth. Figure 8.7c shows ΔS at 5 km depth for a normal connected north-south nodal plane corresponding to the seismicity which lies south of the Kalabsha Fault and is mainly associated with Abu Dirwa fault.

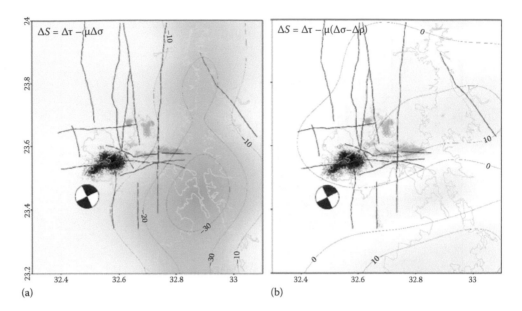

(a)   (b)

FIGURE 8.6   (a) ΔS without pore pressure and (b) ΔS with pore pressure (all in KPa). These calculations are performed at 20 km depth corresponding to the focal depth of the 14 November 1981 earthquake and its time epoch. The epicentre of the earthquake is shown by the red star. The considered focal mechanism (strike, dip and rake as 64°, 90° and 180°, respectively) is also shown. These results represent the seismicity shown by black. All other earthquakes are shown by grey.

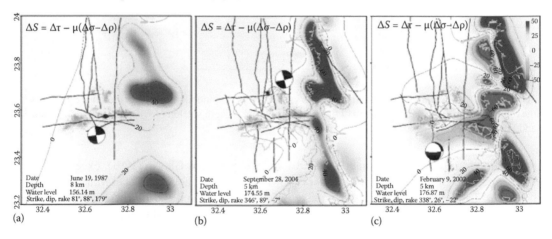

(a)   (b)   (c)

FIGURE 8.7 ΔS with pore pressure (a) at the time epoch of 19 June 1987 at 8 km depth, (b) at the time epoch of 28 September 2004 for the north-south trending nodal planes at 5 km depth and (c) at the time epoch of 9 February 2002 at 5 km depth.

From the analysis, Gahalaut and Hassoup (2012) concluded that the occurrence of earthquakes in the Aswan region is strongly influenced by the change in pore pressure from the reservoir operation. All the active faults of the region associated with the seismicity, which are stabilized when only stresses due to water load are considered, become destabilized when the effect of pore pressure is included.

8.6.3   RTS due to Koyna Reservoir, Western Peninsular India

The Koyna–Warna region of the relatively stable peninsular India is a unique site in the world where the seismicity that reportedly began soon after the impoundment of the reservoir in 1961 has continued for more than 50 years (Talwani 1997; Gupta 1992, 2002 and references therein). The main Koyna Earthquake of 10 December 1967 ($M = 6.3$), the largest near a reservoir ever recorded globally, and the ongoing seismicity in the Koyna–Warna region have been considered the reservoir-triggered earthquakes (Gupta 1992, 2002).

The impoundment of the Koyna reservoir began in 1961, and in 1985 another reservoir was impounded in the Warna valley (Figure 8.8). Knowledge about the seismicity of the region prior to 1962 is very limited due to the absence of seismic stations in the area (Gupta 2002). Since 1963, more than 100,000 earthquakes, including about 170 of $M \geq 4$ and about 17 of $M \geq 5$, have been reported from the Koyna–Warna region, and the frequency of the earthquakes of the past 30 years has been almost steady. Three distinct zones of earthquake clusters were identified (Figure 8.8), namely, the north-northeast-south-southwest trending earthquake cluster, referred to as the Koyna Seismic Zone (KSZ), and two almost parallel clusters trending in the south-southeast-north-northwest direction, collectively referred as the Warna Seismic Zone (WSZ). The majority of earthquakes in these zones occurred between 4 and 8 km depth. Predominant strike-slip faulting on the northeast-southwest to north-northeast-south-southwest trending steep faults in the KSZ and predominant normal faulting on the north-northwest-south-southeast to north-south trending faults in the WSZ have been inferred (Singh 2003).

Figure 8.9 shows the fault stability at 8 km depth on the north-northeast-south-southwest oriented planes with left lateral strike-slip motion and north-northwest-south-southeast oriented planes with normal slip motion (Gahalaut et al. 2004a). ΔS values were calculated after 1 month of maximum water level in the Koyna and Warna reservoirs in 1997, as most of the earthquakes are generally reported to occur with a time lag of about 1 month from the highest water levels (Talwani et al. 1996).

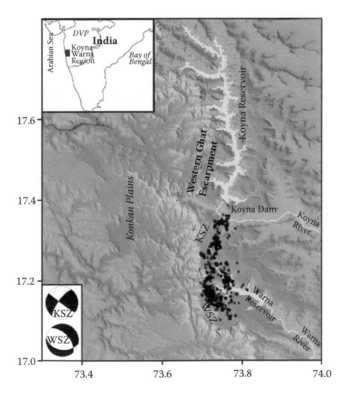

FIGURE 8.8 Epicentral distribution of earthquakes from April 1996 to January 1998 in the Koyna–Warna region over the topography. Well-constrained composite focal mechanisms of earthquake clusters in the KSZ and composite focal mechanisms of earthquake clusters in the WSZ are also shown. DVP, Deccan Volcanic Province.

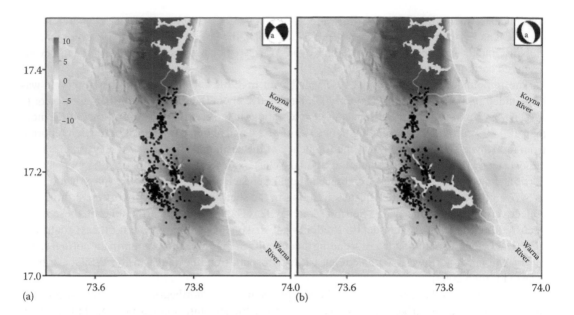

FIGURE 8.9 Stress changes (Δ*S*, in KPa) caused by the Koyna and Warna reservoir operations. (a) Δ*S* on the strike-slip fault planes with strike, dip and rake as 210°, 62° and –10°, respectively. (b) Δ*S* on normal fault planes with strike, dip and rake as 160°, 50° and –90°.

These calculations of Gahalaut et al. (2004a) suggest that the west-northwest dipping strike-slip faults of the KSZ with left lateral motion (Figure 8.9a) and the west-southwest dipping normal faults of the WSZ (Figure 8.9b) are brought closer to failure by the reservoir effects.

However, Gahalaut et al. (2004a) argued that the occurrence of 'normal' earthquakes cannot be attributed to triggering by Koyna–Warna reservoirs alone, despite their favourable influence. According to the mechanism of RTS, only those faults are destabilized by the reservoir effects, which are critically stressed and are close to failure in a manner consistent with the ambient stresses. The stress changes due to reservoirs act only as a trigger and cannot cause an earthquake (Simpson 1986; Chander and Kalpna 1997; McGarr and Simpson 1997). Thus, an earthquake on a fault on which ambient tectonic stresses are not favourably oriented for failure in the desired manner cannot be triggered by the reservoirs, even if the effects of the reservoirs on this fault are to promote slip in the required direction. A similar situation exists in the Koyna–Warna region with regard to normal faults. Although the reservoir effects promote destabilization on the west-southwest dipping normal faults in the southern part of the Koyna–Warna region (Figure 8.9b), these normal faults are unlikely to slip in the compressional regime of peninsular India; rather, they will be stabilized. Thus, the reservoirs cannot trigger earthquakes on these stabilized normal faults. For triggering of normal earthquakes on these faults by the reservoirs, these faults should first be brought to the critical stress level for failure in normal slip by some mechanism. They suggested that the stress transfer on such faults due to the cumulative effect of the left lateral motion on the strike-slip faults in the KSZ, through continuous occurrence of earthquakes, is the mechanism which either causes normal earthquakes on these faults or at least brings these faults to a critically stressed level for failure in normal slip, which may eventually slip under the favourable influence of reservoirs. They also suggested that the fault interaction through stress transfer causes high and persistent seismicity of the Koyna–Warna region.

8.6.4 IMPOUNDMENT OF ZIPINGPU RESERVOIR AND 2008 WENCHUAN EARTHQUAKE

Gahalaut and Gahalaut (2010) carried out a detailed analysis to assess the possible role of the Zipingpu reservoir on the local seismicity and on the occurrence of the 2008 magnitude 7.9

Wenchuan Earthquake. The maximum water level in the reservoir generally occurs during October to December, with a maximum water depth of about 120 m, while the minimum level generally occurs during April to May each year, with a minimum water depth of about 60 m (Lei et al. 2008). The 12 May 2008 Wenchuan Earthquake ($M = 7.9$) occurred on the Longmen Shan thrust-fault system near the eastern margin of the Tibetan plateau (Figure 8.10). The occurrence of the 2008 Wenchuan Earthquake in this region triggered a debate on whether it was influenced by the newly impounded Zipingpu reservoir on Min River, located only about 21 km east of the earthquake epicentre (Klose 2008; Lei et al. 2008; Kerr and Stone 2009). The reservoir impoundment began in March 2004 with a dam height of 156 m.

Gahalaut and Gahalaut (2010) calculated the temporal variation of ΔS to see its variation with change in water level at the earthquake hypocentre, considering the actual water level series (Figure 8.11). Figure 8.12a shows ΔS in a plan view at 19 km depth, which is the focal depth of the earthquake, and Figure 8.12b shows ΔS in a vertical depth section through the reservoir along the strike of the rupture on the same fault plane, at the time of the occurrence of the 2008 Wenchuan Earthquake. Analysis has been carried out on the nodal plane with the strike, dip and rake as 231°, 35° and 138°, respectively, of the 2008 Wenchuan Earthquake focal mechanism. The results of the analysis show that the impoundment of the Zipingpu reservoir had a stabilizing effect on the 2008 Wenchaun earthquake hypocentre region.

The shallow local seismicity near the reservoir before the occurrence of the 2008 Wenchuan Earthquake appeared to be affected by the impoundment of the Zipingpu reservoir (Lei et al. 2008).

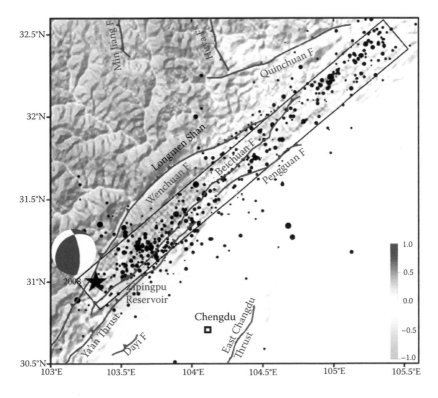

FIGURE 8.10 The 2008 Wenchuan Earthquake, its aftershocks and the rectangular rupture on a topographic map with major faults in the region. The focal mechanism of the 2008 Wenchuan Earthquake is shown by a beach ball and the epicentre by a star. The northwest dipping plane with strike, dip and rake as 231°, 35° and 138°, respectively, corresponds to the fault plane of the earthquake. The Zipingpu reservoir is also shown here. The Coulomb stress change (ΔS, in KPa) at a depth of 19 km on the northwest dipping plane due to the reservoir load and pore pressure is shown by red and green on this map.

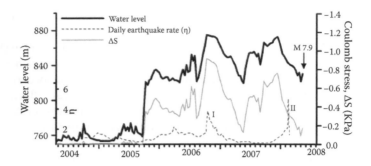

FIGURE 8.11   Water level changes in the reservoir, daily earthquake rate (η) reported by Lei et al. (2008) for earthquakes of $M \geq 0.5$ in the Zipingpu area, along with ΔS on the northwest dipping plane at the 2008 Wenchuan hypocentre. Arrow denotes the time of occurrence of the 2008 Wenchuan Earthquake. Two peaks in the earthquake rate before the 2008 Wenchuan Earthquake are shown by I and II.

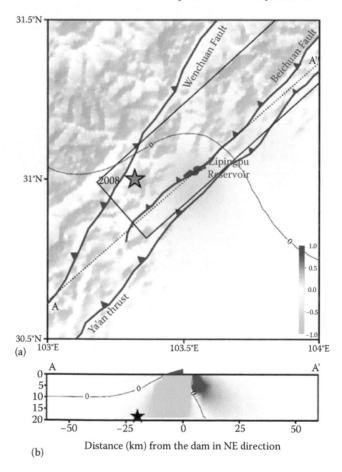

FIGURE 8.12   (a) ΔS on northwest dipping plane at 19 km depth corresponding to the 2008 Wenchuan Earthquake. (b) Corresponding ΔS in the vertical cross section along line A-A'.

They did not find any change in the earthquake frequency in the region after the reservoir impoundment for earthquakes of $M \geq 2.5$. A marginal increase in the frequency of small earthquakes of $M \geq 0.5$ around the reservoir was noticed after September 2005, which coincided with the high water level in the reservoir (Figure 8.11) in October 2006 (marked as I) and February 2008 (marked as II) (Figure 8.11). A higher frequency of earthquakes were observed during February 2008. Thus,

Gahalaut and Gahalaut (2010) calculated ΔS at this time at a depth of 4 km (Figure 8.13), as the majority of these small earthquakes occurred at very shallow depths, ranging between 0 and 8 km (Lei et al. 2008). Most of the low-magnitude local seismicity occurred in the region of positive Coulomb stress, and thus in the region of destabilization (Figure 8.13a). In the depth section of Figure 8.13b, the two main clusters of high seismicity (Lei et al. 2008) lie in the zones of increased ΔS, that is, the region of destabilization. Further, it appears that just beneath the reservoir, the region of negative ΔS (i.e. the region of stabilization) coincides with very low seismic activity. Gahalaut and Gahalaut (2010) concluded that the increase in seismicity horizontally displaced from the reservoir coincided with the region of destabilization, while the lack of seismicity below and in the immediate vicinity of the reservoir coincided with the region of stabilization caused by the reservoir impoundment. Gahalaut and Gahalaut (2010) suggested that the reservoir impoundment probably did not play any role in the occurrence of the 2008 Wenchuan Earthquake at its hypocentre. Earlier studies have found a positive role of the Zipingpu reservoir impoundment on the occurrence of the 2008 Wenchuan Earthquake (Klose 2008; Lei et al. 2008; Kerr and Stone 2009). However, more realistic and robust 3D analysis of Gahalaut and Gahalaut (2010) did not support any significant role of the reservoir in the earthquake occurrence. It has been suggested that the local seismicity

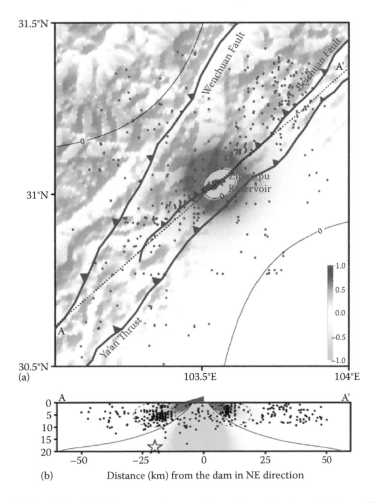

FIGURE 8.13 (a) ΔS at 4 km depth for pure thrust motion with strike, dip and rake as 231°, 35° and 90°, respectively. Earthquakes of $M \geq 0.5$ that occurred before the 2008 Wenchuan Earthquake but after the reservoir impoundment (Lei et al. 2008) are shown by dots. (b) Corresponding ΔS in vertical cross section along line A-A'.

was influenced by the impoundment of the reservoir, as the destabilizing and stabilizing effects of reservoir impoundment led to the increase in seismicity away from the reservoir and low seismicity beneath it, respectively. Deng et al. (2010) arrived at similar conclusions from their 2D analysis.

8.6.5 TARBELA RESERVOIR

Tarbela reservoir on Indus River in the Pakistan Himalaya is a unique case where the occurrence of microearthquakes occurred on low water stands after the impoundment, which started in 1974, of the reservoir (Ibenbrahim et al. 1989). Ibenbrahim et al. (1989) carried out 2D stress analysis and concluded that the Tarbela reservoir exerted a stabilizing influence on nearby seismogenic strike-slip and reverse faults and inferred that unloading would have the opposite effect. They concluded that the elastic effect of reservoir unloading was the main factor in occurrence of these dry season microearthquakes. However, Kalpna and Chander (1997) from their 3D detailed analyses have shown that even at the reduced water levels during dry seasons, the Tarbela reservoir exerted a stabilizing influence on the causative faults of these microearthquakes. And a stabilizing influence could have been overcome through destabilization of the faults due to continued accumulation of horizontal, broadly north-south oriented, compressive stresses of plate tectonic origin. The pore pressure needs to be included in stability analysis before making further inferences.

8.6.6 AÇU RESERVOIR

Açu Dam on Açu River is situated near the active zones of intraplate seismicity in northeast Brazil. Some historical earthquakes and swarms are reported (Ferreira and Assumpção 1983) in the region. Impoundment started after the completion of the dam in May 1983, but the water depth was less than 14 km for the 2 initial years of impoundment and then increased to 35.5 m in 1985. A vertical component seismometer has been continuously operating in the area since August 1987. Three stations with a vertical component campaign from October 1989 to December 1989, four stations with a vertical component campaign from November 1990 to March 1991 and eight stations with a three-component digital seismograph campaign from August 1994 to May 1997 have also been done in parallel (Farreira et al. 1995; Nascimento et al. 2004). Detailed analysis of the data by many workers (Farreira et al. 1995; Nascimento et al. 2004) inferred that seismicity in the region is clustered in several well-defined zones, and there is a temporal and spatial migration. Farreira et al. (1995) correlated the water level of Açu and the seismicity of the area from January 1987 to August 1994, which was clustered directly beneath the reservoir, and concluded that the first 3 years of seismicity was very well correlated with the water level, with a time lag of 3 months due to pore pressure diffusion. Farreira et al. (1995) observed a migration of seismicity to the lake margin from 1990, and also the absence of correlation between the two time series. Nascimento et al. (2005a,b) and Hariri et al. (2010) carried out model studies of pore pressure and hydraulic diffusivity and observed that pore pressure might have played a positive role in triggering the earthquakes after the reservoir impoundment. However, a detailed stability analysis is needed to give a final verdict on the role of reservoir impoundment in triggering earthquakes near the Açu reservoir.

8.7 CONCLUSION

The need to investigate earthquakes near reservoirs stems from two considerations. First, the theme should be pursued for its intrinsic scientific interest. Second, there is the compulsion to develop hydroelectric power in spite of the drawbacks. The key to further progress in understanding the subject lies greatly on an increased database derived from observations at reservoir sites and development in the simulation of this process. The past few years have seen great development on both fronts. Seismic networks have expanded tremendously, and now we have good understanding about the earthquakes. On the theoretical development front, now we understand more about the process of triggered earthquakes. Snow (1972) remarked that no theory about reservoir-triggered earthquakes should be regarded as complete unless it explains both the cases where earthquakes occur after impoundment and the cases where they do not. The use of the porous elastic theory and the concept of reservoir stabilization and destabilization of nearby seismogenic faults are important rationalizations in this regard. It has been possible to see the influence of reservoir operation on various faults in the region, their earthquake-generating potentials, and whether a particular earthquake was actually triggered by reservoir operation. Here, the cases of seismicity near Aswan, Koyna and Rihand were found to be classical and now proven examples where reservoir operations indeed influence the seismicity. On the other hand, the Zipingpu reservoir did not influence the occurrence of the 2008 Wenchuan Earthquake, as thought. There are still several worldwide cases where the role of reservoirs in earthquake occurrence has been debated by the observation of visual correlation between the two, but no rigorous analysis has been undertaken. It is important to analyze such cases. At the same time, there is a need to actually quantify the role of the reservoir in earthquake occurrence. In other words, it is important to quantify the extent of stresses contributed by the reservoir operation in the total strain release during earthquake occurrence and to assess whether there lies any threshold on reservoir size, depth, duration of loading, and so forth. We need to measure in situ rock properties (stresses and hydrologic regime). Further, we need to consider more realistic Earth models by incorporating heterogeneities in the earth crust.

REFERENCES

Bell, M. L., and Nur, A. (1978). Strength changes due to reservoir induced pore pressure and stresses and application to Lake Oroville. *J. Geophys. Res.*, 83, 4469–4483.

Biot, M. A. (1941). General theory of three-dimensional consolidation. *J. Appl. Phys.*, 12, 155–164.

Biot, M. A. (1955). Theory of elasticity and consolidation for a porous anisotropic solid. *J. Appl. Phys.*, 26, 182–185.

Biot, M. A. (1956a). Thermo elasticity and irreversible thermodynamics. *J. Appl. Phys.*, 27, 240–253.

Biot, M. A. (1956b). Theory of propagation of elastic waves in a fluid saturated porous solid. *J. Acoust. Soc. Am.*, 28, 168–191.

Bock, Y. (1980). Load induced stresses and their relation to initial stress field. *J. Geophys.*, 48, 94–100.

Carder, D. S. (1945). Seismic investigations in the Boulder Dam area, 1940–1944, and the influence of reservoir loading on earthquake activity. *Bull. Seis. Soc. Am.*, 35, 175–192.

Castle, R. O., Clark, M. M., Grantz, A., and Savage, J. C. (1975). Tectonics stage: Its significance and characterization in the assessment of seismic effect associated with reservoir impounding. Presented at the 1st International Symposium on Induced Seismicity, Banff, Canada, September 15–19.

Chander, R. (1999). Can dams and reservoirs cause earthquakes? Triggering of earthquakes. *Resonance*, 4, 4–13.

Chander, R., and Kalpna. (1997). On categorising induced and natural tectonic earthquakes near new reservoirs. *Eng. Geol.*, 46, 81–92.

Costain, J. K., Bollinger, G. A., and Speer, J. A. (1987). Hydroseismicity: A hypothesis for the role of water in the generation of intraplate seismicity, Seismol. *Res. Lett.*, 58, 41–64.

Costain, J. K., and Bollinger, G. A. (2010). Review: Research results in hydroseismicity from 1987 to 2009, Bull. Seis. *Soc. Am.*, 100, 1841–1858.

Deng, K., Zhou, S, Wang, R., Robinson, R., Zhao, C., and Cheng, W. (2010). Evidence that the 2008 M_w 7.9 Wenchuan earthquake could not have been induced by the Zipingpu reservoir, Bull Seismol. *Soc. Am.*, 58, 2805–2814.

Ferreira, J. M., and Assumpção, M. (1983). Sismicidade do Nordeste do Brasil. *Rev. Bras. Geofis.*, 1, 67–88.

Ferreira, J. M., Oliveira, R. T. D., Assumpção, M., Moreira, A. M., Pearce, R. G., and Takeya, M. K (1995). Correlation of seismicity and water level in the Açu reservoir—An example from northeast Brazil. Bull. *Seism. Soc. Am.*, 85, 1483–1489.

Gahalaut, K., and Gahalaut, V. K. (2010). Effect of the Zipingpu reservoir impoundment on the occurrence of the 2008 Wenchuan Earthquake and local seismicity. *Geophys. J. Int.*, 183, 277–285.

Gahalaut, K., Gahalaut, V. K., and Pandey, M. R. (2007). A new case of reservoir triggered seismicity: Govind Ballav Pant reservoir (Rihand Dam), central India. *Tectonophysics*, 439, 171–178.

Gahalaut, K., and Gupta, P. K. (2008). An integral equation algorithm for 3-D simulation of pore pressure in a porous elastic medium with heterogeneities. *Geophys. J. Int.*, 175, 1245–1253. doi: 10.1111/j.1365–246X.2008.03940.x.

Gahalaut, K., and Hassoup, A. (2012). Role of fluids in the earthquake occurrence around Aswan reservoir, Egypt. *J. Geophys. Res.*, 117, B02303. doi: 1029/2011JB008796.

Gahalaut, V. K., Kalpna, and Singh, S. K. (2004a). Fault interaction and earthquake triggering in the Koyna-Warna region, India. *Geophys. Res. Lett.*, 31, L11614. doi: 10.1029/2004GL019818.

Gahalaut, V. K., Rao, V. K., and Tewari, H. C. (2004b). On the mechanism and source parameters of the deep crustal Jabalpur earthquake, India, of May 21, 1997: Constraints from aftershocks and change in static stress. *Geophys. J. Int.*, 156, 345–351.

Gough, D. I., and Gough, W. I. (1976). Time dependence and trigger mechanisms for the Kariba (Rhodesia) earthquakes. *Eng. Geol.*, 10, 211–217.

Gowd, T. N., Srirama Rao, S. V., and Gaur, V. K. (1992). Tectonic stress field in the Indian subcontinent. *J. Geophys. Res.*, 97, 11879–11888.

GSI (Geological Survey of India). (2000). *Seismotectonics Atlas of India.* Calcutta, India: GSI.

Gupta, H. K. (1992). Reservoir induced earthquakes. In *Developments in Geotechnical Engineering.* Amsterdam: Elsevier, 355.

Gupta, H. K. (2002). A review of recent studies of triggered earthquakes by artificial water reservoirs with special emphasis on earthquakes in Koyna, India. *Earth Sci. Rev.*, 58, 279–310.

Gupta, H. K., Rastogi, B. K., and Narain, H. (1972). Common features of reservoir associated seismic activities. *Bull. Seis. Soc. Am.*, 62, 481–492.

Hardebeck, J. L., Nazareth, J. J., and Hauksson, E. (1998). The static stress change triggering model: Constraints from two southern California aftershock sequences. *J. Geophys. Res.*, 103, 24427–24437.

Hariri, M. El, Abercrombie, R. E., Rowe, C. A., and Do Nascimento, A. F. (2010). The role of fluids in triggering earthquakes: Observations from reservoir induced seismicity in Brazil. *Geophys. J. Int.*, 181, 1566–1574.

Hassib, G., Hamed, H., Dahy, S., Hassoup, A., and Moustafa, S. (2010). Detection of the seismic quiescence along the seismic activity faults in Kalabsha area, west of Lake Nasser, Aswan. *Acta Geod. Geoph. Hung.*, 45, 210–226. doi: 10.1556/AGeod.45.2010.2.6.

Hassoup A. (2002). Seismicity and water level variations in the Lake Aswan area in Egypt 1982–1997. *J. Seismol.*, 6, 459–467.

Healy, J. H., Rubey, W. W., Griggs, D. T., and Raleigh, C. B. (1968). The Denver earthquakes. *Science*, 161, 1301–1310.

Ibenbrahim, A., Ni, J., Salyards, S., and Ali, I. M. (1989). Induced seismicity of the Tarbela reservoir, Pakistan. *Seismol. Res. Lett.*, 60, 185–197.

Jaeger, J. C., and Cook, N. G. W. (1969). *Fundamentals of Rock Mechanics.* London: Methuen.

Kalpna and Chander, R. (1997). On some microearthquakes near Tarbela reservoir during three low water stands. *Bull Seismol. Soc. Am.*, 87, 265–271.

Kalpna and Chander, R. (2000). Green's function based stress diffusion solution in the porous elastic half space for time varying finite reservoir loads. *Phys. Earth Planet. Interiors*, 120, 93–101.

Kebeasy, R., and Gharib, A. (1991). Active fault and water loading are important factors in triggering earthquake activity around Aswan Lake. *J. Geodyn.*, 14, 73–83.

Kebeasy, R. M., Maamoun, M., Ibrahim, E., Megahed, A., Simpson, D. W., and Leith, W. S. (1987). Earthquake studies at Aswan reservoir. *J. Geodyn.*, 7, 173–193.

Keith, C. M., Simpson, D. W., and Soboleva, O. V. (1982). Induced seismicity and style of deformation at Nurek Reservoir, Tadjik SSR. *J. Geophys. Res.*, 87, 4609–4624.

Kerr, R.A, Stone, R. (2009). A human trigger for the great quake of Sichuan? *Science*, 323, 322.

King, G. C. P., Stein, R. S., and Lin, J. (1994). Static stress changes and the triggering of earthquakes. *Bull. Seismol. Soc. Am.*, 84, 935–953.

Klose, C. D. (2008). The 2008 M7.9 Wenchuan earthquake—Result of local and abnormal mass imbalances? In *AGU Fall Meeting*, U21C-08.

Kuo, J. T. (1969). Static response of a multilayered medium under inclined surface loads. *J. Geophys. Res.*, 74, 3195–3207.

Lei, X., Ma, S., Wen, X.-Z., Su, J.-R., and Du, F. (2008). Integrated analysis of stress and regional seismicity by surface loading—A case study of Zipingpu reservoir. *Seismol. Geol.*, 30, 1046–1064.

Lomnitz, C. (1974). Earthquake and reservoir associated earthquakes. *Eng Geol.*, 10, 197–210.

McGarr, A., and Simpson, D. (1997). Keynote lecture: A broad look at induced and triggered seismicity. In *Rockbursts and Seismicity in Mines*, ed. Gibowicz, S. J. and Lasocki, S. Rotterdam: Balkema, 385–396.

Mead, R. B. (1991). Reservoir and earthquakes. *Eng. Geol.*, 30, 245–262.

Nascimento, A. F., Cowie, P. A., Lunn, R. J., and Pearce, R. G. (2004). Spatio-temporal evolution of induced seismicity at Açu reservoir, NE Brazil. *Geophys. J. Int.*, 158, 1041–1052.

Nascimento, A. F., Lunn, R. J., and Cowie, P. A. (2005a). Numerical modelling of pore pressure diffusion in a reservoir-induced seismicity site in northeast Brazil. *Geophys. J. Int.*, 160, 249–262.

Nascimento, A. F., Lunn, R. J., and Cowie, P. A. (2005b). Modelling the heterogeneous hydraulic properties of faults using constraints from reservoir-induced seismicity. *J. Geophys. Res.*, 110, B09201. doi: 10.1029/2004JB003398.

Rajendran, K., and Talwani, P. (1992). The role of elastic, undrained and drained response in triggering earthquakes at Monticello reservoir, South Carolina, *Bull. Seismol. Soc. Am.*, 82, 1867–1888.

Rao, N. P., Tsukuda, T., Koruga, M., Bhatia, S. C., and Suresh, G. (2002). Deep lower crustal earthquakes in central India: Inferences from analysis of regional broadband data of the 1997 May 21 Jabalpur earthquake. *Geophys. J. Int.*, 148, 132–138.

Rice, J. R., and Cleary, M. P. (1976). Some basic stress diffusion solutions for fluid-saturated elastic porous media with compressible constituents. *Am. Geophys. Union Rev. Geophys. Space Phys.*, 14, 227–242.

Roeloffs, E. A. (1988). Fault stability changes induced beneath a reservoir with cyclic variations in water level. *J. Geophys. Res.*, 93, 2107–2124.

Scholz, C. H. (1990). *The Mechanics of Earthquakes and Faulting*. Cambridge: Cambridge University Press.

Simpson, D. W. (1976). Seismicity changes associated with reservoir loading. *Eng. Geol.*, 10, 123–150.

Simpson, D. W. (1986). Triggered earthquakes. *Annu. Rev. Earth Planet. Sci.*, 14, 21–42.

Simpson, D. W., Gharib, A. A., and Kebeasy, R. M. (1990). Induced seismicity and changes in water level at Aswan reservoir, Egypt. *Gerl. Beitr. Geophysik Leipzog*, 99, 191–204.

Simpson, D. W., Leith, W. S., and Scholz, C. H. (1988). Two types of reservoir induced seismicity. *Bull. Seis. Soc. Am.*, 78, 2025–2040.

Simpson, D. W., and Narasimhan, T. N. (1990). Inhomogeneities in rock properties and their influence on reservoir induced seismicity. *Gerl. Beitr. Geophysik*, 99, 205–219.

Singh, S. K. (2003). Seismic imaging and tectonics of Koyna seismic zone. PhD thesis, Osmania University, Hyderabad, India.

Singh, S. K., Dattatrayam, R. S., Shapiro, N. M., Mandal, P., Pacheco, J. F., and Midha, R. K. (1999). Crustal and upper mantle structure of peninsular India and source parameters of the 21 May 1997 Jabalpur earthquake (Mw=5.8): Results from a new regional broadband network. *Bull. Seismol. Soc. Am.*, 89, 1631–1641.

Snow, D. T. (1972). Geodynamics of seismic reservoirs. In *Proceedings of the Symposium on Percolation through Fissured Rock*, Stuttgart, T2J, 1–9.

Stein, R. S. (2004). Tidal triggering caught in the act. *Science*, 305, 1248–1249.

Stein, S., and Liu, M. (2009). Long aftershock sequences within continents and implications for earthquake hazard assessment. *Nature*, 462, 87–89.

Steinbrugge, V. (1972). Reservoir and earthquake report. Washington, DC: National Research Council.

Talwani, P. (1997). Seismotectonics of the Koyna–Warna area, India. *Pure Appl. Geophys.*, 150, 511–550.

Talwani, P., and Acree, S. (1985). Pore pressure diffusion and the mechanism of reservoir induced seismicity. *PAGEOPH*, 122, 947–965.

Talwani, P., Chen, C., and Gahalaut, K. (2007). Seismogenic permeability, k_s. *J. Geophys. Res.*, 112, B07309. doi: 10.1029/2006JB00466.

Talwani, P., Kumarswamy, S. V., and Sawalwede, C. B. (1996). The re-evaluation of seismicity data in the Koyna-Warna area. Columbia: University of South Carolina.

Ziv, A., and Rubin, A. M. (2000). Static stress transfer and earthquake triggering: No threshold in sight? *J. Geophys. Res.*, 105, 13631–13642.

Zoback, M. D., and Hickman, S. (1982). Physical mechanisms controlling induced seismicity at Monticello Reservoir, South Carolina. *J. Geophys. Res.*, 87, 6959–6974.

Earthquake Precursory Studies in India
An Integrated Approach

Brijesh K. Bansal and Mithila Verma

CONTENTS

9.1 INTRODUCTION

Complex meteorological–hydrological–tectonic interactions lead to various kinds of natural hazards, for example, earthquakes, tsunami, cloud bursts, landslides, floods and avalanches. The Indian subcontinent, with the lofty Himalayan mountain chain in the north and a more than 4500 km long coast line, has repeatedly witnessed widespread occurrences of these extreme events. The 2014 floods in Jammu and Kashmir; the 2013 multiday cloud burst in the Uttarakhand Himalaya; and the 2016 Vardah, 2014 Hudhud and 2013 Phailin cyclonic storms that hit the east coast of India are some recent examples of the meteorological-hydrological hazards, whereas the 2005 Kashmir (Mw 7.6), 2011 Sikkim (Mw 6.9) and 2004 Sumatra (Mw 9.1) earthquakes are examples of extreme events of tectonic-geodynamic origin. Common in all these extreme events is the heavy loss of human lives and large-scale devastation of structures and properties. Earth system science has a major role in developing interlinked science innovations aimed at improving our understanding of the physical processes causing these events and the factors determining their severity, issuing early warning alerts and/or forecasts, making in-depth risk analysis and suggesting technology-based mitigation strategies. These skills are largely attained for meteorological-hydrological hazards as timely forecast and mitigation strategies to minimize the loss of human lives; for example, the 1999 super cyclonic storm that hit the corridor of Odisha on the Indian east coast claimed more than 10,000 lives. However, due to the timely forecast and preventive intervention, thousands of human lives were saved during the 2014 Hudhud and 2013 Phailin cyclonic storms; only 100 lives were lost. Similar emphasis is being placed on finding solutions to the challenge posed by devastating earthquakes. Over the decades, as a result of increased seismic monitoring, the induction of modern geophysical imaging programmes and the introduction of the global positioning system (GPS), our understanding of seismogenesis, seismotectonics and the dynamics controlling the earthquake process has greatly improved. However, despite some promising leads, our scientific innovations are not up to the scale where we can predict earthquakes with respect to time, location and magnitude with stated confidence in any part of the world. To fill these vital gaps, the focus of the National Seismicity and Earthquake Precursor programme is based on two principal components: (1) the seismic hazard assessment and seismic risk evaluation and (2) the systematic earthquake precursory research, with the broad objective to enhance our capabilities towards earthquake prediction.

The seismic hazards assessment programme so far has been largely dependent on a seismic zoning map prepared and published by the Bureau of Indian Standards (BIS). In this map, the entire country has been grouped into four major zones (II–V); zone V is considered to be the most severe, while zone II is the least. However, this map is not sufficient for the assessment of area-specific seismic risks; a detailed hazard zonation or microzonation map is needed for earthquake disaster mitigation and management. The first-level seismic microzonation has already been completed for selected urban centres, including Jabalpur, Guwahati, Bangalore, Ahmadabad and Dehradun, National Capital Territory (NCT) of Delhi and Kolkata. The maps prepared for these cities are being further refined on larger scales with the help of high-resolution satellite data as per the requirements, and a plan has also been firmed up for taking up microzonation of 30 selected cities, which lie in seismic zones V and IV and have a population density of half a million. Verma and Bansal (2013) have reviewed the efforts made so far towards seismic hazard assessment, as well as the future road map for such studies.

In this chapter, we discuss the progress in the area of earthquake precursory research in India. In the first instance, a brief account of the regional tectonics and the spatial distribution of the seismicity in the Indian subcontinent is outlined. A globally recognized class of earthquake precursors and their physical generating mechanisms are introduced before launching on the progression path of precursory research. Here, we first summarize the trend of precursory studies in the early period; the lessons learned from such studies have motivated us to launch many organized multidisciplinary studies. The first results emerging from such integrated studies are discussed to highlight the trend of precursory research in the upcoming period. The progression path of these activities is in two parts, first for the Himalayan interplate region, where working models for earthquake occurrences are quite well established. The induced seismicity in the stable intraplate region is still the subject of debate. Discussion of these results ends with brief details of a major experiment involving a super-deep borehole to study the earthquake process in in situ conditions. Recognizing that early warning before the actual arrival of destructive surface waves to urban cities away from the source region of the earthquake can save human lives and protect major infrastructure, salient features of the programme on earthquake early warning (EEW) initiated in northwest Himalaya are outlined. The second part provides the lead in earthquake precursory research for the Koyna–Warma region, located in the stable continental region of the Indian shield. The region is also a classic case of reservoir-triggered seismicity.

9.2　TECTONICS AND SEISMIC TRACKS IN THE INDIAN SUBCONTINENT

The Indian subcontinent, with its diverse tectonics, including the Himalayan collision zone in the north, the Chaman fault on the west, the Indo-Burmese arc in the northeast, the Andaman Sumatra Trench in the southeast Indian territory and failed rift zones in the Peninsular Indian shield, shows earthquake activity, albeit with varying degrees (Figure 9.1). Reliable instrumental data on earthquakes are available only from the beginning of the twentieth century, when the first seismic observatory was established at Kolkata in northeast India soon after the great Shillong earthquake of

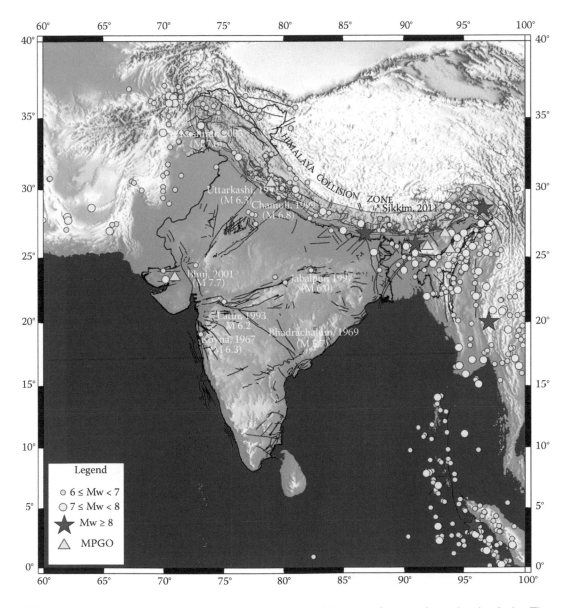

FIGURE 9.1　Seismicity of India and adjoining regions in relation to major tectonics and active faults. The locations of great and some select large and moderate earthquakes (Mw >6) along the Himalayan arc and stable Indian shield region, referred to in the text, are marked by different coloured symbols. The locations of various MPGOs established in different parts of the country for integrated earthquake precursory studies are also shown.

1897 (Mw 8.1). Beginning in 1897, the Himalayan frontal arc has witnessed four great earthquakes, the 1897 Shillong, 1905 Kangra, 1934 Bihar and 1950 Assam earthquakes, in the short interval of 53 years (Figure 9.1). No large earthquake (Mw ≥7.5) has occurred along the arc since 1950. The region, however, remains active and has witnessed several moderate earthquakes. Some recent earthquakes in the Himalaya are the 1991 Uttarkashi (Mw 6.8), 1999 Chamoli (Mw 6.6), 2005 Kashmir (Mw 7.6) and 2011 Sikkim (Mw 6.9) earthquakes. More intriguing is the widespread occurrences of moderate-magnitude earthquakes in the stable continental shield region, the 1967 Koyna (Mw 6.3), 1969 Bhadrachalam (Mw 5.7), 1993 Latur (Mw 6.2), 1997 Jabalpur (Mw 5.8) and 2001 Bhuj (Mw 7.7) Earthquakes. A small pocket of the Koyna–Warna region in the west coast has been active from 1963, following the impoundment of the twin reservoirs of Koyna and Warna in 1962 and 1993, respectively. The recorded seismicity includes the largest triggered earthquake of Mw ~6.3 on 10 December 1967, 22 earthquakes of Mw >5, about 200 earthquakes of Mw ~4 and several thousand smaller earthquakes since 1963 (Figure 9.2). The triggered seismicity started near the southern flank of the Koyna reservoir but migrated south-southeast following the impoundment of the nearby Warna reservoir in 1993. This has turned out to be a case of the longest reservoir triggered seismicity (Gupta 2002).

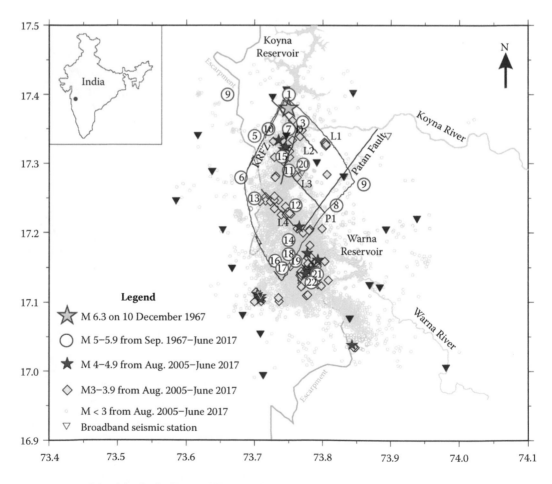

FIGURE 9.2 Seismicity in the Koyna–Warna region after the impoundment of the Koyna and Warna reservoirs. D, Donechiwada fault; P1 fault parallel to Patan fault; L1, L2, L3 and L4, northwest-southeast trending fractures. (Fault geometry from Talwani, P., *Pure Appl. Geophys.*, 150, 473–492, 1997.)

Given the dominance of thrust-type focal mechanisms, earthquake occurrence processes along the Himalayan arc are explained by the underthrusting of the Indian plate beneath the overriding wedge of the Himalaya. GPS measurements corroborate that the convergence due to plate collision is accommodated through a stick and slip manner on seismically active detachment under the Outer and Lesser Himalaya, where great and major earthquakes in the Himalaya are rooted (Ni and Barazangi 1984; Banerjee and Burgmann 2002). Further north, beneath the Higher and Tethys Himalaya, the detachment slips aseismically and is free from earthquakes. The two sections are connected by a steeply dipping ramp, having asperity that accumulates tectonic strain, and accounts for the small- and moderate-magnitude earthquakes, which are confined to a narrow belt, straddling the surface trace of the Main Central Thrust (MCT), referred to as the Himalayan Seismic Belt (HSB).

In contrast to the Himalaya, processes of seismogenesis in the stable continental shield are yet poorly resolved. The seismic activity in the Koyna–Warna region, starting from 1963 to the present, has concentrated on a short 30 km long segment, which is seen as an extension of the $10°$ N–$15°$ E trending Koyna River Fault Zone (KRFZ) to the south in the north-northwest-south-southeast direction (Figure 9.2). This direction closely follows the orientation of major lineament, mega fracture and geomorphic (fissure) features of the region. Due to the thick cover of the Deccan trap, geophysical imaging programmes have limited success in providing detailed structures of the source faults. Our present understanding on the geometry and depth of the source fault relies on seismic data. The introduction of a 10-station digital network has enabled more reliable estimation of a velocity model using seismic tomography, surface wave dispersion, and so forth. Low inferred heat flow of 41 mW/m^2 supports brittle rupture in the 'cold' upper part of the crust, a strong candidate for failure-producing seismicity. Perhaps the anomalous pore fluid pressures on the pre-existing critically stressed faults have a major role in producing failure, triggering the occurrence of continued seismicity in the Koyna–Warna region (Talwani 1997).

9.3 EARTHQUAKE PRECURSORS

Strain built up during the earthquake preparatory cycle and its sudden release during earthquake rupture is likely to produce changes in certain physical and chemical properties of the medium. If these changes could be detected in measured parameters, they could be used as precursors to predict the impending earthquake. For a long time, the search for precursory signals has continued in many active seismic zones, resulting in the identification of a range of potential precursors. In hindsight, some of these parameters have shown characteristic changes which one way or another appear to be related to earthquake occurrences. However, pessimism prevailed, as noted changes were not observed at all earthquake sites or even for different earthquakes in the same region. The lack of a sound physical hypothesis that could explain and validate precursory behaviours further added to the pessimism. The dilatancy–diffusion model based on the behaviour of rocks under stresses in laboratory conditions had some success in explaining some of the noted precursory signals (Scholz et al. 1973). The model hypothesizes that the generation of various precursors is the sequential effect of the opening of minor cracks, the influx of fluids and the material strengthening that rocks exhibit when subjected to accumulating stresses simulating different phases of the earthquake preparatory cycle (Figure 9.3a). The continuing pursuit in different seismic-prone countries has helped to identify a range of precursors. Depending on the physical entities involved, these precursors are grouped into three broad categories, namely, seismological, geomagnetic and geodetic, and geochemical and hydrological (Figure 9.3b), and could be casually linked to the physical processes active during the preparatory phase of the main shock. Despite some important leads, there is still widespread scepticism about their applications to real-time prediction. In the background of a likely physical mechanism, we trace the progress path of these precursory studies in the Indian context to portray the salient nature of precursory signals, as well as factors which

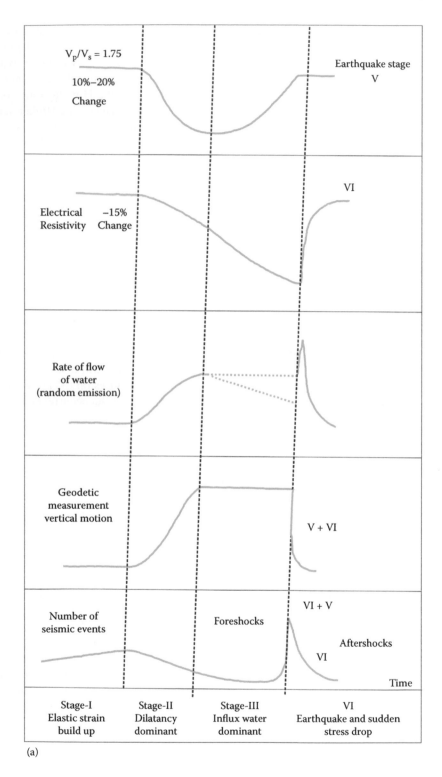

(a)

FIGURE 9.3   (a) Nature of changes in various physical parameters as a function of time as predicted by the laboratory-based dilatancy–diffusion model. (Modified after Scholz, C. H. et al., *Science*, 181, 803–810, 1973.)

(*Continued*)

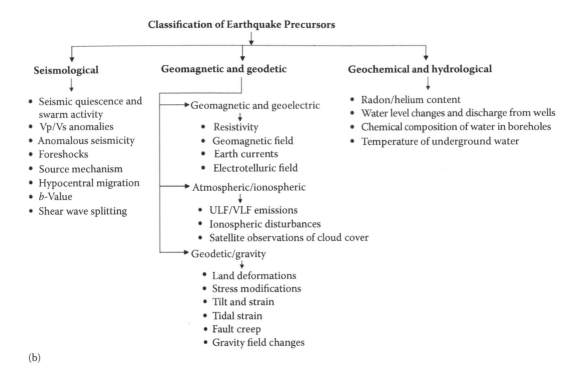

Classification of Earthquake Precursors

Seismological

- Seismic quiescence and swarm activity
- Vp/Vs anomalies
- Anomalous seismicity
- Foreshocks
- Source mechanism
- Hypocentral migration
- *b*-Value
- Shear wave splitting

Geomagnetic and geodetic

Geomagnetic and geoelectric
- Resistivity
- Geomagnetic field
- Earth currents
- Electrotelluric field

Atmospheric/ionospheric
- ULF/VLF emissions
- Ionospheric disturbances
- Satellite observations of cloud cover

Geodetic/gravity
- Land deformations
- Stress modifications
- Tilt and strain
- Tidal strain
- Fault creep
- Gravity field changes

Geochemical and hydrological

- Radon/helium content
- Water level changes and discharge from wells
- Chemical composition of water in boreholes
- Temperature of underground water

(b)

FIGURE 9.3 (CONTINUED)   (b) Group of precursors into three broad categories, namely, seismological, geomagnetic and geodetic, and geochemical and hydrological.

inhibit application of these precursors to real-time prediction. The identification of such factors has enabled us to launch a more organized phase of monitoring where the focus is on integrated studies using multiple parameters.

9.4 PROGRESSION PATH OF EARTHQUAKE PRECURSORY RESEARCH IN INDIA

In India, the initiation of earthquake precursory research closely followed the global quest to isolate and characterize the nature of precursors in different classes. Like elsewhere, in the early stages precursory research was mostly confined to seismological parameters. In the mid-1980s, a national programme on seismicity and seismotectonics provided an umbrella to initiate precursory studies to allied geophysical, geochemical and hydrological parameters. Rationale to induct allied parameters as possible precursors was motivated by the dilatancy–diffusion model, which provided a likely physical mechanism for their generation (Scholz et al. 1973). Despite some convincing field examples, their select appearances, compounded by the lack of any clear relation of the amplitude and duration of the recorded precursors with the magnitude and epicentre distance of the impending earthquake, prevented their application in real-time forecasting. However, a successful forecast of an earthquake in the Koyna–Warna region based on the early identification of a nucleation pattern in the hypocentres of precursory swarms provided a major impetus for the short-term prediction programme (Gupta et al. 2005). With a purpose to improve physical models and capabilities for short-term earthquake forecasting, the Ministry of Earth Sciences, government of India, launched a national programme on earthquake precursors. The establishment of Multi-Parameter Geophysical Observatories (MPGOs) to generate and integrate interdisciplinary parameters is a result of this drive. The growth profile of earthquake precursory studies in India has recently been discussed by Verma and Bansal (2012). In this chapter, we present the synthesis of the results from the early leads and the emerging new developments, to draw a road map for future studies.

9.4.1 Early Leads

9.4.1.1 Seismological Precursors

Reliable earthquake catalogues for the Indian subcontinent are available for a little over the past hundred years. These catalogues have been cautiously used to investigate long-, medium- and short-term seismological precursors.

9.4.1.1.1 Long-Term Precursors

At the plate collision boundaries, where strain energy builds up due to plate motions, the probability of the occurrence of large or great earthquakes is higher in loaded segments, called seismic gaps, than in segments which have unloaded stored strain energy in the form of large or great earthquakes. Demarcation of such seismic gaps is a significant step towards long-term precursors to future large and great earthquakes in a time frame of 10–100 years (Fedotov 1965). Critical analysis based on the energy released, microearthquake activity and spatial distribution of great earthquakes along the Himalayan arc (Figure 9.1) has allowed the identification of three well-known seismic gaps: the Kashmir Gap (west of the 1905 Kangra Earthquake), the Central Himalaya Gap (embedded between the 1905 Kangra and 1934 Nepal–Bihar earthquakes) and the Assam Gap in northeast Himalaya bounded between the 1934 Nepal–Bihar and 1950 Assam earthquakes (Khattri and Tyagi 1983; Srivastava et al. 2013b). But it still remains to be established whether seismic gaps have a higher occurrence probability for great earthquakes than adjoining areas, particularly when the length of the earthquake catalogue used to demarcate gaps is much smaller than the recurrence interval of 300–500 years for great earthquakes in the Himalaya (Molnar and Pandey 1989). This becomes much more critical along the Himalayan arc when all four great earthquakes that formed the rationale for defining the seismic gaps occurred in a very short span of less than 53 years (1897–1950). GPS measurements introduced on the Indian scene have greatly refined our estimates on the plate motion, enabling more precise estimates on the rate of strain accumulation due to ongoing plate convergences. GPS-based estimates of accumulated strains support that numbers of Mw ~8 earthquakes are imminent along the Himalayan arc. However, interpretation of the GPS data does not suggest their size and time of occurrence, and whether an earthquake in a particular segment will occur sooner than that in the neighbouring segments (Gupta and Gahalaut 2014). Emerging GPS data, in combination with improved understanding of the tectonics, are used to address the question, what is the magnitude of the largest earthquake likely to occur along the Himalayan arc, a key parameter in realistic hazard assessment (Srivastava et al. 2013a). The GPS-estimated current state of accumulated slip, coupled with the absence of surface rupture or slip in association with great or large earthquakes in the past millennium, has projected that future earthquake filling of the Kashmir gap could be as large as Mw ≥9 (Bilham and Wallace 2005; Schiffman et al. 2013). However, recognizing the smaller width of the locked section of the plates in the Himalaya in comparison with the active subduction zone, together with the segmented nature of the Himalayan arc, led Srivastava et al. (2013a) to surmise that both factors are not conducive to generate earthquakes with Mw >9 anywhere along the Himalayan arc. The proposition that the Himalayan arc is divided into small segments is corroborated by the observation that the rupture extent of the great and major earthquakes is laterally limited by the transverse subsurface ridges extending from the Indian shield into the Himalaya (Gahalaut and Kundu 2012). In addition to the along-strike segmentation of large-magnitude earthquakes, the transverse structural features on the top of the downgoing Indian plate have also been shown to divide the narrow concentrated HSB, defined by the small- and moderate-magnitude earthquakes (Arora et al. 2012a).

9.4.1.1.2 Medium- and Short-Term Precursors

The potential of the seismic gap hypothesis in providing a long-term forecast of the likely location or time of a future large or great earthquake has still not been realized in the Himalaya. The hypothesis that patterns of seismic swarms and/or quiescence precede large earthquakes

(Evison 1982) has been successfully used as a medium-term precursor for real-time forecasts (Gupta and Singh 1986, 1989). There are spatial and temporal characteristics associated with earthquakes of Mw ≥7.5 in the front-line arc of northeast India, bound by 20° N to 32° N latitude and 87° E to 100° E longitude. Gupta and Singh (1986) observed that large earthquakes are generally preceded by well-defined patterns of increased seismicity in the form of swarms, which is followed by a quiescence just before the main earthquake strikes. They found that the main shock magnitude (*Mm*) has a correspondence with the magnitude of the largest event (*Mp*) in the swarm, and the time interval (*Tp*) between the onset of the swarm and the occurrence of the main shock in days. More significantly, they recognized an area in the vicinity of the India–Burma border where an earthquake swarm has already occurred and the region was experiencing a precursory quiescence. They used these findings to issue a forecast that an earthquake of Mw 8 ± 1/2 could occur in an area bound by 21° N and 25.5° N latitude and 93° E and 96° E longitude well before the end of 1990. The forecast came true with the occurrence of a Mw 7.3 earthquake on 6 August 1988. The epicentral coordinates for this earthquake were 25.149° N, 95.127° E, which is located within the proposed region (Figure 9.4) of forecast (Gupta and Singh 1989). Despite this remarkable success, there is no other example, to our knowledge, where the phenomena of swarm and quiescence have been used for forecasting earthquakes in real time either from northeast India or elsewhere from the Himalaya. However, seismicity data in association with certain Mw >6 earthquakes have been examined to see whether there exist any medium-term precursory patterns. A well-defined couplet of swarm and quiescence extending from 1963 to 1968 and from 1968 to 1975, respectively, was observed before the Mw 6.8 Kinnaur Earthquake, Himachal Himalaya, of 19 January 1975 (Gupta and Gahalaut 2014). Such a prolonged combination of accentuation and quiescence was not conspicuous, but a well-developed short-lived precursory quiescence was observed before the

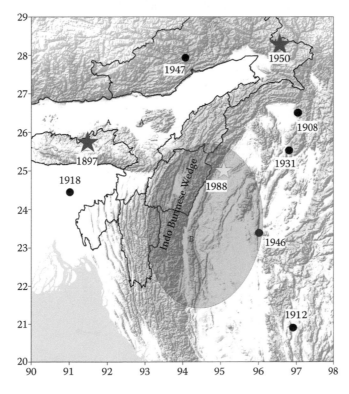

FIGURE 9.4 Map of the northeast India region indicating earthquakes of Mw ≥7.5 since 1897 (filled circles). The elliptical area (dots) shows the preparation zone for a Mw 8 ± 0.5 earthquake identified by Gupta and Singh (1986). The earthquake of Mw 7.3 in the entire region occurred on 6 August 1988 (star).

Mw 6.8 Uttarkashi Earthquake of 19 October 1991 and the Mw 6.8 and 6.6 Chamoli Earthquake of 26 March 1999 in central Himalaya (Lyubushin et al. 2010). Also, a decreasing trend in 'b'-value for minimum ~8 months, as a medium-term precursor, was observed prior to the 1999 Chamoli Earthquake (Lyubushin et al. 2010). Other than these attempts to search long- and medium-term precursors, no clear short-term precursory anomalies were reported from the Himalayan region, but some good success with short-term prediction has been achieved in the Koyna region, Maharashtra.

9.4.2 Geophysical and Geodetic Precursors

9.4.2.1 Geomagnetic and Geoelectric Precursors

Stress-induced changes in the magnetization of rocks, piezomagnetic effect, opening of cracks and stress-induced flow fluids into the dilatant zone of an impending earthquake alter the electrical resistivity of the medium enough to produce geomagnetic and geoelectric changes (Dobrovolsky et al. 1989). The magnetic, electric and electromagnetic (EM) precursors represent a class of precursors that are claimed to have played a key role in successful prediction.

The early geoelectric observations were initiated in the northeast Indian region by Nayak et al. (1983). They measured resistivity, self-potential, earth current, spring discharge and water levels in a dug well at a site near Shillong and observed the precursory changes in resistivity from 7 to 20 days prior to many earthquakes, the 9 March 1980 earthquake (Mw 4.2) in particular (Figure 9.5). The observed precursory changes may be due to the background variations associated with the hydrological effects influenced by rainfall (Guha 1985). However, during the dry season, changes recorded in association with earthquakes could be precursory signals. The changes reported in resistivity were found to be proportional to the distance of hypocentres of earthquakes. Similarly, early attempts to search earthquake-related changes in the geomagnetic field in India were made by Arora and Singh (1979). Reversal in the secular trend in the magnetic field components approximately 3–5 months

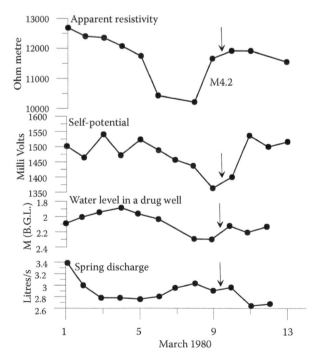

FIGURE 9.5 Temporal changes in apparent resistivity, self-potential, water level in well and rate of spring discharge in relation to the earthquake of 9 March 1980 near Shillong, northeast India. (Modified after Nayak, P. N. et al., *Geoexploration*, 21, 137–157, 1983.)

before the 1963 Koyna Earthquake (Mw 6.3) was observed. Later, as a part of the National Program on Seismicity and Seismotectonics, geomagnetic repeat measurements were carried out along a profile cutting across the major thrusts, as well as high-seismicity zone of the central Himalaya (Arora and Singh 1992). Processing techniques adopted were adequate to isolate anomalous trends exceeding more than 2 nT/year over the regional average, but no significant perturbation was deciphered, probably because even no modest-magnitude earthquake (Mw >5) occurred close to the profile during the period of repeat campaigns between 1988 and 1992. Meanwhile, geomagnetic repeated surveys were also undertaken in the Koyna region in Maharashtra, affected by reservoir-triggered seismicity. Arora (1988) carried out four campaigns of total magnetic measurements at 80 benchmarks around the reservoir between 1978 and 1980. Examination of spatio-temporal evolution of the magnetic field with respect to phase I (May 1978) show a well-developed anomaly pattern for phase III (May 1980) with symmetric positive and negative values; the line of zero value coincided with the Koyna River Fault (Figure 9.6). The spatial distribution maps for the other two phases (January 1979 and October 1980) show a weak anomaly. Demonstrating the lack of any dependence of magnetic patterns on the water level in the reservoir lake, Arora and Singh (1992) suggested that magnetic anomaly patterns for phase III included premonitory magnetic signals of the three earthquakes of Mw >4 that occurred within 20 days in September 1980 in the Koyna region and explained the residual magnetic field pattern in terms of electrokinetic effect. They further suggested that the water diffusion required to initiate electrokinetic phenomena might have occurred along the north-south fault, running parallel to the Koyna River, in response to the high pore pressure existing below the reservoir. The geomagnetic measurements, preferably in continuous mode, could provide an effective mode of detecting precursory signals of earthquakes.

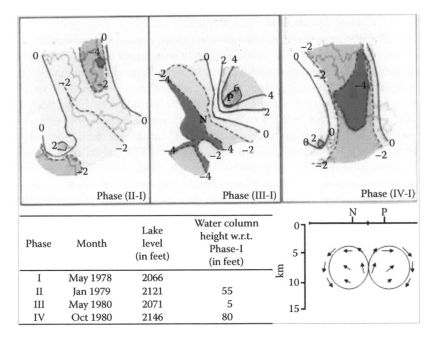

FIGURE 9.6 Contour plots of residual changes in total geomagnetic field around the Koyna reservoir. The plots are shown for three phases with respect to the first phase. The lower panel gives the water level in the reservoir during the four phases. The middle panel depicts the schematic current flow along the KRFZ to explain the symmetric positive–negative anomaly pattern. (Modified after Arora, B. R., in *Earthquake Prediction, Present Status*, ed. S. K. Guha and A. M. Patwardhan, Poona University Press, Pune, 1988, 53–62; Arora, B. R., and Singh, B. P., in *Himalayan Seismicity*, ed. G. D. Gupta, Memoir No. 23, Geological Society of India, Bangalore, 1992, 223–263.)

9.4.2.2 Atmospheric and Ionospheric Precursors

The anomalous EM emission in the ultra-low-frequency (ULF) band (0.001–10 Hz), believed to be emanating from within the focal zones, has received wide attention as a promising short-term precursor from the pioneering work of Hayakawa et al. (1996) and Molchanov et al. (2004). Alternative physical mechanisms, including the deformation or displacement of conductive blocks, fluid flow and the microfracture electrification of crustal materials during elastic straining, are invoked to explain the widely reported pre- and/or co-seismic EM emission (Dudkin et al. 2010 and references therein). The emanating seismoelectromagnetic emissions attenuate very little in the crustal material, and therefore they propagate through the interior of the crust without attenuation and reach the level of ionospheric heights, thereby facilitating their large-scale detection by satellite-based sensors (Hattori et al. 2004; Parrot 2007). This is a unique advantage of EM precursors over the other kinds of precursors. These upward-propagating EM waves are also considered to interpret the often reported anomalies in ionospheric parameters, for example, F-region electron density profiles, GPS-derived total electron contents (TECs) and very low-frequency (VLF) wave recording, using a SoftPAL receiver and automatic whistler detector (Molchanov and Hayakawa 1995; Liu et al. 2002).

On the Indian scene, early ULF magnetic field emissions associated with earthquake events were monitored by Singh et al. (2003, 2005). They examined the amplitude of EM emission in the background of solar flare and magnetic storm activity during the month of October 2005 and found a negative correlation with these events (Kushwah et al. 2007). An interesting result obtained for the 8 October 2005 Kashmir Earthquake (Mw 7.3) included enhanced EM emission, first 10 days before, during 27 and 29 September, and 3 days prior, on 5 October 2005 (Figure 9.7). The precursory characteristics of the signal were confirmed by statistical analysis of the data for a period of 15 days before and 15 days after, from 17 September 2005 to 29 October 2005, by employing a mean (m) and mean and standard deviation ($m = 3$) approach.

Dual-frequency GPS signals at the low-latitude ground station Varanasi were used to study the ionospheric perturbations in TEC associated with nine major earthquakes (Mw >5) occurring within a radius of 2000 km around Varanasi, India. A monthly median of the TEC and associated interquartile range (IQR), upper bound (UB) and lower bound (LB) were utilized as a reference for identifying abnormal signals during all nine earthquakes. It has been observed that the earthquake-related changes in the ionospheric TEC are seen in the time window of 5 days prior to the earthquake. An increase in ionospheric TEC was registered 2 days before the occurrence of the main shock, while a decrease in the ionospheric TEC prevailed 3–5 days prior to the earthquake. Variations in the disturbance storm-time (Dst) index, Ap index (that is daily average level for geomagnetic activity) and auroral electrojets (AE) index during these periods were also examined, which clearly show a geomagnetic quiet period (Priyadarshi et al. 2011). The extremely low-frequency (ELF) and VLF spectra observed from the DEMETER microsatellite above the epicentre of the earthquake lend support to the ionospheric anomalies prior to the earthquakes, which corroborates that such EM emissions are more intense a few days prior to the main seismic shock (Priyadarshi et al. 2011). Pronounced changes in the various surface and atmospheric parameters a few days prior to various earthquakes provide strong evidence of lithosphere–atmosphere–ionosphere coupling.

9.4.2.3 Thermal Anomalies

It has been suggested that stress built up due to seismic activity and associated subsurface degassing might create changes in the thermal regime prior to an earthquake event, and if these changes are detected, they can provide very important information about an impending earthquake (Dey and Singh 2003). The study of thermal anomalies vis-à-vis earthquake occurrence was facilitated by the establishment of the Satellite Earth Station at the Indian Institute of Technology Roorkee (IITR). The thermal regime during the few major earthquakes of January 2001, Bhuj (India); May 2003, Boumerdes (Algeria); and December 2003, Bam (Iran) were investigated using

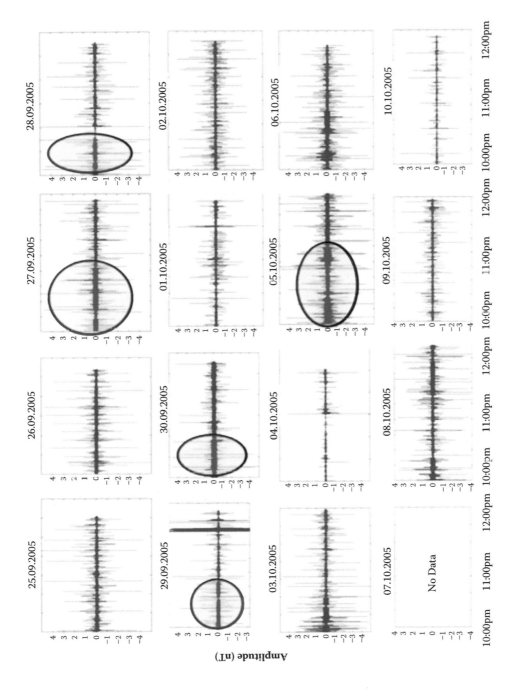

FIGURE 9.7 Time record of ULF emission in vertical magnetic component (Bz) from 25 September 2005 to 10 October 2005 showing the enhancement prior to the 8 October 2005 Kashmir Earthquake in the northwest Himalayan region. (Modified after Kushwah, V. et al., *J. India Geophys. Union*, 11(4), 197–207, 2007.)

National Oceanic and Atmospheric Administration (NOAA) Advanced Very High Resolution Radiometer (AVHRR) thermal datasets. Significant thermal anomalies with a rise in temperature of about 3°C–8°C in the vicinity of the epicentres were observed a few days before the earthquake but started to drop and vanished within a couple of days (Saraf and Choudhury 2003, 2005). In the case of the 26 January 2001 Bhuj Earthquake, the temperature increased to a maximum (28°C–31°C) on 23 January 2001, just 3 days before the Gujarat Earthquake (Figure 9.8). This rise in temperature was about 5°C–7°C higher than the usual temperature of the region. To ascertain that the rise was due to the earthquake on 26 January 2001, the trend of the available weekly average temperature data (1951–1980) was analyzed and found to be contrary to the observed trend for the past 30 years. This study suggested that changes in the thermal regime in the epicentral area, prior to an earthquake event, if detected, can provide very important clues about an impending earthquake event.

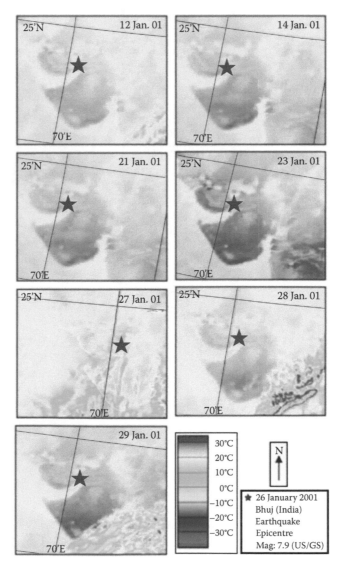

FIGURE 9.8   Geothermal anomalies in the form of the enhanced temperature of 28°C–31°C on 23 January 2001, just 3 days before the 26 January 2001 Bhuj Earthquake in Gujarat, Western India. (Modified after Saraf, A. K. and Choudhury, S., *J. India Geophys. Union*, 9(3), 197–207, 2005.)

9.4.3 Geochemical and Hydrological Precursors

In response to the stress–strain build up during the earthquake preparatory phase, rocks undergo a stage of dilatancy through the development of microcracks, which undergo rapid expansion by opening new fissure zones. The newly opened cracks perturb the steady-state hydrological regime by increased underground water flow into the focal zone of the impending earthquake. This, combined with accompanying modulation of pore pressure, leads to water level changes. The release of gases like radon and helium through the freshly opened surface can explain the anomalous radon concentrations in association with seismic activity. Enhanced water–rock interactions recharge and discharge aquifers with varying chemical constituents. This can explain the geochemical and hydrological precursors, as well as the change in water chemistry in association with earthquake occurrences.

Following the robust field evidence of abrupt variations in radon concentration prior to the 1966 Tashkent Great Earthquake (Ulomov and Mavashev 1967), radon is the most widely searched precursor. In line with global trends, continuous monitoring of radon concentrations in soil samples and groundwater was carried out at selected locations in the northwest Himalaya, and significant precursory changes were observed from a few hours to several days before many earthquakes.

- A radon anomaly associated with the Mw 5.1 Chamba Earthquake of 24 March 1995, in both soil and groundwater, was reported by Virk et al. (1995). The radon concentration abruptly rose to a higher value, up to 7600 counts/day, about 7 days before the event; afterwards, the radon concentration showed a gradual decrease, with pre-earthquake values after the earthquake. In the case of the Mw 6.8 Uttarkashi Earthquake of 20 October 1991, the radon anomalies were recorded about a week before the earthquake occurred (Figure 9.9a). Virk et al. (2001) again observed a helium/radon anomaly 9 days before the Mw 6.6 Chamoli Earthquake that occurred on 29 March 1999 (Figure 9.9b).
- In northeast India, an anomalous large-scale radon emission, a striking high gamma dose rate and a significant increase in He/CH_4 ratio in the thermal spring emanations were observed at Bakreswar, West Bengal, prior to the 2004 great Sumatra Earthquake (Mw 9.3) of 26 December 2004 (Das et al. 2005). They observed the He/CH_4 ratio to be a better index of seismic variations than either He or CH_4 taken separately. On 17 and 18 December 2004, the radon concentration increased sharply to a peak value of 913.61 kBq/m (Figure 9.9c). A remarkable increase in the He/CH_4 ratio to a value of 0.18 and an amplitude swing in He/CH_4 were observed during 20–22 December 2004. They observed that these fluctuations may be due to the change in the thermodynamical state of the fluid reservoir and affect the distribution of the stress field, which in turn leads to enhanced permeation of deep earth volatiles, and helium in particular.

9.5 ORGANIZED APPROACH TO PRECURSORY STUDIES

The successful examples of medium- and short-term predictions in real time based on seismic quiescence, nucleation patterns and the detection of some precursors in radon, geomagnetic activity, resistivity, ULF emissions and geothermal changes, encouraged us to step ahead and adopt an integrated approach for real-time earthquake forecasting. The most significant lead in this direction is the establishment of MPGOs in different seismically active areas of the country (Figure 9.1). Under this scheme, five MPGOs, namely, Ghuttu, Shillong, Imphal, Tejpur and another in Bhuj (by the Institute of Seismological Research), Gujarat, have been established. Here, we discuss results from Ghuttu, the first one to be established and where data of 6–7 years are available for isolating precursory signals. The precursory research at Koyna is not christened under the umbrella of coordinated MPGO studies; however, the data from long-term continuous monitoring of multiparameters have been examined for reliable precursors.

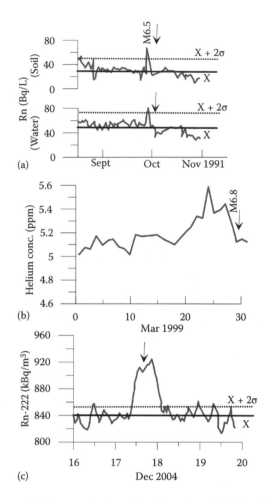

FIGURE 9.9 (a) Radon anomaly in groundwater and soil gas 1 week before the October 1991 Uttarkashi Earthquake (Mw 6.5) recorded in northwest Himalaya. (Modified after Virk, H. S. et al., *J. Geodyn.*, 31, 201–210, 2001.) (b) Helium/radon anomaly 9 days prior to the Chamoli Earthquake (Mw 6.8) of 29 March 1999. (Modified after Virk, H. S. et al., *J. Geodyn.*, 31, 201–210, 2001.) (c) Variations of radon-222 a few days prior to the Great Sumatra Earthquake of 25 December 2004. (Modified after Das, N. K. et al., *Curr. Sci.*, 89(8), 1399–1404, 2005.)

9.5.1 Multi-Parametric Geophysical Observatories: Establishment and Observations

9.5.1.1 Ghuttu Observatory in Northwest Himalaya

The Himalaya is one of the most active seismic intercontinental regions, where devastating earthquakes result due to the continued continent–continent collision between India and Asia. Recognizing the large seismic hazard of the Himalaya, the first Indian MPGO was established at Ghuttu (30.53° N, 78.74° E), Garhwal Himalaya, Uttarakhand (Figure 9.10). Longitudinally, Ghuttu is located in the central Himalaya seismic gap (Khattri and Tyagi 1983), bounded by the 1905 Kangra (Mw ~7.8) Earthquake on the west and the 1934 Bihar–Nepal (Mw ~8.3) earthquake on the east, where accumulated strains are estimated to be large enough to produce great earthquakes (Bilham et al. 2001). The MPGO is located in a narrow HSB, which is best seen as the locked section of the downdip edge of the seismically active detachment (Banerjee and Burgmann 2002), and structurally represents the ramp structure between the seismically active detachment to the south and the aseismically slipping detachment to the north (Pandey et al. 1995; Arora et al. 2012b).

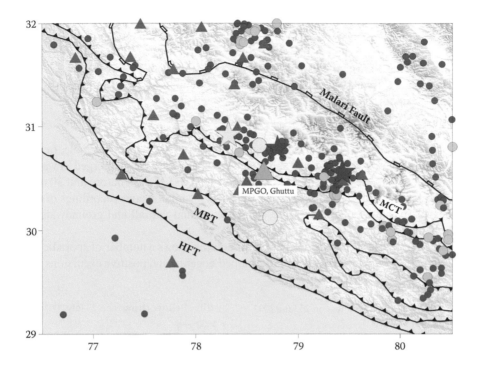

FIGURE 9.10   Location of the MPGO, Ghutu and support magnetic stations (Bhatwari and Pipadali) is shown by peach-coloured triangle and yellow circles, respectively. The 1991 M6.3 Uttarkashi and 1991 M6.8 Chamoli Earthquakes are shown by red stars. Backup seismic network established around the MPGO are represented by pink triangle and distribution of earthquake epicentres recorded between July 2007 and December 2010 are shown by closed circle. Epicentre of *M* 4.9 Kharsali Earthquake (yellow star), largest since the operation of the MPGO in 2007, is also plotted on the map.

In the region, the 1991 Uttarkashi and 1999 Chamoli Earthquakes, both Mw >6, occurred, which exhibited the well-developed pattern of quiescence and accelerated seismicity that invariably precedes the large earthquakes (Lyubushin et al. 2010). Figure 9.10 shows the locations of these two large earthquakes in relation to the background seismic activity in the central seismic gap region of the Himalaya, extracted from the India Meteorological Department (IMD) earthquake catalogue for the period June 1998–June 2011. The Ghuttu observatory is equipped with a superconducting gravimeter (SG), Overhauser magnetometer, triaxial fluxgate magnetometer, ULF band search coil magnetometer, radon data logger and water level recorders, and is also supported by the dense network of GPS and broadband seismographs (BBSs).

The MPGO at Ghuttu became fully functional in April 2007, and during 2007–2014, more than a thousand local earthquakes (1.8 < Mw > 5) were detected within the seismic network (Wadia Institute of Himalayan Geology 2014). Since no earthquake of Mw >5 has occurred within the radius of 200 km, multiple geophysical datasets for some small-magnitude earthquakes (Mw >4) have been processed with different processing strategies to isolate weak precursory changes associated with earthquakes. The Kharsali Earthquake (Mw 5.0) is the largest that has occurred since the establishment of the observatory (Arora et al. 2012b). Details of the observed precursory signals are discussed next.

9.5.1.1.1   Kharsali Earthquake of 22 July 2007 (Mw 5.0)

A moderate earthquake of Mw 5.0 occurred on 22 July 2007 at 23:02:13.22 Universal Time Coordinated (UTC) at an epicentral distance of 60 km from the MPGO, Ghuttu. The earthquake, with its epicentre at 30.91° N, 78.31° E, was estimated to be located at the focal depth of 15 km.

The fault plane solution favoured the role of reverse fault movement with a significant strike-slip component in the generation of the earthquake (Kumar et al. 2012). The focal mechanism and depth support the hypothesis that the earthquake resulted due to thrust movement along the blind basement thrust, marking transition between the downgoing Indian plate and the overriding wedge of the Himalaya. The earthquake was felt in several states of northern India and was also followed by several aftershocks of magnitudes less than 3.3.

9.5.1.1.2   Radon Anomalies

Temporal changes in radon (Figure 9.11) were measured in the 68 m deep borehole, first by a sensor hanging in the air column at 10 m depth and then by a separate sensor submerged in water at a depth of 50 m. The variations are shown for the duration of 61 days, 30 days before and 30 days after the occurrence of the Kharsali Earthquake event. In addition, simultaneous recordings of environmental variables, such as atmospheric pressure, temperature, and rainfall and groundwater level, were made at the sampling interval of 15 minutes.

Examination of radon variations in a closed-air column shows a number of sporadic fluctuations, but two sharp bell changes marked by well-developed negative and positive excursions were distinct

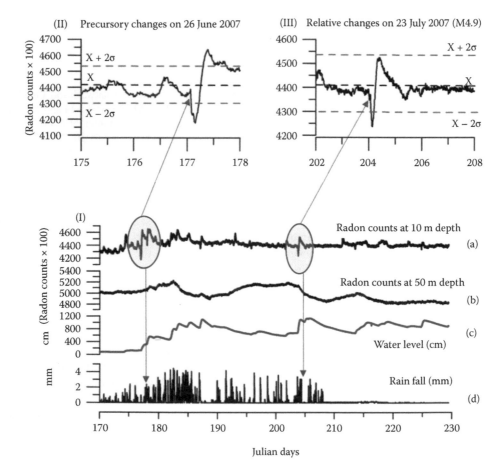

FIGURE 9.11   Radon concentration measured in (a) air column and (b) water column, together with (c) water level in a deep borehole and (d) rainfall recorded at the MPGO, Ghuttu, during a 60-day (170–230 day numbers) interval of 2007. Bell-shaped anomaly in radon with a sharp negative–positive impulse registered 22 days before and on the day of the Kharsali Earthquake on 22 July 2007. (Modified from Choubey, V. M. et al., *Sci. Total Environ.*, 407(22), 5877–5883, 2009.)

features around the Kharsali Earthquake (Choubey et al. 2009). The magnitudes of both positive and negative show an anomalous pattern, statistically significant, as extreme values deviated by more than two standard deviations from the seasonal mean. This was the first anomalous pattern observed on 26 June 2009, approximately 23 days before the occurrence of the Kharsali Earthquake, whereas the second anomalous pattern was recorded a few hours before the event. In comparison, the radon concentration in water shows slow and smooth changes that correlate better with water level fluctuations in the borehole. Considering sharp radon changes in the air column to be characteristic of the precursory signal to an earthquake, an empirical relation incorporating the observed amplitude of the radon peak, decay rate and average prevailing values of radon concentration estimated the magnitude of the impending earthquake as close to 4.6–4.7, in fair agreement with the observed magnitude of the Kharsali Earthquake (Mw 5.0) on 22 July 2007. While this example reinforced the physical rationale that there exists an association between stress build up during an earthquake cycle and radon flux, further validation was emphasized in view of the fact that some of the recorded sharp changes in radon intensity coincided with intense rainfall, groundwater fluctuations and fluctuation in pressure and temperature (Arora et al. 2012b). In a recent communication, Kumar et al. (2017) report anomalous radon gas emission observed in soil and water at the Ghuttu MPGO for the Mw 7.8 Nepal earthquake of 25 April 2016. The earthquake occurred ~600 km to the east of the observatory, and radon emission measured at 50 m depth using a gamma probe showed a prominent pre-seismic temporal change similar to that of soil radon.

9.5.1.1.3   Gravity Field Variations

On the premise that the opening of cracks and the influx of fluids in the dilatant zone of an impending earthquake are expected to alter the mass distribution during the earthquake build up cycle, and hence should be reflected in the gravity field, the measurements of the time-varying gravity field were recorded using an SG. The recorded gravity observations are dominated by solid earth tides and contain the influence of atmospheric pressure, hydrological fluctuations inducted by transient rainfall sequences and accompanying annual cycles in groundwater levels. However, a careful scrutiny of the residual gravity data sampled every second shows an unambiguous co-seismic jump of 5.2 µGal in relation to the Kharsali Earthquake (Mw 4.9) of 22 July 2007 (Figure 9.12). Step gravity jumps are a persistent feature recorded by the network of SGs placed in different parts of the world. The static gravity peak variations recorded in association with a large earthquake could be related to the simple dislocation on the fault (Imanishi et al. 2004). Such changes in gravity fields may correspond to the volumetric strain released during an earthquake. It can be checked that the sign of the observed gravity jump is opposite that of the updip displacement on the source, as evidenced by the fault plane mechanism. Further, based on the observation that observed gravity variations are co-seismic in the sense that they occur when earthquake waves arrive and pass underground where the SG is deployed, the sign and magnitude of the gravity change may be proportional to the volumetric strain released during an earthquake.

9.5.1.1.4   Seismomagnetic Signals

To examine the seismomagnetic signals arising due to the modulation of stress-induced perturbation in the magnetization of the rocks, the total magnetic field (F) is recorded with the help of an Overhauser magnetometer having a higher sensitivity (0.1 nT) and resolution (0.01 nT) than the traditionally used proton precision magnetometers. Even after using a highly sensitive instrument, the isolation of weak seismomagnetic signals remains challenging, as recorded variations are dominated by natural regular quiet daily variations, magnetic storms and substorms, and so forth. Since magnetic variations associated with different sources have different characteristic waveforms, powerful principal component analysis (PCA) has been successfully applied to isolate seismomagnetic signals from the strong variations of solar-terrestrial origin (Hattori et al. 2004; Arora et al. 2012b). The method takes advantage of external variations, especially during the local night hours, having their origin in magneto-spheric processes, at a group of stations separated only by tens of

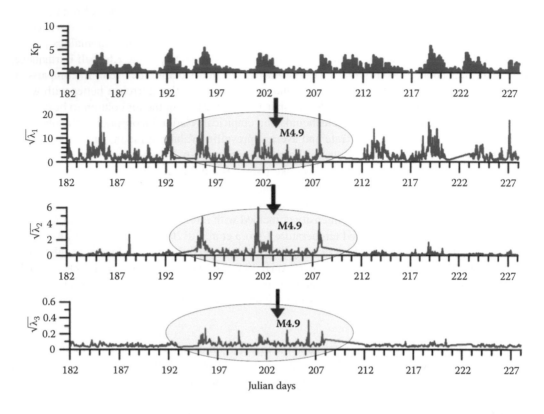

FIGURE 9.12 Co-seismic gravity jump of 5.2 μGal in association with the 22 July 2007 Kharsali Earthquake (Mw 4.9) in Uttaranchal Himalaya. (Modified after Arora, B. R. et al., *Curr. Sci.*, 103(11), 1286–1297, 2012.)

kilometres, which would be identical, and hence global fields with a common waveform can be easily isolated by processing data of two or three stations simultaneously. Stress-induced changes, being local in nature, will produce site-specific waveforms. To exploit this merit of PCA, in addition to Ghuttu, Overhauser magnetometers are being operated at two additional sites, Bhatwari and Pipaldali. Figure 9.13 shows the plot of the square root of magnitude of the first three eigenvalues (λ). Since eigenvalues are a measure of the power of the measured signal, the square root plot shows the temporal variability of the amplitude of the principal waveform. As expected on physical considerations, the time variation of the first eigenvalue (λ1) follows quite closely the global geomagnetic index, *Kp*, indicating control of the magneto-spheric processes. The variability of the second and third eigenvalues is invariably independent of the global geomagnetic activity. It is noted that plots of λ2 and λ3 depict strong variability during the time interval approximately 1 week before and 1 week after the Kharsali Earthquake of 22 July 2007. These higher-order eigenvalues, λ2 and λ3, attain higher amplitude at the Bhatwari station, located on the hanging wall of the source fault. Reverse moments at a depth of 15 km on such a fault are considered to be the source mechanism for this earthquake. The region north of the MCT has also witnessed tectonic history causing granitoid intrusion associated with Tertiary magmatism. The petrologic and magnetic measurements have shown that the paramagnetic minerals biotite and muscovite determine the bulk of the magnetic properties of granitoids, where a single-domain titano-magnetic mineral is the primary carrier of magnetization (Sharma et al. 2011). In the hypocentral depth of the Kharsali Earthquake (~15 km), it has been emphasized that in the thrust domain, the shear heating resulting in response to the accumulating stresses may locally bring the temperature (200°C–400°C) close to the Curie temperature of the titano-magnetic mineral, that is, in the range of 200°C–400°C. It is known that thermal

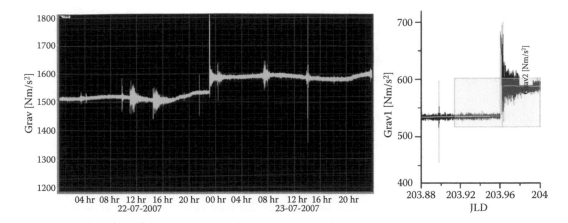

FIGURE 9.13 Time variation of three principal components (λ_1, λ_2, λ_3) in the total geomagnetic field obtained from recordings at the MPGO, Ghuttu, and support stations, and Bhatwari and Pipaldali. The λ_1 control of global geomagnetic activity (Kp) index, where λ_2 and λ_3 depict sharp transient changes associated with the occurrence of the Kharsali Earthquake of 22 July 2007. (Modified after Arora, B. R. et al., *Curr. Sci.*, 103(11), 1286–1297, 2012.)

agitation of magnetic minerals in rocks, close to the Curie temperature, can destroy the alignment of magnetic grains, which may be reflected as small-scale perturbation of the short-period flutuations (Figure 9.13). Thermally excited rocks return to normal conditions after the release of strains following the earthquake (Arora et al. 2012b). That small-scale perturbation in short-period geomagnetic fluctuations is thermally excited can be independently corroborated by the often reported thermal anomalies in the form of abrupt changes in surface temperature of the order of 3°C–7°C occurring around 1–13 days prior to an earthquake and disappearing a few days after the event (Dey and Singh 2003).

9.5.2 Multiparametric Geophysical Observations in the Koyna Region

The multiparameter data, including seismological, hydrological, hydrogeochemical, GPS-based geodesy and EM emission in ULF bands, have been investigated in search of precursory signals. In recent times, near-real-time monitoring of earthquakes has enabled a short-term scientific forecast based on the phenomenon of the nucleation of smaller earthquakes, as well as foreshocks. The proposed method is based on the premise of detecting and observing a nucleation zone of foreshock cluster, which increases over a time window of 50–100 hours prior to a Mw ~4–5 earthquake. Deepening of such a nucleation zone helps to locate a hypocentre of a future moderate-size earthquake at the base of the seismogenic layer (Gupta et al. 2005). Closely spaced seismic networks set up in the Koyna–Warna region have helped in forecasting some of the recent earthquakes of Mw ≥4 in the Koyna region, on 13 November 2005 (Mw 4.0), 26 December 2005 (Mw 4.2), 17 April 2006 (Mw 4.7), 14 October 2007 (Mw 3.4), 2 July 2008 (Mw 3.0) and 12 December 2009 (Mw 5.1), based on the nucleation process. These successful forecasts have been made since 2005, raising hopes for possible earthquake prediction in the future.

Precursory changes of a premonitory nature, for instance, in groundwater level fluctuations in confined aquifers and in hydrogeochemical concentrations, have been observed in the Koyna–Warna region (Chadha et al. 2003; Gavrilenko et al. 2010; Reddy et al. 2010). The hydrological precursory study, which was initiated about 21 years ago, monitors water level fluctuations in 21 boreholes drilled around the seismically active source volume of the Koyna–Warna region. This study resulted in the identification of pre- and co-seismic precursors for earthquakes of Mw >4 occurring in the region. Radon gas measurements are being continuously monitored at three sites in the vicinity of the epicentral region. A sudden increase in radon anomaly by a factor of two was observed about 19 days prior to the 14 March 2005 earthquake (Mw 5.1). The continuous monitoring of data has shown an increase in radon concentration, with three peaks observed during 28 August 2005 to 23 October 2005. These three radon anomalies are correlated with earthquake swarms (Mw ≥2) in the vicinity of Warna during 30 August to 2 September 2005, 7–9 October 2005 and 17–19 October 2005.

9.6   KOYNA DEEP BOREHOLE INVESTIGATIONS

Seismogenesis in reservoir-triggered environments has been studied through numerical modelling of stresses and pore pressure variations in three dimensions (Gahalaut et al. 2010). Moment tensor inversion studies for source mechanism and estimation of seismogenic depth using waveform modelling (Shashidhar et al. 2011), tomography and noise correlation studies for the estimation of detailed crustal structures are some other studies that have been undertaken to understand the reservoir trigger mechanism in conjunction with the crustal structure and local tectonics. Another experiment, involving installation of 97 seismic stations in the Koyna area to delineate the three-dimensional velocity structure of the region using mostly passive sources and two active sources, was completed in early 2010 (Dixit et al. 2011). Based on these studies and also in view of continued seismic activity in smaller seismic volume, the Koyna–Warna region is considered a suitable site for deep borehole investigation to comprehend the mechanism of earthquakes in the region and also to characterize the underground fault geometry, physical properties of rocks, hydraulics, fluid composition, heat flow, in situ stress, pore pressure changes in the 'near-field' region of earthquake occurrence and a window to the upper crustal rocks and Deccan volcanic province.

9.6.1   IDEA OF DEEP BOREHOLE DRILLING IN KOYNA

Existing models, to comprehend the genesis of triggered earthquakes, suffer from a lack of observations in the near field. To investigate further, scientific deep drilling and setting up a fault zone observatory at a depth of 5–7 km are planned in the Koyna area. This is expected to lead to a better understanding of the mechanics of faulting, the physics of reservoir-triggered earthquakes and earthquake hazard assessment. The motivation behind deep drilling down to ~7 km depth resulted from the fact that the Koyna region is a classical RTS site in an intraplate setting where earthquakes are shallow focused with focal depths ranging up to 8–10 km. It was felt that such investigations would provide an opportunity to acquire new information about the properties in the deeper parts of the Earth's crust that will serve to constrain several processes, including seismogenesis in intraplate settings and geologic evolution of cratons, apart from understanding the mechanism of reservoir-triggered earthquake genesis (Gupta et al. 2011).

9.6.2   PRELIMINARY INVESTIGATIONS AND RESULTS

Detailed geophysical investigations, for example, broadband seismology, broadband magneto-tellurics, deep resistivity sounding (DRS), controlled-source audiofrequency magneto-tellurics (CSAMT), land-based gravity–magnetics, airborne gravity–gradiometry and LIDAR surveys at the local and regional scales, have been carried out to constrain the subsurface geology, structure and heat flow regime in the Koyna–Warna area that provide critical inputs for the design of the deep borehole observatory. A few shallow (~1.5 km) exploratory boreholes were drilled (Figure 9.14) for the installation of borehole seismometers to help accurately locate earthquakes, constrain the disposition of fault zones and select the most appropriate site for the deep borehole. Seismometers have been deployed in the granitic basement inside four boreholes and are planned in another set of four boreholes. Details of these studies and their results have been discussed by Gupta et al. (2015). The shallow boreholes have penetrated the Deccan Traps and sampled the granitic basement in the region for the first time. Studies on cores retrieved from the shallow boreholes provide new

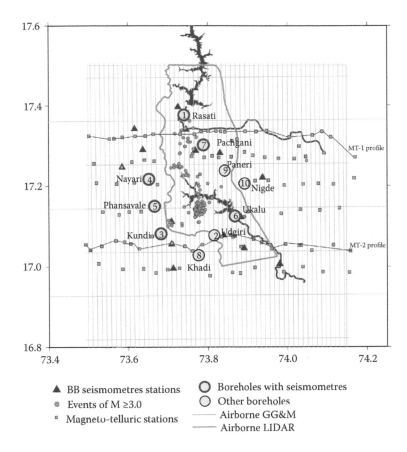

FIGURE 9.14 Deployment plan of the various experiments undertaken in the preparatory phase of the deep drilling programme in the Koyna–Warna region. Earthquakes of Mw ≥3.0 occurred during August 2005 to December 2013. MT-1 and MT-2 indicate the magneto-telluric profiles passing through Rasati in the north and Udgiri in the south, respectively. GG&M represents gravity gradiometry and magnetic data. (After Gupta, H. et al., *Int. J. Earth Sci.*, 104, 1511–1522, 2015.)

and direct information regarding the thickness of the Deccan Traps, the absence of infra-Trappean sediments and the nature of the underlying basement rocks. Temperature estimated at a depth of 6 km in the area, on the basis of heat flow and thermal properties datasets, is not expected to exceed 150°C. Low-elevation airborne gravity gradient and magnetic datasets covering 5012 line km, together with high-quality magneto-telluric data at 100 stations, provide regional information about both the thickness of the Deccan Traps and the occurrence of localized density heterogeneities and anomalous conductive zones in the vicinity of the hypocentral zone. Acquisition of airborne LIDAR data to obtain a high-resolution topographic model of the region has been carried out over an area of 1074 km² centred at the Koyna seismic zone.

Cores recovered from shallow boreholes have provided an opportunity to take up specific geological research problems, including the chronology and palaeoenvironment of the Deccan basalts, the rock magnetics and anisotropy of the Deccan Traps, palaeomagnetic investigations across the 1 km long drill core, the influence of geochemical constraints on microbial diversity, the Ni-Cu sulphide potential of the Deccan basalt, hydraulic continuity in the Deccan Traps and basement rocks, meso- and microstructures in the granitoids, the palaeoenvironmental and palaeoclimatic conditions during Deccan volcanism, petrology, and the compositions and geochronology of the granitic basement. Goswami et al. (2017) measured the rock strength of Archaean basement granitoids recovered from shallow boreholes for the first time. Their main findings include the following:

(1) the rock strength increases linear with increasing confining pressure; (2) the orientations of natural weak planes, such as fractures and fabrics, mostly follow the mineral cleavages; and (3) the large number of low-magnitude earthquakes in the Koyna region during the past few decades may be attributed to low and variable rock strength of the basement granitoids.

9.6.3 DRILLING OF PILOT BOREHOLE

The pilot phase of drilling is a prerequisite for planning the main borehole and setting up the observatory at depth. A pilot bore to a depth of 3000 m, passing through the entire thickness of the Deccan basalt (~1247 m) and ~1753 m into the underlying granitic basement, has been drilled (Roy 2017). The borehole configuration and general litholog are shown in Figure 9.15. It is the deepest borehole drilled in hard crystalline rock formation in India. Cores of basement granite–gneiss from the Koyna pilot borehole (Figure 9.16) show alternate bandings of segregated quartz-ofeldspathic and mafic-rich domains. A suite of downhole geophysical measurements, including the deepest in situ stress tests, have been conducted. These data are critical to designing the fault zone observatory. Repeat measurements over a period of time may provide crucial information on potential temporal changes in rock properties, stress regime and hydrological properties due to ongoing seismic activity in the region (Roy 2017). The core samples recovered from the pilot borehole will also provide an opportunity to examine the microbial life and the processes, in addition to the rock properties.

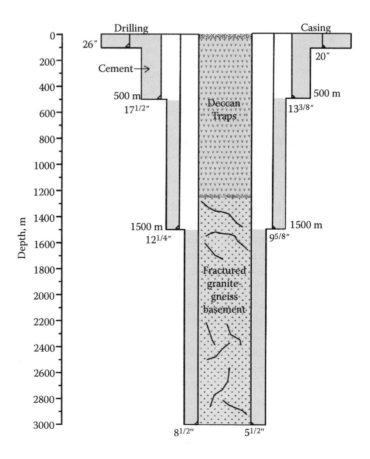

FIGURE 9.15   Schematic design and general litholog of 3 km deep pilot borehole in the Koyna region.

FIGURE 9.16   Cores of basement granite–gneiss from the Koyna pilot borehole showing alternate bandings of segregated quartzofeldspathic and mafic-rich domains.

9.7   EARTHQUAKE EARLY WARNING IN INDIA

9.7.1   PRINCIPLE AND IMPORTANCE OF EEW SYSTEM

The term *Earthquake Early Warning* (EEW) is used to describe a real-time earthquake information system that has the potential to provide warning of approaching strong ground shaking prior to its actual arrival. The principle of EEW is based on the fact that both longitudinal (P) and transverse (shear) waves (S) are generated simultaneously during the earthquake, but P-waves travel at a much higher velocity than the most damaging surface waves. As a consequence, with increasing distance from the place of origin, their arrival times are segregated. If the real-time analysis of the seismometer data can be used to estimate the source parameters of earthquakes, particularly the location and magnitude, a warning with a lead time of a few seconds to a few tens of seconds can be issued on the amplitude and arrival time of more severe ground shakings by the trailing S-waves. Recognizing this critical time lead, seismic switches have been installed in several places, and in the event that the anticipated ground acceleration exceeds a certain prescribed limit, switches turn off critical facilities, for example, nuclear power plants, high-speed trains and gas pipelines. More importantly, the public at large gets a few crucial seconds to come out of buildings, and thus greatly reduce the human loss. Such warning systems, also referred to as earthquake alarm systems, have been implemented in a few select cities and countries, for example, Mexico City, Bucharest, Romania, Istanbul, Taiwan and Japan.

Presently, two types of earthquake early warning systems (EEWSs) are in practice around the world: front detection and regional EEWSs. The former uses observations at a single sensor close to the earthquake source area and estimates the ensuing ground motion at the same site to give early warnings to more distant urban areas (Kanamori 2005). The Seismic Alert System (SAS) in Mexico City, Bucharest (Romania) and Istanbul uses this front detection approach, also referred to as on-site formulation. The regional EEWSs exploit the initial portion of the P-wave at a dense network of seismic stations to estimate the requisite source parameters of the earthquake (location and magnitude) and use this information to predict the more severe ground shakings for the trailing S-wave at given sites of interest (Wu and Kanamori 2005). While regional systems work more accurately, it takes more time to estimate earthquake source parameters as data from several stations are combined. This offset is largely compensated by using modern telecommunication tools, for example, VSat, which transmits data at a much faster speed than the seismic waves. However, both systems have a major drawback of a large 'blind zone' around the epicentre. Since the on-site approach relies on both the P and S phases at the site nearest to the source, the small lead time of the P-wave over the S-wave fails to potentially provide warning at or near the epicentre, that is, blind zone. This drawback can be overcome effectively in the regional approach using the primary wave (P-wave) to estimate source parameters as well as to provide warning. Use of the P-wave from a dense network has the potential to reduce the radius of the blind zone, and potentially provide warning at the epicentre. Many observational parameters have been developed for using the P-wave to assess earthquake hazards (Wu and Kanamori 2005).

9.7.1.1 Experimental EEW System in Northern India

In India, the northern part covering Himalaya is one of the most seismic-prone regions adjoining the Indo-Gangetic Plains (IGP), which has large lateral stretch from west to east and also extends down south to a distance of ~150–200 km reaching up to National Capital Region (NCR) of Delhi. A thick population density, as well as mushrooming industrial houses coupled with poor adherence to earthquake-resistant practices, has substantially increased the seismic vulnerability of the IGP. In the case of a large earthquake in Himalaya, which is rooted some 50–100 km farther north into the Lesser Himalaya, most of places in the IGP can have a lead time of 30–70 seconds before the arrival of damaging seismic waves. Recent studies have shown that the severity of ground motion in several parts of Delhi will be quite high for an Mw 7.5–8 earthquake in Garhwal. Thus, in the event of a large earthquake (Mw ≥7) in Himalaya, the full potential of the EEW system can be attained by giving second life to a large population of northern India, including Delhi. Besides saving lives, the use of EEW in issuing alerts to abort the take-offs and landings of aeroplanes, in shutting down industrial units sensitive to vibration, in warning doctors at the operation table, and so forth, is also gaining importance.

Considering the possible role of EEW system in risk mitigation, the Ministry of Earth Sciences (MoES 2012), government of India, has taken a lead in examining the efficacy of the EEW system in India through the IITR. As a test case, the localized zone of the Garhwal Lesser Himalaya in the northern part of the country has been selected, based on the geodetic data, because it has the

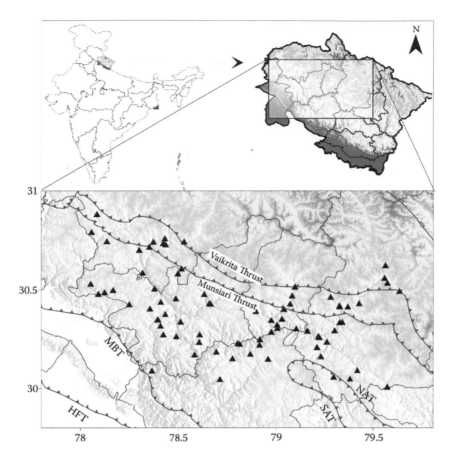

FIGURE 9.17   Location of accelerometers deployed in Uttarkhand Himalaya for the pilot EEW experiment. HFT, Himalaya Frontal Thrust.

potential to generate a great future earthquake. In a selected area of about 100×40 km^2, bounded between the Main Boundary Thrust (MBT) and MCT, 84 state-of-the-art accelerometers were deployed (Figure 9.17). The EEW system is a replica of the one being used in Taiwan. It makes use of the ground motion period parameter τc and the vertical displacement amplitude parameter Pd from the initial 3 seconds of the P waveforms. At a given site, the earthquake magnitude is determined from τc and the peak ground motion velocity (PGV). In this method, incoming ground motion acceleration signals are recursively converted to ground velocity and displacement and a P-wave trigger is constantly monitored. When a trigger occurs, τc and Pd are computed (Wu and Kanamori 2008). The earthquake magnitude and the on-site ground motion intensity are estimated and a warning issued. The system has now been in operation for some time, and smaller events are being used to test and validate it. However, it suffers from poor connectivity and hence still remains a prototype.

9.8   CONCLUSION

Planned science and technology interventions have enhanced our skills to forecast natural hydro-meteorological hazards, saving human lives and properties. Despite some promising leads, the prediction of earthquakes in terms of time, place and magnitude still remains an important goal in minimizing the loss of lives and property. Earthquake precursors that may hold the key to earthquake prediction have been the area of global research for a long time. In the progress path of precursory studies in India, we observe that there are a few success stories where seismic precursors are used to forecast in real time, but their replication to other earthquakes from the same region or to other regions with varying tectonics has inhibited their adoption in a practical prediction programme. There is also growing documentation that precursors do exist in multiple geophysical or geochemical and hydrological parameters, but the sensitivity of each recoded to factors other than geodynamics, restricted their validation and true characterization, and so forth. Recognition of some of these factors has helped us to reorganize and launch a couple of new initiatives, namely, (1) the establishment of the MPGO permits us to cross-validate multiple precursory signals, and thus evaluate the collective value for use in real-time forecasts of earthquakes; (2) drilling a super-deep borehole in Koyna to study the earthquake process in in situ conditions; and (3) the introduction of EEW for the arrival time of destructive surface waves to urban cities to launch an automatic switching off of the hazardous services.

ACKNOWLEDGEMENTS

The authors are thankful to the secretary of the Ministry of Earth Sciences, government of India, for his continued support. A critical review by B. R. Arora helped us to improve the text. Ajeet P. Pandey made some good suggestions. Anup K. Sutar and Renu Bisht helped us modify figures and compile references.

REFERENCES

Arora, B. R. (1988). Tectonomagnetic studies in India. In *Earthquake Prediction, Present Status*, ed. S. K. Guha and A. M. Patwardhan. Pune: Poona University Press, 53–62.

Arora, B. R., Gahalaut, V. K., and Kumar, N. (2012a). Structural control on along-strike variation in the seismicity of the northwest Himalaya. *J. Asian Earth Sci.*, 57, 15–24.

Arora, B. R., Rawat, G., Kumar, N., and Choubey, V. M. (2012b). Multi-parameter geophysical observatory: Gateway to integrated earthquake precursory research. *Curr. Sci.*, 103(11), 1286–1297.

Arora, B. R., and Singh, B. P. (1979). Geomagnetic field precursors associated with earthquakes. *Mausum*, 80, 317–322.

Arora, B. R., and Singh, B. P. (1992). Geomagnetic and geoelectric investigations for seismicity and seismotectonics of the Himalayan region. In *Himalayan Seismicity*, ed. G. D. Gupta. Memoir No. 23. Bangalore: Geological Society of India, 223–263.

Banerjee, P., and Burgmann, R. (2002). Convergence across the northwest Himalaya from GPS measurement. *Geophys. Res. Lett.*, 29, 1652–1655.

Bilham, R., Gaur, V. K., and Molnar, P. (2001). Himalayan seismic hazard. *Science*, 293, 1442–1444.

Bilham, R., and Wallace, K. (2005). Future Mw 8 earthquake in Himalaya: Implication for the 26 December, 2004 M=9 earthquake on eastern margin. *Geol. Surv. India Spec. Publ.*, 85, 1–14.

Chadha, R. K., Pandey, A. P., and Kumpel, H. J. (2003). Search for earthquakes precursors in well water levels in a localized seismically active area of reservoir triggered earthquakes in India. *Geophys. Res. Lett.*, 30, 69-1–69-4.

Choubey, V. M., Kumar, N., and Arora, B. R. (2009). Precursory signatures in the radon and geo-hydrological borehole data for M 4.9 Kharsali earthquake of Garhwal Himalaya. *Sci. Total Environ.*, 407(22), 5877–5883.

Das, N. K., Bhandari, R. K., Ghose, D., Sen, P., and Sinha, B. (2005). Anomalous fluctuation of radon, gamma dose and helium emanating from a thermal spring prior to an earthquake. *Curr. Sci.*, 89(8), 1399–1404.

Dey, S., and Singh, R. P. (2003). Surface latent heat flux as an earthquake precursor. *Nat. Hazards Earth Syst. Sci.*, 3, 749–755.

Dixit, M. M., Kumar, S., Kumar, M. R., Satyanarayana, H. V. S., Anitha, K., Singh, S., Suman, K., Bhagya, K., Peeran, M. M., Vijay, S., and Pandey, A. (2011). High resolution seismic imaging by P-wave and S-wave using passive sources in seismically active zone of Koyna-Warna region of India—Pilot study. Technical report, National Geophysical Research Institute, Hyderabad, India.

Dobrovolsky, I. P., Gershenzon, N. I., and Gokhberg, M. B. (1989). Theory of electrokinetic effects occurring at the final stage in the preparation of a tectonic earthquake. *Phys. Earth Planet. Interiors*, 57, 144–156.

Dudkin, F., Rawat, G., Arora, B. R., Korepanov, V., Leontyeva, O., and Sharma, A. K. (2010). Application of polarization ellipse technique for analysis of ULF magnetic fields from two distant stations in Koyna-Warna seismoactive region, West India, *Nat. Hazards Earth Syst. Sci.*, 10, 1–10.

Evison, F. F. (1982). Generalised precursory swarm hypothesis. *J. Phys. Earth*, 30, 155–170.

Fedotov, S. A. (1965). On distribution patterns for strong earthquake in Kamchataka, the kurile islands and northeastern Japan. *Trudy Inst. Fiz. Zemil Akad. Nauk SSSR*, 36, 66–93.

Gahalaut, K., Gahalaut, V. K., and Chadha, R. K. (2010). Analysis of coseismic water-level changes in the wells in the Koyna-Warna region, Western India. *Bull. Seismol. Soc. Am.*, 100(3), 1389–1394. doi: 10.1785/0120090165.

Gahalaut, V. K., and Kundu, B. (2012). Influence of subducting ridges on the Himalayan arc and on the ruptures of great and major Himalayan earthquakes. *Gondwana Res.*, 21, 1080–1088.

Gavrilenko, P., Singh, C., and Chadha, R. K. (2010). Modelling the hydromechanical response in the vicinity of the Koyna reservoir (India): Results for the initial filing period. *Geophy. J. Int.*, 183(1), 461–477.

Goswami, D., Vyasulu, V. A., Misra, S., Roy, S., Singh, S. K., Sinha, A., Gupta, H. K., Bansal, B., and Nayak, S. (2017). Rock strength measurements on Archaean basement granitoids recovered from scientific drilling in the active Koyna seismogenic zone, Western India. *Tectonophysics*, 712–713, 182–192.

Guha, S. K. (1985). Geoelectrical and geophysical precursors of earthquakes in northeastern India—A discussion. *Geoexploration*, 23, 285–289.

Gupta, H. K. (2002). A review of recent studies of triggered earthquakes by artificial water reservoirs with special emphasis on earthquakes in Koyna, India. *Earth Sci. Rev.*, 58, 279–310.

Gupta, H. K., and Gahalaut, V. K. (2014). Seismotectonics and large earthquake generation in the Himalayan region. *Gondwana Res.*, 25, 204–213.

Gupta, H. K., Mandal, P., Satyanarayana, H. V. S., Shashidhar, D., Sairam, S., Shekhar, M., Singh, A., Uma Devi, E., Kousalya, M., Rao, N. P., and Dimri, V. P. (2005). An earthquake of M~5 may occur at Koyna. *Curr. Sci.*, 89(5), 747–748.

Gupta, H., Nayak, S., Bhaskar Rao, Y. J., Chadha, R. K., Bansal, B. K., Srinagesh, D., Purnachandra Rao, N., Roy, S., Satyanarayana, H. V. S., Shashidhar, D., and Mallika, K. (2011). Deep scientific drilling to study reservoir-triggered earthquakes in Koyna, Western India. *Sci. Dril.*, 12, 53–54. doi: 10.2204/iodp .sd.12.07.2011.

Gupta, H., Rao, N. P., Roy, S., Arora, K., Tiwari, V. M., Patro, P. K., Satyanarayana, H. et al. (2015). Investigations related to scientific deep drilling to study reservoir-triggered earthquakes at Koyna, India. *International Journal of Earth Sciences*, 104, 1511–1522. doi: 10.1007/s00531-014-1128-0.

Gupta, H. K., and Singh, H. N. (1986). Seismicity of the north-east India region. *J. Geol. Soc. India*, 28, 367–406.

Gupta, H. K., and Singh, H. N. (1989). Earthquake swarms precursory to moderate magnitude to great earthquakes in NE India region. *Tectonophysics*, 167, 285–298.

Hattori, K., Serita, A., Gotoh, K., Yoshino, C., Harada, M., Isezaki, N., and Hayakawa, M. (2004). ULF geomagnetic anomaly associated with 2000 Izu islands earthquake swarm, Japan. *Phys. Chem. Earth*, 29, 425–435.

Hayakawa, M., Kawate, R., Molchanov, O. A., and Yumoro, K. (1996). Results of ultra-low-frequency magnetic field measurements during the Guam earthquake of 8 August 1993. *Geophys. Res. Lett.*, 23, 241–244.

Imanishi, Y., Sato, T., Higashi, T., Sun, W., and Okubo, S. (2004). A network of superconducting gravimeters detects submicrogal coseismic gravity changes. *Science*, 306, 476–478.

Kanamori, H. (2005). Real-time seismology and earthquake damage mitigation. *Annu. Rev. Earth Planet. Sci.*, 33, 195–214.

Khattri, K. N., and Tyagi, A. K. (1983). Seismicity patterns in the Himalayan plate boundary and identification of the areas of high seismic potential. *Tectonophysics*, 96, 281–297.

Kumar, N., Chauhan, V., Dhamodharan, S., Rawat, G., Hazarika, D., and Gautam, P. K. R. (2017). Prominent precursory signatures observed in soil and water radon data at multi-parametric geophysical observatory, Ghuttu for Mw 7.8 Nepal earthquake. *Curr. Sci.*, 112(5), 907–909.

Kumar, N., Paul, A., Mahajan, A. K., Yadav, D. K., and Bora, C. (2012). 5.0 Mw Kharsali, Garhwal Himalaya earthquake of July 23, 2007: Source characterisation and tectonic implications. *Curr. Sci.*, 102(12), 1674–1682.

Kushwah, V., Singh, V., and Singh, B. (2007). Ultra low frequency (ULF) amplitude anomalies associated with the recent Pakistan earthquake of 8 October, 2005. *J. India Geophys. Union*, 11(4), 197–207.

Liu, J. Y., Chuo, Y. J., Pulinets, S. A., Tsai, H. F., and Zeng, X. P. (2002). A study on the TEC perturbations prior to the Rei-Li, Chi-Chi and Chia-Yi earthquakes. In *Seismo Electromagnetics: Lithosphere Atmosphere-Ionosphere Coupling*, ed. M. Hayakawa and O. A. Molchanov. Tokyo: Terrapub, 297–301.

Lyubushin, A. A., Arora, B. R., and Kumar, N. (2010). Investigation of seismicity in western Himalaya. *Russ. J. Geophys. Res.*, 11, 27–34.

MoES (Ministry of Earth Sciences). (2012). Scientific deep drilling in the Koyna intraplate seismic zone, Maharashtra. Detailed project report. MoES, Government of India.

Molchanov, O. A., and Hayakawa, M. (1995). Generation of ULF electromagnetic emissions by microfracturing. *Geophys. Res. Lett.*, 22, 3091–3094.

Molchanov, O. A., Schekotov, A. Yu., Fedorov, E., Belyaev, G. G., Solovieva, M. S., and Hayakawa, M. (2004). Preseismic ULF effect and possible interpretation. *Ann. Geophys.*, 47(1), 119–131.

Molnar, P., and Pandey, M. R. (1989). Rupture zones of great earthquakes in the Himalayan region. In *Frontiers of Seismology in India, Proceedings of the Indian Academy of Science*, ed. J. N. Brune. 8 Delhi: Indian Academy of Science, Vol. 98. 61–70.

Nayak, P. N., Saha, S. N., Dutta, S., Rama, Rao, M. S. V., and Sarkar, N. C. (1983). Geoelectrical and geohydrological precursors of earthquakes and northeastern India. *Geoexploration*, 21, 137–157.

Ni, J., and Barazangi, M. (1984). Seismotectonics of the Himalayan collision zone: Geometry of the underthrusting Indian plate beneath the Himalaya. *J. Geophys. Res.*, 89, 1147–1163.

Pandey, M. R., Tandukar, R. P., Avouac, J. P., Lave, L., and Massot, J. P. (1995). Interseismic strain accumulation in the Himalayan crustal ramp in Nepal. *Geophys. Res. Lett.*, 22, 741–754.

Parrot, M. (2007). First results of the DEMETER micro satellite. *Planet Space Sci.*, 54(5), 411–458.

Priyadarshi, S., Kumar, S., and Singh, A. K. (2011). Ionospheric perturbations associated with two recent major earthquakes (M > 5.0). *Phys. Scr.*, 84, 045901. doi: 10.1088/0031 8949/84/04/045901.

Reddy, D. V., Nagabhushanam, P., Sukhija, B. S., and Reddy, R. G. (2010). Continuous radon monitoring in soil gas towards earthquake precursory studies in basaltic region. *Radiat. Meas.*, 45, 35–942.

Roy, S. (2017). Scientific drilling in Koyna region, Maharashtra. *Curr. Sci.*, 112, 11.

Saraf, A. K., and Choudhury, S. (2003). Earthquakes and thermal anomalies. *Geospatial Today*, 2(2), 18–20.

Saraf, A. K., and Choudhury, S. (2005). Thermal remote sensing technique in the study of pre-earthquake thermal anomalies. *J. India Geophys. Union*, 9(3), 197–207.

Schiffman, C., Bali, B. S., Szeliga, W., and Bilham, R. (2013). Seismic slip deficit in the Kashmir Himalaya. *Geophys. Res. Lett.*, 40, 1–4.

Scholz, C. H., Sykes, L. R., and Agarwal, Y. P. (1973). Earthquake prediction: A physical basis. *Science*, 181, 803–810.

Sharma, R., Gupta, V., Arora, B. R., and Sen, K. (2011). Petrophysical properties of the Himalayan granitoids: Implication on composition and source. *Tectonophysics*, 497, 23–33.

Shashidhar, D., Rao, N. P., and Gupta, H. K. (2011). Waveform inversion of local earthquakes using broad band data of Koyna–Warna region, Western India. *Geophys. J. Int.*, 185, 292–304.

Singh, B., Kushwah, V., Singh, V., Tomar, M., and Hayakawa, M. (2005). Simultaneous ULF/VLF amplitude anomalies observed during moderate earthquake in Indian region. *Indian J. Radio Space Phys.*, 34, 221–229.

Singh, R. P., Singh, B., Mishra, P. K., and Hayakawa, M. (2003). On the lithosphere- atmosphere coupling of seismo-electromagnetic signals. *Radio Sci.*, 38, 1–4.

Srivastava, H. N., Bansal, B. K., and Verma, M. (2013a). Largest earthquake in Himalaya: An appraisal. *J. Geol. Soc. India*, 82, 15–22.

Srivastava, H. N., Verma, M., Bansal, B. K., and Sutar, A. (2013b). Discriminatory characteristics of seismic gaps in Himalaya. *Geomatics Nat. Hazards Risk*, 6, 224–242. http://dx.doi.org/10.1080/19475705.2013.839483.

Talwani, P. (1997). On the nature of reservoir-induced seismicity. *Pure Appl. Geophys.*, 150, 473–492.

Ulomov, V. I., and Mavashev, B. Z. (1967). On forerunner of a strong tectonic earthquake. *Dokl. Acad. Sci. USSR*, 176, 319–322.

Verma, M., and Bansal, B. K. (2012). Earthquake precursory studies in India: Scenario and future perspectives. *J. Asian Earth Sci.*, 54–55, 1–8.

Verma, M., and Bansal, B. K. (2013). Seismic hazard assessment and mitigation in India: An overview. *Int. J. Earth Sci.*, 102, 1203–1218.

Virk, H. S., Walia, V., and Sharma, A. K. (1995). Redon precursory signal of Chamba earthquake. *Curr. Sci.*, 69, 452–454.

Virk, H. S., Walia, V., and Kumar, N. (2001). Helium/radon precursory anomalies of Chamoli earthquake, Garhwal Himalaya, India. *J. Geodyn.*, 31, 201–210.

Wadia Institute of Himalayan Geology. (2014). MPGO report on case studies on earthquake events (M≥4.0) occurred between year 2007 and 2014. Dehradun, India: Wadia Institute of Himalayan Geology, 1–100.

Wu, Y. M., and Kanamori, H. (2005). Rapid assessment of damaging potential of earthquakes in Taiwan from the beginning of P waves. *Bull. Seismol. Soc. Am.*, 95(3), 1181–1185.

Wu, Y. M., and Kanamori, H. (2008). Development of an earthquake early warning system using real-time strong motion signals. *Sensors*, 8, 1–9.

10 Geomorphic Features Associated with Erosion

Niki Evelpidou, Isidoros Kampolis and Anna Karkani

CONTENTS

10.1 INTRODUCTION

The Earth experiences alterations upon its relief by exogenous dynamic processes. The most significant of these processes is erosion, which sculpts the superficial layer of the crust. The main processes of erosion are due to moving water, ice, wind and gravity. Erosion depends on various factors, such as climate, rock, soil, morphology and land uses. The climatic conditions determine the wind regime, which is responsible for wind erosion and erosion due to wave action and currents. The amount of rainfall is also dependent on the climate. The more rain, the more surface runoff and erosion that are produced, which induces a higher rill and interrill erosion, as well as riverbed erosion. The rock type determines the erosion type developed on it. For example, a coastal limestone cliff will undergo wave erosion as well as bioerosion, because this lithology is prone to both mechanical and chemical erosion. Lastly, the morphology and land uses pose their own interaction on erosion. Human-made interventions on the plant cover of a slope can enhance rill erosion by reducing the plant volume. Finally, each erosion type produces a variety of distinct geomorphic features.

Erosion is the natural phenomenon of the combined effect of weathering and transportation of the weathered products. It is a two-dimensional process that involves both a static and a dynamic aspect. The Earth surface and subsurface are made of rocks that experience tectonic forces, and also the Earth surface is exposed to winds, heat, rainfall and ocean waves, and as a result, the rocks are weathered and deformed. These forces try to eliminate Earth's relief, leading it towards a peneplain. The static aspect of erosion relies on the weathering process of rocks, which takes place in situ, whereas the dynamic process includes the transportation of the weathered products (rock fragments). This double quality distinguishes erosion from plain weathering.

The first stage of erosion is the detachment of rock fragments, in shapes and sizes that depend on the type of weathering, which might be either mechanical or chemical. The detached particles overlie the parent rock. According to the prevailing climatic and topographic conditions, the transportation of the fragments takes place, comprising the second stage of erosion. This is accomplished via moving water, wind, ice (glaciers) and gravity. These four major factors alter the surface topography and renew it in the course of geologic time. The last stage is the deposition of the transferred material.

Moving water is a significant eroding agent. Large amounts of precipitation produce high quantities of superficial runoff. Increased runoff during rainfall moves notable sediment masses in a drainage basin from its high-relief part towards its low-relief part. This is accomplished through riverbed erosion, embankment erosion and mass movements on the valley sides as streams engrave downward.

Another erosive factor which acts through moving water is wave action impinging on the coast. It is a determinative process that modulates all coastal environments worldwide. Wave energy depends on weather conditions, beach morphology, beach sediment size and tide. During storm events, the produced storm surges alternate beach equilibrium profiles, by eroding beach sediments and provoking their transfer landward.

Wind erosion is the dominant process in arid environments. The wind moves the small sand particles and crashes them, at high velocity, on rocks or human-made constructions, provoking abrasion and particle detachment. The alteration of the desert's landscape, through the movement of dunes, is entirely attributable to the wind. Although the magnitude of wind erosion is difficult to perceive, at places, it in fact generates more soil loss than surface runoff.

In polar regions and high mountains, the geologic processes of erosion and sedimentation are mainly controlled by the presence of ice. Glaciers are huge masses of compacted ice that move under the influence of gravity. It is a very slow movement that goes practically unnoticed, but is actually apparent through its results. For example, when a glacier overcomes an obstacle, the glacier's lower layers go through elastic deformation, whereas the upper and superficial ones deform plastically, resulting in crevasses upon the glacier's surface. As a glacier moves, it draws away any rock fragments located between the glacier's bottom and the landscape's relief; it also carries away any material embedded in its main body.

Gravity is responsible for the shifting of a considerable amount of detached particles. Debris flow, for instance, is a very common process along steep slopes. Generally, gravity forces objects to move towards the zeroing of their dynamic energy.

10.2   GEOMORPHOLOGICAL FACTORS AFFECTING EROSION

Erosion is strictly dependent on geoenvironmental factors which determine its intensity and magnitude. These factors are climate, rock, morphology, land uses, the weathered cap and soil management.

10.2.1   CLIMATE

Climatic conditions have a strong influence on the intensity of erosion. Different climatic parameters affect erosion processes differently on the mainland and on the coastal zone. The wind regime greatly affects both coastal regions and the mainland. The stronger the wind blows, the higher is the height of the wave crushing on the coastline. Wave height is proportional to wind velocity squared. Wind drifts superficial water masses along its direction, through friction. The aforementioned process also leads to the development of wind-induced currents. If wind blows towards a constant direction, then it drifts the molecules of the superficial water layer. This motion is gradually extended in depth, affecting the deeper water layers. The produced waves in their turn impinge on the coastal zone and change its morphology by drifting beach sediments offshore or by eroding coastal cliffs. On another approach, wind abrades rock surfaces and alters landscapes in dry regions by transporting small sand particles and crushing them on rocks. So, erosion phenomena in these regions are in strong correlation to wind regime.

Rainfall is another climatic parameter. This type of precipitation affects mainly terrestrial erosion. Rainfall's erosive ability can be approached through various variables, such as the amount of rain, kinetics, impetus and intensity of the rainfall event (Wischmeier 1959; Foster et al. 1982). The intensity and the amount of the rainfall are the most determinative variables for the erosive ability of precipitation. The intensity describes the erosion per unit of precipitation and can provide an overall estimation of the erosive ability of a storm by multiplying intensity by the amount of rainfall. Important parameters affecting erosion are also the amount and rate of runoff, as well as infiltration. High-intensity rainfalls produce large amounts of water falling on the ground, and surface runoff occurs when the precipitation rate exceeds the soil's infiltration capacity.

The aforementioned equation can be applied for surface runoff erosion. Surface runoff erosion refers to the rate and amount of the superficial runoff. The superficial runoff amount is calculated by the rainfall amount minus the water's infiltration amount. Also, the surface runoff erosion rate is associated with the rainfall intensity minus the infiltration rate.

The erosive ability of rainfall is sufficiently attributed in terms of rainfall intensity and quantity, but a valuable analysis has to be made regarding raindrop size in correlation to erosion. Raindrops impose forces on the soil which depend on their size. The bigger the raindrop that falls on the surface, the greater the forces to be exerted on the soil. The result is that greater imposed forces produce increased values of erosion. This effect is described by the kinetic force of a raindrop impacting the ground, which equals half of the drop's mass times the velocity of the impact squared.

Consequently, a small raindrop presents a small impact velocity and small impact energy. But a storm consisting of thousands of raindrops has a total kinetic energy which represents the sum of the kinetic energies of the individual raindrops.

Experimental research indicates that a reliable index for the erosive ability of precipitation is the result of the total kinetic energy of the storm multiplied by the maximum intensity of 30 minutes (Foster et al. 1982). The mean annual value of this erosive ability can be calculated by the mean kinetic energy of all storms occurring during a year, based on precipitation data (Wischmeier and Smith 1978). The erosive ability varies from year to year, and also within the same year.

Temporary erosive ability is a notable parameter which interacts with changes in vegetation and soil conditions, resulting in significant impacts on the annual erosion. For example, the maximum value of erosion ability corresponds to when greater erosion coincides with the period that soil is most exposed to runoff erosion and raindrop impact. On the other hand, erosion decreases when erosion and soil's exposure status are not at high levels.

The form of precipitation is primarily defined by temperature and presents an important influence on the erosive ability. For example, snowfall produces a low erosion effect; reversely, rainfall provokes greater erosion. Snowfall or rainfall on a frost ground will cause no erosion, but rainfall on a defrosted soil will produce high erosion rates.

The presence of vegetation acts protectively against erosion from raindrop impact and surface runoff. Plant roots, organic material within the soil and decay products reduce soil's vulnerability to erosion. Of course, precipitation, temperature, evapotranspiration and, in some cases, soil humidity determine the amount of the biomass present in the soil profile, as well as the decay within, since the climate defines the growth and the decay of vegetation, which in turn affects erosion. Vegetation or its decay products (fallen leaves and roots) increase the organic material's presence in the soil and the soil cap, reducing soil's susceptibility to erosion. Of course, the volume of the organic material deriving from vegetation, present in the soil, depends on the initial plant mass and its remaining volume after it has been otherwise used. The remaining organic material on the surface of the soil is lost due to decaying processes, which are constant. Notable is the fact that in areas of humid environment and high temperatures, decay occurs at accelerated rates.

10.2.2 ROCK

Rocks undergo mechanical as well as chemical weathering, and their products are transported by erosive agents, such as streams, waves, currents, ice and wind. Along the coastal zone, rocks undergo erosion by the crushing of waves upon them. Erosion takes place in three different processes: (1) the transfer of loose sediments by waves and the induced currents; (2) the abrasion of rock surfaces, that is, on cliffs by waves and the drifted sediments within; and (3) the fluctuation of pressure onto rocks by waves.

Limestone coastal cliffs are resistant to wave erosion. Their lithology renders them difficult to erode but easy to dissolve. Although they undergo mechanical erosion and form coastal caves and coastal platforms, they are mainly subjected to bioerosion and dissolution in the mixing zone. The bioerosion effect includes organisms living in the midlittoral zone, which feed on the vegetation belts, well developed in the coastal zone. Cyanobacteria, patellaceous gastropods (limpets) and chitons graze, erode the underlying rock and abrade the surface with their hard teeth and radulas. This process leads to the formation of coastal notches. On the other hand, dissolution in the mixing zone is a process caused by mixing fresh and sea water in certain parts of the coast, where coastal freshwater discharge takes place. There, the mixing of waters with different pH values affects limestone rocks and dissolves them, forming or broadening present coastal caves.

Rocks in the hinterland are subjected to the erosive action of moving water, ice, wind and gravity. Mechanical and chemical weathering produces various sizes of rock fragments. These particles are transferred by moving water in the form of streams. The development or not of streams depends on the rock type of the surface. For example, on an impermeable lithology the accumulated rainwater will be forced to run off as the infiltration capacity is negligible, whereas in limestone, this water will not be present on the surface. Limestone is permeable and transfers water throughout its mass as its infiltration capacity is high, creating the spectacular underground karst.

Weathering products are transferred by streams and accumulated in sediment traps on the Earth's relief over bedrock. This material is called regolith. Soil scientists define soil as the upper layers of regolith that support plant growth. Generally, the soil is loose material covering the surface and is distinct from solid rocks (Govers and Poesen 1986). The soil performs many functions, like feeding and supporting plant growth and the development of organisms. It is also where all engineering projects are founded. Additionally, soil is the source that provides sediments due to erosion and hosts water circulation.

Certain soils are more prone to erosion than others depending on mineral contents (Bryan 1977; Foster et al. 1985; Agassi 1996). Apart from soil properties, there are external factors that affect its vulnerability, such as the organic material content of the soil.

The most significant factor that defines soil erosion is the soil composition. Soils rich in clays present a low rate of vulnerability to erosion, as they are more resistant to particle detachment, whereas soils with medium-size components are the most vulnerable to erosion. Another factor influencing soil erosion is soil structure, reflecting the sorting of the organic and inorganic materials (Moldenhauer and Wischmeier 1960; Grissenger 1966; Wischmeier et al. 1971).

The erosion of the soils depends on the month and season. Soils are characterized by high erosion vulnerability during the season when the energy and amount of precipitation, as well as the soil's humidity, have their maximum values. Surface runoff increases with high humidity. Additionally, low temperatures facilitate soil humidity, as evapotranspiration is low. Consequently, soil vulnerability tends to be higher during the last months of winter and the first months of spring, rather than at the end of summer. Reversely, soil vulnerability is reduced during the last months of summer, as higher temperatures facilitate biological activity within the soil and produce organic compounds, functioning as composing factors.

Defrosted soil is greatly vulnerable to erosion due to its saturation, which leads to increased surface runoff (Van Klaveren and McCool 1998).

The compaction of the soil and the diffusion of fine particles due to the rainsplash effect can lead to the development of a thick layer on the soil (Bradford et al. 1986; Bradford and Huang 1992; Sumner and Stewart 1992). This layer is called crust when it is dry and sealing when it is humid. The effect of its presence is that it reduces the infiltration rate and, consequently, increases runoff and therefore erosion.

10.2.3 MORPHOLOGY

Morphology is a factor that defines the intensity and magnitude of erosion. The inclination of the landscape facilitates or not the erosive action. For example, high-inclination slopes accelerate the accumulated water's flow on their surface, resulting in streams with high erosive action, whereas high-inclination cliffs reduce erosion provoked by impinging waves on the coast by deflecting wave energy, and in that case, the erosion value depends on the cliff's rock type.

Coastal morphology significantly affects wave action and the induced current regime. Straight coastlines are more prone to erosion, as they are unprotected against open sea waves. If cliffs dominate in a straight coastline, then the presence of weathering materials on the cliff base will determine the erosion value. If the produced loose material on the cliff's base is more than that transferred by waves, then the accumulation of loose material will reduce the erosive action of waves and currents and will stabilize the cliffs. In the case that accumulated loose sediments are less than those drifted by the waves, the cliffs will be subjected to a high erosion rate dependent on their rock type.

In contrast, sheltered coastlines present a more stable morphology, although they can be affected by strong tidal currents. They are well protected against open sea waves and the wave energy coming into the bay is reduced. Of course, in certain sites where high values of tide occur, the fluctuation of tides and especially the flood (high tide) can provoke high tidal waves and enormous erosion on an extended area.

In the case of terrestrial erosion, slopes and their morphology, that is, their length, inclination and shape, are the most important factors affecting runoff erosion. The accumulated sediments at the lower part of a uniform slope vary according to the slope's length and depend on the amount of erosion due to raindrop impact in relation to erosion due to runoff. Additionally, erosion depends on the inclination of the uniform slope (McCool et al. 1987). More particularly, erosion increases linearly with increasing inclination. On the other hand, inclination has a greater effect on runoff erosion than on raindrop impact erosion (Foster 1982).

Erosion in a part of the slope is linked to the distance from the initial point of surface runoff and the inclination in that location (Foster et al. 1977). So, in a downstream location, erosion will be high. For a given location, erosion is proportional to inclination, and this also stands for non-uniform slopes, most common in nature (Figure 10.1). A non-uniform slope presents a varying inclination along its length.

In a convex slope, inclination increases constantly along its length. The maximum erosion rate in a convex slope is much larger than that in a uniform one, and the same goes for the accumulating sediment at the bottom of the slope.

A concave slope presents the opposite geometry as a convex one. Although this type of slope has larger inclination, the induced runoff is minimum (Figure 10.2). The maximum erosion rate is slightly smaller than in a uniform slope, and the sediment production at the bottom of the slope

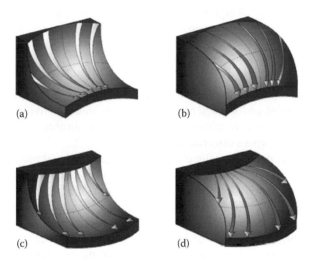

FIGURE 10.1 Various forms occurring in non-uniform slopes, (a) concave–concave form, (b) convex–concave form, (c) concave–convex form, (d) and convex–convex form, and the effect of morphology on surface runoff.

FIGURE 10.2 The slope topography, which defines runoff, is in direct relation to its transfer capability. Consequently, erosion takes place on the upper part of a concave slope, whereas deposition at its lowest part.

follows the same pattern. In a concave slope, erosion occurs at its upper part and deposition is present at the lower part.

Sedimentation on a slope takes place when the produced sediment volume exceeds the transporting ability. The potential sediment for transfer corresponds to the produced sediment volume in the upper part of the slope. In addition, the slope's topography defines runoff, which in turn determines the transporting capacity. Consequently, a slope with a steep upper part and a flat enough lower part satisfies the required conditions for a significant amount of deposition at its end.

Topography has a direct impact on erosion. Soil humidity tends to be greater at the slope bases and in cavities, rather than in the upper parts (Weltz et al. 1998). Consequently, biomass on the soil and within is greater in areas of high humidity. This fact contributes to the spatial fluctuation of erosion, runoff and the impact of the rainsplash effect on the inclination of the slope (Foster 1982).

10.2.4 LAND USE

The term *land use* refers not only to the use of the land but also to the applied management. One land use can characterize an undisturbed area, forbidden to human activities. Forest areas comprise a land use where disturbance is restricted. Also, coastal geomorphic sites of high natural beauty comprise similar land uses where disturbance is forbidden.

Although vegetation is defined by the climatic conditions and soil properties, human interventions control vegetation through the management of cultivated areas and pasture lands. So, vegetation can be greater in regions of controlled grazing, as opposed to overgrazing areas. Consequently, land management affects the erosion processes on a high scale. Land uses have an impact not only on the forces applied on the soil (shear stress, runoff and raindrop impact), but also on the soil's resistance to them. Three parameters of land use are critical for runoff erosion: the vegetation and its role, the surface cap (fallen leaves, stems, etc.) and the mechanical disturbance of the soil (Wischmeier 1975; Renard et al. 1997; Weltz et al. 1998).

Erosion can be eliminated in the portion of the ground that is covered by the broad parts of the plants. In general, branches and the plant's broad parts decelerate runoff and reduce erosivity. Erosion is also reduced by the increased presence of superficial biomass, as well as by the quantity of buried biomass. The foliage of the treetops constitutes the overground part of the plants, which prevents raindrops from reaching the ground. Thick foliage blocks an important percentage of rainfall without raindrops being able to affect surface runoff. The obstruction of the rainwater due to a large extent of thick foliage per soil unit can be significant.

The soil cap is the surface cap which is in direct contact with the soil, protecting it from the rainsplash effect and reducing surface runoff. The soil cap is formed by plant remains coming from fallen leaves, harvest remains, protective layers of hay, artificial material added to the soil and the living parts of plants that are on the ground. The soil cap impact varies according to climate, topography and soil conditions (Meyer et al. 1972; Box 1981).

Generally, soil management preserves soil fertility and structure. Fertile land is associated with high crop yields and good plant cover, resulting in the reduction of raindrop impact, runoff and wind erosion (Morgan 2005). For this reason, mechanical disturbance is applied on soils. This measure enhances superficial coarseness that in turn reduces runoff and the soil's vulnerability to erosion. Additionally, the produced coarseness creates cavities on the soil where water stagnates and therefore produces the reduction of raindrop erodibility and captures the transported sediment. Moreover, the mechanical soil disturbance mixes the superficial plant remains with those buried underground, reducing runoff erosion even further, since this way the infiltration rate increases.

On the other hand, alterations in land uses within the coastal zone or several kilometres away can enhance coastal erosion. For that reason, coastal zone management must be carefully planned. For example, river dams for irrigational purposes can increase erosion in the coastal zone. This human-made construction prevents sediments from moving downstream and reaching the deltaic plain, resulting in the reduction or even the complete lack of sediment reaching the coastal plain.

Consequently, coastal environments cannot counterbalance the loss of beach sediments due to wave and current action, and the retreat of the coastline is imminent.

10.3   TYPES OF EROSION

10.3.1   WATER EROSION

Erosion due to moving water is one of the most significant factors of the relief's alteration. It is produced as a result of water accumulation on the surface, after rainfall events (Figure 10.3). The accumulated water produces the superficial runoff, which flows in specific routes, the stream branches. These branches are parts of a larger hydrologic system called drainage basin. Different types of water erosion take place in different locations, within a single drainage basin (Figure 10.4).

FIGURE 10.3   Areas of surface runoff and concentrated flow. (Photo by Centeri.)

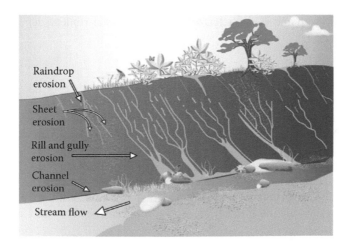

FIGURE 10.4   Types of water-induced erosion.

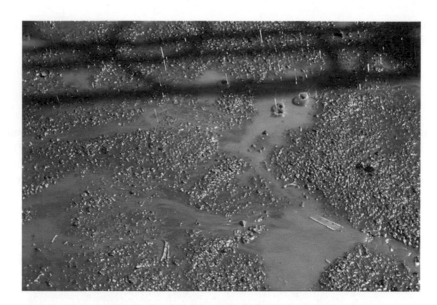

FIGURE 10.5   Splash erosion: The impact of the raindrops. (Photo by Centeri.)

10.3.2   RAINDROP EROSION

This type of erosion is the result of the impact that the raindrop (rainsplash) has on the ground (Figure 10.5). The force applied during this event first condenses the soil and then detaches and dislodges the soil particles. The crushing of the raindrop produces a compaction on the landing point. The dispersing of the raindrop from the landing point provokes flowing jets laterally. These jets carry the detached soil materials entrapped in tiny droplet particles. This combined effect produces a thin surface crust, on the ground, which acts as a sealing cap. This crust reduces the soil's infiltration capacity, resulting in an increase of the surface runoff.

10.3.3   SHEET EROSION

Sheet erosion corresponds to the removal of a uniform layer of the surface ground, due to either the rainsplash effect or the superficial runoff. Water flows in forms of broad sheets, without being restricted into surface depressions. The term *sheet erosion* is not that representative, as the flow is rarely of uniform depth and is characterized by deeper, faster runlets, dependent on microscale variations of the relief.

10.3.4   RILL AND INTERRILL EROSION

Rills are the initial flow paths of runoff, even though erosion is not present in them. These primary flow paths are described as very small branches of a hydrographic network by Emmett (1970), Foster (1971) and Meyer et al. (1975). The critical conditions, under which rills start to form, can be considered in terms of a critical shear stress, after Horton's (1993) theory of slope erosion by overland flow. The interrill areas comprise the terrestrial region between rills (Figure 10.6). Erosion can also be observed on these areas. This is accomplished as the intensity of rainfall increases and rainwater exceeds the infiltration capacity of the soil, so that rain flow on the interrill areas starts (Figure 10.7). Especially for the interrill areas, the rainsplash effect bears a noteworthy impact on erosion.

The location and size of a rill are controlled by the microtopography of the slope and are independent of the macrotopographic features of the wider area.

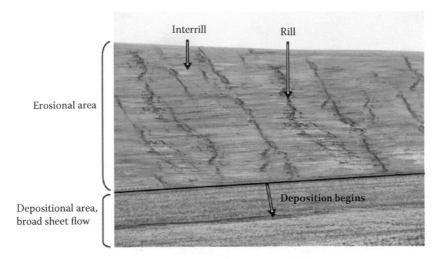

FIGURE 10.6   Rill and interrill erosion on a hill slope. The distinction between the erosion and deposition areas of the slope is obvious.

FIGURE 10.7   Increased rill and interrill erosion on a slope in Tunisia.

10.3.5   EPHEMERAL STREAM EROSION

Ephemeral streams are usually observed in regions where the equilibrium status is disturbed due to agricultural activities. The produced sediment quantity in ephemeral streams is equal to the volume of sediment deriving from rill and interrill erosion in a specific area (Foster 1985; Thomas et al. 1986).

The term *ephemeral* owes its origin to the ephemeral erosion function of the stream during rainfall, which is filled with sediments during the rest of the period (Figure 10.8).

The effect of the cultivation process on erosion, in the adjacent area of ephemeral streams, is of great significance, as it forms a superficial zone on the soil that is prone to erosion. This condition is not present in natural non-cultivated regions.

10.3.6   PERMANENT, INCISED GULLY EROSION

Permanent, incised gullies are characterized by a headcut at the beginning of their channel and continue with steep walls bounding their course (Figure 10.9). They are usually juvenile and are formed within a short period of time. Gullies are usually found in arid and semi-arid environments, where rainfalls

FIGURE 10.8   Blending of ephemeral gully areas with overland flow areas.

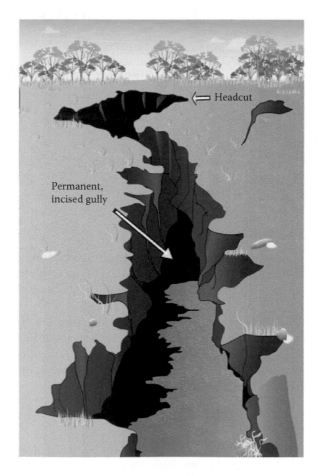

FIGURE 10.9   Schematic view of a permanent, incised gully.

depend greatly on the season and where vegetation is sparse (Charlton 2008). Permanent, incised gullies are present in both cultivated and natural, non-cultivated soils. Their depth is much greater than their width, and this is their major difference from river channels. Erosion in a gully and the consequential transported sediment load do not have a linear relation to flow velocity; on the contrary, gullies carry large amounts of sediments in low flow velocities. Additionally, erosion in permanent, incised gullies is periodical and varies on an annual base (Piest et al. 1975). For example, the weight of the materials, accumulated by the scouring of the lateral walls, stabilizes their base, so that no collapse will take place at this point. Despite this fact, an extreme event of high rainfall can cause significant runoff, resulting in the detachment of soil materials and the destabilization of the gully's lateral walls.

10.3.7 RIVERBED EROSION

Human interventions and alterations in land uses provoke great changes in the fluvial system's equilibrium. The riverbed is a major part of this system, and although it does not depend on human interventions and land use alterations, riverbed erosion can undergo important changes due to perturbations occurring in the river channels and the upstream part of the basin.

Riverbed characteristics, especially its gradient and shape, determine the type of flow and regulate the sediment load that is being deposited in it (Schumm 1977). Consequently, alterations in land uses result in different runoff patterns and sediment volumes which, as a result, trigger changes in the riverbed (Trimble 1977).

In rural areas, intense agriculture can lead to a large increase in upstream runoff, which induces destabilization of the bed and erosion phenomena. As a consequence, the riverbed widens and regression occurs at high rates, resulting in large sediment loads. The most active erosion in the riverbed takes place in the outer part of the meander's inflections, where the river banks retreat by several metres during heavy storm events. Protective measures for the control of riverbed erosion include flow adjustments – by constructing expanded intersections, installing control structures in the channel, expanding protective embankments and deflecting the flow away from the bed walls (Shields et al. 1995).

In non-disturbed forest soils, although river channels are usually permanent, logging and road construction can drastically disturb them (Warrington et al. 1980; Dunne 1998; Elliot 1999; Grace 2000). The effects of this disturbance are increased runoff and sediment load deposition directly on the riverbed. Furthermore, the stability of the riverbed can be affected by major alterations of the sediment load that comes from the upstream part of the drainage basin.

10.3.8 BANK EROSION

In the case of rivers and lakes, erosion is enhanced, due to increased runoff coinciding with intense rainfalls. Increased water flow in river channels can provoke a water level rise, as well as an increase in flow velocity. Flowing water undercuts the river banks, and so their upper part collapses under its weight, provoking alterations, even in the river's course. A potential protective measure against bank erosion is the control of the total runoff through sequential dams along rivers. These dams reduce water momentum and, consequently, its erosive impact on the banks. This measure is also suitable for protection against floods.

10.3.9 EROSION VIA POROSITY

Water also flows within the soil, just below its superficial layer. This flow is accomplished through a 'pipe-like system'. Macropores, interstices and channels comprise the pipes of this system and owe their creation to plant roots, insects and animal activities. Heavy flow can occur in this system, resulting in vast erosion events. Initially, the pipes' diameter is of the order of a few millimetres, but it can widen up to 1 m. As these pipes are close to the surface, a collapse of their upper part towards the formation of open grooves or permanent, engraved flow is likely (Zachar 1982).

Similar flow mechanisms can occur due to the presence of a coarse superficial layer, overlying a dense soil zone. The dense zone acts as an obstacle and prevents water from flowing downwards. So, water flows along the upper boundaries of this dense soil zone, towards the lower part of the slope, making the area vulnerable to erosion (Huang and Laflen 1996). Also, increased soil humidity and water infiltration make soil prone to erosion and result in the acceleration of rill and ephemeral stream erosion (Huang and Laflen 1996).

10.3.10   Erosion due to Snowmelting

Erosion produced by surface runoff can originate from snowmelting; however, it depends on a specific combination of conditions. For example, runoff due to snowmelting after frost can be significant (McCool et al. 1995, 1997; Van Klaveren and McCool 1998). A soil in a defrosting status is highly vulnerable to erosion, and low rainfall of a steady intensity is able to provoke great rill erosion.

10.3.11   Erosion due to Ice Crystal Development

This erosive process occurs mainly in fractured rocks. During the night, as the temperature falls to 0°C or even below 0°C, moisture present in rock cracks starts freezing. As ice crystals form in these tiny voids, enormous stresses are exerted on the fractures' boundaries. These stresses are strong enough to widen the fractures. The intensity of this phenomenon is especially increased in areas of large diurnal temperature range, such as in desserts with a diurnal temperature range of 40°C–50°C (from 40°C to 50°C during the day to 0°C in the night).

10.3.12   Coastal Erosion

This is related to the drifting of beach sediments due to waves and currents (coastal or tidal). Beaches alter their morphology periodically, depending on the climatic conditions that define the wave and current regime. Stronger winds caused by storms increase wave energy, resulting in larger detached amounts of beach sediments. Besides, tidal cycles provoke strong tidal currents, which in some regions deluge areas even several kilometres inland. Consequently, the aforementioned factors contribute to the retreat of beaches consisting of loose sediments. During the retreat of the beach, entire zones of beach ridges or dunes might shift.

10.3.12.1   Erosion due to Wave Action

This type of erosion occurs along the coastline and particularly onshore, where waves impinge on the coast. As waves strike the shoreline at an angle, they drift beach sediment towards their run-up. The return flow in the backwash is, however, influenced by gravity and is forced to follow a direction parallel to the beach slope. As a result, sediment entrained by waves is transported along their direction, advancing in a series of sawtooth motions (Davinson-Arnott 2010). The slope resistance to erosion is related to the wave energy and the cohesion of the slope's rocks (Figure 10.10). The erosive role of the sea waves is double, as not only do they erode the slope's base, but they also remove the weathered material (Figure 10.11). When that material is left to accumulate on the slope's base, the slope retreat will probably slow down or stop, since the material protects the slope from the incoming waves. On a slope's front, two types of pressure are exerted; one is related to the weight of the colliding sea mass (static pressure), and the other depends on the wave type, which is related to the wave's dimensions and the slope's inclination (dynamic pressure). In beaches with loose sediments, the erosion rate is higher during storm events, where waves of significant height impinge on the coast, provoking the drifting of large volumes of sediment offshore, since the beach's resistance to erosion is very low.

FIGURE 10.10 Wave erosion on a limestone cliff in Malta and the development of a sea arch.

10.3.12.2 Erosion due to Currents

Sea currents owe their formation to waves directed towards the coast. As waves break near the shore, water accumulates there, resulting in the formation of longshore currents or rip currents. The type of current is defined by the morphology of the beach, the angle of the wave impinging on the coast and the underwater morphology (Karymbalis 2010). The energy that currents transfer drifts beach sediments and alters beach morphology. Their energy is generally increased during storm events and accelerates the retreat of a beach. In places of specific morphology, tidal currents enhance coastal erosion, especially if the area covered during high tide is considerable.

10.3.12.3 Erosion due to Midlittoral Organisms

Certain organisms living in the midlittoral zone contribute to coastal erosion. A number of vegetation belts are well developed in that zone. Cyanobacteria, patellaceous gastropods (limpets) and chitons (Laborel and Laborel-Deguen 2005) graze, erode the underlying rock and abrade the

FIGURE 10.11   Wave erosion on a cliff, consisting of Neogene sediments, in Cavo Greco, Cyprus. The development of a coastal cave, a sea arch and the absence of loose material at the cliff's base are apparent.

FIGURE 10.12   Tidal notch in Heraion Bay, Perachora Peninsula, Greece.

surface with their hard teeth and radulas. This process results in the formation of notches, which are shaped on carbonate cliffs (Figure 10.12).

10.3.12.4   Erosion due to Sea Level Rise

The sea level has changed repeatedly throughout geologic time, and coastlines have experienced numerous emersions and submersions during Earth's history. During the last 40,000 years, the sea level has oscillated by 150 m, in response to the advance and retreat of glaciers. The rapid sea level rise started 18,000 years ago, and after the Last Glacial Maximum (LGM) about 7,000 years ago, it started to decelerate and has remained virtually stable ever since. Humans began building settlements about 7000 years ago; therefore, civilization has developed during a constant sea level period (Thompson and Turk 1997).

However, the sea level started rising again in the early nineteenth century. Since then, the sea level is estimated by Kemp et al. (2011) to have risen at an average rate of 2.1 mm/year. Two recent studies (Jevrejeva et al. 2008; Kemp et al. 2011) allow for a quantification of the global sea level rise and its rates during the last two millennia and especially during the last two centuries. The second study, by Kemp et al. (2011), is a reconstruction of sea level change over the past 2100 years using salt marsh proxy records from North Carolina. They suggest that the sea level was stable from at least BC 100 to AD 950, and then increased at a rate of 0.6 mm/year for about 400 years (i.e. about 24 cm), followed by an additional period of stability, or slightly dropping sea level, that lasted until the late nineteenth century. According to Jevrejeva et al. (2008), sea level acceleration appears to have started at the end of the eighteenth century, with a rise of 6 cm during the nineteenth century (i.e. at an average rate of 0.6 mm/year), and then 19 cm in the twentieth century (i.e. at a rate of 1.9 mm/year). Other researchers propose that the acceleration in the rate of rise started during the second half of the nineteenth century, around 1865 (Kemp et al. 2011) or 1890 (Woppelmann et al. 2006; Pouvreau 2008; Gouriou 2012). Since 1993, according to altimetric satellites, the rate of the global sea level rise has been estimated to be approximately 3.3 ± 0.4 mm/year (Cazenave and Llovel 2010).

In the Intergovernmental Panel on Climate Change (IPCC) report of 2007 (Bindoff et al. 2007), the rate of sea level rise (as the total of various climatic constituents) has doubled for the most recent period, 1993–2003, in comparison with the period 1961–1993. Even more, it is evident that the contribution of thermal expansion (climatic constituent) in sea level rise has nearly quadrupled, whereas the contribution of ice melting has almost doubled for the most recent period, 1993–2003.

This global sea level rise provokes extensive erosion on the coastal areas, thus destroying the economically thriving regions of the world. All predictions concerning the future of these areas are pessimistic. According to studies posterior to the report of the intergovernmental committee (IPCC) (February 2007), a higher sea level rise is suggested for 2100, reaching 1.5 m. Another study from Pfeffer et al. (2008) predicts an even uglier outcome. This study suggests a possible sea level rise range from 0.8 to 2 m. Despite the discordance, by 2100, a sea level rise up to 1.5–2 m may induce erosion which may be destructive to the coastal zone.

10.3.12.5 Erosion due to Human-Made Constructions

Human interventions on the coastal zone in the form of sea walls, breakwaters and groins, although they are built to protect the coastline from erosion, sometimes end up increasing it. If the planning of these constructions is not accurate, then the implications can be highly destructive (Figure 10.13). On the other hand, other types of human constructions can affect the coastal area harmfully, although located far away, even kilometres away from it. River dams or alteration of land uses in drainage basins can affect the sediment volume reaching the coastal zone. As a consequence, the sediment volume in the coastal zone is decreased progressively, not being able to counterbalance the sediment loss due to wave and current action.

10.3.13 Wind Erosion

The most critical factor in wind erosion is the velocity of the wind. Near the ground, wind velocity has smaller values due to the coarseness provoked by rocks, vegetation and other surface obstacles. Soil accumulated on the land or sand on the beach is unprotected and vulnerable to wind erosion. In some places, soil and sand loss due to wind erosion is higher than other types of erosion. For example, soil loss by wind erosion in Britain, locally called blowing, causes worse problems than

FIGURE 10.13   Erosion induced by human activity along the coastal road of Agios Panteleimon, East Attica, Greece.

water erosion. The light sandy soils of East Anglia, Lincolnshire and east Yorkshire and the light peats of the Fens are the most susceptible to this type of erosion (Huggett 2011). Blows can remove the first 2 cm of the topsoil, including the seeds within it, and damage crops by sandblasting. Also, roads and ditches are blocked by blows. Although this problem in Britain has been known and recorded since the sixteenth century, it was aggravated during the 1960s, probably due to alterations in agricultural techniques (Huggett 2011).

Of course, wind erosion is difficult to perceive, as soil grains are scattered over a large area, making them difficult to become aware of. In urban environments, dust is the most destructive parameter of wind erosion, provoking superficial damages to buildings.

For certain regions on Earth, such as Europe, wind erosion is a less known phenomenon because of its small occurrence in comparison with water erosion. It is a simple process as wind drifts small detached soil grains (deflation) and transfers them along several or thousands of kilometres, until they are deposited. The alterations in environments dominated by wind erosion are significant, such as in the Sahara Desert, where changes in the dunes' position are a constant phenomenon. Also, weathering produced by the particles blown by wind is notable and evident, as, for example, on the Egypt pyramids, which have been abraded through historic times.

10.3.14 HUMAN-MADE EROSION (INCENSEMENT OF THE EROSION RATE)

Human activities usually affect the natural environment in a direct way. Nature knows how to overcome large alterations affecting its equilibrium. But when human appearance imposes stressful changes on natural processes, then things become more complicated and the induced implications are destructive for human-made constructions. Similar alterations (irrigation, cultivation, pasturing and ingathering) on soils induce significant erosion. Also, land levelling, which is the mechanical adjustment of a soil's gradient for the application of agricultural engineering, is an important parameter of human-made erosion. The use of machinery on soils destructs soil aggregates which are determinative for soil's infiltration capacity and the consequently reduced runoff.

The population on Earth has increased rapidly in the last decades. A large part, about 50% of Earth's total population, is located on the coastal zone. This fact imposes stress on the natural processes occurring in coastal areas. The increased inhabitancy on coastal plains triggered the industrial and construction development and significant utilization of the coastal resources. Consequently, man altered the land uses of the coastal zone, and his constructions disturbed the related natural processes, resulting in enormous erosion and landward retreat of the coastline. For instance, the construction of sea walls for the protection of ports alters the longshore current regime, resulting in erosion by removal of the beach sediments on the outer sea wall side and in the accumulation of sediments on the inner sea wall side.

10.3.15 BIOLOGICAL EROSION

Biological erosion leads to phenomena similar to those deriving from erosion via porosity. Biological activity within the soil is important and diverse. Plant roots and insects grow, live and move in the soil. Insects dig and create pipes for their transportation within the soil, and the root system develops pipe systems during its quest for vital nutrients. Both processes lead to extensive erosion under certain conditions.

10.3.15.1 Erosion due to Root System Development

Roots are the means for plants to collect nutrients from the soil and provide them with necessary energy. Roots are spread in a large area around the plant and below the surface. They develop by opening new routes within the soil for nutrient catchment. This process involves the formation of a pipe-like system in the soil that facilitates the underground flow, just below the superficial soil layer. This way, increased subsurface flow can lead to tunnel collapse and the formation of gullies.

10.3.15.2 Erosion due to Underground Living Organisms

In the same way as aforementioned, organisms affect soil erosion. Their activity includes the formation of tiny tunnels within the soil used for their movement, resulting in the facilitation of the subsurface flow. Once increased subsurface flow occurs, these tunnels collapse and transform into gullies.

10.4 EROSION LANDFORMS

10.4.1 WATER EROSION

10.4.1.1 Stream Valleys

Stream valleys owe their creation to the erosive energy of flowing water. As water flows into the stream, it erodes the riverbed and embankments, resulting in certain stream valley profiles indicative of its maturity stage.

10.4.1.2   Gorges

This is a very narrow and deep valley with almost vertical rocky walls, whose height is much greater than its width. Its formation is owed to the erosive activity of water and the existence of faults that initially facilitate weathering and then erosion.

10.4.1.3   Waterfalls

A waterfall corresponds to a broken section of a stream's bed with continuous flow, characterized by an abrupt change of its topographic slope. It owes its formation to intense differential erosion or discontinuities (i.e. faults). The height from which the water flows in a waterfall ranges from a few metres to several hundreds of metres.

10.4.1.4   Meanders

This is a fluvial bed form, characterized by the changing direction of the stream bed with asymmetric banks. The concave section of the stream is steep, whereas its convex section is characterized by a mild inclination. The meandric form is due to the presence of an obstacle in the erosive course of the river. The obstacle may be a rock, more resistant to erosion than the surrounding ones.

10.4.1.5   Knickpoints

This is an abrupt topographic change in the bed of a branch or a part of a drainage network, because of a tectonic line or differential erosion.

10.4.1.6   Pot Holes

These are deep holes present mostly in rocky formations, formed due to turbulent flow erosion.

10.4.1.7   Peneplane

This is a surface characterized by very low topographic inclinations formed as a result of erosion that the whole area has sustained. A peneplane is the last stage in the erosion cycle of the relief.

10.4.2   COASTAL EROSION

10.4.2.1   Coastal Caves

This is a cavity in a rocky coastal area that has been created by the erosive activity of waves. Carbonate rocks are more susceptible to cave formation. Some caves spread out in large areas, penetrate small capes and form impressive arches. The collapse of some caves' roofs brings detached pieces in front of the rocky beaches.

10.4.2.2   Marmites

This is a round ditch created by the turbulent flow of round stones which are transported by waves or turbulent currents.

10.4.2.3   Coastal Platforms

This is a flat rock bench, created by sea erosion between the highest and the lowest sea level. The biological, chemical and mechanical processes, which are considered to be the most important weathering factors, play a primary part in its formation, while the wave energy, which is the main factor of transportation of the erosion products, plays a secondary part. Coastal platforms can also be created by the protracted erosion of the sea notch slope's front. In this case, the platform's form and development are defined by the slope's lithology and stratigraphy. Finally, some coastal platforms are related to eustatic movements, while the presence of a type of terraces is usually attributed to the change of the type of occurring waves.

10.4.2.4   Stacks

This is a rock of pyramidal shape that protrudes in the sea. It is created when the slope retreats, leaving erosion residues at the sea. The sides of a stack are generally steep and vertical, which indicates that the erosion has taken place at the waves' height and not under the sea surface. The term *stack* comes from the word *stakkur*, in the Scandinavian dialect of the Faeroe islands where these particular landforms are very often found in front of high, rocky beaches. Many times, in foreign bibliographies, the terms *pillars*, *chimneys*, *rocks*, *columns*, *skerries* and *needles* are used.

10.4.2.5   Notches

This is a formation located on rocky coasts. Notches are located in places where the sea surface meets the land and are created due to processes of friction, solution or biological erosion. Given that during the last years the sea level has risen, their presence above sea level indicates tectonically active areas. Therefore, by studying the sea fauna in these notches, we collect characteristic data for constant rising periods, for the rising rate and for the earthquake risk of the studied area.

10.4.2.6   Sea Arches

This is a natural opening at the front of a coastal slope and is created by marine processes of erosion. Arches are developed in areas with a lithological and tectonic status that allows the creation of coastal caves by wave action. Their creation is similar to that of coastal caves. Two caves which are created on both sides of a cape may meet after a long time span and form a tunnel at first, and finally an arch as the erosion progresses. The central part of the arch's roof is known as the 'keystone', and it supports the entire structure. The architectonic structure of an arch reflects the hosting lithology. The arch's shape may be arcuate or rectangular, submarine or not, and the height of its opening may reach up to tens of metres above the basic level. Sea arches are considered as ephemeral landforms of differential erosion and exist only for a few decades or centuries.

10.4.3   GLACIER EROSION

10.4.3.1   Glacial Striations

Striations on the transferred rocks are generated by their friction against the glacial valley walls during their transportation by the glacier. Glacial striations are also apparent on the valley walls. The rugged and hard relief of the valley ground is rendered smooth, with striations parallel to the glacier movement.

10.4.3.2   Proglacial Channels

These are drainage channels for the meltwater, formed in front of the glacier's tongue, as meltwater exits the glacier's gate.

10.4.3.3   Gelifluxion

This is a form of ground flow, generated by deglaciation. It is a slow mass movement on a slope of a superficial, water-saturated ground layer that flows on the frozen subsoil.

10.4.3.4   Glacial Debris

This is debris accumulated on a slope by the gravitational fall of glacial round stones.

10.4.3.5   Erratics

These consist of rock blocks located hundreds of kilometres away from the nearest appearance of the respective bedrock (allochthonous origin). The theory that the transportation means of the erratic was ice has gradually prevailed over the previous theory, according to which these large blocks had been moved by water, possibly related to the biblical floods, or that they were detached from large floating icebergs. The residual pieces of the erratics are found abandoned in the border of a regressing glacier.

10.4.3.6   Glacial Gorges

This is a gorge formed by the erosion that glacial meltwater inflicts on rocks.

10.4.3.7   Cirque

This comprises a landform shaped like a bowl, which is actually the starting point of a glacier. In glacial environments, the cirque belongs to the more elevated formations, along with the aretes and horns. The three sides of this depression have escarped walls, and the fourth side is open and descends into the glacial valley comprising the starting point of the glacier. The cirque, before its depression, is a simple irregularity on the side of the mountain, which is later augmented in size as it becomes more and more full of ice. When the glacier starts to heave towards lower altitudes, the open side of the cirque is widened. After the glacier melts, these depressions are usually occupied by small mountain lakes called tarns.

10.4.3.8   Arete

This is a narrow crest with a sharp edge. They are formations occupying higher-elevation points within the glacial environment. They usually separate two parallel glacial valleys, and their composition is that of the bedrock, and therefore must not be confused with the medial moraines that consist of transferred material. Aretes are also formed when the development of two neighbouring 'cirques' erodes the local bedrock until only a narrow ridge is left between them.

10.4.3.9   Horns

This is the meeting point of three or more aretes. It is pointed and has the form of a pyramidal peak with extremely steep sides. Its peak is usually the highest point among the local glacial environment.

10.4.4   Wind Erosion

10.4.4.1   Aeolian Surface

This is an eroded surface where stripes have formed; they are smooth and scratched by the wind that transports sand.

10.4.4.2   Aeolian Sand

This sand is transported and accumulated by wind. It is also characterized by the presence of rounded and dense grains.

10.4.5   Karstic Erosion

10.4.5.1   Fossil Karst

This is the old karst that lies buried within a geological suite (i.e. sedimentary), and which can be uncovered by recent erosion.

10.4.5.2   Exhumation Karst

This is a fossil karst which can be uncovered by recent erosion.

10.4.5.3   Uncovered Karst

This is the karst surface that is continuously exposed to atmospheric processes.

10.4.5.4   Covered Karst

This is a karst surface that lies buried under a cover of alterites and/or under a formation of transported allochthonous material.

10.4.6 EXOKARSTIC FORMS

10.4.6.1 Dolines

This is the most common landform observed in carbonate formations in karst fields. Dolines occur either in isolation or in groups. Their origination may be the collapse of the roof of a subterranean cave, in which case they are called collapse dolines, or the chemical dissolution of the rock, in which case they are called dissolution dolines. Their creation is favoured by the existence of fractures, as happens with the rest of the karst landforms. Usually, the forms of small dolines are funnel shaped with a flat bottom. In the case of a doline with a flat bottom, the landform is considered advanced, since a depth-wise solution, which cannot be perpetual, has stopped due to the presence of resistant formations.

10.4.6.2 Closed Dolines

This is a doline that is not connected to the network of neighbouring valleys.

10.4.6.3 Open Dolines

This doline is generally interconnected in a valley network.

10.4.6.4 Suffusion Dolines

These dolines are formed by soil washing into bedrock fissures, when non-cohesive soil covers a karstified limestone bedrock. As the karstification continues, rock fissures widen. This fact results in the washing of the overlying soil as water drains through these fissures.

10.4.6.5 Uvala

This is a cavity which has been created by the junction of many dolines.

10.4.6.6 Polje

These are large, closed forms, a part of which – at least – is developed in soluble rocks. They give the impression of valleys or basins thanks to their great width and length. The circumferences of these karst plains present a high inclination, their bottom is flat and their drainage is subterranean. Their bottom is covered by fertile soil of polje type.

10.4.6.7 Open Polje

This is a polje which is generally interconnected in a valley's network.

10.4.6.8 Sinkholes

This is an absorbing orifice which is located within a doline or a polje and is the main drainage path for surface waters. It is created as a result of limestone dissolution, particularly in areas where faults exist. Sinkholes advance towards the interior of rocks and form a system of subterranean channels, galleries or caves, usually of labyrinth form.

10.4.6.9 Estavelle

This is a sinkhole which functions alternately as a water absorber or a water supplier.

10.4.6.10 Hum

This is a residual landform which occurs in karst areas, that is, within poljes. Hums are calcareous hummocks, residues of karstificated limestone.

10.4.6.11 Karren, Sculpture

These are small karstic forms which occur in soluble rocks. They are divided into free sculptures, semi-free sculptures and covered sculptures depending on the cover of the rock on which they are developed: if it is naked, partially covered or presents vegetation or soil cover, respectively.

10.4.6.12   Kuppen

This is a relief which is large at its base and arched on top.

10.4.7   ENDOKARSTIC FORMS

10.4.7.1   Caves

The caves are cavities of the ground that have been created in the interior of rocks and communicate with the Earth's surface through small orifices. Caves, in their majority, are underground karstic forms. Caves are the largest category of subterranean karstic forms. For thousands of years, they have accommodated humans, and as a result, the evolution of the human race was closely related to them. Limestone is the most suitable rock for the creation of caves. The flowing water's dissolvent action, through fractures and fissures within limestone, results in the creation of small cavities which form caves when broadened. However, porous limestone is not capable of forming such kind of landforms, because it permits the free intrusion of water towards any direction and its solution takes place in a symmetrical way. Usually, under the entrance of caves a pile of roof material is found, the collapse of which resulted in the communication of the cave with the overlying external surface.

10.4.7.2   Karstic Springs

They are divided into two main categories: headsprings and springs of underground karst. Their creation is caused either by local elevation of the karstic level or by interference of impermeable material (clay and marls), which results in an increase of pressure, due to the closure of the calcareous gaps. Deposits from the precipitation of the crystal sediments of gypsum, dolomite, calcite, and so forth, which occur in the warm periods of the year (in these periods, the concentration of salts in the circulating underground waters is increased), block the calcareous voids.

10.4.7.3   Submarine Karstic Springs

Precipitation water infiltrates in great depths, either through carbonate rock fractures, karst channels or a combination of the aforementioned, and outflows subterraneously, due to the altitudinal difference. Freshwater concentrations floating on seawater are often observed. This effect is caused by the difference in their density. The lenses of freshwater on the seawater are maintained if the speed with which these lenses are supplied with freshwater is higher than the diffusion of the salts of the seawater to the freshwater. Thus, three zones of different water quality can be distinguished: fresh floating waters, subsaline intermediate waters and sea or salty waters.

10.4.7.4   Spring Vauclusienne

This is the reappearance of an underground flow through a siphon, which distributes the water load in a regular way.

10.4.7.5   Karstic Holes

These are karstic forms which have the shape of a semi-circular hole and are associated with the processes of cave creation.

10.5   CONCLUSION

Erosion is a constant process curving the Earth's relief. The main factors controlling erosion are climate, rock type, morphology and land uses. Each one of these poses a greater or lesser influence on the different types of erosion. The various types of water and wind erosion are closely determined by the climatic conditions of an area, and the same stands for biological erosion, but in a minor grade. Especially raindrop erosion, rill and interrill erosion, riverbed erosion, bank erosion, erosion due to waves and currents and wind erosion are the most susceptible types to climate alterations. The rock type affects mainly water erosion and frequently biological erosion, and at a smaller rate the other types. The morphology and land uses play a significant role in the intensity influence of raindrop erosion, rill and interrill erosion, riverbed erosion, wave erosion, wind erosion, erosion due to human-made constructions and biological erosion.

Each type of erosion produces a number of distinct landforms. These forms owe their development to the erosion procedures, which in turn, after a time period, come to an equilibrium status with the geological environment. Human interventions induce this status, provoking alterations in the established balance. As a result, the intensity of erosive phenomena increases, destructing both natural and human building environments.

REFERENCES

Agassi, M., ed. (1996). *Soil Erosion, Conservation, and Rehabilitation*. New York: Marcel Dekker.

Bindoff, N. L., Willebrand, J., Artale, V., Cazenave, A., Gregory, J., Gulev, S., Hanawa, K. et al. (2007). Observations: Oceanic climate change and sea level. In *Climate Change 2007: The Physical Science Basis*, Solomon, S., Qin, D., Manning, M., Chen, Z., Marquis, M., Averyt, K. B., Tignor, M., and Miller, H. L., eds. Cambridge: Cambridge University Press.

Box, J. E. (1981). The effect of surface slaty fragments on soil erosion by water. *Soil Science Society of America Journal*, 45, 111–116.

Bradford, J. M., and Huang, C. (1992). Mechanisms of crust formation: Physical components. In *Soil Crusting: Chemical and Physical Processes*, Sumner, M. D., and Stewart, B. A., eds. Boca Raton, FL: Lewis Publishers, 55–72.

Bradford, J. M., Remley, P. A., Ferris, J. E., and Santini, J. F. (1986). Effects of soil surface sealing on splash from a single waterdrop. *Soil Science Society of America Journal*, 50, 1547–1552.

Bryan, R. B. (1977). Assessment of soil erodibility: New approaches and directions. In *Erosion: Research Techniques, Erodibility and Sediment Delivery*, Toy, T. J., ed. Norwich, England Geo Books, Geo Abstracts Ltd., 57–72.

Cazenave, A., and Llovel, W. (2010). Contemporary sea level rise. *Annual Review of Marine Science*, 2, 145–173.

Charlton, R. (2008). *Fundamentals of Fluvial Geomorphology*. New York: Routledge, 37–51.

Davinson-Arnott, R. (2010). *Introduction to Coastal Processes and Geomorphology*. New York: Cambridge University Press, 139–180.

Dunne, T. (1998). Critical data requirements for prediction of erosion and sedimentation in mountain drainage basins. *Journal of the American Water Resources Association*, 34, 795–808.

Elliot, W. J. (1999). Understanding and modelling erosion from insloping roads. *Journal of Forestry*, 97, 30–34.

Emmett, W. W. (1970). The hydraulics of overland flow on hillslopes. USGS Professional Paper 662A. Washington, DC: U.S. Government Printing Office.

Foster, G. R. (1971). The overland flow processes under natural conditions. In *Biological Effects in the Hydrological Cycle, 3rd International Seminar for Hydrological Professors*, West Lafayette, IN, 18–30 July, 173–185.

Foster, G. R. (1982). Modeling the erosion process. In *Hydrologic Modeling of Small Watersheds*, Haan, C. T., Johnson, H. P., and Brakensiek, D. L., eds. St. Joseph, MI: American Society of Agricultural Engineers, 297–382.

Foster, G. R. (1985). Understanding ephemeral gully erosion (concentrated flow erosion). In *Soil Conservation: Assessing the National Resources Inventory*. Washington, DC: National Academy Press, 90–125.

Foster, G. R., Lombardi, F., and Moldenhauer, W. C. (1982). Evaluation rainfall-runoff erosivity factors for individual storms. *Transactions of the American Society of Agricultural Engineers*, 25(1), 124–129.

Foster, G. R., Meyer, L. D., and Onstad, C. A. (1977). An erosion equation derived from basic erosion principles. *Transactions of the American Society of Agricultural Engineers*, 20(4), 678–682.

Foster, G. R., Young, R. A., and Neibling, W. H. (1985). Sediment composition for nonpoint source pollution analysis. *Transactions of the American Society of Agricultural Engineers*, 28(1), 133–139.

Gouriou, T. (2012). Evolution des composantes du niveau marin a partir d'observations de maregraphie effectuées depuis la fin du 18e siècle en Charente-Maritime. PhD thesis, Université de La Rochelle, 482.

Govers, G., and Poesen, J. (1986). A field-scale study of surface sealing and compaction of loam and sandy loam soils. I. Spatial variability of surface sealing and crusting. In *Assessment of Soil Surface Sealing and Crusting*, Callebaut, F., Gabriëls, D., and De Boodt, M., eds. Gent, Belgium: Flanders Research Centre for Soil Erosion and Soil Conservation, 171–182.

Grace, J. M., III. (2000). Forest road sideslopes and soil conservation techniques. *Journal of Soil and Water Conservation*, 55, 96–101.

Grissenger, E. H. (1966). Resistance of selected clay systems to erosion by water. *Water Resources Research*, 2, 131–138.

Horton, R. E. (1993). The role of infiltration in the hydrologic cycle. *Eos, Transactions, American Geophysical Union*, 20, 693–711.

Huang, C., and Laflen, J. M. (1996). Seepage and soil erosion for a clay loam soil. *Soil Science Society of America Journal*, 60, 408–416.

Huggett, J. R. (2011). *Fundamentals of Geomorphology*. 3rd ed. New York: Routledge Publishing, 314–344.

Jevrejeva, S., Moore, J. C., Grinsted, A., and Woodworth, P. L. (2008). Recent global sea level acceleration started over 200 years ago? *Geophysical Research Letters*, 35, L08715, http://dx.doi.org/10.1029/2008GL033611.

Karymbalis, E. (2010). *Coastal Geomorphology* [in Greek]. Athens: ION Publishing, 49–52.

Kemp, A. C., Horton, B. P., Donnelly, J. P., Mann, M. E., Vermeer, M., and Rahmstorf, S. (2011). Climate related sea-level variations over the past two millennia. *Proceedings of the National Academy of Sciences of the United States of America*, 108, 11017–11022.

Laborel, J., and Laborel-Deguen, F. (2005). Sea-level indicators, biologic. In *Encyclopedia of Coastal Science*, Schwartz, M. L., ed. Dordrecht, The Netherlands: Springer, 833–834.

McCool, D. K., Brown, L. C., Foster, G. R., Mutchler, C. K., and Meyer L. D. (1987). Revised slope steepness factor for the universal soil loss equation. *Transactions of the American Society of Agricultural Engineers*, 30(4), 1387–1396.

McCool, D. K., Saxton, K. E., and Williams, J. D. (1997). Surface cover effects on soil loss from temporally frozen cropland in the Pacific Northwest. In *International Symposium on Physics, Chemistry, and Ecology of Seasonally Frozen Soils*, Fairbanks, AK, 10–12 June, 235–341.

McCool, D. K., Walter, M. T., and King, L. G. (1995). Runoff index values for frozen soil areas of the Pacific Northwest. *Journal of Soil and Water Conservation*, 50, 466–469.

Meyer, L. D., Foster, G. R., and Romkens, M. J. M. (1975). Sources of soil eroded by water from upland slopes. In *Present and Prospective Technology for Predicting Sediment Yields and Sources*. USDA Agricultural Research Service ARS-S-40. Washington, DC: USDA Science and Education Administration, 177–189.

Meyer, L. D., Johnson, C. B., and Foster, G. R. (1972). Stone and woodchip mulches for erosion control on construction sites. *Journal of Soil and Water Conservation*, 27(6), 264–269.

Moldenhauer, W. C., and Wischmeier, W. H. (1960). Soil and water losses and infiltration rates on Ida silt loam as influenced by cropping systems, tillage practices, and rainfall characteristics. *Soil Science Society of America Proceedings*, 24, 409–413.

Morgan, R. P. C. (2005). *Soil Erosion & Conservation*. 3rd ed. Oxford: Blackwell Publishing.

Pfeffer, W. Y., Harper, J. T., and O'Neel, S. (2008). Kinematic constraints on glacier contributions to 21st-century sea-level rise. *Science*, 321(5894), 1340–1343.

Piest, R. F., Bradford, J. M., and Spomer, R. G. (1975). Mechanisms of erosion and sediment movement from gullies. In *Present and Prospective Technology for Predicting Sediment Yields and Sources*. USDA Agricultural Research Service ARS-S-40. Washington, DC: USDA Science and Education Administration, 162–176.

Pouvreau, N. (2008). Trois cents ans de mesures marégraphiques en France: Outils, méthodes et tendances des composantes du niveau de la mer au port de Brest. PhD thesis, Université de La Rochelle, 474.

Renard, K. G., Foster, G. R., Weesies, G. A., McCool, D. K., and Yoder, D. C. (1997). *Predicting Soil Erosion by Water: A Guide to Conservation Planning with the Revised Universal Soil Loss Equation (RUSLE)*. USDA Agricultural Handbook 703. Washington, DC: U.S. Government Printing Office.

Schumm, S. (1977). *The Fluvial System*. Caldwell, NJ: Blackburn Press.

Shields, F. D., Jr., Bowie, A. J., and Cooper, C. M. (1995). Control of streambank erosion due to bed degradation with vegetation and structure. *Journal of the American Water Resources Association*, 31(3), 475–489.

Sumner, M. D., and Stewart, B. A., eds. (1992). *Soil Crusting: Chemical and Physical Processes*. Advances in Soil Science. Boca Raton, FL: Lewis Publishers.

Thomas, A. W., Welch, R., and Jordan, T. R. (1986). Quantifying concentrated-flow erosion on cropland with aerial photogrammetry. *Journal of Soil and Water Conservation*, 41(4), 249–252.

Thompson, R. G., and Turk, J. (1997). *Introduction to Physical Geology.* 2nd ed., Saunders Golden Sunburst Series. Philadelphia: Saunders.

Trimble, S. W. (1977). The fallacy of stream equilibrium in contemporary denudation studies. *American Journal of Science*, 277, 876–887.

Van Klaveren, R. W., and McCool, D. K. (1998). Erodibility and critical shear stress of a previously frozen soil. *Transactions of the American Society of Agricultural Engineers*, 41, 1315–1321.

Warrington, G. E., Knapp, K. L., Klock, G. O., Foster, G. R., and Beasley R. S. (1980). Surface erosion. In *An Approach to Water Resources Evaluation on Non-Profit Silvicultural Sources.* EPA-160018-80-012. Washington, DC: U.S. Environmental Protection Agency.

Weltz, M. A., Kidwell, M. R., and Fox, H. D. (1998). Influence of abiotic and biotic factors in measuring and modelling soil erosion on rangelands: State of knowledge. *Journal of Range Management*, 51, 482–495.

Wischmeier, W. H. (1959). A rainfall erosion index for a universal soil loss equation. *Soil Science Society of America Proceedings*, 23, 246–249.

Wischmeier, W. H. (1975). Estimating the soil loss equation's cover and management factor for undisturbed areas. In *Present and Prospective Technology for Predicting Sediment Yields and Sources.* USDA Agricultural Research Service ARS-S-40. Washington, DC: USDA Science and Education Administration, 118–124.

Wischmeier, W. H., Johnson, C. B., and Cross, B. V. (1971). A soil erodibility monograph for farmland construction sites. *Journal of Soil and Water Conservation*, 26, 189–193.

Wischmeier, W. H., and Smith, D. D. (1978). *Predicting Rainfall-Erosion Losses: A Guide to Conservation Planning.* USDA Agriculture Handbook 537. Washington, DC: U.S. Government Printing Office.

Woppelmann, G., Pouvreau, N., and Simon, B. (2006). Brest sea level record: A time series construction back to the early eighteenth century. *Ocean Dynamics*, 56(5–6), 487–497.

Zachar, D. (1982). *Soil Erosion.* New York: Elsevier Scientific Publishing.

11 Thar Desert
Source for Dust Storm

Priyabrata Santra, Suresh Kumar and M.M. Roy

CONTENTS

11.1 INTRODUCTION

The Indian Thar Desert lies in the western part of India, covering four states, Rajasthan, Gujarat, Punjab and Haryana, with an area of 0.32 million km². Wind erosion is a severe land degradation process in this desert, causing loss of a large amount of nutrient-rich fertile topsoil. Minute soil particles eroded from the desert surface contribute to the dust load in the atmosphere and cause multiple health problems to desert dwellers. Moreover, these suspended dust particles in the atmosphere may be transported over a large distance and cause dust hazards in deposited areas. These suspended dust particles also help to form aerosols in the atmosphere, having a great impact on the regional weather. Field measurements in the Jaisalmer, Rajasthan, region of the Thar Desert have revealed a soil loss of 12.02 t ha^{-1} during mid-June to the end of September. A major portion of these losses are contributed to by a few dust storm events, which generally occur for a short period. Carbon and nitrogen contents in eroded soils were observed as 4 g kg^{-1} and 0.37 g kg^{-1}, respectively. A rapid rate of land conversion from native rangeland to arable farming, along with livestock pressure on rangelands for grazing, may aggravate the problem of wind erosion in the Thar Desert. Control of grazing at the existing rangelands of the desert may greatly reduce the amount of wind-eroded soil loss, and consequently its negative impact on the environment.

Wind erosion is a severe land degradation process in arid and semi-arid areas of the world, which cover 41% of the total land area of the world (Lal 1990). Removal of soil particles by wind is also very active in the rangelands of the Thar Desert. About 13.5% of the total geographical area of India is affected by wind erosion (Sehgal and Abrol 1994). Eroded particles by wind pose severe multifaceted problems in this desert (Dey 1957; Kaul 1959; Bhimaya et al. 1961; Bhimaya and Chowdhary 1961). Loss of nutrient-rich particles from agricultural fields, suspension of fine particles in air and deposition of eroded soil particles on railway tracks, roads, irrigation canals, and so forth, are major wind erosion-related problems in the region (Bittu 1989). During severe dust storm events, the suspended particles may be transported by air over hundreds of kilometres and form a blanket of dust haze over the Indo-Gangetic Plain and surrounding area. Prevailing weather and terrain conditions of this desert are also very congenial for wind erosion. Among climatic factors, wind velocity plays a vital role, and if it exceeds a threshold velocity of 5 m s^{-1} at 0.3 m height from the ground surface, it initiates wind erosion (Schwab et al. 1993). Among terrain properties, soil aggregate distribution, surface roughness, soil moisture, vegetation cover and unsheltered portions of the field are important factors, which have a bearing on the extent of wind erosion. In alliance with the causative factors, indiscriminate grazing in the region further destroys the vegetation and exposes the land surface, thus making it more vulnerable to wind erosion. Minute soil particles (<60 μm) blown by wind in the desert are one of the major sources of particle pollution and cause serious health hazards to people and animals dwelling in this region, especially for children, sensitive persons and old animals (Santra 2006). Introduction of the Indira Gandhi Canal in the desert region from the Harika barrage just after the confluence of the Ravi and Sutlej Rivers was prospected to green the desert, but in contrast, it further accentuated the process of wind erosion through removal of grasses and shrubs from fields due to increased agricultural activity. Combating wind erosion in the desert requires prioritization of different regions according to the severity of the problem to formulate policy and plan accordingly.

11.2 THAR DESERT

The Thar Desert forms the eastern limit of a vast arid tract of an arid region that encompasses the Sahara Desert, the Arabia, Iran and Beluchistan in Africa and Asia. The eastern limit of the Thar is roughly co-terminus with the Aravalli hill ranges – one of the oldest surviving in the world, while its western margin is along the fertile plains of the Indus in Pakistan. In the south, it has a sharp natural boundary with the world's largest saline waste – the Great Rann of Kachh, while in the north, the riparian sub-Himalayan plains define its boundary (Figure 11.1).

Thar Desert is situated between the latitudes of 24°30′ N and the longitudes of 69°30′ N. Within the territory of India, the Thar forms a part of the country's sandy hot arid zone, spread over a 0.32 million km^2 area. Western Rajasthan alone constitutes 61% of the area in the hot arid zone, while 20% of the area lies in the state of Gujarat and 9% lies in the states of Punjab and Haryana. The southern states of Maharashtra, Andhra Pradesh and Karnataka together contribute to the rest of the area (10%). It is one of the most densely

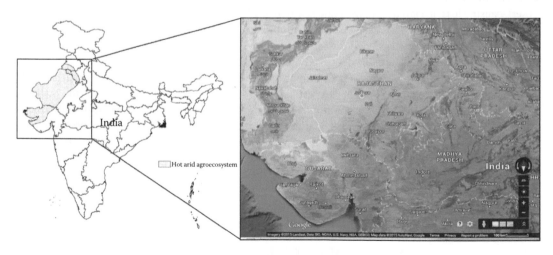

FIGURE 11.1 Location of the Indian Thar Desert.

populated deserts of the world, having an average population density of 61 persons per km^2 as against 3 persons per km^2 in other deserts (Mann et al. 1977). According to 2011 census report by Government of India (http://censusindia.gov.in/), the population density in the hyperarid part of this desert, for example, at Jaisalmer district, is 17 per km^2; Bikaner district, 78 per km^2; and Barmer district, 92 per km^2.

Precipitation in Thar Desert is less than the potential evapotranspiration during most periods in a year and meets less than one-third of the annual water need. A significant portion (91%–96%) of the annual rainfall is generally received during the southwest monsoon period during June through September (Ramakrishna et al. 1992). Another interesting feature of the annual rainfall pattern is the higher coefficient of variability, which often exceeds 50%, and even higher than 70% in extreme arid parts where the annual rainfall is as low as 180 mm. The mean maximum temperature during hot summer months varies from 40°C to 42°C, but in some years the temperature has gone up to 50°C–52°C. During the cold season, the minimum temperature varies from 3°C to 10°C. Frost occurs when the temperature reaches –2°C to –4°C, but the onset varies from 5 to 20 days depending on the western disturbances bringing a cold air mass. Dust storms are frequent during the hot weather season, especially in the months of May and June. The average wind speed during June is more than 25 km h^{-1} in the western part of the Thar Desert. The evaporation rate may be as high as 17.4 mm day^{-1} on a dusty day, which is 18% more than that during a subsequent non-dusty day (Ramakrishna et al. 1987). The frequency of the dust storm and wind erosion activity in the Thar Desert has a good relationship with the seasonal rainfall. Whenever the rainfall in a particular year decreases sharply and is very low, the subsequent summer season experiences a sharp increase in the frequency of occurrence of dust storms (Krishnan 1977).

11.3 FIELD MEASUREMENT OF DUST STORMS

The aeolian sediment load to the atmosphere generated during wind erosion events is generally quantified through different types of sampling devices, which are known as samplers, traps, catchers or collectors. Among wind erosion samplers, the Big Spring Number Eight (BSNE) sampler (Fryrear 1986) and the Modified Wilson and Cook (MWAC) sampler (Wilson and Cook 1980) are mostly used. Other than these two samplers, several indigenously manufactured samplers, for example, the Suspended Sediment Trap (SUSTRA), Pollate Catcher (PULCA), Saltiphone and Cox sand catcher, are available at different arid regions of the world (Goossens et al. 2000). In the Indian Thar Desert, a dust catcher in the design of a Bagnold sampler has been used for a long time for collecting the eroded soil mass from wind erosion events.

11.3.1 DUST CATCHER

A dust catcher contains several samplers at different heights, which are attached to an iron pole. The individual samplers are funnel shaped, with their open top having a diameter of 0.35 m. A metal strip of 0.089 m in height is attached at the top brim of the funnel with an opening area of 0.035 m^2 at one side. A collector is attached at the bottom of the funnel (Figure 11.2). Singh et al. (1992) have suggested that dust storms mostly originate from the Arabia peninsula and travel over the Indian Thar Desert in the south to southwesterly direction (SSW). The meteorological records on wind direction during the last 30 years show a dominant SSW wind direction over the Thar Desert during April–September. Hence, the opening of these samplers is fixed towards SSW. A similar type of sampler was previously used in the Indian Thar Desert by Mann (1985).

However, the dust catcher has a few drawbacks. First, the dust catcher is of a fixed type and does not rotate along with the wind direction. Second, the available dust catcher is more suitable for collecting suspension flow. Third, the top of the sampler is open, and hence suspended soil particles other than eroded ones generated due to wind erosion events may be deposited in the collector. Keeping in mind the constraints of the previously available dust catcher and the present requirement, a new wind erosion sampler for collecting wind-eroded soil mass in the Thar Desert was designed and developed (Santra et al. 2010).

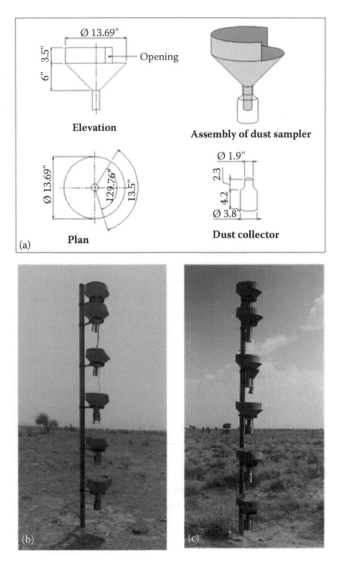

FIGURE 11.2 Dust catcher installed in rangeland sites at Jaisalmer and Chandan. (a) Design of dust catcher, (b) dust catcher installed in an overgrazed rangeland site at Jaisalmer and (c) dust catcher installed in a controlled grazing rangeland site at Chandan. (From Mertia, R. S. et al., *Aeolian Res.*, 2, 135–142, 2010.)

11.3.2 WIND EROSION SAMPLER

The new wind erosion sampler has been designed in such a way that it can collect eroded particles from any direction (Figure 11.3). For this purpose, a vane has been attached to its back. The orifice of the sampler has been kept at 2 cm wide and 5 cm in height. The new erosion sampler has the ability to adjust its orifice towards the wind direction, which is also a feature of the BSNE sampler and many other available samplers outside India. However, in India the dust catcher has been used for the last three decades, and it is a fixed type and thus not able to rotate its orifice towards the wind direction. The newly designed wind erosion sampler overcame the major constraints of the previously available dust catcher in India. Besides these, a few characteristic features of the new erosion sampler are listed below.

FIGURE 11.3 New wind erosion sampler installed at Jaisalmer. (From Santra, P. et al., *Curr. Sci.*, 99(8), 1061–1067, 2010.)

- The top of the sampler is closed and therefore prevents the deposition of fine suspended particles during calm atmospheric conditions after a wind erosion event.
- There is no requirement of a rain hood for the newly designed sampler, and it is able to collect eroded soil particles during dust storm events associated with rainfall.
- It is very cheap to fabricate and needs a maximum of INR 8000 to construct a single assembly with five samplers.
- The vertical gap of the sampler from the centre of the orifice to the bottommost part of the vane is 10 cm, and therefore it is able to collect eroded soil particles starting from 10 cm height from the surface to any desired height.
- It is more suitable to study wind erosion in the form of suspension flow and some portion of saltation flow, but not for surface creep.

11.4 PROCEDURE OF SOIL LOSS CALCULATIONS

In order to assess the soil loss in the field due to wind erosion, samplers were installed at the experimental site by fixing four samplers on an iron pole at 0.25, 0.75, 1.25 and 2.00 m height from the surface. Polythene bags were attached to the bottom of the sampler as indicated in Figure 11.3. The collected eroded masses were recorded after each dust storm event, as well as periodically. Collected aeolian samples in the sampler were carried to the laboratory and weighed by an electrical balance. During the observation periods with rainfall events, collected aeolian soil samples in rain water were decanted and the residual soil left on Whatman filter paper were air-dried to record the sample weight. For each sampler opening, the amounts of collected aeolian sediments were converted to mass flux ($M\,L^{-2}\,T^{-1}$) for the duration of an observation period.

Mass fluxes of aeolian sediments were fitted in a power decay mass–height profile, which has been found to be the best model for the Thar Desert by Mertia et al. (2010) and is shown in Equation 11.1:

$$q(z) = az^{b} \qquad\qquad (11.1)$$

where q is the mass flux ($M\,L^{-2}\,T^{-1}$) of aeolian sediments at height z (L) from the surface, and a and b are empirical constants of the equation. The total aeolian mass transport rate ($M\,L^{-1}\,T^{-1}$) was computed through integration of Equation 11.1 with a lower limit of $z = 0.25$ m to an upper limit of $z = 2$ m. The total aeolian mass transport rate ($M\,L^{-1}\,T^{-1}$) was defined by several researchers as the total mass of aeolian sediments carried over by wind per unit distance across the wind direction per unit time at a downward position of the field. The calculated mass transport rate was converted to soil loss (kg ha^{-1}) by dividing the aeolian mass transport rate by the distance (L) of non-eroding boundaries from the sampling point and multiplying by the duration of the wind erosion event (T). The calculated soil loss following the above procedure represents the soil loss through the suspension mode. To calculate the total soil loss, measurements at the near surface through surface creep samplers need to be integrated into the above procedure. However, total soil loss may be approximated from the above measurement set by integrating Equation 11.1 from the lower limit of $z = 0.01$ m to the upper limit of $z = 2$ m, respectively. The conversion of the aeolian mass transport rate to soil loss as described above depends on the windward distance (L) up to the clear non-eroding boundaries. In the absence of clear non-eroding boundaries, the presence of rocky outcrops or a few sparsely distributed tree barriers is approximated as a semi-eroding boundary at a windward distance.

11.4.1 Field Observations on Soil Loss through Wind Erosion

The extent of soil loss through wind erosion in this desert was quantified by many (Dhir 1961; Gupta and Aggarwal 1978; Gupta et al. 1981; Mann 1985; Gupta 1993; Dhir and Venkateswarlu 1996). Using erosion pins, Gupta and Aggarwal (1978) measured the soil loss and found that within a period of 75 days from April to June, 9 cm of topsoil may be removed from bare sandy plains of desert, which may be as high as 37 cm from bare sand dunes. Gupta et al. (1981) reported a soil loss of 31.2–61.5 kg m^{-2} during April–June. Dust blowing in the zone of 3 m above the soil surface in different parts of the desert was measured by Mann (1985) using a fixed dust catcher and revealed that on a stormy day, the soil loss varies from 50 to 420 kg ha^{-1} and sometimes reached up to 511 kg ha^{-1} for extreme parts of the desert. Using a similar type of dust catcher, Mertia et al. (2010) reported a soil loss of 827 kg ha^{-1} at the overgrazed site at Jaisalmer and 240 kg ha^{-1} at the controlled grazing site at Chandan during the summer (May–July) of 2004.

Using the new wind erosion sampler (Figure 11.3), soil loss has also been successfully measured in the field at Jaisalmer during several dust storm and wind erosion events during 2009 and 2010 (Table 11.1). The most severe dust storm of the year 2009 was observed on 15 June 2009, which was again the most severe dust storm during the last two decades. The average wind speed during this dust storm event was about 50–55 km h^{-1}, as measured from cup anemometer readings before

TABLE 11.1

Wind Erosion and Dust Storm Events Recorded during 2009 and 2010 at Jaisalmer and the Approximate Soil Loss during Each Event

Name of the Observations	Time of Occurrence	Duration	Soil Loss in Suspension Mode (kg ha^{-1})[a]	Total Soil Loss (kg ha^{-1})[b]
		Dust Storm Event (DSE)		
DSE-1	15 June 2009	30 min	389	1166
DSE-2	17 June 2009	20 min	246	466
DSE-3	24 June 2009	15 min	19	58
DSE-4	9 July 2009	25 min	128	1485
DSE-5	14 July 2009	15 min	16	30
DSE-6	5 June 2010	20 min	503	870
		Periodical Observation on Wind Erosion (POWE)		
POWE-1	25 June–2 July 2009	7 days	23	30
POWE-2	15–30 July 2009	16 days	203	3244
POWE-3	31 July–18 August 2009	19 days	255	5287
POWE-4	19 August–3 September 2009	15 days	49	68
POWE-5	4–23 September 2009	20 days	39	188
POWE-6	17 April–5 May 2010	18 days	273	366
POWE-7	6 May–4 June 2010	29 days	953	1434

Source: Santra, P. et al., *J. Agric. Phys.*, 13(1), 13–21, 2013.

[a] Soil loss was computed within the observation heights (0.25–2 m above surface), which generally indicates the soil loss in suspension mode.

[b] Total soil loss was calculated through extrapolation of the mass–height profile up to very near the soil surface (0.01 m).

and after the dust storm events. However, the peak wind gust of the dust storm events might be much higher, which could not be recorded. This severe dust storm started at 4:00 p.m. on 15 June 2009 and lasted for around half an hour. The second-most severe dust storm event of the year 2009 occurred on 17 June 2009, just 2 days after the first severe dust storm of the year. The severity of the second dust storm event in terms of soil loss was comparatively less because of the previous day's rainfall event (15.2 mm), which resulted in the formation of fragile soil aggregates on the desert surface. Besides these two dust storm events, another three events on 24 June, 9 July and 14 July 2009 were also recorded and corresponding aeolian masses were collected. During 2010, the occurrence of wind erosion event was comparatively less because of the good monsoon rainfall during mid-June and onwards. Two periodical wind erosion events and one dust storm event was recorded during April–June; these dust storm events were of short duration (15–30 min) and occurred mostly during the evening time. Another characteristic feature of these dust storm events was that most of them were associated with drizzling at their dissipating stage. A significant amount of eroded masses was also collected periodically during June–September 2009, which mostly occurred due to the sudden occurrence of gusty wind and associated erosion over a small scale. All these periodic observations were also recorded and are presented in Table 11.1 with their duration.

Computed soil loss in suspension mode revealed a maximum soil loss of 389 kg ha^{-1} during the dust storm event on 16 June 2009, which was only 30 min in duration. However, extrapolation of the power decay model (Equation 11.1) resulted in a total soil loss of 1166 kg ha^{-1} during the same dust storm event. On average, the soil loss rate during the dust storm events of the year 2009 was found to be 17 kg ha^{-1} min^{-1}. Periodical observations on eroded soil mass revealed an average soil loss rate of 25 kg ha^{-1} day^{-1}. The cumulative soil loss in suspension mode during mid-June to the end of

September 2009 was 1.36 t ha^{-1}, whereas the total soil loss was 12.02 t ha^{-1}. During 2010, the total amount of soil loss from mid-April to the first week of June was 2.7 t ha^{-1}.

11.5 MAJOR CAUSATIVE FACTORS

11.5.1 Surface Cover Factor

The surface roughness provided by the protruding vegetative surface in the rangelands of the Thar Desert changes the wind velocity profile and increases the threshold wind speed to cause erosion. In the semi-arid zone of the Thar Desert, the rangelands occupy 15%–20% of the lands, whereas they occupy around 45% in the arid region (Mertia 1982). In extreme arid tracts of the Thar Desert, grazing lands occupy 85%–90% of the total area. The growth of vegetation in the rangelands of the Indian desert starts with the onset of monsoon at the end of June or beginning of July, each year. The maximum growth is attained in September–October and afterwards declines upon maturity, as most of the ephemerals have very short life cycles. The above-ground production in the semi-arid regions in Jodhpur (350 mm year^{-1}) has been reported as 1.2–2 t ha^{-1} year^{-1}, whereas in the driest region at Jaisalmer (180 mm year^{-1}), the dry herbage yield may be as high as 2–2.5 t ha^{-1} (Mertia 1982).

The surface cover of *Lasiurus sindicus* (sewan) grass to protect the topsoil from erosion was measured at an experimental site in Chandan during the summer season of 2004 using a quadrate technique. The number of dried grass tussocks per unit area, along with the height and diameter of the tussocks, was measured during hot summer months to calculate the soil loss ratio factor of the revised wind erosion equation (RWEQ) as follows:

$$SLRs = e^{-0.0344(SA^{0.6413})} \tag{11.2}$$

where *SLRs* is the soil loss ratio for the standing plant silhouette, and *SA* is the silhouette area computed by multiplying the number of standing tussocks per square metre area times the average diameter (cm) times the tussock height (cm). The number of dry tussocks of sewan grass, which act as a silhouette, was found to be 0.4 per m^2. The average diameter and height of sewan tussock were 30 and 43 cm, respectively. Accordingly, the soil loss ratio was calculated as 0.15, which indicates that if other factors of RWEQ remain the same, the standing grass tussock of *Lasiurus sindicus* may reduce the soil loss by 85%. After rainfall events, dry tussocks of sewan grass immediately sprout and green foliage growth is started, which further reduces the soil loss.

11.5.2 Weather Factor

Strong wind speed and very scarce rainfall are two key factors that cause wind erosion in the Thar Desert. Strong summer winds blow at an average speed of 20–30 km h^{-1} (daily average), resulting in frequent dust storms. The mean monthly variations of diurnal wind speed are shown in Figure 11.4, which indicates that the daytime average wind speed remains at >3 m s^{-1} for the months of May and June.

The erosivity of strong summer wind may be assessed by calculating wind factor W_f as follows, which is used in the RWEQ for the calculation of weather factors (Fryrear et al. 2000):

$$W_f = \sum_{i=1}^{N} \rho \frac{(U_i - U_t)^2 U_i}{gN} \tag{11.3}$$

where W_f is the wind factor (kg m^{-1} s^{-1}), U_i is the wind speed at 2 m height (m s^{-1}), U_t is the threshold wind speed, N is the number of wind speed observations in a time period, ρ is the air density (1.293 kg m^{-3}) and g is the acceleration due to gravity (9.8 m s^{-2}).

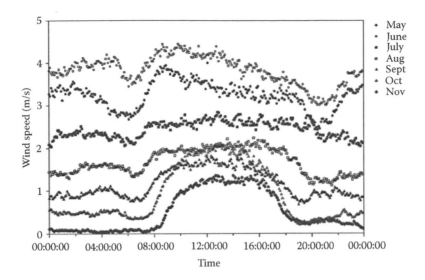

FIGURE 11.4 Diurnal variation of mean monthly wind speed (m s⁻¹) measured at Jaisalmer of the Indian Thar Desert.

Using Equation 11.3, the wind erosivity for four locations in the Thar Desert has been assessed. The threshold wind speed at 2 m height has been found to be about 5 m s⁻¹ (Fryrear et al. 1998), which again depends on soil surface characteristics. Gupta et al. (1981) and Ramakrishna et al. (1990) have found the daily threshold average wind speed for Chandan, Bikaner and Shergarh (Jodhpur) as 2.78, 1.39 and 1.11 m s⁻¹, respectively, at 3 m height. Averaging and conversion of these threshold values to standard the 2 m height resulted in a threshold wind speed of 1.66 m s⁻¹, which has been used in the above calculation. The Jaisalmer and Chandan sites have been found to be more erosive than other sites (Figure 11.5). The highest wind erosivity was observed during the second fortnight of July at Chandan (2.81 kg m⁻¹ s⁻¹).

The weather factor (*WF*) of RWEQ may be calculated from the wind erosivity factor (W_f) and soil wetness (*SW*) factor as follows:

$$WF = W_f \times SW \times SD \tag{11.4}$$

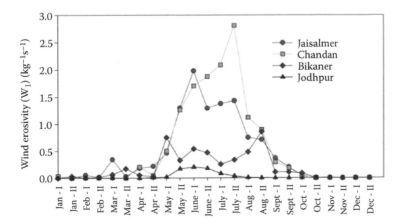

FIGURE 11.5 Wind erosivity (W_f) for four selected locations in western Rajasthan. (Daily wind speed data during 2000–2010 were used.)

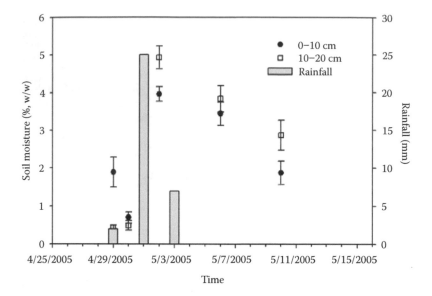

FIGURE 11.6   Soil moisture and its rate of depletion at Jaisalmer.

The snow depth factor may be considered negligible for hot arid situations. Soil moisture content is an important control factor for wind erosion, so a description of the soil moisture status is necessary for the prediction of wind erosion events for either a small region or for a large scale. For the prediction of wind erosion events on a small scale, frequent observations on the soil moisture condition are sufficient, but for those on a large scale, it is necessary to model the soil moisture change.

An attempt has been made to quantify surface as well as subsurface moisture conditions at the wind erosion experimental site of Jaisalmer. For this study, a time period of 2 weeks from 29 April to 10 May 2005 has been considered, when three rainfall events of 2, 25 and 7 mm occurred, respectively, on 29 April and 1 and 3 May. Soil moisture conditions in the soil thickness ranges 0–10, 10–20 and 20–30 cm were quantified gravimetrically for four locations of the experimental site on days 1, 2, 4, 8 and 12 days. Before rainfall events, soil moisture for the surface (0–10 cm) was observed as low as 0.40% (w/w), which was enhanced up to 3.96% (w/w) after three rainfall events, with a total rainfall of 34 mm. An increase in soil moisture (2.87%, w/w) was observed at the soil depth 10–20 cm, although the soil surface had dried, with the soil moisture being 1.88% (w/w) after 7 days from the last rainfall event of 7 mm on 3 May (Figure 11.6).

11.5.3   WIND VELOCITY PROFILE

Simultaneous measurement of wind velocity was also recorded for three heights at 1, 2 and 2.5 m above the surface. The mean wind velocity for three different heights (z) was fitted in the following logarithm equation and is shown in Figure 11.7.

$$U = \sqrt{\left(\frac{\tau}{\rho}\right)} \cdot \left(\frac{1}{k}\right) \cdot \ln\left(\frac{z + z_0}{z_0}\right) \tag{11.5}$$

where τ is the shearing stress, ρ is the density of air, k is the von Kármán constant (0.4) and Z_0 is the aerodynamic roughness length. The daily wind velocities measured at meteorological stations

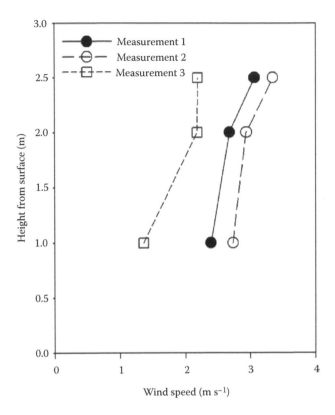

FIGURE 11.7   Wind velocity profile for three different measurements.

for three sites of the Thar Desert were analyzed to explore the seasonal wind velocity distribution. The cumulative distribution pattern of wind velocity for the year 2004 for Jaisalmer, Chandan and Bikaner is shown in Figure 11.8. The wind velocity patterns of Jaisalmer and Chandan coincide with each other. But the pattern for Bikaner is totally different from these two sites. Similarity in the patterns for Jaisalmer and Chandan is due to the short distance between them (~40 km), whereas Chandan is ~250 km away. For 80% of the days in a year, the average daily wind velocity is observed to be <3 m s^{-1} for Bikaner, whereas for Jiaslamer and Chandan, the wind velocity is observed to be greater than >5 m s^{-1} for 30% of the days in a year. This signifies that the wind regime at Jaisalmer and Chandan is more erosive than that in Bikaner.

11.6   POTENTIAL ENVIRONMENTAL HAZARD OF ERODED SOILS

Eroded soil mass, commonly known as dust, emitted during wind erosion events over desert areas remains in the atmosphere for a long time as suspended particles. This suspended particulate matter in the atmosphere has an adverse effect on the respiratory and cardiovascular activity of people, and thus is considered a health hazard (Santra 2006). Furthermore, eroded soil carries high concentrations of carbon and nutrients, thus reducing the soil fertility in source areas and enriching the nutrient contents in unwanted places through dust deposition, like in water bodies (Santra et al. 2006, 2013; Soni et al. 2013). Kayetha et al. (2007) and Singh et al. (2008) have shown an anomalous enhancement of chlorophyll-a concentrations in the Arabian Sea, from a monthly mean value of about 1.5 to >10 mg m^{-3} after the dust storm events, with a time lag of 1–2 or up to 3–4 days in some cases, and thus it strongly influences the marine ecosystem. It has also been reported that dust

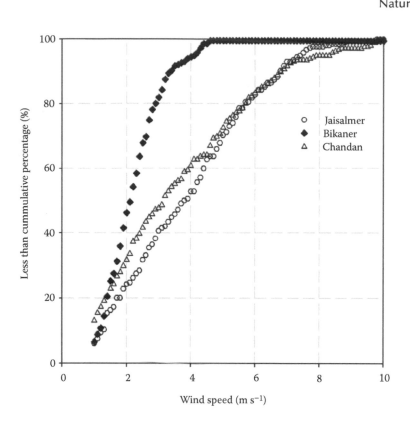

FIGURE 11.8 Cumulative distribution pattern of wind velocity.

deposition significantly reduces the snow albedo over Himalayan glaciers, and sometimes it is difficult to measure snow albedo from remote sensing platforms due to the presence of dust load in the atmosphere over snow glaciers. Moreover, these suspended particulate matters mix with other fine particles of the atmosphere, for example, carbon soot, smoke and salts, and form dense haze, fog and smog. The dust aerosols have an effect on radiative forcing, which is generally defined as the difference of insolation (sunlight) by the Earth and the energy radiated back to space at the top of the atmosphere. A positive forcing warms the system, whereas a negative forcing cools the system. The radiative forcing caused by the mineral dust aerosol depends on the characteristics of aerosol mixtures, whether dominated by absorbing-type particles or scattering-type particles. Dust aerosol also indirectly affects the radiative forcing by acting as condensation nuclei and thereby affecting cloud properties. A detailed review of radiative forcing by aerosol is given by Yu et al. (2006). Recently, the optical and radiative properties of aerosol over the northwestern part of India have been reported by Kant et al. (2015). In general, radiative forcing is governed by the optical parameters of aerosol (aerosol optical density or thickness [AOD/AOT], aerosol size distribution [ASD], angstrom exponent [α], single-scattering albedo [SSA] and imaginary and real parts of the refractive index [RI]). A high value of AOD (>1.0) reflects higher concentrations of aerosol in the atmosphere. The angstrom exponent is inversely related to the aerosol particle size; a small value reflects a larger particle size associated with dust storms. Due to southwesterly winds, dust from the Thar Desert is transported over the Indo-Gangetic Plain (Dey et al. 2004); the extent of dust depends on the meteorological conditions. Thus, a large increase in AOD ($\tau_{a550\,nm}$) during event days (>1) and a corresponding decrease in angstrom exponent ($\alpha_{440-870\,nm}$) (<0.2) was observed over the Indo-Gangetic Plain (Dey et al. 2004; Prasad and Singh 2007), which was also indicated by the presence of coarse particles in dust aerosol. ASD analysis has also shown an increase in the radius of the maxima value of aerosol particles from 1.71 to 2.24 μm with the advancement of the pre-monsoon season over the Indo-Gangetic Plain, implying the presence of higher concentrations of coarse particles in dust aerosol generated from the Thar Desert (Prasad and Singh 2007). Detailed characterization of aerosol and interannual variations over the source area, that is, Thar Desert, as well as over the Indo-Gangetic Plain for the period 2003–2008, shows an average AOD of >0.6 during the premonsoon period (Gautam et al. 2009). An SSA is defined as the ratio of scattering efficiency to the total extinction or attenuation efficiency, a sum of scattering and absorption. A value of SSA close to unity implies that all light extinction is due to scattering, whereas a value close to zero implies that light extinction is mainly due to absorption. The RI of any optical medium is a dimensionless number which describes how light propagates through the medium, and in the case of dust aerosol, it is measured by a complex value of RI, in which the real part accounts for refraction and the imaginary parts account for attenuation. The SSA of dust aerosol was observed to be >0.9 during dusty days over Kanpur located in the centre of the Indo-Gangetic Plain, along with a higher value of the real part of RI (1.5–1.6) and a lower value of the imaginary part of RI (<0.0043), which indicates the presence of a scattering type of aerosol (Prasad et al. 2007; Prasad and Singh 2007). Several studies from different parts of the world have shown a negative net radiative forcing for atmospheric aerosols that show a cooling effect, unlike greenhouse gases. An average change in radiative forcing by -23 W m^{-2} at the surface and by -11 W m^{-2} at the top of the atmosphere was observed by Prasad et al. (2007) during dust event days over the Indo-Gangetic Plain compared with non-dusty days. However, a low positive forcing of 0.7–1.2 K day^{-1} due to dust aerosol over the Thar Desert was reported by Moorthy et al. (2007), which was mainly due to the mixing of dust aerosol with other absorbing types of aerosol. A high degree of correlation between AOD at 500 nm and radiative forcing at the surface was also observed by Prasad et al. (2007). Bhattacharjee et al. (2007) have shown that dust storm events have a strong influence on weather parameters in the atmosphere, and they observed a pronounced enhancement of the CO mixing ratio in the upper atmosphere (~350 hPa) and the water vapour mixing ratio at low altitude (700–850 hPa). Apart from dust hazards, the deposition of eroded dust on roads, railway tracks and irrigation canals demands for regular cleaning operations, which have a large impact on the economy. Different environmental hazards associated with dust from the Thar Desert have been discussed by Narain et al. (2000) and Mertia et al. (2006).

11.6.1 Particulate Matter in Eroded Soil

Particle size analysis of collected aeolian samples during the dust storm event of 15 June 2009 shows that particulate matter having a size of less than 10 μm (PM$_{10}$) was 30% in eroded aeolian mass, whereas the same for desert surface was only 7.5%. Particle size distribution of eroded soil and surface soil is shown in Figure 11.9; fine particles were found to be higher in eroded soils than in the land surface, and therefore have a great role in aerosol loading in the atmosphere.

11.6.2 Nutrient Contents in Eroded Soil

Nutrient contents in wind-eroded soils were determined through laboratory analysis of soil carbon and nitrogen content. In general, it has been observed that eroded soils are rich in nutrients, which might be due to the loss of topsoil through the wind erosion process on which fertilizers or organic manure is applied for cultivation purposes. For example, the soil organic carbon (SOC) content of eroded aeolian mass during the dust storm event on 15 June 2009 was observed to be 2.6 g kg^{-1}, whereas the average SOC content of surface soil was 1.1 g kg^{-1}. This indicates that a large portion of the SOC pool may be lost through a single and short-duration dust storm event.

The average contents of total C and N in eroded soils from the experimental site at the Jaisalmer region of the Thar Desert have been found to be 4 and 0.37 g kg^{-1}, respectively (Santra et al. 2013). In general, both C and N contents were higher during regional dust storm events than during local wind erosion events. The average C and N contents of eroded soils for different sampling heights above the desert surface are given in Table 11.2. The C and N contents of eroded soil do not have

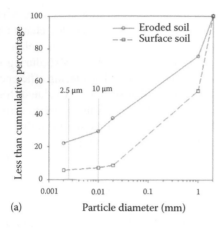

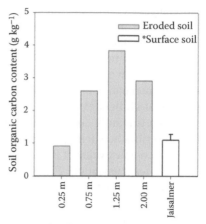

(b) Height of measurements and surface soil

FIGURE 11.9 Particle size distribution and SOC content of eroded soils during the dust storm event on 15 June 2009 and the same for surface soil in the Indian Thar Desert at Jaisalmer. (a) Particle size distribution of eroded soil and (b) SOC content of eroded soil at different heights. (From Santra, P. et al., *Curr. Sci.*, 99(8), 1061–1067, 2010.)

TABLE 11.2

Carbon (C) and Nitrogen (N) Contents of Eroded Soils at Different Sampling Heights during Wind Erosion Events at Jaisalmer

Sampling Heights	C (g kg⁻¹)[a]		§N (g kg⁻¹)	
	DSE	POWE	DSE	POWE
0.25 m	4.25 ± 0.30	3.97 ± 0.52	0.40 ± 0.03	0.34 ± 0.02
0.75 m	4.51 ± 0.33	3.50 ± 0.20	0.42 ± 0.04	0.34 ± 0.02
1.25 m	4.38 ± 0.40	3.47 ± 0.13	0.42 ± 0.05	0.33 ± 0.02
2.00 m	4.40 ± 0.53	3.44 ± 0.41	0.40 ± 0.04	0.34 ± 0.03
Topsoil	5.01 ± 0.18		0.44 ± 0.03	

Source: Santra, P. et al. *J. Agric. Phys.*, 13(1), 13–21, 2013.
Note: DSE, dust storm event; POWE, periodical observation on wind erosion.
[a] Values of C and N content are not significantly different between heights of observations ($p < 0.05$, Tukey test).

much significant ($p < 0.05$) variation with height during both dust storm and local wind erosion events. Due to the loss of nutrient-rich eroded particles from top soil, it reduces the soil's fertility. Sometimes deposition of these nutrient-rich particles at unwanted places can cause environmental hazards (eutrophication in water bodies).

11.7 DUST AEROSOL MONITORING THROUGH REMOTE SENSING

Dust aerosols during wind erosion events in the Thar Desert are transported for longer distances in the Indo-Gangetic Plain, influencing the weather conditions and (poor) air quality, and are considered health hazards (El-Askary et al. 2004, 2006; Deepshikha et al., 2005, 2006; Satheesh and Moorthy 2005; Satheesh et al. 2006; Moorthy et al. 2007). The details on operational remote sensing

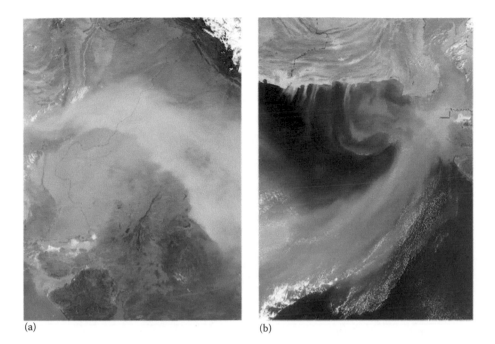

(a)   (b)

FIGURE 11.10   MODIS image of dust transport from source region of the Thar Desert to the Indo-Gangetic
Plain and intercontinental dust transport on (a) 10 June 2005 and (b) 14 December 2003.

for characterization of dust aerosols over land are discussed by Kaufman et al. (1997). El-Askary
et al. (2004, 2006) have also shown the use of multisensor satellite data for characterizing the dust
aerosols. Dust aerosol is generally characterized through satellite remote sensing using either the
Absorbing Aerosol Index (AAI) derived from satellite-measured reflectance in the ultraviolet (UV)
region (Prospero et al. 2002) or the Infrared Difference Dust Index (IDDI) derived from satellite-
measured infrared reflectance (11.5–12.5 µm). A clear view of the regional transport of dust aerosol
on 10 June 2005 is shown in Figure 11.10 using a MODIS satellite image; the intercontinental dust
transport from the Sahara Desert of Africa to the Thar Desert of Asia is clearly visible.

Dust aerosol generated through the wind erosion process was also monitored through MODIS
atmospheric products. The AOT at 0.47 µm over the Thar Desert was mapped during severe dust
storm events. It was found from AOT maps that dust aerosol was more active in the western part of
Jaisalmer and nearby areas (Figure 11.11). A higher value of AOT represents the aerosol concentration content in the atmosphere, which is generally measured by the negative logarithm of the ratio
of spectral irradiance at the top of the atmosphere and on the Earth's surface.

11.8   CONTROL OF GRAZING AND REDUCTION IN WIND EROSION

Control of grazing in the vast rangelands of the Thar Desert may reduce the extent of soil loss
through wind erosion by providing protective grass cover on the surface. In order to quantify the
possible reduction in soil loss through controlled grazing, wind-eroded aeolian mass fluxes (kg m^{-2}
day^{-1}) were measured by a dust catcher at an overgrazed rangeland site and a controlled grazing
rangeland site of the Thar Desert during the summer season of 2004. Most of the erosion-causing
factors at both these sites were similar, except the level of grazing pressure. In the overgrazed site,
3.2 adult cattle units (ACU) ha^{-1} was grazed throughout the year, whereas in the case of controlled
grazing, 2 ACU ha^{-1} was grazed for about 3 months in a year (Mertia et al. 2010). Computation
of mass transport from the observed mass flux data revealed that the average mass transport rate
(kg m^{-1} day^{-1}) during different observation periods was three to five times higher at the overgrazed

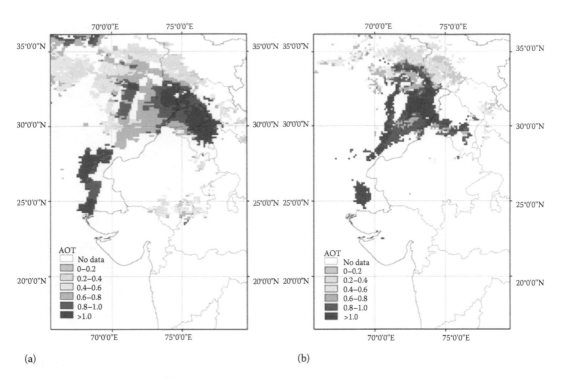

FIGURE 11.11 AOT near Jaisalmer and nearby areas during (a) 8 July 2009 and (b) 9 July 2009. Higher values of AOD show a denser dust aerosol cloud in the atmosphere. (Data from MODIS aerosol level 2 product, MOD04_L2.)

FIGURE 11.12 Wind erosion experimental site at Bhujawar village, Jodhpur. (a) Fallow land for 4–5 years. (b) Continuously cultivated land.

rangeland than at the controlled grazing rangeland site, especially during severe dust storm events. The total amount of soil loss during the summer season (May–July) of 2004 was 827 kg ha^{-1} at the overgrazed site, whereas it was only 240 kg ha^{-1} at the controlled grazing site. These soil loss data from the rangelands of the Thar Desert clearly show that control of grazing at the existing rangelands of the desert may greatly reduce the amount of wind-eroded soil loss, and consequently its negative impact on the environment. Recently, a reduction in loss of fertile soil through keeping the land fallow for 3–5 years, followed by cultivation of 2–3 years in rotation, instead of continuous cultivation, has been demonstrated at Bhujawar village under the UNESCO-aided program Sustainable Management of Marginal Drylands (SUMAMAD – Phase II) (Figure 11.12). Initial observation during 2012 revealed that the wind-eroded mass flux at 0.25 m height from the surface in continuously cultivated land was 0.14 kg m^{-2} day^{-1} during May–June 2012, whereas in grass-covered fallow land, it was almost absent.

11.9 CONCLUSION

Wind erosion in the Thar Desert poses a serious threat to the normal livelihood of all biotic components. The grazing pressure on rangelands is an additional serious problem. Soil loss data measured in the rangeland of the Thar Desert with two grazing situations clearly indicate that simple regulation of grazing through boundary fencing or social fencing may greatly reduce the amount of wind-eroded soil loss, and consequently its negative impact on the environment. At present, a large portion of the geographical area of the Thar Desert in India lies under unprotected rangeland systems, which are therefore overgrazed and thus continue to contribute a significant amount of aeolian sediment load to the atmosphere. Therefore, protection of rangeland sites through fencing and regulation of grazing activity is highly essential to sustain the livelihood of the region, as well as to mitigate future environmental hazards from severe dust storm events. Focus should also be given to developing rangelands in the desert which will conserve biodiversity, as well as minimize the soil loss due to wind erosion. Such goals may be achieved by bringing rules and laws and strictly implementing rangeland policy in the region through community participation.

REFERENCES

Bhattacharjee, P. S., Prasad, A. K., Kafatos, M., and Singh, R. P. (2007). Influence of a dust storm on carbon monoxide and water vapour over the Indo-Gangetic Plains. *Journal of Geophysical Research: Atmosphere*, 112, D18203.

Bhimaya, C. P., and Chowdhary, M. D. (1961). Plantation of windbreaks in the central mechanised farm, Suratgarh. *Indian Forester*, 87, 354–367.

Bhimaya, C. P., Kaul, R. N., and Ganguli, B. N. (1961). Sand dune rehabilitation in western Rajasthan. In *Proceedings of the Fifth World Forestry Congress*, Seatle, USA, 358–363.

Bittu, B. D. (1989). Problem of soil erosion and its control in western Rajasthan. In *Proceedings of the International Symposium on Managing Sandy Soils, Part II*, Jodhpur, India, 551–555.

Deepshikha, S., Satheesh, S. K., and Srinivasan, J. (2005). Regional distribution of absorbing efficiency of dust aerosols over India and adjacent continents. *Geophysical Research Letters*, 32(3), L03810.

Deepshikha, S., Satheesh, S. K., and Srinivasan, J. (2006). Dust aerosols over India and adjacent continents retrieved using METEOSAT infrared radiance: Part I. Sources, regional distribution and radiative effects. *Annals Geophysicae*, 24(1), 37–61.

Dey, B. N. (1957). Reclamation of sand dunes and other shifting sand dunes. West Bengal Forest Bulletin.

Dey, S., Tripathi, S. N., Singh, R. P., and Holben, B. N. (2004). Influence of the dust storms on the aerosol optical properties over the Indo-Gangetic basin. *Journal of Geophysical Research*, 109, D20211.

Dhir, R. P. (1961). Characterization and properties of dune and associated soils. In *Sand Dune Stabilization, Shelterbelts and Afforestation in Dry Zones (FAO Conservation Guide)*. Rome: Food and Agriculture Organization of the United Nations, 41–49.

Dhir, R. P., and Venkateswarlu, J. (1996). Problem of wind erosion in relation to landuse and management and control measures. *Indian Society of Soil Science Bulletin*, 17, 56–65.

El-Askary, H., Gautam, R., and Kafatos, M. (2004). Remote sensing of dust storms over the Indo-Gangetic basin. *Journal of the Indian Society of Remote Sensing*, 32(2), 121–124.

El-Askary, H., Gautam, R., Singh, R. P., and Kafatos, M. (2006). Dust storms detection over the Indo-Gangetic basin using multi sensor data. *Advances in Space Research*, 37(4), 728–733.

Fryrear, D. W. (1986). A field dust sampler. *Journal of Soil Water Conservation*, 41, 117–120.

Fryrear, D. W., Bilbro, J. D., Saleh, A., Schomberg, H. M., Stout, J. E., and Zobeck, T. M. (2000). RWEQ: Improved wind erosion technology. *Journal of Soil and Water Conservation*, 55(2), 183–189.

Fryrear, D. W., Saleh, A., and Bilbro, J. D. (1998). A single event wind erosion model. *Transactions of the American Society of Agricultural Engineers*, 41(5), 1369–1374.

Gautam, R., Liu, Z., Singh, R. P., and Hsu, N. C. (2009). Two contrasting dust-dominant periods over India observed from MODIS and CALIPSO data. *Geophysical Research Letters*, 36, L06813.

Goossens, D., Offer, Z., and London, G. (2000). Wind tunnel and field calibration of five aeolian sand traps. *Geomorphology*, 35(3–4), 233–252.

Gupta, J. P. (1993). Wind erosion of soil in drought-prone areas. In *Desertification and Its Control in the Thar, Sahara and Sahel Regions*, ed. A. K. Sen and A. Kar. Jodhpur, India: Scientific Publishers, 91–105.

Gupta, J. P., and Aggarwal, R. K. (1978). Sand movement studies under different land use conditions of western Rajasthan. Presented at Proceedings of the International Symposium of Arid Zone Research & Development, Jodhpur, India.

Gupta, J. P., Aggarwal, R. K., and Raikhy, N. P. (1981). Soil erosion by wind from bare sandy plains in western Rajasthan, India. *Journal of Arid Environments*, 4, 15–20.

Kant, Y., Singh, A., Mitra, D., Singh, D., Srikanth, P., Madhusudanacharyulu, A. S., and Krishna Murthy, Y. N. V. (2015). Optical and radiative properties of aerosols over two locations in the north-west part of India during premonsoon season. *Advances in Meteorology*, 2015, 517434.

Kaufman, Y. K., Tanre, D., Gordon, H. R., Nakajima, T., Lenoble, J., Frouin, R., Grassl, H., Herman, B. M., King, M. D., and Teillet, P. M. (1997). Passive remote sensing of tropospheric aerosol and atmospheric correction for the aerosol effect. *Journal of Geophysical Research*, 102(D14), 16815–16830.

Kaul, R. N. (1959). Shelterbelt to stop creep of the desert. *Indian Forester*, 85, 191–195.

Kayetha, V. K., Senthilkumar, J., Prasad, A. K., Cervone, G., and Singh, R. P. (2007). Effect of dust storm on ocean color and snow parameters. *Journal of the Indian Society of Remote Sensing*, 35(1), 1–9.

Krishnan, A. (1977). Climatic changes relating to desertification in the arid zone of North West India. *Annals of Arid Zone*, 16(3), 302–309.

Lal, R. (1990). *Soil Erosion in the Tropics: Principles and Management*. New York: McGraw-Hill.

Mann, H. S. (1985). Wind erosion and its control. In *Sand Dune Stabilization, Shelterbelts and Afforestation in Dry Zones (FAO Conservation Guide)*. Rome: Food and Agriculture Organization of the United Nations, 125–132.

Mann, H. S., Malhotra, S. P., and Shankarnarayana, K. A. (1977). Land and resource utilization in arid zone. In Desertification and Its Control. New Delhi: ICAR.

Mertia, R. S. (1982). Studies on improvement and utilization of rangelands of Jaisalmer region. CAZRI Monograph No. 38. Jodhpur, India: Central Arid Zone Research Institute, 1–45.

Mertia, R. S., Prasad, R., Gajja, B. L., Samra, J. S., and Narain, P. (2006). Impact of shelterbelts in arid region of western Rajasthan. Jodhpur, India: Central Arid Zone Research Institute, 76.

Mertia, R. S., Santra, P., Kandpal, B. K., and Prasad, R. (2010). Mass-height profile and total mass transport of wind eroded aeolian sediments from rangelands of Indian Thar Desert. *Aeolian Research*, 2, 135–142.

Moorthy, K. K., Suresh Babu, S., Satheesh, S. K., Srinivasan, J., and Dutt, C. B. S. (2007). Dust absorption over the "Great Indian Desert" inferred using ground-based and satellite remote sensing. *Journal of Geophysical Research*, 112, D09206.

Narain, P., Kar, A., Ram, B., Joshi, D. C., and Singh, R. S. (2000). Wind Erosion in western Rajasthan [Bulletin]. Jodhpur, India: Central Arid Zone Research Institute.

Prasad, A. K., and Singh, R. P. (2007). Changes in aerosol parameters during major dust storm events (2001–2005) over the Indo-Gangetic Plains using AERONET and MODIS data. *Journal of Geophysical Research*, 112, D09208.

Prasad, A. K., Singh, S., Chauhan, S. S., Srivastava, A. K., Singh, R. P., and Singh, R. (2007). Aerosol radiative forcing over the Indo-Gangetic Plains during major dust storms. *Atmospheric Environment*, 41, 6289–6301.

Prospero, J. M., Ginoux, P., Torres, O., Nicholson, S. E., and Gill, T. E. (2002). Environmental characterization of global sources of atmospheric soil dust identified with the NIMBUS 7 total ozone mapping spectrometer (TOMS) absorbing aerosol product. *Review of Geophysics*, 40(1), 2-1–2-31.

Ramakrishna, Y. S., Agnihotri, C. L., and Rao, R. S. (1992). Climatic resources of Thar Desert and adjoining arid areas of India. In *Perspectives of Thar and Karakum*, ed. A. Kar, R. K. Abhichandrani, K. Ananth Ram, and D. C. Joshi. New Delhi: Department of Science and Technology, Government of India, 10–20.

Ramakrishna, Y. S., Rao, A. S., Singh, R. S., Kar, A., and Singh, S. (1990). Moisture, thermal and wind measurements over selected stable and unstable sand dunes in the Indian desert. *Journal of Arid Environments*, 19, 25–38.

Ramakrishna, Y. S., Rao, G. G. S. N., and Ramana Rao, B. V. (1987). Dust storm and associated weather changes in an arid environment. *Contribution to Human Biometeorology, Progress in Biometeorology*, ed. W. Selvamurthy, SPB Academic Publishing, The Hague, The Netherlands 4, 11–15.

Santra, P. (2006). Air pollution through particulate matter and its impact on human health, *Science Reporter*, June, 28–29.

Santra, P., Mertia, R. S., Kumawat, R. N., Sinha, N. K., and Mahla, H. R. (2013). Loss of soil carbon and nitrogen through wind erosion in the Indian Thar Desert. *Journal of Agricultural Physics*, 13(1), 13–21.

Santra, P., Mertia, R. S., and Kushawa, H. L. (2010). A new wind erosion sampler for monitoring dust storm events in the Indian Thar Desert. *Current Science*, 99(8), 1061–1067.

Santra, P., Mertia, R. S., and Narain, P. (2006). Land degradation through wind erosion in Thar Desert – Issues and research priorities. *Indian Journal of Soil Conservation*, 34(3), 214–220.

Satheesh, S. K., and Moorthy, K. K. (2005). Radiative effects of natural aerosols: A review. *Atmospheric Environment*, 39(11), 2089–2110.

Satheesh, S. K., Moorthy, K. K., Kaufman, Y. J., and Takemura, T. (2006). Aerosol optical depth, physical properties and radiative forcing over the Arabian Sea. *Meteorology and Atmospheric Physics*, 91, 45–62.

Schwab, R. O., Fangmeier, D. D., Elliot, W. J., and Frevert, R. K. (1993). *Soil and Water Conservation Engineering*. New York: John Wiley & Sons.

Sehgal, J. L., and Abrol, I. P. (1994). *Soil Degradation in India: Status and Impact*. New Delhi: Oxford & IBH Publishing.

Singh, R. P., Prasad, A. K., Kayetha, V. K., and Kafatos, M. (2008). Enhancement of oceanic parameters associated with dust storms using satellite data. *Journal of Geophysical Research*, 113, C11008.

Singh, R. S., Rao, A. S., Ramakrishna, Y. S., and Prabhu, A. (1992). Vertical distribution of wind and hygrothermal regime during a severe dust storm—A case study. *Annals of Arid Zone*, 31(2), 153–155.

Soni, M. L., Yadava, N. D., Beniwal, R. K., Singh, J. P., Kumar, S., and Meel, B. (2013). Grass based strip cropping systems for controlling soil erosion and enhancing system productivity under drought situations for hot arid western Rajasthan. *International Journal of Agricultural Statistical Science*, 9(2), 685–692.

Wilson, S. J., and Cook, R. U. (1980). Wind erosion. In *Soil Erosion*, ed. M. J. Kirkby and R. P. C. Morgan. Chichester, UK: Wiley, 217–251.

Yu, H., Jaufman, Y. J., Chin, M., Feingold, G., Remer, L. A., Anderson, T. L., Balkanski, Y. et al. (2006). A review of measurement-based assessments of the aerosol direct radiative effect and forcing. *Atmospheric Chemistry and Physics*, 6, 613–666.

12 Coastal Subsidence
Causes, Mapping and Monitoring

*Andrea Taramelli, Ciro Manzo, Emiliana Valentini
and Loreta Cornacchia*

CONTENTS

12.1   INTRODUCTION

Subsidence is a local and widespread phenomenon, appearing as low-lying areas, particularly important along several littorals because the coastal system consists of beaches and coastal wetlands. Ground subsidence has many causes that can be divided into two broad categories: shallow causes due to, for example, peat oxidation, compaction and groundwater extraction, and deep causes due to, for example, tectonic phenomena and the effects of mining and extraction activities (gas and salt). Although subsidence can be a direct hazard (e.g. landslides and sinkholes), more commonly it is associated with hazards that are exacerbated by subsidence, which is a greater threat to land use. Subsidence is generally underacknowledged as a geohazard; indeed, subsidence is generally too slow for human perception, and the effects are often invisible, at least before structural damage appears. This makes subsidence an insidious threat, which may proceed undetected for decades, for example, having significant cumulative effects on flood risk or the integrity of water defences and infrastructure.

Coastal lowland areas are widely recognized as being highly vulnerable to the impacts of climate change, particularly sea level rise (SLR) and changes in runoff, as well as being subject to stresses imposed by the human modification of catchments and delta plain land use (Costanza et al. 2011). The utilization of coastal areas has increased dramatically during the twentieth century, a trend that seems certain to continue through the twenty-first century. It has been estimated that 23% of the world's population lives both within a 100 km distance of the coast and <100 m above sea level, and population densities in coastal regions are about three times higher than the global average (Nicholls et al. 2011). Sixty percent of the world's 39 metropolises with a population of more than 5 million are located within 100 km of the coast, including 12 of the world's 16 cities with populations greater than 10 million (Hallegatte et al. 2013).

The rates of relative sea level rise can greatly exceed the global average in many heavily populated coastal lowland areas due to subsidence (Syvitski et al. 2009). Natural subsidence due to autocompaction of sediment under its own weight is enhanced by subsurface fluid withdrawals and drainage. This increases the potential for inundation, coastal erosion, habitat disruption and saltwater intrusion, especially for the most populated cities in these coastal lowland areas (Bucx et al. 2015). Aside from regional environmental effects of subsidence, there are direct costs related to subsidence and soft soil conditions experienced in coastal lowland areas (Nicholls and Cazenave 2010). The failure of constructions, infrastructure and water defence structures brings high maintenance costs with it and adds up to substantial financial damages in these areas (Chen et al. 2005). Typically, rates of subsidence vary over various spatial scales and depend strongly on local geological conditions and human activity (Nicholls 2004) (Figure 12.1).

In the wake of climate change, the adverse effects can be seen in the coastal lowland areas, and from the perspective of sustainable management of infrastructural assets in these areas, there is an urgent need for adequate data and information to assess

- Spatial variations in subsidence in coastal lowland areas in relation to SLR and its impact on flood risk
- Geographical and temporal variations in subsidence and their impact on the geomorphological, ecological and hydrological systems in coastal lowland areas to anticipate necessary adaptive measures

The holistic view of the environment that scientists apply to study and interpret landscape evolution is based on the principle that the scale of observation of a certain physical process determines the level at which spatial patterns of landscapes can be explained. This means that a physical process, such as subsidence, with all the included drivers, pathways and receptors, may influence ecosystem functions such as coastal vegetation growth, as such processes can bring disturbance and determine perturbation or stress (Taramelli et al. 2015a).

These last three terms, *disturbance*, *perturbation* and *stress*, are widely used to describe factors and phenomena that occur in the evolution of systems, and they each allow the definition of the role of subsidence in landscape analysis and the feedback between landscapes and geomorphology that must be considered in coastal modelling.

Natural hazards are the 'cause' of disturbance when they drive disruption and losses, and their action is simultaneously the 'effect' of disturbance when it produces changes in the main pattern of vegetation after extreme events (Passalacqua et al. 2013; Taramelli et al. 2017). Subsidence in this

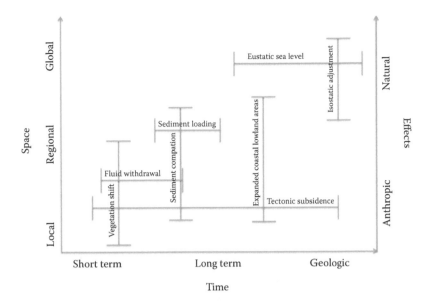

FIGURE 12.1 Temporal and spatial scales of the different processes contributing to the subsidence and relative effects related to the spatial scale. Each process spans a unique distribution of temporal and spatial scales. Further, for each scale, each process may contribute to the net observed subsidence at different rates. At some scales, a specific process may not be relevant at all. The timescales defined in this diagram are approximately quantified as short term (0–1 year), long term (20–400 years) and geologic (>400 years), although the defined timescales would entail overlap. (After Yuill, B. et al., *J. Coastal Res.*, 54, 23–36, 2009.)

context can be considered a complex source of deformations in the steady states of a system; therefore, a reference state of a subsiding ecosystem must be defined spatially and temporally (Taramelli et al. 2014). Although the spatial influence of subsidence can be wider than the specific area of interest due to subsequent flooding and erosion, it can be defined with different levels of precision based on the spatial scale of observation.

Temporal observations include defining the disturbance typology as chronic or continuous and specific conditions of disturbance that generate perturbation or stress. A continuous rate of subsidence determined at large temporal scales shows adapted vegetation with species that are resistant to altimetry variation; however, at shorter temporal ranges, such as groups of rainless years, the intervention of higher rates of subsidence can drive the extinction of several of the most sensitive species (Van Dobben and Slim 2005, 2012; Kaneko and Toyota 2011; Taramelli et al. 2015b).

Moreover, subsidence in dynamic ecosystems, such as coasts, can sometimes represent the continuous inputs required to maintain the viability and organization of living components. Thus, there are ecosystems in a non-equilibrium state that are responding continuously to environmental gradients (rainfall and subsidence) and ecological interactions (competition and extinction) (Taramelli et al. 2015b; Scarelli et al. 2017; Strain et al. 2017). Normal forcing and processes may be considered disturbances when the nominal bounds are exceeded (Pimm 1984; Corenblit et al. 2011). Therefore, the challenge is not only in determining how subsidence is a cause and effect disturbance but also in establishing when a normal force becomes a disturbance by establishing thresholds. Another aspect in analyzing the feedback between subsidence and coastal vegetation is that disturbances are not strictly negative phenomena; when disruptive events free resources or open corridors for cross-species changes, the biodiversity and stability of communities can increase (Reinhardt et al. 2010). Stress is an effect of disturbance and can be considered a specific perturbation characterized by specific variations from the steady state induced by the propagation of disturbance (Antonellini and Mollema 2010; Wang and Temmerman 2013). These phenomena occur for both natural and anthropogenic reasons that may have a synergic effect and increase the total displacement, particularly for heavily populated coastal lowland areas. The main terms to understand the language of feedback at the interface between landscape and physical processes are summarized in Table 12.1.

Natural and anthropogenic pressures may have a synergic effect on the subsidence, thus increasing total displacement (Bucx et al. 2015).

For example, in heavily populated coastal lowland areas, natural subsidence due to the autocompaction of sediment under its own weight is enhanced by subsurface fluid withdrawals, drainage and ground overburden, which increase the potential for inundation, coastal erosion, habitat disruption and saltwater intrusion (Figure 12.2). Coastal subsidence in these depositional zones is a function of five processes: (1) downwarping; (2) tectonic activity; (3) consolidation of Tertiary, Pleistocene and Holocene deposits; (4) shallow subsidence; and (5) underground water and gas extraction (Figure 12.2). When only natural processes are considered, the sum of 1 + 2 + 3 defines deep subsidence. The combination of shallow and deep subsidence equals the total subsidence (Cahoon et al. 1995).

Improvements by new monitoring techniques that produce more accurate results by remote sensing have increased the study of the spatial and temporal evolution of this phenomenon and allowed the construction of a series of simulation models designed to reduce risks to the environment (Bateson et al. 2011), for example, the 'dynamic digital elevation model' (DEM) concept developed by the SUBCOAST FP7 project (Hamersley 2012). As stated by Hung et al. (2010), considerations when selecting a monitoring system are a (1) high spatial sampling density, (2) good measurement accuracy and (3) high temporal frequency. However, the selection is usually made according to the available funding. Various methods have been used for measuring and mapping the spatial gradients and temporal rates of regional and local subsidence and horizontal ground motion (Galloway et al. 1999; Tosi et al. 2013; Crosetto et al. 2016). The methods generally measure relative changes in the position of the land surface, and the observable position is typically a geodetic reference mark that

---

TABLE 12.1

Summary of Concepts and Real Examples of Perturbation, Stress and Disturbance

	Theoretical Definition	Real-World Examples
Disturbance (Bazzaz 1983)	A physical process, force or agent, either abiotic or biotic, causing a perturbation (which can include stress) in an ecological component or system with characteristics relative to a specified reference state and system **Pulse:** When it acts on the system in a pulsing way **Continuous:** When it always acts on the system	Under subsidence effects, biomass can be eliminated, added, reduced or expanded, and matter or energy fluxes can be interrupted or prevented
Perturbation (Odum et al. 1979)	An effect of the disturbance on ecological components relative to a reference condition with a direction, magnitude and persistence **Temporary:** When the system returns to an approximately original steady state **Permanent:** When deviation leads to a different steady state of the system	Water pumping–driven subsidence determines salt intrusion and sensitive species distribution changes with gain and losses of coastal pine wood and dune shrubs
Stress (Barrett and Rosenberg 1981)	A physiological or functional effect relative to a specified reference condition	Where coastal forestry is attacked by salt intrusion, loss of biomass leads to drought and dunes move in the backshores, stressing in a chronic way the perturbed vegetation

Source: Rykiel, E. J., *Aust. J. Ecol.*, 10(3), 361–365, 1985.

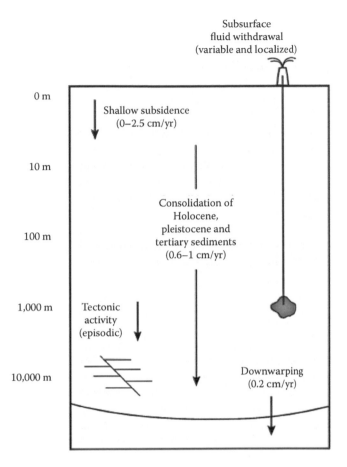

FIGURE 12.2 Processes contributing to total subsidence in coastal regions. Comments and numbers in parentheses indicate the relative magnitude of each process in the Mississippi River Delta, a region where coastal subsidence has been intensely studied. (After Penland, S. et al., Relative sea level rise and delta-plain development in the Terrebonne Parish region, Coastal Geology Technical Report, Louisiana Geological Survey, Baton Rouge, 1988.)

was established so that any movement can be attributed to deep-seated ground movement rather than surficial effects.

Supported by advancements in space-based technologies and combining multisensor space-borne remote sensing (Synthetic Aperture Radar [SAR] and optical) to estimate the variability and associated uncertainties, the aim of this chapter is to review and discuss the causes of coastal subsidence, mapping and monitoring techniques, including satellite-based methods. Different sites will be considered examples of relevant processes and monitoring techniques in subsidence areas. The main objective of this chapter is to develop a structured analysis for the delivery of reliable information about subsidence for relevant local, regional, international and national government managers and scientists, thus covering

- The different causes (both natural and anthropogenic) of the phenomena
- The effects on vegetation as a primary proxy
- The innovative monitoring methodologies
- The 'market' applicability of different real test cases, providing proper follow-up and possibilities to increase the implementation of the new methodologies

12.2 CAUSES

The causes of subsidence are often complex and involve both natural and anthropogenic factors. For this reason, it is important to have different sources of data to monitor these phenomena.

12.2.1 NATURAL CAUSES

Natural subsidence is the result of a series of processes which occur in the ground, causing a vertical displacement of the topographic surface. The natural subsidence can be split into two timescale components: a long-term component controlled by tectonics and geodynamics that is active for periods of millions to hundreds of thousands of years and a short-term component that is most likely controlled by climatic changes (glacial cycles) and active for periods of thousands of years. The main natural triggers are

- Isostasy (glacio-isostasy, sediment load on the substrate)
- Geostatic load (sediment consolidation and/or compaction)
- Tectonics

12.2.2 ISOSTASY

Isostasy is essentially related to the Archimedes principle. It states that continents and high topography are buoyed up by thick continental roots floating in a denser mantle, much like icebergs floating in water. The principle of isostasy states that the elevation of any large segment of crust is directly proportional to the thickness of the crust (Kusky 2005). Lambeck and Purcell (2005) noticed that Scandinavian areas that have recently been glaciated were rising quickly relative to sea level, and they equated this observation with the principle of isostatic rebound. Isostatic rebound is accommodated by the flow of mantle material within the zone of low viscosity beneath the continental crust to compensate for the rising topography. These observations revealed that uplifting is still active and has a rate of 10 mm/year.

Isostatic processes also involve the subsidence process identified as downwarping, which is a crust deformation caused by significant sediment overload. It is active where massive piles of sediments are located in relatively small areas, such as valleys, lakes and deltas (Paola et al. 2001). For example, in the Mississippi River Delta, the sediment pile in geosyncline in the front of the river mouth forms quaternary deposits layered up to 12.000 m in thickness (Penland et al. 1988); the weight of the sedimentary layer and groundwater leads to crust deformation and relative subsidence effects (Coleman and Prior 1982; Blum et al. 2008). If the vertical displacement linked to downwarping is not balanced by a sedimentation rate, subsidence occurs in the area (Figure 12.3) (Yuill et al. 2009).

Crustal flexure from sediment loads is often modelled using elastic or viscoelastic models that predict crustal strain using inputs of imposed loads (i.e. stress), mantle viscosity and lithosphere thickness (Kulp 2000; Watts 2001).

12.2.3 GEOSTATIC LOAD

The volume of the deposited sediments naturally decreases in time as the sediment grains settle through mechanical processes, eventually resulting in compaction and subsidence. Physical sediment compaction occurs in two primary ways: through the expulsion of pore fluid and through the reorientation of sediment grains into a more tightly packed arrangement. Engineers refer to these processes as primary and secondary consolidation, respectively (Figure 12.4) (Yuill et al. 2009). Primary consolidation is a process occurring in low-permeability soils, such as clays in saturated conditions. When clay sediments undergo overloading for an extended period, water is released due to water pressure, and a reduction in the void ratio follows.

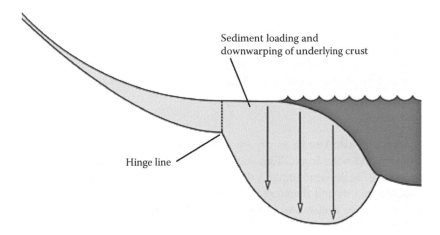

FIGURE 12.3   Downwarping in the delta area: sediment loading deforms the lithosphere, triggering subsidence.

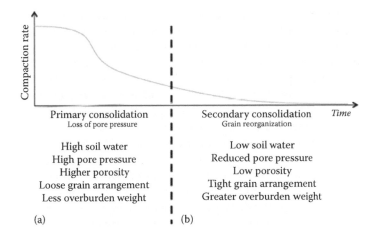

FIGURE 12.4   Plot of the (a) relative compaction rates and (b) sediment properties.

The time process of consolidation is a function of permeability, geostatic load and ground drainage (Hillel 1992). The consolidation of Tertiary and Pleistocene marine and riverine deposits can be a significant component of total subsidence. However, it is the consolidation of recent sediments that is considered the primary cause of background subsidence in coastal regions, with high values near deltas. This process is usually gradual, but it may have high rates in organic soils (the presence of peat).

12.2.4   Shallow Subsidence

The shallow subsurface frequently contains young, compressible deposits. These soft deposits are particularly vulnerable to subsidence caused by natural compaction. The compaction in the upper part is a function of permeability and contributes a shallow component that is often overlooked when calculating total subsidence in coastal regions (Cahoon et al. 1995). Shallow subsidence is the result of primary consolidation and the decomposition of organic matter. It is an especially important process in coastal systems under stress (e.g. due to flooding or a salt wedge moving forward), where the presence of organic matter in the soil, such as roots and rhizomes, and its decomposition leads to rapid subsurface collapse. In the Mississippi River Delta, for example, high rates of relative sea level rise (RSLR) in the coastal marshes have led to increasingly long periods of flooding, plant mortality and rates of shallow subsidence that exceed 2.5 cm/year (Chen and Rybczyk 2005). Natural compaction is causing slow subsidence in numerous deltas, such as in the Mississippi and Po deltas at 5 and 2 mm/year, respectively (Törnqvist et al. 2008; Dokka 2011; Teatini et al. 2011).

12.2.5 TECTONIC SUBSIDENCE

Tectonic subsidence is also known as driving subsidence and is distinguished from the isostatic effects of sediment and water loads. Tectonic forces that affect how continental lithosphere floats on the asthenosphere, as its name implies, drive tectonic subsidence. Three main mechanisms that affect this isostatic balance and therefore drive tectonic subsidence include stretching, cooling and loading. Stretching of the continental lithosphere results in the replacement of the relatively light continental lithosphere with a denser asthenosphere in most situations. The resulting stretched and thinned lithosphere sinks, causing tectonic subsidence. Stretching occurs in several types of sedimentary basins, including rifts, aulacogens, back-arc basins and cratonic basins.

Cooling commonly goes hand in hand with stretching. During stretching, continental lithosphere is heated, becomes less dense and tends to uplift from its decreased density (the net effect in a stretched and heated basin may result in either uplift or subsidence). As continental lithosphere cools, it becomes denser and subsides. Cooling subsidence decreases exponentially with time unless it can cause a significant amount of subsidence hundreds of millions of years following initial cooling. Cooling subsidence is especially important on passive margins and in cratonic basins.

Tectonic loading can also produce subsidence. The additional weight of tectonic loads, such as accretionary wedges or folds and thrust belts, causes continental lithosphere to sink, leading to tectonic subsidence (Bally et al. 1985). Because the lithosphere responds flexurally, the subsidence occurs not only immediately underneath the load but also in broad regions surrounding the load (Sorichetta et al. 2006). Tectonic loading is particularly important in orogenic regions such as foreland basins.

A region where subsidence is dominated by tectonic movement is Bangladesh, in the Bengal basin, formed during the Late Cretaceous when the Indian platform broke away from Gondwana and went towards the Euroasiatic platform (Tapponnier et al. 1986). Since its 2002 launch, Gravity Recovery and Climate Experiment (GRACE) has provided continuous measurements of the global mass flux at seasonal and longer timescales, with spatial resolutions of 300 km or higher (Tapley et al. 2004). The GRACE mission data products are a time sequence of approximately monthly global gravity field estimates (Bettadpur 2007), and these primarily represent the time variability in the distribution of surface and ground water, ice and oceans. The area centred on Bangladesh shows the second largest seasonal anomaly (after the Amazon basin), but the area of large hydrological variability extends beyond Bangladesh. The second largest seasonal anomaly observed in the GRACE gravity field is associated with terrestrial water storage in Southeast Asia centred over the Ganges–Brahmaputra Delta (GBD). The summer monsoon, together with snowmelt from the Himalayas, increases the river flow by greater than a factor of 10, overwhelming the river system. The water table rises up to 12 m, impounding a vast quantity of water in the GBD and producing widespread flooding. The mass of this water produces the large Southeast Asian anomaly in the GRACE gravity field. The weight of the >100 GT of water impounded in Bangladesh acts as a load that depresses the surface of the Earth. A seasonal vertical deflection of 5–6 cm was observed in the global positioning system (GPS) network deployed in Bangladesh (Steckler et al. 2010). Bettinelli et al. (2008) observed similar deformation from GPS stations in Nepal, north of the Gangetic Plain. Time histories of load and deformation are coherent, showing that the response to the load at this timescale is primarily elastic (Wahr et al. 2004; Bevis et al. 2005). The size and shape of this deformation is a function of the elastic properties of the lithosphere. Moreover, SLR in the GBD region has heavy consequences due to shoreline variation; oceanographic and satellite measurements estimate a sea level rising of 1–2 mm/year. We can quantify the interaction of subsidence and sea level, causing an increase of flooding risk in the delta area, particularly during the monsoon season with heavy rain (Syvitski et al. 2009).

However, the role of tectonic subsidence in this dispersal pattern is not well understood. Published and widely cited subsidence rates in excess of 2.5 cm/year are thought to be erroneous, but continued reports of buried artefacts and tree stumps on the lower coastal plain suggest that local, short-term rates of this magnitude may occur. If such high subsidence rates continue, the offset will be filled by sedimentation, as there are no recognized hotspots of land loss or enhanced flooding. In other parts of the delta, from the Sylhet Basin in the northeast to areas of the central Ganges Delta Plain, subsidence rates of 0.1–0.4 mm/year are evident and consistent with modern morphological features (Goodbred et al. 2003).

12.2.6 ACCOMMODATION

Accommodation is the space available for potential sediment to accumulate and is a function of eustasy and subsidence. Sediment influx controls the rate at which this space is filled. The interplay between accommodation and sedimentation rates controls whether the shoreline advances or retreats and the resulting vertical face changes. Changes resulting from uplift or subsidence of the land are known as tectonic movements, whereas those due to an actual rise or fall of sea level are eustatic movements, and relative sea level is the result of these land and sea level changes.

The rates of change of tectonic subsidence, eustatic sea level, sediment thickness and water depth are linked to one another through the accommodation space equation:

$$T + E = S + W \tag{12.1}$$

where T is the rate of tectonic subsidence, E is the rate of eustatic SLR, S is the rate of sedimentation and W is the rate of water depth increase (or deepening). These four variables are defined such that positive values correspond to tectonic subsidence and eustatic SLR (factors that increase accommodation space), as well as sediment accumulation and water depth increase (factors that reflect filling of accommodation space). Reversing the signs of these variables accommodates tectonic uplift, eustatic sea level fall, erosion and decreasing water depth, respectively (Figure 12.5).

Subsidence may be related to the tectonic regime considering constant eustatic sea level at a certain geological time; in this case, there are three main effects on accommodation due to the relationship between subsidence (T) and sedimentation (S):

1. The subsidence rate is lower than sedimentation ($T < S$), the depth decreases and deposition is shallower (Figure 12.6a). The reduction of accommodation causes regression effects, visible by shoreline migration seaward; there is also erosion of sediments under subaerial conditions, which increase the sedimentation. This is the case of compressional tectonic regimes, which are characterized by a maximum horizontal tectonic stress, and sediments are shortened. Shortening generates uplift, which creates a relative sea level fall (RSLF), which decreases the accommodation; however, the RSLF induces erosion, increasing the sediment deposition (Vail et al. 1991).
2. The subsidence rate is equal to sediment deposition ($T = S$). The accommodation is constant, and in the same environmental condition, the increase of space is balanced by terrigenous influx, causing a relatively constant shoreline (Figure 12.6). This case occurs during extensional tectonic regimes, which are characterized by a maximum vertical effective stress, and the sediments are lengthened. This lengthening induces subsidence, which drives RSLR and increases the space available for sediments (accommodation); however, the RSLR induces sedimentation (Vail et al. 1977).
3. The subsidence rate is higher than deposition ($T > S$). The space available for potential sediment increases (Figure 12.6). The shoreline is displaced landward, creating backstepping geometry that is a transgression with a RSLR (Vail et al. 1977).

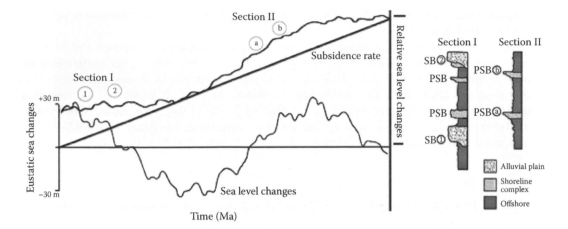

FIGURE 12.5 Interaction between subsidence and multiple orders of eustatic cyclicity, and the effect on sequence stratigraphy. Third-, fourth- and fifth-order eustatic changes combine with subsidence to produce a relative sea level curve, which controls the character of the sequence expression. Section I reflects deposition during a period of lower accommodation related to the third-order sea level fall; points 1 and 2 identify fourth-order sea level falls corresponding to sequence boundaries (SB). Section II reflects deposition during a period of greater accommodation related to the third-order SLR; points a and b designate fourth-order SLRs corresponding to parasequence boundaries (PSB). (Adapted from Van Wagoner, J. C. et al., *Am. Assoc. Petrol. Geol. Methods Explor. Ser.*, 7, 1–55, 1990.)

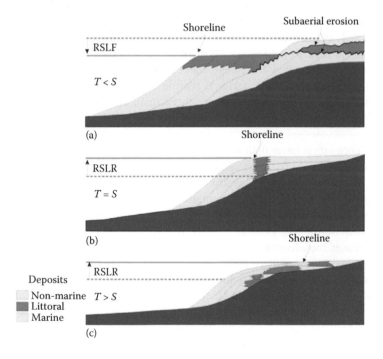

FIGURE 12.6 Subsidence (T) and sedimentation (S) rates scheme. (a) Regression case, $T < S$, shoreline migration seaward. (b) Coastal plain aggradation, $T = S$. (c) Transgression case, $T > S$. (Adapted and modified from Vail, P. R. et al., Seismic stratigraphy and global changes of sea level: Part 3. Relative changes of sea level from coastal onlap: Section 2. Application of seismic reflection configuration to stratigraphic interpretation, Memoir, Texas, 1977.)

When tectonics is combined with eustatic changes, the final accommodation is the sum of the space induced by tectonics and eustasy. In the case in which eustatic SLR increases the accommodation, the amount of water depth created by subsidence is added to the amount created by the eustatic rise, whereas during eustatic sea level fall, there is a decrease in accommodation, and the amount of space created by subsidence must be subtracted from that created by the eustatic sea level fall. Tectonics, eustasy and climate interactions determine the sedimentary process and accommodation. The objective of this section is to understand the principles that lead to the creation, filling and subsequent distribution of accommodation space; these principles, once acquired, are used to divide the sedimentary sequence in units called depositional sequences and systems depositional (systems tracts). Concepts and principles are related to the clastic rocks and silico-clastic rocks, but not in full; the carbonate rocks that have a chance to form in situ (cliffs) respond differently to the formation and changes of accommodation space. A limit of type 1 sequence is characterized by subaerial exposure of the entire area of the former continental shelf, resulting in erosion and rejuvenation of rivers, moving towards the basin facies of coastal and reattachment of coastal onlap farther out and farther down the onlap of the oldest layers.

A limit of the type 2 sequences arises when the relative sea level is lowered along the former continental shelf, without exceeding the old side of the platform. In this limit, there is no significant erosion or displacement towards the basin of coastal onlaps. The discontinuities of the sequences change laterally towards the basin to the correlated surfaces of depositional continuity. Within a cycle of relative variation of the sea level, one can distinguish four phases: a phase of a drop or fall, a phase of low station, a lifting phase and, finally, a phase of high stand. During each stage, it is possible to identify specific associations of depositional system contemporaries called systems tract, characterized by specific boundary conditions (Brown and Fisher 1977; Jervey 1988).

The lowering phase is the formation of groups of depositional systems of fall (falling sea level systems tract) (Hunt and Tucker 1992; Helland-Hansen and Martinsen 1996); during the next phase, the depositional systems of low stationing are deposited (lowstand systems tract), whereas the phases of ascent and the highstand mark the development of the transgressive and highstand systems tract. In the sequences bounded by the limits of type 2, associations of depositional systems of the contemporary margin platform or shelf margin systems tract replace the lowstand systems tract. Each systems tract consists in turn of elementary units, called parasequences, bounded by the surfaces of marine ingression that record cycles of relative variation in the sea level of higher orders of magnitude (Van Wagoner et al. 1990) or fluctuations of input sedimentary, induced by autocyclic processes.

12.2.7 ANTHROPIC CAUSES

Anthropic subsidence is associated with human activities, both at the surface level and at depth. Bioturbation by humans or 'anthroturbation' (Zalasiewicz et al. 2014) fingerprint, comprising phenomena ranging from surface landscaping to boreholes in simple individual structures or complex networks, range from several kilometres' depth to the land surface (compared with animal burrows and vegetation roots that range from centimetres to a few metres in depth).

The natural phenomenon of displacement of lands, coupled with subsidence triggered by human activities like the extraction of material from undergrounds, can lead to topographic modifications or collapse with concomitant consequences in the landscape patterns and serious impacts on the structures and the population (Gernhardt and Bamler 2012; Bucx et al. 2015; Gernhardt et al. 2015).

The anthropic component is linked to exploiting natural resources, such as solid matter (e.g. solid hydrocarbons and metal ores) and fluids (natural gas, liquid hydrocarbons and water), and related to local replacement by other substances (solid waste, drilling mud and water). Moreover, the local shocks due to underground nuclear, military and scientific tests introduce hotspots in the natural displacement rates.

The industrialization and urbanization of a country increases the natural exploitation and consequently increases the subsidence (Carbognin and Gatto 1986) phenomena, inducing changes in the water, gas fluxes and uses of lands where the livelihood of humans can then become threatened or impossible.

There are many examples of human-induced changes in subsidence, most of which are in coastal areas. In the low-lying deltas, the population growth, deforestation, agricultural activities, urbanization, fertilizer, fossil fuel consumption and construction, such as dams and heavy bridges, impact the dynamics of the lands and the water systems. The river channels seen today have migrated from their historically recorded positions for subsurface geotectonic movement and changes in the patterns of water discharge and sediment load (Moors et al. 2011). In the Ganges Plain, on which 250 million people directly depend for agriculture and livelihood, there are now systems to bypass dams that could perhaps maintain sediment flow to the delta, helping to offset subsidence and erosion.

Dixon et al. (2006) analyzed the average rates of subsidence in New Orleans in southern Louisiana by means of radar satellite images detected over the 3-year study period. The researchers point out that the rate observed between 2002 and 2005 is probably at or near the slowest subsidence rate the area has experienced since the 1960s: sinking probably occurred even faster just after the levees were first built. Historically, eastern New Orleans has seen the greatest subsidence. This part of the city was 3–5 m below sea level when the hurricane struck and consequently saw some of the worst flooding. Even if many parts of the city are already metres below, the map indicates areas that sunk up to 28.6 mm each year. Scientists have proposed several causes for subsidence in New Orleans, ranging from natural ones, such as settling of coastal sediments and movement of the Michoud fault, to human ones, such as draining wetlands, diverting sediment-bearing floodwaters from the Mississippi River and pumping groundwater.

In the Italian coasts, the Emilia Romagna coastland south of the Po River Delta, starting from the 1950s, has experienced increases in land settlement and a consequent large groundwater and gas withdrawal. Recent studies on satellite maps show an increase of the subsidence rates obtained for the last decade: the amount of subsidence due only to natural causes is typically a few millimetres per year, while in some areas, the man-induced subsidence reaches values of several millimetres per year (Taramelli et al. 2015b; Fiaschi et al. 2017). Marshland reclamation, groundwater pumping for agricultural and industrial purposes and methane extraction from gas fields near the coastline are the principal anthropogenic causes, and here the subsidence, in combination with SLR, will worsen the inundation risk from the rivers and sea surge exposure. This makes subsidence an insidious threat having significant cumulative effects on SLR, flood, erosion and the integrity of water defences and infrastructures (Nordstrom et al. 2015; Wolff et al. 2016; Cozzolino et al. 2017).

12.2.8 PUMPING ACTIVITIES

The extraction of natural resources is one of the main causes of subsidence for the extension and intensity of the phenomenon (Carbognin and Tosi 2003).

Subsidence due to the groundwater withdrawal is developed where there is intensive abstraction. Such subsidence is attributed to the consolidation of sedimentary deposits in which the groundwater is present, with the consolidation occurring as a result of increasing effective stress (Bell 1988). The total overburden pressure in saturated deposits is borne by their granular structure and the pore water. When groundwater abstraction leads to a reduction in pore water pressure by draining water from the pores, there is a gradual transfer of stress from the pore water to the granular structure. This process occurs in areas where there is a layering of sand, which is more permeable and less compressible, and a clay layer, which has low permeability and high compressibility (Poland 1984; Mishra et al. 1993). The presence of unconsolidated or semi-consolidated soil allows the subsidence; the extent of this process is a function of the compressibility and permeability of sand, the chemical composition of water and the mineralogy of the clays.

The sediments most affected by this type of phenomenon are Quaternary unconsolidated clays with the presence of montmorillonite. The subsidence laws are the same for every type of fluid extracted (Carbognin and Gatto 1986). The decline of the water level in wells causes increases in effective stress, which increases in the part of the overburden load that is supported by the sediments being stressed (Taramelli et al. 2015b). The resulting strain is primarily expressed as a vertical shortening or compaction of the stressed sediments and consequent subsidence of the land surface. Horizontal displacement also occurs, but to a lesser extent. For example, in the Houston–Galveston region of Texas, the highest subsidence rates have been located where the decline of the water level was largest and where the clay thickness in the aquifer system was greatest (Bell 2007). Nevertheless the subsidence rate has slowed in recent years because of controls on groundwater withdrawals (Kearns et al. 2015). Furthermore, the ratio between maximum subsidence and groundwater reservoir consolidation is related to the ratio between the depth of burial and the lateral extent of the reservoir. In other words, small reservoirs that are deeply buried do not give rise to noticeable subsidence, even if subjected to considerable consolidation. By contrast, extremely large underground reservoirs may develop appreciable subsidence (Figure 12.7).

Different techniques can be adopted to control or arrest subsidence, such as the reduction of pumping draft, artificial recharge of aquifers from the land surface, repressuring of aquifers through wells or any combination of these methods. The goal is to manage the overall water supply and distribution in such a way that the water levels in wells tapping the compacting aquifer system, or systems, are stabilized or raised to some degree (Poland 1984).

Anthropic structures are vulnerable to subsidence; for example, urban areas near Tokyo registered a rate of subsidence up to 100 mm/year, for a total vertical displacement of 4 m in a temporal interval of 36 years, from 1930 to 1966 (Nakano et al. 1969), leading to more frequent flooding phenomena. Because the local administration focused on the reduction of groundwater exploitation, the subsidence process decreased, and in some areas, a rebound phenomenon was registered (interval of 36 years, from 1930 to 1966 soils).

Subsidence due to solid extraction drives surfaces to gradual lowering because mine tunnels cannot support the load of the above sediments, causing collapse. Studies in coal mining areas of Germany and The Netherlands can reach decametric displacements (Bekendam and Pattgens 1995; Harnischmacher and Zepp 2014). Such subsidence has a group of geomorphological consequences, including ground fissure formation, disruption of surface drainage and the resultant depressions permanently flooded (Goudie and Viles 2016).

The amount and areal extension of this phenomenon is a function of mineral exploitation, geometry and extraction methods, and the geological and geotechnical characteristics of the materials

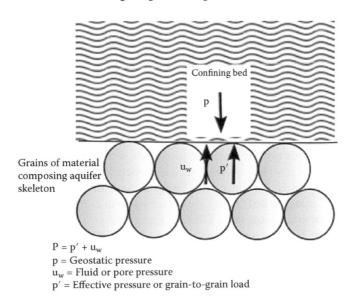

$P = p' + u_w$
p = Geostatic pressure
u_w = Fluid or pore pressure
p' = Effective pressure or grain-to-grain load

FIGURE 12.7 Scheme of the water pressure in an artesian aquifer with a confining bed representation. (From Poland, J., *Guidebook to Studies of Land Subsidence due to Ground-Water Withdrawal*, Michigan Book Crafters, Chelsea, UK, 1984.)

above the mine (Carbognin and Gatto 1986). In this case, it is more difficult to produce models about the evolution and intensity of the phenomenon.

12.2.9 Overloading

The application of loads on the ground, both liquid and solid, in urbanized areas may cause lowering of the topographic surface in particular to recent sediments.

Geological formations, such as loess, debris flow deposits or alluvial sediments above the water table, are subjected to hydrocompaction, which is the compaction and reduction in volume of soils and sediments that occurs when their moisture content is increased (Goudie and Viles 2016). This process is due to the humidification of sediment, which weakens their intergranular strength because of the rearrangement of their particles (Goudie and Viles 2016). It is thought that clay bonding of the particles is responsible for maintaining open textures when the deposits are in a dry condition and for rapid disaggregation and volume loss when submerged in water (Bull 1964). This process produces rapid and irregular subsidence of the ground surface (Carbognin and Gatto 1986), ranging from 1 to nearly 5 m (Poland 1984).

Surface subsidence resulted from wetting without the addition of surcharge load at many sites; at others, a combination of water infiltration and surface loading was required. Lofgren (1969) has reviewed the hydrocompaction process and associated subsidence occurrences in the United States, Europe and Asia.

One of the most famous cases is San Joaquin Valley, where the hydrocompaction of sediment caused by heavy watering is one of the subsidence drivers, with rates up to 3 m in a period of several years (Galloway et al. 1999).

In newer urban areas, the solid loads from the weight of buildings usually have a limited extension. This phenomenon is active in coastal fringe areas where a high level of urbanization, particularly of the partially consolidated sand barriers in the lowlands, causes increases.

12.2.10 Hydraulic Reclaim

Reclaim involves structures and modification of the environment, which allow water to be drained, converting swamps into intensive agriculture or building lands (Carbognin and Tosi 2003). There are different methods to reclaim land: eliminating water permanently (reclaim by channel) or raising the ground level to fill the lowland (reclaim by infilling). In these cases, anthropic actions may cause serious subsidence problems because the drainage of swamps produces compaction, especially for organic materials. Topographic lowering occurs for two main reasons: physical processes, due to the reduction in volume and the increase in density for compaction, and biochemical processes, due to the mass of ground lost due to organic compound oxidation. Generally, after drainage operations, the physical process occurs before the biochemical one (Carbognin and Gatto 1986). In the presence of abundant organic matter, this type of anthropic subsidence can cause lowering rates of 1–2 cm/year. In countries where reclaim is widely adopted, such as Holland, the subsidence rates in the reclaimed area vary from a few centimetres, in the case of inorganic sand, to up to 1 m, in the presence of peat, over several decades (Carbognin and Gatto 1986), reaching up to 2–6 m under sea level in some areas.

The Pontina Plain, an area located in the Lazio region in the centre of Italy, is one of the most important Italian reclaim areas. Here, a pronounced lowering of 5.9 m has occurred during 1811–1994. Since the spring of 1811, the rate of subsidence has reached up to 5 cm/year due to reclaim and increases in the compaction of ground and the oxidation of organic matter (Serva and Brunamonte 1994).

12.3 SUBSIDENCE EFFECTS ON COASTAL VEGETATION

The effects of subsidence in coastal areas are strongly linked to the effects of a worldwide rise in sea level (Wöppelmann et al. 2013; Ablain et al. 2016), and considering that subsidence is often at local scale, it is easy to correlate it to several effects on different coastal vegetation (Dijkema 1997). An estimate of the impact of subsidence on wetlands can be provided to simulate the effects of enhanced SLR. In fact, evidence of high soil subsidence rates in specific coastal test cases can be exploited to infer the effects of future increases in sea level using subsidence as a model. It is particularly interesting to infer how future climate change will influence both vegetation presence and diversity in coastal areas, considering that eustatic rates of SLR have been 1–2 mm/year over the last 100 years (Gornitz 1995), and that the rates of rise relative to the land surface (RSLR) can be higher due to land subsidence and decreases in sediment influx. Moreover, a rising sea level results in a greater frequency and duration of inundation (Titus 1988; Boesch et al. 1994; Wang et al. 2012).

In coastal environments, the elevation with respect to SLR influences a wide range of factors that in turn affect the distribution of organisms. In particular, elevation determines the predominance of either seawater or freshwater influence, leading to different salinization levels; on the contrary, desalinization can occur in relation to weather conditions, namely, high precipitation rates. The balance between saltwater and freshwater can have strong effects on coastal vegetation. Thus, hydrology is the dominant factor controlling the development of spatial variation in wetland plant communities and is responsible for the horizontal zonation of adult plants, seeds and seedlings in a number of different wetland types, including tidal freshwater marshes (Simpson et al. 1983; Abernethy and Willby 1999; Coops et al. 1994; Mitsch and Gosselink 2000). Further, changes in hydrology can affect the specific species composition (Taramelli et al. 2017).

Flooding generally inhibits the emergence of seedlings from the soil seed bank (i.e. the natural storage of viable seeds in the soil of many ecosystems [Baldwin et al. 1996; Baldwin and Mendelssohn 1998]). Because of the primary role of the hydrologic regime in controlling the temporal variation in landscape composition, knowledge about hydrology and erosion or accretion rates is part of the foundation for predicting the potential effects of changes resulting from processes such as subsidence, as well as watershed, land alteration and SLR. Small increases in the frequency and duration of inundation due to natural and anthropogenic processes, such as SLR and coastal subsidence, may change the species composition of tidal freshwater wetlands (Baldwin et al. 2001), at least at specific locations (if communities can migrate inland, species composition of the marsh may remain unchanged). Seedling recruitment and vegetative growth in most species are inhibited by higher water levels, and the abundance of these species would decrease if wetland water levels rise, sediment influx decreases or the sea level relative to the marsh surface increases (i.e. if accretion does not keep pace with eustatic SLR and subsidence). In salt marsh vegetation, the shift from species that are less vulnerable to species that are more vulnerable to increased inundation could decrease the system's capability to respond to changing climate threats, such as SLR (Strain et al. 2017).

12.4   MEASUREMENT TECHNOLOGIES: MAPPING AND MONITORING

Subsidence assessments typically address the spatial changes (magnitude and direction) in the position of the land surface and the process causing the subsidence. Measuring and monitoring subsidence is critical to constrain analyses and forecasts of future subsidence (Galloway and Burbey 2011).

Various methods are used for measuring and mapping spatial gradients, the temporal rates of regional and local subsidence and horizontal ground motion (Galloway et al. 1999). The methods generally measure relative changes in the position of the land surface. The observable position is typically a geodetic reference mark that has been established so that any movement can be attributed to deep-seated ground movement rather than surficial effects. The monitoring network is designed to cover the known or suspected subsidence area and includes stable reference marks that are part of a broader regional geodetic control network (Galloway and Burbey 2011). Land subsidence has been measured using repeat surveys of benchmarks referenced to a known, and presumed stable, reference frame. In many affected areas, an accurate assessment has been hindered or delayed by the lack of a sufficiently stable vertical reference frame (control). The problem of detection in regional land subsidence is compounded by the large aerial scale of the elevation changes and the requirement

TABLE 12.2

Comparison among Different Subsidence Monitoring Systems and Their Precision

Method		Component Displacement	Resolution[a] (mm)	Spatial Density[b] (Samples/Survey)	Spatial Scale (Elements)	Measurement Frequency[d]
Spirit level		Vertical	0.1–1	10–100	Line – network	1 year
Geodimeter		Horizontal	1	10–100	Line – network	1 year
Borehole extensometer		Vertical	0.01–1	1–3	Point	1 month
Horizontal extensometer	Tape	Horizontal	0.3	1–10	Line – array	1 month
	Invar wire	Horizontal	0.0001	1	Line	1 month
	Quartz tube	Horizontal	0.00001	1	Line	1 month
GPS		Vertical	20	10–100	Network	2–3 years[e]
		Horizontal	5			Hourly to daily[f]
InSAR[c]		Range	5–10	100,000–10,000,000	Map pixel[c]	4–35 days
LIDAR		Vertical	Up to 90[g]	1 pt/m²	Point	Yearly
Aerial photogrammetry		Vertical	Up to 100[h]	1 pt/10 cm²	Point	Monthly/ yearly

Source: Modified from Galloway, D. et al., Measuring land subsidence from space, USGS Fact Sheet-051-00, U.S. Geological Survey, Reston, VA, 2000.

[a] Measurement resolution attainable under optimum conditions. Values are given in metric units to conform to standard geodetic guidelines.

[b] Number of measurements generally attainable under good conditions to define the spatial extent of land subsidence at the scale of the survey.

[c] A pixel on an InSAR displacement map is typically from 1 to 80 m² on the ground in function of the sensor (see following paragraphs).

[d] Frequency values for GPS, level and SAR are from Hung et al. (2010).

[e] DGPS (Strozzi et al. 2003a).

[f] Continuous GPS (Strozzi et al. 2003a).

[g] Values from Khan et al. (2014); it depends on the elevation of the device.

[h] Values from Gasperini et al. (2014); it depends on the elevation of the device.

for vertically stable reference marks (benchmarks) located outside the area affected by subsidence. Deformation in the form of vertical and horizontal displacement of the land surface is the principal hazard associated with groundwater extraction. Ground failures, such as earth fissures and surface faults, are associated with areas of differential (vertical) ground displacement (Holzer 1984) and both horizontal and vertical deformation of the aquifer system, which may not be associated with differential (vertical) ground displacement (Helm 1994; Sheng et al. 2003). As given in Table 12.2, there are different measurement and monitoring methods used for subsidence and topographic variations.

Coastal regions are areas sensitive to subsidence, which may occur at different rates and scales. Thus, it is important to use the proper method and note the pros and cons of its application and the scale of phenomena that will be analyzed. Large regional networks may warrant the use of GPS or airborne and space-based geodetic surveys. If more precise and accurate measurements of change are needed on a local scale, extensometers may be used (Galloway and Burbey 2011).

12.4.1 Levelling

Displacement measurements conducted by levelling use a reference point benchmark. Laser and optic devices can collect angular measurements; this method combines easy data collection and a

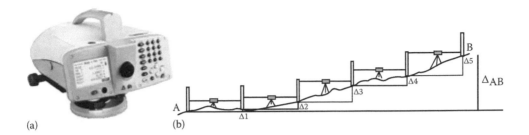

FIGURE 12.8   (a) Levelling device. (b) Levelling scheme.

high level of accuracy and precision, although they are expensive and time consuming (Rappleye 1948; Floyd 1978; Geodetic Survey Division CCRS 1978; Muller 1986; U.S. Army Corps of Engineers 1994). To conduct the survey, it is necessary to create benchmarks, which are a series of elements built to guarantee stability and durability (Figure 12.8).

Levelling using GPS surveying or conventional levelling (spirit level and geodimeter) are alternatives to vertical extensometers, in particular in case of measurements of regional phenomena.

12.4.2   GLOBAL NAVIGATION SATELLITE SYSTEMS

The satellite positioning system determines three-dimensional point positions and can be used to monitor subsidence over large distances or at a regional scale. The advantage of this technique is its good mobility and speedy survey operation; the vertical component accuracy is lower than that obtained by levelling and can reach 1 cm if the satellite signals are not blocked and the record length is sufficiently long.

The first and widely used measuring technique adopts the signal of 24 Navigation Satellite with Time and Ranging (NAVSTAR) satellites that orbit 20 km from the Earth's surface. These satellites transmit binary information of all parameters describing their orbit on two different frequencies (L1 and L2). The GPS measures pseudorange, the pseudodistance between a satellite and navigation satellite receiver.

Karegar et al. (2016) compared the GPS-derived displacement with records of late Holocene RSLR allowing separation of long-term glacial isostatic adjustment-induced displacement and short-term deformation from recent excessive groundwater extraction.

The satellite positioning system allows computing of the coordinates with high precision and accuracy of benchmarks, along a transect or network; by measuring the same points in different surveys throughout years, it is possible to track changes in elevation at the benchmarks and monitor trends over time (Williams et al. 2004). Two main approaches are used with GPS: continuous global positioning system (CGPS) and differential global positioning system (DGPS).

CGPS monitoring consists of constant signal acquisition by a receiver located at the point of interest and by a series of geodetic stations or points whose absolute coordinates are known with high precision (Figure 12.9). The data processing detects the relative displacements in comparison with reference stations that also assess the velocity displacement with an accuracy of millimetres (Bock et al. 2012).

DGPS is another technique adopted to monitor subsidence, and unlike CGPS, it works with one or more reference stations defined as a baseline, whose coordinates are known. A receiver named rover, which measures a series of points identified as benchmarks, creates a series of spatial and temporal baselines at each measurement. Every reference station broadcasts correction of a pseudorange to the receiver, allowing the calculation of positions with better precision.

The time of acquisition is shorter than that of CGPS; the reference stations usually acquire data for the entire day of survey, and the revisiting time of network measurements is yearly (Strozzi et al. 2003a; Gomarasca 2009).

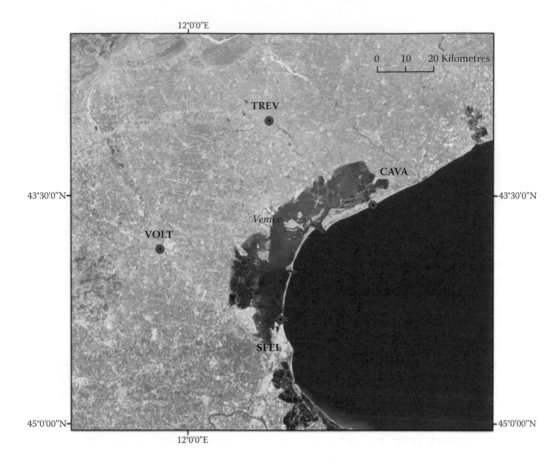

FIGURE 12.9 Two of the stations are located along the lagoon edge (CAVA in the north at Cavallino and SFEL in the south at Forte San Felice near the Chioggia Inlet) and two inland (VOLT at Voltabarozzo near Padua and TREV at Treviso).

In the last years, a dramatic change has been happening to the satellite positioning systems. New constellations are online or upcoming: GLONASS (Russian), Galileo (European) and BeiDou (Chinese). About 120 satellites will be available once all four systems are fully deployed in the next few years (Li et al. 2015). The fusion of the data obtained by these systems will improve the reliability of positioning, optimizing the spatial geometry.

12.4.3 EXTENSOMETERS

Vertical extensometers provide site-specific measurements of subsidence. These instruments consist of a pipe or cable anchored at the bottom of the borehole. The pipe or cable extends from the bottom of the borehole, through the geologic layers that are susceptible to compaction, to the ground surface. The pipe or cable is connected to a recorder that frequently measures the relative distance between the bottom of the borehole and the ground surface. A multilevel compaction monitoring well is a vertical extensometer, which measures compaction of the aquifer system at different depths and is a good tool for understanding the mechanism of subsidence. The advantage is its capability of monitoring multilayer compaction, with high accuracy (approximately 1–5 mm) and stability. The disadvantage is that the device is costly and incapable of determining the subsidence below the depth of the well (Hung et al. 2010). These instruments detect changes in land surface elevation to 1/100 of a foot on a daily basis. The relationship between subsidence, groundwater level fluctuation and the physical properties of the various aquifer systems can also be evaluated by installing a multi-completion monitoring well adjacent to the vertical extensometer.

12.4.4   LIDAR

Airborne light detection and ranging (LIDAR) systems are becoming an increasingly common and effective tool that is essential to measure millimetre-scale ground deformations in relatively large areas. It consists of a laser mounted beneath an airplane or helicopter that follows a predefined path providing a three-dimensional cloud of points. These points are processed to remove non-ground features, producing a DEM (Kraus and Pfeifer 1997; Vosselman 2000). LIDAR can be adopted to assess vertical ground displacements by producing differential DEMs to identify vertical ground displacements over time. This technique was adopted to assess mining subsidence in the Hebi coal mine area in China (Haiyang et al. 2011). Recently, the technique was used in Japan to study changes prior to and after the earthquake of 11 March 2011 (Konagai et al. 2012).

Other authors adopted three different LIDAR DEMs to check subsidence in northern Harris County, revealing changes near the salt domes and surroundings (Khan et al. 2014).

12.4.5   Photogrammetry by UAV

The unnamed aerial vehicle (UAV) is an outstanding tool adopted for coastal and environmental monitoring (Klemas 2015; Turner et al. 2016). This technique consists of the collection of several photos by a camera mounted onboard the aerial platforms with a high overlap rate at a low flight height (<100 m). All these images are then oriented by the structure-from-motion (SfM) photogrammetry (Lowe 2004) and generate a dense point cloud.

Lechner et al. (2012) used a UAV to study subsidence, resulting from underground coal mining that can alter the structure of overlying rock formations. Gasperini et al. (2014) analyzed the UAV potential for monitoring displacement in landfill areas, in particular the evolution of subsidence after the closure, producing a digital surface model (DSM) with a precision of 0.1 m. This level of quality is sufficient to monitor the subsidence by the differential DSM technique (Gasperini et al. 2014). The application in coastal environments for subsidence analysis is still poor due to the high level of error and the consequent high quality of measurements and ancillary data required. However, considering centimetric rates typical of the above-described subsidence causes, it would be possible to apply this methodology during the yearly revisit time.

The integration of UAV data with GPS has provided results as good as those of the dense digital terrain model to support the flooding scenario due to subsidence and SLR in Lipari island measured by GPS (Anzidei et al. 2017).

12.4.6   SAR-Based Techniques

The SAR satellite images play a fundamental role for subsidence monitoring in light of the huge dataset collected in the last 25 years (Sansosti et al. 2015).

The SAR-based techniques are well-proven technology for monitoring the deformation of the Earth's surface (Amelung et al. 1999; Crosetto et al. 2003, 2016; Dixon et al. 2006; Fiaschi et al. 2017).

SAR data for subsidence monitoring can be divided into the following main categories: medium spatial resolution (20–200 m) and low temporal revisit time, such as long-term C-band ERS-1/2 and ENVISAT archives, in addition to the C-band RADARSAT-1/2 data sequences and those provided by the L-band JERS-1 and ALOS-1 systems, whereas Cosmo-SkyMed and TerraSAR-X belong to the second group characterized by better spatial (<20 m) and temporal (<15 days) resolution with the X-band. Finally, the recent generation of satellite missions, such as Sentinel 1, is providing large coverage and short revisit time in the C-band (Torres et al. 2012). These new-generation SAR sensors are impacting the role of Earth observation in research and monitoring activities relevant to surface deformation phenomena in terms of earlier detection with better temporal detail, passing from 35 days of ENVISAT to 6 days of the Sentinel 1 constellation (S1-A/-B) to 4–8 days for the Cosmo-SkyMed constellation case (Sansosti et al. 2015; Tosi et al. 2016; Fiaschi et al. 2017).

Furthermore, the dramatic development of space economy is leading companies to launch or plan to launch in orbit hundreds of new microsatellites with high spatial and temporal resolution (Ban 2016) for both optical (Terra Bella, now acquired by Planet) and SAR (Capella Space 2017) sensors. These changes will deeply transform our way of monitoring the Earth; consequently, there is poor literature on the application of this last generation of SAR sensors in subsidence analysis. SAR emits microwaves that are reflected by the objects present on the Earth, producing multitemporal systematic acquisitions with wide spatial coverage and complex images (amplitude and phase) containing distance information. Interferometric synthetic aperture radar (InSAR) analysis (Bamler and Hartl 1998; Rosen et al. 2000) exploits the phase difference between two SAR passes acquired from the same position at two different times over the same area to generate an interferogram. The occurrence of a displacement between the acquisitions of the two images results in a phase shift proportional to the displacement magnitude. InSAR is a remote sensing technique whose unique features, such as multitemporal systematic acquisitions and wide spatial coverage, make it particularly useful for coastal regional monitoring areas. This technique was widely used to study changes in coastal environments, particularly urban and periurban coasts, such as the Venice Lagoon, Nile River Delta, Taiwan and New Orleans (Bock et al. 2001; Becker and Sultan 2009; Hung et al. 2010; Long and Ding 2010). These environments often have a complex texture due to their spatial variability; consequently, it is fundamental to have data with high sensitivity in order to detect the superficial land changes with accurate resolutions.

To apply the differential interferometric synthetic aperture radar (DInSAR) technique, interferograms were produced from SAR (the band is a function of the sensor) images. DInSAR uses SAR–image pairs collected over a time interval to extract surface displacement by using the phase change of the radar signal in the direction of the radar line of sight (LOS). The use of DInSAR for monitoring the Earth deformation process has received considerable attention because of its great potential for mapping the movement associated with earthquakes, volcanoes and landslides (Zebker et al. 1994; Massonnet and Feigl 1998). DInSAR has also been applied to the monitoring, analysis and interpretation of land subsidence caused by aquifer system compaction (e.g. Galloway et al. 1998; Amelung et al. 1999; Hoffmann et al. 2001, 2003; Bell et al. 2002; Galloway and Hoffmann 2007). InSAR techniques can also detect phase changes related to water level changes in wetlands and produce spatially detailed high-resolution maps of water level variation (Hung et al. 2010); in this case, the combination of InSAR results and other techniques (GPS) can be used as important markers of subsidence phenomena. In coastal areas, SAR-based techniques, such as InSAR, and persistent scatterer interferometry (PSI), for example, permanent scatterer (Ferretti et al. 2001), small-baseline subset (SBAS) (Berardino et al. 2002) and interferometric point target analysis (IPTA) (Werner et al. 2003), are integrated into a Subsidence Integrated Monitoring System (SIMS) to overcome the limits characterizing each technique.

The Venice Lagoon is an area where SIMS efficiently merges the different displacement measurements obtained by high-precision levelling, differential and CGPS data, and SAR-based interferometry, obtaining land deformation data from 1992 to today (Tosi et al. 2009a). Recently, Tosi et al. (2016) integrated Cosmo-SkyMed and PALSAR images to analyze subsidence in the Po River Delta and Venice Lagoon.

12.4.7 INTERFEROMETRIC POINT TARGET ANALYSIS

An important limitation of InSAR is the incomplete spatial coverage due to temporal decorrelation and problems in resolving high-spatial-phase gradients causing gaps in the study area. These limitations can be reduced using sensors with longer wavelengths, such as L-band sensors (Strozzi et al. 2003b). Techniques to interpret the phases of stable reflectors on long-time-series SAR images have been proposed (Ferretti et al. 2001; Werner et al. 2003; Crosetto et al. 2016). These reflectors must satisfy two conditions. The first is that their scattering behaviour corresponds to that of a point target so that there is no geometric decorrelation. This permits phase interpretation even for baselines above the critical baseline, with the advantage that more acquisitions may be included in the analysis. The second condition is that the same reflector remains present over the time period of interest to permit analysis of its phase history (Strozzi et al. 2003b). The IPTA is also known as persistent scatterer (PS) InSAR (Zhang et al. 2011). Phase-stable pointwise targets, called permanent scatterers, can be detected on the basis of a statistical analysis of the amplitudes of their electromagnetic returns (Ferretti et al. 2001). The PSI outputs include the average displacement rates over the observed period and the time series of the deformation per each point with high coherence of the area, which provides information on the temporal evolution of displacements.

12.4.8 SMALL-BASELINE SUBSET

Similar to PSI, the SBAS technique needs a high number of co-registered SAR data to work with differential interferogram series, which are characterized by a small spatial distance between orbital positions (spatial baseline). This geometric constraint limits the spatial decorrelation and topographic errors (Zebker and Villasenor 1992). Through a series of differential interferograms, it is possible to obtain a velocity map of deformation and displacement time series. This technique allows quantifying average displacement per annum per each coherent pixel with an accuracy on the order of millimetres (Berardino et al. 2002; Lanari et al. 2004).

Some of the main advantages of these two techniques are as follows: (1) wide-area coverage, associated with a relatively high spatial resolution, which allows us to capture the overall picture of the deformation phenomena occurring in wide areas while maintaining the capability to measure individual features, such as structures and buildings; (2) relatively low cost compared with conventional methods; (3) all-weather, day–night detection capability; (4) the availability of historical SAR archives, which confers the ability to measure and monitor 'past deformation phenomena' to PSI; for example, it is possible to study ground motion that occurred in the past for which no other data are available; and (5) high sensitivity to terrain motion, on the order of 1 mm/year for deformation velocities.

The sensor adopted for SBAS and land cover influence coherence; for example, the C-band or X-band has a lower coherence than the L-band, as the decorrelation effect is different due to the different response of spatial targets (Tolomei et al. 2013).

In the case of wetland areas, targets change backscatter over time; thus, techniques such as PS may be useless, and SBAS may produce more accurate information.

12.5 EXAMPLE OF STUDY AREAS

12.5.1 RHINE–MEUSE DELTA

The Rhine–Meuse Delta starts at the border between The Netherlands and Germany and flows into the North Sea in The Netherlands (Figure 12.10). The Rhine–Meuse Delta is an area that is particularly vulnerable to climate change and related flood risks (Cuenca et al. 2011). The area is partly below sea level and consists of a thick layer of unconsolidated sand, clay and peat. Localized subsidence contributes to weak zones in natural and man-made water defence structures. Regional subsidence due to drainage, peat oxidation (Cuenca et al. 2013) and oil and gas production (Kooi et al. 1998; Ketelaar 2008) has considerable effects on the regional water management systems and on natural wetland and tidal habitats (http://www.subcoast.eu/rhine.html).

Using PSI technology, surface subsidence parameters of the region of the Dutch (Rhine–Meuse) Delta that is situated below the high water levels of rivers and the sea (which is more than half of The Netherlands) were estimated. This has not yet been performed for such a wide region (~20,000 km^2), and it requires an innovative approach, combining all ERS-1/2 and ENVISAT-ASAR radar datasets,

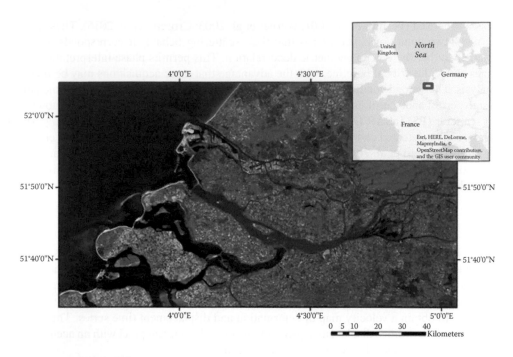

FIGURE 12.10   Rhine–Meuse Delta in The Netherlands. (From Esri, DigitalGlobe, GeoEye, i-cubed, USCS, AEX, Getmapping, Aerogrid, IGN, IGP, swisstopo, and the GIS User Community.)

each of which is defined in a local datum. This overview was based on these two radar data and later expanded using the new high-resolution radar sensors. Cross-heading (ascending and descending) acquisition geometries were used to distinguish between vertical and horizontal motion. It was assimilated with Global Navigation Satellite System (GNSS) positioning data of the continuous Dutch Active GNSS Reference System (AGRS), the basic core net of GPS calibration points ('kernnet') infrastructure, and levelling data. The incorporation of gravity data corrected the map for regional and long-term land rise or subsidence. This task resulted in an absolute wide-scale subsidence map for the Rhine–Meuse Delta. For dyke structures, a complementary detailed radar-based estimation was performed. This new product is different from the wide-area PSI analysis described above because it estimates direct safety-related parameters, for example, related to the stability of the defence system. In a number of specific locations, the added value of integrating sensors with different wavelengths and revisit times (ERS, ENVISAT, RADARSAT and TerraSAR-X) was shown.

In two case studies, the PSI-derived absolute wide-scale subsidence map is combined with gravity measurements, sea level data, geo(hydro)logical data, data on gas extraction or injection and information on buildings and building projects to create the following:

- In two case studies, the PSI-derived absol compaction and gas extraction and injection on subsidence improve the assimilation of PSI-derived subsidence estimates, which multiply the amount of measured data by a factor of 1000. The causes of subsidence, distinguishing between shallow (autocompaction) and deep (gas extraction) effects, are also created. In addition, we distinguish between the subsidence of the radar targets (largely man-made structures) and the actual subsidence of the Earth by adding geomechanical information derived from geological surveys.
- Compaction and gas extraction and injection causing subsidence improve the assimilation of the climatological forecasts for an impact study, in which interpretations of flood risk and the impact on sensitive urban areas are estimated.

12.5.2 Southern Emilia Romagna

The Emilia Romagna region is located in northern Italy on the Adriatic Sea (Figure 12.11). The scope of this pilot service was to address the benefits and critical aspects related to the integration of different types of data for the full three-directional characterization of ground movement, especially taking advantage of the opportunity offered by the new high-resolution satellites TerraSAR-X, Cosmo-SkyMed and RADARSAT-2. A detailed study for a selected area in the southern part of the Emilia Romagna coast was implemented (Strozzi et al. 2012). Draining and reclamation of large valleys and extensive pumping of methane-rich groundwater from shallow depths in southern Emilia Romagna resulted in subsidence and erosion of the coastline, lower efficiency of the waterworks and a significant increase in the hazard of flooding. Several actions have been taken with the aim of monitoring and removing the causes of subsidence and protecting the coastline by sand replenishment. This pilot service extends the conventional PSI processing restricted to displacement estimate in the LOS of the satellite to a full three-directional displacement estimate of the ground-bound radar targets. To measure east-west displacement, two datasets are required, one in descending mode and another in ascending mode (Taramelli et al. 2015b). The PSI analyses of these datasets will reveal a sparse grid of points, some of which will be the same in each dataset. Using LOS values of motion of these common targets, it is possible to measure the east-west values of displacement of those targets. The next stage was to address the third representation, north-south movement, and it will require a third dataset to produce results. In this project, an investigation of the procedure to determine three-directional movement and its testing on real data was necessary. To address the previous issues,

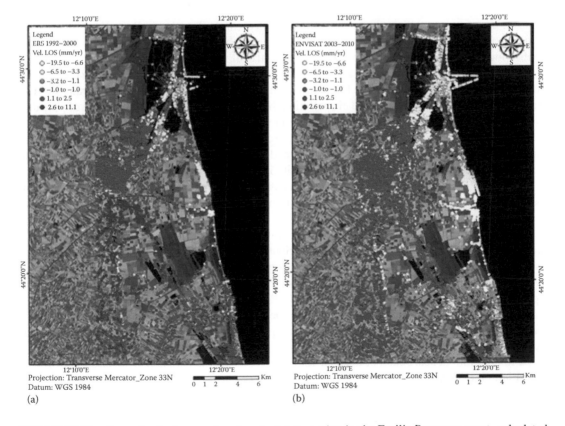

FIGURE 12.11 Ground velocity map (mm/year) of deformation in the Emilia Romagna coast, calculated (a) between 1992 and 2000 using PSInSAR on ERS-1/2 data, and (b) between 2003 and 2010 using PSInSAR on ENVISAT data.

together with the performances of the different satellite SAR data sources, the research on the case study will involve the use of multitemporal, multigeometry, multiresolution and multifrequency SAR data. Satellite radar data will be validated and integrated by information derived from high-accuracy geodetic campaigns (mainly levelling and GNSS techniques) realized in the past and through specific surveys that will be conducted during the project. Long GPS time series will be analyzed for both validation purposes and the establishment of the reference frame for SAR and levelling processing. Specific analysis will be performed on the evaluation of the quality of high-resolution DEMs derived from interferometric processes in different periods on a selected subarea; reference data will be provided by means of available terrestrial, photogrammetric and LIDAR data, eventually integrated by local GPS kinematic surveys. These data represent an important input to the development of flood risk maps. Spatial and temporal resolution of ground control data and remote sensing results are a key issue to reach a quantitative comparison and validation of the applied techniques. Several agencies have set up local topographic networks. In particular, the Emilia Romagna region, with support from IDRODES Ltd. and the Regional Environmental Protection Agency (ARPA) since 1996, has instituted a network aimed at monitoring the entire coastline and interior of the area. Measurements were taken in 1984, 1987, 1993, 1999 and 2005. The last campaign also adopted satellite differential interferometry using the PSInSAR along the coastline and the interior of the area. Measurements were taken in 1984, 1987, 1993, 1999 and 2005. The last campaign was also based on classification techniques to support data validation procedures and PS characterization in a geographic information system (GIS) environment.

12.5.3 Impact on Vegetation Species Richness along the Adriatic Coast

Coastal pine forests of *Pinus pinaster* and *Pinus pinea* can be found near Ravenna (Italy), mostly situated on sand dunes. The Ravenna area is characterized by strong natural (up to 1.5 mm/year) and anthropogenic (up to 35 mm/year) subsidence (Teatini et al. 2006), which is why some of these areas are now at or below sea level. During the past 20 years, a progressive degradation of the coastal pine forests and of the inland class pine forest has been observed. Many possible causes, ranging from groundwater salinity to drought and air pollution, have been proposed for the lowering of pine density and changes in species richness in the adjacent wetlands (Diani and Ferrari 2007). In particular, a trend of groundwater salinization underneath the coastal pine forests of Ravenna has been increasing over time (Marconi et al. 2009): subsidence and salinization will cause serious challenges to the survival of pine forests and plant species diversity in the area.

Antonellini and Mollema (2010) conducted a case study to evaluate if a relationship can be found between groundwater salinity and plant species richness and diversity in the coastal pine forests and wetlands near Ravenna.

Results show that both the depth of the water table from the surface and water salinity are very important factors in controlling pine tree distribution, with a high density where the water table is deep (approximately 1.5 m or more) and salinity is low. Measurements in the dune slacks and wetlands near Ravenna indicate a significant inverse relationship between salinity and vegetation species richness (Antonellini and Mollema 2010).

In conclusion, a dramatic decrease in plant species richness and density occurs in the coastal pine forests and wetlands near Ravenna when a 3 g/L salinity threshold is exceeded (the value can reach 5 g/L for *P. pinaster* and *P. pinea*). Management of the pine forests and wetlands will thus need to take into account each of these factors, based on specific habitat and species richness conservation objectives.

12.5.4 Venice Area

In the last decades, several scientists have studied the city of Venice and its lagoon for the magnitude of the ground surface movement recorded and especially for the conservation and survival of this unique city and the surrounding lagoon, considering the small elevation differences with respect to sea level (Carbognin et al. 2004).

It is well known that during the last century, the relative lowering of Venice is up to 23 cm, consisting of approximately 12 cm of land subsidence, with 3 cm due to natural processes, 9 cm due to anthropogenic processes and 11 cm of SLR (Gatto and Carbognin 1981). The environmental problems induced by SLR are flooding, erosion, fragility of littoral islands and noticeable geomorphological and ecological modifications.

Several measures and actions have been introduced in the lagoon to protect the city and the environment from flooding, such as the regulation of groundwater withdrawal, the reinforcement of littoral islands, the stabilization of coastal protections and the 50 cm elevation of the sidewalks along the Venice canals in the lowest areas. At present, the most important protection project is the Modulo Sperimentale Elettromeccanico (MOSE) (Experimental Electromechanic Module), approved by the Italian government in May 2003, which comprises rows of mobile gates that close the three lagoon inlets of Lido, Malamocco and Chioggia when the Adriatic tide reaches 110 cm above the official Italian datum.

In the framework of project VENEZIA (2001–2003), funded by the European Space Agency (ESA), and project INLET (2006–2009), funded by the Magistrato alle Acque di Venezia, the SIMS, an original procedure that integrates different methods of subsidence monitoring (levelling, CGPS, DGPS, InSAR and PSI), has been developed (Teatini et al. 2005; Tosi et al. 2009a,b).

Until the end of the last century, geometric levelling was a unique method to measure altimetric displacement and evaluate the subsidence rate in the Venice area.

At the end of 1990, other modern techniques based on satellite data have been tested: DGPS and CGPS, the classical interferometry technique (InSAR) (Ferretti et al. 2001) with SAR (Lanari et al. 2004) and IPTA. SIMS is based on the integration of these five techniques to describe the land subsidence of the Venice area on a regional scale (100 × 100 km area). The drawbacks and limitations of each method can be overcome due to the integration of different data in the SIMS; moreover, ground displacement measured by different techniques has been cross-validated using levelling as a reference.

The integration data process has been implemented using the following steps (Teatini et al. 2005; Tosi et al. 2009a): InSAR data resampling on a regular grid; IPTA data filtering to reduce their intrinsic variability by a geostatistic technique based on a spatial autocorrelation model characterized by the presence of a nugget effect; setup of a single database containing the post-processed InSAR and IPTA information, the levelling and GPS data; and interpolation of the overall database over the study region by kriging, assuming that the displacement rate is an isotropic stochastic variable satisfying the intrinsic hypothesis.

The results of the SIMS integration method are summarized and shown in Figure 12.12, in which the analysis of the data from 1992 and 2002 is presented. The ground displacement measures have been interpolated on a 1000 m regular grid, using the stable Treviso area as a reference.

Greater vertical displacements (ranging between 5 and 15 mm/year) have been measured, as shown in Figure 12.12, in the northern and southern coastal areas, whereas the central area of the study site has general stability.

These displacement results are related to the lithostratigrafic structure and to the tectonics of the region, together with a number of anthropogenic activities inducing land subsidence, such as groundwater withdrawal and the oxidation of outcropping peat soils.

More recently, further investigation of the subsidence related to the construction of the mobile barriers of MOSE at the three inlets of Lido, Malamocco and Chioggia has been conducted using high-resolution SAR satellites; in particular, interesting results were obtained by Strozzi et al. (2009) using TerraSAR-X satellites.

The German TerraSAR-X is a new generation of satellites, operating in the X-band (wavelength of 3.11 cm) with a ground spatial resolution of approximately 3 m in StripMap mode and a repeat cycle of 11 days.

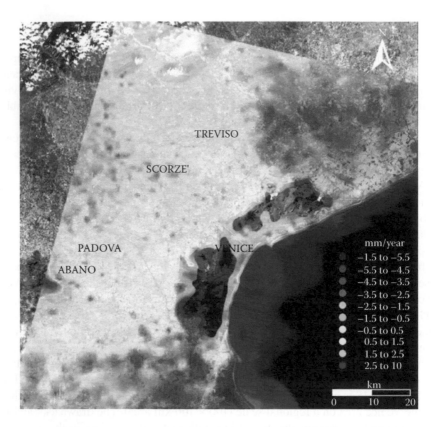

FIGURE 12.12 Displacement rates (mm/year) obtained by the SIMS in the Venice region. (From Teatini, P. et al., *Remote Sens. Environ.*, 98, 403–413, 2005.)

StripMap images have been used for the analysis of PSIs, showing the great potential of high-resolution sensors; in fact, for the entire TerraSAR-X frame, more than 1,400,000 point targets with valuable displacement information have been detected, compared with 50,000 points detected on the same area using ENVISAT ASAR images.

In Figure 12.13, the three inlets of Lido, Malamocco and Chioggia and the city of Venice are presented with the average displacement rates of point targets in the satellite LOS direction derived from TerraSAR-X.

The results clearly show that settlement rates are mainly due to the load of structures under construction or reinforcement.

In the Lido inlet (Figure 12.13a), significant settlements on the order of 20–30 mm/year were detected with TerraSAR-X interferometry at the sides of the central new artificial island and can be explained by consolidation due to the structure load on the shallow lagoon deposits, mainly constituted by unconsolidated sandy–silty layers. Moreover, a lowering on the order of 10–20 mm/year has been detected all along the craft harbour structure in the north side of the inlet, except for the central part, which is uniformly rising by almost 10 mm/year. The uplift is due to the load of the surrounding structures and to the hydraulic overpressure acting on the lock bottom located approximately 10 m below the mean sea level, which is currently drained.

At the Malamocco and Chioggia Inlets, the displacement rates recorded mainly depend on the consolidation of the sea subsurface due to the structure loads. In particular, in the Malamocco inlet (Figure 12.13b), the 1280 m long curved breakwater has settlement rates that rise towards the tip from ~5 to ~25 mm/year.

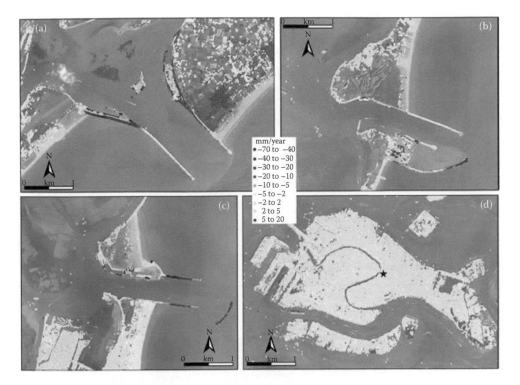

FIGURE 12.13 Mean displacement rates between March 2008 and January 2009 from TerraSAR-X interferometry at the (a) Lido, (b) Malamocco and (c) Chioggia Inlets and at (d) Venice. Negative values indicate settlement, and positive mean uplift. The star indicates the reference point near Rialto Bridge in Venice. (From Strozzi, T. et al., *Remote Sens. Environ.*, 113, 2682–2688, 2009.)

Both the shallow subsoil that is sandier near the beach and the load of the structure that increases seaward can explain this behaviour. Moreover, the lock, completed several years ago, is still showing significant sinking rates on the order of 10–20 mm/year.

In the Chioggia Inlet (Figure 12.13c), large reinforcement works to support the row of gates, and the construction of new harbour structures has been conducted at the northern side, producing a 50 mm/year consolidation rate (Table 12.3).

An influence of mobile barrier construction is detected only at the northern side of the Lido inlet, where groundwater pumping to drain the portion of the lagoon inside of the locks affects a strip of approximately 200 m along the littoral area, with subsidence rates that increase from the previous natural values of approximately 2 mm/year up to 10 mm/year (Figure 12.13a–c). However, the results are very promising and consider a higher level of detail in settlement investigations due to the spatial resolution of 3 m and a revisitation time interval of 11 days.

12.5.5 Wetland Plant Diversity and Subsidence in the North Sea (The Netherlands)

The occurrence of soil subsidence in coastal areas can be used as a case study to mimic SLR and its effects on vegetation. The Wadden Sea in The Netherlands (Figure 12.14) represents an ideal test case because of substantial salt marsh areas and subsidence due to gas extraction activities.

One of the first studies to compare the effects of subsidence on salt marshes to those of SLR was performed by Dijkema (1997). By coupling estimated subsidence rates (in view of planned gas extraction activities in the Wadden Sea) with different hypothetical values for SLR, the effect on salt marsh vegetation was evaluated. The results show that soil subsidence will not influence

TABLE 12.3

Comparison of the Main Characteristics of the Different Techniques Integrated in SIMS

Monitoring Technique	Spatial Resolution	Spatial Characteristics	Vertical Accuracy	Main Limits
Levelling	250–1000 m	Levelling lines	1 mm/year	Time consuming/ expensive
CGPS	Tens to hundreds of kilometres	Few permanent stations	2 mm/year	Expensive/ requires long time series
DGPS	5–10 km	Scattered network	5 mm/year	Accuracy
InSAR	25 m	Builtup areas	2 mm/year	Signal decorrelation in rural areas
IPTA	5–25 m	Scattered point targets	1–2 mm/year	Quite expensive/ requires filtering

Source: Teatini, P. et al., *Remote Sens. Environ.*, 98, 403–413, 2005.

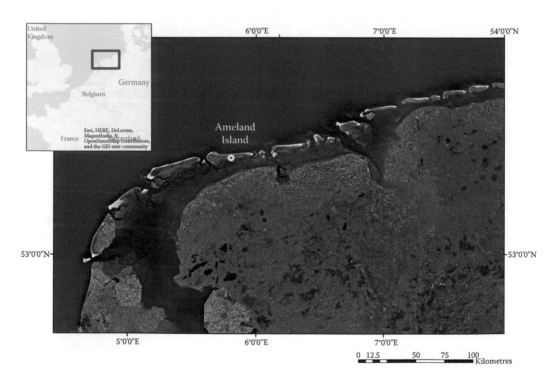

FIGURE 12.14   Wadden Sea in The Netherlands, and location of the Ameland Island case study. (From Esri, DigitalGlobe, GeoEye, i-cubed, USDA, USGS, AEX, Getmapping, Aerogrid, IGN, IGP, swisstopo, and the GIS User Community.)

the accretional balance of salt marshes; however, if estimated subsidence rates are coupled with a 6 mm/year increase in SLR, an accretional deficit is to be expected in some of the Wadden Sea salt marshes.

Simulated subsidence rates result in negative impacts on the mudflat and pioneer zone, where an accretional deficit can lead to cliff and marsh erosion from the sea. While the salt marsh zone itself is expected to compensate and keep up with the predicted ranges of future SLR (up to approximately 10 mm/year), the strongest negative effects of soil subsidence or SLR are expected to affect the pioneer zone.

Van Dobben and Slim (2005, 2012) also conducted a case study on the Ameland Island in the Wadden Sea (The Netherlands), where natural gas extraction has led to a soil subsidence up to 25 cm since 1986. This amount of subsidence was used to infer future changes in vegetation.

The island of Ameland is mostly composed of a sandy dune and salt marsh landscape, for which a monitoring programme was set up when gas extraction started. Because the ultimate effect would be a combination of subsidence and underlying SLR rates, the study concentrates on both soil subsidence effects and its combination with climate change effects.

A statistical analysis was applied to model observed vegetation as a result of measured environmental variables: precipitation, sea level, soil chemistry, phreatic level (as a measure of freshwater influence), flooding frequency (for saltwater influence) and elevation prior to gas extraction and subsidence. To detect the effect of gas withdrawal during 1986–2001, permanent plots were installed and species composition was recorded every 3 years. In 1986, 56 plots were laid out and 10 additional plots were added in 1989. The plots were installed to cover a wide range of habitats and vegetation types (from dunes to salt marshes), as well as the whole range of subsidence values. To evaluate the quantitative change of plant biodiversity, two different measures were used. The first was the number of species per plot, and the second was the compound conservancy value (CCV) (Wamelink et al. 2003; Van Dobben and Wamclink 2009), which attributes a value to each species based on its rarity and rate of decline, in relation to the Red List criteria (IUCN 2001).

The elevation of all plots was determined in relation to the Dutch standard reference level (Normal Amsterdam Ordnance Level [NAP], approximately corresponding to average sea level) by Real Time Kinematic – DGPS in 2001. Subsidence was described as a non-linear function of X and Y coordinates and time, which was parameterized based on geodetic techniques and DGPS measurements conducted between 1985 and 2001. Elevation was determined on the basis of the measured elevation and modelled soil subsidence from 2001. A small temporal change in vegetation occurred and was due to a combination of weather factors, subsidence and eutrophication. The analysis of 15 years of data shows that a generally small loss of diversity has been observed, although in many cases it is non-significant. A non-significant effect of soil subsidence on species number was found, whereas its effect on the CCV is significant. However, a decrease in the CCV was actually observed, and this is the opposite of the change expected based on its relation with elevation (an increase). The authors conclude that the observed loss of diversity is not attributable to soil subsidence, and the same can be inferred for SLR; the biodiversity loss was most likely due to eutrophication.

Despite the absence of an effect on biodiversity, the results support the existence of a relationship between soil subsidence and changes in vegetation. Only small changes in vegetation have occurred during the 15 years of observation, despite the marked subsidence phenomenon; nonetheless, the results show that there is a significant effect of soil subsidence on the shift towards the species of wetter habitats.

In conclusion, the study indicates small loss in biodiversity due to climate change (mostly SLR); however, a small increase in conservancy value is expected. In general, common species are mostly expected to be lost, whereas rarer species (with higher adaptation and resistance to disturbance) may persist. Although a loss of species is to be expected in this scenario, this does not necessarily imply a loss of conservancy value.

12.6   CONCLUSION

Based on the present state of knowledge, no significant analysis of the subsidence system is possible without innovative monitoring technology. The authors would disagree with such a position and would argue that the scientific analysis of individual drivers should proceed with a more integrated assessment of the overall coastal system. This would help to translate new science into policy-relevant forms as it emerges and also help to inform scientists of the most important knowledge gaps that need to be filled.

The analysis of future subsidence trends needs to take into account a range of factors:

- Tectonics and neotectonics (subsidence and uplift)
- Subsidence due to deltaic and/or coastal processes
- Enhanced subsidence due to groundwater withdrawal, drainage, and so forth
- Sediment supply to the coastal system, including climate change effects and human influences within the catchment
- Climate change in coastal areas, especially global SLR, cyclones and storm surges
- Influence of flood management approaches from local scales (e.g. changes in individual farming practices) to national and international initiatives (e.g. the Flood Action Plan)

Given that human influence is likely to continue to intensify, and many of these factors depend on human influence, the best way to organize such an analysis would be to develop a set of scenarios for each of the relevant factors. Thus, innovative monitoring methods for subsidence are effective in analyzing variability and resulting uncertainties in several parameters of relevance to environmental targets. All methods described represent an ongoing scientific challenge, which is becoming more and more important and interesting. The feedback between the climatological, biophysical and morphological parameters illustrated here is an attempt that must be further developed in innovative mapping methods for subsidence studies. The different approaches applied are not only conceptual. They are robust attempts to quantitatively evaluate, at a local to regional scale, different physically based subsidence rate models. The analyses highlight that over time, the morphology of different subsystems represents a balance between inputs (forcing agents such as climate) and responses (related single changes in subsidence rate).

The discussion presented an overview to this complex analysis, and the implications for future development in research science addressing subsidence in coastal areas are profound. Our present understanding suggests that it is quite uncertain how the subsidence rate calculation in coastal areas will develop in the future, as this depends on many different and uncertain factors, including future changes in (1) the sediment supply from the catchments, (2) the net subsidence within the different coastal areas, (3) the management regime within the different coastal environments and (4) the global climate change, especially future sea levels.

REFERENCES

Abernethy, V., and Willby, N. J. (1999). Changes along a disturbance gradient in the density and composition of propagule banks in floodplain aquatic habitats. *Plant Ecology*, 140, 177–190.
Ablain, M., Becker, M., Benveniste, J., Cazenave, A., Champollion, N., Ciccarelli, S. et al. (2016). White paper: Monitoring the evolution of coastal zones under various forcing factors using space-based observing systems. Bern, Switzerland: International Space Science Institute. doi: 10.13140/RG.2.2.14749.31205.
Amelung, F., Galloway, D. L., Bell, J. W., Zebker, H. (1999). Sensing the ups and downs of Las Vegas: InSAR reveals structural control of land subsidence and aquifer-system deformation. *Geology*, 27(6), 483–486.
Antonellini, M., and Mollema, P. N. (2010). Impact of groundwater salinity on vegetation species richness in the coastal pine forests and wetlands of Ravenna, Italy. *Ecological Engineering*, 36(9), 1201–1211.
Anzidei, M., Bosman, A., Carluccio, R., Casalbore, D., D'Ajello Caracciolo, F., Esposito, A. et al. (2017). Flooding scenarios due to land subsidence and sea level rise: A case study for Lipari Island (Italy). *Terra Nova*, 29(1), 44–51.
Baldwin, A. H., Egnotovich, M. S., Clarke, E. (2001). Hydrologic change and vegetation of tidal freshwater marshes: Field, greenhouse, and seed-bank experiments. *Wetlands*, 21(4), 519–531.
Baldwin, A. H., McKee, K. L., Mendelssohn, I. A. (1996). The influence of vegetation, salinity, and inundation on seed banks of oligohaline coastal marshes. *American Journal of Botany*, 83, 470–479.

Baldwin, A. H., and Mendelssohn, I. A. (1998). Response of two oligohaline marsh communities to lethal and nonlethal disturbance. *Oecologia*, 116, 543–555.

Bally, A., Catalano, R., Oldow, J. (1985). *Elementi di Tettonica Regionale*. Bologna, Italy: Editrice Pitagora.

Bamler, R., and Hartl, P. (1998). Synthetic aperture radar interferometry. *Inverse Problems*, 14, R1–R54.

Ban, Y. (2016). Multitemporal remote sensing: Current status, trends and challenges. In: Ban, Y. (eds) *Multitemporal Remote Sensing. Remote Sensing and Digital Image Processing*, 20. Springer, Cham, 1–18. DOI: 10.1007/978-3-319-47037-5_1.

Barrett, G. W., and Rosenberg, R. (1981). *Stress Effects on Natural Ecosystems*. Chichester, UK: Wiley.

Bateson, L., Evans, H., Jordan, C. (2011). GMES-service for assessing and monitoring subsidence hazards in coastal lowland areas around Europe. SubCoast D3.5.1 (OR/11/069). British Geological Survey.

Bazzaz, F. A. (1983). Characteristics of populations in relation to disturbance in natural and man-modified ecosystems. In: Mooney, H. A., Godron, M. (eds) *Disturbance and Ecosystems. Ecological Studies (Analysis and Synthesis)*, 44, 259–275.

Becker, R. H., and Sultan, M. (2009). Land subsidence in the Nile Delta: Inferences from radar interferometry. *The Holocene*, 19, 949–954.

Bekendam, R. F., and Pottgens, J. J. (1995). Ground movements over the coal mines of southern Limburg, The Netherlands, and their relation to rising mine waters. *IAHS Publications-Series of Proceedings and Reports-International Association Hydrological Sciences*, 234, 3–12.

Bell, F. (1988). Subsidence associated with the abstraction of fluids. Geological Society, London, *Engineering Geology Special Publications*, 5.1, 363–376. DOI: 10.1144/GSL.ENG.1988.005.01.40.

Bell, F. (1998). Ground movements associated with the withdrawal of fluids. *Engineering Geology of Underground Movements* (Engineering Geology Special Publication), issue 5, 363–376.

Bell, F. (2007). *Engineering Geology*, 2nd ed. London: Butterworth-Heinemann.

Bell, J. W., Amelung, F., Ramelli, A., Blewitt, G. (2002). Land subsidence in Las Vegas, Nevada, 1935–2000: New geodetic data show evolution, revised spatial patterns, and reduced rates. *Environmental & Engineering Geoscience*, 8(3), 155–174.

Berardino, P., Fornaro, G., Lanari, R., Sansosti, E. (2002). A new algorithm for surface deformation monitoring based on small baseline differential SAR interferograms. *IEEE Transactions on Geoscience and Remote Sensing*, 40(11), 2375–2383.

Bettadpur S. (2007). *Level-2 Gravity Field Product User Handbook*. Report GRACE 327-734, also CSR-GR-03-01, revision 2.3. February 20. Available from GRACE data products archive at http://podaac.jpl.nasa.gov/grace.

Bettinelli, P., Avouac, J. P., Flouzat, M., Bollinger, L., Ramillien, G., Rajaure, S., Sapkota, S. (2008). Seasonal variations of seismicity and geodetic strain in the Himalaya induced by surface hydrology. *Earth and Planetary Science Letters*, 266, 332–334.

Bevis, M., Alsdorf, D., Kendrick, E., Fortes, L., Forsberg, B., Smalley Jr., R., Becker, J. (2005). Seasonal fluctuations in the mass of the Amazon River system and Earth's elastic response. *Geophysical Research Letters*, 32, L16308. doi: 10.1029/2005GL023491.

Blum, M. D., Tomkin, J. H., Purcell, A., Lancaster, R. R. (2008). Ups and downs of the Mississippi Delta. *Geology*, 36, 375–378.

Bock, Y., Wdowinski, S., Ferretti, A., Novali, F., Fumagalli, A. (2012). Recent subsidence of the Venice Lagoon from continuous GPS and interferometric synthetic aperture radar. *Geochemistry, Geophysics, Geosystems*, 13(3), Q03023. DOI:10.1029/2011GC003976.

Bock, Y., Wdowinski, S., Ferretti, A., Novali, F., Rocca, F. (2001). Permanent scatterers in SAR interferometry. *IEEE Transactions on Geoscience and Remote Sensing*, 39, 8–20.

Boesch, D. F., Josselyn, M. N., Mehta, A. J., Morris, J. T., Nuttle, W. K., Simenstad, C. A., Swift, D. J. (1994). Scientific assessment of coastal wetland loss, restoration and management in Louisiana. *Journal of Coastal Research*, 20, 1–103.

Brown, L. F., and Fisher, W. L. Seismic stratigraphic interpretation of depositional systems. Examples from Brazilian rift and pull-apart basins. In: Ch. E. Payton (ed) *Seismic Stratigraphy Applications to Hydrocarbon Exploration*. American Association Petroleum Geologists, Memories, 26, 213–248.

Bucx, T. H. M., van Ruiten, C. J. M., Erkens, G., de Lange, G. (2015). An integrated assessment framework for land subsidence in delta cities. *Proceedings of the International Association of Hydrological Sciences*, 372, 485.

Bull, W. B. (1964). Geomorphology of segmented alluvial fans in western Fresno County, California. *US Government Printing Office*, 437.

Capella Space. (2017). http://www.capellaspace.com.

Carbognin, L., and Gatto, P. (1986). An overview of the subsidence of Venice. *Proceedings of the 3rd International Symposium on Land Subsidence*, 151, 321–328.

Carbognin, L., and Tosi, L. (2003). Il progetto ISES per l'analisi dei processi di intrusione salina e subsidenza nei territori meridionali delle province di Padova e Venezia. Istituto per lo Studio della Dinamica delle Grandi Masse e Consiglio Nazionale delle Ricerche, Venezia (Italy), 63–83.

Carbognin, L., Teatini, P., Tosi, L. (2004). Eustacy and land subsidence in the Venice Lagoon at the beginning of the new millennium. *Journal of Marine Systems*, 51(1), 345–353.

Cahoon, D. R., Reed, D. J., Day, J. W. (1995). Estimating shallow subsidence in microtidal salt marshes of the southeastern United States: Kaye and Barghoorn revisited. *Marine Geology*, 128, 1–9.

Chen, Y., Chen, Y. T., Zhang, G. M. (2005). Forecast and early-warning and preparedness measures for great earthquake disasters in China during the period of the 11th five-year plan. *Journal of Catastrophology*, 20(1), 1–14.

Chen, Z., and Rybczyk, J. (2005). Coastal subsidence. In: Schwartz, M. (ed.), *Encyclopedia of Coastal Science*. Springer Academic Publishers, Netherlands, 302–304.

Coleman, J. M., and Prior, D. B. (1982). Deltaic environments of deposition. *Sandstone Depositional Environments*, 139–178.

Coops, H., Geilen, N., van der Velde, G. (1994). Distribution and growth of the helophyte species *Phragmites australis and Scirpus lacustris* in water depth gradients in relation to wave exposure. *Aquatic Botany*, 48(3–4), 273–284.

Corenblit, D., Baas, A. C. W., Bornette, G., Darrozes, J., Delmotte, S., Francis, R. A., Gurnell, A. M., Julien, F., Naiman, R. J., Steiger, J. (2011). Feedbacks between geomorphology and biota controlling Earth surface processes and landforms: A review of foundation concepts and current understandings. *Earth Science Review*, 106, 307–331.

Costanza, R., Kubiszewski, I., Ervin, D., Bluffstone, R., Boyd, J., Brown, D., Chang, H., Dujon, V., Granek, E., Polasky, S., Shandas, V., Yeakley, A. (2011). Valuing ecological systems and services. *F1000 Biology Report*, 3–14.

Cozzolino, D., Greggio, N., Antonellini, M., Giambastiani, B. M. S. (2017). Natural and anthropogenic factors affecting freshwater lenses in coastal dunes of the Adriatic coast. *Journal of Hydrology*, 551, 804–818.

Crosetto, M., Castillo, M., Arbiol, R. (2003). Urban subsidence monitoring using radar interferometry. *Photogrammetric Engineering & Remote Sensing*, 69(7), 775–783.

Crosetto, M., Monserrat, O., Cuevas-González, M., Devanthéry, N., Crippa, B. (2016). Persistent scatterer interferometry: A review. *ISPRS Journal of Photogrammetry and Remote Sensing*, 115, 78–89.

Cuenca, M. C., Hanssen, R., Hooper, A., Arikan, M. (2011). Surface deformation of the whole Netherlands after PSI analysis. In *Proceedings Fringe 2011 Workshop*, Frascati, Italy, 19–23.

Cuenca, M. C., Hooper, A. J., Hanssen, R. F. (2013). Surface deformation induced by water influx in the abandoned coal mines in Limburg, the Netherlands observed by satellite radar interferometry. *Journal of Applied Geophysics*, 88, 1–11.

Diani, L., Ferrari, C. (2007). La vegetazione della pineta di San Vitale e il pattern spaziale di Pinus Pinca. Monitoraggia e salvaguardia della Pineta di San Vitale e Classe. *Rapporti Tecnici*, Comune di Ravenna.

Dijkema, K. S. (1997). Impact prognosis for salt marshes from subsidence by gas extraction in the Wadden Sea. *Journal of Coastal Research*, 13, 1294–1304.

Dixon, T. H., Amelung, F., Ferretti, A., Novali, F., Rocca, F., Dokka, R. et al. (2006). Space geodesy: Subsidence and flooding in New Orleans. *Nature*, 441(7093), 587–588.

Dokka, R. K. (2011). The role of deep processes in late 20th century subsidence of New Orleans and coastal areas of southern Louisiana and Mississippi. *Journal of Geophysical Research: Solid Earth*, 116(B6).

Ferretti, A., Prati, C., Rocca, F. (2001). Permanent scatterers in SAR interferometry. *IEEE Transactions on Geoscience and Remote Sensing*, 39(1), 8–20.

Fiaschi, S., Tessitore, S., Bona, R., Di Martire, D., Achilli, V., Borgstrom, S., Ibrahim, A., Floris, M., Meisina, C., Ramondini, M., Calcaterra, D. (2017). From ERS-1/2 to Sentinel-1: Two decades of subsidence monitored through A-DInSAR techniques in the Ravenna area (Italy). *GIScience & Remote Sensing*, 54, 3.

Floyd, R. P. (1978). *Geodetic Bench Marks*. NOAA Manual NOS NGS 1. Washington, DC: U.S. Department of Commerce.

Galloway, D., and Burbey, T. (2011). Review: Regional land subsidence accompanying groundwater extraction. *Hydrogeology Journal*, 19, 1459–1486.

Galloway, D., Jones, D., Ingebritsens, S. (1999). Land subsidence. In U.S. Circular 1182. Reston, VA: U.S. Geological Survey.

Galloway, D., Jones, D. R., Ingebritsen, S. (2000). Measuring land subsidence from space. USGS Fact Sheet-051-00. Reston, VA: U.S. Geological Survey.

Galloway, D. L., and Hoffmann, J. (2007). The application of satellite differential SAR interferometry-derived ground displacements in hydrogeology. *Hydrogeology Journal*, 15(1), 133–154. doi: 10.1007/s10040 -006-0121-5.

Galloway, D. L., Hudnut, K. W., Ingebritsen, S. E., Phillips, S. P., Peltzer, G., Rogez, F., Rosen, P. A. (1998). Detection of aquifer system compaction and land subsidence using interferometric synthetic aperture radar, Antelope Valley, Mojave Desert, California. *Water Resources Research*, 34(10), 2573–2585. doi:10.1029/98WR01285.

Gasperini, D., Allemand, P., Delacourt, C., Grandjean, P. (2014). Potential and limitation of UAV for monitoring subsidence in municipal landfills. *International Journal of Environmental Technology and Management*, 17(1), 1–13.

Gatto, P., and Carbognin, L. (1981). The lagoon of Venice: Natural environmental trend and man-induced modification. *Hydrological Sciences—Bulletin des Sciences Hydrologiques*, 26(4), 379–391.

Geodetic Survey Division CCRS. (1978). *Specifications and Recommendations for Control Surveys and Survey Markers*. GS Division, Canada Centre for Remote Sensing.

Gernhardt, S., Auer, S., Eder, K. (2015). Persistent scatterers at building facades – Evaluation of appearance and localization accuracy. *ISPRS Journal*, 100, 92–105.

Gernhardt, S., and Bamler, R. (2012). Deformation monitoring of single buildings using meter-resolution SAR data in PSI. *ISPRS Journal*, 73, 68–79.

Gomarasca, M. A. (2009). *Basics of Geomatics*. Berlin: Springer Science & Business Media.

Gonzalez, P. J., and Fernandez, J. (2007). Drought-driven transient aquifer compaction imaged using multitemporal satellite radar interferometry. *Geology*, 39(6), 551–554.

Goodbred, S. L., Kuehl, S. A., Steckler, M. S., Sarker M. H. (2003). Controls on facies distribution and stratigraphic preservation in the Ganges-Brahmaputra delta sequence. *Sedimentary Geology*, 155, 301–316.

Gornitz, V. (1995). Sea level rise: A review of recent past and near future trends. *Earth Surface Processes and Landforms*, 20, 7–20.

Goudie, A., and Viles, H. (2016). Subsidence in the anthropocene. In: Goudie, A., and Viles, H. (eds) *Geomorphology in the Anthropocene*. Cambridge: Cambridge University Press, 31–56. doi: 10.1017 /CBO9781316498910.004.

Haiyang, Y., Xiaoping, L., Gang, C., Xiaosan, G. (2011). Detection and volume estimation of mining subsidence based on multi-temporal LIDAR data. In *19th International Conference on Geoinformatics*, 24–26 June 2011, Shanghai, China, 1–6. doi: 10.1109/GeoInformatics.2011.5980892.

Hallegatte, S., Green, C., Nicholls, R. J., Corfee-Morlot, J. (2013). Future flood losses in major coastal cities. *Nature Climate Change*, 3, 802–806.

Hamersley, D. (2012). *SubCoast Newsletter*, issue 3.

Harnischmacher, S., and Zepp, H. (2014). Mining and its impact on the earth surface in the Ruhr District (Germany). *Zeitschrift für Geomorphologie, Supplementary Issues*, 58(3), 3–22.

Helland-Hansen, W., and Martinsen, O. J. (1996). Shoreline trajectories and sequences: Description of variable depositional-dip scenarios. *Journal of Sedimentary Research*, 66(4), 670–688.

Helm, D. C. (1994). Horizontal aquifer movement in a Theis-Thiem confined aquifer system. *Water Resources Research*, 30(4), 953–964. doi: 10.1029/94WR00030.

Hillel, D. (1992). *Introduction to Soil Physics*. San Diego: Academic Press.

Hoffmann, J., Galloway, D., Zebker, H. (2003). Inverse modeling of interbed storage parameters using land subsidence observations, Antelope Valley, California. *Water Resources Research*, 39(2), 1031. doi: 10.1029/2001WR001252.

Hoffmann, J., Zebker, H., Galloway, D., Amelung, F. (2001). Seasonal subsidence and rebound in Las Vegas Valley, Nevada, observed by synthetic aperture radar interferometry. *Water Resources Research*, 37(6), 1551–1566.

Holzer, T. L., ed. (1984). *Man-Induced Land Subsidence*. Washington, DC: Geological Society of America, 232.

Hung, W., Hwang, C., Chang, C. P., Yen, J. Y., Liu, C. H., Yang, W. H. (2010). Monitoring severe aquifer-system compaction and land subsidence in Taiwan using multiple sensors: Yunlin, the southern Choushui River alluvial fan. *Environmental Earth Sciences*, 59(7), 1535–1548. doi: 10.1007/s12665-009-0139-9.

Hunt, D., and Tucker, M. E. (1992). Stranded parasequences and the forced regressive wedge systems tract: Deposition during base-level fall. *Sedimentary Geology*, 81(1–2), 1–9. doi: 10.1016/0037-0738(92)90052-S.

IUCN (International Union for Conservation of Nature), Natural Resources, and IUCN Species Survival Commission. (2001). IUCN Red List categories and criteria. Version 3.1. Gland, Switzerland: World Conservation Union.

Jervey, M. T. (1988). Quantitative geological modeling of siliciclastic rock sequences and their seismic expression. *Society of Economic Paleontologists and Mineralogists*, 42, 47–69.

Kaneko, S., and Toyota, T. (2011). Long-term urbanization and land subsidence in Asian megacities: An indicators system approach. In *Groundwater and Subsurface Environments: Human Impacts in Asian Coastal Cities*, ed. M. Taniguchi. Berlin: Springer, 249–270. doi: 10.1007/978-4-431-53904-9_13.

Karegar, M. A., Dixon, T. H., and Engelhart, S. E. (2016). Subsidence along the Atlantic Coast of North America: Insights from GPS and late Holocene relative sea level data. *Geophysical Research Letters*, 43, 3126–3133. doi: 10.1002/2016GL068015.

Kearns, T. J., Wang, G., Bao, Y., Jiang, J., Lee, D. (2015). Current land subsidence and groundwater level changes in the Houston Metropolitan Area (2005–2012). *Journal of Surveying Engineering*, 141(4), 05015002.

Ketelaar, V. B. H. (2008). Monitoring surface deformation induced by hydrocarbon production using satellite radar interferometry. PhD thesis, Delft University of Technology, Delft, the Netherlands.

Khan, S. D., Huang, Z., Karacay, A. (2014). Study of ground subsidence in northwest Harris County using GPS, LIDAR, and InSAR techniques. *Natural Hazards*, 73(3), 1143–1173.

Klemas, V. (2015). Coastal and environmental remote sensing from unmanned aerial vehicles: An overview. *Journal of Coastal Research*, 31(5), 1260–1267.

Konagai, K., Asakura, T., Suyama, S., Kyokawa, H., Kiyota, T., Eto, C., Shibuya, K. (2012). Soil subsidence map of the Tokyo Bay area liquefied in the March 11th Great East Japan Earthquake. In *Proceedings of the International Symposium on Engineering: Lessons Learned from the 2011 Great East Japan Earthquake*, Tokyo, 1–4 March, 855–864.

Kooi, H., Johnston, P., Lambeck, K., Smither, C., Molendijk, R. (1998). Geological causes of recent (~100 yr) vertical land movement in the Netherlands. *Tectonophysics*, 299, 297–316.

Kraus, K., and Pfeifer, N. (1997). A new method for surface reconstruction from laser scanner data. *International Archives of Photogrammetry and Remote Sensing*, 32(Pt. 3), 2W3.

Kulp, M. (2000). *Holocene Stratigraphy, History, and Subsidence. Mississippi Delta Region, North-Central Gulf of Mexico*. Lexington: University of Kentucky.

Kusky, T. (2005). *Encyclopedia of Earth Science*. New York: Facts on File.

Lambeck, K., and Purcell, A. (2005). Sea-level change in the Mediterranean Sea since the LGM: Model predictions for tectonically stable areas. *Quaternary Science Reviews*, 24, 1969–1988.

Lanari, R., Mora, O., Manunta, M., Mallorqui, J., Berardino, P., Sansosti, E. (2004). A small-baseline approach for investigating deformations on full-resolution differential SAR interferograms. *IEEE Transactions on Geoscience and Remote Sensing*, 42, 1377–1386.

Lechner, A. M., Fletcher, A., Johansen, K., Erskine, P. (2012). Characterizing upland swamps using object-based classification methods and hyper-spatial resolution imagery derived from an unmanned aerial vehicle. *Annals of the Photogrammetry, Remote Sensing and Spatial Information Sciences*, 4, 101–106.

Li, X., Zhang, X., Ren, X., Fritsche, M., Wickert, J., Schuh, H. (2015). Precise positioning with current multi-constellation global navigation satellite systems: GPS, GLONASS, Galileo and BeiDou. *Scientific Reports*, 5, 8328.

Lofgren, B. E. (1969). Land subsidence due to the application of water. *Reviews in Engineering Geology*, II, 271–303.

Long, J., and Ding, X. (2010). Monitoring ground subsidence in New Orleans with persistent scatterers. In *International Conference on Multimedia Technology (ICMT)*, 29–31 October, Ningbo, China, 1–5.

Lowe, D. G. (2004). Distinctive image features from scale-invariant keypoints. *Int. J. Comput. Vis.*, 60(2), 91–110.

Marconi, V., Dinelli, E., Antonellini, M., Capaccioni, B., Balugani, E., Gabbianelli, G. (2009). Hydrogeochemical characterization of the phreatic system of the coastal wetland located between Fiumi Uniti and Bevano rivers in the southern Po plain (northern Italy). *Geophysical Research Abstracts*, 11, EGU2009-9771.

Massonnet, D., and Feigl, K. L. (1998). Radar interferometry and its application to changes in the earth's surface. *Reviews of Geophysics*, 36, 441–500.

Mishra, S. K., Singh, R. P., Chandra, S. (1993). Prediction of subsidence in the Indo-Gangetic basin carried by groundwater withdrawal. *Engineering Geology*, 33(3), 227–239.

Mitsch, W. J., and Gosselink, J. G. (2000). *Wetlands*. 3rd ed. New York: John Wiley & Sons.

Moors, E. J., Groot, A., Biemans, H., van Scheltinga, C. T., Siderius, C., Stoffel, M., Huggel, C., Wiltshire, A., Mathison, C., Ridley, J. (2011). Adaptation to changing water resources in the Ganges basin, northern India. *Environmental Science & Policy*, 14, 758–769.

Muller, G. (1986). Appunti di livellazione. Firenze, Italy: IGM (Military Geographical Institute).

Nakano, T., Kadomura, H., Matsuda, T. (1969). Land subsidence in the Tokyo deltaic plain. In *Proceeding of the Tokyo Symposium*, September, 1, IASH/AIHS – Unesco.

Nicholls, R. J. (2004). Coastal flooding and wetland loss in the 21st century: Changes under the SRES climate and socio-economic scenarios. *Global Environmental Change*, 14(1), 69–86.

Nicholls, R. J., and Cazenave, A. (2010). Sea-level rise and its impact on coastal zones. *Science*, 328, 1517. doi: 10.1126/science.1185782.

Nicholls, R. J., Marinova, M., Lowe, J. A., Brown, S., Vellinga, P., De Gusmao, D., Hinkel, J., Tol, R. S. J. (2011). Sea-level rise and its possible impacts given a 'beyond 4°C world' in the twenty-first century. *Philosophical Transactions of the Royal Society A—Mathematical Physical and Engineering Sciences*, 369, 161–181.

Nordstrom, K. F., Armaroli, C., Jackson, N. L., Ciavola, P. (2015). Opportunities and constraints for managed retreat on exposed sandy shores: Examples from Emilia-Romagna, Italy. *Ocean and Coastal Management*, 104, 11–21.

Odum, E. P., Finn, J. T., Franz, E. H. (1979). Perturbation theory and the subsidy-stress gradient. *Bioscience*, 29(6), 349–352.

Paola, C., Mullin, J., Ellis, C., Mohrig, D. C., Swenson, J. B., Parker, G. et al. (2001). Experimental stratigraphy. *GSA Today*, 11(7), 4–9.

Passalacqua, P., Lanzoni, S., Paola, C., Rinaldo, A. (2013). Geomorphic signatures of deltaic processes and vegetation: The Ganges-Brahmaputra-Jamuna case study. *Journal of Geophysical Research: Earth Surface*, 118, 1–12.

Penland, S., Ramsey, K. E., McBride, R. A., Mestayer, J. T., Westphal, K. A. (1988). Relative sea level rise and delta-plain development in the Terrebonne Parish region. Coastal Geology Technical Report. Baton Rouge: Louisiana Geological Survey.

Pimm, S. L. (1984). The complexity and stability of ecosystems. *Nature*, 307(5949), 321–326.

Poland, J. (1984). *Guidebook to Studies of Land Subsidence due to Ground-Water Withdrawal*. Chelsea, UK: Michigan Book Crafters.

Rappleye, H. S. (1948). *Manual of Geodetic Levelling*. Washington, DC: U.S. Government Printing Office.

Reinhardt, L., Jerolmack, D., Cardinale, B. J., Vanacker, V., Wrigh, J. (2010). Dynamic interactions of life and its landscape: Feedbacks at the interface of geomorphology and ecology. *Earth Surface Processes and Landforms*, 35, 78–101.

Rosen, P., Hensley, S., Joughin, I., Li, F., Madsen, S., Rodríguez, E., Goldstein, R. (2000). Synthetic aperture radar interferometry. *Proceedings of the IEEE*, 88(3), 333–382.

Rykiel, E. J. (1985). Towards a definition of ecological disturbance. *Australian Journal of Ecology*, 10(3), 361–365.

Sansosti, E., Manunta, M., Casu, F., Bonano M., Ojha C., Marsella, M., Lanari, R. (2015). Radar remote sensing from space for surface deformation analysis: Present and future opportunities from the new SAR sensor generation. *Rendiconti Lincei*, 26(1), 75–84.

Scarelli, F. M., Barboza, E. G., Cantelli, L., Gabbianelli, G. (2017). Surface and subsurface data integration and geological modelling from the Little Ice Age to the present, in the Ravenna coastal plain, northwest Adriatic Sea (Emilia-Romagna, Italy). *Catena*, 151, 1–15.

Serva, L., and Brunamonte, F. (1994). L'abbassamento del suolo nella Pianura Pontina: Un caso eccezionale di interferenza tra evoluzione naturale ed effetti della bonifica idraulica. Prima monografia sulla difesa del suolo s.l. (G.N.D.C.I.). Comitato Nazionale Difesa del Suolo.

Simpson, R. L., Good, R. E., Leck, M. A., Whigham, D. F. (1983). The ecology of freshwater tidal wetlands. *Bioscience*, 33, 255–259.

Sheng, Z., Helm, D. C., Li, J. (2003). Mechanisms of earth fissuring caused by groundwater withdrawal. *Environmental & Engineering Geoscience*, 9(4), 313–324. doi: 10.2113/9.4.351.

Sorichetta, A., Seeber, L., McHugh, C. M. G., Cormier, M.-H., Taramelli, A. (2006). Crustal unloading along a continental transform: Flexural response to the North Anatolian Fault in Izmit Gulf, NW Turkey. Presented at International Workshop on Comparative Studies of the North Anatolian Fault (Northwest Turkey) and the San Andreas Fault (Southern California), Istanbul, 14–18 August.

Steckler, M. S., Nooner, S. L., Akhter, S. H., Chowdhury, S. K., Bettadpur, S., Seeber, L., Kogan, M. G. (2010). Modeling Earth deformation from monsoonal flooding in Bangladesh using hydrographic, GPS, and Gravity Recovery and Climate Experiment (GRACE) data. *Journal of Geophysical Research: Solid Earth*, 115(B8). doi: 10.1029/2009JB007018.

Strain, E., Belzen, J., Comandini, P., Wong, J., Bouma, T. J., Airoldi, L. (2017). The role of changing climate in driving the shift from perennial grasses to annual succulents in a Mediterranean salt marsh. *Journal of Ecology*, 105(5), 1374–1385. doi: 10.1111/1365-2745.1279.

Strozzi, T., Teatini, P., Tosi, L. (2009). TerraSAR-X reveals the impact of the mobile barrier works on Venice coastland stability. *Remote Sensing of Environment*, 113, 2682–2688.

Strozzi, T., Tosi, L., Wegmuller, U., Werner, C., Teatini, P., Carbognin, L. (2003a). Land subsidence monitoring service in the Lagoon of Venice. In *2003 IEEE International Geoscience and Remote Sensing Symposium 2003 (IGARSS '03)*, 21–25 July, Toulouse, France, vol. 1, 212–214. doi: 10.1109 /IGARSS.2003.1293727.

Strozzi, T., Wegmuller, U., Werner, C., Wiesmann, A., Spreckel, V. (2003b). JERS SAR interferometry for land subsidence monitoring. *IEEE Transactions on Geoscience and Remote Sensing*, 41(7), 1702–1708.

Strozzi, T., Werner, C., Wiesmann, A., Wegmuller, U. (2012). Topography mapping with a portable real-aperture radar interferometer. *IEEE Geoscience and Remote Sensing Letters*, 9(2), 277–281.

Syvitski, J. P. M., Kettner, A. J., Overeem, I., Hutton, E. W. H., Hannon, M. T., Brakenridge, G. R., Day, J., Vörösmarty, C., Saito, Y., Giosan, L., Nicholls R. J. (2009). Sinking deltas due to human activities. *Nature Geoscience*, 2, 681–686. doi: 10.1038/ngeo629.

Tapley, B. D., Bettadpur, S., Watkins, M., Reigber, C. (2004). The Gravity Recovery and Climate Experiment: Mission overview and early results. *Geophysical Research Letters*, 31, L09607. doi: 10.1029/2004GL019920.

Tapponnier, P., Peltzer, G., Armijo, R. (1986). On the mechanics of the collision between India and Asia. *Geological Society Special Publication*, 19(1), 113–157.

Taramelli, A., Di Matteo, L., Ciavola, P., Guadagnano, F., Tolomei, C. (2015b). Temporal evolution of patterns and processes related to subsidence of the coastal area surrounding the Bevano River mouth (Northern Adriatic)—Italy. *Ocean and Coastal Management*, 108, 74–88.

Taramelli, A., Valentini, E., Cornacchia, L. (2015a). Remote sensing solutions to monitor biotic and abiotic dynamics in coastal ecosystems. In *Coastal Zones: Solutions for the 21st Century*, ed. J. Baztan, O. Chouinard, B. Jorgensen, P. Tett, J.-P. Vanderlinden, and L. Vasseur. Amsterdam: Elsevier, 125–135. doi: 10.1016/B978-0-12-802748-6.00009-7.

Taramelli, A., Valentini, E., Cornacchia, L., and Bozzeda, F. (2017). A hybrid power law approach for spatial and temporal pattern analysis of salt marsh evolution. *Journal of Coastal Research* (Special Issue 77: Coastal Resilience: Exploring the Many Challenges from Different Viewpoints), 62–72.

Taramelli, A., Valentini, E., Cornacchia, L., Mandrone, S., Monbaliu, J., Thompson, R., Hogart, S., Zanuttigh, B. (2014). Modelling uncertainty in estuarine system by means of combined approach of optical and radar remote sensing. *Coastal Engineering*, 87, 77–96.

Teatini, P., Ferronato, M., Gambolati, G., Gonella, M. (2006). Groundwater pumping and land subsidence in the Emilia-Romagna coastland, Italy: Modeling the past occurrence and the future trend. *Water Resources Research*, 42(1), 1–19.

Teatini, P., Tosi, L., Strozzi, T. (2011). Quantitative evidence that compaction of Holocene sediments drives the present land subsidence of the Po Delta, Italy. *Journal of Geophysical Research: Atmospheres*, 116, B08407.

Teatini, P., Tosi, L., Strozzi, T., Carbognin, L., Wegmuller, U., Rizzetto, F. (2005). Mapping regional land displacements in the Venice coastland by an integrated monitoring system. *Remote Sensing of Environment*, 98, 403–413.

Tolomei, C., Taramelli, A., Moro, M., Saroli, M., Salvi, S. (2013). Analysis of DGSD impending over the Fiastra lake (central Italy), by geomorphological assessment and deformation monitoring using satellite SAR interferometry. *Geomorphology*, 201, 281–292. doi: org/10.1016/j.geomorph.2013.07.002.

Titus, J. G. (1988). Greenhouse effect, sea level rise and coastal wetlands. EPA-230-05-86-013. Washington, DC: U.S. Environmental Protection Agency.

Törnqvist, T. E., Wallace, D. J., Storms, J. E. A., Wallinga, J., Van Dam, R. L., Blaaum, M., Derksen, M. S., Klerks, C. J. W., Meijneken, C., Snijders, E. M. A. (2008). Mississippi delta subsidence primarily caused by compaction of Holocene strata. *Nature Geoscience*, 1, 173–176.

Torres, R., Snoeij, P., Geudtner, D., Bibby, D., Davidson, M., Attema, E. et al. (2012). GMES Sentinel-1 mission. *Remote Sensing of Environment*, 120, 9–24.

Tosi, L., Teatini, P., Carbognin, L., Brancolini, G. (2009a). Using high resolution data to reveal depth-dependent mechanisms that drive land subsidence: The Venice coast, Italy. *Tectonophysics*, 474, 271–284.

Tosi, L., Teatini, P., Carbognin, L., Brancolini, G., Rizzetto, F. (2009b). Variabilità spaziale della subsidenza attuale nell'area veneziana. *Memorie Descrittive Della Carta Geologica d'Italia Servizio Geologico d'Italia*, 88, 27–30.

Tosi, L., Da Lio, C., Strozzi, T., Teatini, P. (2016). Combining L- and X-band SAR interferometry to assess ground displacements in heterogeneous coastal environments: The Po River Delta and Venice Lagoon, Italy. *Remote Sensing*, 8(4), 308.

Turner, I., Harley, M., Drummond, C. (2016). UAVs for coastal surveying. *Coastal Engineering*, 114, 19–24.

U.S. Army Corps of Engineers. (1994). *Topographic Surveying*. U.S. Army Corps of Engineers.

Vail, P. R., Audemard, F., Bowman, S. A., Eisner, P., Peres-Cruz, C. (1991). The stratigraphic signatures of tectonics, eustasy and sedimentology – An overview. Cycles and events in stratigraphy. Berlin: Springer-Verlag, 617–659.

Vail, P. R., Mitchum Jr., R. M., Thompson, S. (1977). Seismic stratigraphy and global changes of sea level: Part 3. Relative changes of sea level from coastal onlap: Section 2. Application of seismic reflection configuration to stratigraphic interpretation. In *M26: Seismic Stratigraphy—Applications to Hydrocarbon Exploration*, Memoir, Texas, 63–81.

Van Dobben, H. F., and Slim, P. A. (2005). Evaluation of changes in permanent plots in the dunes and upper salt marsh at Ameland East: Ecological effects of gas extraction. *Monitoring effecten van bodemdaling op Ameland-Oost*, 1–36.

Van Dobben, H. F., and Slim, P. A. (2012). Past and future plant diversity of a coastal wetland driven by soil subsidence and climate change. *Climatic Change*, 110(3–4), 597–618.

Van Dobben, H. F., and Wamelink, G. W. W. (2009). A red-list-based biodiversity indicator and its application in model studies in the Netherlands. In *Progress in the Modelling of Critical Thresholds, Impacts to Plant Species Diversity and Ecosystem Services in Europe*, ed. J. P. Hettelingh, M. Posch, J. Slootweg. CCE Status Report 2009. Bilthoven, the Netherlands: Coordination Centre for Effects, 77–81.

Van Wagoner, J. C., Mitchum, R. M., Campion, K. M., Rahmanian, V. D. (1990). Siliciclastic sequence stratigraphy in well logs, cores, and outcrops. *American Association of Petroleum Geologists Methods in Exploration Series*, 7, 1–55.

Vosselman, G. (2000). Slope based filtering of laser altimetry data. *International Archives of Photogrammetry and Remote Sensing*, 33(B4), 958–964.

Wahr, J., Swenson, S., Zlotnicki, V., Velicogna, I. (2004). Time-variable gravity from GRACE: First results. *Geophysical Research Letters*, 31, L11501. doi: 10.1029/2004GL019779.

Wamelink, G. W. W., Ter Braak, C. J. F., Van Dobben, H. F. (2003). Changes in large-scale patterns of plant biodiversity predicted from environmental economic scenarios. *Landscape Ecology*, 18(5), 513–527.

Wang, C., and Temmerman, S. (2013). Does biogeomorphic feedback lead to abrupt shifts between alternative landscape states: An empirical study on intertidal flats and marshes. *Journal of Geophysical Research: Earth Surface*, 118, 229–240.

Wang, J., Gao, W., Xu, S., Yu, L. (2012). Evaluation of the combined risk of sea level rise, land subsidence, and storm surges on the coastal areas of Shanghai, China. *Climatic Change*, 115(3–4), 537–558.

Watts, A. B. (2001). *Isostasy and Flexure of the Lithosphere*. Cambridge: Cambridge University Press.

Werner, C., Wegmuller, U., Strozzi, T., Wiesmann, A. (2003). Interferometric point target analysis for deformation mapping. In *Proceedings of IEEE Geoscience and Remote Sensing Symposium, IGARSS*, 21–25 July, Toulouse, France, vol. 7, 4362–4364.

Williams, S. D. P., Bock, Y., Fang, P., Jamason, P., Nikolaidis, R. M., Prawirodirdjo, L., Miller, M., Johnson, D. J. (2004). Error analysis of continuous GPS position time series. *Journal of Geophysical Research: Solid Earth*, 109(B3).

Wolff, C., Vafeidis, A. T., Lincke, D., Marasmi, C., Hinkel, J. (2016). Effects of scale and input data on assessing the future impacts of coastal flooding: An application of DIVA for the Emilia-Romagna coast. *Frontiers in Marine Science*, 3, 41.

Wöppelmann, G., Le Cozannet, G., Michele, M., Raucoules, D., Cazenave, A., Garcin, M. et al. (2013). Is land subsidence increasing the exposure to sea level rise in Alexandria, Egypt? *Geophysical Research Letters*, 40(12), 2953–2957.

Yuill, B., Lavoie, D., Reed, D. (2009). Understanding subsidence processes in coastal Louisiana. *Journal of Coastal Research*, 54, 23–36.

Zalasiewicz, J., Waters, C. N., Williams, M. (2014). Human bioturbation, and the subterranean landscape of the Anthropocene. *Anthropocene*, 6, 3–9.

Zhang, Y., Zhang, J., Wu, H., Lu, Z., Guangtong, S. (2011). Monitoring of urban subsidence with SAR interferometric point target analysis: A case study in Suzhou, China. *International Journal of Applied Earth Observation and Geoinformation*, 13(5), 812–818. doi: 10.1016/j.jag.2011.05.003.

Zebker, H., and Villasenor, J. (1992). Decorrelation in interferometric radar echoes. *IEEE Transactions on Geoscience and Remote Sensing*, 30, 950–959.

Zebker, H. A., Rosen, P. A., Goldstein, R. M., Gabriel, A., Werner, C. L. (1994). On the derivation of coseismic displacement fields using differential radar interferometry: The Landers earthquake. *Journal of Geophysical Research*, 99(19), 617–634.

13 Subsidence Mapping Using InSAR

Mukesh Gupta

CONTENTS

13.1 INTRODUCTION

Subsidence, a common hazard on the Earth's surface, can occur due to natural or anthropogenic causes and affect life and building structures in the areas where subsidence occurs. Mapping and monitoring of subsidence is of utmost importance, as it can greatly help in the safety, mitigation and management of life, property and natural resources. In subsidence, the vertical downward displacement (sudden or gradual) of the Earth's surface occurs with respect to a local datum (Figure 13.1), caused by continuous or episodic natural and anthropogenic activities. It can be due to mining, extraction of hydrocarbon and groundwater, natural compaction, geological faults, urban settlement, thawing permafrost, sinkholes, drainage of organic soils and isostatic and tectonic movements. The subsidence and the resulting surface deformation can cause serious environmental hazards, especially in the coal mining areas (Singh and Yadav 1995), where coal fire may further aggravate the disaster (Prakash et al. 2001). Surface subsidence due to reservoir depletion and induced seismicity may also occur as a result of geothermal exploitation (Carnec and Fabriol 1999). Land subsidence can also trigger a chain of catastrophic events, such as damage to underground pipelines and the transport system, severe flooding due to destruction of dams, snow avalanches, tsunamis (due to ocean floor subsidence caused by earthquakes), building collapse and road damage.

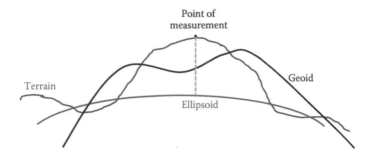

FIGURE 13.1 Illustration of different reference surfaces: ellipsoid, geoid and terrain, in relation to the point of measurement.

Mapping of subsidence is a graphical representation of geophysical information related to vertical displacement of the Earth's surface, horizontal movements, topographic or crustal deformation and the rate of change of the above, if persistent through a long duration of time (Gupta et al. 2014). Subsidence is a time-varying three-dimensional phenomenon; its mapping (two-dimensional) and modelling pose some technological challenges (Liu and Huang 2013). Different mapping and modelling techniques provide different accuracies to adequately map the nature of subsidence. Several network-based surveying methods, for example, levelling, total station (electronic distance metre [EDM] surveying) and global positioning system (GPS) surveying (in static and Real Time Kinematic [RTK] mode) (Schofield and Breach 2007), are in use for mapping the subsidence. Levelling and total station methods can deliver 0.1 mm height change resolution, while the GPS can provide a height determination accuracy of 5 mm in static and 2–3 cm in RTK mode. However, despite good capability of providing sufficient accuracy, these methods are point based, time-consuming, costly and weather dependent.

The synthetic aperture radar interferometry (InSAR) (space or airborne) technique provides highly accurate sub-millimetre-level displacements of any ground motion (Massonnet et al. 1993). It exploits the phase difference from two SAR images acquired over an area. The phase differences between the two images are calculated to generate the interferogram. The phase can be converted into height information using the geometry. Satellite-based microwave techniques of subsidence mapping provide all-weather, day and night, and large-swath coverage on a high temporal basis. Multitemporal differential InSAR (DInSAR) can provide ~1 cm vertical resolution for subsidence occurring at shorter time intervals (Cascini et al. 2006). However, atmospheric disturbances (tropospheric delay) can lead to misinterpretation of DInSAR results. The corrections due to atmospheric disturbances can be made using GPS observations (Ge et al. 2007). InSAR, assisted with geographic information systems (GIS) and various subsidence modelling techniques, is capable of mapping ground subsidence to an unprecedented accuracy.

In recent years, a newly developed technique, permanent or persistent scatterer InSAR (PSInSAR), has replaced the conventional InSAR (Musson et al. 2004; Meisina et al. 2008). PSInSAR is a tool for mapping very precise ground deformation on a sparse grid of permanent scatterers (or radar targets) best suited for high-resolution mapping of extremely slow ground deformation. The persistent scatterers (PSs) can be permanent scattering points or stable reflectors, such as rooftops, bridges, dams, antennas, towers and other prominent natural features. PSInSAR is a cost-effective method, in which no ground equipment is needed for mapping, and it is also successful in vegetated areas. The problem of loss of coherency in the multitemporal InSAR method is overcome in PSInSAR due to a large number of permanent scatterers that do not change despite changing vegetation. Ferretti et al. (2000) have developed a methodology for the detection and analysis of permanent scatterers. It can be used in a wide range of natural hazards associated with surface deformation, for example, subsidence, landslides (Farina et al. 2006) and tectonic motion (Colesanti et al. 2003).

13.2   InSAR AND PSInSAR METHODS

Conventional subsidence techniques involved point observations at selected locations that were easily accessible for repeat visits, provided sufficient coverage of the vulnerable area and were cost-effective. The network of such points, however, is a crucial geodetic design, which requires a controlled optimization. A good network design is important to ensure sufficient precision and reliability of the estimated parameters. An optimal design considers positioning of benchmarks for reasonable spatial extent and density. In addition to the above, there can be other considerations: the requirement of benchmark installation, limitations in installation and having a sufficient density of benchmarks, a check on time sensitivity such that the rate of deformation is detectable and selection of a reference location. Here, the details of InSAR and PSInSAR are discussed.

13.2.1   InSAR

InSAR is a powerful tool to map subsidence over large areas with accuracy as low as a few millimetres per year covering several kilometres. While conventional geodetic surveys are conducted for studying limited areas of subsidence and surface deformation, the satellite-based observations cover larger areas on a frequent basis, highly suitable for detailed mapping of land subsidence. Because it is based on remote measurements using satellite- or aircraft-based SAR antenna, it does not require any specific field measurements. This ability of InSAR can be considered a major advantage over conventional methods. The InSAR uses the advantage of the phase component of the reflected microwave to measure the changes in the range distance (Bamler and Hartl 1998). The patterns of interference between phase components called inter-ferograms are the key for mapping surface deformation. InSAR is a technique for extracting topographic information from complex radar signals. It exploits the phase differences of two complex SAR images acquired from different orbits and/or at different times. Topographic information about the scene is obtained by differencing the phase of each pixel. The vertical separation between the two receiving antennas enables phase difference to be computed which is related to the pixel height by a geometrical relationship given by Li and Goldstein (1990) (Equations 13.1 and 13.2):

$$\phi = \frac{4\pi}{\lambda}(B_H \sin\theta - B_V \cos\theta) \tag{13.1}$$

$$h = H - r\cos\theta \tag{13.2}$$

where λ is the radar wavelength; h is pixel height; B_H and B_V are horizontal and vertical base-lines, respectively; and ϕ is the phase difference. Parameter H is the satellite altitude, r is the slant distance from the base of the target to the satellite antenna and θ is the look angle (Figure 13.2).

The phase difference of the two SAR images is calculated for an interferogram where fringes represent the whole range of the phase from 0 to 2π in a full colour cycle. The correlation of the phase information of two SAR images is measured as coherence in the range of 0–1. The phase of the signal contains information about coherent displacements of all scatterers imaged by the radar. The degree of coherence can be used as a quality measure because it significantly influences the accuracy of the phase difference and height measurements. The phase information is directly related to the topography. This phase information contains an ambiguity of 2π. It is necessary to solve this ambiguity to calculate the elevation at each point. This ambiguity solution is known as phase unwrapping and leads to topographic information. The unwrapped phase consists of the phase difference from the two viewing geometries. This is proportional to the slant range two-way path difference of the ground resolution cells from two viewing geometries.

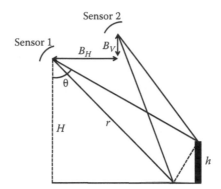

FIGURE 13.2   Geometry of terrain height calculation from InSAR sensors.

This is converted to terrain height using baseline and imaging geometries. The InSAR technique has been extensively used in land subsidence investigation (Massonnet et al. 1997; Fielding et al. 1998; Galloway et al. 1998).

However, a major limitation of InSAR is the decorrelation of signal at different spatial and temporal levels as acquired from two SAR antennas, and the atmospheric perturbations in the interferometric phase (Zebker and Villasenor 1992). This makes it difficult to map surface displacement if the surface area is highly vegetated, snow covered, flooded, and so forth, which enhances the decorrelation of signals. This problem can be circumvented to a great extent using numerous multi-temporal SAR images. Sequential InSAR processing steps are described in the next section.

13.2.2   DIGITAL ELEVATION MODEL GENERATION

Figure 13.3 provides complete processing steps for computing digital elevation model (DEM) from InSAR. The steps involve master-to-slave registration, resampling and filtering, interferogram generation, coherence image generation, phase unwrapping, slant-to-height conversion and geocoding.

13.2.3   MASTER-TO-SLAVE REGISTRATION

Registration of a master SAR image to a slave SAR image begins with estimation of approximate offsets in the line and pixel direction using precise orbits. The initial offsets in the pixel and line direction are estimated visually using tie points from the slave and master amplitude images. These initial offsets are further refined to single-pixel-level accuracy using image-to-image coarse correlation. Image-to-image coarse correlation is estimated using a number of windows distributed across the amplitude images. Coarse correlation does not use interpolation to compute sub-pixel-level accuracy. Both spatial and frequency domain coarse correlation yield similar results, but frequency domain coarse correlation is computationally more efficient. It is sufficient to use 10–20 windows of size 32 × 32 for coarse correlation.

Fine correlation estimates sub-pixel-level offset between the amplitude images using the interpolation technique in either the frequency or spatial domain. Since these offsets are used to model slave-to-master translation, the fine correlation needs to be estimated over a large number of windows (typically 1000). Only the matched points having a correlation coefficient above a threshold are used to compute the polynomial function for the master-to-slave translation.

13.2.4   RESAMPLING AND FILTERING

Resampling of the slave image to match the master image is the most computation-intensive among all tasks. The relative strength of the interpolation schemes needs to be evaluated. The resampled

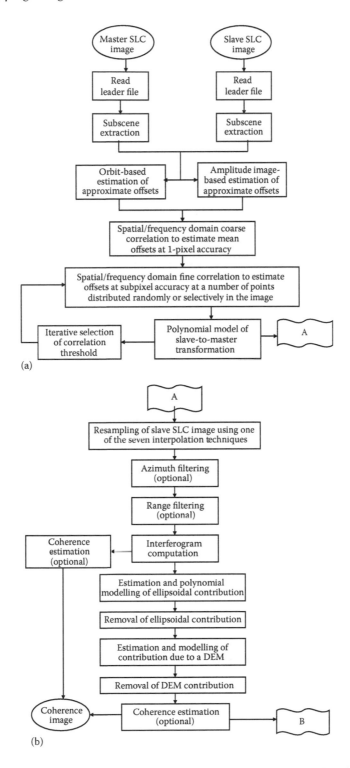

FIGURE 13.3 (a–c) Complete processing steps for generating a DEM using InSAR single look complex (SLC). *(Continued)*

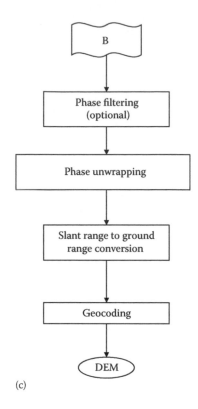

(c)

FIGURE 13.3 (CONTINUED) (a–c) Complete processing steps for generating a DEM using InSAR single look complex (SLC).

image is also filtered in the azimuth and range direction to improve the quality of the interferogram. Even though filtering is an optional step, it is strongly advised. The main cost of filtering has been a requirement for additional disk space rather than computational power.

13.2.5 INTERFEROGRAM GENERATION

Interferogram generation is a trivial task of multiplying the complex conjugate of the slave image by the master image. However, the phase of the complex raw interferogram needs to go through a number of additional processing steps before it can be used to extract terrain parameters. Figure 13.4 shows a decimated overview of the amplitude and phase of a complex raw interferogram of the Jharia mining area in northern India (SAC Report 2001). The systematic contribution due to an Earth ellipsoid is computed and removed from the raw interferogram. Figure 13.5 shows the decimated overview of the synthetic interferogram phase corresponding to an Earth ellipsoid model, and the flattened interferogram in a contour-like pattern. The phase of the complex interferogram is filtered to reduce the number of residues for more efficient unwrapping to be described in the following steps.

13.2.6 COHERENCE IMAGE GENERATION

The degree of correlation between the two complex images used in interferogram generation is a measure of interferogram quality. It can be used in both path-following and region-growing unwrapping techniques as a quality map. The critical issue in a coherence study is identifying the ideal temporal baseline and spatial window size for coherence computation. Figure 13.6 shows the coherence image of the Jharia mining area, India, computed using a 3 × 15 window (15 is in the azimuth

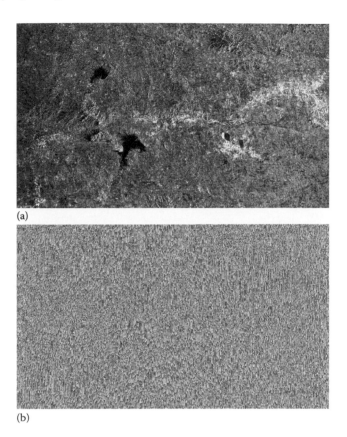

(a)

(b)

FIGURE 13.4   Decimated overview of (a) amplitude and (b) raw phase image of the complex interferogram (using European Remote Sensing Satellite Synthetic Aperture Radar [ERS SAR]) of the Jharia mining area, India.

direction). The water bodies are the areas of minimum coherence. The hilly areas also show a lower coherence due to layover and foreshortening.

13.2.7   Phase Unwrapping

Phase unwrapping solves for 2π ambiguities in phase and is essential for height computation and surface deformation studies. In principle, only terrain slope can be computed without phase unwrapping. Various algorithms are available, for example, the classic Goldstein's phase unwrapping algorithm, quality-guided phase unwrapping algorithm, mask cut algorithm, Flynn's minimum discontinuity algorithm, unweighted least-square algorithm, preconditioned conjugate gradient (PCG) algorithm, weighted multigrid algorithm and minimum L_p-norm algorithm. However, a quality-guided algorithm with a minimum gradient quality map option is found to be quite successful. Least-square algorithms result in smoothly unwrapped images, but break up the unwrapped image into a number of regions. Figure 13.7a shows the unwrapped image of the Jharia mining area generated using a quality-guided algorithm with a minimum gradient quality map (Ghiglia and Pritt 1998). Unwrapping hilly and other low-coherence regions has always been difficult. Figure 13.7b shows the phase residue map of the area computed using the Goldstein algorithm. Most of the residues occur in low-coherence regions occupied by water bodies and hills.

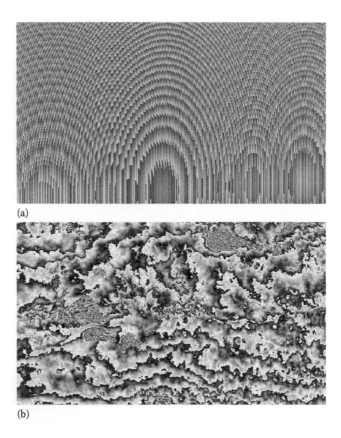

(a)

(b)

FIGURE 13.5   (a) Synthetically generated ellipsoidal phase. (b) Phase of the complex interferogram after removal of the ellipsoidal phase.

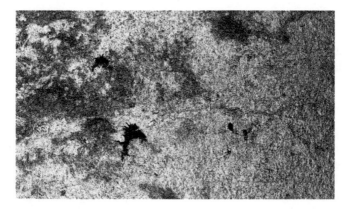

FIGURE 13.6   Coherence image computed over 3 × 15 windows. Lighter shades represent high-coherence regions.

13.2.8   SLANT-TO-HEIGHT CONVERSION

The unwrapped phase images consist of the phase difference from the two viewing geometries. This is proportional to the slant range two-way path difference of the ground resolution cells from the two viewing geometries. This is converted to terrain height using baseline and imaging geometries. The height obtained by this process contains constant bias, which is removed using a single ground control point (GCP) with a known height value.

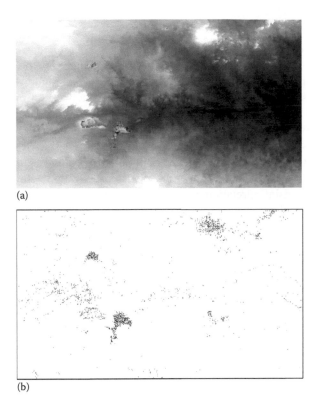

(a)

(b)

FIGURE 13.7   (a) Unwrapped phase image of the Jharia mining area, generated using quality-guided algorithm (darker shade represents lower phase values). (b) Phase residues in the wrapped phase image obtained using Goldstein's algorithm.

13.2.9   Geocoding

The rectification of the height image can be carried out using any image processing software. A grid of coordinates, extracted from the latitude and longitude images, has been used as control points to rectify and resample the height image into a ground range image. The accuracy of geocoding can be evaluated using more control points derived from the map. Figure 13.8 shows the amplitude image of the complex interferogram draped over the height image generated using InSAR.

FIGURE 13.8   Topographic contour map of the Jharia mining area generated using InSAR.

13.2.10 PSInSAR

The issue of loss of coherence or the phase instability or temporal decorrelation in a subsidence-prone area remains a challenge when using the InSAR technique. The method of PSInSAR provides tremendous improvement over the conventional InSAR technique. A large number of PSs, for example, rooftops and bridges, act as a natural geodetic network of radar targets. This technique can be applied not only in subsidence studies, but also in monitoring landslides, and tectonic motions (Meisina et al. 2008). Ferretti et al. (2000) were the first to propose the PSInSAR technique, which was further modified in the past decade. Unlike InSAR, PSInSAR uses a large number of (at least 20 or more) interferometric data pairs for better results and accuracy. It also calculates the motion of scatterers and the atmospheric propagation delay. PSInSAR eliminates possible errors due to misinterpretation caused by atmospheric factors, and the inaccuracies due to DEM are also removed.

The initial processing steps of the PSInSAR technique are similar to those of the conventional InSAR up to the phase computation for each pixel. The phase is influenced by inaccuracies of external DEM, atmospheric phase delay (Higgins et al. 2014), displacement of scatterer, speckle and decorrelation noise. Because the scatterers exist to remain coherent over several years, they can be identified in the sequential SAR images. The contributions to the interferometric phase of each PS can be computed using spatio-temporal analysis of images. The relative motion, elevation and atmospheric delay can be estimated for each PS; thus, subsidence can be mapped to a greater accuracy using the following steps (Ferretti et al. 2000):

- Various interferograms pertaining to all images are generated from the same master image.
- Differential interferograms are created using a reference DEM and the satellite ephemeris data.
- Elevation error and atmospheric propagation delay are estimated. These estimates contain errors due to limited accuracy of the DEM and local topographic effects.
- Refinement to remove errors due to atmospheric propagation delay. As the accuracy improves, more PSs can be identified.

Table 13.1 provides the major advantages and disadvantages of conventional DInSAR and PSInSAR techniques. The small spatial and temporal baselines are a requirement in DInSAR so that best coherence can be achieved; however, PSInSAR can make use of larger baselines (both spatial and temporal). A coarse DEM can be used in the PSInSAR since accurate elevation is computed for each PS, whereas in DInSAR, high-resolution DEM is selected for better results.

13.2.11 PERMANENT SCATTERER CANDIDATES

The coherence maps associated with various interferograms can be exploited to identify PSs. A threshold of correlation can be set for such an identification of targets. If a target exhibits coherence

TABLE 13.1

Comparison between PSInSAR and Conventional DInSAR Techniques

	PSInSAR	Conventional DInSAR
Baseline	No limits	<200 m
Atmospheric artefacts	Strongly reduced	No reduction
Coherence	>0.7 on a single pixel	>0.3 on several pixels
DEM accuracy	100 m	Baseline dependent
Number of SAR images	>30	≥2

always greater than a certain value, it would qualify as a PS candidate. The selection of PS solely based on coherence can be erroneous due to varying baselines and limited DEM accuracy. The PS can be identified by optimizing the coherence threshold and the size of estimation window, that is, averaging the data over larger areas (Ferretti et al. 2001).

13.3 EXAMPLES OF SUBSIDENCE MAPPING

13.3.1 Mapping in Ganges–Brahmaputra Delta

Ground subsidence caused by sediment uploading, compaction and tectonics has been mapped in the sinking Ganges–Brahmaputra Delta. Higgins et al. (2014) have derived (using ALOS PALSAR) a subsidence map of the Ganges–Brahmaputra Delta in Bangladesh using InSAR. A land subsidence rate of 0 to >18 mm per year is observed in Dhaka and surrounding areas.

13.3.2 Mapping in Barcelona, Spain

Subsidence and uplift caused by water extraction and pumping have been mapped showing rates of 2–14 mm/year. Devanthéry et al. (2014) have generated a deformation velocity map (using TerraSAR-X) of the airport and port of Barcelona, derived using the PSInSAR technique. It includes more than 5.4 million PSs in an area of 1019 km² (Figure 13.9).

13.3.3 Mapping of Tungurahua Volcano, Ecuador

Champenois et al. (2014) have derived (using ENVISAT SAR) a mean line-of-sight velocity map superimposed on a Shuttle Radar Topographic Mission (SRTM) DEM, and related profiles of the Tungurahua volcano in Ecuador between 2003 and 2009 using PSInSAR. The generated map used 190,000 PS pixels with a spatial density of 25 PS/km². Champenois et al. (2014) measured continuous large-scale uplift with a maximum line-of-sight displacement rate of about 8 mm/year.

FIGURE 13.9 Deformation velocity map of airport and port of Barcelona, derived using PSInSAR technique (using TerraSAR-X). (From Devanthéry, N. et al., *Remote Sens.*, 6 (7), 6662–6679, 2014. Copyright © Devanthéry et al. (2014); Open Access article distributed under the terms and conditions of the Creative Commons Attribution License.)

13.4 CONCLUSION

Our understanding of subsidence of the ground has enormously improved over the last 20 years since the development of satellite- and aircraft-based InSAR techniques. Before the development of the InSAR technique, subsidence was estimated using ground surveys and modelling methods; however, the conventional geodetic surveying methods lacked detailed mapping of large areas at high temporal resolution. Depending on the level of accuracy and requirements, a plethora of methods and techniques are available today to study the land subsidence, caused by any natural or anthropogenic activities, in unprecedented detail. Especially in the past decade, the combined use of a large number of SAR images and the InSAR technique has resulted in the development of the PSInSAR technique, which can provide exceedingly improved accuracy of subsidence mapping of large areas. Satellite-based InSAR techniques have the edge over conventional ground-based surveying techniques, that the former can be applied in areas of harder accessibility (e.g. volcanoes, marshes, landslides and coalfields) with an unprecedented accuracy at the millimetre level. For reconnaissance surveys of large areas, GPS surveying (both static and RTK) has proved to be extremely useful, as the desired accuracy is only at the centimetre level. Levelling and total station are very reliable for use mainly in construction sites (e.g. roads and bridges).

For scientific research, development and its applications in various fields, the InSAR and PSInSAR techniques provide the basis for high-quality subsidence mapping of any part of the Earth's surface. Based on the rate of subsidence and its mapping using SAR interferometry, the platform for secondary applications, such as water management, disaster management, mitigation, rehabilitation, agricultural management and urban planning, can be set. As the satellite technology enters into the next generation, with the constellation of SAR satellites (e.g. Sentinel series and Radar Constellation Mission) slated to be launched, even better techniques of subsidence mapping can be developed in the future.

ACKNOWLEDGEMENTS

The author is thankful to Shailesh R. Nayak, Ministry for Earth Sciences, government of India, for encouragement and guidance in promoting InSAR work. Thanks are due to Kamini K. Mohanty for InSAR processing using Delft Object-oriented Radar Interferometric Software (DORIS).

REFERENCES

Bamler, R., and Hartl, P. (1998). Synthetic aperture radar interferometry. *Inverse Probl.*, *14* (4), R1–R54.

Carnec, C., and Fabriol, H. (1999). Monitoring and modeling land subsidence at the Cerro Prieto Geothermal Field, Baja California, Mexico, using SAR interferometry. *Geophys. Res. Lett.*, *26* (9), 1211–1214.

Cascini, L., Ferlisi, S., Fornaro, G., Lanari, R., Peduto, D., and Zeni G. (2006). Subsidence monitoring in Sarno urban area via multi-temporal DInSAR technique. *Int. J. Remote Sens.*, *27* (8), 1709–1716.

Champenois, J., Pinel, V., Baize, S., Audin, L., Jomard, H., Hooper, A., Alvarado, A., and Yepes, H. (2014). Large-scale inflation of Tungurahua volcano (Ecuador) revealed by persistent scatterers SAR interferometry. *Geophys. Res. Lett.*, *41*, 5821–5828.

Colesanti, C., Ferretti, A., Prati, C., and Rocca, F. (2003). Monitoring landslides and tectonic motion with the permanent scatterers technique. *Eng. Geol.*, *68* (1), 3–14.

Devanthéry, N., Crosetto, M., Monserrat, O., Cuevas-González, M., and Crippa, B. (2014). An approach to persistent scatterer interferometry. *Remote Sens.*, *6* (7), 6662–6679.

Farina, P., Colombo, D., Fumagalli, A., Marks, F., and Moretti, S. (2006). Permanent scatterers for landslide investigations: Outcomes from ESA-SLAM project. *Eng. Geol.*, *88* (3), 200–217.

Ferretti, A., Prati, C., and Rocca, F. (2000). Nonlinear subsidence rate estimation using permanent scatterers in differential SAR interferometry. *IEEE Trans. Geosci. Remote Sens.*, *38* (5), 2202–2212.

Ferretti, A., Prati, C., and Rocca, F. (2001). Permanent scatterers in SAR interferometry. *IEEE Trans. Geosci. Remote Sens.*, *39* (1), 8–20.

Fielding, E. J., Blom, R. G., and Goldstein, R. M. (1998). Rapid subsidence over oil fields measured by SAR interferometry. *Geophys. Res. Lett.*, *25* (17), 3215–3218.

Galloway, D. L., Hudnut, K. W., Ingebritsen, S. E., Phillips, S. P., Peltzer, G., Rogez, F., and Rosen, P. A. (1998). Detection of aquifer system compaction and land subsidence using interferometric synthetic aperture radar, Antelope Valley, Mojave Desert, California. *Water Resour. Res.*, *34* (10), 2573–2585.

Ge, L., Chang, H. C., and Rizos, C. (2007). Mine subsidence monitoring using multi-source satellite SAR images. *Photogramm. Eng. Remote Sens.*, *73* (3), 259–266.

Ghiglia, D. C., and Pritt, M. D. (1998). *Two-Dimensional Phase Unwrapping: Theory, Algorithms, and Software.* New York: Wiley, 1–6.

Gupta, M., Mohanty, K. K., Kumar, D., and Banerjee, R. (2014). Monitoring surface elevation changes in Jharia coalfield, India using synthetic aperture radar interferometry. *Environ. Earth Sci.*, *71* (6), 2875–2883.

Higgins, S. A., Overeem, I., Steckler, M. S., Syvitski, J. P., Seeber, L., and Akhter, S. H. (2014). InSAR measurements of compaction and subsidence in the Ganges-rahmaputra Delta, Bangladesh. *J. Geophys. Res. Earth Surf.*, *119* (8), 1768–1781.

Li, F. K., and Goldstein, R. M. (1990). Studies of multibaseline spaceborne interferometric synthetic aperture radars. *IEEE Trans. Geosci. Remote Sens.*, *28* (1), 88–97.

Liu, Y., and Huang, H. J. (2013). Characterization and mechanism of regional land subsidence in the Yellow River Delta, China. *Nat. Hazards*, *68* (2), 687–709.

Massonnet, D., Holzer, T., and Vadon, H. (1997). Land subsidence caused by the East Mesa geothermal field, California, observed using SAR interferometry. *Geophys. Res. Lett.*, *24* (8), 901–904.

Massonnet, D., Rossi, M., Carmona, C., Adragna, F., Peltzer, G., Feigl, K., and Rabaute, T. (1993). The displacement field of the Landers earthquake mapped by radar interferometry. *Nature*, *364* (6433), 138–142.

Meisina, C., Zucca, F., Notti, D., Colombo, A., Cucchi, A., Savio, G., Giannico, C., and Bianchi, M. (2008). Geological interpretation of PSInSAR data at regional scale. *Sensors*, *8* (11), 7469–7492.

Musson, R. M. W., Haynes, M., and Ferretti, A. (2004). Space-based tectonic modeling in subduction areas using PSInSAR. *Seismol. Res. Lett.*, *75* (5), 598–606.

Prakash, A., Fielding, E. J., Gens, R., Van Genderen, J. L., and Evans, D. L. (2001). Data fusion for investigating land subsidence and coal fire hazards in a coal mining area. *Int. J. Remote Sens.*, *22* (6), 921–932.

SAC Report. (2001). SAR interferometry using Delft Object-oriented Radar Interferometry Software (DORIS). SAC/RESA/MWRG/ESHD/TR/07/2001. Ahmedabad, Gujarat, India: Space Applications Centre (ISRO), 1–22.

Schofield, W., and Breach, M. (2007). *Engineering Surveying.* 6th ed. London: Taylor & Francis.

Singh, R. P., and Yadav, R. N. (1995). Prediction of subsidence due to coal mining in Raniganj coalfield, West Bengal, India. *Eng. Geol.*, *39* (1–2), 103–111.

Zebker, H. A., and Villasenor, J. (1992). Decorrelation in interferometric radar echoes. *IEEE Trans. Geosci. Remote Sens.*, *30* (5), 950–959.

14 Earthquakes and Associated Landslides in Pakistan

Shah F. Khan, Muhammad Asif Khan,
Ulrich Kamp and Lewis A. Owen

CONTENTS

14.1 INTRODUCTION

Earthquakes are among the most devastating natural hazards, threatening life and properties in many parts of the world. They are particularly abundant along the boundaries between lithospheric plates. About 85% of the fatalities have occurred due to earthquakes in the Alpine–Himalaya chain stretching from Spain in western Europe to Indonesia in East Asia (Bilham 2009). Many fatalities result from the damage and collapse of buildings, since often their construction did not follow appropriate safety codes. During major earthquakes, secondary types of damage, such as landslides and snow avalanches, are common, particularly in mountains. In the 1556 Shensi Earthquake in China, about 830,000 casualties occurred during the ground rupture and hillside collapse (Gates and Ritchie 2007; Bilham 2009).

Pakistan is a country with contrasting landforms, from the highest mountains on Earth, characterized by great vertical relief and steep slopes, to the low-lying flat regions of the Indus–Gangetic Plains. Owing to its active tectonic history, the region provides an excellent natural laboratory to understand the dynamism of earthquake-associated landslides. This chapter gives an overview of these earthquake-triggered landslides in Pakistan in the past ~200 years. *Landslide* describes a geological phenomenon that includes a wide range of ground movements, for example, rockfalls, shallow debris flows, earthflows, mudflows and sturzströme. Often, the process of movement is also known as *landsliding*. Additionally, *landslide* also describes the resulting landform itself.

14.2   TECTONIC SETTING

Pakistan is located astride the Indian and Eurasian continental–continental lithospheric plate boundary that has produced major mountain ranges, including the Waziristan–Sulaiman–Kirthar (Axial Belt), Karakoram–Himalaya, Pamir–Hindu Kush and Makran–Chagai–Ras Koh regions (Figure 14.1).

The Hindu Kush, Karakoram and Pamir Ranges owe their origin to a Middle Jurassic tectonic amalgamation of microcontinents of Gondwana affinity at the southern margin of Eurasia (Şengör 1984; Gaetani et al. 1990; Gaetani 1997; Angiolini et al. 2013). The Neotethys ocean that separated the Karakoram from the Indian plate became a site for two major north dipping subduction zones: one at the immediate southern margin of the Karakoram plate and the other to the south in an intraoceanic setting. Both subduction zones consumed the Neotethys and triggered the Late Cretaceous northward drift of the Indian plate that ultimately collided with the southern margin of the Eurasia plate that was defined by the Karakoram plate (Burg 2011). The Indus suture zone is the main Himalayan structure that demarcates the Indian plate to its south from the subduction complexes and Gondwanaic microcontinents at the southern margin of Eurasia in the north. The deformed northern edge of the Indian plate (including the obliterated lithosphere of the Neotethys) that was emplaced inward onto the Indian plate craton during the Late Palaeocene–Recent India–Eurasia collision and the accompanying Himalayan orogeny define the Himalaya (Figure 14.2). Prior to the Eocene, the northward subduction of the Neotethys resulted in the formation of an Andean-type continental arc also known as Trans-Himalaya, including parts of the southern margin of the Eurasian plate north of the Indus suture zone (Hodges 2000).

The Himalaya, where it enters Pakistan from the east, undergoes a major orogenic inflection from its overall northwestern trend to stretch west-southwest and then southwest-south around what is known as the western Himalayan syntaxis that is defined by several crustal-scale north trending fold structures and syntaxial bends (Zeitler et al. 2001). The first major Himalayan orographic inflection is caused by the Nanga Parbat and the Hazara–Kashmir syntaxes (Figure 14.2), which cause the northwest trending Kashmir Himalaya at its eastern limbs to attain a west-southwest trend to its

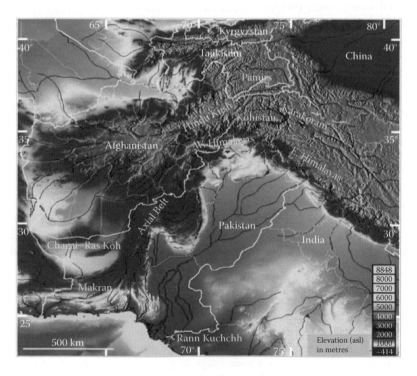

FIGURE 14.1   Physiographic map of South-Central Asia depicting major mountain ranges in Pakistan and adjacent countries. (Modified map downloaded from www.maps-for-free.com.)

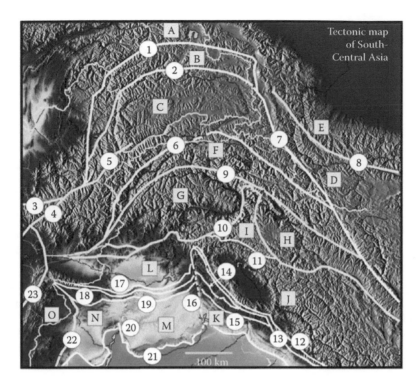

FIGURE 14.2 Tectonic map of northern Pakistan and South-Central Asia depicting major tectonic blocks (A–O), suture zones (yellow lines) and boundary faults (white lines). Tectonic blocks include (A) north Pamir; (B) central Pamir; (C) south Pamir; (D) Qiangtang; (E) Kunlun; (F) Karakoram–Hindu Kush; (G) Kohistan Island arc; (H) Ladakh Island arc; (I) Nanga Parbat syntaxis; (J) Kashmir Himalayas (hinterlands); (K) Kashmir Himalayas (forelands) with Hazara–Kashmir syntaxis at their northwest termination; (L) western Himalayas (hinterlands); (M) western Himalayas (foreland), including Hill Ranges, Potwar Plateau and Salt Range; (N) western Himalayas (foreland), including Kohat Plateau and Trans-Indus Ranges; and (O) Axial Belt (Kurram–Waziristan Ranges). Major faults include (1) Jinsha suture, (2) Rushan–Pshart suture, (3) Heart Fault, (4 and 5) Eastern Hindu Kush Fault, (6) Tirich–Kilik Fault, (7) Karakoram Fault, (8) Karakas Fault, (9) Karakoram–Kohistan suture, (10) Indus suture (Main Mantle Thrust), (11) Battal Fault, (12) Panjal Fault, (13) eastern MBT, (14) Balakot–Bagh Fault, (15) Raisi Fault, (16) Balakot–Jhelum Fault, (17) Khairabad Fault, (18) Nathiagali–Hisartang Fault, (19) western MBT, (20) Kalabagh Fault, (21) Main Frontal Thrust (Salt Range Thrust), (22) Kurram boundary mélange and (23) Gardez Fault. (Data from Beck, R. A. et al., *Acta Geodyn. Geomater.*, 9, 114–144, 1996; Searle, M. P., and Khan, M. A., *Geological Map of North Pakistan and Adjacent Areas of Northern Ladakh and Western Tibet (Western Himalaya, Salt Ranges, Kohistan, Karakoram, Hindu Kush), Scale: 1:650,000*, Oxford University, Oxford, 1996; Gaetani, M., *Sediment. Geol.*, 109, 339–359, 1997; Hodges, K. V., *Geol. Soc. Am. Bullet.*, 112, 324–350, 2000; Angiolini, L. et al., *Terra Nova*, 25, 352–360, 2013.)

west defined by the western Himalaya in northern Pakistan (Searle and Khan 1996; DiPietro and Pogue 2004). A second major orographic inflection occurs in the extreme northwestern border areas of Pakistan with Afghanistan at Kurram, where the northwest Himalaya of Pakistan abuts against the Waziristan–Sulaiman Ranges (Figures 14.2 and 14.3). The Waziristan–Sulaiman Ranges trend parallel to and lie to the east of the Chaman transform fault that defines the western plate boundary of the Indian plate against the Afghan block of Eurasia (Beck et al. 1996; Khan et al. 2003; Ul-Hadi et al. 2013). Internally, this is a typical fold-thrust belt derived from the shelf–slope sequence at the northwestern edge of the Indian plate, and emplaced onto the Indian plate to the southeast (Figure 14.3). The Waziristan–Muslim Bagh–Zhob ophiolite and melange sequences in the hinterlands of the Waziristan–Sulaiman Ranges represent the obliterated Neotethyan oceanic crust between the India and Afghan blocks. The Waziristan–Sulaiman Ranges loop around yet another syntaxis, the Quetta

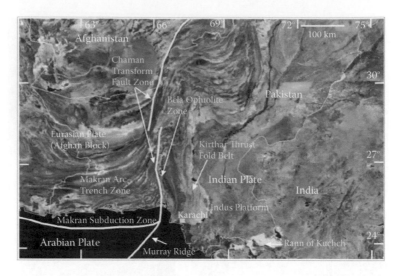

FIGURE 14.3 Tectonic map of south-central Pakistan showing the western plate boundary defined by the Chaman transform fault and the Sulaiman–Kirthar Thrust-Fold Belt. The active Makran subduction zone and associated Makran–Chagai Trench–Arc system define the plate–boundary link between the Himalayas in the east and Zagros in the west. (Data from Farhoudi, G., and Karig, D. E., *Geology*, 5, 664–668, 1977; Kazmi, A. H., and Jan, M. Q., *Geology and Tectonics of Pakistan*, Graphic Publishers, Karachi, Pakistan, 1997; Schelling, D. D., *Geol. Soc. Am. Spec. Papers*, 328, 287–302, 1999.)

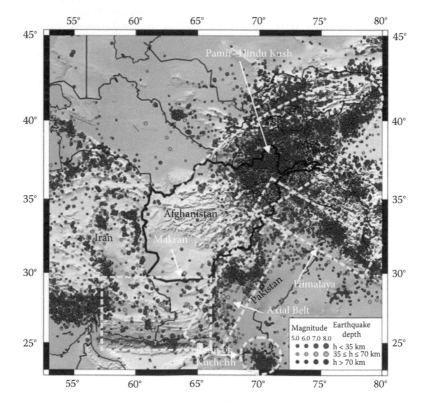

FIGURE 14.4 Seismicity map of South-Central Asia showing major seismotectonic provinces in Pakistan and immediate surroundings. (From Owen, L. A., *Geol. Soc. Spec. Publ.*, 338, 389–407, 2010.)

syntaxis, and from Quetta southward to Karachi extend as the Kirthar Range. The Kirthar Range has a tectonic setting similar to that of the Waziristan–Sulaiman Ranges, with the Ornach-Nal segment of the Chaman fault and the Bela ophiolite forming the western hinterlands, and the fold-thrust belt tectonically emplaced eastwards onto the Indus basin (Kazmi and Jan 1997; Schelling 1999).

The Makran subduction zone in the Gulf of Oman is the western extension of the Himalayan plate boundary into the Zagros Range (Figure 14.3). This zone subducts northwards underneath the Pakistani–Iranian Makran coast at the southern margin of the Afghan (Helmond) and Lut blocks of Eurasia (Farhoudi and Karig 1977).

According to this general tectonic setting, Pakistan can be divided into five seismotectonic provinces: (1) Pamir–Hindu Kush, (2) Karakoram–Himalaya, (3) Axial Belt (Waziristan–Sulaiman–Kirthar Ranges), (4) Makran and (5) Rann of Kuchchh (Figure 14.4). All these are part of the Indian-Eurasian plate boundary except for the Rann of Kuchchh, which is an intraplate seismotectonic province within the Indian plate. Each of these seismotectonic provinces has generated major earthquakes in the past two centuries, and with the exception of the Rann of Kuchchh province, all have experienced earthquake-triggered landslides. Rann of Kuchchh, despite being dominantly a low-lying gentle terrain, is susceptible, like the other four provinces, to earthquake-triggered surface disturbances, such as surface ruptures, differential uplift and subsidence, lateral spreading and liquefaction (Rastogi 2001; Biswas 2005).

14.3 PAMIR–HINDU KUSH SEISMOTECTONIC PROVINCE

The western part of the Karakoram–Kohistan suture zone (also termed the Main Karakoram Thrust [MKT]) in northern Pakistan extends southwest as the Kunar fault in Afghanistan, where it merges with the Gardez Fault – a splay of the Chaman fault (Figure 14.2). This fault represents the south-eastern fringes of the Pamir–Hindu Kush seismic zone. The major earthquakes associated with the Kunar fault are those from 19 February 1842 (moment magnitude [M_w] 7.6), 7 July 1909 (M_w 7.5) and 16 September 1956 (M_w 6.7), of which the earliest caused widespread damage in northwest Pakistan, including the collapse of more than 40% of the adobe houses in Peshawar (Ambraseys and Bilham 2003a, 2014). All three earthquakes triggered numerous rockfalls, landslides, debris flows and snow avalanches in northwest Pakistan and across the border in Afghanistan. The latest significant earthquake occurred on 25 April 2013 (M_w 5.6) and had a focal depth of 62 km, resulting in 18 fatalities and injuring 70 people in villages near Jalalabad in Afghanistan.

14.4 KARAKORAM–HIMALAYA SEISMOTECTONIC PROVINCE

The Karakoram–Himalaya seismotectonic province encompasses the entire area of northern Pakistan, from the Salt Range in the hanging wall of the Himalayan Frontal Thrust to the Karakoram–Hindu Kush Ranges extending across the northern borders of Pakistan (Figure 14.2). A clustering of historical seismicity and associated landsliding identified three seismic zones: (1) Kashmir Himalayas–Indus–Kohistan, (2) Nanga Parbat and (3) Darel–Hamran Kohistan.

14.4.1 KASHMIR HIMALAYAS–INDUS–KOHISTAN SEISMIC ZONE

Based on 1973–1975 telemetric seismic networks in northern Pakistan that recorded ~10,000 local earthquakes (Seeber and Armbruster 1979, 1984), a wedge-shaped subsurface structure is defined in the upper Hazara region south of the Indus suture zone, known as the Indus–Kohistan Seismic Zone. The structure consists of a horizontal upper planar surface at a depth of ~11 km and a lower planar surface that dips northeast, defined by concentrated hypocentres ranging in depth from ~11 to 40 km (Figure 14.5). The upper planar surface is considered a northward continuation of the décollement underlying the Salt Range, Potwar and Hazara foreland ranges of outer western Himalaya in Pakistan (Seeber 1979). The fault plane closely resembles the frontal faults of the Kashmir Himalaya in trend and geographical position, which is considered a northwestern extension of the latter. These basement faults,

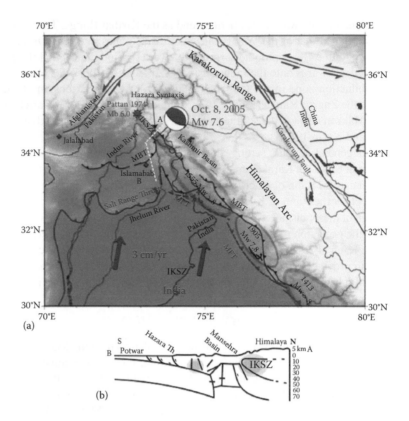

FIGURE 14.5   (a) Tectonic map of the northwest Himalayas showing location of the Indus–Kohistan Seismic Zone (IKSZ). Note the possible continuation of the IKSZ southeast, with earthquake ruptures associated with earthquakes of the thrust system at the front of the Kashmir Himalayas (Himalayan Arc). (b) Inset shows the wedge-shaped geometry of the IKSZ in a cross section across the Hazara–Potwar Himalayas. (Modified after Bilham et al., *Seismological Research Letters*, 78, 601–613, 2007. Profile from Bendick, R. et al., *Geology*, 35, 267–270, 2007.)

defined on the basis of seismicity, are discordant with the arcuate structural trends in the overlying sedimentary wedge in the Hazara arc and in the Hazara–Kashmir syntaxis (Seeber 1979; Stack 1982).

14.4.1.1   The 1974 Pattan Earthquake

The 28 December 1974 Pattan Earthquake (body wave magnitude [M_b] 6.0) is one of the most infamous earthquakes associated with the Indus–Kohistan Seismic Zone (Figure 14.5). The earthquake's epicentre was located at the village of Roniala near the town of Dubair (35.09° N, 72.88° E), ~10 km southwest of the town of Pattan, with a focal depth of ~15 km (Pennington 1979). The main shock was followed by a series of aftershocks, of which more than 200 were recorded by the Tarbela observatory and associated portable seismic stations deployed after the earthquake. The rupture plane defined by aftershocks was 10 km long, through the main shock, and was oriented N 113° E. No earthquake surface ruptures were identified in the field (Aslam and Ghazanvi 1975; Pennington 1979; Ambraseys et al. 1981), although several cracks, some tens of metres long, were seen near Pattan and Pallas. An isoseismal intensity map for the epicentral region was produced, noting a Modified Mercalli Intensity (MMI) of VIII at Dubair Qala and Jijal, VII at Pattan, VI at Kayal, V at Besham and IV at Kamila and Thakot (Ambraseys et al. 1975).

Rockfalls and landslides caused the major damage in the epicentral area in and around Dubair. The village of Shorgarah, located 6.5 km west of Dubair, and the Roniala village upstream in the Dubair Nala were almost totally destroyed by massive rockfalls that killed more than 60 people. The earthquake triggered a large landslide between Shorgarha and Roniala, producing 200 m long scarps on the east-facing mountain slopes. The Karakoram Highway (KKH) was extensively damaged between Dubair and Jijal, where a retaining wall collapsed and rockfalls buried the road. Rockfalls, debris flows and landslides caused deadly damage near Pattan.

14.4.1.2 The 2005 Kashmir Earthquake

The 8 October 2005 M_w 7.6 Kashmir Earthquake was one of the most disastrous earthquakes in the Himalaya. It caused >80,000 fatalities and >69,000 injuries, destroyed >32,000 buildings and displaced ~2.8 million people, with losses amounting to $2.3 billion (Peiris et al. 2006; Kamp et al. 2008). A long surface rupture (>75 km) traversed the major population centres of Balakot, Muzaffarabad and Bagh. Displacement on the rupture was ≤7 m and generated a shaking equivalent to an MMI of X (Ali et al. 2009; Anggraeni et al. 2010). The resultant ground motions caused extensive landslides on soil-covered high-relief slopes in the mountainous regions (elevations between 1500 and 4500 m) that damaged numerous buildings and resulted directly or indirectly in many human fatalities (Petley et al. 2006).

The earthquake occurred on a rupture plane ~75 km long and ~35 km wide, striking 331° and dipping 29°. The epicentre was at the northwestern tip of the northwest trending Kashmir Himalaya, in the core of the north-northwest trending Hazara–Kashmir syntaxis (Figures 14.6 and 14.7). The Hazara–Kashmir syntaxis is cored by the structurally lowest (Sub-Himalayan) thrust sheet comprising the Miocene Murree Formation (composed of maroon shale and sandstone) and the underlying Palaeocene Patala (composed of limestone with shale intercalations) and Precambrian Muzaffarabad (composed of limestone and dolomite) Formations. The first indications of the causative fault responsible for the earthquake were (1) the linear concentration of building damage and slope failures and landslides, and (2) the location of the aftershocks following the main event.

Muzaffarabad, the district capital, was the closest town to the epicentre that was severely damaged. However, the cities of Balakot, located 15 km northwest of the epicentre, and Bagh, located 35 km southeast of the epicentre, were almost totally destroyed. Initial assessment of the destruction showed that the damage was concentrated in a line, stretching from the Allai Kohistan in the northwest to Bagh in the southeast. Between Balakot and Muzaffarabad, the earthquake fault was likely the Main Boundary Thrust (MBT) at the western limb of the Hazara–Kashmir

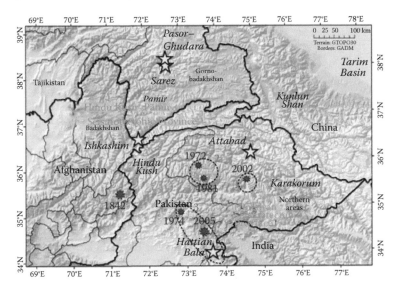

FIGURE 14.6 Locations of major earthquakes (red stars) and earthquake-triggered landslides that resulted in the formation of landslide-dammed lakes (yellow stars) in Pakistan and South-Central Asia: 1842 Kunar (M_w ~7.6), 1972 Hamran (M_b 6.3), 1974 Pattan (M_b 6.0), 1981 Darel (M_b 6.1), 2002 Nanga Parbat (M_b 6.2) and 2005 Kashmir (M_w 7.6). (Modified after Schneider, J. F. et al., Recent cases and geomorphic evidence of landslide-dammed lakes and related hazards in the mountains of Central Asia, presented at *Proceedings of the Second World Landslide Forum*, Rome, 3–7 October 2011.)

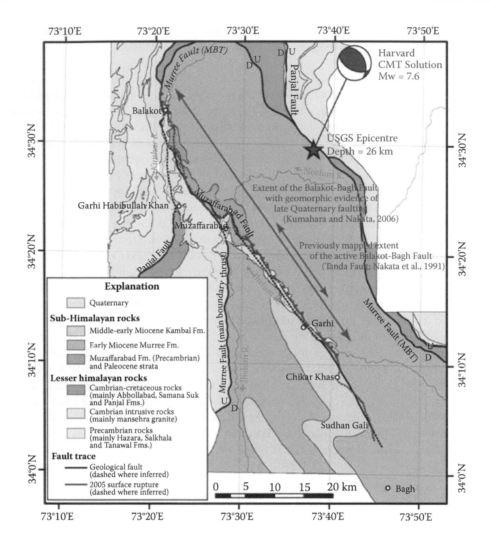

FIGURE 14.7 Geological map of Hazara–Kashmir syntaxis showing the position of the Kashmir Earthquake epicentre and associated surface rupture, causative fault (Balakot–Bagh Fault) and geological setup. (From Hughes, R., Astor Valley Earthquake, northern areas Pakistan: A field assessment, 1, Earthquake Engineering Field Investigation Team (EEFIT) Short Report, Institution of Structural Engineers, London, 2003.)

syntaxis, while between Muzaffarabad and Bagh, the earthquake fault was likely the Jhelum thrust (Tapponnier et al. 2006). Parts of the Jhelum thrust had been previously mapped as the active Tanda fault by Nakata et al. (1991). Northwest of Balakot, the coincidence of damage and aftershocks indicates involvement of the Indus–Kohistan Seismic Zone (Seeber and Armbruster 1979). These observations show that in the broader sense, the northwest trending Kashmir–Himalayan deformation front served as the causative fault of the 2005 Kashmir Earthquake.

The earthquake rupture for the Kashmir Earthquake has extensively been studied both in the field (Gaetani et al. 1990; Avouac et al. 2006; Harp et al. 2006; Pathier et al. 2006; Bendick et al. 2007; Hewitt 2009 and Chini et al. 2011) and by using satellite data (Hughes 2003). These studies successfully identified a ~90 km long regional deformation strip extending southeastward from Balakot to Bagh. Displacement field maps showed a sharp discontinuity at the southwestern edge of this deformation strip that corresponds well with the pre-earthquake mapped active faults (Nakata et al. 1991; Hewitt 2009). Additionally, the hanging wall and footwall of the causative fault were identified with prominent west- or south-directed horizontal displacement in the hanging wall and some east-directed displacement in the footwall. Field-based studies described the geomorphic expressions of the rupture: its typical surface expression included a northeast side-up pressure ridge and bulging in the form of a monoclonal scarp characterized by tension cracks at the crest of the scarp and compressional features, such as roll-ups, at the scarp base (Sayab and Khan 2010). Tension cracks locally exhibited a left-stepping en échelon pattern indicating a minor right-lateral strike-slip component. A common observation was that all the standing features (trees, poles, walls, etc.) showed opposite tilting: back-tilting in the hanging wall and fore-tilting in the footwall of the monoclonal crests. Both satellite-based and field studies documented a maximum average vertical separation (slip) of ~4 m on the northern segment between Muzaffarabad and Balakot, with a maximum slip of ~7 m observed some 15 km northwest of Muzaffarabad (Figure 14.8). The central segment, between Muzaffarabad and the point where the rupture crossed the Jhelum River towards Chikkar, had moderate amounts of average slip, with greater slip on the northwestern half of the fault segment (~2.9 m) compared with its southern half (~0.7–2.0 m).

The earthquake triggered thousands of mass movement events, including landslides, rockslides, rockfalls and debris falls or flows across an area of 7500 km² centred over the earthquake's epicentre. Some landslides partially or wholly dammed rivers and streams, while others caused roads to collapse, slide and/or be blocked by debris, and bridges and other infrastructure to be damaged (Harp and Crone 2006;

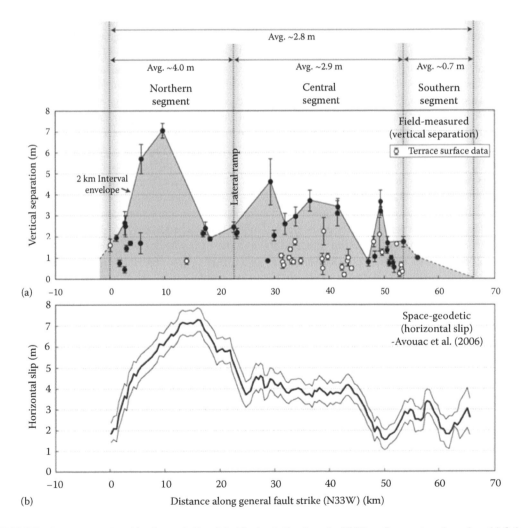

FIGURE 14.8 Vertical and horizontal slip of the Kashmir Earthquake 2005 surface rupture based on (a) field and (b) geodetic measurements. (From Fujiwara, S. et al., *Eos*, 87, 73, 2006.)

Schneider 2006; Kamp et al. 2008, 2010; Kaneda et al. 2008). The earthquake attracted numerous studies that covered many aspects of earthquake-triggered landslides, including inventorying, classification, typology, size, mechanics, controlling factors, temporal evolution and susceptibility analysis (Fujiwara et al. 2006; Harp and Crone 2006; Kumar et al. 2006; Dunning et al. 2007; Sato et al. 2007; Owen et al. 2008; Schneider 2009; Khattak et al. 2010; Saba et al. 2010; Khan et al. 2013).

Immediately after the earthquake, there was a general perception that the earthquake had triggered all the thousands of landslides present in the epicentral region. However, the first release of pre- and post-earthquake ASTER satellite images by the National Aeronautics and Space Administration (NASA) clearly showed that many landslides pre-dated the 2005 earthquake, and that a large number had been reactivated during the earthquake. For example, Saba et al. (2010) found that 117 of their mapped 158 landslides in a 36 km^2 area around Muzaffarabad pre-dated the earthquake. Moreover, the 80 million m^3 Hattian debris avalanche triggered by the earthquake rupture at the crest of Dana Hill near Chikkar was at least partially reactivated on a large historic landslide (Harp and Crone 2006; Kumar et al. 2006; Kamp et al. 2010).

According to Owen et al. (2008), more than 90% of the landslides were coalesced overhanging bedrocks or slope debris in the form of rock and debris falls or topples that formed talus slopes and debris cones. Approximately 40 rockfalls covering 1.5 km^2 occurred in the surroundings of Muzaffarabad (Owen et al. 2008), which is an increase in area covered by rockfalls of almost 100% compared with pre-earthquake conditions. Translational and rotational slides were the second-most common landslide type, and the majority of landslides occurred on pre-existing sites and were reactivated by the earthquake (Saba et al. 2010). However, the number of translational and rotational landslides doubled within a year of the earthquake event, especially following the monsoon during the summer of 2006. Co-seismic surface disturbance was manifested in large-scale fissuring along ridge crests in response to lateral spreading, as well as at slopes owing to gravity-induced slope failures (Lawrence and Ghauri 1983; Khan et al. 2013). Eight co-seismic rock avalanches – one of the most common co-seismic mass movement types during this earthquake – were identified around Muzaffarabad (Saba et al. 2010). Rock avalanches were instrumental in partial or complete blockage of rivers and streams in the earthquake-affected area, for example, at the Nisar Camp on the Neelam River and at Hattian Bala on a tributary of the Jhelum River (Figure 14.9).

Landslide numbers were mainly determined using satellite imagery supplemented by fieldwork in accessible areas, for example, 1293 landslides at 174 locations to produce a landslide inventory (Owen et al. 2008). From ASTER satellite imagery, the number of landslides was found to increase from 369 in 2001 to 2252 in October 2005 in an area of 2250 km^2 between Hattian in the southeast and Balakot in the northwest (Kamp et al. 2010). Submetre-resolution IKONOS and QuickBird satellite images were used in a selected study area of 36 km^2 around Muzaffarabad to identify 158 earthquake-triggered landslides (Saba et al. 2010). Sato et al. (2007) used 2.5 m resolution SPOT 5 stereo images to identify 2424 landslides in a study area of 2805 km^2. Other regional studies showed landside densities of between 0.86 and 1 landslides per square kilometre (Kamp et al. 2008, 2010), while more local studies generated greater landslide densities, approaching 5 landslides per square kilometre immediately after the earthquake (Owen et al. 2008; Saba et al. 2010); this number increased to 11 landslides per square kilometre in the subsequent years due to heavy monsoonal rains.

Landslides of varying sizes were observed across the affected surrounding region (Figure 14.10), with the maximum concentration on valley slopes associated with major rivers, such as Kunhar (between Kaghan through Balakot and Muzaffarabad), Neelam (between Titwal and Muzaffarabad) and Upper Jhelum (between Hattian and Muzaffarabad) (Sato et al. 2007; Kamp et al. 2008). However, major landslides (area >1 ha) (about 9%) mainly occurred along the eastern slopes of the Kunhar River, the lower reaches of the Neelam River near Muzaffarabad and the northeastern slopes of the Upper Jhelum River close to the earthquake rupture, especially on the uplifted hanging wall of the earthquake fault (Sato et al. 2007).

Earthquake-triggered landslides generated by the Kashmir Earthquake have been extensively analyzed in terms of controlling factors such as earthquake rupture, topography, geomorphology,

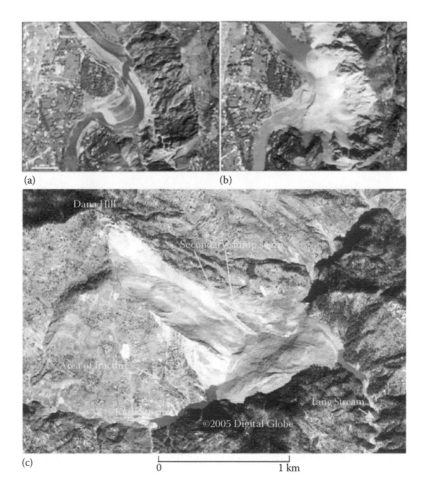

(a)

(b)

(c)

0   1 km

FIGURE 14.9 During the 2005 Kashmir Earthquake, event-triggered mass movements partially or completely blocked rivers. (a) Pre-earthquake IKONOS image of Neelam River at Nisar Camp, 2 km upstream from Muzaffarabad. (b) Post-earthquake IKONOS image showing partial blockage of the Neelam River at Nisar Camp. (c) ASTER image of Hattian Bala rock debris avalanche blocking two tributaries (Karli and Tang) of the Jhelum River, which resulted in the formation of two lakes. (From Schneider J. F., in Kausar, A. B., Karim, T., Khan, T., eds., *International Conference on 8 October 2005 Earthquake in Pakistan: Its Implications and Hazard Mitigation*, Islamabad, Pakistan, 18–19 January 2006, extended abstracts, 76–78.)

lithology and anthropogenic factors (Aydan 2006; Harp and Crone 2006; Kumahara and Nakata 2006; Sato et al. 2007; Kaneda et al. 2008; Ali et al. 2009; Konagai and Sattar 2012). The studies have shown that

- A great majority of the landslides, especially those with areas of >10,000 m², occurred at the hanging wall side of the Balakot–Bagh Fault. More than one-third of the landslides occurred within 1 km of the active fault (Sato et al. 2007).
- There was a strong lithological control. The greatest number (1147) of landslides, with a distribution density of 3.2 per square kilometre, occurred within the Precambrian dolomite and limestone (Muzaffarabad Formation). Miocene sandstone and siltstone (Murree Formation), Precambrian schist and quartzite (Salkhala Formation and Hazara Slates) and Eocene and Palaeocene limestone and shale (Patala–Margala Formations) were noticed to be favourable for earthquake-triggered landslides (Sato et al. 2007; Schneider 2009).

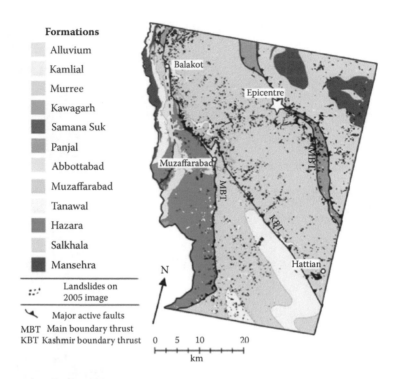

FIGURE 14.10 Geological map of Hazara–Kashmir syntaxis showing location of earthquake rupture, epicentre of the 2005 Kashmir Earthquake and earthquake-triggered landside distribution. (From Kamp, U. et al., *Geomorphology*, 101, 631–642, 2008.)

- More than 53% of the earthquake-triggered landslides occurred at steep slopes destabilized by pre-earthquake developmental activities, such as excavation for roads (Owen et al. 2008).
- Large landslides occurred on convex slopes rather than on concave slopes; furthermore, large landslides occurred on steeper (≥30°) slopes than on gentler slopes (Sato et al. 2007).
- Many large landslides occurred on south- and southwest-facing slopes, which is consistent with the interferometric synthetic aperture radar (InSAR)-detected horizontal-dominant direction of crustal deformation on the hanging wall (Sato et al. 2007).

Extensive co-seismic ground fissuring in a region governed by monsoonal climatic conditions was a major concern for long-term slope stability of the epicentral region of the Kashmir Earthquake. This attracted many studies on landslide change detection (Khattak et al. 2010; Khan et al. 2013). Saba et al. (2010) showed that in a 36 km² large area around Muzaffarabad, the number of landslides increased from 158 to 391 by the end of the monsoon season in 2006, with a corresponding increase in landslide area from 2.9 to 3.7 km²; this was followed by 2 years without significant changes. Khan et al. (2013) showed that between 2007 and 2010, 45% of the landslides showed no changes, 44% showed vegetation recovery and 11% showed an increase in area. All these change detection studies suggested that earthquake-triggered landslides continued to grow within the first 2 years of the event and then became stable in the subsequent years.

The Hattian Bala rock and debris avalanche located ~30 km southeast of the epicentre was the largest landslide triggered by the earthquake (Figures 14.9c and 14.10). The landslide formed on Dana Hill at an altitude of 2038 m on the southeast-facing valley slope of the Karli River, a tributary of the Jhelum River (Figures 14.11 and 14.12a). The landslide produced a scar of about 2 km length from the hill's summit to the foot of the slopes and 0.5 km width. An estimated 82 million m³ of debris moved downslope and eventually up the opposite valley slope to a height of up to 300 m above the river (Harp

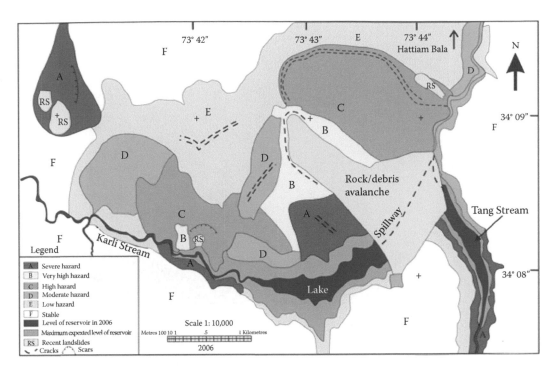

FIGURE 14.11 Map of Hattian Bala rock debris avalanche and surroundings showing existing slope stabilities and landslide-dammed tributary streams with lakes. Dotted lines mark the location of unlined spillways constructed across the landslide body to drain the lakes. (From Geological Survey of Pakistan, National Collaborative Project 'Potential Hazards of Landslides and Mitigation Measures at Hattian Bala, 2006–09'.)

and Crone 2006; Dunning et al. 2007; Owen et al. 2008; Schneider 2009). The avalanche debris was derived almost exclusively from the Miocene Murree Formation and consists of reddish sandstone, mudstone and shale. Although the landslide debris included large boulders of 7–8 m in diameter, much of it was composed of sand, silt and clay-sized particles. The co-seismic nature of the Hattian Bala rock avalanche is clear as the northeast dipping earthquake rupture traverses across the Dana Hill: it was apparently a downdip oblique-slip collapse of the valley slope in the hanging wall of the rupture. The debris that filled the Karli Valley formed a 1.3 km long landslide dam (Figure 14.12a) that also blocked the drainage of the Tang River at its confluence with the Karli River, resulting in the formation of two lakes: Karli Lake, with a maximum volume of 86 million m^3, and Tang Lake, with a maximum volume of 5 million m^3 (Sattar et al. 2011; Konagai and Sattar 2012). The hazard of overtopping or breaching of the Hattian Bala landslide dam was a major concern for the safety of the downstream population. To reduce the possibility of a dam burst and flash flooding, two spillways of 450 and 130 m in length, respectively, were constructed across the landslide dam to help slowly drain the lakes. Nevertheless, Tang Lake had filled up by February 2006 and started overflowing, while Karli Lake started to overtop its dam by April 2007. However, the spillways helped reduce the volume of Karli Lake to 62 million m^3 and of Tang Lake to 3.5 million m^3 (Sattar et al. 2011). They also helped regulate the outflow from the lakes without any significant flood for about 4 years after the earthquake (Figure 14.12b). However, throughout this period the unlined spillways suffered gradual toe- and head-cutting erosion on the downstream face of the landslide dam. Ultimately, by 9 February 2010, the water of Karli Lake breached the debris dam after 5 days of incessant rainfall. The peak outflow was ~5500 m^3/s, ~36 million m^3 of water was drained from the lake, and 7.78 million m^3 of debris was eroded from the landslide dam. The flash flood and accompanied debris flow destroyed 24 houses and a bridge. There were several landslides at the peripheries of the lakes; one on the right bank destroyed 174 houses, displacing more than 1000 residents (Konagai and Sattar 2012; Sattar et al. 2011). Dunning and Armitage (2011)

(a)

(b)

FIGURE 14.12 (a) During the 2005 Kashmir Earthquake, the Hattian Bala landslide blocked the Karli and Tang streams (view looking downstream). (From Harp, E. L., and Crone, A. J., Landslides triggered by the October 8, 2005 Pakistan earthquake and associated landslide-dammed reservoirs, U.S. Geological Survey Open-File Report 2006-1052 13, U.S. Geological Survey, Reston, VA, 2006.) (b) Breach of Karli Lake in 2008 through unlined spillway (view looking upstream). (From Schneider, J. F., *J. Seismol.*, 13, 387–398, 2009.)

suggested that landsliding into Karli Lake caused the overtopping and breaching, while Konagai and Sattar (2012) concluded that landsliding at the lake margins had been caused by the drawdown during the emptying of the lake. The landslide debris was never completely removed from the flooding, which resulted in the formation of a new Zalzal Lake that still poses a threat to downstream villages.

14.4.2 NANGA PARBAT SEISMIC ZONE

At Nanga Parbat, owing to exceptionally high exhumation rates of up to ~20 mm per year (Shroder and Bishop 2000), deep crustal Indian plate rocks have been exposed in just ~10 Ma (Zeitler et al. 1982, 2001; Treloar et al. 2000). Despite these spectacular neotectonics, the

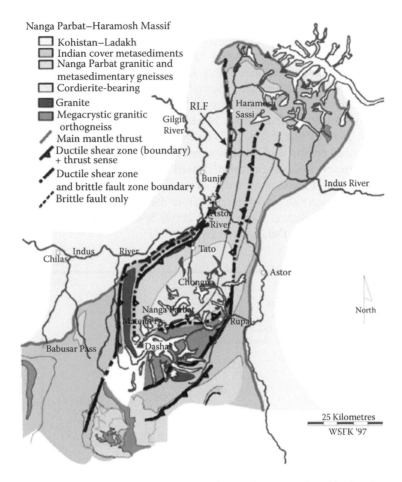

FIGURE 14.13 Geological map of Nanga Parbat syntaxis showing the Raikot–Liachar–Bunar Fault (RLF). Note the hypothetical position of the 1840 earthquake's epicentre (red star) and the instrumental epicentre location of the 2002 earthquake pair (yellow star). (Map after Pctley, D. et al., in Marui, H., ed., *Disaster Mitigation of Debris Flows, Slope Failures and Landslides*, Universal Academy Press, Tokyo, 2006, 47–55. Instrumental earthquake location after Shroder, J. F., *Geomorphology*, 26, 81–105, 1998.)

Nanga Parbat syntaxis is not prominent on seismicity maps of Pakistan. However, a record of 15 earthquakes of $M_b \geq 4$ occurred in a time window from 1970 to 1988 and suggests active seismicity (Sverdrup et al. 1994). Of 1500 seismic events recorded in the Nanga Parbat area between 1995 and 1997, 380 were clearly associated with the active Raikot–Liachar–Bunar Fault (Meltzer et al. 2001). The largest (M_b 6.2) recorded earthquake occurred on 20 November 2002. So far, more than 350 large rock avalanches and landslides have been identified in the upper Indus River basin (Hewitt et al. 2011). Of these, around 20 major events involving $>10^5$ m^3 of debris have occurred at the western margin of the Nanga Parbat syntaxis (Shroder 1993; Hewitt 2009) (Figure 14.13).

14.4.2.1 The 1840 Nanga Parbat Earthquake

The northern areas of Pakistan (Gilgit–Baltistan region) have a poor record of historical earthquakes. However, a detailed geomorphic appraisal of the Indus Valley at the western margin of the Nanga Parbat syntaxis shows evidence, in the form of large rock avalanches and landslides, for both prehistoric and historic seismic events. The prehistoric Hattu Pir, Thelichi and Liachar landslides may attest to tectonic instability, particularly along the Liachar thrust at the eastern bank of the Indus River (Shroder 1993, 1998). A series of landslides and rock avalanches have been identified along the Bunji–Raikot segment of the Indus River (Figure 14.13). At least three of these occurred in December 1840 and may have created a major dam that blocked the Indus River for more than 6 months (Shroder 1989). Landslides in the Tatta Pani area, some 6 km downstream from the Liachar thrust, are believed to have occurred at the same time as those in the Bunji–Raikot segment, and they also dammed the Indus (Shroder 1998). There is a general consensus that the mega-landslide and river blockage that occurred in December 1840 (Shroder 1998; Hewitt 2009; Delaney and Evans 2011) dates the causative earthquake event. Mason et al. (1930) noted, '[There is] not the slightest doubt that the dam was directly caused by a great fall of rock from the western spurs of Nanga Parbat'. Based on detailed field studies, the damming was likely caused by collapse of the Gor ridge across the river, which was subsequently compounded by slope failure on the eastern slopes of the river from the Liachar ridge (Figures 14.14 and 14.15) (Shroder 1998, Shroder and Bishop 2000; Hewitt 2009). Specifics of this event had been reported as follows (Shroder 1993, 1998; Delaney and Evans 2011):

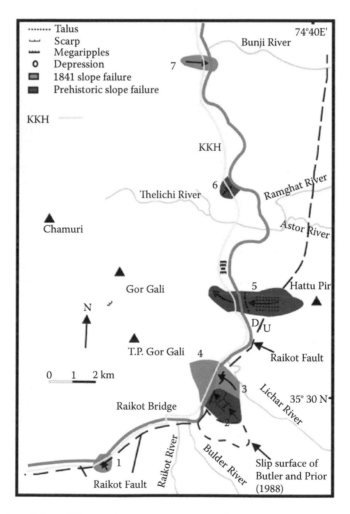

FIGURE 14.14 Map of slope failure and mass movements in the Indus Valley between Raikot Bridge and Bunji. (1) Tatta Pani debris falls and slides (1841 and recent). (2) Prehistoric rockslides. (3) Rock slip of 1841. (4) Gor Gali debris slide of 1841. (5) Prehistoric Hattu Pir rockslide reactivated in 1841. (6) Prehistoric Thelichi debris slide. (7) Bunji rockslide of 1841. (From Shroder, J. F., in Shroder, J. F., ed., *Himalaya to the Sea: Geology, Geomorphology, and the Quaternary*, Routledge, London, 1993, 1–42.)

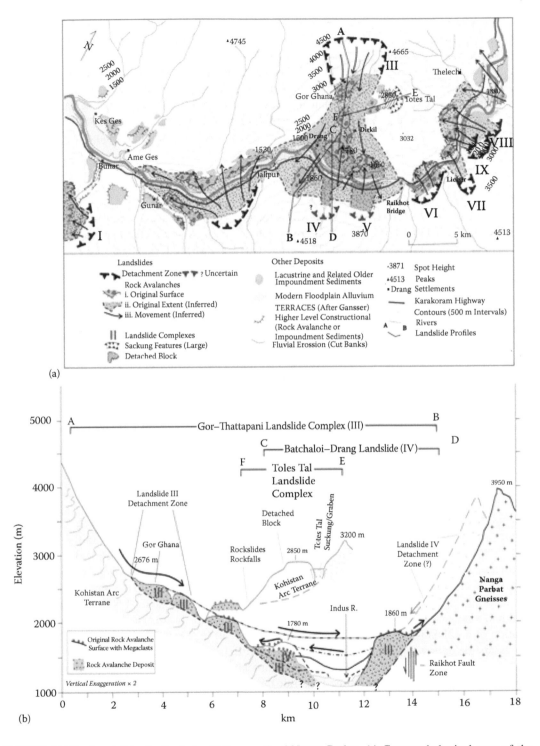

FIGURE 14.15 Landslides in the Indus Valley north of Nanga Parbat. (a) Geomorphological map of the western margin of the Nanga Parbat syntaxis between Bunar and Thelichi. (b) Related cross sections. (From Hewitt, K., *Quat. Sci. Rev.*, 28, 1055–1069, 2009.)

- The dammed lake backed up behind the confluence of the Hunza and Gilgit Rivers near Danyur to a height of ~1415 m, that is, 65 km upstream from Liachar, which suggests the dam was ~300 m high.
- The outburst flood in June 1841 devastated a large tract of the Indus River down to Kalabagh. According to recorded eyewitness accounts, a ~25 m high wall of mud and discoloured water gushed out into Peshawar basin from the Indus Gorge at Tarbela and wiped out a 500-men-strong Sikh Army that camped at the riverbanks near Attock. A thick blanket of black sand from the flood covered vast tracks of agricultural land on the banks of the Indus River between Tarbela and Attock.

14.4.2.2   The 2002 Nanga Parbat Earthquake

The Nanga Parbat syntaxis experienced two earthquakes of moderate magnitude in November 2002 (Figure 14.13): The first occurred on 2 November (M_b 5.3), while the main shock occurred on 20 November (M_b 6.2) (Mahmood et al. 2002). The latter one was followed by a series of aftershocks during the next 40 days. The epicentres for these earthquakes were located close to the western margin of the Nanga Parbat syntaxis that is delimited by the active Riakot–Liachar–Bunar fault (Lawrence and Ghauri 1983; Butler and Prior 1988).

The foreshock of the 2 November event was accompanied by at least three tremors with widespread damage to villages such as Tato in the Raikot and Muttath Valleys. Eleven people were killed, 400 injured and 4000 rendered homeless when 1000–1500 houses were destroyed. Blockages from earthquake-triggered landslides occurred at the Tatta Pani ~1 km downstream from the Raikot Bridge, and also all along the KKH from Bunji to Chilas, a strip that was partially cleared by 11 November for light traffic. The main earthquake of 20 November had its epicentre north of the village of Liachar. Much of its associated damage and aftershocks were in the densely populated Astor Valley, where the villages of Mushkin, Dushkin, Turbaling and Harchu were destroyed (Riaz and Khattak 2002). At least 23 people were killed, more than 100 people were seriously injured, and an estimated additional 7000 people were left homeless. Many pre-existing landslides, and among these the two largest ones at Doian and Mushkin (Butler et al. 1989), were reactivated and caused closure of the Astor Valley road and the KKH between Bunji and Chilas.

14.4.2.3   The 2010 Attabad Landslide

On 4 January 2010, a major landslide occurred at Attabad in the Hunza Valley in the Karakoram (36°18′50″ N, 74°49′13″ E) without any signs of an earthquake (Petley 2011; Schneider et al. 2011). A rock mass of ~45 million m³ advanced down from the northern valley side over a vertical distance of 1.2 km and horizontal distance of 1.3 km. The landslide mass moved up the opposite valley slope by 150–180 m and resulted in a dam that measured ~300 m across and ~1.5 km along the valley, and had a height of 120–200 m above the Hunza River (Figure 14.16). The blockage led to the formation of a ~21 km long lake with a volume of ~450 × 10⁶ m³. Several villages were flooded, and the landslide dam was overtopped on 28–29 May 2010. Based on the experiences from the Hattian Bala landslide in Kashmir from 2005, the government of Pakistan excavated an unlined 14 m deep spillway along the lowest edge of the landslide dam. Chances of initial outburst through headward erosion at the downstream toe of the landslide were minimized by the presence of an extraordinary large granite boulder underneath the spillway. By May 2012, the lake level was lowered by 20 m, and by June 2013, a 7 km long stretch of the KKH, the Shishkat Bridge, 466 ha of land and 327 buildings were reclaimed (FWO 2013).

The Attabad landslide also had an impact on the pre-existing sediments at the valley bottom. A dark silt–clay deposit, the remnant of a lake that had formed during the 1858 landslide dam at Salamanabad some 3 km downstream of the present site, was squeezed out in three major components. Part of this sediment was deposited ~150 m up the opposite side of the valley and subsequently slid down to spread onto the main landslide body. The remaining part of the sediment

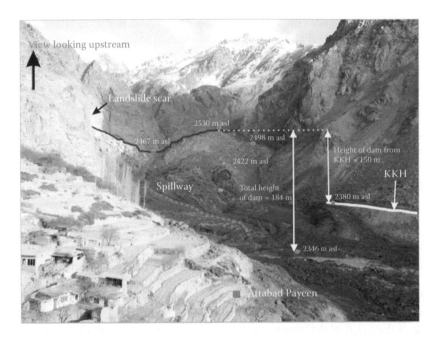

FIGURE 14.16   Upstream view of the 2010 Attabad landslide blocking the Hunza River, above sea level (asl). (Courtesy of G. A. Khatak, pers. comm.)

turned into mudflows at both ends of the landslide; one component travelled ~1.5 km upstream, and the other 3 km downstream. The latter killed 19 people in a hamlet located close to the valley floor at Sarat. Overall, ~140 houses were destroyed and more than 1600 people were displaced from their homes (Petley 2010).

Although the Attabad landslide did not occur during an earthquake, it is considered to be the result of the 20 November 2002 Astor earthquake located ~50 km to the south (35.52° N, 74.68° E) (Mahmood et al. 2002; Hughes 2003; Petley 2011) (Figure 14.13). Petley (2010) concluded that '[the Attabad landslide] has all the signs of being an incipient large landslide 300 m from bottom to top and 600 m across and perhaps triggered by the earthquake and topographic amplification effect–, so in total more than 5.5 million m³ of scree could collapse off the mountainside, down 1000 m, and across the Hunza Valley'. Investigators discovered cracks in the landslide that originally had been found by locals in early February 2003 at slopes above Attabad village (Hussain and Awan 2009; Petley 2010). These cracks were up to 1.5 m wide, particularly at the uppermost reaches of the rock face, and progressively widened and lengthened over the years until significant movements were reported subsequent to the M_w 7.6 Kashmir Earthquake on 8 October 2005. By then, the phenomenon had attracted the attention of the Pakistani government, as well as several non-governmental organizations. The Geological Survey of Pakistan reported extensive development of the cracks by 2009, which led to the evacuation of people from the slopes (Hussain and Awan 2009). In addition to the human fatalities and loss of homes, the landslide had serious social and economic impacts; for example, the KKH was closed, stalling Pakistan–China bilateral trade and cutting off the upstream population from access to the rest of the country.

14.4.3   DAREL–HAMRAN KOHISTAN SEISMIC ZONE

The Darel–Hamron Kohistan seismic zone is characterized by patches of seismicity (Khurshid et al. 1984; Ambraseys and Bilham 2003b). It straddles the drainage divide between the Kohistan and Ghizer Rivers and has an average altitude of ~4000 m, with deeply carved narrow valleys. The 1972 Hamran (M_b 6.3) and 1981 Darel (M_b 6.1) earthquakes are the main seismic events and have triggered hundreds of landslides on valley slopes in east-central Kohistan (Khurshid et al. 1984; Anbraseys and Bilham 2003b) (Figure 14.16). A detailed field account of the 1981 Darel earthquake disaster provides the eyewitness accounts gathered from locals in the upper reaches of Darel Valley: houses collapsed as a result of the intense shaking, followed by the roar of landslides and boulders as they crashed down from the mountainsides. The villages of Yashot (population ~1000), Junishall (population ~500) and Ghabar (population ~500) in the upper reaches of the Darel Valley were completely destroyed (Stack 1982). Massive rockfalls smashed into trees and spread into the ter-raced fields. The tributary stream that passes through the villages acted as a conduit for boulders to advance down. The ~41 km jeep track that starts at Shatial at the KKH and leads in the Darel Valley was blocked at several points by rockfalls, landslides and debris flows. The fact that the Pakistani Army engineers were able to reopen the track within 2 weeks of the disaster suggests that none of these mass movements were particularly large.

14.5 AXIAL BELT SEISMOTECTONIC PROVINCE

Despite the mountainous terrain and common occurrence of earthquakes of $M_w > 6$ that produce co-seismic slips as large as 4 m (Figure 14.17), earthquake-triggered landslides in the Axial Belt (Chaman–Sulaiman–Kirthar) seismic province are rare. The 1931 Mach earthquake (M_w 7.3) caused massive landslides, particularly in the Bolan Pass region, where several tunnels were blocked as a consequence of the uplift of the hanging wall along the earthquake fault (Yeats et al. 2007; Hands Pakistan 2013). Likewise, the road between Quetta and Sibi, especially a 2 km long stretch south of Mach, was seriously damaged by landsliding, which caused transportation to stall for 9 days. The 29 October 2008 earthquake near Pishin, ~100 km northeast of Quetta, damaged many villages in the

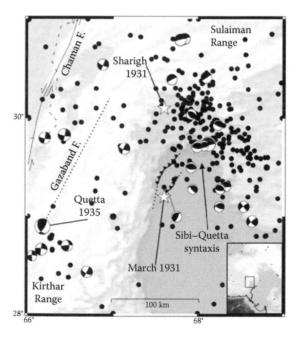

FIGURE 14.17 Part of the Axial Belt (Sulaiman–Kirthar thrust-fold belt) depicting principal tectonic fea-tures (Chaman fault and Sibbi–Quetta syntaxis), recent seismicity ($M_w > 5$) and locations of the Sharigh, Mach (stars) and Quetta earthquakes (focal mechanism beach ball). Focal mechanisms are scaled according to magnitude: the largest focal mechanism is M_w 7.7, while the smallest is M_w 5. (From Baig, M. S., in Kausar, A. B. et al., eds., *International Conference on 8 October 2005 Earthquake in Pakistan: Its Implications and Hazard Mitigation*, Islamabad, Pakistan, 18–19 January 2006, extended abstracts, 21–22.)

Ziarat and Pishin areas and triggered landsliding on steep slopes. At least one village, Wam Killy, was completely destroyed and partially buried. There are no reports of triggered landsliding for the 24 September 2013 Awaran earthquake (M_w 7.8). However, several roads approaching Awaran from Karachi via the Kirthar Range were blocked by landslides, which forced authorities to deliver much of the relief supplies by helicopter for at least a week (Martin and Kakar 2012).

14.6 MAKRAN SEISMOTECTONIC PROVINCE

Earthquakes along the subduction zone characterize the Makran seismotectonic province (Figure 14.18). The focal depths progressively increase northward from the coastal region inland, essentially defining the edge of the subducting plate. The northern intermediate deep-focused interslab earthquakes have generated MMIs of I–IV in the epicentral region. However, they have rarely triggered any significant landsliding. In case of the 2011 Dalbandin earthquake (M_w 7.2), rockfalls were reported in the Chagai Hills to the northeast of Dalbandin (Byrne et al. 1992). To the south in the Ras Koh Hills, rocks were dislodged from the summit of a ridge overlooking a manganese mine in the Khargoshkan area. In contrast, earthquakes with a shallow focus in the Makran coastal areas are well known to have caused considerable surface disturbances.

14.6.1 THE 1945 MAKRAN EARTHQUAKE

The 27 November 1945 Makran earthquake (M_w 8.1) is reported to have caused widespread surface damage along the Makran coast and several surface disturbances (Rajendran et al. 2008). These included ground rupturing at Pasni and Ormara, resulting in oozing out of warm water, terrace and beach uplift, and shoreline recession near Ormara, and reactivation of extinct and emergence of

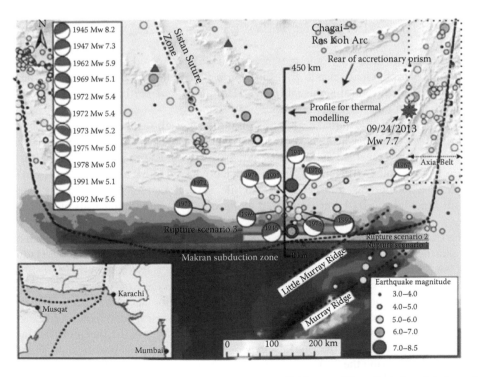

FIGURE 14.18 Tectonic map of the Makran region of western Pakistan and Iran depicting plate boundaries, accretionary prism and Chagai–Ras Koh Arc, together with historical seismicity. Note the location of the 24 September 2013 earthquake (star). (Modified after Smith, G. L. et al., *Geophys. Res. Lett.*, 40, 1528–1533, 2013.)

new mud volcanoes in the sea between Ormara and Pasni and at Gawadar (Rajendran et al. 2008). Extensive landslides were triggered along the entire coast. The principal artefact was a major seaward landslide involving part of the coastal area at Pasni (including part of the village), resulting in the creation of ~1000 m of new land along the coast (Rajendran et al. 2008). The earthquake generated a major tsunami in the Gulf of Oman, with waves estimated to be 5–10 m high along the Makran coast and up to 2 m high at Karachi. The initial wave shortly after the earthquake was followed by a second destructive tsunami that swept over the Pasni village 90–120 minutes after the earthquake. This delayed tsunami was attributed to a submarine landslide in the Gulf of Oman (Ambraseys and Melville 1982; Bilham et al. 2007; Owen 2010).

14.7 CONCLUSION

Of the five seismotectonic provinces in Pakistan, the Karakoram–Himalaya seismotectonic province has had the greatest number of earthquake-triggered landslides. The Pamir–Hindu Kush seismotectonic province, with a mean elevation and relief only slightly less than that of the Karakoram–Himalaya seismotectonic province, is also significantly susceptible to earthquake-triggered landslides, but here the record is far smaller. This may be partly owing to a lack of knowledge, since the historical record of earthquakes and associated landsliding from this region is poor. However, there are other possibilities accounting for this discrepancy. First, northwestern Pakistan – including the Chitral, Dir and Peshawar districts and the Bajaur, Mohmand, Khyber and Kurram tribal agencies – is within the extreme southeastern fringes of the Pamir–Hindu Kush seismotectonic province and, as such, mostly outside the main earthquake zone. Second, the earthquakes in the Pamir–Hindu Kush seismotectonic province are deep (~300 km) focused, and thus generate much lower surface intensities than those associated with shallow earthquakes. Although the Axial Belt (Western Highlands) and Makran seismotectonic provinces are characterized by greater seismic activity than the Himalayan seismotectonic province, earthquake-triggered landsliding is less common, which is mainly caused by lower mean elevations and less vertical relief.

In the Himalayan seismotectonic province, earthquake-triggered landslides are mainly associated with the two principal seismic zones of Nanga Parbat and Kashmir–Himalaya. In the third seismic zone of north-central Kohistan (Darel–Hamran region), earthquakes and associated landsliding are less common.

The 2005 Kashmir Earthquake, in terms of human causalities, population displacement and economic losses, was the most devastating earthquake experienced in the entire Himalaya. It was characterized by surface disturbances along a ~100 km rupture with vertical slips of up to 7 m, fissures associated with lateral spreading, selective land uplift and subsidence. Earthquake-triggered landsliding was unprecedented. While several factors controlled the landsliding, almost all landslides (>0.01 km²) occurred along the uplifted hanging wall of the earthquake rupture.

Of the earthquake-triggered landslides in Pakistan, the most dramatic are mega-landslides and rock avalanches that blocked rivers and created enormous lakes. These include the Hattu Pir and Liachar landslides on the Indus River. Their proximity to the active Raikot–Liachar–Bunar Fault at the western margin of the Nanga Parbat syntaxis attests to their triggering likely being due to earthquake rupturing or associated ground motions. Three major landslides occurred in 1840 along the same segment of the Indus River, of which one blocked the river for 6 months. In 2005, the co-seismic Hattian Bala landslide in Kashmir dammed two tributaries of the Jhelum River. In contrast, the 2010 Attabad landslide in the Hunza Valley of the Karakoram was post-seismic: it is the result of ground shaking during the 2002 Nanga Parbat earthquake and the 2005 Kashmir Earthquake, in combination with hydrometeorological processes in subsequent years.

Landslide-dammed lakes, especially on the large rivers of the scale of Indus, Jhelum and Hunza, can have catastrophic consequences for areas downstream. Once the lakes are filled to their maximum potential, breaching of the natural dam can result in flood waves as high as 30 m to downstream distances on the scale of hundreds of kilometres. The catastrophic flood of June 1841 associated with the breach of the mega-landslide dammed lake at Liachar on the Indus River is an infamous example of this hazard. The 2005 Hattian Bala and 2010 Attabad landslide-dammed lakes carried the potential to be catastrophic flood hazards. However, Pakistan successfully thwarted catastrophic breaches of these natural dams by controlled slow draining through low-cost temporary spillways across the landslide bodies. This experience serves as a useful case study for tackling such situations in comparable high-relief, seismically active mountainous terrains prone to landslide–dam outburst hazards.

ACKNOWLEDGEMENTS

S.F.K. acknowledges support by the National Centre of Excellence in Geology (NCEG) University of Peshawar and the University of Montana. M.A.K. is grateful to the Higher Education Commission (HEC) of Pakistan and the NCEG for supporting his projects on earthquake research in Pakistan since 2001. U.K. and L.A.O. thank the National Science Foundation for financial support (EAR-0602675) for their research on the 2005 Kashmir Earthquake.

REFERENCES

Ali, Z., Qaisar, M., Mahmood, T., Shah, M. A., Iqbal, T., Serva, L., Burton, P. W. (2009). The Muzaffarabad, Pakistan, earthquake of 8 October 2005: Surface faulting, environmental effects and macroseismic intensity. *Geological Society of London, Special Publication*, 316, 155–172.

Ambraseys, N., Bilham, R. (2003a). Earthquakes in Afghanistan. *Seismological Research Letters*, 74, 107–123.

Ambraseys, N., Bilham, R. (2003b). Earthquakes and associated deformation in northern Baluchistan 1892–2001. *Bulletin of the Seismological Society of America*, 93, 1573–1605.

Ambraseys, N., Bilham, R. (2014). The tectonic setting of Bamiyan and the seismicity in and near Afghanistan for the past twelve centuries. In Margottini, C., ed., *After the Destruction of the Giant Buddha Statues in Bamiyan (Afghanistan) in 2001*. Berlin: Springer, 67–94.

Ambraseys, N., Lensen, G., Moinfar, A. (1975). The Pattan Earthquake of 28 December 1974. UNESCO Technical Report, RP/1975-76/2.222.3. Paris: UNESCO.

Ambraseys, N., Lensen, G., Moinfar, A., Pennington, W. (1981). The Pattan (Pakistan) earthquake of 28 December 1974: Field observations. *Quarterly Journal of Engineering Geology and Hydrogeology*, 14, 1–16.

Angiolini, L., Zanchi, A., Zanchetta, S., Nicora, A., Vezzoli, G. (2013). The Cimmerian geopuzzle: New data from south Pamir. *Terra Nova*, 25, 352–360.

Anggraeni, D., Van der Meijde, M., Shafique, M. (2010). Modelling the impact of topography on seismic amplification at regional scale. Enschede, the Netherlands: UT-ITC, 46.

Aslam, M., Ghazanvi, M. (1975). A note on the Pattan Earthquake (Hazara-Swat Kohistan), N.W.F.P. 28 December 1974. Report of the Geological Survey of Pakistan, NWFP Circle, Peshawar. Quetta, Pakistan: Geological Survey of Pakistan.

Avouac, J. P., Ayoub, F., Leprince, S., Konca, O., Helmberger, D. V. (2006). The 2005, Mw 7.6 Kashmir Earthquake: Sub-pixel correlation of ASTER images and seismic waveforms analysis. *Earth and Planetary Science Letters*, 249, 514–528.

Aydan, O. E. (2006). Geological and seismological aspects of Kashmir Earthquake of October 8, 2005 and a geotechnical evaluation of induced failures of natural and cut slopes. *Journal of the School of Marine Science and Technology*, 4(1), 25–44.

Baig, M. S. (2006). Active faulting and earthquake deformation in Hazara-Kashmir syntaxis, Azad Kashmir, northwest Himalaya, Pakistan. In Kausar, A. B., Karim, T., Khan, T., eds., *International Conference on 8 October 2005 Earthquake in Pakistan: Its Implications and Hazard Mitigation*, Islamabad, Pakistan, 18–19 January, extended abstracts, 21–22.

Beck, R. A., Burbank, D. W., Sercombe, W. J., Khan, A. M., Lawrence, R. D. (1996). Late Cretaceous ophiolite obduction and Palaeocene India–Asia collision in the westernmost Himalaya. *Acta Geodynamica et Geomaterialia*, 9, 114–144.

Bendick, R., Bilham, R., Khan, M. A., Khan, S. F. (2007). Slip on an active wedge thrust from geodetic observations of the 8 October 2005 Kashmir earthquake. *Geology*, 35, 267–270.

Bilham, R., Lodi, S., Hough, S., Bukhary, S., Khan, A. M., Rafeegi, S. F. A. (2007). Seismic hazard in Karachi, Pakistan: Uncertain past, uncertain future. *Seismological Research Letters*, 78(6), 601–631.

Bilham, R. (2009). The seismic future of cities. *Bulletin of Earthquake Engineering*, 7, 839–887.

Biswas, S. K. (2005). A review of structure and tectonics of Kutch Basin, western India, with special reference to earthquakes. *Current Science*, 88, 1592–1600.

Burg, J. P. (2011). The Asia–Kohistan–India collision: Review and discussion. In Brown, D., Ryan, P. D., eds., *Arc-Continent Collision*. Berlin: Springer, 279–309.

Butler, R. W., Prior, D. J. (1988). Tectonic controls on the uplift of the Nanga Parbat massif, Pakistan Himalayas. *Nature*, 333, 247–250.

Butler, R. W., Prior, D. J., Knipe, R. J. (1989). Neotectonics of the Nanga Parbat syntaxis, Pakistan, and crustal stacking in the northwest Himalayas. *Earth and Planetary Science Letters*, 94, 329–343.

Byrne, D. E., Sykes, L. R., Davis, D. M. (1992). Great thrust earthquakes and aseismic slip along the plate boundary of the Makran subduction zone. *Journal of Geophysical Research*, 97, 449–478.

Chini, M., Cinti, F. R., Stramondo, S. (2011). Co-seismic surface effects from very high resolution panchromatic images: The case of the 2005 Kashmir (Pakistan) earthquake. *Natural Hazards and Earth System Science*, 11, 931–943.

Delaney, K. B., Evans, S. G. (2011). Rockslide dams in the northwest Himalayas (Pakistan, India) and the adjacent Pamir Mountains (Afghanistan, Tajikistan), Central Asia. In Evans, S. G., Scarascia-Mugnozza, G., Hermanns, R. L., Strom, A., eds., *Natural and Artificial Rockslide Dams*. Berlin: Springer, 205–242.

DiPietro, J. A., Pogue, K. R. (2004). Tectonostratigraphic subdivisions of the Himalaya: A view from the west. *Tectonics*, 23, TC5001. doi: 10.1029/2003TC001554.

Dunning, S. A., Mitchell, W. A., Rosser, N. J., Petley, D. N. (2007). The Hattian Bala rock avalanche and associated landslides triggered by the Kashmir earthquake of 8 October 2005. *Engineering Geology*, 93, 130–144.

Dunning, S. A., Armitage, P. (2011). The grain-size distribution of rock-avalanche deposits: Implications for natural dam stability. In: Evans S., Hermanns R., Strom A., Scarascia-Mugnozza G. (eds) *Natural and Artificial Rockslide Dams. Lecture Notes in Earth Sciences*, vol 133. Springer, Berlin, Heidelberg.

Farhoudi, G., Karig, D. E. (1977). Makran of Iran and Pakistan as an active arc system. *Geology*, 5, 664–668.

Fujiwara, S., Tobita, M., Sato, H. P., Ozawa, S., Une, H., Koarai, M., Hiroyuki, N., Midori, F., Hiroshi, Y., Takuya, N., Hayashi, F. (2006). Satellite data gives snapshot of the 2005 Pakistan earthquake. *Eos*, 87, 73.

Frontier Works Organization (FWO), Newsletter January–December 2013, Complied by: PR section, HQ FWO, Designed & Produced by ASTRAL HATCH INC. 051-8730659, Frontier Works Organizations (Headquarters) 509, Kashmir Road, R.A. Bazar, Rawalpindi-Pakistan. https://www.fwo.com.pk /news-info/newsletters.

Gaetani, M. (1997). The Karakorum block in Central Asia, from Ordovician to Cretaceous. *Sedimentary Geology*, 109, 339–359.

Gaetani, M., Garzanti, E., Jadoul, F., Nicora, A., Tintori, A., Pasini, M., Khan, K. S. A. (1990). The north Karakorum side of the Central Asia geopuzzle. *Geological Society of America Bulletin*, 102, 54–62.

Gates, A. E., Ritchie, D. (2007). *Encyclopedia of Earthquakes and Volcanoes*. New York: Checkmark Books.

Hands Pakistan. (2013). Pakistan earthquake 2013 updates report. Karachi, Pakistan: HANDS International. http://www.hands.org.pk.

Harp, E. L., Crone, A. J. (2006). Landslides triggered by the October 8, 2005 Pakistan earthquake and associated landslide-dammed reservoirs. U.S. Geological Survey Open-File Report 2006-1052 13. Reston, VA: U.S. Geological Survey.

Hewitt, K. (2009). Catastrophic rock slope failures and late Quaternary developments in the Nanga Parbat–Haramosh massif, Upper Indus basin, northern Pakistan. *Quaternary Science Reviews*, 28, 1055–1069.

Hewitt, K., Gosse, J., Clague, J. J. (2011). Rock avalanches and the pace of late Quaternary development of river valleys in the Karakoram Himalaya. *Geological Society of America Bulletin*, 123, 1836–1850.

Hodges, K. V. (2000). Tectonics of the Himalaya and southern Tibet from two perspectives. *Geological Society of America Bulletin*, 112, 324–350.

Hughes, R. (2003). Astor Valley earthquake, northern areas Pakistan: A field assessment, 1. Earthquake Engineering Field Investigation Team (EEFIT) Short Report. London: Institution of Structural Engineers.

Hussain, S. H., Awan, A. A. (2009). Causative mechanisms of terrain movement in Hunza Valley. Quetta, Pakistan: Geological Survey of Pakistan, 19.

Kamp, U., Growley, B. J., Khattak, G. A., Owen, L. A. (2008). GIS-based landslide susceptibility mapping for the 2005 Kashmir earthquake region. *Geomorphology*, 101, 631–642.

Kamp, U., Owen, L. A., Growley, B. J., Khattak, G. A. (2010). Back analysis of landslide susceptibility zonation mapping for the 2005 Kashmir earthquake: An assessment of the reliability of susceptibility zoning maps. *Natural Hazards*, 54, 1–25.

Kaneda, H., Nakata, T., Tsutsumi, H., Kondo, H., Sugito, N., Awata, Y., Akhtar, S. S. et al. (2008). Surface rupture of the 2005 Kashmir, Pakistan, earthquake and its active tectonic implications. *Bulletin of the Seismological Society of America*, 98, 521–557.

Kazmi, A. H., Jan, M. Q. (1997). *Geology and Tectonics of Pakistan*. Karachi, Pakistan: Graphic Publishers.

Khan, M. A., Abbasi, I. A., Qureshi, A. W., Khan, S. R. (2003). Tectonics of Afghan-India collision zone, Kurram-Waziristan region, N. Pakistan. In *Conference Proceedings (ATC 2003), Pakistan Association of Petroleum Geologists*, October 3–5, 2003, Islamabad, Pakistan, 219–234.

Khan, S. F., Kamp, U., Owen, L. A. (2013). Documenting five years of landsliding after the 2005 Kashmir earthquake, using repeat photography. *Geomorphology*, 197, 45–55.

Khattak, G. A., Owen, L. A., Kamp, U., Harp, E. L. (2010). Evolution of earthquake-triggered landslides in the Kashmir Himalaya, northern Pakistan. *Geomorphology*, 115, 102–108.

Khurshid, A., Yielding, G., Ahmad, S., Davison, I., Jackson, J. A., King, G. C. P., Zuo, L. B. (1984). The seismicity of northernmost Pakistan. *Tectonophysics*, 109, 209–226.

Konagai, K., Sattar, A. (2012). Partial breaching of Hattian Bala landslide dam formed in the 8th October 2005 Kashmir earthquake, Pakistan. *Landslides*, 9, 1–11.

Kumar, K. V., Martha, T. R., Roy, P. S. (2006). Mapping damage in the Jammu and Kashmir caused by 8 October 2005 Mw 7.3 earthquake from the Cartosat–1 and Resourcesat–1 imagery. *International Journal of Remote Sensing*, 27, 4449–4459.

Kumahara, Y., Nakata, T. (2006). Active faults in the epicenter area of the 2005 Pakistan earthquake. Special Publication 41. Hiroshima: Research Centre for Regional Geography, Hiroshima University, 54.

Lawrence, R. D., Ghauri, A. A. K. (1983). Evidence of active faulting in Chilas district, northern Pakistan. *Geological Bulletin of the University of Peshawar*, 16, 185–186.

Mahmood, T., Qaisar, M., Ali, Z. (2002). Source mechanism of Astor Valley earthquake of November 20, 2002 inferred from teleseismic body waves. *Geological Bulletin of the University of Peshawar*, 35, 151–161.

Martin, S. S., Kakar, D. M. (2012). The 19 January 2011 Mw 7.2 Dalbandin earthquake, Balochistan. *Bulletin of the Seismological Society of America*, 102, 1810–1819.

Mason, K., Gunn, J. P., Todd, H. J. (1930). The Shyok flood in 1929. *Himalayan Journal*, 2, 35–47.

Meltzer, A., Sarker, G., Beaudoin, B., Seeber, L., Armbruster, J. (2001). Seismic characterization of an active metamorphic massif, Nanga Parbat, Pakistan Himalaya. *Geology*, 29, 651–654.

Nakata, T., Tsutsumi, H., Khan, S. H., Lawrence R. D. (1991). Active faults of Pakistan: Map sheets and inventories. Special Publication 21. Hiroshima: Research Centre for Regional Geography, Hiroshima University, 141.

Owen, L. A. (2010). Landscape development of the Himalayan–Tibetan orogen: A review. *Geological Society of London, Special Publication*, 338, 389–407.

Owen, L. A., Kamp, U., Khattak, G. A., Harp, F. L., Keefer, D. K., Bauer, M. A. (2008). Landslides triggered by the 8 October 2005 Kashmir earthquake. *Geomorphology*, 94, 1–9.

Pathier, E., Fielding, E. J., Wright, T. J., Walker, R., Parsons, B. E., Hensley, S. (2006). Displacement field and slip distribution of the 2005 Kashmir Earthquake from SAR imagery. *Geophysical Research Letters*, 33, L20310.

Peiris, N., Rossetto, T., Burton, P., Mahmood, S. (2006). EEFIT mission: October 8, 2005 Kashmir Earthquake. London: Report of the Institution of Structural Engineers.

Pennington, W. D. (1979). A summary of field and seismic observations of the Pattan Earthquake – 28 December 1974. In Farah, A., DeJong, K. A., eds., *Geodynamics of Pakistan*. Quetta, Pakistan: Geological Survey of Pakistan, 143–147.

Petley D. (2010). The landslide at Attabad in Hunza, Gilgit/Baltistan: Current situation and hazard management needs. Report prepared for Focus Humanitarian Assistance, Pakistan, based upon a rapid field assessment on 26th February–4th March 2010.

Petley, D. (2011). The Attabad landslide crisis in Hunza, Pakistan: Lessons for the management of valley blocking landslides. Technical report. Durham, UK: Institute of Hazard, Risk and Resilience, and International Landslide Centre in the Department of Geography, Durham University.

Petley, D., Dunning, S., Rosser, N., Kausar, A. B. (2006). Incipient landslides in the Jhelum Valley, Pakistan following the 8th October 2005 earthquake. In Marui, H., ed., *Disaster Mitigation of Debris Flows, Slope Failures and Landslides*. Tokyo: Universal Academy Press, 47–55.

Rajendran, C. P., Ramanamurthy, M. V., Reddy, N. T., Rajendran, K. (2008). Hazard implications of the late arrival of the 1945 Makran tsunami. *Current Science*, 95, 1739–1743.

Rastogi, B. K. (2001). Ground deformation study of Mw 7.7 Bhuj earthquake of 2001. *Episodes*, 24, 160–165.

Riaz, M., Khattak, G. A. (2002). Earthquakes in Nanga Parbat area, north Pakistan: Causes and concerns with particular reference to Basha Dam. *Geological Bulletin of the University of Peshawar*, 35, 163–173.

Saba, S. B., Van der Meijde, M., Van der Werff, H. (2010). Spatiotemporal landslide detection for the 2005 Kashmir earthquake region. *Geomorphology*, 124, 17–25.

Sato, H. P., Hasegawa, H., Fujiwara, S., Tobita, M., Koarai, M., Une, H., Iwahashi, J. (2007). Interpretation of landslide distribution triggered by the 2005 northern Pakistan earthquake using SPOT 5 imagery. *Landslides*, 4, 113–122.

Sattar, A., Konagai, K., Kiyota, T., Ikeda, T., Johansson, J. (2011). Measurement of debris mass changes and assessment of the dam-break flood potential of earthquake-triggered Hattian landslide dam. *Landslides*, 8, 171–182.

Sayab, M., Khan, M. A. (2010). Temporal evolution of surface rupture deduced from coseismic multi-mode secondary fractures: Insights from the October 8, 2005 (Mw 7.6) Kashmir Earthquake, NW Himalaya. *Tectonophysics*, 493, 58–73.

Schelling, D. D. (1999). Frontal structural geometries and detachment tectonics of the northeastern Karachi arc, southern Kirthar Range, Pakistan. *Geological Society of America, Special Papers*, 328, 287–302.

Schneider J. F. (2006). Earthquake-triggered mass movements in northern Pakistan with special reference to Hattian landslide. In Kausar, A. B., Karim, T., Khan, T., eds., *International Conference on 8 October 2005 Earthquake in Pakistan: Its Implications and Hazard Mitigation*, Islamabad, Pakistan, 18–19 January, extended abstracts, 76–78.

Schneider, J. F. (2009). Seismically reactivated Hattian slide in Kashmir, Northern Pakistan. *Journal of Seismology*, 13, 387–398.

Schneider, J. F., Gruber, F. E., Mergili, M. (2011). Recent cases and geomorphic evidence of landslide-dammed lakes and related hazards in the mountains of Central Asia. Presented at Proceedings of the Second World Landslide Forum, Rome, 3–7 October.

Searle, M. P., Khan, M. A. (1996). *Geological Map of North Pakistan and Adjacent Areas of Northern Ladakh and Western Tibet (Western Himalaya, Salt Ranges, Kohistan, Karakoram, Hindu Kush), Scale: 1:650,000.* Oxford: Oxford University.

Seeber, L., Armbruster, J. (1979). Seismicity of the Hazara arc in northern Pakistan: Decollement vs. basement faulting. In Farah, D., DeJong, K. D., eds., Geodynamics of Pakistan. Quetta, Pakistan: Geological Survey of Pakistan, 131–147.

Seeber, L., Armbruster, J. (1984). Some elements of continental subduction along the Himalayan front. Tectonophysics, 105, 263–278.

Şengör, A. C. (1984). The Cimmeride orogenic system and the tectonics of Eurasia. *Geological Society of America, Special Papers*, 195, 1–74.

Shroder, J. F. (1989). Hazards of the Himalaya. *American Scientist*, 77, 564–573.

Shroder, J. F. (1993). Himalaya to the sea: Geomorphology and the Quaternary of Pakistan in the regional context. In Shroder, J. F., ed., *Himalaya to the Sea: Geology, Geomorphology, and the Quaternary.* London: Routledge, 1–42.

Shroder, J. F. (1998). Slope failure and denudation in the western Himalaya. *Geomorphology*, 26, 81–105.

Shroder, J. F., Bishop, M. P. (2000). Unroofing of the Nanga Parbat Himalaya. *Geological Society of London, Special Publication*, 170, 163–179.

Smith, G. L., McNeill, L. C., Wang, K., He, J., Henstock, T. J. (2013). Thermal structure and megathrust seismogenic potential of the Makran subduction zone. *Geophysical Research Letters*, 40, 1528–1533.

Stack, T. (1982). The Darel Valley earthquake: An overlooked disaster in northern Pakistan. *Disasters*, 6, 169–176.

Sverdrup, K. A., Schurter, G. J., Cronin, V. S. (1994). Relocation analysis of earthquakes near Nanga Parbat-Haramosh Massif, northwest Himalaya, Pakistan. *Geophysical Research Letters*, 21, 2331–2334.

Tapponnier, P., King, G., Bollinger, L. (2006). Active faulting and seismic hazard in the western Himalayan Syntaxis, Pakistan. In Kausar, A. B., Karim, T., Khan, T., eds., *International Conference on 8 October 2005 Earthquake in Pakistan: Its Implications and Hazard Mitigation*, Islamabad, Pakistan, 18–19 January, extended abstracts, 6–7.

Treloar, P. J., Rex, D. C., Guise, P. G., Wheeler, J., Hurford, A. J., Carter, A. (2000). Geochronological constraints on the evolution of the Nanga Parbat syntaxis, Pakistan Himalaya. *Geological Society of London, Special Publication*, 170, 137–162.

Ul-Hadi, S., Khan, S. D., Owen, L. A., Khan, A. S., Hedrick, K. A., Caffee, M. W. (2013). Slip-rates along the Chaman fault: Implication for transient strain accumulation and strain partitioning along the western Indian plate margin. *Tectonophysics*, 608, 389–400.

Yeats, R. S., Parsons, T., Hussain, A., Yuji, Y. (2007). Stress changes with the 08 October 2005 Kashmir earthquake: Lessons for future. In Kausar, A. B., Karim, T., Khan, T., eds., *International Conference on 8 October 2005 Earthquake in Pakistan: Its Implications and Hazard Mitigation*, Islamabad, Pakistan, 18–19 January, extended abstracts, 16–17.

Zeitler, P. K., Johnson, N. M., Naeser, C. W., Tahirkheli, R. A. (1982). Fission-track evidence for Quaternary uplift of the Nanga Parbat region, Pakistan. *Nature*, 298, 255–257.

Zeitler, P. K., Koons, P. O., Bishop, M. P., Chamberlain, C. P., Craw, D., Edwards, M. A., Hamidullah, S. et al. (2001). Crustal reworking at Nanga Parbat, Pakistan: Metamorphic consequences of thermal-mechanical coupling facilitated by erosion. *Tectonics*, 20, 712–728.

15 Landslides in Jamaica
Distribution, Cause, Impact and Management

Servel Miller, Anestoria Shalkowski, Norman Harris,
Dionne Richards and Lyndon Brown

CONTENTS

15.1 INTRODUCTION

Jamaica has one of the highest natural hazard risk exposures in the world, with more than 90% of the population exposed to two or more natural hazards. The island of Jamaica is particularly prone to multiple hazards, including hurricanes, earthquakes and slope instability, due to its geographical position (within the track of Atlantic hurricanes and its location on the Caribbean 'tectonic' plate) and its topography and geology (steep slopes with highly weathered material). Of these hazards, slope instability is the most common, affecting not only mountainous areas but also the coastal plains, where submarine landslides have been known to generate tsunamis. One such tsunami contributed to the destruction of the then capital city of Port Royal in 1692.

Landslides are predominantly triggered by seismic activities and heavy rainfall associated with hurricanes and tropical depressions. These landslides have caused loss of lives, widespread destruction to the built and natural environment and long-term damage to the socioeconomic development of the country. The slope instability problem is compounded by the lack of awareness of the impact

FIGURE 15.1   Damage to road, Tangle River.

by the general public, developers and planners, as well as uncontrolled and unplanned urbanization on marginal lands susceptible to slope failure.

Landslides are one of the most significant threats to Jamaica's development, and it needs more investment in resources to deal with them. If a more holistic approach to slope instability is adopted, this could result in significant reduction in the societal and economic losses that the island experiences as a result of slope instability.

Between 1980 and 2012, Jamaica experienced more than 30 significant natural hazardous events that resulted in major disasters (EM-DAT 2012). These disasters between 2008 and 2010 resulted in almost $3 billion worth of damage (World Bank and GFDRR 2010). According to Jamaica's umbrella government office that has responsibility for natural hazards, disaster mitigation and management, the Office of Disaster Preparedness and Emergency Management (ODPEM), 'landslides are the most common natural hazard in Jamaica' (ODPEM 2012). Each year, landslides result in significant damage to infrastructure (Ahmad et al. 1993) and the built environment (Figures 15.1 and 15.2), the blocking of roads and damage to agricultural land, generally causing millions of dollars' worth of damage when they occur (ECLAC 2001; ODPEM/CDB 2003) and in some instances loss of lives (Zans 1959; Jamaica Observer 2010). The problem is compounded by uncontrolled and unplanned urbanization on marginal lands susceptible to slope failure, therefore leading to an increased level of vulnerability among the population (Jackson 2005).

This chapter provides the spatial distribution of landslide damage in Jamaica on both a local and a national scale. It examines the physical, social and economic impacts of slope instability, the causes, and the management strategies being utilized to better mitigate the impact of this natural phenomenon.

15.2   LANDSLIDE DISTRIBUTION AT A NATIONAL SCALE

Jamaica is a small country located in the Caribbean (Figure 15.3). It has a mountainous interior which is fringed by a coastal plain. The topography of Jamaica is characterized by rugged steep hills and mountains surrounded by low-lying alluvial plains. The mountainous regions with their steep slopes, covered by deeply weathered soils, are particularly prone to slope instability. It is this

FIGURE 15.2   Houses fully and partially buried under debris, Gordon Town.

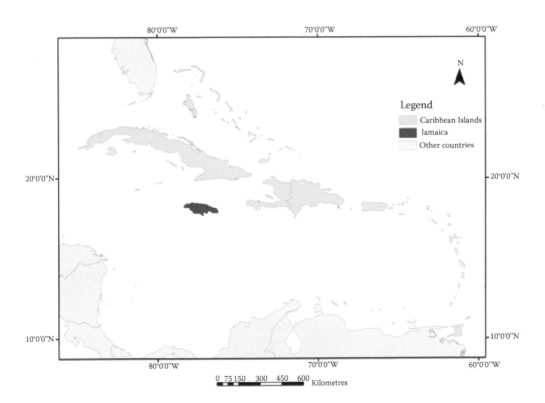

FIGURE 15.3   Location of Jamaica in relation to the Caribbean.

that accounts for the numerous landslides found around the island (Figure 15.4). While most areas around the entire island are vulnerable to slope movements, Figure 15.5 highlights that most of the known slope failures occur in areas where the lithologies of volcanics, volcaniclastics and metasediment shale and sandstone sequences conglomerate and limestones dominate. Structurally, these occur within inliers and the Wagwater Belt.

Jamaica is located within the track of the North Atlantic hurricanes passing through the Caribbean. These hurricanes and/or storms are normally accompanied by intense and island-wide rainfall. Historically, rainfall is the most effective mechanism for triggering widespread slope movements in

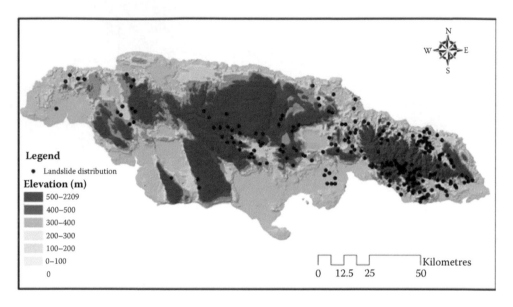

FIGURE 15.4   National distribution of landslides in relation to topography.

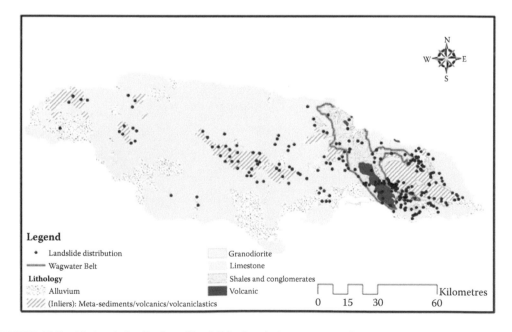

FIGURE 15.5   National distribution of landslides in relation to topography.

Jamaica (Barker and McGregor 1995). The incidence of landslides increases particularly in those areas which have a high annual rainfall (see Section 15.5 for more detail). Although the most recent landslides in Jamaica are associated with periods of heavy rainfall, numerous landslides have been recorded that were associated with earthquakes. For example, after the 1993 earthquake, more than 40 landslides were mapped in the eastern half of the island (Ahmed 1995a). One of the first recorded and most infamous landslides in Jamaica's history, the Judgement Cliff landslide (see also Section 15.4 for more detail), is associated with the earthquake of 1692 that destroyed the city of Port Royal. As stated by Zans (1959), 'At Yellows a great mountain split, and fell into the level land, and covered several settlements, and killed 19 people. One of the persons, whose name was Hopkins, had his plantation removed half a mile from the place where it formerly stood'. More than 85 million cubic metres of material was displaced during this event.

15.3 LANDSLIDES IN JAMAICA: HISTORY AND IMPACT

Most people tend to regard landslides as accidents, as something that only happens to others because, in comparison with earthquakes, hurricanes and floods, landslides affect relatively small and sharply delineated areas (ODPEM 2012), and hence in the past little attention was given to slope instability. While there seems to be increasing awareness of the hazards, due to coverage in both print and electronic media of recent occurrences, slope instability is not a new phenomenon to Jamaica. There are records of landslide impacts and occurrences dating back to 1692, the year in which an earthquake destroyed Port Royal. The 1692 earthquake is thought to have triggered submarine landslides that caused a tsunami which inundated and devastated the city (Table 15.1). Active landslides, as mentioned before, become evident particularly after heavy rainfall where the debris blocks and/or destroys roads. During periods of heavy rainfall, most of the roads within the country, particularly in the eastern regions, are blocked and/or damaged by landslides, affecting buildings particularly in urban areas (Figure 15.6). In the last 30 years, such extreme incidences have occurred in, for example, 1988, associated with Hurricane Gilbert (Eyre 1989); 1993, associated with a 5.4 earthquake (Ahmad 1995a); 1998, associated with Hurricane Mitch; May–June 2002, associated with a tropical depression (ReliefWeb 2002); and 2012, associated with Hurricane Sandy (Wignall 2012) (Table 15.1). The consequences were that after each event, several villages and towns were isolated and cut off from essential services and supplies; farmers were unable to transport food produce to market and for processing. On an annual basis, the cost from the damage from these landslides runs into millions of dollars. There are no concrete figures that can be given for the damage caused by landslides over the years (Jamaica Gleaner 1998a,b). Landslides also result in the damming of rivers, as was the case along the Cascade River in St. Thomas parish, where the river rose more than 50 ft above its normal level during November 1909 (ODPEM 2002). The landslide dams have proven to cause significant damage. When the dams are breached, this results in flooding downstream, often with little warning, damaging agricultural fields, livestock and houses on the plains. A very good example is the 1937 Millbank landslides (see Section 15.4), which dammed the Rio Grande for 6 months. When the dam was breached, it destroyed two bridges and agricultural lands (Table 15.1).

In addition to the active landslides during periods of intense rainfall, there are many relict slides which may be more widespread than previously believed (Miller et al. 2009). Recent studies (Miller et al. 2007; Bhalai 2010) demonstrated that the distribution of inland landslides has been grossly underestimated in the parishes of St. Mary, St. Thomas and Portland. Increasingly, there is the need to recognize and map existing landslides and evaluate the slope instability on the island. This has taken on added importance with the rapid infrastructure development, for example, the 'highway 2000' housing project and the building of hotels. Road cuttings in particular have triggered movement along relict landslides, and this has significantly delayed the project and, as a result, led to increasing construction costs (Miller et al. 2009).

TABLE 15.1

Some Major Slope Movements in Jamaica, Including Fatalities, Injuries and Economic Costs

Year and Event	Slope Movement Type	No. of Persons Killed/ Injured/Missing	Economic/Social Impact
Great earthquake of 1 June 1692	Widespread liquefaction, rockfalls and avalanches; landslide dammed river	Very high; 3000+ killed	Port Royal destroyed
Judgement Cliff landslide induced by rainfall of October 1692	Complex rockslide slump; Yallahs River dammed, causing flooding	At least 19 killed	Homes and plantation destroyed
Kingston Earthquake of 14 January 1907	Liquefaction, widespread landslide activities in eastern Jamaica	About 1000 killed, catastrophic events	90,000 homeless, submarine cables damaged; property damage around £2 million
Whitfield Hall landslide, Blue Mountain, torrential rains of 1909	Complex landslide		Hundreds of hectares of cultivated land and coffee estate destroyed
Millbank landslide, Portland, 27 November 1937; torrential rain	Complex rockslide slump; Rio Grande River dammed for 6 months, causing flooding	5 killed; missing and injured unknown	Extensive damage to agricultural lands, livestock and housing; roads blocked; 2 bridges destroyed
Chelsea landslide, Portland, November 1940	Complex rockslide slump; Swift River dammed, causing flooding	10 killed	Extensive damage to agricultural property, roads and bridges
Island-wide debris flows associated with Hurricane Gilbert	Extensive debris flows		More than 450 landslides mapped in above rock region; mostly in weather volcanic rocks
Rockfall in Bog Walk Gorge, widespread debris flow in western, central and eastern parishes	Debris flows		4 killed, 10 injured, river dammed and cause flooding, pipeline damaged in Portland; road blocked/damaged in Cedar Valley, Morgans Pass, Killet, Main Ridge, Bog Walk and Dias in Hanover
May/June 2002 rainfall induced island-wide debris flows	Extensive debris flow	1 killed, 1 injured	Extensive damage to agriculture, roads, bridge and properties; the Bog Walk Gorge blocked for 6 months
Widespread debris flow in eastern Jamaica, 2012 – associated with Hurricane Sandy	Debris flow		Damage to agriculture fields, roads blocked in St. Thomas (White Hall to Hillside, Hall Head to Morant River Bridge, Coley to Low Mountain, Yallahs to Petersfield), bridge damaged

15.4   MAJOR LANDSLIDES AND SLOPE INSTABILITY EVENTS

During the late seventeenth century, Port Royal, Jamaica, was the second largest (Boston was the largest) English town in the Americas. Port Royal was destroyed in 1692 by an earthquake and tsunami generated by slumping of part of the land mass into the sea. The earthquake triggered the massive slump along the seafront, which led to one-third of the existing town of Port Royal being subducted below sea level (Figure 15.7). These landslides subsequently generated a tsunami that inundated the area and led to further destruction of the town after the earthquake (Lander et al. 2002).

FIGURE 15.6   Landslide debris impacting the built environment and burying cars.

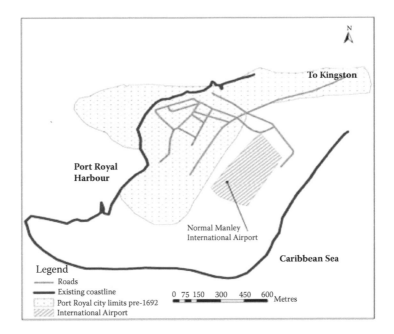

FIGURE 15.7   Map showing the change in the land area of Port Royal following the 1692 landslide.

The slope instability that occurred with the earthquake of 1692 is the most significant landslide event in Jamaica's history. The liquefaction associated with this earthquake resulted in the subduction of buildings below the Earth's surface (Figure 15.8), thus contributing to the further demise of the city of Port Royal.

The 1692 earthquake that destroyed Port Royal also triggered the most conspicuous landslide on the Jamaican landscape, the Judgement Cliff landslide (Figure 15.9). The landslide resulted in the loss of 19 lives and the destruction of a village and plantation (Zans 1959). The Judgement Cliff

FIGURE 15.8   Top of armoury (previously 10 m above Earth surface) at current land surface due to liquefaction and subsidence.

FIGURE 15.9   Judgement Cliff landslide looking east across the Yallahs River.

landslide was a 'classic' rotational landslide with three distinctive benches and scarps. Geomorphic mapping indicates that Judgment Cliff is the northern-most extension of the landslide complex (Figure 15.10). While most of the debris from the older landslides have disappeared due to the fast-flowing Yallahs River, the scarps are still very much evident. The Judgement Cliff landslides formed within white micritic limestone, which is highly lithified and compacted. The presence of extensive faulting at the top of the scarp, the steep slopes and the erosive nature of the Yallahs

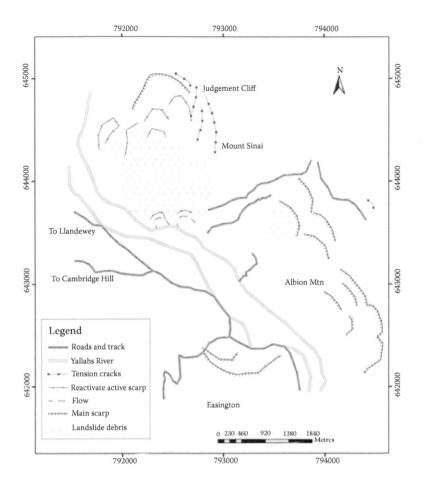

FIGURE 15.10   Geomorphic map of Judgement Cliff.

channel highlight the combination of factors that are important in the generation of landslides in Jamaica. These factors are discussed in more detail in Section 15.5.

The Preston landslides in the rural community of St. Mary are another example of large-scale landslides, but unlike the previous one mentioned, these were slow-moving landslides, with movement occurring over several months. These landslides displaced a significant portion of the village of Preston. The landslide was initially triggered by heavy rainfall, with the most active movement taking place over a 10-month period between 1986 and 1987 (Harris 1991). The Preston landslide occurred in competent limestone in gentle to moderate slopes. Borehole data reveal that the more pervious limestone sits over impervious shale. Between these two lithologies, there is a thin layer of soft clay (Harris 1991) (Figure 15.11). The build up of pore water pressure in the overlying limestone as a result of the rainfall during 1986 and the incompetent clay layer was the primary cause of slope failure in the Preston area.

The landslide resulted in 17 families having to be relocated at a cost of $1.5 million (US$1 = J$5.5 in 1986). As agriculture was the main livelihood of the residents, destruction of agricultural land and farms meant living in the community became unsustainable for most of the residents. This was compounded by the loss of essential services such as water (broken pipes) and electricity supply as a result of the destruction of pylons and poles, as well as the loss of roads leading to the community. Although not all houses in the Preston lands were affected by landslides, the impact on livelihood and essential services and infrastructure had a much more long-lasting effect than just the physical destruction of the built environment (Merriman and Browitt 1993).

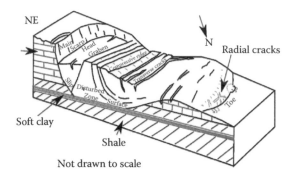

FIGURE 15.11 Diagrammatic representation of Preston Lands landslide.

15.4.1 MILLBANK LANDSLIDE AND SLOPE INSTABILITY IN THE RIO GRANDE VALLEY

The Rio Grande Valley in the parish of Portland is one of the most landslide-prone areas in Jamaica, and it has a long history of landslide events severely affecting communities within the valley (Bhalai 2006). For example, in 2005, following heavy rains associated with Hurricane Dennis, the communities of Berrydale and Millbank were severely affected by landslides. The landslides damaged residential buildings and destroyed infrastructure, such as roads, water mains and critical facilities and services, for example, health centres and schools (PIOJ 2007). The Rio Grande Valley is also the location of several destructive and large deep-seated landslides (Figure 15.12). These landslides have left permanent scars on the Jamaican landscape. These include the Unity landslide (Bhalai 2006),

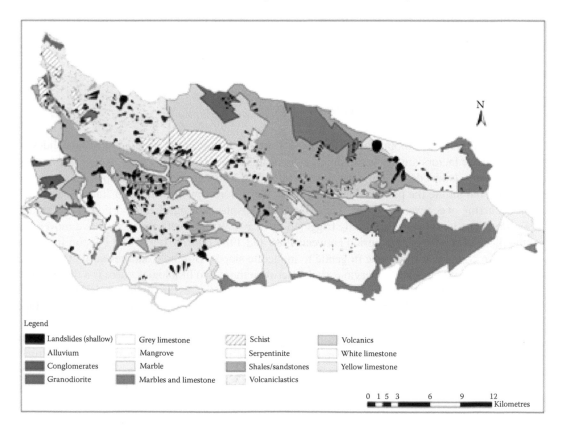

FIGURE 15.12 Geomorphic map of landslides in the Rio Grande Valley. This map highlights the Millbank and Jupiter landslides.

FIGURE 15.13   Reactivate scarp within the relict landslide at Millbank following the rains of 2005.

the Jupiter Fording landslide (Robinson et al. 1997), the Burlington landslides (Bhalai 2010) and the Millbank landslides (Harris and Rammelaere 1985). Of these, the Millbank was the most significant in terms of impact and damage to the built and natural environment (Robinson et al. 1997).

The 1937 Millbank landslides occurred after 2 months of rain, killing eight people and destroying a section of the community of Millbank. The debris formed a landslide dam in the Rio Grande, which when breached caused flooding of the surrounding agricultural lands, washing away of houses and the destruction of two bridges and other infrastructure. The scarp of the landslides is still clearly visible on the landscape and is continuously being reactivated (Figure 15.13), resulting in a continuous threat to the community of Millbank. During the 1937 rainfall event that triggered the Millbank landslides, another 70 slope failures occurred within a 8 km radius which affected communities along the Rio Grande (Daily Gleaner 1937a,b).

The landslides at Millbank are evident by two distinct main scarps (Figure 15.14). According to the Daily Gleaner (1937b), these landslides occurred 5 days apart on 24 and 29 November 1937. The landslides merged in the alley of the Rio Grande, blocking the river and forming a landslide dam. The landslides extend from an elevation of 826 m and continue to the river, for an overall length of 1842 m (Richards 2006). Harris and Rammelaere (1985) gave a possible depth of 18–30 m and calculated a volume of 2×106 m³ for the displaced material of the main landslide. The Millbank landslide is a complex slide with movement along shale and sandstone boundaries, as well as along faulted planes (Figure 15.14).

15.4.2 JUPITER LANDSLIDE

The Jupiter landslide is the largest known active slide on the island, covering an area of 246 ha. This landslide is located between the communities of Ginger House and Comfort Castle, approximately 8 km from the head of the Rio Grande River, 3.5 km in a northwesterly direction from the community of Millbank. The Jupiter landslide is a complex slide whose movement appears to have initiated along faulted boundaries, around the crown and main scarp of the landslide, where highly permeable limestone is in contact with the differentially permeable shale and sandstone sequence of rock. Movement is also evident along the shale–sandstone interface, where most of the movement appears to be taking place.

The crown of the landslide is at an elevation of 660 m, along the Blue Mountain Ridge, and travels a length of 2269 m down to Jupiter Fording, crossing the Rio Grande and across to the Dry River. The landslide material is widest at the base where it crosses the Rio Grande. At this point, it is more than 1541 km wide, with limestone debris extending from Ginger House to Comfort Castle (Figure 15.15). The average depth of the landslide is 46 m, with an estimated 114 million m³ of material displaced (Richards 2006).

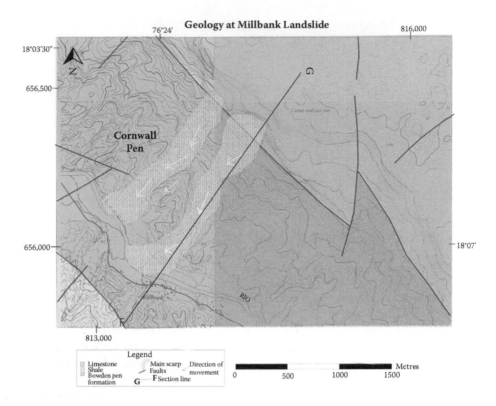

FIGURE 15.14   Millbank landslides in relation to the geological structures and lithology.

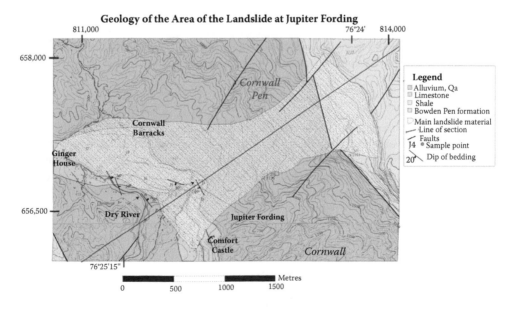

FIGURE 15.15   Geology of the area in and around the Jupiter fording landslide.

Cross Section of Landslide and Surrounding Geology

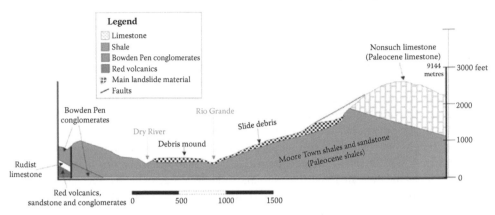

FIGURE 15.16   Cross section of the Jupiter landslide showing associated geology.

The trigger for this landslide has not been determined. As evident from Figures 15.15 and 15.16, the area is highly faulted and seismic data indicate that this area is seismically active. However, it is more likely that the slides were triggered by a combination of factors, including rainfall, river undercutting and earthquakes, a scenario similar to that of the Judgement Cliff landslides.

While the Jupiter landslide's direct impact on the built environment is negligible, as the area around it is undeveloped, the landslide is located 3.5 km downstream from the Millbank landslide mentioned above (Figure 15.14) and has the potential to form a landslide dam (landslide debris extending to the bank of the river opposite the slide), which when breached could cause flood-related damage in communities along the bank of the Rio Grande. The age of the landslides has not been determined, but major movement is believed to have occurred on a number of occasions based on the composition of the landslide debris, which shows evidence of rework and redistribution (Richards 2006).

15.5   CAUSES OF LANDSLIDES: PREPARATORY AND TRIGGERS

On any slope there are two forces acting antagonistically. There is the shear stress, which promotes movement, and the shear strength, which tries to resist movement. Generally, when a slope is stable, the shear strength tends to exceed the shear stress. In slopes that are at the point of movement, the shear stress and the shear strength are approximately equal; in this case, there is no margin of stability (Crozier 1986). Whenever the shear stress exceeds the shear strength, slope movement may occur. This increase in shear stress may be due to internal forces and/or external transient forces, for example, rainfall and/or earthquakes. Slope movement is not always due directly to factors increasing the shear stress. If there is a decrease in shear strength, the shear stress increases automatically, and whenever it exceeds the shear strength, movement occurs. Some of the factors that may contribute to decreased shear strength are weathering, change in water regime and organic decay.

Crozier (1986) argues that there are three factors contributing to slope instability: preparatory, triggering and controlling factors. The preparatory factors generally contribute to making a slope marginally stable. They may best be described as those factors that make a slope susceptible to slope movement, without initiating any form of movement (Crozier 1986). Triggering factors, on the other hand, initiate movement and change the state of the slope from being marginally stable to actively unstable (Crozier 1986). The controlling factors come into play once movement has been initiated. These factors dictate the conditions of movement as it takes place. In essence, they control the rate of movement, the form of movement and the period of time over which the slope movement takes place. This section explores factors that condition slopes for failure, as well as those that trigger landslides and maintain their movement.

15.6 PREPARATORY FACTORS

There are a number of factors commonly cited which are believed to contribute to the distribution of landslides throughout Jamaica (Maharaj 1993; Ahmad 2001; Miller et al. 2009). These include geology, slope gradient, proximity of fault lines to drainage channel, land use and slope aspect. These factors are believed to play a significant role in the distribution of landslides across Jamaica. However, they do not act unilaterally, but normally in conjunction with each other. Not all factors contribute in the same manner, and the degree to which each contributes differs.

Within a geographical information system (GIS), a landslide inventory of all known landslides was spatially analyzed to evaluate their relationship to lithology, proximity to major fault lines, proximity to fluvial channels, land use, slope gradient and slope aspect. All maps were rasterized (pixel of 5*5 m = size of smallest landslide used). The landslide inventory map was reclassified (coded), with 1 = landslide area and 0 = non-landslide area. Factor maps were also recoded (e.g. on the lithological map, alluvium = 1, granodiorite = 2, shale = 5, etc.). One pixel was randomly selected within each landslide, and the equivalent pixel on the lithology, proximity to major fault, proximity to fluvial channel, land use, slope gradient and slope aspect was also selected (Table 15.2). These data were then inputted in a statistical programme SPSS to analyze the combination of factors contributing to slope instability. Bivariate correlation analysis indicated that there is a significant relation between lithology, slope gradient and proximity to fault lines. Backward stepwise analysis and further analysis using multivariate analysis confirmed this, and that slope gradient lithology and land use can be effectively used to predict the location of existing landslides. The three factors that the statistical analysis indicates are most likely to prepare the slope for failure are discussed in more detail below. Statistical analysis was undertaken to analyze the combination of factors contributing to slope instability.

Seven variables were selected to be used in the landslide susceptibility analysis: Geology, distance from fault lines, slope, land use, rainfall, aspect and distance from river-head. There was a need to identify systematically the relationship between the variables before the analysis could proceed and to decide which variable would be useful. First, bivariate correlation analysis indicated that there is a significant relation between lithology, slope gradient and proximity to fault lines. Second, a backward stepwise discriminant analysis was used to determine what the significant variables are and which would contribute least to the model. The F-value for a variable indicates its statistical significance in the discrimination between groups; that is, it is a measure of the extent to

TABLE 15.2
Example of Pixel Sample for Statistical Analysis

Lithology	Proximity of Fault	Rainfall	Slope Gradient	Land Use	Proximity of Fluvial	Slope Aspect	Landslide
6	3	2	4	1	3	4	1
8	2	2	4	2	2	3	1
6	4	2	4	2	1	1	1
8	5	2	5	1	2	2	1
3	1	3	3	1	4	4	1
10	2	3	2	2	3	2	1
8	3	3	5	2	6	5	1
10	3	3	2	2	4	6	1
3	1	3	3	2	2	4	1
10	2	3	3	2	3	2	1
10	1	3	4	2	3	3	1
10	1	3	4	2	4	5	1
3	1	3	5	1	3	3	1

TABLE 15.3

Selection of Variable Based on Discriminant Analysis

Step No.	No. of Variables in Model	Approx. *F*-Value	Variable Removed
0	6	187.8	
1	5	218.3	Aspect
2	3	259.5	Proximity to fluvial channel and slope gradient
Final	3	259.5	

Note: Variable entered were geology, proximity to fault lines, slope gradient, land use, slope aspect and proximity to fluvial channel.

which a variable makes a unique contribution to the prediction of group membership. The smaller the *F*-value of the model, the better it is. The *F*-value was then used to eliminate those variables which were contributing the least to the model. Based on this discriminant analysis, slope aspect was removed first, and then proximity to fluvial channels and slope gradient (Table 15.3). The variables left in the model were geology, proximity to fault line, land use and rainfall.

15.6.1 GEOLOGY

Geology in most instances is the major factor influencing susceptibility to slope failure. The geotechnical properties of a rock, superficial deposit or regolith dictate the susceptibility of that material to slope failure and also the type of mass movement that may result (Yilmaz and Karacan 2002; Miller et al. 2009). The mode of occurrence of slope failure in cohesive rock is different from that in less cohesive deposits and regolith (Cooke and Doorknamp 1974). Jones and Lee (1994) emphasized that as the clay content of rocks increases, they become more prone to sliding. As such, sliding in clayey bedrocks (shale, mudrock and interbedded argillaceous rocks) appears to be more common than in any other rock type. These influencing factors are discussed in more detail in the following sections.

Previous research on landslides in Jamaica has highlighted certain geological sequences as being more susceptible to slope instability than others (Miller et al. 2007; Bhalai 2010). For example, the Wagwater conglomerates and Richmond shale and sandstone sequences (Wagwater Formation) that are widespread throughout the Wagwater Belt in eastern Jamaica, the schist and volcaniclastic rocks of the Cretaceous age inliers and the 'yellow' limestone rocks (of the Eocene–Miocene age) around the flanks of the inliers have a very high density of landslides (Miller et al. 2007; Shalkowski 2009; Bhalai 2010). This relationship is highlighted by Figure 15.5 (national level), as well as mapping and analysis being done at a detailed scale in the eastern section of the island (Figures 15.17 and 15.18); there is a much higher density of landslides on the lithologies highlighted above. However, one feature that is evident from field mapping is that although the density of landslides in the white limestones (biomicrites and dolomitic limestones) is not very high compared with the more deeply weathered volcaniclastics and shale–sandstone conglomerate, the landslides that occur within this lithology are much larger and deep seated (+5 m). Examples of this are the Jupiter and Judgement Cliff landslides discussed in Section 15.3. During periods of heavy rainfall, the debris particularly within the main body and 'toes' gets reactivated to form debris flows. In general terms, these large landslides in the white limestones are relatively stable but could easily become reactivated as a result of changing land-use patterns (oversteepening during road cuts) and/or during an earthquake, as happened during the earthquake of 1993 (Ahmad and Robinson 1994).

Some of the most significant landslides affecting infrastructure are located in the Wagwater conglomerate and the Richmond shale and sandstone sequences that collectively form the Wagwater Formation (Figure 15.5). Their impacts are particularly evident, for example, along the Junction

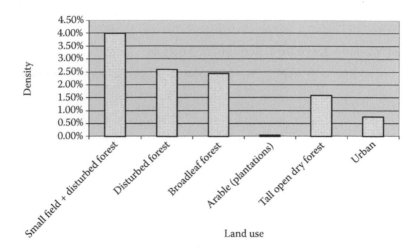

FIGURE 15.17   Landslide inventory and lithology map for St. Thomas.

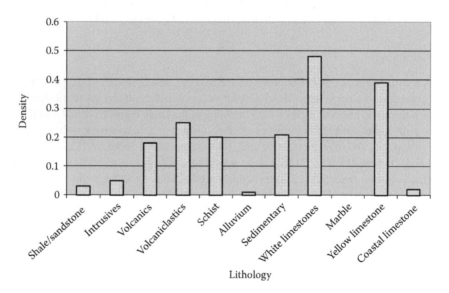

FIGURE 15.18   Landslide density within lithology classes.

Main road in St. Mary parish and Gordon Town road to Mavis Bank in St. Andrew. Rock within the Wagwater Formation is of a terrigenous origin, composed mainly of eroded and redeposited andesitic lava flows and localized beds of limestone. These reddish-purple coloured rocks are deeply weathered and generally exist in faulted contact with the Richmond (shale and sandstone) Formation (Harris 2002).

15.6.2   ROLE OF PORE WATER PRESSURE WITHIN GEOLOGICAL SEQUENCES

One of the major reasons for slope failure is the build up of pore water pressure. Pore water pressure may build up when water enters a permeable rock or soil mass, and as a result of an underlying impervious rock, it cannot escape. This leads to a decrease in the shear strength of the slope material and an increase in the shear stress acting on it, which may lead to failure (Terzaghi 1950; Hillel 1998). The stratigraphy of a slope may therefore act as a pre-condition for rock failure. Landslides are common in sequence of sandstone and shale (Figure 15.19) where differential pore

FIGURE 15.19   Slope failure of Richmond Formation, sequence of shale and sandstone (Stanton Harcourt Road, Portland).

water pressure is more likely to develop. In a simplistic model, water enters the more permeable sandy layers. However, the impervious shale layers prevent the water from readily passing through, resulting in the build up of pore water pressure that may contribute to slope failure. Where the upper section of the sequences is fractured due to weathering and some cases faulting, water is likely to accumulate in the upper fractures' layer of sandstone and shale, but may encounter a thick layer of impervious shale that may result in the build up of pore water pressure and subsequent slope failure. This problem is compounded if the upper permeable layer is deeply weathered and composed of clays susceptible to slope failure.

Impervious and permeable sequences are not exclusive to shale and sandstone sequences, but may also exist with a combination of rocks and regolith, where the rock, for example, the volcanic, conglomerated and volcanics, is deeply weathered and/or fractured. The regolith is permeable to water, whereas the underlying 'solid' and unweathered rock is impervious to water. It is therefore possible that impervious rock underlying a thick layer of regolith may form some sort of confined aquifer or perched water table. Such a relationship will increase the pore water pressure, which may lead to slope failure.

15.6.3   ROLE OF ROCK WEATHERING

Weathering of the rocks forms a key factor in making them susceptible to slope failure. It is evident from field investigation that deeply weathered rock and a thick layer of regoliths on moderate to steep slopes are primary contributors to slope instability. While some lithologies are relatively stable on steep slopes when 'fresh', weathering has the potential to significantly change their physical strength, through, for example, chemical alteration of their composition to form clays. Research has shown that changes in the shear strength of slopes can be directly linked to the weathering of rocks to form clay minerals (Wen and Chen 2007). Where clays are major components of slope material, changes in shear strength occur when clays become saturated with water as it dissociates and behaves in a plastic manner (Velde 1995). The level of dissociation is dependent on the type of clay. Clay minerals generally have a strong interaction with water due to their high specific surface area, their platy structure mineral and the polarity of the water molecule. The weak ionic bond between phyllosilicate layers in the structure of the clay has the potential to allow bipolar water molecules to dissolve them, resulting in dissociation and loss of share strength of the material (Yilmaz and Karacan 2002).

Failed surface material taken from several landslides within the slope instability-prone region of eastern Jamaica was analyzed to determine the clay composition using x-ray diffraction (XRD) and optical microscopy. The analysis indicated that predominant clays were a mixture of primary (in situ) and secondary (residual) clays, in addition to mixed-layer clays, such as chlorite-vermiculite and illite-smectite (Figures 15.20 and 15.21).

The XRD shows a distinctive smectite peak of about 6 nm that shifts to approximately 7 nm when heat treated (Figure 15.20). Distinctive peaks are also evident for chlorite and unweathered plagioclase and quartz. The steplike changes observed from glycerol treatment prove the existence of mixed interlayer minerals within these samples (Forlati et al. 1996), suggesting that the presence of smectite is a decisive factor in paving the way to sliding phenomena, which are then triggered by a sudden increase in rainfall water and the ensuing hydration processes. Gillot (1986) determined that material rich in smectite content has increased plasticity, compressibility and swelling pressure, which may contribute to slope failure. In the case of smectites, the specific surface area is approximately 760 m^2/g. Thus, smectite-rich clays softened by increased water content can exhibit the properties of a lubricant and can be critical for slope stability. This assertion has also been made by Yilmaz and Karacan (2002), who showed that smectite-rich soils drastically change their geotechnical characteristics when saturated with water. For example, clay soils with 25% smectite have an average cohesion of 160 kPa and an average internal friction angle of 20° when dry, but these change to around 37 kPa and an internal angle of 9°.

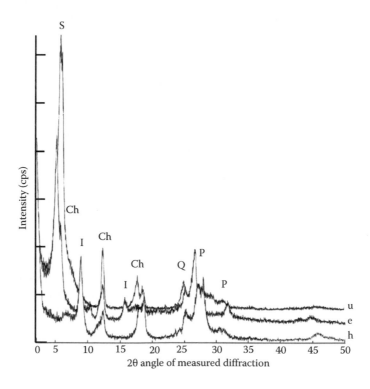

FIGURE 15.20   XRD patterns of failed material from Portland. The diagram shows the changes that occur when the samples are ethylene glycol (e) and heat (h) treated. u, untreated samples; Ch, chlorite; I, illite; S, smectite; P, plagioclase; Q, quartz.

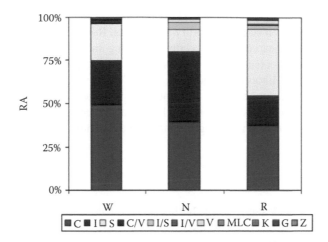

FIGURE 15.21 Relative abundance of clay minerals per lithological unit. W, Wagwater Formation; N, Newcastle Volcanics; R, Richmond Formation; C, chlorite; I, illite; S, smectite; C/V, chlorite-vermiculite; I/S, illite-smectite; I/V, illite-vermiculite; V, vermiculite; MLC, mixed-layer clay; K, kaolinite; G, gibbsite; Z, zeolite.

The weathering of parent rock to form clays like smectite and illite in Jamaica's landslide-prone area plays a significant role in slope stability. Rainfall has the potential to induce significant changes in the mechanics and behaviour of clay-dominant slope material. As such, it is of critical importance to understand the controlling role that, for example, smectite-rich layers can represent in the generation of slope failure in areas with slopes with gradients as low as 9°.

15.6.4 PROXIMITY TO FAULTS

Major landslides, for example, Jupiter, Judgement Cliff and Millbank, as discussed above (see Section 15.3), occurred in areas with mapped faults. The relationship between landslide occurrences and faults was undertaken by determining the quantities and density of landslides with classes of 'proximity to fault lines'. The results from the analysis indicate that there is an association between proximity to fault and the occurrence of slope failure. Landslides were more likely to occur in close proximity to fault lines than farther away (Figure 15.22). This relationship may be the result of a combination of factors. First, Jamaica is a seismically active region, so movement along these

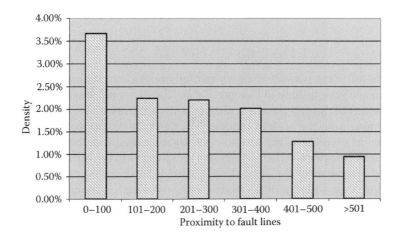

FIGURE 15.22 Comparison of landslides with proximity of faults.

faults has the potential to trigger landslides. However, it is possible that fault lines and zones have a more long-term influence on slope instability in the area. Rocks in close proximity to fault lines and within fault zones are sheared and brecciated. Shearing and brecciation create planes of weakness, and as shear strength decreases, less shear stress is required before failure occurs. Within the fault zone, there is more likelihood that the movement of water will contribute to the weathering of rocks and formation of clay minerals that contribute to slope movement, as seen in Section 15.6.3. The percolation of water along fissures in the fault zone also increases the pore pressure within the rock, which also decreases the shear strength, rendering them more susceptible to slope failure (see Section 15.6.2).

15.6.5 LAND USE

Over the last 50 years, there has been a marked increase in the number of recorded landslides (Miller et al. 2007). This may be due to increased awareness and the recording system. However, evidence does point to changing land-use patterns having a very strong influence on their occurrences. Land-use changes for example; road construction that oversteepens slopes or traverses existing relict landslides, commercial agricultural farms (e.g., coffee and forestry), and quarrying activities that destabilize these slopes making them prone to landslides. The growing population and resulting urbanization of marginal land use means increased exposure to landslide hazards by the population. Within the Jamaican setting, the pressure for houses also means that a vulnerable population, usually those that are below the poverty line, with large families, very limited education and limited resources to construct sound buildings, has occupied hillsides that are already marginally stable. The deforestation, oversteepening and increased surface rainfall have also significantly increased the frequency and magnitude of slope instability events to the economic and social detriment of these communities.

The results obtained from the comparison of landslide occurrence with land-use type indicate that most of the landslides occur within broadleaf forests which were cleared for farming and other purposes. Landslide occurrence also correlates strongly with the location of small fields (Figure 15.22). There is generally a low correlation of landslide occurrences and urban area. This is not surprising, as large urban centres are mostly located on the low-lying alluvium plains, which generally are not susceptible to land slippage. The small villages located in the hilly terrain that is more prone to land slippage were not included within the land-use classification, but rather encompass the 'disturbed broadleaf forest' category. In the future, classification may be worthwhile to disaggregate the small urban areas as a class by themselves in order to assess their impact. The low correlation with the plantation is a direct result of the plantation being located at low elevation, on flat alluvium plains.

15.7 CAUSES OF SLOPE INSTABILITY TRIGGERING

Landslide events following periods of heavy rainfall are evident on an annual basis, particularly during the hurricane season, June to October. Table 15.4 shows the long-term mean rainfall since 1951 for the island. The rainfall pattern is distinct with two defined wet periods, April to June and August to November, with the peak in October. The first period, with peak rainfall in May, is due to an annual tropical depression. The second is mostly associated with the North Atlantic hurricane season. By and large, the North Atlantic hurricane season tends to bring the most rainfall, triggering widespread rainfall throughout the island. During a single hurricane event, these storms are capable of 'dumping' more than 400 mm in landslide-prone areas like St. Thomas in 3 days (Table 15.5), causing widespread damage and indicating that most slope movements occur during the second half of the North Atlantic hurricane season, but more importantly during the month of October. Although rainfall is the predominant trigger, the earliest recorded events, for example, the submarine landslides that contributed to the destruction of Port Royal and the Bog Walk Gorge landslide

TABLE 15.4

Long-Term Rainfall Averages (mm) Since 1951 for Parishes in Jamaica

Administrative Areas	Monthly Rainfall (mm)												Total
	Jan	Feb	Mar	Apr	May	Jun	Jul	Aug	Sep	Oct	Nov	Dec	
Clarendon	54	39	58	83	181	149	87	119	180	257	104	67	**1378**
St. Catherine	53	50	57	91	171	139	108	138	174	238	121	88	**1428**
Kingston/St. Andrew	53	49	56	103	180	123	50	168	215	287	187	112	**1583**
St. Ann	145	90	78	117	164	115	50	97	130	177	214	219	**1596**
Trelawny	99	76	69	115	181	130	96	154	166	222	167	131	**1606**
Manchester	60	52	85	134	237	175	102	169	213	291	118	70	**1706**
St. James	91	77	62	111	223	203	145	182	202	253	138	104	**1791**
St. Elizabeth	69	60	84	182	243	163	145	204	211	273	133	71	**1838**
St. Mary	181	129	106	148	175	122	81	116	110	209	263	268	**1908**
Westmoreland	64	70	91	164	302	262	261	275	245	290	122	70	**2216**
St. Thomas	121	91	65	120	251	219	150	213	281	368	232	177	**2288**
Hanover	88	91	87	146	294	309	237	275	264	291	133	87	**2302**
Portland	321	236	185	273	321	278	231	245	273	373	477	457	**3670**
Total	108	85	83	137	225	184	134	181	205	271	185	148	1946

TABLE 15.5

Typical Precipitation Amount during a Hurricane (Hurricane Wilma 2005) for the Parish of St. Thomas

Station	October 2005			Total Rainfall (mm)
	15th	16th	17th	
Cedar Valley	137.1	175.0	110.0	422.1
Serge Island	50.0	100.0	100.0	250.0
Sunning Hill	124.8	57.6	81.6	264
Batchelor's Hall	76.2	168.0		244.2

following the 1692 earthquake, are evidence of the impact of seismic activity on the distribution of earthquakes throughout the island. One notable trend, as highlighted above, is that landslide events associated with rainfall are more widespread and common, whereas those triggered by earthquakes are less rare but occur on a larger scale and are more catastrophic. In some instances, for example, in the case of the Judgement Cliff landslide, it may be a combination of both rainfall and seismic events. The following section explores some of the significant slope instability events and landslides themselves triggered by rainfall and seismic activity.

15.7.1 Rainfall Triggering

Meteorological events can bring prolonged and/or intense periods of rainfall to the island. These events include tropical depressions, cold fronts, upper-level jets, cold fronts interacting with a low- to midlevel trough, prefrontal troughs and hurricanes. Since 1980, there have been more than 30 such events that triggered major slope instability events (Table 15.6). As evident from Table 15.6, most of these slope instability events impact the eastern section of the country (St. Thomas, Portland, St. Mary and Kingston) but have also affected south and central parishes (parishes of St. Catherine and Manchester).

TABLE 15.6
Significant Landslides for the Period 1980–2012

Year	Trigger Event	Location
1985 November 19	Hurricane Kate	Portland, KSA
1986 June 11–17	No record	Eastern Jamaica
1988 September 10–12	Hurricane Gilbert	Portland, St. Mary, KSA
1991 November 9–13	Tropical depression	Portland, St. Mary
1993 January 25–26	Cold front, low-pressure area, upper-level jet	Portland, St. Mary
1993 May 21–23	Low-pressure system	Portland, KSA, St. Mary
1993 April 21–30	Low-pressure system	Portland
1994 January 22–25	Surface trough	KSA, Portland
1995 May 1–3	Tropical wave	KSA
1995 December 12	Cold front interacting with a low- to midlevel trough	Portland, St. Mary
1995 December 29–30	Frontal trough	Portland
1997 June 4–5	Surface trough, upper-level divergent flow	Portland
1998 September 15–October 1	Hurricane Georges	St. Andrew
1998 October 24–26	Hurricane Mitch	St. Andrew
1998 November 4	Cold front	Eastern Jamaica
1998 December 18	Prefrontal trough	KSA, St. Mary, western Portland
1999 March 15–19	Cold front	St. Mary
1999 October 25	Low-level trough	Portland
1999 December 30–31	Surface trough	Portland
2001 January 1–3	Stationary front	Eastern Portland
2001 February 19–20	No record	Portland
2001 October 21–November 5	Hurricane Michelle	KSA, Portland, St. Mary
2001 November 19	No record	Portland
2003 May 24–26	Surface trough	Portland
2003 December 18	Frontal system	Portland
2004 September	Hurricane Ivan	Predominantly eastern Jamaica
2005	Hurricanes Dennis and Emily	Predominantly eastern Jamaica
2012	Hurricane Sandy	Widespread damage in St. Thomas parish

Note: KSA, Kingston and St. Andrew.

Most of the landslides outlined in Section 15.3 are large landslides with distinctive scars on the landscape. However, the more frequent landslides, mostly debris flow and translation slides associated with seasonal rainfall, can prove to be more disruptive, devastating and costly. Except for 1993, when a 5.3 earthquake triggered landslides around eastern Jamaica, all the major slope movement events in the last 30 years are associated with rainfall.

Following the rainfall associated with Hurricane Gilbert in 1998 (Manning et al. 1992), 478 landslides were recorded, mostly debris flows and translational landslides. The debris flows were particularly extensive, with some having tracks extending in length from just over 50 m to up to 214 m. While the cost of the damage was not quantified, the landslide caused extensive damage to roads, buildings and agriculture (Manning et al. 1992).

In January 1998, more than 400 mm of rain fell over a 2-day period in the Fellowship area of Portland, triggering mud and debris flows. Approximately 4000 m² of debris buried a shop where

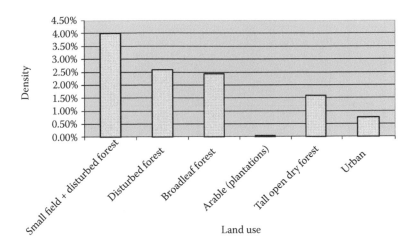

FIGURE 15.23   Comparison of landslide scarp within land-use classes.

13 people were taking shelter from the rain. The landslide resulted in the loss of four lives and injuries to eight others who had to be excavated from the debris (Harris 2002). Throughout the Rio Grande Valley area (Figure 15.23), main roads were blocked, often collapsed due to undermining by landslides. Water pipelines, telephone lines and electricity transmission lines were damaged or destroyed. Agricultural farms and orchards were seriously affected. In total, 270 landslides were mapped, which damaged 35 houses and destroyed 25 of them. A total 1270 m of road suffered some degree of damage, with 806 m of that total being destroyed. It was estimated that the landslide causes J\$231 million in damage (1998 figures) (Harris 2002).

Between 20 May and 12 June 2002, the eastern and central section of the island received significant rainfall that triggered widespread landslides. In the Manchester Highlands over the 10-day period from May 22 to 31, more than 1500 mm of rain fell. In St. Thomas, the Cedar Valley rainfall station recorded 1182 mm. This was 4.4 times the 30-year monthly mean. In St. Catherine, the Innswood station recorded 788 mm. This was 6.3 times the 30-year monthly mean. In Clarendon, the Osbourne Store station recorded 1058 mm. This was 9.8 times the 30-year monthly mean. The 24-hour rainfall of 246 mm recorded on 22 May was 2.3 times the 108 mm normally received in 1 month (ODPEM/CDB 2003).

Landslides blocked roads in 523 recorded locations and damaged roads in 18 locations. Throughout the island, some 18 communities were affected due to the damage to transport infrastructure. The blocked roads left hundreds of people stranded and/or isolated. In extreme cases, emergency services were severely affected, for example, preventing the injured from being transported to the hospital. In St. Thomas, five houses, as well as water mains in five locations were severely damaged (ReliefWeb 2002; ODPEM/CDB 2003). In total, 523 landslides were recorded; however, it should be noted that mapping took place along main roads and where access could be gained. Numerous landslides in inaccessible areas may have gone reported. Most of the landslides (375) mapped were in the Trout Hall to Frankfield in Clarendon and the Bath and Hayfield (144) areas of St. Thomas.

15.7.2   Rainfall Intensity and Duration

Analysis of these landslide events since 1980 in relationship to rainfall duration and intensity demonstrates that landslides may be triggered by short high-intensity rainfall events. However, the majority of landslides are triggered by low- to midintensity rainfall events which occur continuously over longer (+24-hour) time periods (Figure 15.24). It may likely be the case that during short

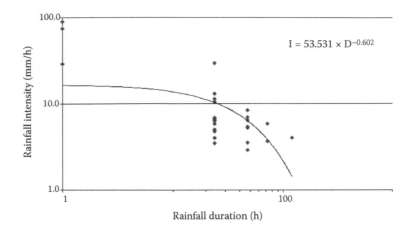

FIGURE 15.24 Rainfall intensity and duration in relation to landslide-triggered events between 1950 and 2010.

intense periods, surface runoff and erosion may be dominant and not mass movement. Moreover, even for a low-intensity event, if it occurs over a long time period, oversaturation and build up of pore water pressure are likely to trigger landslides (Miller et al. 2009).

15.7.3 TRIGGERING BY EARTHQUAKES

Compared with rainfall, there is less recorded evidence that landslides are triggered by earthquakes. The exact quantity and distribution of landslides related to the 1962 and 1907 earthquakes (destroyed Kingston) has not being verified. However, historical records mention widespread occurrences of slope failures and landslide-related damage in Kingston and the parish of St. Thomas related to these two seismic events. Ahmad (1995a) believes that historical records and scientific reports between 1667 and 1993 reveal approximately 13 earthquake-triggered landslide-forming events in Jamaica. In more recent times, following the 1993 earthquake (5.3 on the Richter scale), 40 landslides were triggered throughout eastern Jamaica. Most of these were small debris flows and rockfall, but they were still able to cause damage to buildings and agricultural fields (Eyre 1989).

Large earthquakes generate landslides that affect a significant area even away from the epicentre (Keefer 1984; Crosta 2004). As indicated above, earthquakes trigger not only surface slides but also submarine landslides that may also generate tsunamis. The earthquake events of 1692 and 1907 are known to have generated tsunamis triggered by landslides, with wave heights reaching as high as 2 m in Kingston and across the eastern parishes. Chirp sonar profiles in the Kingston Harbour have shown that steep submarine slopes still exist in the harbour, with slopes sitting at precariously high angles above the angle of repose. Some profiles show slopes reaching angles as high as 35°, some of which are within the very busy shipping channel (Kingston Harbour is the seventh largest natural harbor in the world). Sonar mapping in the harbour was done to identify the major historic earthquakes' epicentre, correlating escarpment and recent earthquake epicentres as the locations of the faults responsible for generating earthquakes that are poorly known. Preliminary results of the Chirp study reveal evidence of at least one potentially active fault beneath Kingston Harbour, showing patterns of a submarine landslide (Hornbach et al. 2011) (Figure 15.25). According to Hornbach et al. (2011), the slide is directly adjacent to an area in Kingston Harbour that reported a local tsunami during the 1907 earthquake, and may be a result of dislocation of steep submarine slopes, resulting in tsunamis for those events. The Chirp mapping also indicates that steep failure-prone slopes exist along the edge of the Palisadoes in the vicinity of Port Royal, the city destroyed by a tsunami and earthquake in 1692.

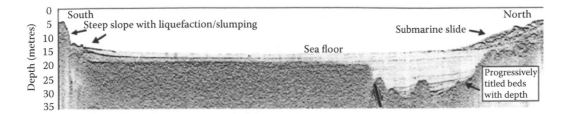

FIGURE 15.25 A Chirp profile that was collected at the eastern end of the harbour shows tilted beds at the Palisadoes to the south and evidence of recent slumping along the north shore. Steep slopes along the Palisadoes highlight the potential for liquefaction near the southern drop-off that pervades across the south side of the harbour. (From Hornbach et al. 2011.)

15.8 RESPONSE, MANAGEMENT STRATEGIES AND CRITIQUE

The damage caused by landslides is of economic, societal and environmental concern (Ahmad et al. 1993; Ahmad 1995b; Mines and Geology Division 1998). Although it is difficult to prevent most of these landslides from occurring, we can certainly mitigate their impact. Effective management requires a more holistic approach than that currently being employed. In order to do this, there first needs to be an inventory of existing landslides or the development of a model indicating where landslides are likely to occur. In so doing, and by undertaking an all-island risk assessment, it would be possible to improve land zonation policies and raise awareness of various stakeholders (developers, homeowners, planners and engineers), as well as the general public, and inform slope instability management strategies.

Until 1980, there was no particular organization responsible for coordinating the management of natural hazards in Jamaica. An active managerial approach to natural hazards on the island started in the 1980s with the establishment of the Office of Disaster Preparedness and Emergency Relief Coordination (ODIPREC), as a response to the June 1979 rains and Hurricane Allen (1980). This agency was later institutionalized by the Disaster Preparedness and Emergency Act (1993). This act *inter alia* resulted in a name change from ODIPREC to ODPEM, and it was later revised in the early 2000s.

Until the early 2000s, the landslide management strategy was mostly responsive rather than being proactive. This problem has been recognized by local geoscientists, mainly from the University of the West Indies and the Mines and Geology Division, government of Jamaica, and ODPEM. These organizations are trying to be more proactive by undertaking research into landslide mapping and risk evaluation. However, while there is a more coordinated strategy to slope investigation in Jamaica, there is still a tremendous amount of work that needs to be undertaken, particularly in terms of the engineering approach in response to slope failure.

The engineering approach is normally by use of retaining walls and gabion baskets (Figure 15.26). While this method may be useful for small landslips, it is generally not suitable for most of the landslides that occur on the island, and in most instances, it has not been tremendously effective (Figure 15.27). These measures are normally undertaken with little geotechnical investigation to determine the mechanics, dimensions and factors contributing to movement; the walls are normally inadequate to stabilize the landslides. In most cases, the landslides are deep seated and extend beyond the dimension of the obvious (scarp) landslip. The results show that a high percentage of these fail and collapse (Figure 15.27), resulting in significant damage to the buildings, land and infrastructure that they were there to protect.

Increasingly, the government agencies responsible for landslide management have realized that more research in landslide risk assessment is needed to complement the structural response. Increasingly more is being invested in landslide hazard mapping, assessment and risk modelling. There has also been a mean increase in the trend to more non-structural mitigative strategies, including community loss reduction and capacity building projects, encouragement of sustainable land management practices and training for stakeholders (e.g. farmers, engineers, planners

FIGURE 15.26   Retaining wall and Gabion basket built in response to slope failure.

FIGURE 15.27   Retaining wall collapses.

and developers). For example, there are projects that aid farmers in developing effective hillside farming practices that reduce slope instability and train local masons to build gabion baskets, the development and execution of community outreach programmes designed to teach local residents about hazard recognition and hazard mapping, and technical workshops for engineers, technocrats, management officials and informal construction personnel. Landslide risk reduction control mechanisms and all disaster risk reduction operations within the island are summarized in Table 15.7.

TABLE 15.7

Summary of Landslide Management Strategy in Jamaica (as of 2012)

	Loss Reduction	Policy Planning	Sustainable Development and Capacity Building	Training
Responsible agencies	National Works Agency, National Environmental Planning Agency, Town Planning Authority, Ministry of Local Development, Ministry of Agriculture	ODPEM, Mines and Geology Division, Ministry of Agriculture and Fisheries, Ministry of Housing and Water, Ministry of Science and the Environment, Ministry of Local Development, Town Planning Authority, National Works Agency	UWI, Mines and Geology Division, Ministry of Agriculture and Fisheries, National Works Agency, Ministry of Local Development, Ministry of Housing and Water, Ministry of Science and Environment, Water Resources Authority, National Environmental Planning Agency, Town Planning Authority	UWI; secondary, tertiary and vocational institutions; all relevant agencies and government ministries
Works completed	Retaining walls and the widening of river valleys, e.g. near the mouth of the Yallahs River	Development of a national hazard mitigation policy	Landslide inventory; preliminary landslide high-risk maps; vulnerability maps for the parishes of Portland, St. Catherine, St. Mary, St. Thomas, Kingston and St. Andrew	Summer courses by the Unit for Disaster Studies
Current status	Design of a hillside development manual	Development of a national hazard mitigation strategy, execution of the MOSSAIC model within a few communities by ODPEM	Ongoing	The UDS has ceased its summer training course at this time; the Environmental Management Unit of the UWI now offers a master's programme in disaster risk reduction

Note: UWI, University of the West Indies.

Submarine landslides pose a significant risk, as they can generate a near-shore tsunami, with little time to respond. There is an urgent need for a greater scientific understanding of the nature and distribution of these landslides. Work similar to that of Hornbach et al. (2011) using Chirp sonar mapping needs to be urgently done along the continental shelf and harbours particularly close to coastal urban settlements. This sonar mapping should be complemented by geotechnical analysis, LIDAR and earthquake ground acceleration modelling, which will help to address the concern of liquefaction and tsunami occurrence due to submarine slope failure.

15.9 CONCLUSION

Landslides have continued to cause significant damage to the built and natural environment, and loss of lives and livelihood. In the long term, they pose a threat to the socioeconomic development of the island of Jamaica. These landslides are predominantly triggered by rainfall and earthquakes. Because Jamaica is a seismically active country and experiences intense and high-frequency rainfall associated with hurricanes and tropical depressions, triggering of mass movements in the deeply weathered material on steep slopes will continue if effective management strategies are not more widely implemented. These strategies have to be multidisciplinary and adapt a multiagency approach. The continuous predominant use of retaining walls and gabion baskets to respond to failed surfaces needs to be re-examined, and more novel and effective techniques (soil nails, rock bolts, erosional controls and biotechnical slope protection) explored to assess their effectiveness in this setting. There is also more need for research in understanding both landslide processes on the island and the socioeconomic impact of landslides to better inform policy development. If a more holistic approach to slope instability is adopted, this could result in a significant reduction in the societal and economic losses the island experiences as a result of landslides.

REFERENCES

Ahmad, R. (1995a). Earthquake-induced landslides in the Jamaican Neogene plate boundary zone. Presented at the 3rd Geological Conference of the Geological Society of Trinidad and Tobago—1995.

Ahmad, R. (1995b). Landslides in Jamaica: Extent, significance and geological zonation. In D. Barker and D. M. McGregor, eds., *Environment and Development in Small Island States: The Caribbean*. Kingston: The Press, University of the West Indies, pp. 147–169.

Ahmad, R. (2001). Damaging landslides related to the intense rainstorm of October 27–November 2, 2001, Portland, Jamaica. In *8th December Field Guide*. Kingston: Jamaican Geographical Society, pp. 1–10.

Ahmad, R., Earle, A., Hughes, P., Maharaj, R., and Robinson, E. (1993). Landslide damage to the Boar River water supply pipeline, Bromley Hill, Jamaica: Case study of a landslide caused by Hurricane Gilbert (1988). *Bulletin of the International Association of Engineering Geology*, 47(1), 59–70.

Ahmad, R., and Robinson, E. (1994). Geological evolution of the Liguanea Plain—The landslide connection. In M. Greenfield and R. Robinson, eds., *Proceedings of the First Conference, Faculty of Natural Sciences*, Mona, Jamaica, pp. 22–23.

Barker, D., and McGregor, D. F. M., eds. (1995). *Environment and Development in the Caribbean: Geographical Perspectives*. Kingston: The Press, University of the West Indies.

Bhalai, S. (2006). Landslide damage assessment, Rio Grande Valley, Portland. Kingston: Mines and Geology Division, Ministry of Land and Environment.

Bhalai, S. (2010). Landslide susceptibility of Portland, Jamaica: Assessment and zonation. *Caribbean Journal of Earth Science*, 41, 39–54.

Cooke, R. U., and Doorknamp, J. C. (1974). *Geomorphology in Environmental Management: An Introduction*. Oxford: Clarendon Press.

Crosta, G. B. (2004). Introduction to the special issue on rainfall triggered landslides and debris flows. *Engineering Geology*, 73, 191–192.

Crozier, M. (1986). *Landslides: Causes, Consequences and Environment*. London: Croom Helm.

Daily Gleaner. (1937a). Huge landslide raises Portland's death toll. Gleaner Company of Jamaica.

Daily Gleaner. (1937b). Mighty landslide at Millbank kills children and destroys homes. Gleaner Company of Jamaica.

ECLAC (Economic Commission for Latin America and the Caribbean). (2001). Jamaica: Assessment of the damage caused by flood rains and landslides in association with Hurricane Michelle, October 2001—Implications for economic, social and environmental development. General LC/CAR/G.672. ECLAC Subregional Headquarters for the Caribbean, Caribbean Development and Cooperation Committee, 7 December.

EM-DAT (Emergency Events Database). (2012). The international disaster database—Jamaica: Country profile natural disaster. http://www.emdat.be/search/node/jamaica.

Eyre, L. A. (1989). Hurricane Gilbert: Caribbean record-breaker. *Weather*, March.

Forlati, F., Lancellotta, R., and Osella, A. (1996). The role of swelling marl in planar in the Langhe region. In *Proceedings of the 7th International Symposium on Landslides*, Trondheim, Norway, vol. I, pp. 721–725.

Gillot, E. J. (1986). Some clay-related problems in engineering geology in North America. Clay Minerals, 21, 261–278.

Harris, N. (1991). Landslides in Jamaica: A case study of the Preston Lands landslide, St. Mary. Kingston: Mines and Geology Division.

Harris, N. (2002). Rain-induced slope failures and damage assessment in Portland, Jamaica, related to the flood event of 3–4 January 1998. In T. Jacskon, ed., *Caribbean Geology: Into the Third Millennium*. Kingston: University Press, pp. 235–255.

Harris, N., and Rammelaere, M. (1985). Landslide disaster—Millbank and Chelsea areas, Portland. Kingston: Mines and Geology Division.

Hillel, D. (1998). *Environmental Soil Physics*. San Diego: Academic Press.

Hornbach, M., Mann, P., Frohlich, C., Ellins, K., and Brown, L. (2011). Assessing geohazards near Kingston, Jamaica: Initial results from Chirp profiling: Special section: Geoscientist without borders. *Leading Edge*, pp. 410–413.

Jackson, R. (2005). Managing natural hazards in Jamaica. Office of Disaster Preparedness and Emergency Management. Kingston: ODPEM. https://www.mona.uwi.edu/cardin/virtual_library/docs/1091/1091.pdf.

Jamaica Gleaner. (1998a). Experts target landslides. Gleaner Company of Jamaica.

Jamaica Gleaner. (1998b). The island takes a beating. Gleaner Company of Jamaica.

Jamaica Observer. (2010). Four more bodies found. *Jamaica Observer*, http://www.jamaicaobserver.com/news/Four-more-bodies-found_8015958.

Jones, D. K., and Lee, E. M. (1994). *Landsliding in Great Britain*. London: Stationery Office Books (TSO).

Keefer, D. (1984). Landslides caused by earthquakes. *Geological Society of America Bulletin*, 95, 406–421.

Lander, J. F., Whiteside, L. S., and Lockridge, P. A. (2002). A brief history of tsunamis in the Caribbean. *Science of Tsunami Hazards*, 20(2), 57–94.

Maharaj, R. (1993). Landslide processes and landslide susceptibility analysis from an upland watershed: A case study from St. Andre, Jamaica, West Indies. *Engineering Geology*, 34, 53–79.

Manning, P., McCain, T., and Ahmad, R. (1992). Landslides triggered by 1988 Hurricane Gilbert along roads in the Above Rocks area, Jamaica. *Journal of the Geological Society of Jamaica*, 12, 34–53.

Merriman, A., and Browitt, P. (1993). *Natural Disasters: Protecting Vulnerable Communities: Proceedings of the Conference Held in London, 13–15 October 1993*. London: T. Telford.

Miller, S., Brewer, T., and Harris, N. (2009). Rainfall thresholding and susceptibility assessment of rainfall-induced landslides: Application to landslide management in St Thomas, Jamaica. *Bulletin of Engineering Geology and the Environment*, 68(4), 539–550.

Miller, S., Harris, N., Williams, L., and Bhalai, S. (2007). Landslide susceptibility assessment for St. Thomas, Jamaica, using geographical information system and remote sensing methods. In R. Teeuw, ed., *Remote Sensing Hazardous Terrain*. Geological Society Special Publication 283. London: Geological Society, 77–91. doi: 10.1144/SP283.7.

Mines and Geology Division. (1998). Preliminary landslide damage assessment for Portland, Jamaica; January 3–4, 1998. Kingston: Mines and Geology Division.

ODPEM (Office of Disaster Preparedness and Emergency Management). (2002). Draft national hazard mitigation policy/programme. Kingston: ODPEM.

ODPEM (Office of Disaster Preparedness and Emergency Management). (2012). Landslides. Kingston: ODPEM. http://www.odpem.org.jm/DisastersDoHappen/TypesofHazardsDisasters/Landslides/tabid/253/Default.aspx.

ODPEM/CDB (Office of Disaster Preparedness and Emergency Management/Caribbean Development Bank). (2003). Assessment of 2003 May–June rainfall event. Internal report. Camp Road, Jamaica: ODPEM.

PIOJ (Planning Institute of Jamaica). (2007). The poverty-environment nexus: Establishing an approach for determining special development areas in Jamaica. Prepared by Sustainable Development and Regional Planning Division, PIOJ.

ReliefWeb. (2002). OAS responds to Jamaica's flood emergency. https://reliefweb.int/report/jamaica/oas-responds-jamaicas-flood-emergency.

Richards, D. V. (2006). Landslide dams along the Rio Grande in Portland, Jamaica. Unpublished MSc thesis, University of the West Indies.

Robinson, E., Laughton, D. V., and Ahmad, R. (1997). Burlington and Jupiter landslides, Rio Grande Valley: Comparison with Millbank landslide as examples of processes producing natural dams. In R. Ahmad, ed., *Natural Hazards and Hazard Management in the Greater Caribbean and Latin America: Proceedings of the Second Caribbean Conference on Natural Hazards and Disasters.* Jamaica: UDS, University of the West Indies, pp. 56–61.

Shalkowski, A. (2009). Clay mineralogy of altered rocks from failed slopes in the Wagwater Belt. Unpublished PhD thesis, Hiroshima University.

Terzaghi, T. (1950). Mechanism of landslides. In S. Paige, ed., *The Application of Geology to Engineering Pratice.* Berkey volume. Baltimore, MD: Geological Society of America.

Velde, B. (1995). Composition and mineralogy of clay minerals. In B. Velde, ed., *Origin and Mineralogy of Clays.* New York: Springer-Verlag, pp. 8–42.

Wen, B., and Chen, H. (2007). Mineral composition and element composition as indicators for the role of groundwater in the development of landslip zones: A case study of large-scale landslides in the Three Gorges area—China. *Earth Science Frontiers,* 14, 98–106.

Wignall, M. (2012). Sandy was no Gilbert, but it was also dangerous. *Jamaica Observer.* http://www.jamaica observer.com/columns/Sandy-was-no-Gilbert--but-it-was-also-dangerous_12843396.

World Bank and GFDRR [Global Facility for Disaster Reduction and Recovery]. (2010). Disaster risk management in Latin America and the Caribbean region: GFDRR country notes. https://www.gfdrr.org/sites /default/files/publication/drm-country-note-2010-jamaica.pdf.

Yilmaz, I., and Karacan, E. (2002). A landslide in clayey soils: An example from the Kizildag region of the Sivas-Erzincan highway (Sivas, Turkey). Environmental Geosciences, 9, 35–42.

Zans, V. A. (1959). Judgement Cliff landslide in the Yallahs Valley. *Geonotes,* 2, 43–48.

16 Landslides
Causes, Mapping and Monitoring – Examples from Malaysia

Omar F. Althuwaynee and Biswajeet Pradhan

CONTENTS

16.1 INTRODUCTION

Landslides are one of the most disastrous natural hazards to occur worldwide, causing catastrophic damage to both lives and property every year. Landslides can be of many types (e.g. toppling, sliding, flowing or spreading) depending on the types of mechanisms involved, occurring at different scales (local, medium or large) and propagation times (varying from a few minutes to a few days). Generally, the mechanisms of landslides are dependent on independent environmental parameters (conditioning factors), such as slope material, geomorphic conditions (the same as rocks, soil or artificial fill) and the triggering factors. This process creates a downward and outward movement of slope material (Sidle and Ochiai 2006). In recent years, a number of studies have been carried out related to the development of advanced modelling approaches in landslide hazard and risk modelling (Hutchinson 1995; Guzzetti et al. 1996, 2012; Leroi 1996; Aleotti and Chowdhury 1999; Angeli et al. 2000; Huabin et al. 2005; Van Westen et al. 2006; Sharma et al. 2012). These predictive modelling approaches have become a major interest for engineering professionals, as well as for the community and local administrations.

16.2 LANDSLIDES IN MALAYSIA

In the beginning of the 1980s, Varnes (1984) introduced a technical report on landslide assessment dealing with sources for landslide classification. Technically, landslides are the result of environmental or anthropogenic activities or a combination of both, and landslide materials vary from rock and soil to mixed material which moves downwards on slopes due to gravity (Van Westen et al. 2008).

In tropical countries like Malaysia, most of the slope failure events occur due to prolonged spells of heavy rainfall. This condition causes high pore water pressure on the slope face, which leads to increased strain. Chigira and Oyama (2000) stated that pyrite is a common mineral contained in rocks that is oxidized by oxygen, and sulphuric acid is generated from this oxidized zone. The produced sulphuric acid, which can dissolve underlying materials (e.g. calcite, zeolite and volcanic glass), makes the structure of the rock very weak and initiates slope instability. Through analysis of the thematic environmental variables (conditioning factors) of the previously occurring landslides, it is possible to forecast and predict future landslides for the areas which have similar geographical, geological and environmental conditions. However, landslide forecasting is not a simple task.

Landslide hazard zonation (LHZ) divides the land surface into susceptible areas and ranks them according to the degree of actual hazard impact (Varnes 1984). Landslide susceptibility mapping is generally performed by statistical, conventional, heuristic or deterministic-based models. Analysis of these methods depends on some key factors, that is, data availability, spatial accuracy, data scale and complexity. Hartlen and Viberg (1988) identified and elaborated on some of these analytic methods and categorized them based on (1) relative hazard analysis, which compares current hazards with previous similar hazards; (2) absolute hazards, that is, factors of safety; (3) empirical hazards, that is, curve analysis based on several parameters, such as slope height and angle, mainly depending on geotechnical analysis and lab work conclusions; and (4) monitoring hazards, which compares the triggering and erosion factors using historical landslide location data. In addition, the employed model or data must be checked in terms of accuracy and reliability.

To start an effective landslide susceptibility analysis, various data are required: (1) a complete landslide inventory containing details about the spatial location, date and time of occurrence and types; (2) thematic environmental variables (e.g. geology, hydrogeology and geomorphological factors); (3) the durability of geomaterials; (4) reports and field surveys; and (5) triggering factors, such as rainfall records and historical seismicity; for instance, Sidle and Ochiai (2006) explained many mass movement scenarios triggered by heavy rain and earthquakes on either a natural geomorphologic slope or artificial fill.

Predictions are an essential outcome of all models and should be evaluated scientifically. Generally, there are two methods adopted to validate the prediction results: time robustness and space robustness. Results may be plotted graphically using the receiver operating characteristic (ROC) curve, consisting of both success rate and prediction rate curves.

In Malaysia, continuous development, urbanization, deforestation and weathering erosion of soil masses cause serious threats to slope stability and loss of lives and properties (Pradhan and Lee 2007). In tropical Malaysia, torrential rainfall attenuates the stability of metamorphosed rocks, which is the main bedrock in these areas. Moreover, in recent years the degree of deforestation has severely increased due to constructional activities which have caused severe soil erosion downstream. To overcome the landslide problem, urbanization is being avoided and constructions planned in gentle terrains. In this context, a national landslide hazard zoning map is a prerequisite to assist in decision making. To further strengthen these initiatives, landslide detection and monitoring systems are also being developed for timely mitigation measures.

16.3 LANDSLIDE TYPES

Landslides are classified on the basis of movement and estimated mass volume, which is calculated based on the morphology and geometry of the detachment area and the deposition zone (Guzzetti et al. 2012). Guzzetti et al. (2005) pointed out the significance of satellite imagery that provides a wealth of information about the morphological characteristics and appearance related to age, activity, depth and velocity.

Landslides types are categorized according to the various types of mass movements or slopes failures:

- Shallow rapid landslides (debris slides, avalanches and flow): Triggered by severe rain from a single huge storm as well as long and rapid periods of snowmelt.
- Rapid deep slides and flows: Quick failure caused by an individual instance of severe rain or an extended rain event with a large storm, such as a sudden cloudburst which reaches 100 mm/h precipitation or more, in a few minutes, enough to create flood conditions.
- Slow, deep-seated sedimentation process of landslides (slumps, earthflows and lateral spreads), also named creep process; the process starts due to extended rainfalls and continues for several days' or weeks' lag time, until the movement of slope movement activity.
- Slow flows and deformations (soil creep and solifluction): Is the downward movement of water-saturated soil, caused by waterlogged soil slowly moving downhill on top of an impermeable layer.

16.3.1 TRIGGERING FACTOR ASSESSMENT

A landslide is the result of the interplay of two important factors: predisposing and triggering factors that determine the probability of the landslide's occurrence. Predisposing factors can cause slope failure by very low speed and long duration. They are considered terrain attributes, and they are used in landslide susceptibility assessment. These factors, through processes such as stress release, weathering or erosion, might lead to a slope failure situation (Corominas and Moya 2008). Triggering factors (i.e. rainfall events, snowmelting, earthquakes, sea storms and cloudbursts) are responsible for rapid slope failure, with various intensity levels within a very short time (Althuwaynee et al. 2014a; Pellicani et al. 2014).

Caine (1980) proposed a well-known rainfall threshold which was derived through analyzing a collection of worldwide datasets of rainfall intensities and durations. It is expressed as

$$I = 14.82D^{-0.39} \tag{16.1}$$

where I is the critical rainfall intensity in millimetres per hour and D is the duration (h) of the storm.

Crozier (1997) concluded that the longer the duration of the rain, the smaller the average rainfall intensity. Unfortunately, this expression has some conflicts with the findings of others since it was not validated on a universal scale. In a case study of the estimated rainfall threshold of Kuala Lumpur's metropolitan and surrounding areas, Althuwaynee et al. (2014b) confirmed that not all high-intensity rainfall produces landslides, but it depends on the permeability of the soil materials that build up and the dissipation of positive pore pressures in a short time. The landslide mechanism represents a degree of response to rainfall, as high-intensity and short-lasting rains often trigger shallow landslides and debris flows. Rainstorms and snowmelting are usually reported as being rockfall-trigger factors, and after studying eight worldwide earthquakes, Lei (2012) found a strong relationship between rock mass fracture density and earthquakes.

16.3.2 MONITORING

Monitoring systems are dominant tools for understanding the kinematic aspects of slope movement. Different site conditions (hilly, mountainous or densely populated areas) have different monitoring techniques. Remote sensing (RS) and geographic information systems (GIS) techniques are now commonly being used for evaluation, prediction and mitigation of landslides with high precision.

16.3.2.1 Remote Sensing-Based Monitoring

RS-based approaches gather basic information about landslides, which can be used in spatial analysis within the GIS environment (Joyce et al. 2014). There are two types of RS-based landslide monitoring approaches: direct and indirect methods. The direct methods have some drawbacks:

- It is difficult to figure out the exact movement time of large-scale and high-velocity slides.
- Small mass movement needs very high-resolution images.
- There are high costs and discontinuities in the data because of low temporal resolution.

The indirect RS monitoring approach describes the environmental changes which are related to the slope failure activities, for instance, regional hazard identification and infrastructure defects caused by severe slope failure. Zhihua (2007) discussed the Tiantaixiang and Qianjiangpin landslides that occurred in three locations.

Active RS, which produces its own source of energy, is able to detect and collect the reflected backscatter from the target through many techniques, such as differential interferometric synthetic aperture radar (DInSAR), the least-squares method (LS), light detection and ranging (LIDAR), the permanent scatterer method (PS), and the small-baseline subset method (SBAS) (Joyce et al. 2014). All these techniques are suitable for large-scale areas, and laser scanning is one of the most accepted and powerful techniques, among others. Laser scanning consists of two main technologies: airborne laser scanning (ALS) and ground-based terrestrial laser scanning (TLS). ALS plays an important role in landslide mapping by increasing the resolution of the landslide contours to recognize the scarps and displaced materials. ALS combined with a high-resolution digital elevation model (HRDEM) can be used for the purpose of landslide mapping validation (Jaboyedoff et al. 2010). TLS was developed in the 1990s, made by the development of the electronic distance metre (EDM) and total station instruments. TLS is able to provide accuracy up to centimetres and can be used for crack opening detection, acoustic disturbances, microseismicity and/or pre-failure deformation. Jaboyedoff et al. (2010) suggested the integration of both ALS and TLS technologies to overcome some drawbacks of ALS related to shadow-covered areas. Major indicators of landslide monitoring are macrogeological features, ground displacement, displacement in boreholes and groundwater (pore pressure, groundwater table and groundwater chemistry).

16.3.2.2 Field Monitoring

Many real-time monitoring methods have been used to detect the initial landslide movements and evaluate the displacement. These methods are generally based on the mechanical principles of mass movement before the failure occurrence, such as the increment of shear stress that causes a low shear strength in debris mass materials due to the trapped water level (Manchao 2009). Subsequently, Manchao (2009) examined four different RS methods for landslide forecasting, and found that the landslides occur when the sliding force is greater than the antisliding force.

There are many useful techniques to monitor the Earth surface and detect the initial slope movements and displacement. These techniques vary between surveying and continuous monitoring systems, such as precision tape, fixed wire extensometers, offsets from baselines, surveying triangulation, surveying traverses, geometrical levelling, EDM, terrestrial photogrammetry, rod of crack openings, precision theodolites, aerial photogrammetry and global positioning system (GPS). Each technique has its degree of accuracy depending on the area that can be covered (Gili et al. 2000).

GPS acts as an efficient system for monitoring the mass movements, which can be used in the frame of an individual tool since it provides highly accurate results. Mora et al. (2003) discussed the many GPS monitoring techniques that are available, such as (1) differential GPS (DGPS) (Squarzoni et al. 2005), (2) fast static or rapid static (FS) and Real Time Kinematics (RTK) with a centimetre level of accuracy (Huang et al. 2017), (3) EDM and GPS (Pradhan and Sameen 2017b), (4) GPS and InSAR (Yin et al. 2010), (5) GPS stations and inclinometers (Bayer et al. 2016) and (6) time domain reflectometry (TDR) (Bayer et al. 2016). GPS tools provide higher accuracy and productivity than very long-baseline interferometry (VLBI), satellite laser ranging (SLR), precise levelling or geodetic surveying (Mora et al. 2003).

The subsurface geometry, particularly the surface of rupture, has an important role in landslide analysis (Bruno and Martillier 2000). Surface observations are capable of producing only qualitative estimations, but boreholes and geophysical analysis are able to provide more quantitative models of the geometry. Borehole methods are more costly and require uncertain interpolations between the holes compared with geophysical analysis. Also, in some cases landscape ruggedness makes the geophysical analysis practically not possible. As a result, more research can be seen on surface observation than with geophysical analysis.

16.3.2.3 Landslide Inventory Mapping

The past is the key to the future occurrence of landslides (Varnes 1984). Current and historical data are required for landslide analysis (Guzzetti et al. 2000). A landslide inventory is generally prepared by interpreting orthophoto maps and aerial photographs, and using standard field-based geomorphologic methods.

Guzzetti et al. (2012) discussed various scopes in landslide inventory mapping, such as (1) shaping the extent of landslide phenomena in areas with watersheds varies, from a local (Cardinali et al. 2001), regional and national or global scale (Van Den Eeckhaut and Hervás 2012); (2) initial step towards landslide risk assessment (Althuwaynee and Pradhan 2016); (3) distribution between landslide types with morphological and geological characteristics (Guzzetti et al. 1996); and (d) monitoring the development of landscapes dominated by mass-wasting processes (Guzzetti et al. 2012) from large (1:5,000) to small (1:500,000) scales.

Generally, a landslide inventory map is prepared by either a single method or a combination of some techniques. The selection of the method is based on the objectives, geological environment of the study area, experience and availability of RS data, such as aerial photographs, high-resolution satellite and LIDAR elevation data (Guzzetti et al. 2000; Van Westen et al. 2006). Fiorucci et al. (2011) used satellite imagery to prepare the landslide inventory, which contained information about the morphological characteristics, type and appearance of the movement, such as age, activity, depth and velocity. Vegetation coverage is an indicator and one of the important criteria to detect the landslides, but it is not always true depending on the area and soil characteristics and vegetation conditions; sometimes the quick growth of vegetation may mask the landslide scars (especially in tropical areas). Recently, many new techniques have been developed which can be grouped into three broad categories:

1. Visual (heuristic) interpretation of optical images, including panchromatic, composite, false colour and pan-sharpened ('fused') images as visual interpretation of monoscopic satellite images (Mondini et al. 2011a)
2. Analysis of the multispectral images (Lin et al. 2017), including image classification, semiautomatic detection of landslides (Mondini et al. 2011b), change detection methods, index thresholding methods and clustering methods (Cheng et al. 2013)
3. Analysis of SAR images (Del Ventisette et al. 2014) to measure the surface deformations, three-dimensional visualization of stereoscopic satellite images, semi-automatic detection of landslide features from the analysis of HRDEMs (Pradhan and Sameen 2017a), object-oriented image classification (Martha et al. 2010) and multiple change detection techniques for the semi-automatic detection of landslides.

Santangelo et al. (2010) has shown the capability of high-quality digital cameras to simplify the documentation of landslides with low cost. However, since the landslides occur in different geoenvironments (geology, weather conditions and slopes), no standard methodology and protocols exist for landslide inventory mapping for the preparation and updating of the landslide maps, which reduce the credibility of the inventory map (Guzzetti et al. 2012).

16.4 LANDSLIDE THEMATIC ENVIRONMENTAL VARIABLES (CONDITIONING FACTORS)

To predict the future hazard map of landslide-prone areas, proper spatial representations of conditioning factors are required. One must understand the failure type and characteristics of the study area in order to have an accurate and optimized selection of conditioning factors (Van Westen et al. 2008). A large number of conditioning factors have been used for the hazard mapping (Pradhan and Lee 2010b; Nefeslioglu and Gokceoglu 2011; Felicísimo et al. 2012), whereas others have made efforts to use few conditioning factors (Ohlmacher and Davis 2003; Manzo et al. 2012). Some of the main conditioning factors which were used widely in landslide studies are discussed:

- DEM maps are generally used to extract various geomorphological factors, such as slope angle, slope aspect, surface curvature, surface roughness, flow direction and flow accumulation (Wilson and Gallant 2000). The slope angle has a direct effect on slope stability. Slope aspect strongly affects the hydrological process through evapotranspiration. In the Northern Hemisphere, a north-facing slope experiences more frequent periods of wetting and drying; therefore, a higher rate of landslide is expected than with a south-facing slope (Sidle and Ochiai 2006). Elevation is one of the most important conditioning factors in the generation of mass movements, but it cannot be used individually to detect landslides (Lo et al. 2017).
- Geological maps are used to extract the rock types, depth of the weathering profile, discontinuities, geological structure, slope angle relationship, distance from active faults and fault parameters (fault length). Other factors, such as frost friction, salt weathering and thermal stress, can also be derived using this conditioning factor (Preisig et al. 2016).
- Soil maps are used to extract the soil types, soil depth, grain size distribution and hydrological properties, such as pore volume.
- Hydrological parameters, such as the spatial distribution of the water table, soil moisture, evapotranspiration, infiltration, precipitation (spatial and temporal distribution of rain), stream network and distance from drainage. The main attributes of the rainfall that can affect landslide are total rainfall, short-term intensity, antecedent storm precipitation (spatial and temporal distribution of rain) and storm duration. According to Zhai et al. (2003), the majority of a real-time warning system is based on a telemetered network of recording rain gauges.
- Geomorphology maps are used to extract the geomorphological units, genetic classification of main landform building processes and slope facets.
- Land-use maps are used for mapping the vegetation types or normalized difference vegetation index (NDVI), and distance from the roads (Althuwaynee et al. 2012a). The dense vegetation increases the hills' stability in two ways; first, it helps to remove the soil moisture through the evapotranspiration process, and second, it provides a rooting network, which adds a cohesion bond for soil particles.
- Combination of two or more maps: A stream power index (SPI) is made by using a combination of the catchment area and the slope gradient information. A topographic wetness index (TWI) has been used extensively to describe the effect of topography on the location and size of saturated source areas of runoff generation (Bui et al. 2016).

Recent studies (Pradhan and Lee 2010a; Floris et al. 2011) have shown that using a large number of conditioning factors may not produce accurate and reliable results (Manzo et al. 2012).

16.5 GENERAL CLASSIFICATION

Among all the available methods, qualitative and quantitative analysis are the most common approaches in landslide susceptibility mapping (Aleotti and Chowdhury 1999). Qualitative methods, such as the analytic hierarchy process (AHP) (Kayastha et al. 2012) and weighted linear combination (Akgun et al. 2008), are widely used by geologists and geomorphologists and engineers for regional studies.

In recent decades, due to the development of computing tools, the quantitative methods have become very popular; they are built based on the numerical expressions of the relationship between conditioning factors and landslides (Figure 16.1).

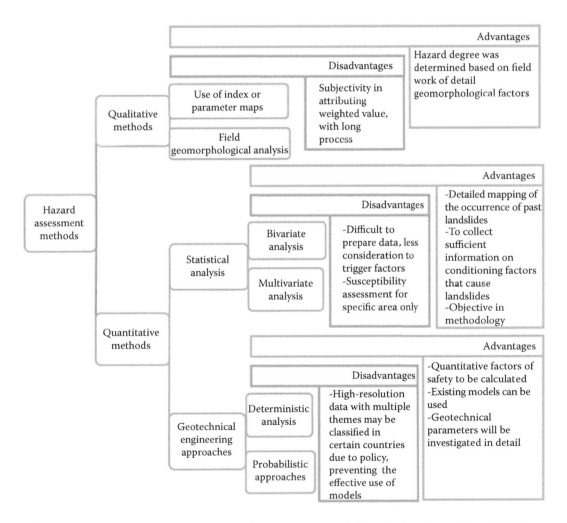

FIGURE 16.1 Comparison of landslide hazard assessment method classifications. (From Huabin, W. et al., *Prog. Phys. Geogr.*, 29(4), 548–567, 2005; Aleotti, P., and Chowdhury, R., *Bull. Eng. Geol. Environ.*, 58(1), 21–44, 1999.)

16.5.1   QUALITATIVE APPROACHES

For qualitative analysis (probability and losses expressed in qualitative terms are based entirely on the judgement of the person or persons carrying out the susceptibility or hazard assessment), aerial photo and fieldwork are important (Aleotti and Chowdhury 1999), which requires geomorphological analysis and the overlaying of index maps with or without weighting (Aleotti and Chowdhury 1999; Haugen 2016).

- Field geomorphological analysis is carried out by geologists without considering the standard assessment rules. This method contains a few drawbacks in data selection since it uses implicit results rather than explicit analytical rules. Also, another disadvantage of this method is related to the field survey, which is often time consuming and expensive.
- On the contrary, the overlaying of index maps utilizes the personal experience of the expert by assigning weights for each factor.

16.5.2   QUANTITATIVE APPROACHES

The quantitative approaches are represented by deterministic and statistical analysis methods (Mihir et al. 2014). The deterministic quantitative method utilizes the engineering principles of slope instability in terms of the safety factor. These methods are often used to map the small area due to the exhaustive data requirements. The statistical analysis governs the slope stability by assigning a contribution weight to each conditioning factor (Bui et al. 2013). The statistical analysis can be either bivariate or multivariate. Sidle and Ochiai (2006) stated that 'in bivariate statistical analysis, each conditioning factor is evaluated separately in conjunction with landslide density or volume, and bivariate analysis is used to identify factors that are significant related to landslide occurrence, than the assigned relative weight'.

Multivariate approaches are empirical in nature and consider the relationships among the conditioning factors and landslides. After preparing all the conditioning factors, the presence or absence of a landslide can be determined through multivariate analysis. A multicollinearity test can be used to detect if there are any interdependent spatial factors. The method should be carried out before the multivariate analysis using analysis of variance (ANOVA) of the inflation factor, tolerance and the co-linearity index method.

Bivariate statistical analysis considers landslide density in each individual conditioning factor separately. This process can make for some uncertainties, as it is possible that the same conditioning factor can have different consequences depending on the process types (Constantin et al. 2011). Some other limitations are related to the tendency to simplify the conditioning factors and generalize all landslide inventories. Unfortunately, this method heavily relies on the GIS expert's knowledge instead of the earth scientist expert's opinion. Finally, each mass movement type should have its own set of conditioning factors and should be analyzed separately (Kirschbaum 2009).

Multivariate statistical analysis, such as logistic regression (LR) and decision tree (DT), is very robust in landslide susceptibility mapping (Althuwaynee et al. 2014a). Some of the drawbacks of bivariate statistic methods can be amended in multivariate analysis. Table 16.1 shows some approaches with some case studies from Malaysia.

16.6   LANDSLIDE SUSCEPTIBILITY MODELLING APPROACHES

Landslides can be analyzed through qualitatively or quantitatively statistical approaches (Lu et al. 2014). Machine learning, bivariate statistical analysis and multivariate analysis are highly recommended for landslide studies, which are discussed briefly in this section.

TABLE 16.1

Review Articles of Predictive Modelling and Evaluation Approaches Used in Landslide Modelling (Malaysia Study Area)

Year	Title	Location	Technique	Data	Evaluation
2006	Probabilistic landslide hazards and risk mapping on Penang Island	Penang Island	Probability–FR	Slope, aspect, curvature and distance from drainage, geology and distance from lineament, land use, soil, precipitation, vegetation	FR = 81%
2006	Landslide hazard mapping at Selangor using FR and LR models	Selangor	FR and LR models	Slope, aspect, curvature and distance from drainage, lithology, distance from lineaments, land cover, vegetation index, precipitation	FR = 93.04% LR = 90.34%
2007	Utilization of optical RS data and GIS tools for regional landslide hazard analysis using an ANN model	Selangor	Artificial neural network	Slope, aspect, curvature and distance from drainage, lithology and distance from lineament, land cover, vegetation index	ANN = 83%
2008	Landslide risk analysis using an ANN model focusing on different training sites	Penang Island	Artificial neural network	Slope, aspect, curvature distance from drainage, distance from lineament geology, land cover, NDVI, soil, precipitation	ANN = 93%
2009	Delineation of landslide hazard areas on Penang Island by using FR, LR and ANN models	Penang Island	FR, LR, artificial neural	Slope, aspect, curvature, distance from drainage geology, distance from lineament, land use, soil, precipitation, vegetation index	FR = 86.41 LR = 89.59 ANN = 83.55%
2009	Use of geospatial data and fuzzy algebraic operators for landslide hazard mapping	Penang Island	Data-derived model (FR) and a knowledge-derived model (fuzzy operator)	Slope angle, slope aspect, slope curvature, distance from drainage, lithology map, distance from lineament, land use, NDVI	FR = 80% Fuzzy or operator = 56%
2009	Manifestation of RS data and GIS on landslide hazard analysis using spatial-based statistical models	Cameron area	FR model, bivariate, LR	Slope, aspect, curvature, distance to rivers, lithology and distance to faults, land cover, vegetation index, precipitation	FR = 89.25% LR = 85.73%

(Continued)

TABLE 16.1 (CONTINUED)

Review Articles of Predictive Modelling and Evaluation Approaches Used in Landslide Modelling (Malaysia Study Area)

Year	Title	Location	Technique	Data	Evaluation
2010	Landslide susceptibility mapping by neurofuzzy approach in a landslide-prone area (Cameron Highlands)	Cameron Highlands	Neurofuzzy	Altitude, slope gradient, curvature, distance from the drainage, distance from the road, lithology, distance from the faults, NDVI	Neurofuzzy = 97% for model 5
2010	Manifestation of an advanced fuzzy logic model coupled with geoinformation techniques to landslide susceptibility mapping and their comparison with LR modelling	Selangor	Fuzzy logic	Slope gradient, slope exposure, plan curvature, altitude, SPI, TWI, distance from drainage, distance from road, lithology, distance from faults, soil, land cover, NDVI	Fuzzy logic = 94%
2010	Regional landslide susceptibility analysis using backpropagation neural network model at Cameron Highlands	Cameron Highlands	Backpropagation neural network	Slope, aspect, curvature and distance from drainage, lithology and distance from lineament, soil type, rainfall, land cover, vegetation index value	Backpropagation neural network = 83%
2010	Landslide susceptibility mapping of a catchment area using FR, fuzzy logic and multivariate LR approaches	Cameron catchment area	FR, fuzzy logic and multivariate regression models	Aspect, curvature, distance from drainage, lithology and distance from lineament, land cover, vegetation index, precipitation	FR model = 89% Fuzzy logic = 84% LR = 85%
2010	Use of GIS-based fuzzy logic relations and their cross-application to produce landslide susceptibility maps in three test areas in Malaysia	Penang Hill, Bertam Valley and Putrajaya areas	Fuzzy logic relations and their cross-application of membership	Slope, slope exposure, plan curvature, topography, distance from the drainage, distance from the road, land cover, soil texture and types, lithology, distance from the fault lines, NDVI	Fuzzy logic = 82%
2010	Weight-of-evidence model applied to landslide susceptibility mapping in a tropical hilly area	Cameron Highlands	Weight-of-evidence model (a Bayesian probability model)	Slope aspect, plan curvature and distance from drainage, lithology and distance from the lineament, soil texture, land cover, NDVI	WoE = 97%

(Continued)

TABLE 16.1 (CONTINUED)

Review Articles of Predictive Modelling and Evaluation Approaches Used in Landslide Modelling (Malaysia Study Area)

Year	Title	Location	Technique	Data	Evaluation
2011	Manifestation of an adaptive neurofuzzy model on landslide susceptibility mapping in Klang Valley	Klang Valley areas, Selangor area	Neurofuzzy model	Altitude, slope angle, plan curvature, distance from drainage, soil type, distance from faults, NDVI	ANFIS = 98%
2011	Soil erosion assessment and its correlation with landslide events using RS data and GIS: a case study at Penang Island	Penang Island	Universal soil loss equation (USLE) method	L and S, terrain factors; K, soil erodibility factor; R, rainfall runoff erosive factor; C, land cover and management factor; P, conservation practices factor	
2012	Application of an EBF model in landslide susceptibility mapping	Selangor	EBF	Slope, aspect, curvature, altitude, surface roughness, lithology, distance from faults, NDVI, land cover, distance from drainage, distance from road, SPI, soil type, precipitation	EBF = 82%

Note: WoE, weight of evidence; ANFIS, adaptive neurofuzzy inference system.

16.6.1 Frequency Ratio

The frequency ratio (FR) is a simple probability model being used in many landslide analyses (Pradhan and Lee 2010b). Theoretically, the probabilities might be defined through the relationship of the landslide occurrence to each conditioning factor. The FR model answers a conditional probability question for a certain factor, like a slope, that is, how to find the probability of a randomly selected pixel area being a landslide-susceptible area and also between a given slope angle range. All the conditioning factors must be carried in the same number of pixel size before FR (Equation 16.2) is applied (Pradhan and Lee 2010b).

$$FR = \frac{PLO}{PIF} \tag{16.2}$$

where *PLO* is the percentage of landslide occurrence in each sub-category and *PIF* is the percentage of each category of a factor that affects the landslide.

Values greater than 1 reflect high correlation with landslide occurrence, and values smaller than 1 represent a lower correlation. Using FR, maps are produced for different landslide-prone areas and validated through field survey assessment (Pradhan et al. 2009; Pradhan 2010a; Pradhan and Youssef 2010).

16.6.2 Evidential Belief Function

The evidential belief function (EBF) method is considered a mathematical model which represents the spatial integration based on the rule of combination and is able to provide a framework to estimate the weights for conditioning factors (Carranza et al. 2005). The EBF method is capable of handling the analysis even with an incomplete dataset (Carranza et al. 2005). The predictive landslide susceptibility map is used for quantitative knowledge of the spatial relationship between landslides and the landslide conditioning factors. Thus, the analysis is started by extracting and building the conditioning factors and categorizing and reclassifying each of them.

The main part of the Dempster–Shafer theory is represented by Bel (lower probability) and Pls (upper probability) (Likkason et al. 1997). There are three important functions in the Dempster–Shafer theory, as shown below.

$$M: 2^{\odot} = \{\varnothing, Tp, Tp^-, \odot\} \quad \odot = \{Tp, Tp^-\} \tag{16.3}$$

where *Tp* is the class pixels affected by the landslide and *Tp⁻* is the class pixels not affected by the landslide.

To get the Bel results, Equation 16.4 is used:

$$\begin{aligned} \lambda(Tp)Eij &= \left[N(L \cap Eij) / N(L) \right] / \left[N(Eij) - N(L \cap Eij) / \left(N(A) - N(L) \right) \right] \\ &= N/D \end{aligned} \tag{16.4}$$

where $N(L \cap Eij)$ is the number of landslide pixels in the domain; $N(L)$ is the total number of landslide, or $\sum N(L \cap Eij)$; $N(Eij)$ is the number of pixels in the domain; $N(A)$ is the total number of pixels in the domain, or $\sum N(Eij)$; N is the proration that a landslide will occur; and D is the proportion of the non-landslide area.

Bel can be calculated using Equation 16.5:

$$\text{Bel} = \lambda(Tp)Eij / \sum \lambda(Tp)Eij \tag{16.5}$$

Similarly, the Dis value can be calculated using Equation 16.6:

$$\lambda(Tp^-)Eij = \left[\left(N(L)-N(L\cap Eij)\right)/N(L)\right]\Big/\left[\left(N(A)-N(L)-N(Eij)+N(L\cap Eij)\right)/\left(N(A)-N(L)\right)\right]$$
$$= K/H \tag{16.6}$$

where K is the proportion of landslides that do not occur and H is the proportion of non-landslide areas in other attributes outside class.

Dis results are calculated using Equations 16.7 through 16.9:

$$\text{Dis} = \lambda(Tp^-)Eij/\Sigma\lambda(Tp^-)Eij \tag{16.7}$$

$$Unc = 1-\text{Dis}-\text{Bel} \tag{16.8}$$

$$Pls = 1-\text{Dis} \tag{16.9}$$

Another significant advantage of the EBF model, in addition to the predictive map, is its ability to show the degree of uncertainty of the same zone (Althuwaynee et al. 2012a).

Model capability may be concluded as

- Degrees of belief, which show susceptible areas.
- Degrees of disbelief, which show non-susceptible areas.
- Degrees of uncertainty, which assess the quality of input conditioning factors, by indicating the current evidences and proofs for landslide causative information.
- Degrees of plausibility, which show the cases in which more spatial evidence is required. Furthermore, it shows all the integrated evidence except the disbelief. The degree of plausibility also reveals where spatial evidence is sufficient and efficient or inefficient to prove that a landslide-triggering factor will affect the dependent factors.

Data-driven EBFs can produce acceptable results even with a gross input dataset.

16.6.3 INDEX OF ENTROPY

The index of entropy is a bivariate statistical analysis which was proposed by Vlcko et al. (1980) based on the relationship between the landslide inventory map as a dependent variable and several conditioning factors. The hazard contribution weight for each conditioning factor is expressed as a level of entropy index. The entropy index takes two main considerations in its process: first, finding the most predominant conditioning factors that initiate the mass movements, and second, elaborating the extent of disorder in the environment. Equations 16.10 through 16.16 represent the steps of index of entropy analysis which can be used for producing the landslide susceptibility map (Bednarik et al. 2010).

$$P_{ij} = \frac{P_{sd}}{P_C} \tag{16.10}$$

$$(P_{ij}) = \frac{P_{ij}}{\sum_{j=1}^{S_j} P_{ij}} \tag{16.11}$$

where P_c is the area of the category after primary reclassification, P_{sd} is the area of landslides within the given category and (P_{ij}) is the probability density.

H_j and $H_{j\,max}$ are the entropy values of Equations 16.12 and 16.13, and they are written as

$$H_j = -\sum_{i=1}^{S_j}(P_{ij})\log_2(P_{ij}), \quad j = 1,\ldots\ldots,n \tag{16.12}$$

$$H_{j\,max} = \log_2 S_j, \tag{16.13}$$

where S_j is the number of classes and I_j is the information coefficient.

$$I_j = \frac{H_{j\,max} - H_j}{H_{j\,max}} \tag{16.14}$$

$$W_j = I_j \times P_j \tag{16.15}$$

where W_j represents the resultant weight value for the parameter.

The final susceptibility map is prepared by summation of weighted multiplications for the secondary reclassified parametric maps (Equation 16.16).

$$y = \sum_{i=1}^{n}(\text{factor\_recl2}) \times W_j \tag{16.16}$$

where factor_recl2 is an integer number and represents each pixel contribution weight after reclassifying the parametric map of the conditioning factors. y is the value of landslide susceptibility in the final map.

Using the calculated probability P_{ij} (Equation 16.10) and probability density (P_{ij}) (Equation 16.11), each input parametric map is reclassified (column recl_2). The weight index of conditioning factors is the nucleus of a regional landslide hazard degree assessment and zonation. However, the lack of data can always cause the results of the final susceptibility map to be misleading.

16.6.4 ARTIFICIAL NEURAL NETWORKS

The artificial neural network (ANN) is the most popular machine learning method, applied by many particularly during the last 5 years (Lee et al. 2004; Lee and Evangelista 2006; Pradhan and Lee 2007, 2010c). ANN is defined as a model of reasoning based on the human brain that can be considered a simplified reproduction or mirror of the highly complicated system (Ercanoglu 2005). Garrett (1994) defined ANN as a computational mechanism to acquire, represent and compute a mapping from one multivariate space of information to another, given a set of data representing that mapping.

ANNs are a family of biological learning models in machine learning. The ANN model comprises interconnected neurons or nodes, which are structured into layers with random or full interconnections among successive layers (Rumelhart et al. 1986). The ANN model comprises input, hidden and output layers that are responsible for receiving, processing and presenting results, respectively (Pradhan and Lee 2010c). Each layer contains nodes connected by numeric weights and output signals. The weights are the functions of the sum of the inputs to the node modified by a simple activation function (Bishop 1995). The possibility of learning is the most important feature that attracts researchers to use ANNs.

The most common type of ANN is the multilayered backpropagation algorithm (Rumelhart et al. 1986; Bishop 1995). A typical backpropagation network contains an input layer, an output layer and at least one hidden layer. 'The hidden and output layer nodes process their inputs by multiplying them by a corresponding weight, summing the product, and processing the sum using a nonlinear transfer function' (Pradhan and Lee 2010c). The transfer function and the net connectivity are held constant. Initially, the network is provided with random connection weights, which represent the input to the network and the expected output. Error signals are calculated by measuring the distance between network output results and the desired output results (training data). Through an iterative procedure of backpropagation of errors, weights are automatically modified to reduce the output error and the network is trained to classify the training data properly. In a successive phase, the trained ANN is able to classify data samples with an unknown classification (Figure 16.2).

Some of the advantages of ANN are

- ANNs allow us to have a different view of problems which cannot be solved by statistical methods due to their theoretical limitations (Perus and Krainc 1996).
- The ANN method is independent of the statistical distribution of the data, and there is no need for specific statistical variables.
- ANN allows the target classes to be defined with consideration to their distribution in the corresponding domain of each data source (Lee et al. 2003).
- ANN uses less training data to perform accurate analysis than the statistical methods (Paola and Schowengerdt 1995).

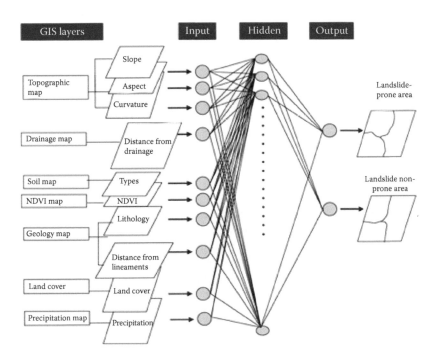

FIGURE 16.2   Architecture of the ANN. (From Pradhan, B., and Lee, S., *Environ. Earth Sci.*, 60(5), 1037–1054, 2010.)

16.6.5 LOGISTIC REGRESSION

LR is one of the multivariate statistical approaches, which is often cited as one of the most efficient data-driven techniques. LR has many advantages, such as the conditioning factor can be either continuous or discrete, or any combination of both types, and it does not necessarily have a normal distribution (Althuwaynee et al. 2014a). The LR model has been widely applied in landslide susceptibility mapping (Devkota et al. 2013). In this technique, a dependent variable is considered a binary variable representing the absence or presence of landslides. Moreover, the LR method represents a probability in the range of [0,1] on an S-shaped curve.

The probability (p) of the occurrence of a landslide in each pixel can be calculated using Equation 16.17.

$$p = \frac{1}{1 + s^{-z}} \qquad (16.17)$$

where Z represents a linear combination of conditioning factors with the presence or absence of the landslide, and its value varies from $-\infty$ to $+\infty$. To apply the LR model, each class of the contributing factors is considered an independent variable for the statistical regression. It is recommended to normalize the conditioning factor dataset by creating layers of binary values for each class (Althuwaynee et al. 2014a).

To prepare the training and testing landslide data, Nefeslioglu and Gokceoglu (2011) recommended using an equal number of pixels which show the occurrence of landslide and no landslide. After the data is produced, it can be split into training and testing data (Brenning 2005). The next step is to change the conditioning factors into a continuous format. After preparing the conditioning factors, the relationship of each of them with landslide occurrence is evaluated. The LR method utilizes the classes of each conditioning factor as an indicator and assigns the weights to other classes based on that indicator. z is a linear combination, and it follows that LR involves fitting an equation of the following form to the data:

$$z = b_o + b_2 x_2 + \cdots + b_p x_p \qquad (16.18)$$

where b_o is the intercept of the model, b_i ($i = 0, 1, 2, \ldots, n$) represents the coefficients of the LR model and x_i ($i = 0, 1, 2, \ldots, n$) denotes the conditioning factors. The final map is constructed using Equation 16.17, which represents the results in the range of 0–1. The pre-process steps mentioned earlier should be applied before running the LR model.

16.6.6 VALIDATION PROCESS

Generally, time robustness and space robustness are the main validation methods for landslide susceptibility and hazard mapping. In time robustness validation, landslide inventory is divided into two time periods: past occurrence (model training data) and future occurrence (model testing data). Testing data are used to examine how well the rest of the inventory can be predicted using the model. For space robustness, landslide inventory is divided into two groups randomly: group 1 of training data and group 2 of prediction data. A third method is a combination of space robustness and time robustness, which can be used whenever a complete landslide inventory is available to produce a comprehensive approach (Van Westen et al. 2006).

Another method which can be used to evaluate the model performance is the ROC curve, which can measure the success and prediction rate. The ROC curve has been adopted in different scientific fields, such as medical diagnostic testing (Swets 1979) and machine learning modelling (Nandi and Shakoor 2010). The area under the curve (AUC) can be used as a base

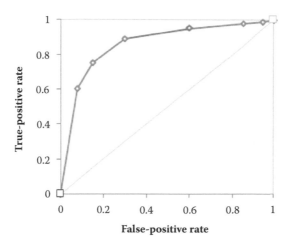

FIGURE 16.3 ROC curve.

metric to assess the overall quality of a model (Hanley and McNeil 1983). The higher the area under the fitting curve (value >50%), the better the validity of the performance. The main axes on the ROC curve represent the false-positive (FP) and true-positive (TP) rates, which are pairs and are derived from different contingency tables (Figure 16.3). Points which are closer to the upper-right corner correspond to the lower values, whereas points in the upper-left corner represent higher values (Frattini et al. 2010).

In the case of grid cell units where landslides correspond to single grid cells and all the terrain units have the same area, the y-axis corresponds to TP, analogous with the ROC space, and the x-axis corresponds to the number of units classified as positive (Swets 1979). The main advantage of this method is related to its simple structure, which is understandable for many landslide researchers.

Success and prediction rate curves are frequently used to validate and evaluate the performance of the model. In Figure 16.4, the y-axis represents the percentage of correctly classified objects and the x-axis represents the percentage of area classified as positive (Bui et al. 2014). The success rate curve is obtained by plotting the respective sensitivities against the total proportions of the dataset classified as landslide. As the training landslide data are used to generate the model, the validation process using the training data does not represent the real efficiency of the developed model. So the prediction capability of the model cannot be achieved using success rate. The prediction rate shows how well the model can predict the landslide in an area (Bui et al. 2012).

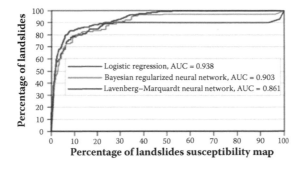

FIGURE 16.4 Prediction rate curves and success rate curve, AUC. (From Bui, D. T. et al., *Geomorphology*, 171–172, 12–29, 2012; Tien Bui, D. et al., *Catena*, 96, 28–40, 2012.)

16.7   LANDSLIDES IN MALAYSIA: SELANGOR CASE STUDY

Malaysia lies in tropical regions; the region suffers with rainfall and related debris flows, which leads to the erosion of slopes and scouring of channels, which results in a disastrous impact on humans and infrastructure. Since the 1980s, due to the rapid development in highlands, many landslides were recorded in densely populated cities, which happened due to the cutting slopes and other factors (Weng Chan 1998) (Table 16.2 and Figure 16.5).

16.7.1   APPLICATION OF EBF MODEL IN LANDSLIDE SUSCEPTIBILITY MAPPING IN SELANGOR

Kuala Lumpur and its vicinity areas have a major role in economic and social development in Malaysia (Figure 16.6). These areas are classified as landslide-prone areas which have faced frequent landslides in recent years (Pradhan 2010b). The study area covers approximately 1975 km^2 of settlement, with a smaller percentage of peat swamp forest, abandoned mines, grassland and a

---

TABLE 16.2

List of Most Serious Landslides That Have Occurred around Selangor

Date	Event	Fatalities	Remarks
December 1993	Highland Towers collapsed, Taman Hill view, Ulu Klang, Selangor	48	Shallow rotational slide/cut slope in granitic formation
June 1995	Genting Highlands slip road	22	Debris flow, natural slope in meta-sediment formation
January 1999	Squatter's settlement, Sandakan Town, Sabah	13	Shallow rotational/natural slope in meta-sediment formation
May 1999	Bukit Antarabangsa, Ulu Klang, Selangor	Injured and trapped	Debris flow, natural slope in metasediment formation
November 2002	Bungalow of the Affin Bank chairman general (RtD) Tan Sri Ismail Omar collapse, causing landslide in Taman Hillview, Ulu Klang, Selangor	8	Debris flow, sliding/flowing of debris soil of abandoned projects during heavy rain
December 2003	New Klang Valley Expressway (NKVE) near the Bukit Lanjan	Closed the expressway for 6 months	
November 2004	Flowing of debris soil from uphill bungalow project – toppled the back portion of neighbouring downslope bungalow, Gombak, Kuala Lumpur	1	Debris flow, sliding/ metasediment formation
December 2004	Rockfall – buried back portion of illegal factory at the foot of limestone hill, Bercham, Ipoh City, Perak	2	Natural limestone cliff
May 2006	Kampung Pasir, Ulu Klang, Selangor	4	
November 2007	Functional damage on PLUS Expressway for 8 hours/Butterworth to Ipoh	Highway damage	Natural limestone cliff
December 2007	12 houses damaged by slope failure at Bukit Cina, Kapit, Sarawak	4	
February 2009	Bukit Ceylon, Kuala Lumpur	1	
May 2011	Orphanage home in FELCRA Semungkis, Hulu Langat, Selangor	16	

Source: Public Works Department, Malaysia; Althuwaynee, O. F. et al., *Computers & Geosciences*, 44, 0, 120–135, 2012a.

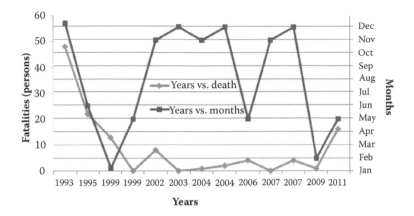

FIGURE 16.5   Relationship between years, months and fatalities. Kuala Lumpur and surrounding areas, Malaysia case study.

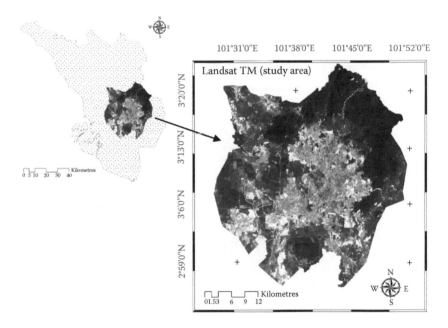

FIGURE 16.6   Study area location map.

few scrub areas. Temperature varies from 29°C to 32°C and precipitation varies from 58 to 240 (mm/month). This condition is suitable to trap a large amount of water within the steep terrain and increase the pore water pressure (Corominas and Moya 2008).

In the last decades, in the study area massive deforestation activities have been observed due to the construction of new residential and industrial zones. However, this action carries negative effects in terms of reducing the slope stability of hilly areas (Evett et al. 2006). The geology and detailed geomorphological characteristics of the study area are discussed by Althuwaynee et al. (2012a).

In this study, a total of 14 landslide conditioning factors were constructed: slope, aspect, curvature, altitude, surface roughness, lithology, distance from faults, NDVI, land cover, distance from drainage, distance from road, SPI, soil type and precipitation. All the conditioning factors were converted to raster grid format in a GIS environment.

16.7.2 SPATIAL DATABASE USED

The quality of the input datasets has a direct effect on the precision of the results. Thus, many data sources were utilized, such as an archived scale of 1:5,000 to 1:50,000 of aerial photographs, SPOT 5 panchromatic satellite images and landslide reports of 1981–2003. A total number of 220 landslides were mapped, and 70% of the landslide locations (representing 153 landslide cells) were randomly selected for the purpose of training the model (Figure 16.7).

A topographic map with a scale of 1:25,000 was used to extract the altitude, slope angle, slope aspect, roughness and curvature. The SPI was computed from the slope and catchment area. Also, the distance from the road and the distance from the drainage were constructed from the DEM using ArcGIS v. 9.3. A geological map with a scale of 1:63,000 was used to produce a lithological map and distance from faults. A soil map with a scale of 1:100,000 was used to extract the soil properties of the study area. The Landsat TM was used to extract the land cover types and NDVI maps. A record of the last 29 years (1981–2010) of historical rainfall data was collected via two rain gauge stations. These data were used to construct the precipitation 100 × 100 cell size map (Figure 16.8).

Using the EBFs (Figure 16.9), a landslide hazard map was produced (Equation 16.19). Figure 16.10 shows (a) Bel (degree of belief) and (b) Dis (degree of disbelief).

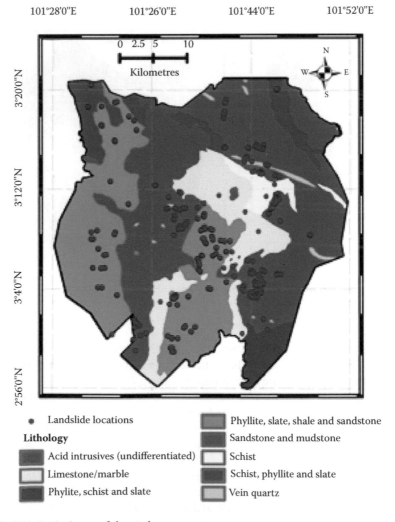

FIGURE 16.7   Lithological map of the study area.

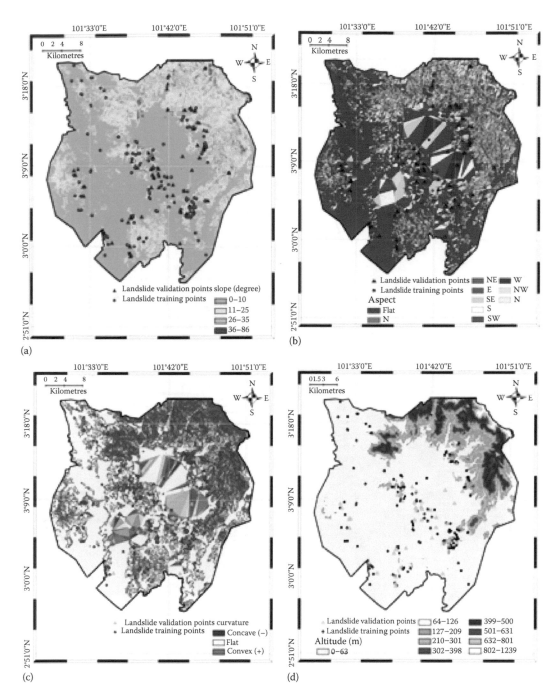

FIGURE 16.8 Landslide conditioning factors used for EBF: (a) slope, (b) aspect, (c) curvature, (d) altitude.

(*Continued*)

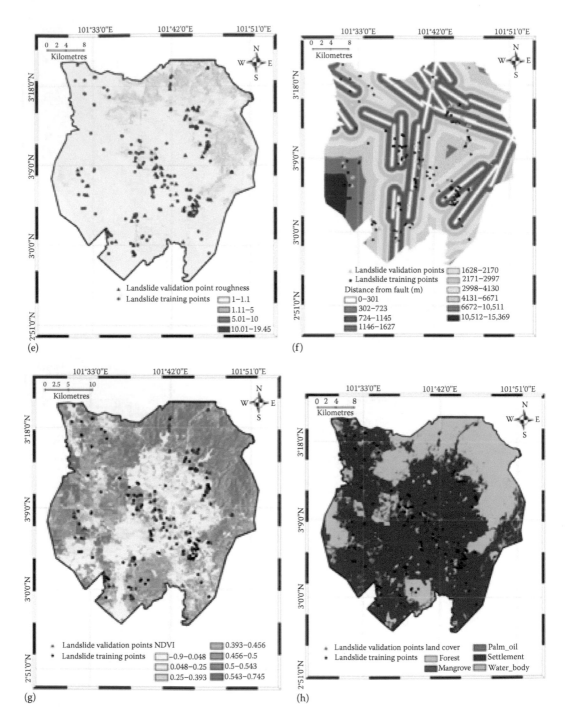

FIGURE 16.8 (CONTINUED)   Landslide conditioning factors used for EBF: (e) surface roughness, (f) distance to faults, (g) NDVI, (h) land cover.   (*Continued*)

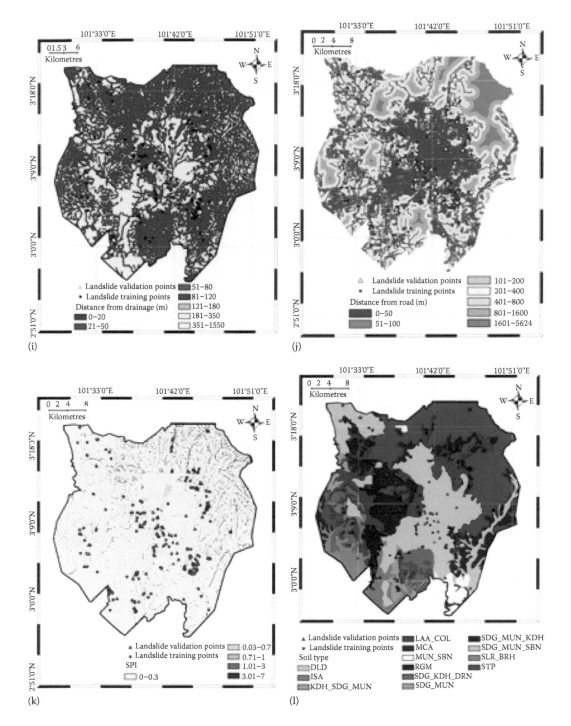

FIGURE 16.8 (CONTINUED) Landslide conditioning factors used for EBF: (i) distance from drainage, (j) distance from road, (k) SPI, (l) soil type.   *(Continued)*

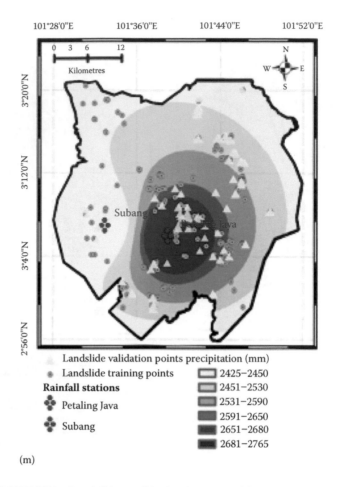

FIGURE 16.8 (CONTINUED)   Landslide conditioning factors used for EBF: (m) precipitation.

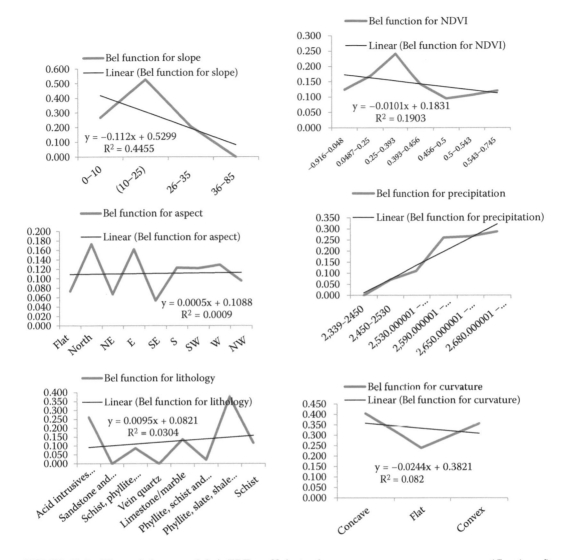

FIGURE 16.9   Thematic layers and their EBF coefficient values.   (*Continued*)

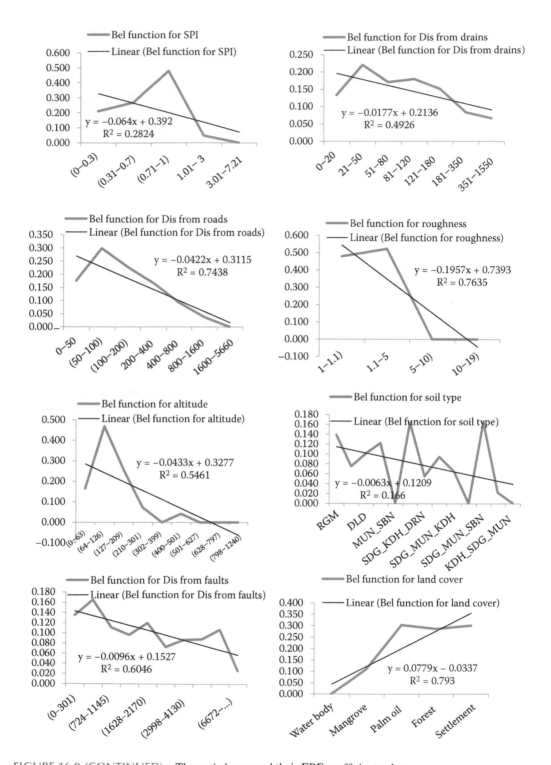

FIGURE 16.9 (CONTINUED)   Thematic layers and their EBF coefficient values.

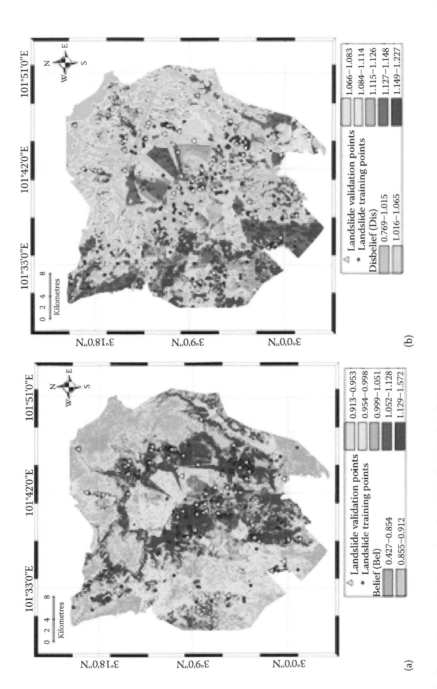

(Continued)

FIGURE 16.10   Integration result of mass function: (a) belief, (b) disbelief.

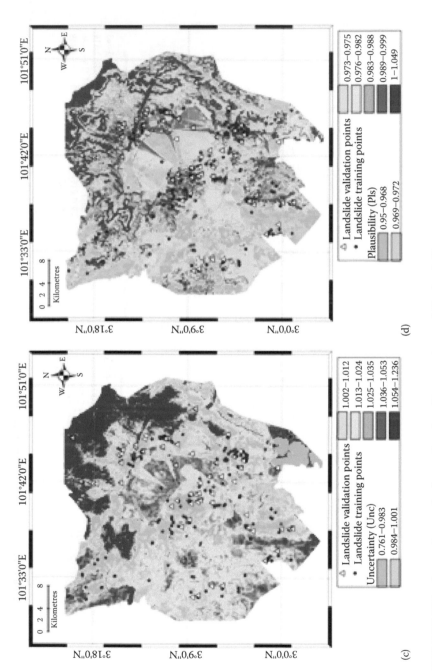

FIGURE 16.10 (CONTINUED) Integration result of mass function: (c) uncertainty and (d) plausibility.

$$Pls \geq Bel$$
$$Unc\,(Ignorance\ or\ Doubt) = Pls - Bel$$
$$If\ Unc = 0,\ then\ Bel = Pls \qquad\qquad (16.19)$$
$$Dis = (1 - Pls)\ or\ (1 - Unc)$$
$$Bel + Unc + Dis = 1$$

The spatial relationship between landslides and conditioning factors was recognized, and the results were transformed into evidential data layers. All covariates were integrated by the overlying technique as the dominant approach, and then a landslide susceptibility map was generated. According to Park (2010), the EBF model is able to represent the degree of uncertainty associated with datasets and target propositions considered.

Finally, the map was divided into five classes, using the quantile-based classification approach, in order to produce a susceptibility map (Pradhan and Lee 2010b; Pradhan et al. 2010; Tien Bui et al. 2012). The integrated uncertainty map Unc (degree of uncertainty) characterizes the areas where the spatial evidence is insufficient to provide real proof for a landslide existing. In this case, additional field assessment for validation and exploration purposes is needed (Althuwaynee et al. 2012a). The integrated map of plausibility Pls (degree of plausibility) represents mass movement location and detects the areas that require further evidence for hazard occurrence. Also, it is able to assess the inefficient conditioning factors that may negate the results.

16.7.3 ACCURACY ASSESSMENT

A total number of 220 landslide locations were divided randomly into two groups for the purpose of training and testing; 70% (153 cells) of the locations were used to train the model and 30% (67 cells) were selected to validate the efficiency of the model. For validation, success rate and prediction rate methods were used. The success rate curve compares the training dataset locations with the landslide susceptibility map. The susceptibility map was divided into equal quantile classes from the high- to the low-susceptibility index values to be used in the validation process. Prediction accuracy was carried out using the data that were not used to build the model (Brenning 2005; Pradhan and Lee 2010b). Perfect prediction accuracy can be achieved when AUC is equal to 1. In this study, AUC shows 0.82 and 0.745 for the success and prediction rate accuracy, respectively (Figure 16.11).

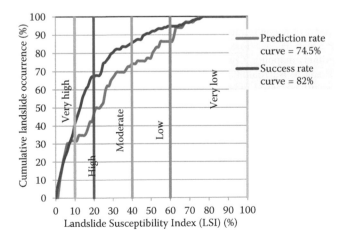

FIGURE 16.11   Prediction and success rate curves for the landslide susceptibility map.

16.8 CONCLUSION

Landslides are a complex natural phenomenon varying in scale from the local to the regional, and they have direct and indirect impacts. Many aspects of landslides, causes, mapping and monitoring are reviewed in this chapter. The main forms of landslides are shallow rapid landslides, rapid and slow deep slides, deep-seated landslides and slow flows and deformations. Landslides are the result of the interplay of two factors which are predisposing (slope angle, slope aspect, elevation, vegetation cover, etc.) and triggering factors (rainfall event, snowmelting, earthquake, sea storm, etc.).

The monitoring techniques consist of RS and field-based methods. As mentioned earlier, the first stage of landslide mapping is the construction of landslide inventory, which links to the distribution of different types of landslides with several morphological and geological characteristics. Many methods were reviewed, such as qualitative approaches that are mostly used by geologists and geomorphologists, and quantitative approaches which utilize the engineering principles of slope instability in terms of the safety factor.

To perform the validation, it is recommended to use different time-series data. The most common technique to validate the performance of the modelling is ROC curves, which can produce success and prediction rate curves, which have been discussed briefly. Also, a case study from Malaysia was used to perform landslide susceptibility mapping using the data-driven EBF method.

Having sufficient knowledge of landslides and their spatial relation to conditioning factors will lead to better analysis with high precision. The results discussed in this chapter deal with appropriate modelling tools and data and can provide very useful information for decision making and policy planning in landslide-prone areas. This can assist the local government, especially in developing countries, to control and mitigate future landslides and reduce the damage and loss of lives they cause.

REFERENCES

Akgun, A., Dag, S., and Bulut, F. (2008). Landslide susceptibility mapping for a landslide-prone area (Findikli, NE of Turkey) by likelihood-frequency ratio and weighted linear combination models. *Environmental Geology*, 54, 1127–1143.

Aleotti, P., and Chowdhury, R. (1999). Landslide hazard assessment: Summary review and new perspectives. *Bulletin of Engineering Geology and the Environment*, 58(1), 21–44.

Althuwaynee, O., Pradhan, B., Park, H.-J., and Lee, J. (2014a). A novel ensemble decision tree-based CHi-squared Automatic Interaction Detection (CHAID) and multivariate logistic regression models in landslide susceptibility mapping. *Landslides*, 11, 1063–1078.

Althuwaynee, O. F., and Pradhan, B. (2016). Semi-quantitative landslide risk assessment using GIS-based exposure analysis in Kuala Lumpur City. *Geomatics, Natural Hazards and Risk*, 1–27. doi: 10.1080/19475705.2016.1255670.

Althuwaynee, O. F., Pradhan, B., and Ahmad, N. (2014b). Estimation of rainfall threshold and its use in landslide hazard mapping of Kuala Lumpur metropolitan and surrounding areas. *Landslides*, 12, 861–875.

Althuwaynee, O. F., Pradhan, B., and Lee, S. (2012a). Application of an evidential belief function model in landslide susceptibility mapping. *Computers & Geosciences*, 44(0), 120–135.

Althuwaynee, O. F., Pradhan, B., Mahmud, A. R., and Yusoff, Z. M. (2012b). Prediction of slope failures using bivariate statistical based index of entropy model. Presented at the 2012 IEEE Colloquium on Humanities, Science and Engineering (CHUSER), Kota Kinabalu, Sabah, Malaysia.

Angeli, M.-G., Pasuto, A., and Silvano, S. (2000). A critical review of landslide monitoring experiences. *Engineering Geology*, 55(3), 133–147.

Bayer, B., Bertello, L., Simoni, A., Berti, M., Schmidt, D., Generali, M., and Pizziolo, M. (2016). Ground surface deformations induced by tunneling under deep-seated landslides in the northern Apennines of Italy imaged using advanced InSAR techniques. Presented at the Proceedings of the 12th International Symposium on Landslides: Landslides and Engineered Slopes: Experience, Theory and Practice, Napoli, Italy, 12–19 June.

Bednarik, M., Magulová, B., Matys, M., and Marschalko, M. (2010). Landslide susceptibility assessment of the Kralovany-Liptovskı Mikulás railway case study. *Physics and Chemistry of the Earth, Parts A/B/C*, 35(3–5), 162–171.

Bishop, C. M. (1995). *Neural Networks for Pattern Recognition*. Oxford: Oxford University Press.

Brenning, A. (2005). Spatial prediction models for landslide hazards: Review, comparison and evaluation. *Natural Hazards and Earth System Sciences*, 5, 853–862.

Bruno, F., and Martillier, F. (2000). Test of high-resolution seismic reflection and other geophysical techniques on the Boup landslide in the Swiss Alps. *Surveys in Geophysics*, 21(4), 335–350.

Bui, D. T., Ho, T. C., Revhaug, I., Pradhan, B., and Nguyen, D. B. (2014). *Landslide Susceptibility Mapping Along the National Road 32 of Vietnam Using GIS-Based J48 Decision Tree Classifier and Its Ensembles Cartography from Pole to Pole*. Berlin: Springer, 303–317.

Bui, D. T., Pradhan, B., Lofman, O., Revhaug, I., and Dick, O. B. (2012). Landslide susceptibility assessment in the Hoa Binh province of Vietnam: A comparison of the Levenberg-Marquardt and Bayesian regularized neural networks. *Geomorphology*, 171–172, 12–29.

Bui, D. T., Pradhan, B., Lofman, O., Revhaug, I., and Dick, Ø. B. (2013). Regional prediction of landslide hazard using probability analysis of intense rainfall in the Hoa Binh province, Vietnam. *Natural Hazards*, 66(2), 707–730.

Bui, D. T., Tuan, T. A., Klempe, H., Pradhan, B., and Revhaug, I. (2016). Spatial prediction models for shallow landslide hazards: A comparative assessment of the efficacy of support vector machines, artificial neural networks, kernel logistic regression, and logistic model tree. *Landslides*, 13(2), 361.

Caine, N. (1980). The rainfall intensity: Duration control of shallow landslides and debris flows. *Geografiska Annaler: Series A: Physical Geography*, 62, 23–27.

Cardinali, M., Antonini, G., Reichenbach, P., and Guzzetti, F. (2001). Photo-geological and landslide inventory map for the Upper Tiber River basin. CNR Gruppo Nazionale per la Difesa dalle Catastrofi Idrogeologiche Publication n. 2154.

Carranza, E. J. M., Woldai, T., and Chikambwe, E. M. (2005). Application of data-driven evidential belief functions to prospectivity mapping for aquamarine-bearing pegmatites, Lundazi District, Zambia. *Natural Resources Research*, 14(1), 47–63. doi: 10.1007/s11053-005-4678-9.

Cheng, G., Guo, L., Zhao, T., Han, J., Li, H., and Fang, J. (2013). Automatic landslide detection from remote-sensing imagery using a scene classification method based on BoVW and pLSA. *International Journal of Remote Sensing*, 34(1), 45–59.

Chigira, M., and Oyama, T. (2000). Mechanism and effect of chemical weathering of sedimentary rocks. *Developments in Geotechnical Engineering*, 84, 267–278.

Constantin, M., Bednarik, M., Jurchescu, M. C., and Vlaicu, M. (2011). Landslide susceptibility assessment using the bivariate statistical analysis and the index of entropy in the Sibiciu Basin (Romania). *Environmental Earth Sciences*, 63(2), 397–406.

Corominas, J., and Moya, J. (2008). A review of assessing landslide frequency for hazard zoning purposes. *Engineering Geology*, 102(3–4), 193–213.

Crozier, M. (1997). The climate-landslide couple: A Southern Hemisphere perspective. Rapid mass movement as a source of climatic evidence for the Holocene. Stuttgart: Gustav Fischer Verlag, 333–354.

Del Ventisette, C., Righini, G., Moretti, S., and Casagli, N. (2014). Multitemporal landslides inventory map updating using spaceborne SAR analysis. *International Journal of Applied Earth Observation and Geoinformation*, 30(0), 238–246. http://dx.doi.org/10.1016/j.jag.2014.02.008.

Devkota, K., Regmi, A., Pourghasemi, H., Yoshida, K., Pradhan, B., Ryu, I. et al. (2013). Landslide susceptibility mapping using certainty factor, index of entropy and logistic regression models in GIS and their comparison at Mugling–Narayanghat road section in Nepal Himalaya. *Natural Hazards*, 65(1), 135–165. doi: 10.1007/s11069-012-0347-6.

Ercanoglu, M. (2005). Landslide susceptibility assessment of SE Bartin (West Black Sea region, Turkey) by artificial neural networks. *Natural Hazards and Earth System Sciences*, 5, 979–992.

Evett, S. R., Tolk, J. A., and Howell, T. A. (2006). Soil profile water content determination: Sensor accuracy, axial response, calibration, temperature dependence, and precision. *Vadose Zone Journal*, 5(3), 894.

Felicísimo, Á. M., Cuartero, A., Remondo, J., and Quirós, E. (2012). Mapping landslide susceptibility with logistic regression, multiple adaptive regression splines, classification and regression trees, and maximum entropy methods: A comparative study. *Landslides*, 10, 175–189.

Fiorucci, F., Cardinali, M., Carlà, R., Rossi, M., Mondini, A., Santurri, L. et al. (2011). Seasonal landslide mapping and estimation of landslide mobilization rates using aerial and satellite images. *Geomorphology*, 129, 59–70.

Floris, M., Iafelice, M., Squarzoni, C., Zorzi, L., De Agostini, A., and Genevois, R. (2011). Using online databases for landslide susceptibility assessment: An example from the Veneto region (northeastern Italy). *Natural Hazards and Earth System Sciences*, 11, 1915–1925.

Frattini, P., Crosta, G., and Carrara, A. (2010). Techniques for evaluating the performance of landslide suscep-tibility models. *Engineering Geology*, 111(1–4), 62–72.

Garrett, J. (1994). Where and why artificial neural networks are applicable in civil engineering. *Journal of Computing in Civil Engineering, ASCE*, 8(2), 129–130.

Gili, J. A., Corominas, J., and Rius, J. (2000). Using global positioning system techniques in landslide monitor-ing. *Engineering Geology*, 55(3), 167–192.

Guzzetti, F., Cardinali, M., and Reichenbach, P. (1996). The influence of structural setting and lithology on landslide type and pattern. *Environmental and Engineering Geoscience*, 2(4), 531–555.

Guzzetti, F., Cardinali, M., Reichenbach, P., and Carrara, A. (2000). Comparing landslide maps: A case study in the upper Tiber River Basin, central Italy. *Environmental Management*, 25(3), 247–263.

Guzzetti, F., Mondini, A. C., Cardinali, M., Fiorucci, F., Santangelo, M., and Chang, K. T. (2012). Landslide inventory maps: New tools for an old problem. *Earth-Science Reviews*, 112(1), 42–66.

Guzzetti, F., Reichenbach, P., Cardinali, M., Galli, M., and Ardizzone, F. (2005). Probabilistic landslide hazard assessment at the basin scale. *Geomorphology*, 72(1), 272–299.

Hanley, J. A., and McNeil, B. J. (1983). A method of comparing the areas under receiver operating characteris-tic curves derived from the same cases. *Radiology*, 148(3), 839.

Hartlen, J., and Viberg, L. (1988). General report: Evaluation of landslide hazard. In *Proc. 5th Int. Symp. on Landslides*, Lausanne, 2, 1037–1058.

Haugen, B. D. (2016). Qualitative and quantitative comparative analyses of 3D LIDAR landslide displacement field measurements. Golden: Colorado School of Mines.

Huabin, W., Gangjun, L., Weiya, X., and Gonghui, W. (2005). GIS-based landslide hazard assessment: An overview. *Progress in Physical Geography*, 29(4), 548–567.

Huang, G., Du, Y., Meng, L., Huang, G., Wang, J., and Han, J. (2017). Application performance analysis of three GNSS precise positioning technology in landslide monitoring. In *China Satellite Navigation Conference* (139–150). Springer, Singapore.

Hutchinson, J. (1995). Landslide hazard assessment. Presented at the Proceedings of the VI International Symposium on Landslides, Christchurch.

Jaboyedoff, M., Oppikofer, T., Abellán, A., Derron, M. H., Loye, A., Metzger, R., and Pedrazzini, A. (2010). Use of LIDAR in landslide investigations: A review. *Natural Hazards*, 61(1), 5–28.

Joyce, K., Samsonov, S., Levick, S., Engelbrecht, J., and Belliss, S. (2014). Mapping and monitoring geologi-cal hazards using optical, LIDAR, and synthetic aperture RADAR image data. *Natural Hazards*, 73(2), 137–163.

Kayastha, P., Dhital, M., and De Smedt, F. (2012). Application of the analytical hierarchy process (AHP) for landslide susceptibility mapping: A case study from the Tinau watershed, west Nepal. *Computers & Geosciences*, 52, 398–408.

Kirschbaum, D. B. (2009). Multi-scale landslide hazard and risk assessment: A modeling and multivariate sta-tistical approach. PhD thesis, Columbia University. http://adsabs.harvard.edu/abs/2009PhDT.......189K.

Lee, S., and Evangelista, D. (2006). Earthquake-induced landslide-susceptibility mapping using an artificial neural network. *Natural Hazards and Earth System Sciences*, 6, 687–695.

Lee, S., Ryu, J. H., Min, K., and Won, J. S. (2003). Landslide susceptibility analysis using GIS and artificial neural network. *Earth Surface Processes and Landforms*, 28(12), 1361–1376.

Lee, S., Ryu, J. H., Won, J. S., and Park, H. J. (2004). Determination and application of the weights for land-slide susceptibility mapping using an artificial neural network. *Engineering Geology*, 71(3), 289–302.

Lei, C. (2012). Earthquake-triggered landslides. Presented at the 1st Civil and Environmental Engineering Student Conference, London.

Leroi, E. (1996). Landslide hazard-risk maps at different scales: Objectives, tools and developments *Landslides*, 1, 35–51.

Likkason, O. K., Shemang, E. M., and Suh, C. E. (1997). The application of evidential belief function in the integration of regional geochemical and geological data over the Ife-Ilesha Goldfield, Nigeria. *Journal of African Earth Sciences*, 25(3), 491–501.

Lin, J., Wang, M., Yang, J., and Yang, Q. (2017). Landslide identification and information extraction based on optical and multispectral UAV remote sensing imagery. In *IOP Conference Series: Earth and Environmental Science*, 57(1), 012017. Beijing, China.

Lo, C.-M., Lee, C.-F., and Keck, J. (2017). Application of sky view factor technique to the interpretation and reactivation assessment of landslide activity. *Environmental Earth Sciences*, 10(76), 1–14.

Lu, P., Catani, F., Tofani, V., and Casagli, N. (2014). Quantitative hazard and risk assessment for slow-moving landslides from persistent scatterer interferometry. *Landslides*, 11, 685–696.

Manchao, H. (2009). Real-time remote monitoring and forecasting system for geological disasters of landslides and its engineering application. *Chinese Journal of Rock Mechanics and Engineering*, 6, 003.

Manzo, G., Tofani, V., Segoni, S., Battistini, A., and Catani, F. (2012). GIS techniques for regional-scale landslide susceptibility assessment: The Sicily (Italy) case study. *International Journal of Geographical Information Science*, 27(7), 1433–1452.

Martha, T. R., Kerle, N., Jetten, V., van Westen, C. J., and Kumar, K. V. (2010). Characterising spectral, spatial and morphometric properties of landslides for semi-automatic detection using object-oriented methods. *Geomorphology*, 116(1), 24–36.

Mihir, M., Malamud, B., Rossi, M., Reichenbach, P., and Ardizzone, F. (2014). Landslide susceptibility statistical methods: A critical and systematic literature review. In *EGU General Assembly Conference Abstracts* (16). Vienna, Austria.

Mondini, A., Guzzetti, F., Reichenbach, P., Rossi, M., Cardinali, M., and Ardizzone, F. (2011a). Semi-automatic recognition and mapping of rainfall induced shallow landslides using optical satellite images. *Remote Sensing of Environment*, 115(7), 1743–1757.

Mondini, A. C., Chang, K.-T., and Yin, H.-Y. (2011b). Combining multiple change detection indices for mapping landslides triggered by typhoons. *Geomorphology*, 134(3), 440–451.

Mora, P., Baldi, P., Casula, G., Fabris, M., Ghirotti, M., Mazzini, E., and Pesci, A. (2003). Global positioning systems and digital photogrammetry for the monitoring of mass movements: Application to the Ca'di Malta landslide (Northern Apennines, Italy). *Engineering Geology*, 68(1), 103–121.

Nandi, A., and Shakoor, A. (2010). A GIS-based landslide susceptibility evaluation using bivariate and multivariate statistical analyses. *Engineering Geology*, 110(1), 11–20.

Nefeslioglu, H., and Gokceoglu, C. (2011). Probabilistic risk assessment in medium scale for rainfall-induced earthflows: Catakli catchment area (Cayeli, Rize, Turkey). *Mathematical Problems in Engineering*, 2011, 280431.

Ohlmacher, G. C., and Davis, J. C. (2003). Using multiple logistic regression and GIS technology to predict landslide hazard in northeast Kansas, USA. *Engineering Geology*, 69(3), 331–343.

Paola, J. D., and Schowengerdt, R. A. (1995). A detailed comparison of backpropagation neural network and maximum-likelihood classifiers for urban land use classification. *IEEE Transactions on Geoscience and Remote Sensing*, 33(4), 981–996.

Park, N.-W. (2010). Application of Dempster-Shafer theory of evidence to GIS-based landslide susceptibility analysis. *Environmental Earth Sciences*, 62(2), 367–376. doi: 10.1007/s12665-010-0531-5.

Pellicani, R., Frattini, P., and Spilotro, G. (2014). Landslide susceptibility assessment in Apulian Southern Apennine: Heuristic vs. statistical methods. *Environmental Earth Sciences*, 72, 1097–1108.

Perus, I., and Krainc, A. (1996). aiNet: A neural network application for 32bit Windows environment. User's manual. Version 1.

Pradhan, B. (2010a). Landslide susceptibility mapping of a catchment area using frequency ratio, fuzzy logic and multivariate logistic regression approaches. *Journal of the Indian Society of Remote Sensing*, 38(2), 301–320.

Pradhan, B. (2010b). Remote sensing and GIS-based landslide hazard analysis and cross-validation using multivariate logistic regression model on three test areas in Malaysia. *Advances in Space Research*, 45, 1244–1256.

Pradhan, B., and Lee, S. (2007). Utilization of optical remote sensing data and GIS tools for regional landslide hazard analysis using an artificial neural network model. *Earth Science Frontiers*, 14(6), 143–151.

Pradhan, B., and Lee, S. (2010a). Delineation of landslide hazard areas on Penang Island, Malaysia, by using frequency ratio, logistic regression, and artificial neural network models. *Environmental Earth Sciences*, 60(5), 1037–1054.

Pradhan, B., and Lee, S. (2010b). Landslide susceptibility assessment and factor effect analysis: Backpropagation artificial neural networks and their comparison with frequency ratio and bivariate logistic regression modelling. *Environmental Modelling and Software*, 25(6), 747–759.

Pradhan, B., and Lee, S. (2010c). Regional landslide susceptibility analysis using back-propagation neural network model at Cameron Highland, Malaysia. *Landslides*, 7(1), 13–30.

Pradhan, B., Lee, S., and Buchroithner, M. F. (2009). Use of geospatial data and fuzzy algebraic operators to landslide-hazard mapping. *Applied Geomatics*, 1(1), 3–15.

Pradhan, B., Lee, S., and Buchroithner, M. F. (2010). Remote sensing and GIS-based landslide susceptibility analysis and its cross-validation in three test areas using a frequency ratio model. *Photogrammetrie, Fernerkundung, Geoinformation*, 2010(1), 17–32.

Pradhan, B., and Sameen, M. I. (2017a). *Effects of the Spatial Resolution of Digital Elevation Models and Their Products on Landslide Susceptibility Mapping Laser Scanning Applications in Landslide Assessment.* Berlin: Springer, 133–150.

Pradhan, B., and Sameen, M. I. (2017b). *Laser Scanning Systems in Landslide Studies Laser Scanning Applications in Landslide Assessment.* Berlin: Springer, 3–19.

Pradhan, B., and Youssef, A. M. (2010). Manifestation of remote sensing data and GIS on landslide hazard analysis using spatial-based statistical models. *Arabian Journal of Geosciences*, 3(3), 319–326.

Preisig, G., Eberhardt, E., Smithyman, M., Preh, A., and Bonzanigo, L. (2016). Hydromechanical rock mass fatigue in deep-seated landslides accompanying seasonal variations in pore pressures. *Rock Mechanics and Rock Engineering*, 49(6), 2333.

Rumelhart, D. E., Hintont, G. E., and Williams, R. J. (1986). Learning representations by back-propagating errors. *Nature*, 323(6088), 533–536.

Santangelo, M., Cardinali, M., Rossi, M., Mondini, A., and Guzzetti, F. (2010). Remote landslide mapping using a laser rangefinder binocular and GPS. *Natural Hazards and Earth System Sciences*, 10, 2539–2546.

Sharma, L., Patel, N., Debnath, P., and Ghose, M. (2012). Assessing landslide vulnerability from soil characteristics—A GIS-based analysis. *Arabian Journal of Geosciences*, 5, 789–796.

Sidle, R. C., and Ochiai, H. (2006). *Landslides: Processes, Prediction, and Land Use.* Washington, DC: American Geophysical Union.

Squarzoni, C., Delacourt, C., and Allemand, P. (2005). Differential single-frequency GPS monitoring of the La Valette landslide (French Alps). *Engineering Geology*, 79(3), 215–229.

Swets, J. A. (1979). ROC analysis applied to the evaluation of medical imaging techniques. *Investigative Radiology*, 14(2), 109–121.

Tien Bui, D., Pradhan, B., Lofman, O., Revhaug, I., and Dick, O. B. (2012). Spatial prediction of landslide hazards in Hoa Binh province (Vietnam): A comparative assessment of the efficacy of evidential belief functions and fuzzy logic models. *Catena*, 96, 28–40.

Van Den Eeckhaut, M., and Hervás, J. (2012). State of the art of national landslide databases in Europe and their potential for assessing landslide susceptibility, hazard and risk. *Geomorphology*, 139, 545–558. http://dx.doi.org/10.1016/j.geomorph.2011.12.006.

Van Westen, C. J., Castellanos, E., and Kuriakose, S. L. (2008). Spatial data for landslide susceptibility, hazard, and vulnerability assessment: An overview. *Engineering Geology*, 102(3–4), 112–131.

Van Westen, C. J., van Asch, T. W. J., and Soeters, R. (2006). Landslide hazard and risk zonation—Why is it still so difficult? *Bulletin of Engineering Geology and the Environment*, 65(2), 167–184. doi: 10.1007/s10064-005-0023-0.

Varnes, D. J. (1984). Landslide hazard zonation—A review of principles and practice. IAEG Commission on Landslides. Paris: UNESCO.

Vlcko, J., Wagner, P., and Rychlikova, Z. (1980). Evaluation of regional slope stability. *Mineralia Slovaca*, 12(3), 275–283.

Weng Chan, N. (1998). Responding to landslide hazards in rapidly developing Malaysia: A case of economics versus environmental protection. *Disaster Prevention and Management: An International Journal*, 7(1), 14–27.

Wilson, J. P., and Gallant, J. C. (2000). Digital terrain analysis. In *Terrain Analysis: Principles and Applications*, ed. J. P. Wilson and J. C. Gallant. Hoboken, NJ: Wiley, 1–27.

Yin, Y., Zheng, W., Liu, Y., Zhang, J., and Li, X. (2010). Integration of GPS with InSAR to monitoring of the Jiaju landslide in Sichuan, China. *Landslides*, 7(3), 359–365. doi: 10.1007/s10346-010-0225-9.

Zhai, G., Fukuzono, T., and Ikeda, S. (2003). Effect of flooding on megalopolitan land prices: A case study of the 2000 Tokai flood in Japan. *Journal of Natural Disaster Science*, 25(1), 23–36.

Zhihua, W. (2007). Remote sensing for landslide survey, monitoring and evaluation. *Remote Sensing for Land & Resources*, 1, 10–15.

17 Mapping and Monitoring of Landslides Using LIDAR

Michel Jaboyedoff, Antonio Abellán, Dario Carrea, Marc-Henri Derron, Battista Matasci and Clément Michoud

CONTENTS

17.1 INTRODUCTION

During the last decades, the research on surface processes has taken its lead from two main developments: (1) faster and more portable remote sensing devices and (2) more powerful computer processors and graphic capabilities. One of these remote sensing techniques – the light detection and ranging (LIDAR, all acronyms are found in Table 17.1), also known as laser scanner – has provided detailed and accurate topographic surveys since the beginning of the twenty-first century (Carter et al. 2001). LIDAR allows us to acquire more than 10,000 points per second from fixed or moving positions, leading to huge sets of data in a three-dimensional (3D) Cartesian coordinate system (i.e. *X*, *Y* and *Z* coordinates), normally referred to as 'point clouds'. This detailed and accurate representation of the terrain surface has been extensively used to characterize, monitor and model landslides

TABLE 17.1

List of Acronyms

Acronym	Meaning
ALS	Airborne laser scanner
DEM	Digital elevation model
EDM	Electronic distance metre
FOV	Field of view
GNSS	Global Navigation Satellite System
GPS	Global positioning system
HRDEM	High-resolution digital elevation model
INS	Inertial navigation system
InSAR	Interferometric Synthetic Aperture Radar
IMU	Inertial measurement unit
LASER	Light amplification by stimulated emission of radiation
LIDAR	Light detection and ranging
LOS	Line of sight
MASER	Microwave amplification by stimulated emission of radiation
SLR	Single-lens reflex
TIN	Triangular irregular network
TLS	Terrestrial laser scanner
TOF	Time of flight

(Rowlands et al. 2003; McKean and Roering 2004; SafeLand Deliverable 4.1 2010; Jaboyedoff et al. 2012; Abellán et al. 2014).

The two main categories of LIDAR for mapping and monitoring landslides are static and mobile devices. Static devices are commonly named ground-based LIDAR or terrestrial laser scanners (TLSs). Regarding mobile devices, the different instruments are normally classified depending on the platform configuration: either boats (Alho et al. 2009; Michoud et al. 2015), cars (Lato et al. 2009) or flying devices (Shan and Toth 2008). The latter, also known as airborne laser scanners (ALSs), include LIDAR assembled on airplanes, but the device can also be mounted over unnamed aerial vehicles (UAVs) and helicopters (Vallet and Skaloud 2004; Li et al. 2011). Examples of point cloud density that can be derived from TLS systems normally range from 100 to 500 points/m^2, whereas ALS systems range from 0.5 to 50 points/m^2.

Point clouds can be converted to surfaces through different surface reconstruction processes, usually involving Delaunay triangulation. For instance, the point cloud produced by an ALS is normally transformed into digital elevation models (DEMs) or high-resolution DEMs (HRDEMs), which provide very detailed geomorphological information (Chigira et al. 2004; Ardizzone et al. 2007).

The main achievements on the application of the LIDAR technique to landslide investigations include (1) the improvement of landslide mapping and inventories using HRDEM (Waleed and Biswajeet 2017); (2) a complete characterization of the 3D slope deformation and signs of precursory activity, allowing the forecasting of landslide failures (Rosser et al. 2007; Collins and Sitar 2008; Abellan et al. 2010; Pedrazzini et al. 2011); (3) an accurate and detailed monitoring of rock failures (Rosser et al. 2005; Teza et al. 2007; Collins and Sitar 2008; Oppikofer et al. 2008; Prokop and Panholzer 2009; Abellan et al. 2010; Kromer et al. 2017; Caudal et al. 2017); (4) the improvement in rock slope characterization in inaccessible areas thanks to the automatic extraction of discontinuity orientation (Bornaz et al. 2002; Kemeny and Turner 2008; Gigli and Casagli 2011; Riquelme et al. 2014); and (5) the identification of geomorphic signatures of debris flows based on an HRDEM and accurate erosion and deposit quantification (Cavalli and Marchi 2008; Bremer and Sass 2012).

17.2 LIDAR AND LASER SCANNING TECHNIQUES

17.2.1 History

The history of LIDAR technology is directly linked to the Light Amplification by Stimulated Emission of Radiation (LASER) evolution. In 1957, the physicists Charles Townes and Arthur Shawlow at Bell Labs investigated the development of optical Microwave Amplification by Stimulated Emission of Radiation (MASER), which later became LASER in 1959 (Brooker 2009). In May 1960, the first instrument which successfully produced a series of pulsed lasers was designed (Hecht 1994). Less than 1 year later, the first LIDAR was manufactured for military purposes (Brooker 2009). In the 1970s, laser technologies became generalized for civilian applications, such as surveying. Some examples include the use of the electronic distance metre (EDM) in quarries and tunnel environments (Dallaire 1974; Petrie 1990; Petrie and Toth 2009a).

At the same time, the first ALS systems were able to measure the range of the emitted pulse with a precision of 1 m (Miller 1965; Krabill et al. 1984), for altimetry and bathymetry purposes (Sheperd 1965; Hoge et al. 1980). Nevertheless, this technology was limited by the poor accuracy on the position and orientation of the airplane given by the inertial navigation systems (INSs). Afterwards, laser devices became more accurate, less heavy, less expensive and eye-safe. Moreover, in the early nineties the improvements on the navigation satellite constellations GPS-NAVSTAR (United States) and GLONASS (USSR) allowed for the overcoming of the limitations of the former INS, considerably improving the ALS accuracy (Petrie and Toth 2009b; Beraldin et al. 2010). In the late nineties, theoretical and practical aspects of ALS techniques were well known and several important works were published (Baltsavias 1999a; Wehr and Lohr 1999). Early applications of ALSs consisted of the production of the DEM (Baltsavias 1999b; Carter et al. 2001), the mapping of topographical changes of Greenland's ice sheet (Krabill et al. 1995) and the creation of a 3D model of an urban environment (Haala and Brenner 1999). The mapping and modelling of landslides from an HRDEM at the regional scale started in the beginning of the twenty-first century (Haugerud et al. 2003; Schulz 2004, 2007; Van Den Eeckhaut et al. 2007; Haneberg et al. 2009).

During the last decade, the Geoscience Laser Altimeter System (GLAS) sensor illuminated the Earth with three lasers. One of these sensors was especially designed to monitor different environmental variables, such as polar ice sheet mass balances, vegetation canopy and land elevation with a vertical accuracy of 3 cm (Zwally et al. 2002; Abshire et al. 2005; Schutz et al. 2005). One of the most relevant topographic products derived from the ICESat mission was the 500 m pixel resolution DEM of Antarctica (DiMarzio 2007; Shan and Toth 2008). Meanwhile, a space-borne EDM device named Mars Orbiter Laser Altimeter (MOLA) was developed to observe Mars from 1999 to 2001 in order to survey its topography and atmosphere (Smith et al. 2001). This sensor measured the surface of Mars for 15 months, obtaining a DEM of Mars with a resolution of a few square kilometres and a vertical accuracy of 1 m (Smith et al. 2001).

At the same time, pioneering triangulation-based TLS systems were designed and used for archeological purposes (Beraldin et al. 2000). Later, ground-based LIDAR sensors were used for surveying and monitoring deformations of urban structures (Gordon et al. 2001; Lichti et al. 2002). Geological applications of TLSs have been experiencing fast developments since 2002, including (1) the extraction of orientations and roughness of rock slope discontinuities (Slob et al. 2002; Kemeny and Post 2003; Fardin et al. 2004; Feng and Röschoff 2004; Jaboyedoff et al. 2007; Sturzenegger and Stead 2009; Garcia-Selles et al. 2011; Riquelme et al. 2014), (2) the monitoring of volcanic activity (Hunter et al. 2003), (3) the litho-stratigraphic modelling (Bellian et al. 2005; Buckley et al. 2008; Kurz et al. 2008; Humair et al. 2015), (4) the detection of rockfalls (Lim et al. 2005; Rosser et al. 2005; Collins and Sitar 2008; Abellan et al. 2010), and (5) the monitoring of landslides (Teza et al. 2007; Collins and Sitar 2008; Monserrat and Crosetto 2008; Fey et al. 2015).

17.2.2 Instrument Principle

17.2.2.1 LIDAR Functioning

A LIDAR device consists of a combination of a laser rangefinder and a scanning mechanism which allows for the precise measurement of a distance to a target and the orientation of the laser beam. The scanner device works through the internal rotation of one or two mirrors and/or the rotation of the whole device. Additional components normally include an electronic unit, an imaging device (e.g. digital camera) and specific software to control the whole system.

There are three main types of LIDAR, corresponding to three different ways of measuring the range, for instance, the distance along the line of sight (LOS) between the sensor and the terrain (Figure 17.1): (1) pulse LIDAR measures the time of flight (TOF) of a laser pulse, (2) phase LIDAR uses the phase shift between the emitted and received signals, and (3) triangulation LIDAR uses a camera to locate the laser spot on the scanned surface (Vosselman and Maas 2010).

Pulse LIDAR is based on the measurement of the time delay (TOF) of a laser pulse travelling from the source (e.g. TLSs) to a reflective target (e.g. ground surface) and back to the source, as follows (Petrie and Toth 2009a):

$$d = \frac{1}{2} c \times \Delta t \tag{17.1}$$

where d is the range, c is the speed of light in the air (3×10^8 m/s) and Δt is the TOF. This technique allows for centimetre accuracy measurements at several kilometres' distance. Phase LIDAR allows for more precise acquisition (some millimetres), but it only works for short ranges (less than 100 m). Phase LIDAR uses a continuous amplitude and/or frequency-modulated beam instead of pulses. The range is deduced from the phase shifts between emitted and backscattered signal for a couple of frequencies (Petrie and Toth 2009a). Triangulation LIDAR estimates the range from two angles: the illumination angle of the laser beam and the observation angle of the camera (Beraldin et al. 2010). With a high accuracy (around 0.1 mm), triangulation LIDAR has very short ranges (less than 5 m). Pulse (or TOF) LIDAR is the main device used for landslide monitoring and mapping, since it is the only one that makes possible acquisitions for ranges longer than 100 m (up to a couple of kilometres for the most advanced sensors). There are few applications of phase and triangulation LIDAR for

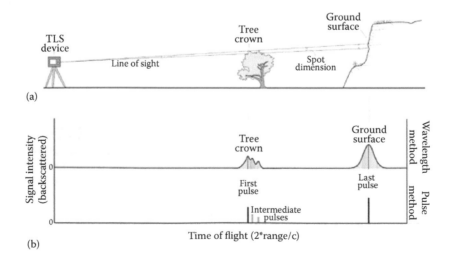

FIGURE 17.1 (a) Sketch of LIDAR functioning principle showing TLS device, LOS and spot dimension. (b) Signal intensity vs. TOF for two different methods of measurement: full-wavelength method vs. pulse method.

landslides, such as very detailed characterization of sliding surfaces or analogue modelling in the lab (Carrea et al. 2012).

The footprint size depends on the range and the beam divergence angle. The shape of the footprint on the surface depends on the local incidence angle, that is the angle between the LOS and the normal to the surface (Figure 17.2). The range is measured at the peak of intensity of the backscattered pulse. If the surface is normal to the laser beam, it corresponds to the centre of the footprint (point P in Figure 17.2). If the incidence angle is high (more than 70°) or the local relief is rough, the recorded point of location (point P′ in Figure 17.2) can lie anywhere within the spot (Schaer et al. 2007; Lichti and Skaloud 2010).

17.2.2.2 Multiple Echoes

Even if strongly collimated, laser beams of LIDAR are affected by divergence; that is, the beam width increases with the range. Typical diameters of beam footprints are 30–50 cm for ALSs at 1 km range, and 10 cm for TLSs at 500 m range (Lichti and Skaloud 2010). As a result, the consequences are that several backscattered pulses can correspond to a single emitted pulse (multiple echoes). For instance, when a pulse hits a tree, part of the pulse can be reflected by the top of the canopy, another part by branches and the last one by the ground (Figure 17.1). This characteristic of the beam to 'penetrate' the vegetation is a key advantage of LIDAR compared with photogrammetry, in order to get ground elevation models of vegetated areas. However, a dense forest may prevent the beam from reaching the ground surface. Full-wave LIDAR devices, usually ALS, provide a continuous record of the backscattered signal (full waveform). Most TLSs record only one or several discrete pulses. In landslide studies, the last pulse (i.e. the pulse backscattered by the ground surface) is generally used. But it may be convenient to have other return pulses for vegetation removal purposes (Petrie and Toth 2009a).

17.2.2.3 Other Parameters: Intensity + Colour

In addition to 3D surface point locations, LIDAR devices measure the signal intensity of the backscattered laser signal. Intensity mainly depends on the type of material, orientation of the slope, range and laser wavelength. Additional external cameras can be coupled with LIDAR during acquisition, allowing the assignment of a colour attribute (i.e. RGB) to each point of the LIDAR point

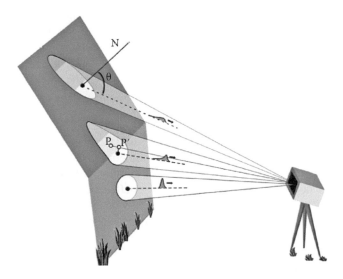

FIGURE 17.2 LIDAR footprint deformation (θ, incidence angle; N, normal to the surface). The point position is attributed to the centre of the beam (point P), but the range is measured at the maximum intensity of the backscattered pulse. Then, the actually measured point P′ is located slightly before the rock face on the LOS.

cloud. During ALS acquisitions, natural colours and/or infrared images are usually taken simultaneously. A single-lens reflex (SLR) digital camera is frequently used to provide natural colours to TLS point clouds, but more advanced imaging devices, such as hyperspectral cameras, can be used too (Kurz et al. 2011). Some of the main uses of intensity values or colour attributes for landslide studies include bedrock mapping, vegetation removal and interactive visualization of point clouds (Alba et al. 2011).

17.2.2.4 ALS versus TLS

ALSs are widely used, by either national topography agencies or private companies, to cover large areas for topographical surveying or forest management (Shan and Toth 2008). The plane usually flies 1000 m above the ground, and the ALS records 1–2 points by ground square metre (recent developments increase these values by one order of magnitude), with a position accuracy of 5–30 cm (Petrie and Toth 2009b). Mobile techniques require a precise recording of the location and altitude of the scanning device during the acquisition, which is actually carried out by coupling the LIDAR with a Global Navigation Satellite System (GNSS) (usually a differential global positioning system [GPS]) and with an inertial measurement unit (IMU). Then the location of each point of a scan is recalculated as a function of the LIDAR position and orientation during a post-processing stage. Due to its heavy logistics, an ALS is operated by professional surveying companies. Other smaller mobile platforms are also occasionally used for laser scanning of landslides. For instance, LIDAR sensors can be assembled to helicopters, taking advantage of their relatively low flying speed and flying heights close to the terrain surface, which allows acquiring 3D information on cliffs from an oblique view. This was used to scan a 2 km² area around the Åknes rockslide (Norway), providing 20 measurement points per square metre with a position precision of 5 cm (Blikra et al. 2006; Jaboyedoff et al. 2011). Also, offshore laser scanning from a boat is used for coastal cliff and river bank stability analysis, with a typical average point spacing of 10–20 cm (Alho et al. 2009). Finally, a fast-growing technique is the setup of laser scanners on cars or trains in order to acquire high-resolution ground models along transport corridors. This technique has been used to assess rock wall stability along roads and railways in Ontario (Canada), with an average point density of 100 points/m² (Lato et al. 2009).

A TLS is a fixed-position scanning technique that only requires the device to be set up on a tripod opposite the surface to scan within the maximum range limit. Because of this relatively easy implementation, a TLS is usually operated directly by the geoscientists in charge of the landslide investigation. With a high point density (up to 10,000 points/m²) and accuracy (about 1 cm), it is mostly limited by the availability of a good point of view, the maximum range of acquisition and the relatively small areas that can be covered in a single acquisition due to a limited field of view (FOV) angle (for instance, 40° in TLS Ilris3D). Usually, in order to cover the whole area of interest, several scans have to be acquired from different points of view.

17.2.2.5 Spacing, Accuracy, Resolution and Data Types

The major differences between the two most commonly used techniques (i.e. TLSs and ALSs) are the (1) average point spacing, metre scale for ALSs and centimetre scale for TLSs; (2) position and range accuracy, some tens of centimetres for ALSs and about 1 cm for TLSs; and (3) point cloud geometry, as ALSs are usually shooting subvertically, they are very efficient in order to produce a horizontally projected surface, such as gridded DEMs. But it only provides a poor representation of steep areas like rock cliffs. In addition to this, and opposite TLS devices, an ALS is not able to have several ground surface elevation values of ground elevation (Z) for each horizontal position (X, Y), and thus it is unable to represent overhanging walls. Properly speaking, ALS products are 2.5-dimensional (2.5D) elevation surfaces and not full 3D representations of the terrain surface. As a result, most ALS processing software is inefficient to work properly with point clouds from TLS acquisitions.

Measurement accuracy on single points depends on range, surface reflectivity and incident angle (Ingensand 2006; Lichti 2007; Abellán et al. 2009; Lichti and Skaloud 2010). Inaccuracy of surface location and geometry is due not only to single measurement errors, but also to alignment and merging processes. This is particularly critical when several acquisitions are merged in one dataset. The quality of the final dataset must be carefully checked, because even very small inaccuracies can produce large uncertainties in change detection as discussed by Schürch et al. (2011).

Point cloud resolution determines the level of detail that can be observed from a scanned point cloud (Pesci et al. 2011), where the laser beam width plays an important role (Lichti and Jamtsho 2006). However, TLS resolution and point spacing are normally mixed up (Lichti and Jamtsho 2006). It is recommended to place the optimal point spacing as 86% of the laser beam width at a given distance.

17.2.2.6 Data Acquisition Issues: Occlusion and Biases

17.2.2.6.1 Occlusion

Due to relief roughness and natural variability, the acquisition of rock slope geometry from a single station is normally affected by the existence of shadow areas (also called occluded areas). In order to minimize occlusion issues (Kemeny and Turner 2008), different series of best practices and protocols are proposed for survey planning. For instance, occlusion can be minimized using data acquisition from multiple locations (Lato et al. 2010; Stock et al. 2011).

17.2.2.6.2 Biases

Discontinuity characterization may be affected by biases depending on the intersection angle between discontinuity sets and the sensor's LOS (Sturzenegger and Stead 2009; Lato et al. 2010): the density of those families with a normal vector parallel to the LIDAR's LOS may be overrepresented compared with the density of families with a normal vector perpendicular to the LOS of the instrument. Indeed, this is a similar effect observed by Terzaghi (1962), in which the density of families with direction parallel to the slope orientation may be overrepresented compared with the density of those families perpendicular to the slope.

17.2.3 DATA TREATMENT

A series of pre-processing steps are necessary in order to transform the RAW LIDAR data (unfiltered point cloud in both ALSs and TLSs) into real terrain points, as follows (Figure 17.3).

17.2.3.1 Full-Waveform and Automatic Filtering

The first truly-operational full-waveform (FW) LIDAR systems for airborne applications appeared in 1999 (Mallet and Bretar 2009). This new generation of instruments is able to capture not only the discrete signal return but also the entire waveform. The first experimental full-waveform LIDAR system was originally developed for forestry canopy and building segmentation. At the present time, full-waveform data are mainly used for ALS applications in order to discriminate the surface terrain from forestry, which is of great interest for vegetation removal and DEM generation. Current research on full-waveform LIDAR data includes signal decomposition for object classification: terrain, vegetation, wires, buildings, and so forth.

The integration of LIDAR-derived 3D data with complementary information registered by other remote sensing methods is also in development. For instance, the coupling of LIDAR and InSAR data was successfully explored for displacement monitoring (Farr et al. 2008; Lingua et al. 2008; Roering et al. 2009). Other spectral attributes provided by multispectral imaging can also help improve interpretation and segmentation of 3D LIDAR data (Kurz et al. 2011).

17.2.3.2 Non-Ground-Point Filtering (Including Vegetation Removal)

Non-ground-point (e.g. vegetation, cables, birds and dust) filtering is a time-consuming task that is commonly performed manually. Although this approach may be straightforward when non-ground

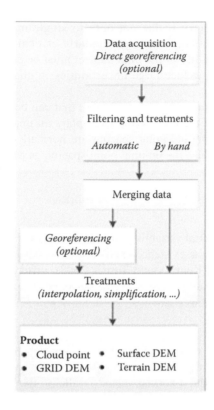

FIGURE 17.3   Workflow showing the main steps of data treatment.

points appear at small portions of the terrain, new semi-automatic methods aiming to classify ground points are being currently developed. These methods of classification, which are normally supervised by the user, are based on the identification of different values of the point cloud that allow us to separate ground from non-ground points. For instance, while some methods use the colour-coded value (RGB) or intensity of the laser return (Lichti 2005; Franceschi et al. 2009), other approaches include the study of some geometrical parameters as single point statistics (Huang et al. 2000), local 3D point cloud statistics (e.g. Lalonde et al. 2006) and multiscale dimensionality criteria (Brodu and Lague 2012), or use classifiers such as support vector machine (Cristianini and Shawe-Taylor 2000), expectation–maximization (EM) algorithms and/or Gaussian mixture models (Xu and Jordan 1996). Although these methods are highly promising, manual vegetation removal is still currently necessary in most of the cases.

17.2.3.3   Co-Registration and Georeferencing

Georeferencing (also known as alignment) is the process of coordinate transformation from an internal referential system relative to the LIDAR device into a real-world system of reference (e.g. Universal Transverse Mercator [UTM], latitude and longitude and national grid coordinates). This process is typically a rigid-based body transformation consisting of a combination of translation and rotation (Oppikofer et al. 2009). Two methods are often used for georeferencing:

1. *Target-based registration*: The alignment of the TLS point cloud is classically carried out with ground control points (GCPs) whose coordinates are previously acquired using differential GPS, total station, and so forth. Complementarily, some modern techniques include the alignment of the 3D point cloud using co-registration of the scanned well-known geometries (i.e. cubes, spheres, cylinders, and so forth) (Rabbani et al. 2007).

2. *Iterative closest point (ICP) methods*: This is aimed at minimizing the Euclidian distance between different point clouds, as proposed by Besl and McKay (1992) and Chen and Medioni (1992). For a fast convergence of the alignment using this technique, an *a priori* alignment is normally carried out using either GCPs or already oriented and referenced commercial DEMs.

Co-registration aims at (1) reducing occlusion and shadow areas by merging multiple scans from the same epoch and different viewpoints into one single point cloud and (2) comparing different datasets and detecting changes on the topographical surface by aligning scans acquired sequentially, for instance, by performing a multitemporal TLS survey during a certain time lag in order to study slope evolution.

Since the final accuracy of single point position is unknown, error quantification is not always straightforward (Barbarella et al. 2017). Co-registration errors mainly depend on instrumental accuracy, point spacing and percentage of overlapping area. Techniques for error propagation control must be performed when dealing with the overlapping of several scans, especially in long linear features, as when studying landslides along transportation corridors, marine cliffs, and so forth (Lim et al. 2010; Lato et al. 2015). Although it is evident in classical surveying techniques, the points used for alignment must be restricted to stable areas, for instance, the surroundings or the stable parts of a given landslide.

17.2.3.4 Point Cloud Comparison

Point cloud comparison between sequential LIDAR datasets allows the detection of temporal differences in terrain morphology at multiple scales (Lim et al. 2005; Kromer et al. 2015). In order to make a comparison, several steps are required: (1) acquisition of a reference point cloud at initial time (t_0), (2) construction of a reference surface at t_0, (3) acquisition of successive data point clouds at different time lapses (t_1, t_2, etc.), (4) co-registration between reference surface and successive data point clouds (Besl and McKay 1992; Chen and Medioni 1992) and (5) change detection through the calculation of the differences for each period of comparison.

The single point distances between reference and successive point clouds are normally computed along a user-defined vector in one dimension. This vector can be defined by both using a fixed orientation (Rosser et al. 2005; Girardeau-Montaut 2006) or using different local vectors defined perpendicular to the rock face at each section of the slope, as described in Brodu and Lague (2012). Another method of comparison includes the use of the 'shortest distance' between the data point cloud and the reference model. The real 3D deformation of single blocks in terms of rotations and translations can be quantified by using the roto-translational (RT) matrix technique (Collins and Sitar 2008; Monserrat and Crosetto 2008; Caudal et al. 2017). More recently, a series of innovative approaches for surface change detection utilize semi-automatic techniques for 4D investigation of slope failures (Kromer et al. 2015; Williams et al. under review).

Regarding sign criteria, a positive value is normally interpreted as an increase of material (e.g. deposition) or a displacement towards the external part of the slope (pre-failure deformation). Similarly, negative values are normally interpreted as a loss of material in a given period (e.g. soil erosion and rockfalls).

17.3 LANDSLIDE APPLICATIONS

17.3.1 LANDSLIDES

Since the beginning of the 2000s, the ALS system has been widely used to produce HRDEMs at regional scale (Petrie and Toth 2009b). This technique quickly became an essential document of a regional landslide inventory map (McKean and Roering 2004; Jaboyedoff et al. 2012; Petschko et al. 2016). By creating a hill-shaded surface, it is possible to highlight structures that cannot

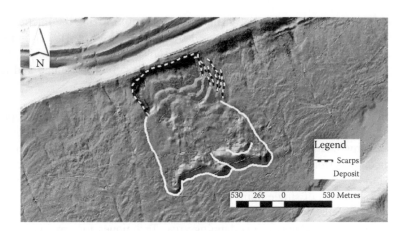

FIGURE 17.4 Mapping of a large-scale landslide detected using a LIDAR-based HRDEM. Bern Canton, northwest Switzerland. (Data from Géodonnées. © Swisstopo, DV084371.)

be seen from aerial photos or in field observation (Schulz 2004; Van Den Eeckhaut et al. 2007; Haneberg et al. 2009; Burns et al. 2010) either due to landslide scale (from metric to kilometric) or location (in inaccessible or densely vegetated areas) (Figure 17.4). Different approaches are usually employed for landslide mapping and identification, with various degrees of automation: (1) a manual or expert-based approach, which focuses on the identification of landslides based on geomorphological evidence of crowns, main scarps, flanks, foot or toe and vegetation cover (Zaruba and Mencl 1982; Cruden and Varnes 1996; Soeters and Van Westen 1996; Agliardi et al. 2001; Braathen et al. 2004), and (2) automatic geomorphological feature recognition, which extracts different information derived from HRDEMs, such as slope angle, curvature or roughness (Glenn et al. 2006).

When several generations of ALS scans are available, change detection between HRDEMs can be used to characterize topographic changes. This technique has been employed to detect slope failures, soil erosion, slope displacements and terrain deformation (Burns et al. 2010; DeLong et al. 2012; Fey et al. 2015). Some examples of ALSs for coastal cliff monitoring and erosion analysis are discussed in Young and Ashford (2008) and Reineman et al. (2009). Furthermore, high-resolution topography from airborne LIDAR can also be compared with less accurate historical aerial photographs, in order to analyze the kinematics of the landslide over a long-term period.

17.3.2 ROCK SLOPES

17.3.2.1 Structural Analysis

The aim of the structural study is the detection and characterization of the main joint sets responsible for the fracturing of the rock mass. The analysis of the network of discontinuities is an essential step to study a rockfall scar or to assess the potential instable areas of a rock slope. The measurements of the orientation (Figure 17.5), undulation, persistence, spacing, opening, roughness and water presence of the main joint sets are usually taken in the field with classical methods, such as photography, direct observations and compass measurements (ISRM 1978). A TLS allows us to perform most of these tasks faster, with high accuracy, and even in inaccessible vertical areas (Kemeny and Post 2003; Jaboyedoff et al. 2009; Lato et al. 2010; Gigli and Casagli 2011; Longoni et al. 2012). Georeferenced 3D TLS point clouds allow us to extract the orientation of structures and perform distance or volume measurements (Oppikofer et al. 2009; Olsen et al. 2015). Similar results can be obtained from an HRDEM generated from aerial LIDAR, but with significant limitations imposed by the vertical shooting direction and by the size of the cells. The representation of structural features like planes or lines, which is classically done with stereo-plots (ISRM 1978), can also be

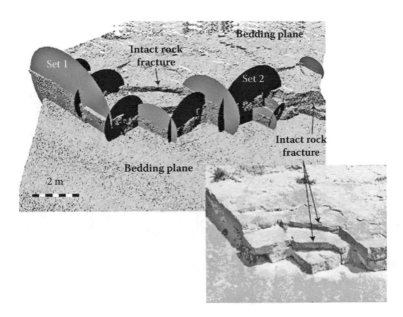

FIGURE 17.5 Discontinuity mapping on a terrestrial LIDAR point cloud, rendered using greyscale intensity values. (From Sturzenegger, M., and Stead, D., *Nat. Hazards Earth Syst. Sci.*, 9, 267–287, 2009. With permission.)

automatically extracted using modern algorithms. Early studies used a manually selected section of the point cloud and calculated the best-fitting plane (Oppikofer et al. 2009; Sturzenegger and Stead 2009), which can be laborious and time consuming. Later, some authors proposed the construction of 2.5D surfaces, such as triangular irregular networks (TINs), for the automatic calculation of the discontinuity orientation (Slob et al. 2002; Kemeny and Turner 2008). More recently, techniques allow automatically computing local plane orientations directly from the point cloud (Ferrero et al. 2009; Garcia-Selles et al. 2011; Gigli and Casagli 2011; Riquelme et al. 2014). In order to facilitate structural interpretations, the orientation and inclination of each point of a TLS point cloud, or of a DEM cell, can be represented using a unique colour (Jaboyedoff et al. 2007). The resolution of the datasets plays an important role in the analysis of discontinuity orientation: as discussed in Sturzenegger and Stead (2009), low-resolution datasets may lead to (1) a truncation of non-persistent discontinuity sets due to a lack of information and (2) a shifting of discontinuity orientation due to a non-realistic smoothening of the surface geometry.

17.3.2.2 Monitoring of Fragmental Rockfalls

Due to the steepness of rock slopes, terrestrial LIDAR is more extensively used than airborne LIDAR for the monitoring of fragmental rockfalls. Moreover, terrestrial LIDAR provides an unprecedented level of detail at a site-specific slope, with both resolution and accuracy ranging from several millimetres to a few centimetres. A pioneer approach for monitoring fragmental rockfalls along a section of marine cliffs in northeast England was developed by Lim et al. (2005) and Rosser et al. (2005). Since then, TLSs have become a widely used tool for automatic rockfall detection and volume calculation (Abellan et al. 2010; Stock et al. 2011; Collins and Stock 2012; Carrea et al. 2014; Tonini and Abellan 2014; Micheletti et al. 2017; Williams et al. under review). Using the TLS technique, Stock et al. (2012) detected a series of progressive rockfalls that occurred in exfoliated granitic rocks, analyzing how a given fracture event initiated subsequent adjacent failures. A common application of TLSs for rockfall monitoring includes the analysis of magnitude–frequency (MF) characterization of fragmental rockfalls. Indeed, several authors have observed a scale-invariant pattern in an MF dimension under log-log scale (Rosser et al. 2005; Dewez et al. 2009;

FIGURE 17.6   Model comparison showing the changes (i.e. rockfalls) along time on La Cornalle study area (Vaud, Switzerland).

Lim et al. 2010; Santana et al. 2012) (Figure 17.6), suggesting that a self-organized criticality (SOC) may rule rockfall phenomena (Hergarten 2003). Some other applications include the monitoring of several million cubic metres of rockslides (Figure 17.7) in order to quantify deformation rates and failure mechanisms (Collins and Sitar 2008; Avian et al. 2009; Oppikofer et al. 2009; Ravanel et al. 2010), and the failure forecasting based on a given precursory indicator: either minor-scale rockfalls or tertiary creep deformation (Rosser et al. 2007; Abellan et al. 2010; Royan et al. 2013; Kromer et al. 2017). More recently, the combination of web-retrieved images and LIDAR data for rock slope monitoring has been explored (Guerin et al. 2017; Voumard et al. 2017).

17.3.2.3   RockFall Susceptibility Assessment

The simplest way to detect the potential rockfall source areas is by defining a slope angle threshold above which rockfalls are more susceptible to occur (Guzzetti et al. 2003). This method can be optionally coupled with other criteria, such as the presence of cliff areas (Jaboyedoff and Labiouse 2003). The slope threshold can be gathered from a detailed statistical analysis of slope angles, thus allowing us to identify the cliff area (Strahler 1954; Baillifard et al. 2004; McKean and Roering 2004; Loye et al. 2009). A simple GIS approach developed by Baillifard et al. (2003) allows identifying potential rockfall source areas along roads according to the presence or absence of four parameters: faults, scree slopes, rock cliffs and steep slopes. A great improvement related to the use of DEMs for the detection of rockfall-prone areas consists of the automatic realization of a kinematic analysis along the whole area (Wagner et al. 1988; Gokceoglu et al. 2000). This analysis allows the detection of the topography areas where a discontinuity of a given orientation can lead to an instability. Using a standard classic stability criterion (Norrish and Wyllie 1996), together with a statistical analysis of the kinematic tests, Gokceoglu et al. (2000) were able to produce probability maps of different failure mechanisms, such as planar sliding, toppling and wedge sliding. The kinematic analysis criterion was applied to accurately detect the most unstable areas from TLS data (Fanti et al. 2013). Spacing and persistence of the joint sets, which can be determined both in the field and on LIDAR point clouds (Sturzenegger and Stead 2009; Lato et al. 2010; Riquelme et al. 2015), should also be taken into account in the stability calculations, as suggested by Jaboyedoff et al. (2004). The number of potential failures per cliff area can be calculated based on the spacing and on the trace length values, which considerably improves the rockfall susceptibility maps (Brideau et al. 2011; Matasci et al. accepted).

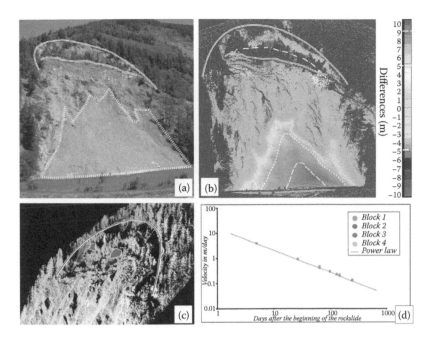

FIGURE 17.7   Different techniques for the quantification of a mass movement located on Montatuay mountain (Valais, Switzerland) using a ground-based LIDAR. (a) Picture of the study area. (b) Filtered point cloud of the study area; each point is coloured according to the differences calculated between the 2009 ALS DEM and the TLS point cloud captured in April 2011. (c) Rotated perspective of the LIDAR point cloud, including the manual selection of different blocks for subsequent tracking. (d) Velocity computation for each of the previously selected blocks; the straight line in log-log scale shows a progressive stabilization of the slope.

17.3.3 DEBRIS FLOWS

LIDAR is used for various tasks related to debris flow mapping, characterization and modelling. HRDEMs from ALSs are frequently used to identify and characterize initiation and propagation zones of debris flows (Staley et al. 2006; Conway et al. 2010; Lopez Saez et al. 2011; Lancaster et al. 2012) and deposition fans (Staley et al. 2006), or to estimate the eroded and deposited volumes (Breien et al. 2008; Scheidl et al. 2008; Bull et al. 2010). Remarkably, Schürch et al. (2011) have used TLSs to detect surface changes along a debris flow channel in Switzerland, and Cavalli and Marchi (2008) have combined TLSs and ALSs to quantify sediment transport for debris flow events in the Austrian Alps, observing several metre changes between the two epochs (Figure 17.8). ALS data are also used to estimate the volume of sediments that can be potentially entrained along the channels by a future event (Abancó and Hürlimann 2014). Apart from geomorphic analysis, a DEM from laser scanning is also used in debris flow propagation modelling. For instance, Stolz and Huggel (2008) have assessed the impact of HRDEM cell sizes (1, 4 and 25 m) on modelled propagations, in particular the effect of the resolution on debris flow confinement along recent channels.

17.3.4 INPUT FOR MODELLING

Mass movement modelling is the process of simulating its stability and/or its propagation across time. This analysis can be carried out using conventional representations of a terrain's surface, either a DEM or an HRDEM. Note: Current models do not utilize real 3D point clouds but 2.5D projections of the topography.

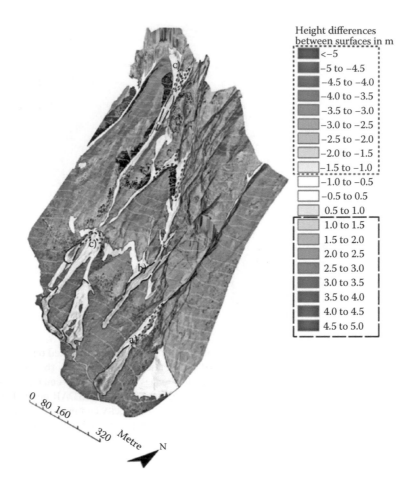

Height differences
between surfaces in m
<-5
-5 to -4.5
-4.5 to -4.0
-4.0 to -3.5
-3.5 to -3.0
-3.0 to -2.5
-2.5 to -2.0
-2.0 to -1.5
-1.5 to -1.0
-1.0 to -0.5
-0.5 to 0.5
0.5 to 1.0
1.0 to 1.5
1.5 to 2.0
2.0 to 2.5
2.5 to 3.0
3.0 to 3.5
3.5 to 4.0
4.0 to 4.5
4.5 to 5.0

0 80 160   320   Metre   N

FIGURE 17.8 Surface differences between pre-event (ALS) and post-event (TLS) datasets (negative values for erosion). (From Bremer, M., and Sass, O., *Geomorphology*, 138(1), 49–60, 2012. With permission.)

One of the first applications of LIDAR topographic data for landslide modelling was presented by Dietrich et al. (2001) using the SHALSTAB model. These authors showed that susceptibility mapping of shallow landslides can be considerably improved when using an HRDEM of 1 m cell size obtained from an ALS. The results of this simulation demonstrated that the distribution of potentially unstable areas was closer to the inventoried landslides than when using a low-resolution DEM. In addition, the total surface of the affected terrain was smaller. The same trend has also been observed in other available models, such as TRIGRS, which includes more sophisticated infiltration models (Savage et al. 2003; Godt et al. 2008).

The optimal accuracy and resolution of the DEM for modelling should be carefully considered. Although an HRDEM usually improves the modelling of the phenomena, most of the conventional modelling packages do not necessarily require more than a 5–10 m of DEM resolution (Liao et al. 2011). Modelling using a higher-resolution DEM can even introduce artefacts and nonrealistic noises that can mask the main terrain features (Tarolli and Tarboton 2006). For instance, ALS-derived DEMs used for mapping debris flows kinematics have a strong dependency on lateral spreading on the DEM resolution (Stolz and Huggel 2008; Horton et al. 2013). It was found that the fine resolutions produced geometric artefacts, degrading the quality of the modelling results. Similarly, the use of an HRDEM for the simulation of rockfall propagation can lead to significant differences when compared with lower-resolution DEMs (Agliardi and Crosta 2003; Janeras et al. 2004). For instance, the lateral dispersion of the trajectories and the kinetic energy vary with DEM resolution that controls the roughness of the terrain surface (Agliardi and Crosta 2003). According to Agliardi and Crosta (2003), the maximum kinetic energy normally increases with the DEM resolution. However, the opposite result is presented in Jaboyedoff et al. (2006). This discrepancy may be explained by the fact that a higher-resolution DEM has little impact on the accuracy of the results when other parameters (geology, vegetation, etc.) are poorly described (Fuchs et al. 2014). The resolutions of the LIDAR DEM are often too fine compared with the other required input data (Brideau et al. 2011).

Erosion and hillslope evolution modelling has been greatly enhanced by using an HRDEM (Roering 2008; Neugirg et al. 2016). However, most of the current packages for landslide modelling are not able to make the best use of high-resolution LIDAR data. The increasing power of computers will certainly lead to extensive use of HRDEMs as input for stability and run-out modelling. One can expect that the exploitation of full 3D data will be attained in the near future.

17.4 CONCLUSION

The current knowledge concerning the mapping and monitoring of landslides is now far beyond what was expected by the end of the nineties. This is mainly due to the recent advances in remote sensing techniques (including LIDAR), together with the exponential growth of computational capabilities. For instance, laser scanning has allowed for the development of new techniques to monitor slope changes and slope movements, providing innovative ways for hazard qualification and a better understanding of the surface processes, such as rockfall failures.

Dense 3D data have become popular thanks to LIDAR. Nevertheless, one of the present-day key issues is the scarce software to extract all the landslide-relevant information from the unprocessed

FIGURE 17.9 (a) Example of discontinuity mapping along a highway at Pontarlier (east France) using an Optech Lynx Mobile Mapper LiDAR. (b) Picture of the study area.

LIDAR data, especially from full-waveform data. At present, different modelling techniques benefit only partially from the full resolution of the LIDAR-derived DEMs. This is linked to the fact that the density and quality of the geometrical information are much higher than the other parameters required for hazard assessment, such as geological maps. Despite this, the recent advances achieved on the study of rockfalls and rock slopes indicate that HRDEMs possess great potential for the development of new methods and techniques for the analysis of landslide phenomena.

New applications and techniques are now emerging and will be used routinely in the near future: (1) the development of new algorithms for the analysis of complex slope deformations in space and time (Kromer et al. 2015; Williams et al. under review); (2) the detection of slope movement in real time and its integration on early warning systems (Olsen et al. 2012; Kromer et al. 2015; Williams et al. under review); (3) the possibility to rapidly acquire accurate 3D data along transport corridors using mobile mappers (Figure 17.9); (4) the improvement of TLS technical capabilities, including maximum range, accuracy and point density; (5) the merging of web-retrieved images and terrestrial LIDAR data for slope monitoring (Guerin et al. 2017; Voumard et al. 2017) and (5) the acquisition of Earth surface topography by a new generation of satellite LIDAR such as ICESat-2 planned for 2018 and (6) the releasing of LIDAR data to the public domain in open repositories, including 'Open Topography' and 'Point Cloud Library' is helping raising awareness on the importance of 3D and 4D data treatment.

In conclusion, we can say that LIDAR technology has completely revolutionized the characterization, mapping and monitoring of landslides during the last decades. However, a series of great challenges still remain on the development of innovative tools to rapidly extract geohazard-related information from LIDAR data.

ACKNOWLEDGEMENTS

This work has been supported by the Swiss National Science Foundation, Project numbers 138015 (understanding landslide precursory deformation from superficial 3D data), 127132 (rockfall sources: towards accurate frequency estimation and detection) and 144040 (characterizing and analyzing 3D temporal slope evolution). In addition, the second author was funded by the H2020 Program of the European Commission during manuscript review [Ref: MSCA-IF-2015-705215].

REFERENCES

Abancó, C., and Hürlimann, M. (2014). Estimate of the debris-flow entrainment using field and topographical data. *Nat. Hazards*, 71(1), 363–383.
Abdulwahid, W. M., and Pradhan, B. (2017). Landslide vulnerability and risk assessment for multi-hazard scenarios using airborne laser scanning data (LIDAR). *Landslides*, 14(3), 1057–1076.
Abellan A., Vilaplana J. M., Calvet J., and Blanchard, J. (2010). Detection and spatial prediction of rock falls by means of terrestrial laser scanning modelling. *Geomorphology*, 119, 162–171.
Abellán, A., Jaboyedoff, M., Oppikofer, T., and Vilaplana, J. M. (2009). Detection of millimetric deformation using a terrestrial laser scanner: Experiment and application to a rockfall event. *Nat. Hazards Earth Syst. Sci.*, 9, 365–372.
Abellán, A., Jaboyedoff, M., Oppikofer, T., Rosser, N. J., Lim, M., and Lato, M. (2014). State of science: Terrestrial laser scanner on rock slopes instabilities. *Earth Surf. Proc. Landforms*, 39, 1, 80–97. DOI: 10.1002/esp.3493.
Abshire, J. B., Sun, X., Riris, H., Sirota, J. M., McGarry, J. F., Palm, S., Yi, D., and Liiva, P. (2005). Geoscience Laser Altimeter System (GLAS) on the ICESat Mission: On-orbit measurement performance. *Geophys. Res. Lett.*, 32(4) L21S02, DOI:10.1029/2005GL024028.
Agliardi, F., and Crosta, G. B. (2003). High resolution three-dimensional numerical modelling of rockfalls. *Int. J. Rock Mech. Min. Sci.*, 40, 455–471. DOI: 10.1016/S1365-1609(03)00021-2.
Agliardi, F., Crosta, G., and Zanchi, A. (2001). Structural constraints on deep-seated slope deformation kinematics. Eng. Geol., 59, 83–102.
Alba, M., Barazzetti, L., Roncoroni, F., and Scaioni, M. (2011). Filtering vegetation from terrestrial point clouds with low-cost near infrared cameras. *Ital. J. Remote Sens.*, 43, 55–75.

Alho, P., Kukko, A., Hyyppä, H., Kaartinen, H., Hyyppä, J., and Jaakkola, A. (2009). Application of boat-based laser scanning for river survey. *Earth Surf. Proc. Landforms*, 34, 1831–1838.

Ardizzone, F., Cardinali, M., and Galli, M. (2007). Identification and mapping of recent rainfall-induced landslides using elevation data collected by airborne LIDAR. *Nat. Hazards Earth Syst. Sci.*, 7, 637–650.

Avian, M., Kellerer-Pirklbauer, A., and Bauer, A. (2009). LIDAR for monitoring mass movements in permafrost environments at the cirque Hinteres Langtal, Austria, between 2000 and 2008. *Nat. Hazards Earth Syst. Sci.*, 9, 1087–1094. DOI: 10.5194/nhess-9-1087-2009.

Baillifard, F., Jaboyedoff, M., and Sartori, M. (2003). Rockfall hazard mapping along a mountainous road in Switzerland using a GIS-based parameter rating approach. *Nat. Hazards Earth Syst. Sci.*, 3, 435–442.

Baillifard, F., Jaboyedoff, M., Rouiller, J.-D., Couture, R., Locat, J., Robichaud, G., and Gamel, G. (2004). Towards a GIS-based hazard assessment along the Quebec City Promontory, Quebec, Canada. In *Landslides Evaluation and Stabilization*, Lacerda, W., Ehrlich, M., Fontoura, A., and Sayão, A. A., eds. Rotterdam: Balkema, 207–213.

Baltsavias, E. P. (1999a). Airborne laser scanning: Basic relations and formulas. *ISPRS J. Photogramm. Remote Sens.*, 54, 199–214.

Baltsavias, E. P. (1999b). A comparison between photogrammetry and laser scanning. *ISPRS J. Photogramm. Remote Sens.*, 54, 83–94.

Barbarella, M., Fiani, M., and Lugli, A. (2017). Uncertainty in terrestrial laser scanner surveys of landslides. *Remote Sensing*, 9(2), 113.

Bellian, J. A., Kerans, C., and Jennette, D. C. (2005). Digital outcrop models: Applications of terrestrial scanning LIDAR technology in stratigraphic modeling. *J. Sediment. Res.*, 75 (2), 166–176.

Beraldin, J.-A., Blais, F., and Boulanger, P. (2000). Real world modelling through high resolution digital 3D imaging of objects and structures. *ISPRS J. Photogramm. Remote Sens.*, 55, 230–250.

Beraldin, J.-A., Blais, F., and Lohr, U. (2010). Laser scanning technology. In *Airborne and Terrestrial Laser Scanning*, Vosselman, G., and Maas, H.-G., eds. Dunbeath, Scotland: Whittles Publishing, 1–42.

Besl, P., and McKay, N. (1992). A method for registration of 3-D shapes. *IEEE Trans. Pattern Anal. Mach. Intell.*, 14, 239–256.

Blikra, L. H., Anda, E., Belsby, S., Jogerud, K., and Klempe, Ø. (2006). Åknes/Tafjord prosjektet: Statusrapport for Arbeidsgruppe 1 (Undersøking og overvaking). Åknes/Tafjord project. Stranda, Norway.

Bornaz, L., Lingua, A., and Rinaudo, F. (2002). Engineering and environmental applications of laser scanner techniques. *Int. Arch. Photogramm. Remote Sens.*, 34(3B), 40–43.

Braathen, A., Blikra, L. H., Berg, S. S., and Karlsen, F. (2004). Rock-slope failures in Norway, type, geometry and hazard. *Norwegian J. Geol.*, 84, 67–88.

Breien, H., De Blasio, F., Elverhøi, A., and Høeg, K. (2008). Erosion and morphology of a debris flow caused by a glacial lake outburst flood, western Norway. *Landslides*, 5(3), 271–280.

Bremer, M., and Sass, O. (2012). Combining airborne and terrestrial laser scanning for quantifying erosion and deposition by a debris flow event. *Geomorphology*, 138(1), 49–60.

Brideau, M. A., Pedrazzini, A., Stead, D., Froese, C., Jaboyedoff, J., and van Zeyl, D. (2011). Three-dimensional slope stability analysis of South Peak, Crowsnest Pass, Alberta, Canada. *Landslides*, 8, 139–158. DOI: 10.1007/s10346-010-0242-8.

Brodu, N., and Lague, D. (2012). 3D terrestrial LIDAR data classification of complex natural scenes using a multi-scale dimensionality criterion: Applications in geomorphology. *ISPRS J. Photogramm. Remote Sens.*, 68, 121–134.

Brooker, G. (2009). Introduction to sensing. In *Introduction to Sensors for Ranging and Imaging*. Raleigh, NC: The Institution of Engineering and Technology, 743p.

Buckley, S. J., Howell, J. A., Enge, H. D., and Kurz, T. H. (2008). Terrestrial laser scanning in geology: Data acquisition, processing and accuracy considerations. *J. Geol. Soc.*, 165, 625–638.

Bull, J. M., Miller, H., Gravley, D. M., Costello, D., Hikuroa, D. C. H., and Dix, J. K. (2010). Assessing debris flows using LIDAR differencing: 18 May 2005 Matata event, New Zealand. *Geomorphology*, 124, 75–84.

Burns, W. J., Coe, J. A., Kaya, B. S., and Ma, L. (2010). Analysis of elevation changes detected from multi-temporal LIDAR surveys in forested landslide terrain in western Oregon. *Environ. Eng. Geosci.*, 16(4), 315–341.

Carrea, D., Abellán, A., Derron, M. H., Gauvin, N., and Jaboyedoff, M. (2012). Using 3D surface datasets to understand landslide evolution: From analogue models to real case study. Presented at the Proceedings of the 11th International Symposium on Landslides and 2nd North American Symposium on Landslides, Banff, Alberta, Canada, 2–8 June.

Carrea, D., Abellán, A., Derron, M.-H., and Jaboyedoff, M. (2014). Automatic rockfalls volume estimation based on terrestrial laser scanning data. Presented at the International Association Engineering Geology Conference, Torino, Italy, 15–19 September.

Carter, W., Shrestha, R., Tuell, G., Bloomquist, D., and Sartori, M. (2001). Airborne laser swath mapping shines new light on earth's topography. *EOS*, 82(46), 549–564.

Caudal, P., Grenon, M., Turmel, D., and Locat, J. (2017). Analysis of a large rock slope failure on the east wall of the LAB Chrysotile Mine in Canada: LIDAR Monitoring and Displacement Analyses. *Rock Mech. Rock Eng.*, 50(4), 807–824.

Cavalli, M., and Marchi, L. (2008). Characterisation of the surface morphology of an alpine alluvial fan using airborne LIDAR. *Nat. Hazards Earth Syst. Sci.*, 8, 323–333.

Chen, Y., and Medioni, G. (1992). Object modelling by registration of multiple range images. *Image Vis. Comput.*, 10, 145–155.

Chigira, M., Duan, F., Yagi, H., and Furuya, F. (2004). Using an airborne laser scanner for the identification of shallow landslides and susceptibility assessment in an area of ignimbrite overlain by permeable pyroclastics. *Landslides*, 1, 203–209.

Collins, B. D., and Stock, G. M. (2012). LIDAR-based rock-fall hazard characterization of cliffs. In *Geotechnical Engineering State of the Art and Practice*, Hryciw, R. D., Athanasopoulos-Zekkos, A., and Yesiller, N., eds. ASCE Geotechnical Special Publication No. 225. Reston, VA: American Society of Civil Engineers, 3021–3030.

Collins, B., and Sitar, N. (2008). Processes of coastal bluff erosion in weakly lithified sands, Pacifica, California, USA. *Geomorphology*, 97, 483–501.

Conway, S. J., Decaulne, A., Balme, M. R., Murray, J. B., and Towner, M. C. (2010). A new approach to estimating hazard posed by debris flows in the Westfjords of Iceland. *Geomorphology*, 114(4), 556–572.

Cristianini, N., and Shawe-Taylor J. (2000). *An Introduction to Support Vector Machines: And Other Kernel-Based Learning Methods*. Cambridge: Cambridge University Press.

Cruden, D. M., and Varnes, D. J. (1996). Landslide types and process. In *Landslides: Investigation and Mitigation*, Turner, A. K., and Schuster, R. L., eds. Transportation Research Board Special Report 247. Washington, DC: National Academy Press, 36–75.

Dallaire, E. E. (1974). Electronic distance measuring revolution well under way. *Civ. Eng.*, 44, 66–71.

DeLong, S. B., Prentice, C. S., Hilley, G. E., and Ebert, Y. (2012). Multitemporal ALSM change detection, sediment delivery, and process mapping at an active earthflow. *Earth Surf. Proc. Landforms*, 37(3), 262–272. DOI: 10.1002/esp.2234.

Dewez, T., Gebrayel, D., Lhomme, D., and Robin, Y. (2009). Quantifying morphological changes of sandy coasts by photogrammetry and cliff coasts by lasergrammetry. *La Houille Blanche*, 1, 32–37. DOI: 10.1051/lhb:2009002.

Dietrich, W. E., Bellugi, D., and Real de Asua, R. (2001). Validation of the shallow landslide model, SHALSTAB, for forest management. In *Land Use and Watersheds: Human Influence on Hydrology and Geomorphology in Urban and Forest Areas*, Wigmosta, M. S., and Burges, S. J., eds. Water Science and Application, vol. 2. Hoboken, NJ: Wiley, 195–227.

DiMarzio, J. P. (2007). GLAS/ICESat 500 m laser altimetry digital elevation model of Antarctica. Boulder, CO: National Snow and Ice Data Center.

Fanti, R., Gigli, G., Lombardi, L., Tapete, D., and Canuti, P. (2013). Terrestrial laser scanning for rockfall stability analysis in the cultural heritage site of Pitigliano (Italy). *Landslides*, 10, 409–420. DOI: 10.1007/s10346-012-0329-5.

Fardin, N., Feng, Q., and Stephansson, O. (2004). Application of a new in situ 3D laser scanner to study the scale effect on the rock joint surface roughness. *Int. J. Rock Mech. Min. Sci.*, 41, 329–335.

Farr, T. G., Rignot, E., Saatchi, S., Simard, M., and Treuhaft, R. (2008). LIDAR/InSAR synergism for earth science applications. In *Synthetic Aperture Radar (EUSAR), 2008 7th European Conference*, Tralee, Country Kerry, Ireland, 2–5 June, 1–3.

Feng, Q. H., and Röschoff, K. (2004). In-situ mapping and documentation of rock faces using a full-coverage 3-D laser scanning technique. *Int. J. Rock Mech. Min. Sci.*, 41(3) 139–144.

Ferrero, A. M., Forlani, G., Roncella, R., and Voyat, H. I. (2009). Advanced geostructural survey methods applied to rock mass characterization. *Rock Mech. Rock Eng.*, 42(1), 631–665. DOI: 10.1007/s00603-008-0010-4.

Fey, C., Rutzinger, M., Wichmann, V., Prager, C., Bremer, M., and Zangerl, C. (2015). Deriving 3D displacement vectors from multi-temporal airborne laser scanning data for landslide activity analyses. *GIScience & Remote Sensing*, 52(4), 437–461.

Franceschi, M., Teza, G., Preto, N., Pesci, A., Galgaro, A., and Girardi, S. (2009). Discrimination between marls and limestones using intensity data from terrestrial laser scanner. *ISPRS J. Photogramm. Remote Sens.*, 64(6), 522–528. DOI: 10.1016/j.isprsjprs.2009.03.003.

Fuchs, M., Torizin, J., and Kühn, F. (2014). The effect of DEM resolution on the computation of the factor of safety using an infinite slope model. *Geomorphology*, 224, 16–26. DOI: 10.1016/j.geomorph.2014.07.015.

Garcia-Selles, D., Falivene, O., Arbues, P., Gratacos, O., Tavani, S., and Munoz, J. A. (2011). Supervised identification and reconstruction of near-planar geological surfaces from terrestrial laser scanning. *Comput. Geosci.*, 37, 1584–1594. DOI: 10.1016/j.cageo.2011.03.007.

Gigli, G., and Casagli, N. (2011). Semi-automatic extraction of rock mass structural data from high resolution LIDAR point clouds. *Int. J. Rock Mech. Min. Sci.*, 48, 187–198.

Girardeau-Montaut, D. (2006). Détection de changement sur des données géométriques tridimensionnelles. Doctorat Traitement du Signal et des Images, TSI/TII, ENST. https://tel.archives-ouvertes.fr/pastel-00001745/.

Glenn, N. F., Streutker, D. R., Chadwick, D. J., Thackray, G. D., and Dorsch, S. J. (2006). Analysis of LIDAR-derived topographic information for characterizing and differentiating landslide morphology and activity. *Geomorphology*, 73(1–2), 131–148.

Godt, J. W., Baum, R. L., Savage, W. Z., Salciarini, D., Schulz, W. H., and Harp, E. L. (2008). Transient deterministic shallow landslide modeling: Requirements for susceptibility and hazard assessments in a GIS framework. *Eng. Geol.*, 102, 214–226. DOI: 10.1016/j.enggeo.2008.03.019.

Gokceoglu, C., Sonmez, H., and Ercanoglu, M. (2000). Discontinuity controlled probabilistic slope failure risk maps of the Altindag (settlement) region in Turkey. *Eng. Geol.*, 55, 277–296.

Gordon, S., Litchi, D., and Stewart, M. (2001). Application of a high-resolution, ground-based laser scanner for deformation measurements. In *New Techniques in Monitoring Surveys I, Proceedings of the 10th FIG International Symposium on Deformation Measurements*, Orange, California, 23–32, 19–22 March 2001.

Guerin, A., Abellán, A., Matasci, B., Jaboyedoff, M., Derron, M. H., and Ravanel, L. (2017). Brief communication: 3-D reconstruction of a collapsed rock pillar from Web-retrieved images and terrestrial LIDAR data—The 2005 event of the west face of the Drus (Mont Blanc massif). *Nat. Hazards Earth Syst. Sci.*, 17(7), 1207.

Guzzetti, F., Reichenbach, P., and Wieczorek, G. F. (2003). Rockfall hazard and risk assessment in the Yosemite Valley, California, USA. *Nat. Hazards Earth Syst. Sci.*, 3, 491–503.

Haala, N., and Brenner, C. (1999). Extraction of buildings and trees in urban environments. *ISPRS J. Photogramm. Remote Sens.*, 54, 130–137.

Haneberg, W. C., Cole, W. F., and Kasali, G. (2009). High-resolution LIDAR-based landslide hazard mapping and modeling, UCSF Parnassus Campus, San Francisco, USA. *Bull. Eng. Geol. Environ.*, 68, 263–276.

Haugerud, R. A., Harding, D. J., Johnson, S. Y., Harless, J. L., Weaver, C. S., and Sherrod, B. L. (2003). High-resolution LIDAR topography of the Puget Lowland, Washington—A Bonanza for earth science. *GSA Today*, 13, 4–10.

Hecht, J. (1994). *Understanding Lasers: An Entry Level Guide*. 2nd ed. New York: IEEE Press.

Hergarten, S. (2003). Landslides, sandpiles, and self-organized criticality. *Nat. Hazards Earth Syst. Sci.*, 3, 505–514. DOI: 10.5194/nhess-3-505-2003.

Hoge, F. E., Swift, R. N., and Frederick, E. B. (1980). Water depth measurement using an airborne pulsed neon laser system. *Appl. Optics*, 19, 871–883.

Horton, P., Jaboyedoff, M., Rudaz, B., and Zimmermann, M. (2013). Flow-R, a model for susceptibility mapping of debris flows and other gravitational hazards at a regional scale. *Nat. Hazards Earth Syst. Sci.*, 13, 869–885. DOI: 10.5194/nhess-13-869-2013.

Huang, J., Lee, A., and Mumford, D. (2000). Statistics of range images. Presented at the IEEE International Conference on Computer Vision and Pattern Recognition 2000 (CVPR'00), Los Alamitos, CA, 13–15 June 2000.

Humair, F., Carrea, D., Matasci, B., Jaboyedoff, M., and Epard, J. L. (2015). Stratigraphic layers detection and characterization using Terrestrial Laser Scanning point clouds: Example of a box-fold. *Eur. J. of Remote Sens.*, 48, 541–568.

Hunter, G., Pinkerton, H., Airey, R., and Calvari, S. (2003). The application of a long-range laser scanner for monitoring volcanic activity on Mount Etna. *J. Volcanol. Geotherm. Res.*, 123, 203–210.

Ingensand, H. (2006). Metrological aspects in terrestrial laser-scanning technology. Presented at the Proceedings of the 3rd IAG/12th FIG Symposium 2006, Baden, Austria, 22–24 May 2006.

ISRM (International Society for Rock Mechanics). (1978). Suggested methods for the quantitative description of discontinuities in rock masses. *Int. J. Rock Mech. Min. Sci. Geomechan. Abstr.*, 15, 319–368.

Jaboyedoff, M., and Labiouse, V. (2003). Preliminary assessment of rockfall hazard based on GIS data. In *10th International Congress on Rock Mechanics ISRM 2003—Technology Roadmap for Rock Mechanics*, Johannesburg, South Africa, 575–578, 8–12 September 2003.

Jaboyedoff, M., Baillifard, F., Philippossian, F., and Rouiller, J.-D. (2004). Assessing the fracture occurrence using the "weighted fracturing density": A step towards estimating rock instability hazard. *Nat. Hazards Earth Syst. Sci.*, 4, 83–93.

Jaboyedoff, M., Couture, R., and Locat, P. (2009). Structural analysis of Turtle Mountain (Alberta) using digital elevation model: Toward a progressive failure. *Geomorphology*, 103, 5–16. DOI: 10.106/j.geomorph .2008. 14.012.

Jaboyedoff, M., Giorgis, D., and Riedo, M. (2006). Apports des modèles numériques d'altitude pour la géologie et l'étude des mouvements de versant. *Bull. Soc. Vaud. Sc. Nat.*, 90(1), 1–21.

Jaboyedoff, M., Metzger, R., Oppikofer, T., Couture, R., Derron, M.-H., Locat, J., and Turmel, D. (2007). New insight techniques to analyze rock-slope relief using DEM and 3D-imaging cloud points: COLTOP-3D software. In *Rock Mechanics: Meeting Society's Challenges and Demands, Proceedings of the 1st Canada-US Rock Mechanics Symposium*, Vancouver, Canada, 27–31 May, 61–68.

Jaboyedoff, M., Oppikofer, T., Abellan, A., Derron, M.-H., Loye, A., Metzger, R., and Pedrazzini, A. (2012). Use of LIDAR in landslide investigations: A review. *Nat. Hazards*, 61, 5–28. DOI: 10.1007/s11069 -010-9634-2.

Jaboyedoff, M., Oppikofer, T., Derron, M.-H., Blikra, L. H., Böhme, M., and Saintot A. (2011). Complex landslide behaviour and structural control: A three-dimensional conceptual model of Åknes rockslide, Norway. *Geol. Soc. Spec. Publ.*, 351, 147–161.

Janeras, M., Navarro, M., Arnó, G., Ruiz, A., Kornus, W., Talaya, J., Barberà, M., and López, F. (2004). LIDAR applications to rock fall hazard assessment in Vall de Núria. In *Proceedings of the 4th ICA Mountain Cartography Workshop*, Vall de Núria, Spain, 1–14, 30 September–2 October 2004.

Kemeny, J., and Post, R. (2003). Estimating three-dimensional rock discontinuity orientation from digital images of fracture traces. *Comput. Geosci.*, 29, 65–77.

Kemeny, J., and Turner, K. (2008). Ground-based LIDAR: Rock slope mapping and assessment. FHWA-CFL/TD-08-006. Washington, DC: Federal Highway Administration. www.ctiponline.org/publications /view_file.ashx?fileID=98.

Krabill, W. B., Collins, J. G., Link, L. E., Swift, R. N., and Butler, M. L. (1984). Airborne laser topographic mapping results. *Photogramm. Eng. Remote Sens.*, 50, 685–694.

Krabill, W. B., Thomas, R. H., Martin, C. F., Swift, R. N., and Frederick, E. B. (1995). Accuracy of airborne laser altimetry over Greenland ice sheet. *Int. J. Remote Sens.*, 16, 1211–1222.

Kromer, R. A., Abellán, A., Hutchinson, D. J., Lato, M., Edwards, T., and Jaboyedoff, M. (2015). A 4D filtering and calibration technique for small-scale point cloud change detection with a terrestrial laser scanner. *Remote Sensing*, 7(10), 13029–13052.

Kromer, R. A., Abellán, A., Hutchinson, D. J., Lato, M., Chanut, M. A., Dubois, L., and Jaboyedoff, M. (2017). Automated terrestrial laser scanning with near-real-time change detection—Monitoring of the Séchilienne landslide. *Earth Surf. Dynam.*, 5(2), 293.

Kurz, T. H., Buckley, S. J., Howell, J. A., and Schneider, D. (2008). Geological outcrop modeling and interpretation using ground based hyperspectral and laser scanning data fusion. In *WG VIII/12: Geological Mapping, Geomorphology and Geomorphometry, Proceedings of the International Society for Photogrammetry and Remote Sensing Conference*, Beijing, 3–11 July, 1229–1234.

Kurz, T., Buckley, S., Howell, J., and Schneider, D. (2011). Integration of panoramic hyperspectral imaging with terrestrial LIDAR. *Photogramm. Rec.*, 26(134), 212–228.

Lalonde, J.-F., Vandapel, N., Huber, D., and Hebert, M. (2006). Natural terrain classification using three-dimensional ladar data for ground robot mobility. *J. Field Robot.*, 23(1), 839–861.

Lancaster, S. T., Nolin, A. W., Copeland, E. A., and Grant, G. E. (2012). Periglacial debris-flow initiation and susceptibility and glacier recession from imagery, airborne LIDAR, and ground-based mapping. *Geosphere*, 8(2), 417–430.

Lato, M. J., Diederichs, M. S., and Hutchinson, D. J. (2010). Bias correction for view-limited LIDAR scanning of rock outcrops for structural characterization. *Rock Mech. Rock Eng.*, 43(5), 615–628. DOI: 10.1007 /s00603-010-0086-5.

Lato, M., Hutchinson, J., Diederichs, M., Ball, D., and Harrap, R. (2009). Engineering monitoring of rockfall hazards along transportation corridors: Using mobile terrestrial LIDAR. *Nat. Hazards Earth Syst. Sci.*, 9, 935–946.

Liao, Z., Hong, Y., Kirschbaum, D., Adler, R., Gourley, J. J., and Wooten, R. (2011). Evaluation of TRIGRS (transient rainfall infiltration and grid-based regional slope-stability analysis)'s predictive skill for hurricane-triggered landslides: A case study in Macon County, North Carolina. *Nat. Hazards*, 58, 325–339. dx.doi.org/10.1007/s11069-010-9670-y.

Lichti, D. D. (2005). Spectral filtering and classification of terrestrial laser scanner point clouds. *Photogramm. Rec.*, 20, 218–240.

Lichti, D. D. (2007). Error modelling, calibration and analysis of an AM-CW terrestrial laser scanner system. *ISPRS J. Photogramm. Remote Sens.*, 61(5), 307–324.

Lichti, D. D., and Jamtsho, M. (2006). Angular resolution of terrestrial laser scanners. *Photogramm. Rec.*, 21(114), 141–160.

Lichti, D. D., and Skaloud, J. (2010). Registration and calibration. In *Airborne and Terrestrial Laser Scanning*, Vosselman, G., and Maas, H.-G., eds. Caithness, UK: Whittles Publishing, 83–133.

Lichti, D., Gordon, S., and Stewart, M. (2002). Ground-based laser scanners: Operation, systems and applications. *Geomatica*, 56(1), 21–33.

Lim, M., Petley, D. N., Rosser, N. J., Allison, R. J., Long, A. J., and Pybus, D. (2005). Combined digital photogrammetry and time-of-flight laser scanning for monitoring cliff evolution. *Photogramm. Rec.*, 20(1), 109–129.

Lim, M., Rosser, N. J., Allison, R. J., and Petley, D. N. (2010). Erosional processes in the hard rock coastal cliffs at Staithes, North Yorkshire. *Geomorphology*, 114, 12–21.

Lin, Y., Hyyppa, J., and Jaakkola, A. (2011). Mini-UAV-borne LIDAR for fine-scale mapping. *IEEE Geosci. Remote S.*, 8(3), 426–430.

Lingua, A., Piatti, D., and Rinaudo, F. (2008). Remote monitoring of a landslide using an integration of GB-InSAR and LIDAR techniques. In *International Archives of the Photogrammetry, Remote Sensing and Spatial Information Sciences*, Beijing, vol. XXXVII, pt. B1, 361–366, 3–11 July 2008.

Longoni, L., Arosio, D., Scaioni, M., Papini, M., Zanzi, L., Roncella, R., and Brambilla, D. (2012). Surface and subsurface non-invasive investigations to improve the characterization of a fractured rock mass. *J. Geophys. Eng.*, 9(5), 461.

Lopez Saez, J., Corona, C., Stoffel, M., Gotteland, A., Berger, F., and Liébault, F. (2011). Debris-flow activity in abandoned channels of the Manival torrent reconstructed with LIDAR and tree-ring data. *Nat. Hazards Earth Syst. Sci.*, 5, 1247–1257.

Loye, A., Jaboyedoff, M., and Pedrazzini, A. (2009). Identification of potential rockfall source areas at regional scale using a DEM-based quantitative geomorphometric analysis. *Nat. Hazards Earth Syst. Sci.*, 9, 1643–1653. DOI: 10.5194/nhess-9-1643-2009.

Mackey, B. H., Roering, J. J., and McKean, J. A. (2009). Long-term kinematics and sediment flux of an active earthflow, Eel River, California. *Geology*, 37(9), 803–806.

Mallet, C., and Bretar, F. (2009). Full-waveform topographic LIDAR: State-of-the-art. *ISPRS J. Photogramm. Remote Sens.*, 64(1), 1–16.

Matasci, B., Stock, G. M., Carrea, D., Collins, B. D., Guerin, A., Matasci, G., and Ravanel, L. (accepted). Assessing rockfall susceptibility in steep and overhanging slopes using three-dimensional analysis of failure mechanisms. *Landslides.*

McKean, J., and Roering J. J. (2004). Landslide detection and surface morphology mapping with airborne laser altimetry. *Geomorphology*, 57, 331–351.

Micheletti, N., Tonini, M., and Lane, S. N. (2017). Geomorphological activity at a rock glacier front detected with a 3D density-based clustering algorithm. Geomorphology, 278, 287–297.

Michoud, C., Carrea, D., Costa, S., Derron, M., Jaboyedoff, M., Delacourt, C., Maquaire, O., Letortu, P., and Davidson, R. (2015). Landslide detection and monitoring capability of boat based mobile laser scanning along Dieppe coastal cliffs, Normandy. *Landslides*, 12(2), 403–418.

Miller, B. (1965). Laser altimeter may aid photo mapping. *Aviat. Week Space Technol.*, 88, 60–64.

Monserrat, O., and Crosetto, M. (2008). Deformation measurement using terrestrial laser scanning data and least squares 3D surface matching. *ISPRS J. Photogramm. Remote Sens.*, 63(1), 142–154.

Neugirg, F., Stark, M., Kaiser, A., Vlacilova, M., Della Seta, M., Vergari, F., Schmidt, J., Becht, M., and Haas, F. (2016). Erosion processes in calanchi in the Upper Orcia Valley, Southern Tuscany, Italy based on multitemporal high-resolution terrestrial LIDAR and UAV surveys. *Geomorphology*, 269, 8–22.

Norrish, N., and Wyllie, D. (1996). Rock slope stability analysis. In *Landslides: Investigation and Mitigation*, Turner, A. K., and Schuster, R. L., eds. Transportation Research Board Special Report 247. Washington, DC: National Academy Press, 673.

Olsen, J. M., Butcher, S., and Silvia, E. P. (2012). Real-time change and damage detection of landslides and other earth movements threatening public infrastructure. Final Report SR 500-500, OTREC-RR-11-23. Oregon Department of Transportation Research Section. http://www.oregon.gov/ODOT/TD/TP_RES /docs/Reports/2012/SR500_500_Landslides.pdf.

Olsen, M. J., Wartman, J., McAlister, M., Mahmoudabadi, H., O'Banion, M. S., Dunham, L., and Cunningham, K. (2015). To fill or not to fill: Sensitivity analysis of the influence of resolution and hole filling on point cloud surface modeling and individual rockfall event detection. *Remote Sensing*, 7(9), 12103–12134.

Oppikofer, T., Jaboyedoff, M., and Keusen, H. R. (2008). Collapse at the eastern Eiger flank in the Swiss Alps. *Nat. Geosci.*, 1, 531–535.

Oppikofer, T., Jaboyedoff, M., Blikra, L., Derron, M.-H., and Metzger, R. (2009). Characterization and monitoring of the Aknes rockslide using terrestrial laser scanning. *Nat. Hazards Earth Syst. Sci.*, 9, 1003–1019.

Pedrazzini, A., Jaboyedoff, M., Froese, C. R., Langenberg, C. W., and Moreno, F. (2011). Structural analysis of Turtle Mountain: Origin and influence of fractures in the development of rock slope failures. *Geol. Soc. Spec. Publ.*, 351, 163–183.

Pesci A., Teza, G., and Bonali, E. (2011). Terrestrial laser scanner resolution: Numerical simulations and experiments on spatial sampling optimization. *Remote Sens.*, 3(1), 167–184.

Petrie, G. (1990). Laser-based surveying instrumentation and methods. In *Engineering Surveying Technology*, Kennie, T. J. M., and Petrie, G., eds. London: Halsted Press, 48–83.

Petrie, G., and Toth, C. (2009a). Introduction to laser ranging, profiling, and scanning. In *Topographic Laser Ranging and Scanning*, Shan, J., and Toth, C.K., eds. Boca Raton, FL: CRC Press, 1–27.

Petrie, G., and Toth, C. (2009b). Airborne and spaceborne laser profilers and scanners. In *Topographic Laser Ranging and Scanning*, Shan, J., and Toth, C. K., eds. Boca Raton, FL: CRC Press, 29–85.

Petschko, Bell, R., and Glade, T. (2016). Effectiveness of visually analyzing LIDAR DTM derivatives for earth and debris slide inventory mapping for statistical susceptibility modeling. *Landslides*, 13(5), 857–872.

Prokop, A., and Panholzer, H. (2009). Assessing the capability of terrestrial laser scanning for monitoring slow moving landslides. *Nat. Hazards Earth Syst. Sci.*, 9, 1921–1928.

Rabbani, T., Dijkman, S., van den Heuvel, F., and Vosselman, G. (2007). An integrated approach for modelling and global registration of point clouds. *ISPRS J. Photogramm. Remote Sens.*, 61(6), 355–370.

Ravanel, L., Allignol, F., Deline, P., Gruber, S., and Ravello, M. (2010). Rock falls in the Mont-Blanc massif in 2007 and 2008. *Landslides*, 7(4), 493–501.

Reineman, B. D., Lenain, L., Castel, D., and Melville, W. K. (2009). A portable airborne scanning LIDAR system for ocean and coastal applications. *J. Atmos. Ocean. Technol.*, 26(12), 2626–2641.

Riquelme, A., Abellan, A., Tomás, R., and Jaboyedoff, M. (2014). A new approach for semi-automatic rock mass joints recognition from 3D point clouds. *Comput. Geosci.*, 68(0), 38–52. DOI: 10.1016/j.cageo.2014.03.014.

Riquelme, A., Abellan, A., and Tomás, R. (2015). Discontinuity spacing analysis in rock masses using 3D point clouds. *Eng. Geol.* 195, 185–195. DOI:10.1016/j.enggeo.2015.06.009.

Roering, J. J. (2008). How well can hillslope evolution models 'explain' topography? Simulating soil production and transport using high-resolution topographic data. *Geol. Soc. Am. Bull.*, 120, 1248–1262.

Roering, J. J., Stimely, L. L., Mackey, B. H., and Schmidt, D. A. (2009). Using DInSAR, airborne LIDAR, and archival air photos to quantify landsliding and sediment transport, *Geophys. Res. Lett.*, 36, L19402. DOI: 10.1029/2009GL040374.

Rosser, N. J., Lim, N., Petley, D. N., Dunning, S., and Allison, R. J. (2007). Patterns of precursory rockfall prior to slope failure. *J. Geophys. Res.*, 112(F4), F04014, 14 p, doi:10.1029/2006JF000642.

Rosser, N. J., Petley, D. N., Lim, M., Dunning, S. A., and Allison, R. J. (2005). Terrestrial laser scanning for monitoring the process of hard rock coastal cliff erosion. *Q. J. Eng. Geol. Hydrogeol.*, 38, 363–375.

Rowlands, K. A., Jones, L. D., and Whitworth, M. (2003). Landslide laser scanning: A new look at an old problem. *Q. J. Eng. Geol. Hydrogeol.*, 36(2), 155–157.

Royan, M. J., Abellán, A., Jaboyedoff, M., Vilaplana, J. M., and Calvet, J. (2013). Spatio-temporal analysis of rockfall pre-failure deformation using terrestrial LIDAR. *Landslides*, 11, 697–709. DOI: 10.1007/s10346-013-0442-0.

SafeLand Deliverable 4.1. (2010). Review of techniques for landslide detection, fast characterization, rapid mapping and long-term monitoring. Edited for the SafeLand European project by Michoud, C., Abellán, A., Derron, M.-H., and Jaboyedoff, M. https://www.ngi.no/eng/Projects/SafeLand.

Santana, D., Corominas, J., Mavrouli, O., and Garcia-Sellés, D. (2012). Magnitude–frequency relation for rockfall scars using a terrestrial laser scanner. *Eng. Geol.*, 145, 50–64.

Savage, W. Z., Godt, J. W., and Baum, R. L. (2003). A model for spatially and temporally distributed shallow landslide initiation by rainfall infiltration. In *Debris-Flow Hazards Mitigation: Mechanics, Prediction, and Assessment*, Rickenmann, D., and Chen, C., eds. Rotterdam: Millpress.

Schaer, P., Skaloud, J., Landtwing, S., and Legat, K. (2007). Accuracy estimation for laser point cloud including scanning geometry. Presented at the 5th International Symposium on Mobile Mapping Technology, MMT '07, Padua, Italy.

Scheidl, C., Rickenmann, D., and Chiari, M. (2008). The use of airborne LIDAR data for the analysis of debris flow events in Switzerland. *Nat. Hazards Earth Syst. Sci.*, 5, 1113–1127.

Schulz, W. H. (2004). Landslides mapped using LIDAR imagery, Seattle, Washington. U.S. Geological Survey Open-File Report 2004-1396. Seattle, WA: U.S. Department of the Interior, 1–11.

Schulz, W. H. (2007). Landslide susceptibility revealed by LIDAR imagery and historical records, Seattle, Washington. *Eng. Geol.*, 89, 67–87.

Schutz, B. E., Zwally, H. J., Shuman, C. A., Hancock, D., and DiMarzio, J. P. (2005). Overview of the ICESat mission. *Geophys. Res. Lett.*, 32 L21S01, DOI 10.1029/2005GL021009.

Schürch, P., Densmore, A., Rosse, N., Lim, M., and McArdell, B. W. (2011). Detection of surface change in complex topography using terrestrial laser scanning: Application to the Illgraben debris-flow channel. *Earth Surf. Proc. Landforms*, 36, 1847–1859.

Shan J., and Toth, K. (2008). *Topographic Laser Ranging and Scanning: Principles and Processing.* Boca Raton, FL: CRC Press, Taylor & Francis Group.

Sheperd, E. C. (1965). Laser to watch height. *New Sci.*, 1, 33–37.

Slob, S., and Hack, R. (2004). 3-D terrestrial laser scanning as a new field measurement and monitoring technique. In *Engineering Geology for Infrastructure Planning in Europe: A European Perspective*, Hack, R., Azzam, R., and Charlier, R., eds. Lecture Notes in Earth Sciences 104. Berlin: Springer, 179–190.

Slob, S., Hack, R., and Keith, T. (2002). An approach to automate discontinuity measurements of rock faces using laser scanning techniques. In *Proceedings of the International Symposium on Rock Engineering for Mountainous Regions—Eurock 2002*, Funchal, Portugal, 25–28 November, 87–94.

Smith, D. E., Muhleman, D. O., and Ivanov, A. B. (2001). Mars Orbiter Laser Altimeter: Experiment summary after the first year of global mapping of Mars. *J. Geophys. Res.*, 106(E10), 23689–23722.

Soeters, R., and Van Westen, C. J. (1996). Slope instability recognition, analysis, and zonation. In *Landslides: Investigation and Mitigation*, Turner, A. K., and Schuster, R. L., eds. Transportation Research Board Special Report 247. Washington, DC: National Academy Press, 129–177.

Staley, D. M., Wasklewicz, T. A., and Blaszczynski, J. S. (2006). Surficial patterns of debris flow deposition on alluvial fans in Death Valley, CA using airborne laser swath mapping data. *Geomorphology*, 74(1–4), 152–163.

Stock, G. M., Bawden, G. W., Green, J. K., Hanson, F., Downing, G., Collins, B. D., Bond, S., and Leslar, M. (2011). High-resolution three-dimensional imaging and analysis of rock falls in Yosemite Valley, California. *Geosphere*, 7, 573–581.

Stock, G. M., Martel, S. J., Collins, B. D., and Harp, E. L. (2012). Progressive failure of sheeted rock slopes: The 2009–2010 Rhombus Wall rock falls in Yosemite Valley, California, USA. *Earth Surf. Proc. Landforms*, 37, 546–561.

Stolz, A., and Huggel, C. (2008). Debris flows in the Swiss National Park: The influence of different flow models and varying DEM grid size on modeling results. *Landslides*, 5(3), 311–319.

Strahler, A. N. (1954). Quantitative geomorphology of erosional landscapes. *Compt. Rend. 19th Intern. Geol. Cong. Sec.*, 13, 341–354.

Sturzenegger, M., and Stead, D. (2009). Quantifying discontinuity orientation and persistence on high mountain rock slopes and large landslides using terrestrial remote sensing techniques. *Nat. Hazards Earth Syst. Sci.*, 9, 267–287.

Tarolli, P., and Tarboton, D. G. (2006). A new method for determination of most likely landslide initiation points and the evaluation of digital terrain model scale in terrain stability mapping. *Hydrol. Earth Syst. Sci.*, 10, 663–677. DOI: 10.5194/hess-10-663-2006.

Terzaghi, K. (1962). Stability of slopes on hard unweathered rock. *Geotechnique*, 12, 251–263.

Teza, G., Galgaro, A., Zaltron, N., and Genevois. R. (2007). Terrestrial laser scanner to detect landslide displacement fields: A new approach. *Int. J. Remote Sens.*, 28(16), 3425–3446.

Tonini, M., and Abellan A. (2014). Rockfall detection from terrestrial LIDAR point clouds: A clustering approach using R. *J. Spatial Inf. Sci.*, 8, 95–110. DOI: 5311/JOSIS.2014.8.123.

Vallet, J., and Skaloud, J. (2004). Development and experiences with a fully-digital handheld mapping system operated from a helicopter. *Int. Arch. Photogramm Remote Sens.*, 35(B5), 791–796.

Van Den Eeckhaut, M., Poesen, J., Govers, G., Verstraeten, G., and Demoulin, A. (2007). Characteristics of the size distribution of recent and historical landslides in a populated hilly region. *Earth Planet. Sci. Lett.*, 256, 588–603.

Vosselman, G., and Maas, H. (2010). *Airborne and Terrestrial Laser Scanning.* Boca Raton, FL: CRC Press.

Voumard, J., Abellán, A., Nicolet, P., Penna, I., Chanut, M.A., Derron, M.H., and Jaboyedoff, M. (2017). Using street view imagery for 3-D survey of rock slope failures. *Nat. Hazards Earth Syst. Sci.*, 17, 2093–2107, https://doi.org/10.5194/nhess-17-2093-2017.

Wagner, A., Leite, E., and Olivier, R. (1988). Rock and debris-slides risk mapping in Nepal—A user-friendly PC system for risk mapping. In *Proceedings of the 5th International Symposium on Landslides*, Lausanne, Switzerland, 10–15 July, vol. 2, 1251–1258.

Wehr, A., and Lohr, U. (1999). Airborne laser scanning—An introduction and overview. *ISPRS J. Photogramm. Remote Sens.*, 54, 68–82.

Williams, J. G., Rosser, N. J., Hardy, R. J., Brain, M. J., and Afana, A. A. (in review). Optimising 4D approaches to surface change detection: Improving understanding of rockfall magnitude-frequency, *Earth Surf. Dynam. Discuss.*, https://doi.org/10.5194/esurf-2017-43.

Xu, L., and Jordan, M. I. (1996). On convergence properties of the EM algorithm for Gaussian mixtures. *Neural Comput.*, 8(1), 129–151.

Young, A. P., and Ashford, S. A. (2008). Instability investigation of cantilevered seacliffs. *Earth Surf. Proc. Landforms*, 33(11), 1661–1677. DOI: 10.1002/esp.1636.

Zaruba, Q., and Mencl, V. (1982). *Landslides and Their Control*. Amsterdam: Elsevier.

Zwally, H. J., Schutz, B., Abdalati, W., Abshire, J., Bentley, C., Brenner, A., Bufton, J. et al. (2002). ICESat's laser measurements of polar ice, atmosphere, ocean, and land. *J. Geodyn.*, 34, 405–445.

18 Radar Monitoring of Volcanic Activities

Zhong Lu and Daniel Dzurisin

CONTENTS

18.1 INTRODUCTION

Earth is home to about 1500 volcanoes that have erupted in the past 10,000 years, and today volcanic activity affects the lives and livelihoods of a rapidly growing number of people around the globe. About 20 volcanoes are erupting on Earth at any given time; 50–70 erupt each year, and about 160 erupt each decade. Impressive as these statistics are, they do not include a large but unspecified number of volcanic vents along submarine midocean ridges that girdle the globe (Smithsonian Institution, Global Volcanism Program, http://www.volcano.si.edu/faq.cfm#q3).

The nature of eruptive activity ranges from the quiet outpouring of fluid lava on the ocean floor and in places like Hawaii (http://hvo.wr.usgs.gov/) to the explosive ejection of volcanic ash, pumice and other fragmental material at volcanoes like Mount Fuji (Japan), Mount St. Helens (United States), Chaitén (Chile) and others along the Pacific Ring of Fire and elsewhere. Less frequent, larger events, like the 1912 eruption of Novarupta–Mount Katmai (Alaska) (Hildreth and Fierstein 2012) and the 1991 eruption of Mount Pinatubo (Philippines) (Newhall and Punongbayan 1996), produce local and regional impacts that can last for decades, and shorter-term effects on the average global temperature. At the extreme end of the spectrum, it has been suggested that a catastrophic eruption at Lake Toba on the Indonesian island of Sumatra about 74,000 years ago caused a decade-long 'volcanic winter', resulting in a genetic bottleneck that profoundly affected the course of human evolution—an idea that remains plausible but controversial (Robock 2013 and references therein).

The products of volcanic eruptions also vary widely, giving rise to a large range of associated hazards (Myers et al. 2008). Explosive eruptions produce ballistic ejecta (solid and molten rock fragments) that can impact the surface up to several kilometres away from the vent. Smaller fragments are carried upward in eruption columns that sometimes reach the stratosphere, forming eruption clouds that pose a serious hazard to aircraft. Large eruption clouds can extend hundreds to thousands of kilometres downwind, resulting in ash fall over large areas. Heavy ash fall can collapse buildings, and even minor amounts can cause significant damage and disruption to everyday life. Volcanic gases in high concentrations can be deadly. In lower concentrations, they contribute to health problems and acid rain, which causes corrosion and harms vegetation. Lava flows and domes extruded during mostly non-explosive eruptions can inundate property and infrastructure, and create flood hazards by damming streams or rivers. Pyroclastic flows – high-speed avalanches of hot pumice, ash, rock fragments and gas – can move at speeds in excess of 100 km/h and destroy everything in their path. In some cases, gravitational collapse of an unstable volcanic edifice results in a devastating debris avalanche; the most famous example is the 1980 debris avalanche at Mount St. Helens, which extended more than 20 km down the North Fork Toutle River Valley. Debris flows and lahars (volcanic mudflows) triggered by eruptions inundate valleys for distances approaching 100 km, causing long-term ecological impacts and increased flood hazards.

Assessment, monitoring and preparedness are three keys to mitigating the adverse impacts of volcanic activity. Radar can play a direct role in helping to monitor volcanoes and assess hazards, both during periods of unrest and during ensuing eruptions. For example, interferometric synthetic aperture radar (InSAR) images can be used to distinguish between deep and shallow sources of volcano deformation, and between the deformation pattern caused by magma accumulation in a subsurface reservoir and that caused by upward intrusion of a magma-filled dyke from the reservoir towards the surface. During the course of an eruption when the volcano is obscured by clouds or darkness, Synthetic Aperture Radar (SAR) intensity images might be the only means available to track hazardous developments, such as the emergence of a gravitationally unstable lava dome. Ground-based Doppler radars can track volcanic ash clouds and provide short-term warnings to aircraft and to areas downwind that are likely to receive ash fall. Insights gained from radar studies also can contribute to improved public awareness and preparedness for volcanic activity through proactive public information programmes. For additional information about volcano hazards, hazard assessments and eruption preparedness, see http://volcanoes.usgs.gov/.

18.2   RADAR

The term *radar* is derived from 'radio detection and ranging', a phrase that encapsulates some of radar's essential characteristics and capabilities. Radar systems make use of the radio and microwave portion of the electromagnetic spectrum, with wavelengths ranging from a few millimetres to 100 m or more. Most volcano applications, including SAR and InSAR, make use of wavelengths ranging from a few centimetres to a few tens of centimetres. All radar systems employ a radio transmitter that sends out a beam of microwaves either continuously or in pulses. By measuring the time it takes radio waves travelling at the speed of light to make the round trip from the radar to a target and back, a tracking radar system can determine the distance to the target. If the target is moving with respect to the radar, its velocity can be determined from the frequency of the return signal, which differs from that of the transmitted signal as a result of the Doppler effect. The distance to the target, strength of the return signal and Doppler shift are three fundamental parameters provided by tracking radars. Because of these capabilities, tracking radars are essential tools for air traffic control and weather monitoring.

A typical tracking radar employs a scanning strategy in which the beam sweeps through a range of azimuth and elevation angles in order to map a volume of interest. For example, the radar might transmit pulses while rotating 360° in azimuth at a fixed elevation angle, and then repeat the scan at progressively higher or lower elevation angles. Return echoes from targets are received by the radar antenna and processed by the receiver. Once the radar sweeps through all elevation slices, a volume scan is complete, which provides a three-dimensional view of the airspace around the radar site. Tracking radars equipped with Doppler capability, such as those used for air traffic control, can determine both the location and speed of targets (aircraft) within their range. Weather radars take advantage of the fact that the strength of the return signal depends on the size, density, state (e.g. solid hail and liquid rain) and shape of scatterers in the beam's path. Based on empirical relationships, the approximate rainfall rate at the ground can be estimated from observations made by weather radar (e.g. https://radar.weather.gov). Weather radars equipped with Doppler capability can peer inside thunderstorms and determine if there is rotation in the cloud, which often is a precursor to the development of tornadoes.

Two characteristics of radar that are important for volcano monitoring are (1) unlike optical and infrared systems that are inherently passive (i.e. they rely on natural reflected energy or radiated energy originating at the source), radar is an active sensor that provides its own illumination, and (2) owing to their longer wavelength, radar signals penetrate water clouds, diffuse ash clouds and sparse to moderate vegetation better than visible light, enabling limited 'see-through' capability for objects that are opaque at optical wavelengths. Because radar is an active microwave system, it is equally effective in darkness and daylight, and during bad weather or good. This is a tremendous advantage for volcano monitoring, which requires round-the-clock operations during periods of unrest.

Ground-based Doppler radars have been utilized to detect and track volcanic ash clouds (Harris and Rose 1983; Rose et al. 1995; Dubosclard et al. 1999; Lacasse et al. 2004; Houlié et al. 2005; Tupper et al. 2005; Marzano et al. 2006), which can pose a hazard to buildings, infrastructure, human health and aviation systems (Rose 1977; Miller and Casadevall 2000). Figure 18.1 shows time-series images of a developing ash cloud during the 2009 eruption at Redoubt volcano, Alaska

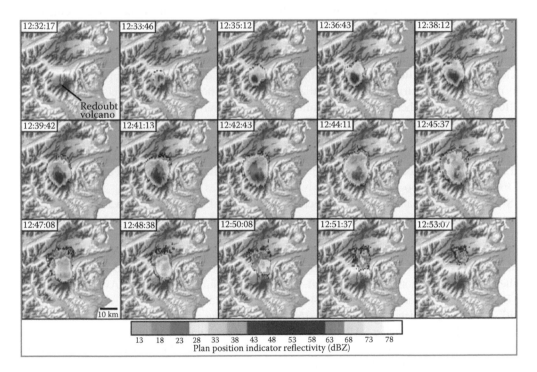

FIGURE 18.1 Sequence of radar reflectivity images at an altitude of 7.9 km above sea level over Redoubt volcano, Alaska, from a Doppler weather radar located about 82 km east of the volcano. The images show the growth and decline of an eruption cloud on 23 March 2009. Times are in Universal Time Coordinated for starts of volume scans, each of which take 90 seconds to complete. The colour bar at the bottom shows reflectivity values in decibels relative to Z (dBZ), a unit commonly used in weather radar to compare the equivalent reflectivity (Z) of a radar signal scattered from a remote object (volcanic ash, in this case) with the return from a droplet of rain with a diameter of 1 mm. For weather clouds, dBZ values can be converted to estimates of rainfall rate using an empirical formula. In this case, the greater dBZ values (warmer colours) correspond to denser parts of the ash cloud. (Modified from Schneider, D., and Hoblitt, R., *J. Volcanol. Geotherm. Res.*, 259, 133–144, 2013.)

(Schneider and Hoblitt 2013). The images show the extent and radar reflectivity of the cloud at an altitude of 7.9 km above sea level. The nearly circular cloud was characterized by a high reflectivity core and lasted at detectable levels for about 20 minutes (Schneider and Hoblitt 2013).

An imaging radar aboard the German Space Agency's TerraSAR-X satellite was used to track the growth, destruction and regrowth of a lava dome at Mount Cleveland volcano in the central Aleutian Islands, Alaska, during 2011–2013 (Lu and Dzurisin 2014, chap. 6, sect. 6.15.5). Mount Cleveland is remote, difficult to access and often obscured by clouds, so the SAR observations provided timely information about the eruption that would not have been available otherwise. Satellite SAR imagery provided similar information that aided hazard assessments during recent eruptions at the Merapi (2013) and Sinabung (2013–2014 ongoing) volcanoes in Indonesia (Smithsonian Institution Global Volcanism Program, http://volcano.si.edu/).

18.3 SYNTHETIC APERTURE RADAR

SAR is an imaging radar system designed, as the name implies, to take advantage of a large 'synthetic' antenna to produce images of much better resolution than would be possible otherwise. SAR systems operate on the same principles as Doppler radars, but have additional capability to distinguish among return signals from individual resolution elements within a target footprint. SARs are side looking, that is, they direct signals to the side of their path across the surface rather than straight down. As a result, the arrival path of the radar signal is oblique to the surface being imaged. Return signals from near-range parts of the target (the part closest to the ground track of the radar) generally arrive back at the radar sooner than return signals from far-range areas, so the relationship between round-trip travel time and range can be used to organize return signals in the across-track, or range, direction. In the along-track, or azimuth, direction, the Doppler principle comes into play. Signals returned from areas that are ahead of the radar as it travels along its path are shifted to slightly higher frequencies, while returns from trailing areas are shifted to slightly lower frequencies. An imaging radar uses the relationship between return signal frequency and relative velocity between radar and target to organize return signals in the azimuth direction. In this way, the returns from each resolution element on the ground can be assigned unique coordinates in range and azimuth. The resulting data can be processed into an image of the target area, which contains information about topography and radar reflective properties of the surface.

SAR systems take advantage of the fact that each point along the ground swath is illuminated for an extended period of time while the footprint of the radar beam moves across it. The resolution of an imaging radar is inversely proportional to the size (aperture) of the antenna, so a SAR is capable of much better resolution than is possible with a real aperture radar. Conceptually, a SAR image processor makes use of this fact to 'synthesize' a large virtual antenna, and thus achieves much higher spatial resolution than is practical with a real aperture radar. Most SAR systems designed for Earth orbit use an antenna that is 1–4 m wide and 10–15 m long, with a look angle in the range of 10°–60°, to illuminate a footprint 50–150 km wide in the range direction and 5–15 km wide in the azimuth direction. Such a SAR system is capable of producing a ground resolution of 1–10 m in azimuth and 1–20 m in range, which is an improvement by about three orders of magnitude over a comparable real aperture system. Because a SAR actively transmits and receives signals backscattered from the target area, and because radar wavelengths are mostly unaffected by weather clouds, a SAR can operate effectively during day and night under most weather conditions to produce images at times and under conditions that render most optical imaging systems useless for surface observations.

Using a sophisticated image processing technique called SAR processing (Curlander and McDonough 1991; Bamler and Hartl 1998; Henderson and Lewis 1998; Rosen et al. 2000; Massonnet and Souyris 2008), both the intensity and phase of the signal backscattered from each ground resolution element can be calculated and portrayed as part of a complex-valued SAR image. The intensity of a resulting single-look complex (SLC) image is controlled primarily by terrain slope, surface roughness and surface relative permittivity. Note that *dielectric constant* is the historical term often used to describe this property, but *surface relative permittivity* is more precise and currently accepted by the Institute of Electrical and Electronics Engineers (IEEE) Standards Board (IEEE Standard Definitions of Terms for Radio Wave Propagation 1997). The phase component is controlled mainly by the round-trip travel time from SAR to ground, which is affected by atmospheric conditions (water vapour in the troposphere slows the speed of electromagnetic waves, and electron density in the ionosphere shortens the propagation path) and by interaction of the radar signal with the ground surface.

18.4 INTERFEROMETRIC SYNTHETIC APERTURE RADAR

InSAR involves the use of two or more SAR images of the same area to extract the land surface topography plus any surface deformation that might have occurred during the interval between image acquisitions. The images can be created by spatially or temporally separated SARs (i.e. two SARs operating at the same time at slightly different locations, or a single SAR that images the same target area from similar vantage points at two different times). The spatial separation between two SAR antennas is called the baseline. The two antennas can be mounted on a single platform for simultaneous interferometry. This is the usual implementation for aircraft and space-borne systems such as the Topographic SAR (TOPSAR) and Shuttle Radar Topography Mission (SRTM) systems (Farr et al. 2007), which are used to generate digital elevation models (DEMs). Alternatively, InSAR images can be created by using a single antenna on an airborne or space-borne platform in nearly identical repeating flight paths or orbits for repeat-pass interferometry (Gray and Farris-Manning 1993; Massonnet and Feigl 1998). For the latter case, even though the antennas do not illuminate the same area at the same time, the two sets of signals recorded during the two passes will be highly correlated if the scattering properties of the ground surface remain undisturbed during the time between image acquisitions. This is the typical implementation for past and present space-borne sensors, such as the U.S. Seasat and Shuttle Imaging Radar-C (SIR-C); European Remote Sensing Satellites (ERS-1 and ERS-2), Environmental Satellite (Envisat) and Sentinel-1A/B; Canadian Radar Satellite (Radarsat-1 and Radarsat-2); and Japanese Earth Resources Satellite (JERS-1) and Advanced Land Observing Satellite (ALOS) and ALOS-2 – all of which operate at wavelengths ranging from a few centimetres (X-band and C-band) to tens of centimetres (L-band) (Table 18.1). This configuration enables InSAR measurements of surface deformation with millimetre to centimetre precision at a spatial resolution of a few tens of metres over a large region.

18.4.1 InSAR Processing Flow

A SAR image represents the intensity and phase of the reflected (or backscattered) signal from each ground resolution element in the form of a complex-valued data matrix (Figure 18.2). Generating an interferogram requires two SLC SAR images. Neglecting phase shifts induced by the transmitting and receiving antenna and SAR processing algorithms, the phase value of a pixel in an SLC SAR image (Figure 18.2b) can be represented as

$$\phi_1 = W\left\{ -\frac{4\pi}{\lambda} r_1 + \varepsilon_1 \right\} \tag{18.1}$$

where r_1 is the apparent range distance (including possible atmospheric delay) from the antenna to the ground target, λ is the radar wavelength, ε_1 is the sum of phase shifts due to the interaction between the incident radar wave and scatterers within the resolution cell and $W\{\}$ is a wrapping operator so that the observed ϕ_1 is wrapped into the interval of $(-\pi, \pi)$. Because the backscattering phase (ε_1) is a randomly distributed (unknown) variable, the phase value (ϕ_1) in a single SAR image

TABLE 18.1

Satellite SAR Sensors Capable of InSAR Mapping

Mission	Agency	Period of Operation	Orbit Repeat Cycle (days)	Band/ Frequency (GHz)	Wavelength (cm)	Incidence Angle (°) at Swath Centre	Resolution (m)
Seasat	NASA	June 1978–October 1978	17	L-band/1.275	23.5	23	25
ERS-1	ESA	July 1991–March 2000	3, 168 and 35[a]	C-band/5.3	5.66	23	30
JERS-1	JAXA	February 1992–October 1998	44	L-band/1.275	23.5	39	20
ERS-2	ESA	April 1995–July 2011	35 and 3[b]	C-band/5.3	5.66	23	30
Radarsat-1	CSA	November 1995–2013	24	C-band/5.3	5.66	10–60	10–100
Envisat	ESA	March 2002–April 2012	35 and 30[c]	C-band/5.331	5.63	15–45	20–100
ALOS	JAXA	January 2006–May 2011	46	L-band/1.270	23.6	8–60	10–100
TerraSAR-X	DLR	June 2007–present	11	X-band/9.65	3.11	20–55	0.24–260
Radarsat-2	CSA	December 2007–present	24	C-band/5.405	5.55	10–60	3–100
COSMO-SkyMed	ASI	June 2007–present	1, 4, 5, 7, 8, 9, 12 and 16[d]	X-band/9.6	3.12	20–60	1–100
RISAT-2	ISRO	April 2009–present	14	X-band/9.59	3.13	20–45	1–8
TanDEM-X[e]	DLR	June 2010–present	11	X-band/9.65	3.11	20–55	1–16
RISAT-1	ISRO	April 2012–present	25	C-band/5.35	5.61	15–50	3–50
Sentinel-1A	ESA	April 2014–present	12	C-band/5.405	5.55	20–47	5–40
ALOS-2	JAXA	May 2014–present	14	L-band/1.2575 (and more)	23.9 (and more)	8–70	1–100
Sentinel-1B	ESA	April 2016–present	12	C-band/5.405	5.55	20–47	5–40

Note: ASI, Italian Space Agency; CSA, Canadian Space Agency; ESA, European Space Agency; ISRO, Indian Space Research Organization; JAXA, Japan Aerospace Exploration Agency; RISAT-1, Radar Imaging Satellite-1; RISAT-2, Radar Imaging Satellite-2.

[a] To accomplish various mission objectives, the ERS-1 repeat cycle was 3 days from 25 July 1991 to 1 April 1992 and from 13 December 1993 to 9 April 1994; 168 days from 10 April 1994 to 20 March 1995; and 35 days at other times.

[b] The ERS-2 repeat cycle was mainly 35 days. During the few months before the end of the mission, the ERS-2 repeat cycle was changed to 3 days to match the 3-day-repeat ERS-1 phases in 1991–1992 and 1993–1994.

[c] The Envisat repeat cycle was 35 days from March 2002 to October 2010, and 30 days from November 2010 to April 2012.

[d] A constellation of four satellites, each of which has a repeat cycle of 16 days, can collectively produce repeat-pass InSAR images at intervals of 1, 4, 5, 7, 8, 9 and 12 days, respectively.

[e] TerraSAR add-on for digital elevation measurements.

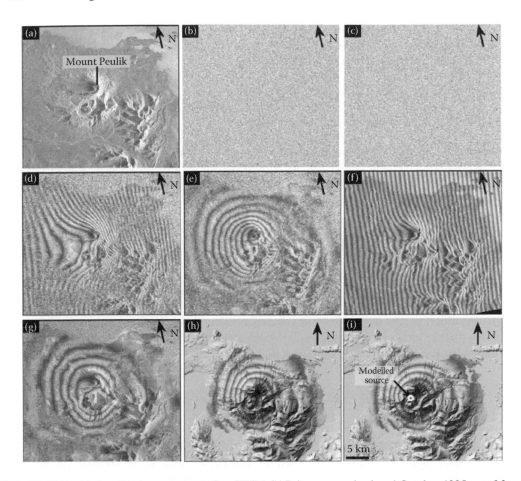

FIGURE 18.2 (a) Amplitude component of an ERS-1 SAR image acquired on 4 October 1995 over Mount Peulik volcano, Alaska. (b) Phase component of the SLC SAR image corresponding to the amplitude image in (a). (c) Phase component of an ERS-2 SAR image of Mount Peulik acquired on 9 October 1997. The amplitude component is similar to that in (a) and therefore is not shown. The phase values represented in (b) and (c) look spatially random but nonetheless contain useful information after InSAR processing. (d) Original interferogram formed by differencing the phase values of two co-registered SAR images, (b) and (c). The resulting InSAR image contains fringes produced by the differing viewing geometries, topography, any atmospheric delays, surface deformation and noise. The perpendicular component of the InSAR baseline is 35 m in this case. (e) Flattened interferogram produced by removing the effect of a flat Earth surface from the original interferogram (d). (f) Simulated interferogram representing the contribution of topography in the original interferogram (d) using knowledge of the InSAR imaging geometry and a DEM. (g) Topography-removed interferogram produced by subtracting the simulated interferogram (f) from the original interferogram (d). The resulting interferogram contains fringes produced by surface deformation, any atmospheric delays and noise. (h) Georeferenced topography-removed interferogram overlaid on a shaded relief image produced from a DEM. The concentric pattern of fringes indicates ~17 cm of uplift centred on the volcano, which occurred during an aseismic inflation episode between 1996 and 1998 prior to a strong earthquake swarm ~30 km to the northwest (Lu et al. 2002b). (i) Model interferogram produced using a best-fit inflationary point source at ~6.5 km depth with a volume change of ~0.043 km^3 overlaid on the shaded relief image (compare to (h)). Each interferometric fringe (full-colour cycle or band) represents 360° of phase change (b–f), or 2.83 cm of range change (g–i) between the ground and the satellite along the satellite look direction. Areas of loss of InSAR coherence are uncoloured in (h) and (i).

cannot be used to calculate the range (r_1) and is of no practical use. However, assume that a second SLC SAR image of the same area (with the phase image shown in Figure 18.2c) is obtained at a different time with a phase value represented by

$$\phi_2 = W\left\{-\frac{4\pi}{\lambda}r_2 + \varepsilon_2\right\} \tag{18.2}$$

Note that, by itself, the second SAR image cannot provide useful range information (r_2) either.

An interferogram (Figure 18.2d) is created by co-registering two SAR images and differencing the corresponding phase values (Figure 18.2b,c) on a pixel-by-pixel basis. The phase value of the resulting interferogram (Figure 18.2d) is

$$\phi = \phi_1 - \phi_2 = W\left\{-\frac{4\pi(r_1 - r_2)}{\lambda} + (\varepsilon_1 - \varepsilon_2)\right\} \tag{18.3}$$

The fundamental assumption in repeat-pass InSAR is that the scattering characteristics of the ground surface do not change during the interval between image acquisitions. The degree of change can be quantified by the interferometric coherence value, which is discussed in Section 18.5.3. Assuming that the interactions between radar waves and scatterers remain the same (i.e. $\varepsilon_1 = \varepsilon_2$), the interferometric phase value can be expressed as

$$\phi = W\left\{-\frac{4\pi(r_1 - r_2)}{\lambda}\right\} \tag{18.4}$$

Typical values for the range difference, ($r_1 - r_2$), are from a few metres to several hundred metres. The SAR wavelength (λ) is of the order of several centimetres. Because the measured interferometric phase value (ϕ) is modulated by 2π, ranging from $-\pi$ to π, there is an ambiguity of many cycles (i.e. numerous 2π values) in the interferometric phase value. Therefore, the phase value of a single pixel in an interferogram is of no practical use. However, the change in range difference, $\delta(r_1 - r_2)$, between two neighbouring pixels that are a few metres apart is normally much smaller than the SAR wavelength. So the phase difference between two nearby pixels, $\delta\phi$, can be used to infer the range difference ($r_1 - r_2$) to a precision that is a small fraction of the radar wavelength. This explains how the InSAR technique can determine range changes to within a few millimetres or centimetres based on observed phase differences between two co-registered images.

The phase (or range distance difference) in the original interferogram (Figure 18.2d) contains contributions from both the topography and any possible ground surface deformation. Therefore, the topographic contribution needs to be removed from the original interferogram in order to derive a deformation map. The most common procedure is to use an existing DEM and knowledge of the InSAR imaging geometry to produce a synthetic interferogram that represents the topographic effect and subtract it from the interferogram to be studied (Massonnet and Feigl 1998; Rosen et al. 2000). This is the so-called two-pass InSAR technique. Alternatively, a synthetic interferogram that represents the topographic contribution can be produced from a different interferogram of the same area that is either insensitive to deformation or does not span the deformation episode (if known by some other means). The procedures are then called three-pass or four-pass InSAR (Zebker et al. 1994). Because the two-pass InSAR method is commonly used for deformation mapping, we explain briefly how to simulate the effect of topography in an InSAR image based on an existing DEM.

Two steps are required to simulate a topography-only interferogram based on a DEM. In the first step, the DEM needs to be resampled to project heights from a map coordinate into the appropriate radar geometry via geometric simulation of the imaging process. The InSAR imaging geometry is shown in Figure 18.3. The InSAR system acquires two images of the same scene with SAR platforms located at A_1 and A_2. The baseline, defined as the vector from A_1 to A_2, has a length B and is tilted with respect to the horizontal by angle α. The slant range r from the SAR to a ground target T with an elevation value h is linearly related to the measured phase values in the SAR images by Equations 18.1 and 18.2. The look angle from A_1 to the ground point T is θ_1. For each ground resolution cell at ground range r_g with elevation h, the slant range value (r_1) should satisfy

$$r_1 = \sqrt{(H+R)^2 + (R+h)^2 - 2(H+R)(R+h)\cos\left(\frac{r_g}{R}\right)} \qquad (18.5)$$

where H is the SAR altitude above a reference Earth surface, which is assumed to be a sphere with radius R. The radar slant range and azimuth coordinates are calculated for each point in the DEM. This set of coordinates forms a non-uniformly sampled grid in SAR coordinate space. The DEM

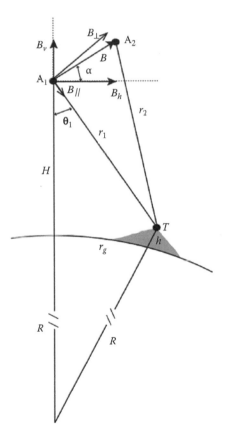

FIGURE 18.3 Schematic showing InSAR imaging geometry. Two SAR images of the same target area are acquired from vantage points A_1 and A_2. The baseline B (the spatial distance between SAR antennas A_1 and A_2) is tilted with respect to the horizontal by angle α, and can be represented by a pair of horizontal (B_h) and vertical (B_v) components, or by a pair of parallel $(B_{//})$ and perpendicular $(B_\perp)$ components. The slant range distances from A_1 and A_2 to a ground target T with elevation above the ground surface h are r_1 and r_2, respectively. The altitude of A_1 is H, and the ground range from A_1 to T is r_g. The look angle from A_1 to T is θ_1. The radius of the spherical Earth is R.

height data are then resampled into a uniform grid in the radar coordinates using the values from the non-uniform grid.

In the second step, the precise look angle from A_1 to ground target T at ground range r_g, slant range r_1 and elevation h is calculated:

$$\theta_1 = \arccos\left[\frac{(H+R)^2 + r_1^2 - (R+h)^2}{2(H+R)r_1}\right] \tag{18.6}$$

Finally, the interferometric phase value due to the topographic effect at target T can be calculated:

$$\phi_{dem} = -\frac{4\pi}{\lambda}(r_1 - r_2) = \frac{4\pi}{\lambda}\left(\sqrt{r_1^2 - 2(B_h \sin\theta_1 - B_v \cos\theta_1)r_1 + B^2} - r_1\right) \tag{18.7}$$

where B_h and B_v are horizontal and vertical components of the baseline B (Figure 18.3).

Figure 18.2e shows the simulated topographic effect in the original interferogram (Figure 18.2d) calculated using an existing DEM and the InSAR imaging geometry (Figure 18.3). Removing the topographic effect (Figure 18.2e) from the original interferogram (Figure 18.2d) results in an interferogram that represents ground surface deformation during the time interval between image acquisitions, plus measurement noise (Figure 18.2f). The resulting phase value can be written as

$$\phi_{def} = W\{\phi - \phi_{dem}\} \tag{18.8}$$

In common practice, an ellipsoidal Earth surface characterized by its major axis, e_{maj}, and minor axis, e_{min}, is used instead of a spherical Earth model. The radius of the Earth at the imaged area is then

$$R = \sqrt{(e_{min} \sin\beta)^2 + (e_{maj} \cos\beta)^2} \tag{18.9}$$

where β is the latitude of the centre of the imaged region.

If h is taken as zero, the procedure outlined in Equations 18.5 through 18.9 will remove the effect of an ellipsoidal Earth surface on the interferogram. This results in a flattened interferogram, in which phase values can be approximated as

$$\phi_{flat} = -\frac{4\pi}{\lambda}\frac{B\cos(\theta_1 - \alpha)}{r_1 \sin\theta_1}h + \phi_{def} = -\frac{4\pi}{\lambda}\frac{B_\perp}{H \tan\theta_1}h + \phi_{def} \tag{18.10}$$

where $B_\perp$ is the perpendicular component of the baseline with respect to the incidence angle θ_1 (Figure 18.3). Removing the effect of an ellipsoidal Earth surface from the original interferogram (Figure 18.2d) results in the flattened interferogram shown in Figure 18.2g.

If ϕ_{def} in Equation 18.10 is negligible (i.e. no deformation) or can be removed from an independent source (Lu et al. 2013), the phase value in Equation 18.10 can be used to calculate the surface height h. This explains how InSAR can be used to produce an accurate, high-resolution DEM for a large region. If the primary goal is to produce a DEM but the interferogram is also affected by ground deformation (i.e. ϕ_{def} is not negligible), the deformation effect can be calculated from a second interferogram that is less sensitive to topography, and then removed from the first interferogram (Lu and Dzurisin 2014).

For the ERS-1/2 satellites, H is about 800 km, θ_1 is about $23° \pm 3°$, λ is 5.66 cm and $B_\perp$ should be less than 1100 m for a coherent interferogram. Therefore, Equation 18.10 can be approximated as

$$\phi_{flat} \approx -\frac{2\pi}{9600} B_\perp h + \phi_{def} \tag{18.11}$$

For an interferogram with $B_\perp$ of 100 m, 1 m of topographic relief produces a phase value of about 4°. However, producing the same phase value requires only 0.3 mm of surface deformation. ϕ_{flat} in Equation 18.11 can be considered a function of two variables, h and ϕ_{def}. The coefficient (i.e. $2\pi B_\perp/9600$) for h is much less than 1, while the coefficient for ϕ_{def} is equal to 1. So for a given imaging geometry, the interferogram phase value is much more sensitive to changes in topography (i.e. surface deformation ϕ_{def}) than to the topography itself (h). This explains why repeat-pass InSAR is capable of mapping surface deformation with millimetre to centimetre precision.

With the two-pass InSAR technique, DEM errors can be incorrectly mapped into apparent surface deformation. The effect is characterized by the so-called 'altitude of ambiguity', which is the amount of DEM error required to generate one interferometric fringe in a topography-removed interferogram (Massonnet and Feigl 1998). Because the altitude of ambiguity is inversely proportional to the perpendicular baseline $B_\perp$, interferometric pairs with small baselines are better suited for deformation analysis. Conversely, pairs with larger baselines (within the constraint imposed by coherence; see Section 18.5.2) are preferable for DEM generation.

One significant error source in repeat-pass InSAR deformation measurements is inhomogeneity in the atmosphere that results in path-delay anomalies (Lu and Dzurisin 2014). Differences in atmospheric water vapour content (and temperature and pressure to a lesser extent) at two observation times can cause differing path delays and consequent anomalies in an InSAR deformation image. Atmospheric delay anomalies can reduce the accuracy of InSAR-derived deformation measurements from several millimetres under ideal conditions to a few centimetres under more typical conditions, thus obscuring subtle changes that could hold clues to the cause of the deformation. The difficulty with estimating water vapour conditions with the needed accuracy and spatial density is an important limiting factor in deformation monitoring with InSAR.

Four methods have been proposed to estimate the water vapour content and remove its effect from deformation interferograms. The first method is to estimate water vapour concentrations in the target area at the times of SAR image acquisitions using short-term predictions from operational weather models (Foster et al. 2006). The problem with this approach is that current weather models have much coarser resolution (a few kilometres) than InSAR measurements (tens of metres). This deficiency can be remedied to some extent by integrating weather models with high-resolution atmospheric measurements, but this approach requires intensive computation. The second method is to estimate water vapour concentration from continuous global positioning system (CGPS) observations in the target area (Li et al. 2005). The spatial resolution (i.e. station spacing) of local or regional CGPS networks at volcanoes is typically a few kilometres to tens of kilometres, which renders this method ineffective in most cases. The third approach to correcting atmospheric delay anomalies in InSAR observations is to utilize water vapour measurements from optical satellite sensors such as the Moderate Resolution Imaging Spectroradiometer (MODIS), Advanced Spaceborne Thermal Emission and Reflection Radiometer (ASTER) and European Medium Resolution Imaging Spectrometer (MERIS) (Li et al. 2003). The disadvantage of this method is the requirement of nearly simultaneous acquisitions of SAR and cloud-free optical images. The fourth and most promising technique is to correct atmospheric delay anomalies using a multitemporal InSAR technique (Section 18.6) (Ferretti et al. 2001; Lu and Dzurisin 2014). Because the atmospheric artefacts are generally spatially correlated and temporally random, they can be mitigated through temporal high-pass and spatial low-pass filtering of multitemporal interferograms.

Another significant error source in two-pass InSAR processing is baseline uncertainty due to inaccurate determination of the SAR antenna positions at the times of image acquisitions. Therefore, baseline refinement during InSAR processing is recommended. A commonly used method is to determine the baseline vector based on an existing DEM via a least-squares approach (Rosen et al. 1996). For this method, areas of the interferogram that are used to refine the baseline should have negligible deformation or deformation that is well characterized by an independent data source.

The final procedure in two-pass InSAR is to rectify the SAR images and interferograms into a geographic coordinate system, which is a backward transformation of Equation 18.5. The georeferenced interferogram (Figure 18.2h) and derived products can be readily overlaid with other data layers to enhance the utility of the interferograms and facilitate data interpretation. Figure 18.2h shows six concentric fringes that represent about 17 cm of range decrease (mostly uplift) centred on the southwest flank of Mount Peulik volcano, Alaska. The volcano inflated aseismically from October 1996 to September 1998, a period that included an intense earthquake swarm that started in May 1998 more than 30 km northwest of Peulik (Lu et al. 2002; Lu 2007; Lu and Dzurisin 2014).

Interferometric phase values need to be unwrapped to remove the modulo 2π ambiguity before estimating the topography or deformation source parameters (Goldstein et al. 1988; Costantini 1998; Chen and Zebker 2000). Phase unwrapping is the process of restoring the correct multiple of 2π to each pixel of the interferometric phase image. Interferograms are often spatially filtered before phase unwrapping (Goldstein and Werner 1998). Two popular phase unwrapping methods utilize branch cut (Goldstein et al. 1998) and minimum cost flow (Costantini 1998; Chen and Zebker 2000) algorithms.

In-depth descriptions of InSAR processing techniques are given by many (Zebker et al. 1994; Bamler and Hartl 1998; Henderson and Lewis 1998; Massonnet and Feigl 1998; Rosen et al. 2000; Hanssen 2001; Hensley et al. 2001; Lu and Dzurisin 2014).

18.5   InSAR PRODUCTS AND THEIR APPLICATIONS TO VOLCANOES

The InSAR processing techniques include a number of steps, precise registration of an InSAR image pair, interferogram generation, removal of the curved Earth phase trend, adaptive filtering, phase unwrapping, precise estimation of the interferometric baseline, generation of a surface deformation image (or a DEM map), estimation of interferometric correlation and rectification of interferometric products. Using a single pair of SAR images as input, a typical InSAR processing chain outputs two SAR intensity images, a deformation map or DEM and an interferometric correlation map.

18.5.1   SAR INTENSITY IMAGE

Volcanic surfaces do not scatter microwaves uniformly. The strength of the return signal at the SAR is controlled primarily by surface roughness and relative permittivity of the target. Surface roughness refers to the SAR wavelength-scale variation in the surface relief. Surfaces that are rough at the scale of the radar wavelength generally are brighter in radar images than smooth ones, because some of the roughness elements are oriented perpendicular to the incoming signal and reflect energy back towards the source. With smooth surfaces, most of the energy is deflected forward, away from the source, which causes them to appear dark. For this reason, blocky lava flows tend to exhibit stronger backscattering returns than pyroclastic flows, which in turn produce higher backscattering than ash deposits. Therefore, SAR intensity images are useful for distinguishing and mapping volcanic ash deposits, lava flows and pyroclastic flows (Lu et al. 2004; Lu and Dzurisin 2014). In cloud-prone volcanic areas (such as the Aleutian volcanic arc), all-weather SAR intensity imagery can be one of the most useful data sources available to track the course of volcanic eruptions in this way. Relative permittivity is an electric property of material that influences radar return strength and is controlled primarily by moisture content of the imaged surface. The effect of relative permittivity variations on radar images is of secondary importance to surface roughness variations, as most natural, dry rocks and soils have a narrow range of values of relative permittivity. Limited laboratory results have indicated that relative permittivity values of volcanic rocks of similar mineralogy and composition tend to increase with the bulk density but decrease with the porosity of volcanic rocks (Russ et al. 1999). Obviously, mapping volcanic deposits based on relative permittivity can generally be complicated by moisture content, mineralogy, composition and other parameters.

Figure 18.4 shows an example in which time-series SAR intensity images were used to track eruptive activity at Mount Cleveland volcano in the central Aleutian Arc, Alaska. Thermal anomalies at Mount Cleveland were noted in satellite data starting on 19 July 2011, and a small lava dome in the summit crater was first observed on 2 August 2011 (Lu and Dzurisin 2014). Time-series TerraSAR-X images revealed that the new dome grew rapidly until 29 December 2011, when it

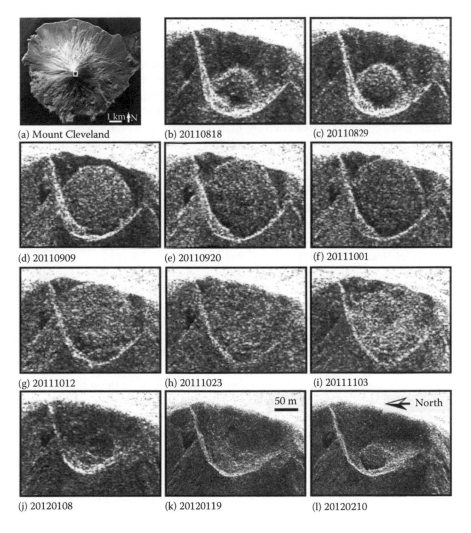

(a) Mount Cleveland   (b) 20110818   (c) 20110829

(d) 20110909   (e) 20110920   (f) 20111001

(g) 20111012   (h) 20111023   (i) 20111103

(j) 20120108   (k) 20120119   (l) 20120210

FIGURE 18.4   (a) Airborne L-band SAR intensity image of Mount Cleveland volcano, Alaska, acquired in 2009 by NASA's Uninhabited Aerial Vehicle Synthetic Aperture Radar (UAVSAR). Lava flows of various ages are apparent in the SAR image. The white rectangle in the summit area shows the extent of the SAR intensity images shown in (b–l). (b–l) Time-series X-band TerraSAR-X intensity images of the Mount Cleveland summit crater showing lava dome growth (b–i), destruction (j–k) and regrowth (l) during the period from 18 August 2011 to 10 February 2012. The initial dome was destroyed by an explosive eruption on 29 December 2011, and a second dome was clearly visible in the SAR image acquired on 10 February 2012. Cycles of dome growth and destruction continued into 2013.

was destroyed by an explosion. A new dome was visible in a TerraSAR-X image acquired on 10 February 2012 (Figure 18.4). That dome likely was destroyed by a series of three explosions during 8–13 March 2012. A third dome, which was first seen in satellite imagery on 28 March 2012, was destroyed by an explosive eruption on 4 April 2012. Additional small explosions occurred during April–June 2012, and a fourth dome was observed in the crater on 26 June 2012. Multiple explosions were detected and a small lava flow was extruded in May 2013 (http://avo.alaska.edu /volcanoes/volcact.php?volcname=Cleveland). In this case, morphological changes at the summit of Mount Cleveland that could be discerned in a series of SAR intensity images, but were otherwise obscured from view, played a key role in monitoring activity throughout the course of the eruption (Wang et al. 2015).

18.5.2 InSAR Deformation Image and Source Parameters Derived from Modelling

Unlike a SAR intensity image, an InSAR deformation image is derived from phase components of two overlapping SAR images. SAR is a side-looking sensor, so an InSAR deformation image depicts ground surface displacements in the SAR line-of-sight (LOS) direction, which generally include both vertical and horizontal components. InSAR deformation images have an advantage for modelling purposes over point measurements made with GPS, for example, because InSAR images provide more complete spatial coverage than is possible with even a dense network of CGPS stations. On the other hand, CGPS stations provide better precision and much better temporal resolution than is possible with InSAR images. The temporal resolution of InSAR measurements is constrained by the orbit repeat times of SAR satellites, that is, typically several days to weeks for currently operational satellites. For hazards monitoring, a combination of periodic areal InSAR observations and continuous data streams from networks of in situ deformation sensors (e.g. CGPS, tiltmeters and strainmeters), integrated with seismic, gas emission and other remote sensing information, is highly desirable (Poland et al. 2006b; Dzurisin et al. 2009; Biggs et al. 2010b; Currenti et al. 2011, 2012; Del Negro et al. 2013).

For understanding volcanic processes, numerical models are often employed to estimate physical parameters of the deformation source based on observations. The high spatial resolution of surface deformation data provided by InSAR makes it possible to constrain models with various geometries, such as the spherical point pressure source (Mogi 1958), dislocation source (sill or dyke source) (Okada 1985), ellipsoid source (Davis 1986; Yang et al. 1988) and penny-crack source (Fialko et al. 2001). Among the physical parameters of interest, the location and volume change of the source usually are the most important.

The most widely used source in volcano deformation modelling is the spherical point pressure source (widely referred to as the Mogi source) embedded in an elastic homogeneous half space (Mogi 1958). In a Cartesian coordinate system, the predicted displacement u at the free surface due to a change in volume ΔV or pressure ΔP of an embedded sphere is

$$u_i\left(x_1 - x_1', x_2 - x_2', 0 - x_3'\right) = \Delta P(1-v)\frac{r_s^3}{G}\frac{x_i - x_i'}{|R^3|} = \Delta V\frac{(1-v)}{\pi}\frac{x_i - x_i'}{|R^3|} \tag{18.12}$$

where x_1', x_2' and x_3' are the horizontal coordinates and depth of the centre of the sphere, R is the distance between the centre of the sphere and the observation point (x_1, x_2 and 0), ΔP and ΔV are the pressure and volume changes in the sphere, v is Poisson's ratio of the host rock (typical value is 0.25), r_s is the radius of the sphere and G is the shear modulus of the host rock (Johnson 1987; Delaney and McTigue 1994).

A non-linear least-squares inversion approach is often used to optimize the source parameters (Press et al. 2007). Inverting the observed interferogram in Figure 18.2h using a Mogi source results in a best-fit source located at a depth of 6.5 ± 0.2 km. The calculated volume change is

0.043 ± 0.002 km^3. Figure 18.2i shows the modelled interferogram based on the best-fit source parameters, which agrees very well with the observed deformation field shown in Figure 18.1h.

Because many volcanic eruptions are preceded by pronounced ground deformation in response to increasing pressure in a magma reservoir or to upward intrusion of magma, surface deformation patterns can provide important insights into the structure, plumbing and state of restless volcanoes (Dvorak and Dzurisin 1997; Dzurisin 2003, 2007). Numerous studies have shown that in some cases, surface deformation is the first detectable sign of volcanic unrest, preceding seismicity or other precursors to an impending intrusion or eruption (Lu and Dzurisin 2014). Therefore, mapping surface deformation and deriving source characteristics is a primary focus of most InSAR studies of volcanoes (Massonnet et al. 1995; Lu et al. 1997, 2000a,b,c, 2002, 2005, 2007, 2010; Wicks et al. 1998, 2002, 2006, 2011; Dzurisin et al. 1999, 2005; Amelung et al. 2000, 2007; Zebker et al. 2000; Mann et al. 2002; Pritchard and Simons 2002, 2004a,b; Masterlark and Lu 2004; Fukushima et al. 2005; Lundgren and Lu 2006; Poland et al. 2006a; Wright et al. 2006; Yun et al. 2006; Hooper et al. 2007; Calais et al. 2008; Biggs et al. 2009, 2010a,b; Fournier et al. 2010; Lu and Dzurisin 2010; Ji et al. 2013; Lee et al. 2013; Parker et al. 2014).

Figure 18.5 shows several interferograms of Mount Okmok, a dominantly basaltic volcano in the central Aleutian volcanic arc, Alaska; each has a temporal separation of 1 year, and collectively they span from 1997 to 2008. Okmok erupted during February–April 1997 and again during July–August 2008. The inter-eruption deformation interferograms suggest that Okmok began to reinflate soon after its 1997 eruption, but the inflation rate generally decreased with time during 1997–2001: from about 10 cm/year during 1997–1998 to about 8 cm/year during 1998–2000, and further to about 4 cm/year during 2000–2001 (Figure 18.5b–e). The rate increased again during 2001–2003 (Figure 18.5f and g), reaching a maximum of about 20 cm/year during 2002–2003 (Figure 18.5g), before slowing to about 10 cm/year during 2003–2004 (Figure 18.5h). The caldera floor subsided 3–5 cm during 2004–2005 (Figure 18.5i), rose a similar amount during 2005–2006 (Figure 18.5j) and then did not move appreciably during 2006–2007 (Figure 18.5k). About 15 cm of uplift occurred from summer 2007 to 10 July 2008, shortly before the 12 July 2008 eruption (Figure 18.5l). This remarkable series of interferograms was interpreted as indicative of a variable rate of magma supply to a shallow storage zone beneath Okmok during the inter-eruption period of 1997–2008 (Lu et al. 2010; Lu and Dzurisin 2014).

Modelling these interferograms using a Mogi source suggests that a magma storage zone centred about ~3.5 km beneath the centre of the 10 km diameter caldera floor was responsible for the observed deformation at Okmok. The InSAR deformation images can be used to track the accumulation of magma beneath Okmok as a function of time. The total volume of magma added to the shallow storage zone from the end of the 1997 eruption to a few days before the 2008 eruption was 85%–100% of the amount that was extruded during the 1997 eruption (Lu and Dzurisin 2014).

Because InSAR is an imaging technique with good spatial resolution, it is also highly effective for mapping localized deformation associated with volcanic flows. For example, Figure 18.5k shows that the 1997 lava flow at Okmok subsided about 3 cm/year during 2006–2007, nearly a decade after it was emplaced. Lu et al. (2005) constructed two-dimensional finite element models of the localized deformation field and concluded that the subsidence likely was caused by thermoelastic cooling of the 1997 flow. They also reported that a significant amount of subsidence (1–2 cm/year) could be observed with InSAR even 50 years after emplacement of the 1958 lava flows at Okmok. This has implications for positioning geodetic markers and deformation sensors at Okmok and other similar volcanoes, and for interpretation of resulting point measurement data (e.g. GPS, tilt and borehole strain). InSAR images can provide an important spatial context for such endeavours, thus helping to avoid misinterpretations caused by unrecognized deformation sources, such as young flows, localized faulting or hydrothermal activity.

18.5.3 InSAR Coherence Image

An InSAR coherence image is a cross-correlation product derived from two co-registered complex-valued (both intensity and phase components) SAR images (Zebker and Villasenor 1992; Lu and

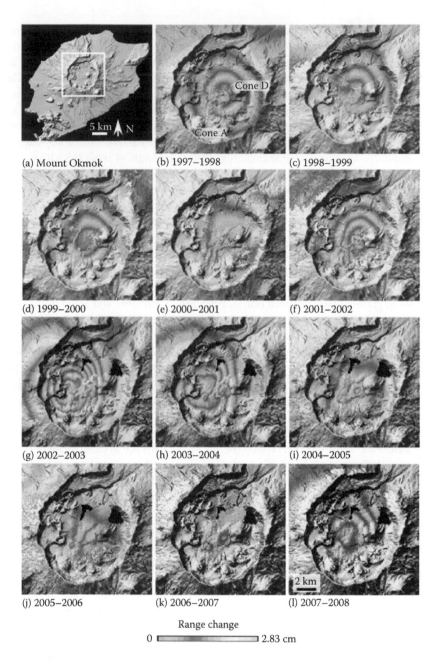

(a) Mount Okmok   (b) 1997–1998   (c) 1998–1999

(d) 1999–2000   (e) 2000–2001   (f) 2001–2002

(g) 2002–2003   (h) 2003–2004   (i) 2004–2005

(j) 2005–2006   (k) 2006–2007   (l) 2007–2008

Range change

0 ▭▭▭▭▭▭▭▭ 2.83 cm

FIGURE 18.5 (a) Shaded relief image of Mount Okmok volcano in the central Aleutian Arc, Alaska. The white square shows the extent of interferograms in (b–l). (b–l) Multitemporal 1-year InSAR images showing the intereruption deformation of Mount Okmok from 1997 (after the end of the 1997 eruption) to 2008 (before the 2008 eruption). InSAR deformation phase values are draped over the corresponding portion of the shaded relief image. Each fringe (full colour cycle) represents 2.83 cm of range change between the ground and satellite along the satellite LOS direction. Areas that lack interferometric coherence are uncoloured.

Freymueller 1998). It depicts changes in backscattering characteristics on the scale of the radar wavelength. Constructing a coherent interferogram requires that SAR images correlate with each other; that is, the backscattering spectrum must be substantially similar over the observation period. Physically, this translates into a requirement that the ground scattering surface be relatively undisturbed at the scale of the radar wavelength during the time between measurements. Loss of InSAR coherence is often referred to as decorrelation. Decorrelation can be caused by the combined effects of (1) thermal decorrelation caused by uncorrelated noise sources in radar instruments, (2) geometric decorrelation resulting from imaging a target from very different look angles, (3) volume decorrelation caused by volume backscattering effects and (4) temporal decorrelation due to surface changes over time (Lu and Kwoun 2008).

InSAR coherence is estimated by cross-correlation of the SAR image pair within a small window of pixels. An InSAR coherence map is generated by computing the cross-correlation in a moving window over the entire image. The reliability of a deformation image or InSAR-derived DEM map can be assessed based on the InSAR coherence map. On the one hand, loss of InSAR coherence renders an InSAR image useless for measuring ground surface deformation. So for this application, the greater the coherence shown by a coherence map, the more reliable is the associated deformation image. Geometric and temporal decorrelation can be mitigated by choosing an image pair with a short baseline and brief temporal separation, respectively, so choosing such a pair is recommended when the goal is to measure surface deformation.

On the other hand, the pattern of decorrelation within a coherence image can provide useful information about surface modifications caused by volcanic activities, such as heavy ash fall or various types of flows. These phenomena modify the surface to a degree that coherence is lost, providing an efficient means to delineate the impacted areas without detailed fieldwork. Even though useful deformation measurements cannot be retrieved over areas of decorrelation, time-sequential InSAR coherence maps can be used to map the extent and progression of eruptive products, such as active lava flows. As an example, Figure 18.6 shows two TerraSAR-X InSAR coherence images along the East Rift Zone and south flank of Kīlauea volcano on the Big Island of Hawaii. The X-band images do not maintain coherence in areas of dense rainforest outside the lava flow field from the 1983 to the present Puʻu ʻŌʻō–Kupaianaha eruption, nor on an active lava flow in the central part of the flow field (dark areas in the images). Elsewhere in the flow field, where young but inactive flows have cooled and stabilized, coherence is generally maintained. As a result of these differences, we can see that (1) flow activity extended all the way to the ocean during 15 September–23 December 2011, and (2) from 25 January to 25 May 2012, the flow expanded laterally but did not reach the ocean. Time-series images such as these can aid in mapping the extent and progress of volcanic flows, and thus also in assessing the inundation threat to nearby areas.

18.5.4 DIGITAL ELEVATION MODEL

A precise DEM can be a very important dataset for characterizing and monitoring man-made and natural hazards, including those posed by volcanic activity. For example, a DEM is necessary to simulate potential mudflows (lahars) that are commonly associated with volcanic eruptions, large earthquakes and heavy rainfall in steep terrain. The ideal SAR configuration for DEM production is a single-pass (simultaneous) two-antenna system (e.g. SRTM). However, repeat-pass single-antenna InSAR also can be used to produce useful DEMs. Either technique is advantageous in areas where the traditional photogrammetric approach to DEM generation is hindered by persistent clouds or other factors (Lu et al. 2003; Lu and Dzurisin 2014).

There are many sources of error in DEM construction from repeat-pass SAR images, including inaccurate determination of the InSAR baseline, atmospheric delay anomalies and possible surface deformation due to tectonic, volcanic or other sources during the time interval spanned by the images. To generate a high-quality DEM, these errors must be identified and corrected using a multi-interferogram approach (Lu et al. 2003, 2013; Lu and Dzurisin 2014). A data fusion technique, such as the wavelet method, can be used to combine DEMs from several interferograms with different spatial resolution, coherence and vertical accuracy to generate the final DEM product (Ferretti et al. 1999). One example of the utility of precise InSAR-derived DEMs is illustrated in Figure 18.7, which shows the extent and thickness of a lava flow extruded during the 1997 Okmok eruption. The flow's three-dimensional distribution was derived by differencing two DEMs that represent the

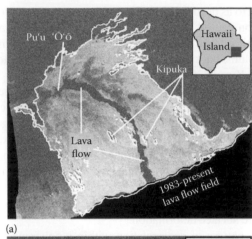

(a)

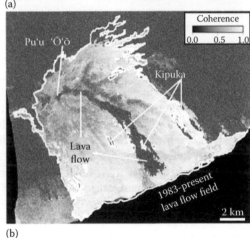

(b)

FIGURE 18.6 TerraSAR-X InSAR coherence images showing a portion of the East Rift Zone and south flank of Kīlauea volcano, Hawaii (inset). The images span (a) 15 September–23 December 2011 and (b) 25 January–25 May 2012. The extent of lava flows from the ongoing Pu'u 'Ō'ō–Kupaianaha eruption, which began in 1983, is outlined in white. Areas outside the lava field are covered by dense rain forest, which results in coherence loss (dark areas). The same is true for a flow in the central part of the field that was active while the images were acquired. (Images were processed and provided by Michael Poland, USGS Hawaiian Volcano Observatory.)

surface topography before and after the eruption. Multiple repeat-pass interferograms were used to correct various error sources and generate the two high-quality DEMs (Lu et al. 2003).

The TerraSAR-X tandem mission for DEM measurements (TanDEM-X) was launched by the German Aerospace Center (DLR) in 2010 (http://www.dlr.de/hr/en/desktopdefault.aspx/tabid-2317/). TanDEM-X is a high-resolution InSAR mission that relies on an innovative flight formation of two tandem TerraSAR-X satellites to produce InSAR-derived DEMs on a global scale with accuracy better than that of SRTM (Krieger et al. 2007). X-band SARs on the two satellites record data synchronously with a closely controlled baseline separation of 200–500 m. Precise baseline information and simultaneous data acquisitions result in InSAR images that are nearly immune to the baseline errors, atmospheric contamination and temporal decorrelation that sometimes plague DEMs derived from repeat-pass InSAR. Thus, the TanDEM-X mission enables the production of a global DEM of unprecedented accuracy, coverage and quality: TanDEM-X DEMs have a specified relative vertical accuracy of 2 m and an absolute vertical accuracy of 10 m at a horizontal resolution of 12 m (Krieger et al. 2007).

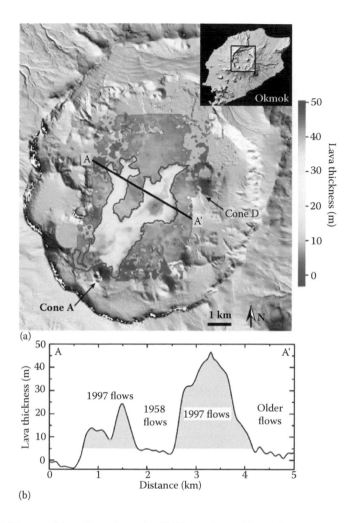

(a)

(b)

FIGURE 18.7 Thickness of lava flows from the 1997 eruption at Mount Okmok produced by differencing pre- and post-eruption DEMs derived from InSAR. (a) Map view of the 1997 lava flows. The red line represents the flow perimeter based on field mapping in August 2001 (Lu et al. 2003). The inset is a shaded relief image of Okmok; the black rectangle shows the extent of (a). (b) Lava thickness along the profile A-A′ across the 1997 flows and a small portion of the underlying 1958 flows that are not covered by 1997 flows. The locations of A-A′ are shown in (a).

18.6 MULTI-INTERFEROGRAM InSAR

When more than two SAR images are available for a given study area, multi-interferogram InSAR processing can be employed to improve the accuracy of deformation maps (or other InSAR products) (Ferretti et al. 2001, 2007; Berardino et al. 2002; Hooper et al. 2007; Rocca 2007; Zhang et al. 2011, 2012; Lu and Dzurisin 2014; Lu and Zhang 2014). A goal of multi-interferogram InSAR processing is to characterize the spatial and temporal behaviours of the deformation signal plus various artefacts and noise sources (e.g. atmospheric delay anomalies, including radar frequency-dependent ionosphere refraction and non-dispersive troposphere delay of the radar signals; orbit errors; and DEM-induced artefacts) in individual interferograms, and then to remove the artefacts to retrieve time-series deformation measurements at the SAR pixel level.

Among several approaches to multi-interferogram analysis, persistent scatterer InSAR (PSInSAR) is one of the newest and most promising. PSInSAR exploits the distinctive backscattering characteristics of certain ground targets (PS; examples include buildings, houses, bridges, dams, large boulders or rock outcrops) and the unique characteristics of atmospheric delay anomalies to improve the accuracy of conventional InSAR deformation measurements (Ferretti et al. 2001). The SAR backscattering signal of a PS target has a broadband spectrum in the frequency domain, which implies that the radar phase of a PS target correlates over much longer time intervals and over much longer baselines than that of other targets. As a result, if the backscatter signal from a given pixel is dominated by return from one or more PSs, the pixel remains coherent over longer time intervals and longer baselines than it would in the absence of the PS pixels. Therefore, at PS pixels, the limitation imposed by loss of coherence in conventional InSAR analysis can be overcome. Because InSAR coherence is maintained at PS pixels, the atmospheric contribution to the

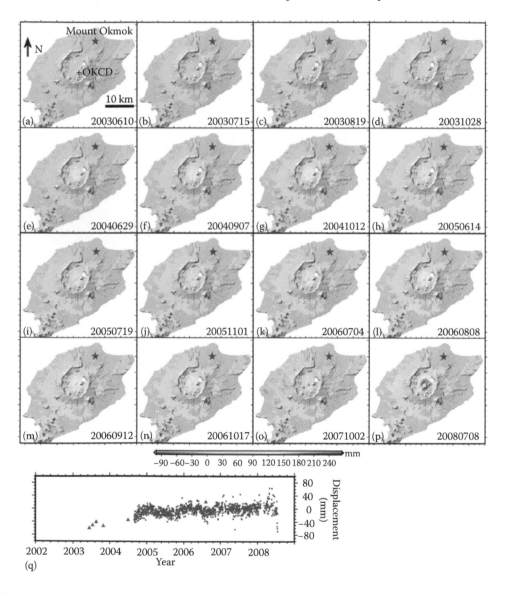

FIGURE 18.8 (a–p) Time-series deformation maps for Mount Okmok based on PSInSAR processing of 19 Envisat SAR images acquired during 2003–2008. The red star in the northeast quadrant represents the pixel used for PSInSAR processing. The location of the CGPS station OKCD is indicated by a black cross (+) in (a). (q) Comparison of time-series PSInSAR measurements (red triangles) with CGPS observations (blue dots) at OKCD. PSInSAR displacements are with respect to the reference pixel; the start time of the PSInSAR time series is 10 June 2003.

backscattered signal, DEM error and orbit error can be identified and removed from the data using a multi-interferogram iterative approach. After these errors are removed, displacement histories at PS pixels can be resolved with millimetre accuracy. If a sufficient number of PS pixels exist in a series of interferograms, relative displacements among them can provide a detailed picture of the surface deformation field.

Figure 18.8a–p shows time-series deformation maps for the period 2003–2008 at Mount Okmok based on PSInSAR processing of a stack of 19 Envisat SAR images. The average inflation rate near the centre of the caldera is slightly less than 50 mm/year. The subsidence of 1997 lava flows on parts of the caldera floor is also discernible in some of the images. The PSInSAR-derived time-series displacements match CGPS measurements at nearby points (Figure 18.8q), demonstrating that PSInSAR can be useful either as a stand-alone tool or in conjunction with other techniques to track volcanic deformation.

18.7 CONCLUSION

Radar in various forms can provide timely observations of volcanic ash clouds, eruptive flows and ground surface deformation before, during and after eruptions. SAR and InSAR products can be used to (1) characterize changing volcanic landscapes that might otherwise be unmonitored or hidden from view, (2) map and measure the deformation of volcanic flows that can persist for decades, (3) estimate physical parameters of subsurface magma reservoirs and conduit systems, (4) monitor changes in reservoir volume and magma migration pathways and (5) contribute to eruption forecasts and volcanic hazard assessments. With more satellite radar platforms either operational or in the planning stages, SAR and InSAR are becoming increasingly important tools for studying volcanoes and associated hazards.

ACKNOWLEDGEMENTS

ERS-1, ERS-2 and Envisat, and TerraSAR-X SAR images are Copyright © 1992–2011 European Space Agency and © 2011–2012 DLR, respectively. ERS-1 and ERS-2 SAR images were provided by Alaska Satellite Facility. We thank Lei Zhang, Michael Poland, and David Schneider and Richard Hoblitt for providing the materials used in Figures 18.8, 18.6 and 18.1, respectively. This work was supported by the National Aeronautics and Space Administration's (NASA) Earth Surface and Interior Program (NNX14AQ95G), the U.S. Geological Survey (USGS) Volcano Hazards Program and the Shuler–Foscue Endowment at Southern Methodist University.

REFERENCES

Amelung, F., Jonsson, S., Zebker, H., and Segall, P. (2000). Widespread uplift and 'trapdoor' faulting on Galapagos volcanoes observed with radar interferometry. *Nature*, 407, 993–996.

Amelung, F., Yun, S. H., Walter, T. R., Segall, P., and Kim, S. W. (2007). Stress control of deep rift intrusion at Mauna Loa volcano, Hawaii. *Science*, 316, 1026–1030.

Bamler, R., and Hart, P. (1998). Synthetic aperture radar interferometry. *Inverse Problems*, 14, R1–R54.

Berardino, P., Fornaro, G., Lanari, R., and Sansosti, E. (2002). A new algorithm for surface deformation monitoring based on small baseline differential SAR interferograms. *IEEE Transactions on Geoscience and Remote Sensing*, 40, 2375–2383.

Biggs, J., Amelung, F., Gourmelen, N., Dixon, T., and Kim, S. W. (2009). InSAR observations of 2007 Tanzania rifting episode reveal mixed fault and dyke extension in an immature continental rift. *Geophysical Journal International*, 179, 549–558.

Biggs, J., Anthony, E. Y., and Ebinger, C. J. (2010a). Multiple inflation and deflation events at Kenyan volcanoes, East African Rift. *Geology*, 37, 979–982.

Biggs, J., Lu, Z., Fournier, T., and Freymueller, J. (2010b). Magma flux at Okmok Volcano, Alaska from a joint inversion of continuous GPS, campaign GPS and InSAR. *Journal of Geophysical Research*, 115, B12401. doi: 10.1029/2010JB007577.

Calais, E., d'Oreye, N., Albaric, J., Deschamps, A., Delvaux, D., Déverchére, J., Ebinger, C. et al. (2008). Aseismic strain accommodation by dyking in a youthful continental rift, East Africa. *Nature*, 456, 783–787. doi: 10.1038/nature07478.

Chen, C., and Zebker, H. (2000). Network approaches to two-dimensional phase unwrapping: Intractability and two new algorithms. *Journal of the Optical Society of America A*, 17, 401–414.

Costantini, M. (1998). A novel phase unwrapping method based on network programming. *IEEE Transactions on Geoscience and Remote Sensing*, 36, 813–818.

Curlander, J. C., and McDonough, R. N. (1991). *Synthetic Aperture Radar: Systems and Signal Processing.* 1st ed., Wiley Series in Remote Sensing and Image Processing. Hoboken, NJ: Wiley-Interscience.

Currenti, G., Napoli, R., and Del Negro, C. (2011). Toward a realistic deformation model of the 2008 magmatic intrusion at Etna from combined DInSAR and GPS observations. *Earth and Planetary Science Letters*, 312, 22–27.

Currenti, G., Solaro, G., Napoli, R., Pepe, A., Bonaccorso, A., Del Negro, C., and Sansosti, E. (2012). Modelling of ALOS and COSMO-SkyMed satellite data at Mt Etna: Implications on relation between seismic activation of the Pernicana fault system and volcanic unrest. *Remote Sensing of Environment*, 125, 64–72.

Davis, P. M. (1986). Surface deformation due to inflation of an arbitrarily oriented triaxial ellipsoidal cavity in an elastic half-space, with reference to Kīlauea Volcano, Hawaii. *Journal of Geophysical Research*, 91, 7429–7438.

Delaney, P. T., and McTigue, D. F. (1994). Volume of magma accumulation or withdrawal estimated from surface uplift or subsidence, with application to the 1960 collapse of Kīlauea volcano. *Bulletin of Volcanology*, 56, 417–424.

Del Negro, C., Currenti, G., Solaro, G., Greco, F., Pepe, A., Napoli, R., Pepe, S., Casu, F., and Sansosti, E. (2013). Capturing the fingerprint of Etna volcano activity in gravity and satellite radar data. *Scientific Reports*, 3, 3089. doi: 10.1038/srep03089.

Dubosclard, G., Cordesses, R., Alard, P., Hervier, C., Coltelli, M., and Kornprobst, J. (1999). First testing of a volcano Doppler radar (Voldorad) at Mt. Etna. *Journal of Geophysical Research*, 26, 3389–3392.

Dvorak, J., and Dzurisin, D. (1997). Volcano geodesy—The search for magma reservoirs and the formation of eruptive vents. *Reviews of Geophysics*, 35, 343–384.

Dzurisin, D. (2003). A comprehensive approach to monitoring volcano deformation as a window on eruption cycle. *Reviews of Geophysics*, 41, 1–29. doi: 10.1029/2001RG000107.

Dzurisin, D. (2007). *Volcano Deformation—Geodetic Monitoring Techniques.* Vol. 7, Springer-Praxis Books in Geophysical Sciences. Berlin: Springer.

Dzurisin, D., Lisowski, M., and Wicks, C. W. (2009). Continuing inflation at Three Sisters volcanic center, central Oregon Cascade Range, USA, from GPS, leveling, and InSAR observations. *Bulletin of Volcanology*, 71, 1091–1110.

Dzurisin, D., Lisowski, M., Wicks, C. W., Poland, M. P., and Endo, E. T. (2005). Geodetic observations and modeling of magmatic inflation at the Three Sisters volcanic center, central Oregon Cascade Range, USA. *Journal of Volcanology and Geothermal Research*, 150, 35–54.

Dzurisin, D., Wicks, C., Jr., and Thatcher, W. (1999). Renewed uplift at the Yellowstone caldera measured by leveling surveys and satellite radar interferometry. *Bulletin of Volcanology*, 61, 349–355.

Farr, T. G., Rosen, P. A., Caro, E., Crippen, R., Duren, R., Hensley, S., Kobrick, M. et al. (2007). The Shuttle Radar Topography Mission. *Reviews of Geophysics*, 45, RG2004. doi: 10.1029/2005RG000183.

Ferretti, A., Prati, C., and Rocca, F. (1999). Multibaseline InSAR DEM reconstruction—The wavelet approach. *IEEE Transactions on Geoscience and Remote Sensing*, 37, 705–715.

Ferretti, A., Prati, C., and Rocca, F. (2001). Permanent scatterers in SAR interferometry. *IEEE Transactions on Geoscience and Remote Sensing*, 39, 8–20.

Ferretti, A., Savio, G., Barzaghi, R., Borghi, A., Musazzi, S., Novali, F., Prati, C., and Rocca, F. (2007). Submillimeter accuracy of InSAR time series: Experimental validation. *IEEE Transactions on Geoscience and Remote Sensing*, 45, 1142–1153.

Fialko, Y., Khazan, Y., and Simons, M. (2001). Deformation due to a pressurized horizontal circular crack in an elastic half-space, with applications to volcano geodesy. *Geophysical Journal International*, 146, 181–190.

Foster, J., Brooks, B., Cherubini, T., Shacat, C., Businger, S., and Werner, C. L. (2006). Mitigating atmospheric noise for InSAR using a high resolution weather model. *Geophysical Research Letters*, 33, L16304. doi: 10.1029/2006GL026781.

Fournier, T. J., Pritchard, M. E., and Riddick, S. N. (2010). Duration, magnitude, and frequency of subaerial volcano deformation events: New results from Latin America using InSAR and a global synthesis. *Geochemistry, Geophysics, Geosystems*, 11, Q01003. doi: 10.1029/2009GC002558.

Fukushima, Y., Cayol, V., and Durand, P. (2005). Finding realistic disk models from interferometric synthetic aperture radar data: The February 2000 eruption at Piton de la Fournaise. *Journal of Geophysical Research*, 110, B03206. doi: 10.1029/2004JG0003268.

Gray, L., and Farris-Manning, P. (1993). Repeat-pass interferometry with airborne synthetic aperture radar. *IEEE Transactions on Geoscience and Remote Sensing*, 31, 180–191.

Goldstein, R., Zebker, H., and Werner, C. (1998). Satellite radar interferometry: Two-dimensional phase unwrapping. *Radio Science*, 23, 713–720.

Goldstein, R. M., and Werner, C. L. (1998). Radar interferogram filtering for geophysical applications. *Geophysical Research Letters*, 25, 4035–4038.

Hanssen, R. (2001). *Radar Interferometry—Data Interpretation and Error Analysis*. Dordrecht, the Netherlands: Kluwer Academic Publishers.

Harris, D. M., and Rose, W. I. (1983). Estimating particle sizes, concentrations and total mass of ash in volcanic clouds using weather radar. *Journal of Geophysical Research*, 88, 10969–10983.

Henderson, F., and Lewis, A. (1998). *Principles and Applications of Imaging Radar: Manual of Remote Sensing*. 3rd ed., vol. 2. Hoboken, NJ: John Wiley & Sons.

Hensley, S., Munjy, R., and Rosen, P. (2001). Interferometric synthetic aperture radar (IFSAR). In *Digital Elevation Model Technologies and Applications—The DEM Users Manual*, ed. D. F. Maune. 2nd ed. Bethesda, MD: American Society for Photogrammetry and Remote Sensing, 142–206.

Hildreth, W., and Fierstein, J. (2012). The Novarupta-Katmai eruption of 1912—Largest eruption of the twentieth century; centennial perspectives. U.S. Geological Survey Professional Paper 1791. http://pubs.usgs .gov/pp/1791/.

Hooper, A., Segall, P., and Zebker, H. (2007). Persistent scatterer interferometric synthetic aperture radar for crustal deformation analysis, with application to Volcán Alcedo, Galápagos. *Journal of Geophysical Research*, 112, B07407. doi: 10.1029/2006JB004763.

Houlié, N., Briole, P., Nercessian, A., and Murakami, M. (2005). Sounding the plume of the 18 August 2000 eruption of Miyakejima volcano (Japan) using GPS. *Geophysical Research Letters*, 32, L05302.

Ji, L. Y., Lu, Z., Dzurisin, D., and Senyukov, S. (2013). Pre-eruption deformation caused by dike intrusion beneath Kizimen volcano, Kamchatka, Russia, observed by InSAR. *Journal of Volcanology and Geothermal Research*, 256, 87–95.

Johnson, D. J. (1987). Elastic and inelastic magma storage at Kīlauea volcano. In *Volcanism in Hawaii*, ed. R. W. Decker, T. L. Wright, and P. H. Stauffer. U.S. Geological Survey Professional Paper 1350. Washington, DC: U.S. Government Printing Office, 1297–1306.

Krieger, G., Moreira, A., Fiedler, H., Hajnsek, I., Werner, M., Younis, M., and Zink, M. (2007). TanDEM-X—A satellite formation for high-resolution SAR interferometry. *IEEE Transactions on Geoscience and Remote Sensing*, 45, 3317–3341.

Lacasse, C., Karlsdóttir, S., Larsen, G., Soosalu, H., Rose, W. I., and Ernst, G. G. J. (2004). Weather radar observations of the Hekla 2000 eruption cloud, Iceland. *Bulletin of Volcanology*, 66, 457–473.

Lee, C. W., Lu, Z., Won, J. S., Jung, H. S., and Dzurisin, D. (2013). Dynamic deformation of Seguam Island, Alaska, 1992–2008, from multi-interferogram InSAR processing. *Journal of Volcanology and Geothermal Research*, 260, 43–51.

Li, Z., Fielding, E., Cross, P., and Muller, J.-P. (2005). InSAR atmospheric correction—GPS topography-dependent turbulence model (GTTM). *Journal of Geophysical Research*, 110, B02404. doi: 10.1029/2005JB003711.

Li, Z., Muller, J.-P., and Cross, P. (2003). Comparison of precipitable water vapor derived from radiosonde, GPS, and moderate-resolution imaging spectroradiometer measurements. *Journal of Geophysical Research*, 108, 4651. doi: 10.1029/2003JD003372.

Lu, Z. (2007). InSAR imaging of volcanic deformation over cloud-prone areas—Aleutian Islands. *Photogrammatric Engineering & Remote Sensing*, 73, 245–257.

Lu, Z., and Dzurisin, D. (2010). Ground surface deformation patterns, magma supply, and magma storage at Okmok volcano, Alaska, inferred from InSAR analysis: II. Co-eruptive deflation, July-August 2008. *Journal of Geophysical Research*, 115, B00B02. doi: 10.1029/2009JB006970.

Lu, Z., and Dzurisin, D. (2014). *InSAR Imaging of Aleutian Volcanoes—Monitoring a Volcanic Arc from Space*. Springer-Praxis Books in Geophysical Sciences. Berlin: Springer.

Lu, Z., Dzurisin, D., Biggs, J., Wicks, C., Jr., and McNutt, S. (2010). Ground surface deformation patterns, magma supply, and magma storage at Okmok volcano, Alaska, from InSAR analysis: I. Inter-eruption deformation, 1997–2008. *Journal of Geophysical Research*, 115, B00B03. doi: 10.1029/2009 JB006969.

Lu, Z., Dzurisin, D., Wicks, C., Power, J., Kwoun, O., and Rykhus, R. (2007). Diverse deformation patterns of Aleutian volcanoes from satellite interferometric synthetic aperture radar (InSAR). In *Volcanism and Subduction: The Kamchatka Region*, ed. J. Eichelberger, E. Gordeev, P. Izbekov, M. Kasahara, and J. Lees. American Geophysical Union Geophysical Monograph Series 172. Washington, DC: American Geophysical Union, 249–261.

Lu, Z., Fatland, R., Wyss, M., Li, S., Eichelberger, J., Dean, K., and Freymueller, J. T. (1997). Deformation of volcanic vents detected by ERS 1 SAR interferometry, Katmai National Park, Alaska. *Geophysical Research Letters*, 24, 695–698.

Lu, Z., and Freymueller, J. (1998). Synthetic aperture radar interferometry coherence analysis over Katmai volcano group, Alaska. *Journal of Geophysical Research*, 103, 29887–29894.

Lu, Z., Fielding, E., Patrick, M., and Trautwein, C. (2003). Estimating lava volume by precision combination of multiple baseline spaceborne and airborne interferometric synthetic aperture radar: The 1997 eruption of Okmok volcano, Alaska. *IEEE Transactions on Geoscience and Remote Sensing*, 41, 1428–1436.

Lu, Z., Jung, H. S., Zhang, L., Lee, W. J., Lee, C. W., and Dzurisin, D. (2013). DEM generation from satellite InSAR. In *Advances in Mapping from Aerospace Imagery: Techniques and Applications*, ed. X. Yang and J. Li. Boca Raton, FL: CRC Press, 119–144.

Lu, Z., Mann, D., Freymueller, J., and Meyer, D. (2000c). Synthetic aperture radar interferometry of Okmok volcano, Alaska: Radar observations. *Journal of Geophysical Research*, 105, 10791–10806.

Lu, Z., Masterlark, T., and Dzurisin, D. (2005). Interferometric synthetic aperture radar (InSAR) study of Okmok volcano, Alaska, 1992–2003: Magma supply dynamics and post-emplacement lava flow deformation. *Journal of Geophysical Research*, 110, B02403. doi: 10.1029/2004JB003148.

Lu, Z., and Kwoun, O. (2008). Radarsat-1 and ERS interferometric analysis over southeastern coastal Louisiana: Implication for mapping water-level changes beneath swamp forests. *IEEE Transactions on Geoscience and Remote Sensing*, 46, 2167–2184.

Lu, Z., Power, J., McConnell, V., Wicks, C., and Dzurisin, D. (2002). Pre-eruptive inflation and surface interferometric coherence characteristics revealed by satellite radar interferometry at Makushin volcano, Alaska: 1993–2000. *Journal of Geophysical Research*, 107, 2266. doi: 10.1029/2001JB000970.

Lu, Z., Rykhus, R., Masterlark, T., and Dean, K. (2004). Mapping recent lava flows at Westdahl volcano, Alaska, using radar and optical satellite imagery. *Remote Sensing of Environment*, 91, 345–353.

Lu, Z., Wicks, C., Dzurisin, D., Power, J., Moran, S., and Thatcher, W. (2002). Magmatic inflation at a dormant stratovolcano—1996–98 activity at Mount Peulik volcano, Alaska, revealed by satellite radar interferometry. *Journal of Geophysical Research*, 107, 2134. doi: 10.1029/2001JB000471.

Lu, Z., Wicks, C., Dzurisin, D., Thatcher, W., Freymueller, J., McNutt, S., and Mann, D. (2000b). Aseismic inflation of Westdahl volcano, Alaska, revealed by satellite radar interferometry. *Geophysical Research Letters*, 27, 1567–1570.

Lu, Z., Wicks, C., Power, J., and Dzurisin, D. (2000a). Ground deformation associated with the March 1996 earthquake swarm at Akutan volcano, Alaska, revealed by satellite radar interferometry. *Journal of Geophysical Research*, 105, 21483–21496.

Lu, Z., and Zhang, L. (2014). Frontiers of Radar Remote Sensing. *Photogrammetric Engineering & Remote Sensing*, 80, 5–13.

Lundgren, P., and Lu, Z. (2006). Inflation model of Uzon caldera, Kamchatka, constrained by satellite radar interferometry observations. *Geophysical Research Letters*, 33, L063012. doi: 10.1029/2005GL025181.

Mann, D., Freymueller, J., and Lu, Z. (2002). Deformation associated with the 1997 eruption of Okmok volcano, Alaska. *Journal of Geophysical Research*, 107, 10.1029/2001JB000163.

Marzano, F. S., Barbieri, S., Vulpiani, G., and Rose, W. I. (2006). Volcanic ash cloud retrieval by ground-based microwave weather radar. *IEEE Transactions on Geoscience and Remote Sensing*, 44, 3235–3246.

Massonnet, D., Briole, P., and Arnaud, A. (1995). Deflation of Mount Etna monitored by spaceborne radar interferometry. *Nature*, 375, 567–570.

Massonnet, D., and Feigl, K. (1998). Radar interferometry and its application to changes in the earth's surface. *Reviews of Geophysics*, 36, 441–500.

Massonnet, D., and Souyris, J. S. (2008). *Imaging with Synthetic Aperture Radar*. Lausanne, Switzerland: EPFL Press.

Masterlark, T., and Lu, Z. (2004). Transient volcano deformation sources imaged with InSAR: Application to Seguam island. *Journal of Geophysical Research*, 109, B01401. doi: 10.1029/2003JB002568.

Miller, T. P., and Casadevall, T. J. (2000). Volcanic ash hazards to aviation. In *Encyclopaedia of Volcanolcanology*, ed. H. Sigurdsson. New York: Academic, 915–930.

Mogi, K. (1958). Relations between the eruptions of various volcanoes and the deformation of the ground surfaces around them. *Bulletin of the Earthquake Research Institute, University of Tokyo*, 36, 99–134.

Myers, B., Brantley, S. R., Stauffer, P., and Hendley, J. W., II. (2008). What are volcano hazards? USGS Fact Sheet 002-97. http://pubs.usgs.gov/fs/fs002-97/.

Newhall, C. G., and Punongbayan, R. S. (1996). Eruptive history of Mount Pinatubo. In *Fire and Mud: Eruptions and Lahars of Mount Pinatubo, Philippines.* Seattle: University of Washington Press, 165–195.

Okada, Y. (1985). Surface deformation due to shear and tensile faults in a half-space. *Bulletin of the Seismological Society of America,* 75, 1135–1154.

Parker, A., Biggs, J., and Lu, Z. (2014). Investigating long-term subsidence at Medicine Lake Volcano, CA, using multi-temporal InSAR. *Geophysical Journal International,* 199, 844–859.

Poland, M. P., Bürgmann, R., Dzurisin, D., Lisowski, M., Masterlark, T., Owen, S., and Fink, J. (2006a). Constraints on the mechanism of long-term, steady subsidence at Medicine Lake volcano, northern California, from GPS, leveling, and InSAR. *Journal of Volcanology and Geothermal Research,* 150, 55–78.

Poland, M., Hamburger, M., and Newman, A. (2006b). The changing shapes of active volcanoes: History, evolution, and future changes for volcano geodesy. *Journal of Volcanology and Geothermal Research,* 150, 1–13. doi: 10.1016/j.jvolgeores.2005.11.005.

Press, W. H., Teukolsky, S. A., Vetterling, W. T., and Flannery, B. P. (2007). *Numerical Recipes in C—The Art of Scientific Computing.* 3rd ed. Cambridge: Cambridge University Press.

Pritchard, M. E., and Simons, M. (2002). A satellite geodetic survey of large-scale deformation of volcanic centres in the central Andes. *Nature,* 418, 167–171.

Pritchard, M. E., and Simons, M. (2004a). An InSAR-based survey of volcanic deformation in the central Andes. *Geochemistry, Geophysics, Geosystems,* 5. doi: 10.1029/2003GC000610.

Pritchard, M. E., and Simons, M. (2004b). Surveying volcanic arcs with satellite radar interferometry: The central Andes, Kamchatka, and beyond. *GSA Today,* 14, 4–11.

Robock, A. (2013). The latest on volcanic eruptions and climate. *Eos, Transactions, American Geophysical Union,* 94, 305–312.

Rocca, F. (2007). Modeling interferogram stacks. *IEEE Transactions on Geoscience and Remote Sensing,* 45, 3289–3299.

Rose, W. I. (1977). Scavenging of volcanic aerosol by ash: Atmospheric and volcanological implications. *Geology,* 5, 621–624.

Rose, W. I., Kostinski, A. B., and Kelley, L. (1995). Real time C band radar observations of 1992 eruption clouds from Crater Peak/Spurr volcano, Alaska. *U.S. Geological Survey Bulletin,* 2139, 19–26.

Rosen, P., Hensley, S., Joughin, I. R., Li, F. K., Madsen, S. N., Rodriguez, E., and Goldstein, R. M. (2000). Synthetic aperture radar interferometry. *Proceedings of the IEEE,* 88, 333–380.

Rosen, P., Hensley, S., Zebker, H., Webb, F. H., and Fielding, E. J. (1996). Surface deformation and coherence measurements of Kīlauea volcano, Hawaii, from SIR-C radar interferometry. *Journal of Geophysical Research,* 101, 23109–23125.

Rust, A. C., Russell, J. K., and Knight, R. J. (1999). Dielectric constant as a predictor of porosity in dry volcanic rocks. *Journal of Volcanology and Geothermal Research,* 91, 79–96.

Schneider, D., and Hoblitt, R. (2013). Doppler weather radar observations of the 2009 eruption of Redoubt volcano, Alaska. *Journal of Volcanology and Geothermal Research,* 259, 133–144.

Tupper, A., Oswalt, J. S., and Rosenfield, D. (2005). Satellite and radar analysis of volcanic-cumulonimbi at Mount Pinatubo, Philippines, 1991. *Journal of Geophysical Research,* 110, D9204, http://dx.doi.org/10.1029/2004JD005499.

Wang, T., Poland, M. P., and Lu, Z. (2015). Dome growth at Mount Cleveland, Aleutian Arc, quantified by time series TerraSAR-X imagery. *Geophysical Research Letters,* 42, 1–8. doi: 10.1002/2015GL066784.

Wicks, C., de la Llera, J. C., Lara, L., and Lowenstern, J. (2011). The role of dyking and fault control in the rapid onset of eruption at Chaitén volcano, Chile. *Nature,* 478, 374–377.

Wicks, C., Dzurisin, D., Ingebritsen, S., Thatcher, W., Lu, Z., and Iverson, J. (2002). Magma intrusion beneath the Three Sisters Volcanic Center in the Cascade Range of Oregon, USA, from interferometric radar measurements. *Geophysical Research Letters,* 29. doi: 10.1029/2001GL014205.

Wicks, C., Thatcher, W., and Dzurisin, D. (1998). Migration of fluids beneath Yellowstone caldera inferred from satellite radar interferometry. *Science,* 282, 458–462.

Wicks, C., Thatcher, W., Dzurisin, D., and Svarc, J. (2006). Uplift, thermal unrest and magma intrusion at Yellowstone caldera. *Nature,* 440, 72–75.

Wright, T., Ebinger, C., Biggs, J., Ayele, A., Yirgu, G., Keir, D., and Stork, A. (2006). Magma-maintained rift segmentation at continental rupture in the 2005 Afar dyking episode. *Nature,* 442, 291–294.

Yang, X.-M., Davis, P. M., and Dieterich, J. H. (1988). Deformation from inflation of a dipping finite prolate spheroid in an elastic half-space as a model for volcanic stressing. *Journal of Geophysical Research,* 93, 4249–4257.

Yun, S., Segall, P., and Zebker, H. (2006). Constraints on magma chamber geometry at Sierra Negra Volcano, Galapagos Islands, based on InSAR observations. *Journal of Volcanology and Geothermal Research*, 150, 232–243.

Zebker, H., Amelung, F., and Jonsson, S. (2000). Remote sensing of volcano surface and internal processing using radar interferometry. In *Remote Sensing of Active Volcanism*, ed. P. Mouginis-Mark, J. A. Crisp, and J. H. Fink. AGU Monograph. Washington, DC: American Geophysical Union, 179–205.

Zebker, H. A., Rosen, P. A., Goldstein, R. M., Gabriel, A., and Werner, C. L. (1994). On the derivation of coseismic displacement fields using differential radar interferometry—The Landers earthquake. *Journal of Geophysical Research*, 99, 19617–19634.

Zebker, H., and Villasenor, J. (1992). Decorrelation in interferometric radar echoes. *IEEE Transactions on Geoscience and Remote Sensing*, 30, 950–959.

Zhang, L., Ding, X. L., and Lu, Z. (2011). Modeling PSInSAR time-series without phase unwrapping. *IEEE Transactions on Geoscience and Remote Sensing*, 49, 547–556. doi: 10.1109/TGRS.2010.2052625.

Zhang, L., Lu, Z., Ding, X. L., Jung, H. S., Feng, G. C., and Lee, C. W. (2012). Mapping ground surface deformation using temporarily coherent point SAR interferometry—Application to Los Angeles Basin. *Remote Sensing of Environment*, 117, 429–439.

19 Active Volcanoes
Satellite Remote Sensing

Nicola Pergola, Eugenio Sansosti and Francesco Marchese

CONTENTS

19.1 INTRODUCTION

Active volcanism represents a serious threat for society and the environment. A continuous, efficient and accurate surveillance of active volcanoes is the proper strategy to manage and reduce risks for both the environment and population. A range of observations is generally required, not only for monitoring purposes but also for improving our understanding of processes underpinning volcanic eruptions. Some parameters and phenomena can be observed from space, taking advantage of Earth observation (EO) satellites in orbit, offering synoptic coverage, timely and frequent sampling, multispectral capabilities and significant cost savings compared with other observing systems. Sometimes, satellite-based observations are the unique opportunity to get systematic and accurate information about volcanic phenomena that, as in the case of eruptive volcanic clouds, may rapidly disperse from hundreds to thousands of kilometres from the source.

According to the Global Volcanism Program (2017), there are more than 1500 active volcanoes worldwide, continuously emitting a large amount of ash, gases and hot material, menacing the security of neighbouring people, the environment, infrastructure and air traffic. The United Nations Environment Programme (UNEP) reports that more than 26,000 people have died in volcanic disasters between 1975 and 2000 (UNEP 2004). Volcanic eruptions have both social (Baxter 2005) and economic impacts. As an example, the refined estimate of the total costs of the Mount St. Helens (Washington State) eruption in 1980 was $1.1 billion (Tilling et al. 1990). As a further example, the estimated economic impact of the 1989–1990 eruption of Redoubt volcano (United States), only on the aviation industry, was more than $101 million (Tuck and Huskey 1994). More recently, the economic losses of aviation companies due to the Eyjafjallajökull (Iceland) eruption in 2010 were estimated to be approximately U.S.\$1.7 billion over a 6-day period (IATA 2010). Furthermore, because of the increase in population worldwide, the number of people and the total cost of social infrastructure close to active volcanoes are increasing. Evidence for increased exposure to volcanic hazards includes a steady increase in the number of fatal eruptions over the last 500 years (Simkin et al. 2001).

Although significant successes have been achieved in understanding volcanoes, many crucial aspects of volcanic activity and processes still remain not fully understood. Systematic observations and continuous monitoring of volcanic activity might help in improving our knowledge of volcanic processes, in better identifying possible signs of volcanic unrest and in better managing and assessing the risk when an eruptive event is in progress. However, many active volcanoes around the world are still inadequately or poorly monitored or not monitored at all. A recent study carried out on 439 active volcanoes in 16 developing countries reveals that more than 60% have rudimentary or no monitoring, including several volcanoes identified as posing a high risk to large populations (Tilling et al. 1990). Another recent report reveals that globally, only about 50% of volcanoes are currently monitored by dedicated ground-based systems (ESA-EUMETSAT 2010).

For decades, satellite remote sensing has played a significant role in this context, offering synoptic coverage, timely and frequent sampling on a large scale, multispectral capabilities and significant cost savings compared with other observing systems and sources of information. Moreover, satellite-based observations represent, sometimes, the unique tool capable of getting systematic and accurate information about volcanic phenomena that, as for volcanic emissions in the atmosphere (ash clouds), may rapidly disperse from hundreds to thousands of kilometres from the source. One recent example is the 2010 Eyjafjallajökull (Iceland) eruption that had a large-scale impact (i.e. affecting several European Union [EU] countries) on social and economic human activities and confirmed that the global perspective offered by satellite systems is crucial (ESA-EUMETSAT 2010).

Satellite remote sensing, besides providing an actual and enhanced monitoring capability, may contribute to better understanding of volcanic processes, providing a huge amount of data that can be routinely used by experts and volcanic observatories. EO satellites, since the 1980s and 1990s, have contributed to achieve significant advances in volcanology, helping the scientific community address major volcanological questions and challenges, and allowed for a periodic or systematic monitoring of volcanism at a global scale. Here, the present capabilities and the main achievements of satellite remote sensing for active volcanoes are reviewed, from a scientific and an operational point of view. The most employed space-based technologies will be mentioned, with the aim of providing a first-level overview of the current wide usage of satellite observations to provide relevant (i.e. qualitative and sometimes quantitative) volcanological information at a global scale.

19.2 SATELLITE REMOTE SENSING OF ACTIVE VOLCANOES

New developments in remote sensing techniques have expanded the capability of scientists worldwide to monitor and understand volcanoes using satellites. Although none of the present operational EO satellite missions were specifically designed or intended for volcanological studies, data from EO satellites have been increasingly and successfully applied to study a wide range of volcanological issues (Oppenheimer 1998; Francis and Rothery 2000). The synoptic coverage, the multispectral capability, the long-term availability and the high temporal frequency of observations offered by sensors flying aboard satellites give, in fact, unprecedented opportunities for active volcanism monitoring and investigation. Both geostationary (GEO) and low Earth orbit (LEO) satellites are used to provide relevant observations and useful information at least for (1) detecting and monitoring eruptive events; (2) assessing thermal energy emitted from the volcano; (3) detecting, tracking and characterizing volcanic plumes and emissions in the atmosphere; (4) measuring volcano topography and topographic changes; and (5) mapping volcano surface deformation.

The spectral capabilities of current satellite technology allow for both qualitative information of ongoing processes at volcanoes and quantitative retrievals of a number of volcanic parameters. Moreover, the generally high sampling frequency (up to an image every 15 minutes) and the time continuity of observations from space (the typical lifetime of a single satellite mission is 5–7 years, but a 'suite' of EO satellites ensures decades of observation records) allow for periodic and recursive measurements, suitable for mapping and monitoring changes globally. Both passive (i.e. sensors measuring signals produced by natural sources) and active (i.e. systems producing electromagnetic signals to be directed towards the Earth's surface and measuring the component reflected back to the sensor by the observing target) satellite technologies are employed, exploiting information contained in the reflected and/or emitted radiation within a wide range of the electromagnetic spectrum (i.e. from the ultraviolet [UV] up to the microwave spectral regions). A list of main satellite systems so far used for investigating and monitoring active volcanoes is reported in Table 19.1.

Here, we review some of the significant applications and advances achieved in volcanology by exploiting the past and present generation of EO satellites. Among all the possible volcanic phenomena, satellite observations are demonstrated to be useful for investigating a wide range of processes during eruptive phases, as well as in pre-eruptive and unrest periods. We have organized this review by focusing on three main topics: (1) thermal activity, (2) eruption plumes and (3) topography and deformations.

19.3 THERMAL ACTIVITY

Several studies have shown the potential of infrared satellite observations for identifying and quantifying volcanic thermal activities, such as lava flows, lava fountains, lava lakes, fumarole fields and pyroclastic deposits (Rothery et al. 1988; Glaze et al. 1989; Harris et al. 1995, 1997, 1999, 2000; Wooster 2001; Wright et al. 2005). Such a potential has significantly increased the monitoring capabilities of volcanoes, especially those located in very remote areas, generally not routinely monitored by traditional geophysical devices, which are difficult to use because of obvious logistical reasons. In addition, in the areas well monitored by ground-based systems, satellite observations have demonstrated usefulness in supporting decision makers, for example, providing rapid alerts on eruption onset, for mapping lava flows and for estimating heat and mass fluxes.

The first observations of volcanic thermal anomalies from space were performed by the National Aeronautics and Space Administration's (NASA) Nimbus 1 satellite in 1964, which was used to study the Hawaiian volcanoes Mauna Loa and Kilauea (Gawarecki et al. 1965). Nimbus satellites carried the high-resolution infrared radiometer (HRIR) sensor, having a spectral channel in the range of 3.5–4.1 μm, which was employed 2 years later to observe lava on Surtsey volcano (Iceland) (Williams and Friedman 1970).

In recent decades, a number of satellite sensors having channels in the infrared bands of the electromagnetic spectrum have been used for volcanological applications. Sensors like the Advanced Very High Resolution Radiometer (AVHRR) and Advanced Along-Track Scanning Radiometer (AATSR) have largely been used for detecting active volcanic surfaces (Harris et al. 1995; Wooster and Rothery 1997a; Kaneko et al. 2002; Pergola et al. 2004a). In particular, the AVHRR, thanks to channels in the medium infrared (MIR) and thermal infrared (TIR) bands and to a good compromise between spatial and temporal resolution (Table 19.1), enabled both identification and quantitative characterization of lava flows (Harris et al. 1997). Other polar orbiting satellite sensors, like the Enhanced Thematic Mapper Plus (ETM+), the Advanced Spaceborne Thermal Emission and Reflection Radiometer (ASTER) and the Advanced Land Imager (ALI), thanks to their spectral capability in the short-wave infrared (SWIR) region, at higher spatial resolution (i.e. tens of metres compared with 1 km of AVHRR-like medium- or coarse-resolution sensors) demonstrated their efficiency for mapping lava flows (Figure 19.1). Moreover, these sensors were profitably used for studying the thermal structure of hot magmatic surfaces, like lava domes and crater lakes (Flynn et al. 2001; Trunk and Bernard 2008; Rose and Ramsey 2009). Geostationary satellite platforms, such as

TABLE 19.1

Active and Passive Satellite EO Missions Used in Volcanology

Spectral Range	Satellite	Sensor/Mode	Orbit/Altitude	Spatial Resolution	Temporal Resolution (Passes per Day)	Spectral Range (Wavelength)	Lifespan	Main Application
				Passive Systems				
VIS/NIR/SWIR/MIR/TIR	NOAA	AVHRR suite	LEO/830 km	1 km	4	0.5–12.5 μm	1979–today	Ash, thermal
	GOES	VISSR/VAS	GEO/35, 790 km	1–4 km/10 km	24–96	0.5–13.7 μm; 0.7–14.7 μm	1980–today	Ash, thermal
	ERS-1/2	ATSR suite	LEO/780 km	1 km	1/3	0.55–12.0 μm	1991–2011	Thermal
	ENVISAT	AATSR	LEO/800 km	1 km	1/3	0.55–12.0 μm	2002–2012	Thermal
	GMS-5	VISSR	GEO/ 36,000 km	1.25–5 km	24	0.5–12.5 μm	1995–2003	Ash, thermal
	HIMAWARI-8/9	AHI	GEO/ 36,000 km	0.5 km VIS; 1 km VNIR; 2 km SWIR/MIR/TIR	144	0.4–13.4 μm	2014–today	Ash, thermal
	Terra/Aqua	MODIS	LEO/705 km	0.25–1 km	4	0.4–14.4 μm	1999–today	Ash, thermal, SO$_2$
	Aqua	AIRS	LEO/705 km	15 km	2	0.4–15.4 μm	2002–today	SO$_2$, ash
	MetOp-A/B	IASI	LEO/817 km	12 km	2	3.7–15.5 μm	2007–today	SO$_2$, ash
	MetOp-A/B	AVHRR	LEO/830 km	1 km	4	0.5–12.5 μm	2006–today	Ash, thermal
	Suomi/NPP	VIIRS	LEO/830 km	0.35–0.75 km	2	0.4–12 μm	2011–today	Ash, thermal
	MSG	SEVIRI	GEO/35, 800 km	3 km IR; 1 km HRV (VIS)	96	0.6–13.40 μm	2006–today	Ash, thermal, SO$_2$
	Terra	ASTER	LEO/705 km	15 m VNIR; 30 m SWIR; 90 m TIR	1/16	0.52–11.65 μm	1999–today	Thermal
	MTSAT-1R/2	Imager	GEO/ 35,800 km	1–4 km	48	0.55–12.5 μm	2005–today	Ash, thermal
	LANDSAT LDCM	TM/ETM+OLI/ TIRS	LEO/705 km	30 m/15 m VNIR; 120 m/60 m/30 m TIR	1/16	0.45–12.5 μm	1982–today	Thermal
	Sentinel-2	MSI	LEO/786 km	10 m VNIR; 20 m SWIR	1/5	0.44–2.2 μm	2015–today	Thermal

(Continued)

TABLE 19.1 (CONTINUED)

Active and Passive Satellite EO Missions Used in Volcanology

Spectral Range	Satellite	Sensor/Mode	Orbit/Altitude	Spatial Resolution		Temporal Resolution (Passes per Day)	Spectral Range (Wavelength)	Lifespan	Main Application
VIS/NIR/SWIR/MIR/TIR	Sentinel-3	SLSTR	LEO/815 km	0.5 km VNIR/SWIR 1 km MIR/TIR		1	0.55–12 μm	2016–today	Ash, thermal
	EO-1	ALI	LEO/705 km	30 m		1/16	0.4–2.35 μm	2000–today	Thermal
	Aura	TES	LEO/705 km	5 × 8 km		1	3.2–15.4 μm	2004–today	SO_2
	NOAA	HIRS suite	LEO/830 km	~18 km		2	3–15 μm 0.69 μm	1979–today	SO_2
UV	Aura	OMI[a]	LEO/705 km	24 × 13 km		1	270–500 nm	2004–today	SO_2
	ERS-2	GOME[a]	LEO/780 km	320 × 40 km		1/3	240–790 nm	1995–2011	SO_2
	ENVISAT	SCIAMACHY[b]	LEO/800 km	30 × 120 km		1/6	240–2380 nm	2002–2012	SO_2
	MetOp	GOME-2[a]	LEO/817 km	80 × 40 km		2	250–790 nm	2006–today	SO_2
	Nimbus 7—Meteor 3 Adeos—Earth Probe	TOMS	LEO/variable	39–64 km		1	312–380 nm	1978–2005	SO_2
MW	NOAA	SBUV/2	LEO/830 km	170 km		1	160–400 nm	1985–today	SO_2
	Aura/UARS	MLS	LEO/705 km	5 × 500 km		1	118–2250 GHz	1991–today	SO_2

Active Systems

Spectral Range	Satellite	Sensor/Mode	Orbit/Altitude	Ground Range	Azimuth	Temporal Resolution (Passes per Day)	Spectral Range (Wavelength)	Lifespan	Main Application
MW	SeaSAT	SAR/Stripmap	LEO/800 km	20 m	6 m	1/24	23.5 cm (L-band)	1978–1978	Topography deformation
	ERS-1	SAR/Stripmap	LEO/800 km	25 m	5 m	1/35	5.66 cm (C-band)	1991–2001	Topography deformation
	ERS-2	SAR/Stripmap	LEO/800 km	25 m	5 m	1/35	5.66 cm (C-band)	1995–2011	Topography deformation
	JERS-1	SAR/Stripmap	LEO/570 km	16 m	7.5 m	1/44	23.5 cm (L-band)	1991–1998	Topography deformation
	ENVISAT	SAR/Stripmap	LEO/800 km	25–50 m	5 m	1/35	5.63 cm (C-band)	2002–2012	Topography deformation
		SAR/Scan		25–50 m	100 m				

(Continued)

TABLE 19.1 (CONTINUED)

Active and Passive Satellite EO Missions Used in Volcanology

Spectral Range	Satellite	Sensor/Mode	Orbit/Altitude	Spatial Resolution Ground Range	Azimuth	Temporal Resolution (Passes per Day)	Spectral Range (Wavelength)	Lifespan	Main Application
MW	RADARSAT-1	SAR/Stripmap SAR/Scan	LEO/800 km	20–30 m 25–40 m	7.5 m 25–35 m	1/24	5.66 cm (C-band)	1995–today	Topography deformation
	RADARSAT-2	SAR/Stripmap SAR/Scan SAR/Spotlight	LEO/800 km	20–30 m 25–40 m 2–5 m	7.5 m 25–35 m 1 m	1/24	5.55 cm (C-band)	2007–today	Topography deformation
	Sentinel-1	SAR/Stripmap SAR/TOPS	LEO/690 km	5 m 5–20 m	5 m 20–40 m	1/6	5.55 cm (C-band)	2014–today	Topography deformation
	COSMO-SkyMed	SAR/Stripmap SAR/Scan SAR/Spotlight	LEO/620 km	3–15 m 7–30 m 1 m	3 m 16–20 m 1 m	1/4–1/8	3.1 cm (X-band)	2007–today	Topography deformation
	TERRASAR-X/ TanDEM-X	SAR/Stripmap SAR/Scan SAR/Spotlight	LEO/510 km	1–3 m 2–3 m 1–3 m	2.4 m 16 m 1 m	1/11	3.11 cm (X-band)	2007–today	Topography deformation
	ALOS-1	SAR/Stripmap SAR/Scan	LEO/690 km	9–30 m 15–75 m	5 m 50 m	1/46	23.6 cm (L-band)	2006–2011	Topography deformation
	ALOS-2	SAR/Stripmap SAR/Scan SAR/Spotlight	LEO/630 km	6–10 m 10–30 m 3 m	5 m 50 m 1 m	1/14	22.9 cm (L-band)	2014–today	Topography deformation
VIS/NIR	CALIPSO	CALIOP	LEO/705 km	0.330 km		1/16	532 nm 1,064 nm	2006–today	Aerosol

Note: VIS, visible; VNIR, visible to near infrared; MW, microwave; NIR, near infrared.

[a] UV + VIS capability.

[b] UV + VIS + SWIR capability.

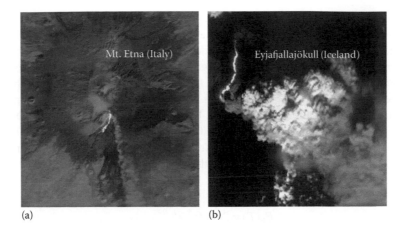

(a)   (b)

FIGURE 19.1   High-resolution infrared satellite imagery for lava flow mapping: (a) ASTER composite image of 30 December 2002 showing Mt. Etna (Italy) lava flows and (b) EO-1/ALI composite image of infrared channels showing Eyjafjallajökull (Iceland) eruptive activity of 4 May 2010. (Courtesy of M. Abrams, Jet Propulsion Laboratory, Pasadena, CA.)

the Geostationary Operational Environmental Satellite–Visible-Infrared Spin-Scan Radiometer (GOES-VISSR), the Meteosat Second Generation–Spinning Enhanced Visible and Infrared Imager (MSG-SEVIRI) and the Multifunctional Transport Satellites (MTSAT), providing data at high temporal resolution (from 30 to 15 minutes), have significantly increased the monitoring capabilities of active volcanoes. Indeed, they were suited for timely detecting eruption onsets (Harris et al. 2002; Pergola et al. 2008, 2015), as well as for providing continuous estimations of the lava discharge rates (Gouhier et al. 2011; Ganci et al. 2012).

Therefore, especially sensors offering improved spectral resolution, high dynamic range and enhanced spatial and/or temporal sampling, like the Moderate Resolution Imaging Spectroradiometer (MODIS) onboard the Terra and Aqua satellites, have opened new and exciting scenarios for the surveillance of active volcanoes also in operational contexts.

In the following, various applications of satellite remote sensing for thermal monitoring of volcanoes are discussed.

19.3.1   PRINCIPLES OF HOTSPOT DETECTION FROM SPACE

The identification of thermal anomalies (i.e. hotspots) from space is a key application in the general context of the thermal monitoring of volcanoes. In this context, high temporal resolution satellite systems (AVHRR, MODIS, GOES and SEVIRI) may be very useful, especially in pre-eruptive phases, timely identifying, for instance, abrupt changes in the thermal state of volcanoes which may announce new impending activities and in co-eruptive scenarios (i.e. during volcanic crisis), providing quantitative estimations of the thermal flux and mass eruption rate. The latter may in turn be used by numerical models simulating lava flow paths and predicting their space–time evolution for operational purposes.

Several algorithms have been up to now developed to detect volcanic hotspots by using polar and/or geostationary satellite data (Higgins and Harris 1997; Dean et al. 1998; Wright et al. 2002; Kervyn et al. 2008; Koeppen et al. 2011; Marchese et al. 2011; Piscini and Lombardo 2014; Coppola et al. 2016). These methods are commonly based on the measurement of the Earth's emitted radiation in the infrared region of the electromagnetic spectrum, generally exploiting properties of hot magmatic surfaces to be more radiant in the spectral range of 2–5 µm compared with environmental and natural surfaces. Wien's displacement law (Equation 19.1) describes this spectral behaviour:

$$\lambda_{max} \cong 2898/T \qquad (19.1)$$

where T is the temperature (K) of the emitting blackbody and λ_{max} is the peak emission wavelength (μm) (i.e. the wavelength where the emitted radiance is maximum). On the basis of Equation 19.1, volcanic surfaces having temperatures on the order of 1200 K emit the maximum of their thermal radiation at around 2.4 μm, in the SWIR band. Magmatic surfaces at lower temperatures (800–1000 K) emit the maximum of their thermal radiation in the MIR band, around 3–4 μm. Therefore, as the temperature decreases, the peak of thermal emission moves towards longer wavelengths, reaching around 10 μm, in the TIR band, for surfaces at background temperatures (i.e. at around 300 K). Equation 19.1 is derived by Planck's law, which describes the spectral radiance $B_\lambda(T)$ (W m^{-3} sr^{-1}) emitted by a blackbody in thermal equilibrium (representing a schematization of an ideal thermal radiator emitting more energy at every wavelength than any other body at the same temperature), at temperature T, as a function of the wavelength:

$$B_\lambda(T) = 2hc^2 / \lambda^5 (e^{hc/\lambda kT} - 1) \tag{19.2}$$

where h is Planck's constant (6.6256 × 10^{-34} J s), c is the light speed (2.9979 × 10^8 m s^{-1}), k is Boltzmann's gas constant (1.3806 × 10^{-23} J K^{-1}) and λ is the wavelength. An example of Planck curves for blackbodies at different temperatures (i.e. from background up to magmatic sources) is displayed in Figure 19.2. The inverse of Planck's function is used for converting the spectral radiance to the correspondent brightness temperature (BT) value (i.e. representing the blackbody temperature). As far as real surfaces are considered, emitted radiance is always less than the blackbody radiance (Equation 19.2), so that $BT < T$. The spectral emissivity $\varepsilon_\lambda = R_\lambda(T)/B_\lambda(T)$ describes the capability of a real surface to emit thermal radiation compared with a blackbody having the same temperature T, at wavelength λ. Spectral emissivity varies from 0 to 1 for real surfaces, while for a blackbody, $\varepsilon_\lambda = 1$. Emissivity estimations of high-temperature magmatic surfaces, as lava bodies, can be found in some published works (Burgi et al. 2002).

However, a satellite sensor at the top of the atmosphere, operating in the infrared region, measures not only the thermal radiation emitted by the Earth's surface, but also other contributions.

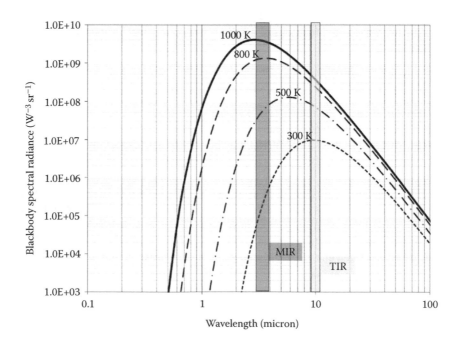

FIGURE 19.2 Planck curves of blackbodies at different, both ambient and magmatic, temperatures.

In particular, the total measured radiance in the infrared region is described by the following radiative transfer equation:

$$R_\lambda = \tau_\lambda \varepsilon_\lambda B_\lambda(T_s) + \rho_\lambda \tau_\lambda R_{\lambda D} + \int_0^{z_{sat}} B_\lambda\big(T(z)\big)\frac{\partial \tau}{\partial z}dz \qquad (19.3)$$

In this equation, R_λ is the total radiance measured at the top of the atmosphere at wavelength λ. The term $\tau_\lambda \varepsilon_\lambda B_\lambda(T_s)$ represents the thermal radiation emitted by the surface, having a temperature of T_s and emissivity of ε_λ, which undergoes attenuation passing through the atmosphere having transmittance τ_λ. The term $\rho_\lambda \tau_\lambda R_{\lambda D}$ represents instead the radiation reaching the sensor after the interaction with the Earth's surface (i.e. by reflection, modulated by the superficial reflectance ρ_λ) and the atmosphere (i.e. by absorption, modulated by the atmospheric transmittance τ_λ), where $R_{\lambda D}$ represents the downwelling (i.e. directed to the Earth's surface) radiation. Finally, the term $\int_0^{z_{sat}} B\big(T(z)\big)\frac{\partial \tau}{\partial z}dz$ is the upwelling radiation (i.e. directed towards the satellite sensor) coming from all the atmospheric layers, from the surface (i.e. at height $z = 0$) up to the top of the atmosphere (i.e. at satellite height $z = z_{sat}$). Although Equation 19.3 is generally true and always represents the actual situation, it should be stressed that when very hot surface sources are concerned (as in the case of active volcanic features), atmospheric contributions to the measured radiance can often be considered negligible.

19.3.1.1   Hotspot Detection Algorithms

Volcanic hotspot detection from space is generally a matter of automatically identifying highly radiative pixels over the satellite scenes with a reasonable trade-off between reliability (i.e. keeping the false-positive rate low) and sensitivity (i.e. ability to detect any thermal changes, including low- to midlevel signals).

The first algorithms developed to detect surface thermal anomalies (forest fires) applied fixed thresholds to single MIR images (Langaas 1992; Arino and Melinotte 1995) exploiting, as before mentioned, the properties of hot surfaces to significantly increase the pixel-integrated temperature even when they occupy only a small portion of the pixel area. In particular, a thermal anomaly ranging from 900 to 1300 K must occupy between 10^{-5} and 10^{-6} of the 1 km^2 pixel to be easily identified (Oppenheimer 1998). In Figure 19.3, we show an example of hotspots over Shinmocdake volcano (Japan) (white pixels in the inset of the figure) potentially detectable on the TERRA-MODIS MIR channel on 3 February 2011 at 1350 Greenwich mean time (GMT). The aforementioned methods do not consider, however, both seasonal (emissivity variations) and environmental (high reflectance of sparsely vegetated areas in daytime, and local warming effects) factors that, by strongly affecting MIR radiances, represented a significant cause of omission (i.e. false-negative) and/or commission (i.e. false-positive) errors (Di Bello et al. 2004). These limitations stimulated researchers to develop more advanced and refined algorithms, capable of improving hotspot identification by satellite under different observational conditions.

Multichannel, fixed-threshold techniques, exploiting the different spectral behaviour of high-temperature surfaces in the MIR and TIR bands compared with natural and environmental ones, showed the first significant improvements compared with single-channel algorithms (Kennedy et al. 1994; Li et al. 2001). Some of these methods were specifically tailored to volcanological applications (Wright et al. 2002). Among them, MODVOLC uses thresholds set sufficiently high to perform globally (Wright 2016). However, some limitations (low sensitivity to subtle hotspots) still affect the performance of these methods, encouraging the development of more advanced algorithms, like the so-called 'contextual' approaches (Giglio et al. 1999, 2003). The latter generally compare the $\Delta BT = BT_{MIR} - BT_{TIR}$ spectral difference (where BT_{MIR} and BT_{TIR} represent the brightness temperatures measured in the MIR and TIR bands, respectively) of each pixel under investigation with the

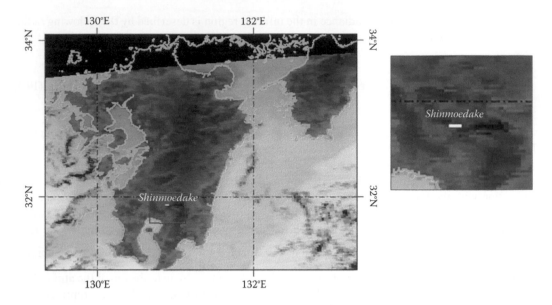

FIGURE 19.3   Volcanic hotspot (white pixels on the image and in the inset) detected by the TERRA-MODIS, mid-infrared image (i.e. channel 22 image, centred at 3.9 μm), during the Shinmoedake (Japan) eruption on 3 February 2011 at 1350 GMT (image reprojected in LAT-LONG WGS 84).

ΔBT of neighbouring pixels (representing the background signals) in terms of spatial mean and standard deviation (Giglio et al. 1999; Kaneko et al. 2002). Contextual algorithms used to detect volcanic hotspots (VAST) (Higgins and Harris 1997) have been shown to perform well, especially when applied on a local or regional scale (Kaneko et al. 2002; Steffke and Harris 2011), despite the dependence on background (presence of cloudy pixels), generally representing a non-negligible issue for these techniques.

Multitemporal algorithms (Pergola et al. 2004a; Marchese et al. 2011), analyzing signal in the space–time domain and using self-adaptive thresholds, specific for each place and time of observation, have demonstrated that active magmatic surfaces may be identified by satellite with a high trade-off between reliability and sensitivity. In particular, algorithms like RST_{VOLC} (Marchese et al. 2011) have shown that low-level hotspots, which are sometimes associated with pre-eruptive phases (Dehn et al. 2000, 2002; Kaneko et al. 2006; Pergola et al. 2009; Van Manen et al. 2010; Girina 2012), may be detected without generating significant false detections (Marchese et al. 2012). Finally, a number of hybrid approaches have also been proposed to detect hot volcanic features, combining one or more features of the above-mentioned methods (Kervyn et al. 2008; Koeppen et al. 2011).

Automated monitoring systems, implementing some of the above-mentioned techniques, currently perform an operational surveillance of active volcanoes from space (MIROVA 2017; MODVOLC 2017; REALVOLC 2017), also providing estimates of volcanic heat flux which is retrieved by satellite (using infrared MODIS data) starting from detected hotspot pixels.

19.3.2   TIME-SERIES ANALYSES OF VOLCANIC HOTSPOTS

As mentioned earlier, the first satellite observations of thermal volcanic activities dated back to the 1960s. Moreover, some EO space missions have provided data with continuity since the 1970s (LANDSAT and National Oceanic and Atmospheric Administration [NOAA]), ensuring a continuous data stream and long-term time series of observations, that can be used to better understand volcanic processes and interpret possible changes at volcanic sources. Several works have shown that time-series analysis of emitted radiances at volcanoes, measured by satellite, may be an important contribution for studying the thermal behaviour of volcanoes over long-term time periods. Among the studies performed to investigate the thermal state of active volcanoes, a time series of infrared nighttime satellite records, provided by the Along-Track Scanning Radiometer (ATSR) sensor, was analyzed to study eruptions of the Lascar (Chile) volcano of 1992–1995 (Wooster and Rothery 1997a). A cycle of increasing and decreasing SWIR radiance was also recognized, and interpreted as a variation in gas flow through fumaroles on the crater floor, which was followed by a major explosive event. Other works analyzing the time series of satellite-derived SWIR radiances have been more recently performed, assessing and quantifying thermal emissions at volcanoes (Blackett and Wooster 2011).

Several studies investigating the temporal trend of MIR signals have demonstrated the high potential of temporally resolved space observations in providing timely alerts about eruption onsets (Harris et al. 2000; Marchese et al. 2010, 2014), as well as about new phases of unrest (Dehn et al. 2002; Pieri and Abrams 2005; Marchese et al. 2012). One example is reported in Figure 19.4, where a 3-day time series (between 18 and 20 February 2013) of 15-minute SEVIRI observations in the MIR spectral band, recorded over the Mt. Etna area (Sicily, Italy), is reported. It is evident that the satellite sensor was capable of promptly identifying the first paroxysmal event of 2013, which occurred on 19 February. In particular, the MIR signal collected over the Mt. Etna pixel started to significantly deviate from the coincident background level (recorded over a nearby, non-volcanic area) from 0330 GMT on 19 February 2013, thus accurately setting the exact moment of the event onset.

Together with the aforementioned studies, even advanced statistical approaches were used for dynamically characterizing thermal signals over active volcanic areas in time domain (Marchese et al. 2006;

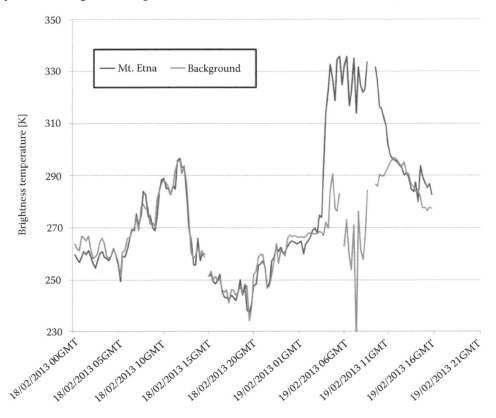

FIGURE 19.4 Time series of SEVIRI MIR measurements recorded every 15 minutes over the Mt. Etna area from 18 to 20 February 2013. The red line refers to Mt. Etna pixels, whereas the blue line is related to a background (i.e. non-volcanic) nearby pixels. The strong deviation between the two curves, observed starting from 19 February at 0315 GMT, was due to the first paroxistic event that occurred at Etna in 2013.

Lovallo et al. 2007, 2009). Finally, similar works have shown that when time series of infrared satellite signals are analyzed, the different thermal phases of volcanoes (lava dome growth periods and lava effusions) may be well discriminated and investigated from space (Pergola et al. 2009; Van Manen et al. 2012).

19.3.3 THERMAL ANOMALY CHARACTERIZATION AND QUANTIFICATION

Once volcanic thermal anomalies are detected, they may be characterized in terms of temperature and/or extent, providing information aiming at better investigating the ongoing volcanic processes.

The 'dual-band' method, originally proposed to analyze the thermal structure of forest fires, starting from radiance measured in two spectral bands of the AVHRR sensor (Dozier 1981), can also be used for characterizing thermal volcanic surfaces (Rothery et al. 1988).

This method assumes that two thermal components occupy the pixel area in the presence of a hot surface: the hot part (i.e. magmatic source), having temperature T_h and covering a fraction p of the pixel, and the cool part (i.e. background) at temperature T_b over the remaining pixel fraction $(1-p)$. If one of these parameters is assumed (generally T_h), or estimated from independent information (field data), the remaining two terms may be derived by solving the following equation system:

$$R_{\lambda_1} = pB_{\lambda_1}(T_h) + (1-p)B_{\lambda 1}(T_b)$$
$$R_{\lambda_2} = pB_{\lambda_2}(T_h) + (1-p)B_{\lambda 2}(T_b) \tag{19.4}$$

where R_{λ_1} and R_{λ_2} are the radiances measured in two different spectral bands, λ_1 and λ_2, of the used satellite sensor (MIR and TIR bands of AVHRR), while $B_{\lambda_1}(T_h)$, $B_{\lambda_1}(T_b)$, $B_{\lambda_2}(T_h)$ and $B_{\lambda_2}(T_b)$ represent the Planck functions at correspondent surface temperatures and wavelengths λ_1 and λ_2. Improved versions of the dual-band method have been proposed (Oppenheimer et al. 1993; Wooster and Rothery 1997b; Flynn et al. 2000; Lombardo and Buongiorno 2006) taking into account its limitations, widely reported and discussed in literature (Rothery et al. 1988). As an example, only two thermal components are insufficient to correctly describe the thermal structure of the pixel area in the presence of a lava flow. Quantitative parameters derived by this method in its different formulations are generally used for estimating the heat flux by means of the Stefan–Boltzmann law:

$$Q_{rad} = \sigma\varepsilon A \left| pT_h^4 + (1-p)T_b^4 \right| \tag{19.5}$$

where Q_{rad} is the heat flux (W), σ is Stefan–Boltzmann's constant (5.67×10^{-8} W m^{-2} K^{-4}) and ε is the surface emissivity, while T_h, T_b and p have the same meanings as before. The heat flux gives an estimation of the energy released by the volcanic features, such as lava dome, lava fountains and lava flows (Wright et al. 2005; Coppola et al. 2009). Its time variations, better derivable by using high-temporal-resolution data, describe changes in the intensity of thermal emissions as a consequence of increases or decreases in the thermal activity in progress at volcanoes. Some approximations of the heat flux were also proposed for estimating this parameter directly from data measured in the MIR band by sensors like MODIS, quantifying the relative errors (Kaufman et al. 1998; Wooster et al. 2003). Indeed, reliable estimations of this parameter are very important for retrieving the effusion rate (i.e. the rate at which lava is erupted, generally expressed in cubic metres per second), which is an important parameter for volcanologists (Harris et al. 2007). In addition, the effusion rate (E_r) is a critical parameter for numerical models generally employed to simulate lava flow paths, forecasting areas that may be invaded by lava flows. The effusion rate can be estimated using the formulation proposed by some authors (i.e. Pieri and Baloga 1986; Crisp and Baloga 1990), adapted to infrared satellite data in the following studies (Harris et al. 1998) and referring to a specific flow model in steady state.

In particular, considering the total heat flux, as the sum of the radiative flux (Q_{rad}) and of heat losses from lava flow both by convection (Q_{conv}) and conduction (Q_{cond}), the effusion rate may be calculated as

$$E_r = Q_{rad} + Q_{conv} + Q_{cond}/\rho(C_p \delta T + C_L \Delta\phi) \quad (m^3 s^{-1}) \tag{19.6}$$

where Q_{rad} is the radiant heat flux measured following Equation 19.5, or using different approximations; ρ is the lava density; C_p is the specific heat capacity; and δT is the eruption temperature minus the solidus, while $\Delta\Phi$ and C_L respectively represent the crystallization in cooling through δT and the latent heat of crystallization. The terms Q_{conv} and Q_{cond} are calculated by means of relative equations reported in the literature (Harris et al. 1998, 2000). Satellite measurements of effusion or the eruption rate have been shown to be compatible with field data when available (Harris et al. 1998, 2000), although the assumption of the model underlying Equation 19.6 is still the object of an important scientific debate (Dragoni and Tallarico 2009; Harris and Baloga 2009).

19.4 ERUPTION PLUMES

During eruptive phases, volcanoes generally emit several gases (H_2O, CO_2 and SO_2) and solid materials (ash minerals and glass) into the atmosphere. Atmospheric emissions of volcanoes may have an impact on the local environment and on the health of the population living in the surrounding areas (Delmelle 2003). Volcanic plumes (especially ash) may also pose a serious threat for aviation (Casadevall 1994, 1996; Miller and Casadevall 2000; Robock 2000; Duffell et al. 2003) and may affect climate, even on a global scale (Robock 2000). The characteristics, intensity and typology of atmospheric emissions also reflect the status of volcanoes, with changes in gas composition and/or flux which, for instance, may possibly announce incoming eruptions (Oppenheimer 1998; Duffell et al. 2003). Therefore, continuous and accurate observations of volcanic emissions in the atmosphere are necessary for hazard mitigation as well as for improving our knowledge about eruption dynamics and properties. The present satellite technology offers a wide range of spatial, temporal and spectral resolutions adequate to investigate and monitor volcanic plumes. Furthermore, satellites, thanks to their synoptic view, are often the only means of getting information on very rapid and large-scale phenomena, like volcanic plume dispersions. In fact, due to wind speed and wind shear, the transport of volcanic plumes over long distances in a short amount of time is generally possible (Durant et al. 2011).

Regarding volcanic ash clouds, the first objective is their accurate and timely detection and their discrimination from conventional weather clouds. Continuous tracking of ash clouds is also required, to avoid encounters with airplanes (because ash may seriously damage aircraft engines) and to mitigate aviation hazards. Besides the spatial (i.e. two-dimensional [2-D]) extent and distribution of volcanic clouds, a three-dimensional (3-D) reconstruction is also important, providing information about the height reached by the plumes. Finally, gas monitoring and mapping is also required and achievable from space, although present technological and observational limits actually restrict applications to sulphur dioxide, when injected in large quantities into the atmosphere.

19.4.1 Ash Cloud Detection and Tracking

Volcanic ash means all the particulate emissions (mostly silicate materials) emitted by a volcano during an eruption (Stix and Gaonac'h 2000). Generally, the fine (1–15 μm in size) component of ash can be detected and tracked by satellite as coarse particles rapidly fall out (Rose et al. 1995).

Since the 1980s, weather satellites have been recognized as a potential source of valuable information about volcanic ash clouds (Sawada 1987). Initially, satellite images were only used to study eruption cloud dynamics, by visual inspection, without exploiting their multispectral capability

(Hanstrum and Watson 1983; Malingreau and Kaswanda 1986; Sparks et al. 1987; Glaze et al. 1999). The first studies suggesting a physically based approach to distinguish andesitic ash from meteorological clouds were performed by Prata (1989a), who identified a spectral signature of ash in the TIR region of the electromagnetic spectrum. In particular, it was recognized as a 'reverse absorption' of ash (Prata 1989b; Holasek and Rose 1991), in comparison with weather clouds, in satellite images acquired in AVHRR channel 4 (10.3–11.3 m) and channel 5 (11.5–12.5 μm). Specifically, the brightness temperature difference (BTD) in these two spectral bands was used to enhance discrimination, as the BTD should be positive for water or ice (contained in meteorological clouds) and negative for ash (present in volcanic plumes):

$$BTD = BT(11\,\mu m) - BT(12\,\mu m) < 0 \ ash$$
$$BTD = BT(11\,\mu m) - BT(12\,\mu m) > 0 \ water/ice \tag{19.7}$$

Although there are some limitations and drawbacks of this method, extensively discussed in the scientific literature (Simpson et al. 2000, 2001; Prata et al. 2001), it represented pioneering work in this field, and under specific conditions, it works well. As the real world is always more complex than theoretic models, advanced detection and discrimination methods were afterwards proposed by several authors, to improve the identification of volcanic ash (Mosher 2000; Hillger and Clark 2002; Yu et al. 2002; Ellrod et al. 2003; Pergola et al. 2004b; Pavolonis et al. 2006; Picchiani et al. 2011; Francis et al. 2012; Guéhenneux et al. 2015; Taylor et al. 2015). These approaches, mostly employing infrared radiances, use principal component analysis, implement multitemporal processing schemes, perform water vapour corrections and, occasionally, introduce short-wave channels (visible and mid-infrared) information or exploit information provided by spectral channels centred in the SO_2 absorption bands (at 8.6 μm wavelengths) to achieve improved results in terms of sensitivity of detection and reliability of ash discrimination.

Once detected, ash clouds need to be continuously monitored and tracked, in space and time domains, in order to mitigate aviation hazards posed by fine ash particles which are often not detectable by onboard radar systems (Holasek and Self 1995). High-temporal-resolution satellite imagery is generally preferred in this aim, by exploiting the near-real-time (NRT) observation rate offered by present weather EO platforms (Figure 19.5). In particular, space systems orbiting with geostationary altitude (MSG and GOES) presently guarantee observations in a 15- to 30-minute interval, providing an unprecedented source of information in terms of timeliness and continuity (Holasek and Self 1995; Davies and Rose 1998; Mayberry et al. 2003; Francis et al. 2012).

Ash detection by satellite, as mentioned, is generally a matter of discrimination. Thus, qualitative products (ash or no ash digital masks) are generally provided. However, multispectral satellite observations may provide further, more quantitative information. In particular, a pioneer work (Wen and Rose 1994) demonstrated that volcanic ash masses can be retrieved from infrared satellite radiances by using a microphysical model of the ash particles together with a detailed radiative transfer model. This retrieving method (named 'split window'), based on a series of assumptions, is capable of revealing particle size (i.e. effective radius) and optical depth. Integrating these parameters over the ash-affected area and fixing ash density leads to the inference that the total mass and the mass loading of the volcanic cloud can be quantified as input parameters for dispersion models.

Although other, more advanced retrieval methods have been developed (Prata and Grant 2001), uncertainties still remain high (e.g. due to difficulties in precisely estimating cloud thickness), with mass loading errors of about 40%–50% (Wen and Rose 1994). However, there are many examples in the literature of quantitative satellite-based ash maps (Schneider et al. 1995; Watson et al. 2004; Kearney and Watson 2009) which have contributed to investigate the space–time evolution of volcanic atmospheric emissions and to better assess aviation hazards posed by volcanic ash.

Finally, high-spectral-resolution infrared satellite sensors, such as the Atmospheric Infrared Sounder (AIRS) and the Infrared Atmospheric Sounding Interferometer (IASI), have shown good

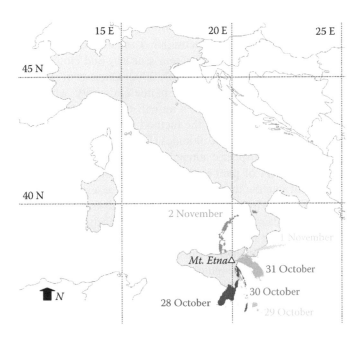

FIGURE 19.5 Detection and tracking of volcanic ash by frequent satellite images: AVHRR-based ash maps achieved during Mt. Etna eruption in October–November 2002. (Adapted from Pergola N. et al. *Annals of Geophysics*, 51, 1, 187–301, 2008.)

potential for investigating ash features (Chahine et al. 2006), better taking into account, for instance, the variable mineralogy of the volcanic plumes (Klüser et al. 2012). Examples of the profitable use of these sensors for detecting and retrieving ash properties can be found in several published studies (Carn et al. 2005; Clarisse et al. 2008; Gangale et al. 2010). The *AIRS concavity index* correctly observed, for instance, the transition from basaltic to magmatic glass dominating the fine ash during the Eyjafjallajökull 2010 (Iceland) eruption (IATA 2010).

19.4.2 PLUME HEIGHT

The height reached by an eruption cloud is another important piece of information, from both a volcanological and an operational point of view. The information about eruptive cloud altitude, in fact, can be used to infer size, character and potential impact of the eruption which is in progress, as well as to provide more accurate, 3-D information about ash-affected areas, which is definitely more useful for aviation hazard assessment and mitigation (Tupper et al. 2012). Although there is no standard satellite product able to provide volcanic cloud altitude, several methods and techniques have been proposed and used for inferring plume heights by satellite observations (Sparks et al. 1987). One approach exploits the stereoscopic view of some satellite systems, such as the Along Track Scanning Radiometer (ATSR), the Multi-angle Imaging SpectroRadiometer (MISR) and the ASTER, which, by furnishing a pair of observations at nadir and forward viewing, are capable of determining cloud-top height by parallax analysis with relatively good accuracy (Prata and Turner 1997). Unfortunately, only a limited number of present satellite sensors have stereoscopic capability, together with resolutions (spatial, spectral and temporal) adequate to investigate volcanic eruptive clouds.

Another satellite method to derive cloud-top height is based on a simple radiometric temperature measurement (Glaze et al. 1989). This approach is based on three main assumptions: (1) volcanic plumes behave as blackbodies (i.e. they have emissivity close to unity), (2) volcanic cloud top is in thermodynamic equilibrium with surrounding air and (3) volcanic clouds are thick and opaque.

Cloud-top altitude can be then assessed by selecting the coldest pixel over the plume, generally looking at radiance measured in the 11 μm spectral band and, for accurate estimations, by using an (independent) atmospheric temperature profile (from a nearby and close-in-time radiosonde measurement) to obtain cloud-top altitude (Sawada 1987, 2002; Glaze et al. 1989; Holasek et al. 1996; Tupper et al. 2004). Inaccuracies arise because of plume emissivity, which is poorly known, and because the thermodynamic equilibrium is often unrealistic (Woods and Self 1992). Moreover, if volcanic clouds are close to the tropopause, the method is more prone to large errors, due to the small rate of change of temperature at these altitudes (Prata and Grant 2001). Another issue can be the availability of independent atmospheric temperature profiles.

During daylight, measurement of plume shadow lengths on the Earth's surface, achieved by exploring visible channels of satellite imagery, can be used to assess plume height by trigonometric computations (Glaze et al. 1989; Holasek and Self 1995). Analysis of shadows over sea surfaces is simpler than that over land (because of reduced topography effects) and provides better accuracies. Drawbacks are due to its limited use in daylight. Furthermore, the 'shadow' method provides a better measure of the height of cloud edges than of the central part of the plume; thus, the method generally furnishes a lower bound of the plume height (Glaze et al. 1999).

The very high temporal sampling of geostationary satellites might also be exploited to derive plume speed from consecutive (very close in time) observations (Sparks et al. 1987; Lynch and Stephens 1996; Simkin et al. 2001; Tupper et al. 2004). The plume is assumed to be driven by atmospheric winds; thus, plume speed is an indirect measure of wind speed. Once wind speed is derived, height estimates might be inferred by analyzing nearly coincident (in space and time) wind speed profiles, obtained by independent sources. This method generally furnishes rough estimations (error range of several hundred to thousands of metres), with major uncertainties coming from ambiguous wind vectors and speeds at different elevations (Steffke et al. 2010). The availability and spatial or temporal resolution of ancillary wind fields should also be carefully taken into account.

Finally, more accurate estimations of plume height can be achieved by the 'CO$_2$ slicing' method. It exploits the differential absorption of atmospheric CO$_2$ in contiguous infrared spectral channels, selected within the CO$_2$ absorption band (i.e. around 13.5–14 μm), to retrieve cloud-top pressure and, consequently, cloud-top height (Menzel et al. 2002). As the CO$_2$ slicing algorithm determines the height of the radiative centre of the cloud, it works better for optically thick than for optically thin clouds. Moreover, problems in estimating cloud pressure may arise when the difference in radiance between the cloud and clear sky is low, as occurs in a relatively isothermal atmosphere (Menzel et al. 2002; Richards et al. 2006). In addition, CO$_2$ channels are presently available only on a few space-based instruments and will not be present on most of the new-generation satellite missions. In conclusion, despite several methods and techniques having been so far developed and proposed, an accurate and consistent estimate of the plume height still represents an issue in the monitoring of volcanic ash clouds from space (Webley and Mastin 2009).

19.4.3 Plume Gas Measurements

Among the several gases emitted by an active volcano, the first three, ranked for their relative abundance in emissions, are water vapour, CO$_2$ and SO$_2$. Unfortunately, water vapour and carbon dioxide are normally present with high concentrations in the atmosphere, resulting in definitively hard-to-discriminate volcanogenic contributions from background levels of these two atmospheric components. On the other hand, SO$_2$ is relatively scarce in the background atmosphere, with only limited emissions coming from anthropogenic sources. Therefore, current satellite applications for volcanogenic gases are essentially restricted to the detection and mapping of sulphur dioxide, not only when emitted in large quantities during eruptive phases but also for passive degassing (Carn et al. 2008). Accurately measuring and monitoring volcanic SO$_2$ emissions and fluxes improves our knowledge of volcano behaviour and may have an impact on forecasting hazardous events (Schneider et al. 1999). Both decreases and increases in gas emissions may occur before main eruptive events.

Only to provide a couple of examples, some eruptions of Mt. Etna were preceded by significant drops in SO_2 emission rates (Schneider et al. 1999), whereas main events at St. Helens during 1980–1988 occurred after increases in gas emission (Caltabiano et al. 1994). Moreover, the identification of sulphur dioxide is particularly important for its potential impact on aircraft engines and because it may give a good indication about possible ash presence within volcanic plumes (McGee and Sutton 1994), although SO_2 and ash may separate in distinct clouds under vertical wind shear conditions (Constantine et al. 2000; Carn et al. 2009). Sulphur dioxide has spectral signatures in both infrared (7.3 and 8.6 µm) and UV (300 and 350 nm) regions of the electromagnetic spectrum (Constantine et al. 2000). Both absorption regions can then be used for retrieving SO_2 from space, with relative advantages and drawbacks.

Concerning UV, the ability to map SO_2 in volcanic clouds from EO satellites was first demonstrated in 1983 (Thomas and Watson 2010), studying the eruption of El Chichon (Mexico) in 1982 by using Total Ozone Mapping Spectrometer (TOMS) data, a UV spectrometer measuring solar reflected radiance in six different bands, operational onboard various satellite platforms since 1979 (Table 19.1). The algorithm assumes that solar radiation is attenuated by absorbing species in the atmosphere according to the Beer–Bouguer–Lambert law:

$$I(\lambda) = I_0(\lambda)\exp\left(-\tau_a(\lambda)\right) \tag{19.8}$$

where $I(\lambda)$ is the measured radiance at the top of the atmosphere and at wavelength λ, while $\tau_a(\lambda)$ is the atmospheric transmittance and $I_0(\lambda)$ is the expected radiance for a clear (i.e. totally transparent) atmosphere. The transmittance $\tau_a(\lambda)$, with negligible scattering effects, is directly proportional to the gas concentration. The simultaneous retrieval of TOMS radiances at four wavelengths (312.5, 317.5, 331.3 and 339 nm) was used to calculate the SO_2 columnar content (Krueger 1983; Krotkov et al. 1997; Constantine et al. 2000). Afterwards, other SO_2 retrieval schemes were developed by using TOMS and new-generation hyperspectral instruments. They include the Solar Backscatter Ultraviolet/2 (SBUV/2), the Scanning Imaging Absorption Spectrometer for Atmospheric Cartography (SCIAMACHY), the Global Ozone Monitoring Experiment (GOME) and the Ozone Monitoring Instrument (OMI), which have further increased the sensitivity to volcanic SO_2 (McPeters 1993; Bovensmann et al. 1999; Krueger et al. 2000). As an example, the differential optical absorption spectroscopy (DOAS) algorithm, also implemented on data provided by GOME-1/2 and SCIAMACHY, offering extended spectral range capabilities (Burrows et al. 1999; Krotkov et al. 2006), has become a standard technique for identifying and quantifying trace gases in the atmosphere, including volcanic sulphur dioxide (Vogel et al. 2012). However, although UV sensors allow for a reliable identification (i.e. suitable for tracking volcanic plumes for long distances) and retrieval of volcanic SO_2 (Prata and Grant 2001), some limitations affect their usage. They may be summarized in the low spatial resolution of UV data and in satellite observations limited to daytime conditions.

The retrieval of volcanic SO_2 at infrared wavelengths offers, on the other hand, the advantages of performing measurements under different illumination conditions (i.e. in both day and night) by using higher-spatial-resolution data than UV records (Clarisse et al. 2012). The infrared retrievals are generally performed in two main absorption bands of sulphur dioxide, at 7.3 and 8.6 µm, respectively. A wide description of the 7.3 µm retrieval scheme, which also analyzes satellite signal in regions unaffected by SO_2 to estimate the contribution of the unperturbed atmosphere to the radiance measured by satellite (Clarisse et al. 2012), can be found in the literature (Realmuto et al. 1994; Prata and Bernardo 2007; Thomas and Prata 2011). In general, at 7.3 µm the signal absorption because of sulphur dioxide is particularly strong. However, since atmospheric water vapour significantly absorbs signal at the same wavelength, this retrieval analysis works well only for volcanic plumes extending over 3 km in altitude (i.e. above most of the water vapour content) and for volcanic plumes with a low water vapour loading. Basically, the SO_2 retrieving scheme in this region relies on the difference of the measured (i.e. SO_2-affected) and the normal (i.e. SO_2-free) radiance at 7.3 μm. Normal radiance is generally estimated by using different spectral bands (6.7 and 11 μm), not affected by the presence of SO_2. Besides water vapour, thin clouds and low thermal contrasts between background and cloud radiances generally represent the main factors affecting performances of this method (Prata and Bernardo 2007). At 8.6 μm, the signal absorption because of atmospheric water vapour and other gas species is less significant than at 7.3 μm. The correspondent retrieval scheme is then generally used for retrieving SO_2 in the lower troposphere (Realmuto et al. 1994). The main factors affecting the performance of this method, which is based on a least-squares fit procedure applied to the satellite sensor measurements and to the simulated radiances, are ash and sulphate aerosols (capable of absorbing signal at 8.6 μm, causing an overestimation of the retrieved SO_2 columnar abundance) and the low thermal contrast between volcanic clouds and the ground (Realmuto et al. 1994; Prata et al. 2003).

Finally, the sulphur dioxide may also be retrieved in the microwave region of the electromagnetic spectrum by using the Microwave Limb Sounder (MLS) measuring the microwave emission from the limb (edge) of the Earth's atmosphere (Read et al. 2003; Livesey et al. 2006). Different gas species in the upper troposphere, including volcanic SO_2 (McGee and Sutton 1994), may be studied by means of this instrument, although its poor spatial resolution (30 km across-track and 150 km along-track) generally impacts the accuracy of retrieval analysis (Stohl et al. 2001; Livesey et al. 2006).

19.5 VOLCANO TOPOGRAPHY AND DEFORMATION

The topography of active volcanoes can change rapidly and quite significantly. Emplacement of massive lava flows, opening of new vents or large collapses associated with volcanic eruptions may significantly modify the surface of a volcano, thus making the available topography information readily outdated. The availability of tools capable of generating in a rapid way a digital elevation model (DEM) of the volcano is of extreme importance since this information is useful in volcanic risk reduction. For example, updated topography is fundamental for forecasting the direction of possible lava flows through numerical simulation (Costa and Macedonio 2005; Carranza and Castro 2006).

Similarly, and more importantly, subtle changes to volcanic surfaces (ground deformation) can be related to several volcanic hazards. Phenomena such as recharging of magma in the volcano plumbing system, overpressurization of the magma chambers and movements due to gravitative effects, such as flank instability, may cause measurable ground deformation which is worth monitoring both for volcanic risk mitigation and for scientific investigation of volcano deep structure.

Synthetic Aperture Radar (SAR) is an active remote sensing imaging technology that has proven to be very powerful in measuring topography and ground deformation (Bamler and Hartl 1998; Massonet and Feigl 1998; Burgmann et al. 2000; Rosen et al. 2000). A SAR system is usually mounted onboard a satellite (or an aircraft) with a side-looking configuration with respect to the flight direction; this causes the image to have a peculiar geometry, as sketched in Figure 19.6: one dimension (i.e. *azimuth* or along-track) represents the position of an imaged ground target along the satellite flight track. The other dimension (i.e. *range* or across-track) accounts for its distance from the sensor position in the across-track plane. This distance is computed along the so-called *line of sight* (LOS) of the radar, which is tilted with respect to the vertical by an angle known as the *look angle* (Figure 19.6). For this reason, the range direction is often called the *slant range*.

Being an active sensor operating in the microwave region of the electromagnetic spectrum, a SAR system is characterized by a continuous (night and day and through cloud) imaging capability (Franceschetti and Lanari 1999). For this reason, SAR systems can play a key role in monitoring active volcanoes that are very often covered by meteorological clouds, ash clouds and/or volcanic plumes, which may strongly limit the use of optical sensors.

Topography and ground deformation mapping require at least the use of two SAR images. While for topographic application they should ideally be acquired at the same time (or over a short

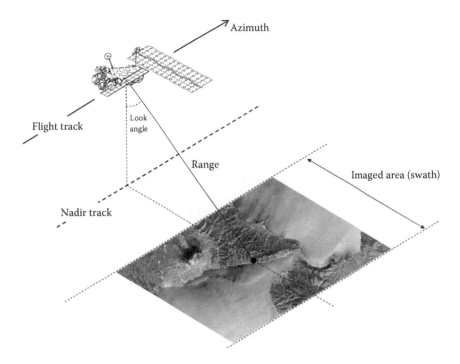

FIGURE 19.6   Side-looking geometry of an SAR system.

interval), in the ground deformation case the acquisition interval must at least cover the deformative episode under investigation.

For decades, a number of SAR systems have been orbiting around Earth, operating onboard several LEO satellite missions, and collecting a huge amount of valuable data (Table 19.1). The global coverage and continuous acquisition policy of some satellites (e.g. ERS-1/2, ENVISAT and ALOS) allowed the collection of data over most active volcanic areas. Those data have been applied to several case studies, including, for instance, the investigation of long-term ground deformation at volcanoes, such as Etna, by using ERS and ENVISAT data (Palano et al. 2007; Neri et al. 2009), and the monitoring of large areas, such as the entire Indonesian archipelago (Philibosian and Simons 2011) and the central Andes volcanic arc (Pritchard and Simons 2002). In the latter case, the background activity level of about 900 volcanoes over a period of time as long as 8 years (from 1992 to 2000) was studied.

The technological trend in EO radar satellites is oriented to a continuous improvement of performances (Sansosti et al. 2014), particularly in terms of spatial and temporal resolutions (Table 19.1), thus clearly posing the basis for an operational use of SAR interferometry for active volcanoes (Stevens and Wadge 2004; Sansosti et al. 2010).

19.5.1   TOPOGRAPHY MEASUREMENT

The first experimental study on the use of SAR images to measure terrain elevation dates back to 1974 (Graham 1974). However, it was the development of phase-preserving radar sensors that led to the promotion, in the late 1980s, of the interferometric SAR (InSAR) technique as an effective tool for topographic mapping (Zebker and Goldstein 1986; Li and Goldstein 1990). This was possible by exploiting the phase difference between two SAR images of the same area acquired by slightly different orbits separated in the across-track direction. The corresponding images, called *interferograms*, show phase cycles of 2π radians, usually represented by colour *fringes*, that may represent either topography or topography variation. As an example, Figure 19.7 shows an interferogram of

FIGURE 19.7   Interferogram of Nabro volcano (Ethiopia) from COSMO-SkyMed data. Each colour cycle (fringe) represents a subsidence of about 1.5 cm that occurred between 16 October 2011 and 28 May 2012.

the Nabro volcano (Ethiopia) obtained from data acquired by COSMO-SkyMed satellites between 16 October 2011 and 28 May 2012. In this case, each colour cycle (fringe), for example, from blue to blue, represents a subsidence of about 1.5 cm. This phenomenon affects the whole volcanic edifice and is a consequence of the degassing activity associated with a very large eruption that began on 12 June 2011.

The capability of SAR interferometry to measure topography and ground deformation can be easily understood in terms of simple geometrical considerations. Indeed, the phase of the radar signal is related to both the travel path of the electromagnetic wave and the backscattering characteristics of the ground. Interferometric approaches exploit the phase difference of two SAR images (Elachi and van Zyl 2006). Therefore, assuming that electromagnetic properties of the ground do not change significantly between the two acquisitions, the backscatter contribution is essentially cancelled out by the interference operation and the interferometric phase, say φ, is solely related to the travel path difference, δR, through the following simple equation:

$$\delta R = \frac{\lambda}{4\pi}\varphi \qquad (19.9)$$

where λ is the wavelength of the transmitted electromagnetic wave. However, in practical cases, slight changes of the backscatter contribution between the two acquisitions appear as noise (i.e. decorrelation noise) associated with the range difference measurement (Zebker and Villasenor 1992). Change of the electromagnetic characteristics with time (temporal decorrelation) is related, for instance, to abrupt variations of the groundwater content (i.e. from dry to wet), or to the presence of a vegetation or snow coverage that causes random dislocation of elementary scatterers within the resolution cell. This may represent a significant drawback for application to volcanoes, which are often covered by heavy vegetation and snow. In fact, experiments on Mt. Etna showed that a careful selection of images that excludes winter acquisitions strongly reduces the decorrelation noise on the volcano summit, seasonally capped by snow (Ruch et al. 2012). On the other hand, decorrelation noise is usually very low on cold and non-vegetated lava flows; this property can be used for tracking lava flow emplacements, as demonstrated on Kilauea volcano, Hawaii (Dietterich et al. 2012).

In addition to temporal changes, backscattering characteristics are also subject to variation with the incidence angle of the electromagnetic wave: the larger the distance between the two acquisition orbits tracks (i.e. *baseline*), the larger the (spatial decorrelation) noise, thus establishing an upper limit for the baseline value in interferometric applications.

Once the measurement of the range difference is available, the scene's topography is easily retrieved. The two range measurements (or, equivalently, one range and the range difference) allow the determination of the target's absolute position as the intersection of the two equirange circles in the across-track plane (Rosen et al. 2000) (Figure 19.8a). It is evident that, in this way, the terrain height is univocally determined. On the other hand, this is not possible when only one range measurement is available. Indeed, with reference to Figure 19.8b, all the targets lying on the equirange

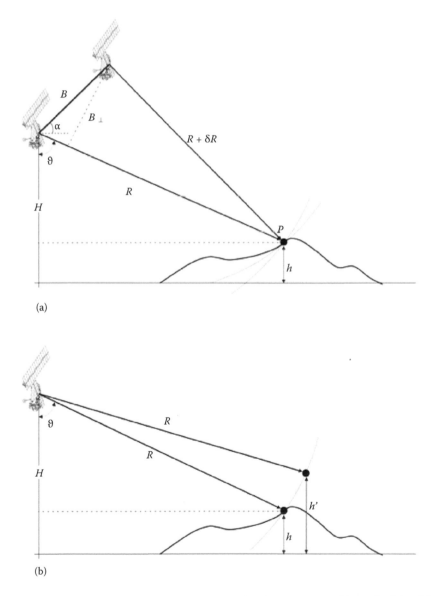

(a)

(b)

FIGURE 19.8 Topography and ground deformation mapping from space: working principle of InSAR systems. (a) Two range measurement case: The height of the target is univocally determined. (b) Single range measurement case: Height is not resolved.

circle, which have different heights, are imaged as the same pixel because they are at the same distance from the radar, thus demonstrating that a single radar is not able to solve for topography.

As for any stereoscopic system, the SAR interferometry capability of measuring topography can be translated in formulas by using simple trigonometry (Sansosti 2004). Indeed, the height, say h, of a resolution cell can be computed as follows (Figure 19.8a):

$$h = H - R\cos\vartheta \qquad (19.10)$$

where H is the altitude of the satellite orbit, and R is the range of the first image. The angle ϑ is the look angle, which can be computed, as a derivation of the cosine law, from the following equation (Rosen et al. 2000; Sansosti 2004):

$$\sin(\vartheta - \alpha) = \frac{R^2 + B^2 - (R + \delta R)^2}{2R^2B} \qquad (19.11)$$

where B is the separation between the two orbit tracks (referred to as baseline), α is the baseline tilt angle and δR is the range difference, which is related to the interferometric phase through Equation 19.9.

Sensitivity analysis of this system (Rosen et al. 2000; Sansosti 2004) shows that for a given precision of the range difference measurements (interferometric phase), the error on the computed height increases for larger ranges and/or for larger look angles, while it decreases when the component of the baseline perpendicular to the LOS (*perpendicular baseline*) enlarges. For this reason, the perpendicular baseline, identified by $B_\perp$ in Figure 19.8a, is a key parameter in SAR interferometry.

Moreover, it can be demonstrated that the range difference must be measured with a very high precision in order to obtain a reasonable accuracy of the derived topography information (Sansosti 2004). As an example, by using typical C-band system parameters (such as ERS or ENVISAT satellites), that is, $B_\perp = 200$ m, $\vartheta = 23°$ and an orbit altitude of about 800 km, we need a range difference as precise as 5 mm to obtain a height precision of 10 m. This becomes a feasible task only when range differences are computed via interferometric procedures (Franceschetti and Lanari 1999), unless very high-resolution radar systems are used.

In practical cases, the precision of DEMs derived from the satellite InSAR techniques is in the range of metres, or a few tens of metres. Although more precise methods exist, the InSAR technique has the advantage of being cheaper and frequently updatable thanks to the satellite's relatively short repeat time.

DEMs have been widely used for studying and monitoring active volcanoes. For example, InSAR was useful for updating the topography of the Valle del Bove area at Etna after the large lava flows occurred during the 1991–1993 eruption, whose thickness reached 80–100 m, thus strongly altering the local topography (Sansosti et al. 1999). Moreover, comparison of a sequence of several InSAR topography measurements with existing DEMs has been used to measure increases in lava thickness (up to 140 m) at Santiaguito volcano, Guatemala, between 2000 and 2009, thus allowing the estimation of the total volume of erupted lava (Ebmeier et al. 2012). A similar approach has also been applied to the measurement of the topographic and volume changes during the 2010 eruption of the Merapi volcano (Kubanek et al. 2015) by using high-precision DEMs computed from TanDEM-X data.

19.5.2 Deformation Measurements

In principle, ground displacements could be measured by comparing two DEMs, as for the mentioned case of lava flow thickness measurements. However, this approach would reach a precision in the order of metres (or a fraction of a metre, with newer SAR systems), while subtle deformation in the centimetre range is usually of higher interest in volcanic areas, being associated with magma movements at depth, and therefore pre-eruptive activity.

To this end, the high sensitivity of the phase information to range variations can be exploited by applying a technique nowadays known as differential SAR interferometry (DInSAR). In this case, a simulated phase signal (synthetic interferogram) is first computed by using both an available DEM and knowledge of the system acquisition geometry (orbit tracks and look angle); this is then subtracted from a measured interferogram spanning the deformative event, thus obtaining the so-called *differential interferogram*, which accounts for the deformation only. In this way, the topography compensation becomes possible. Indeed, with reference to the typical C-band configuration discussed earlier, 10 m of variation (error) in the height would cause less than a 5 mm change to the range difference if the perpendicular baseline is short enough (less than 200 m); shorter baselines would imply even less range sensitivity to DEM errors, with the DEM being unnecessary in the limit of zero baseline. This simple analysis makes clear that, at least in principle, millimetric precision can be reached for ground displacement measurement.

In the differential interferometry case, the phase, say φ_d, is related to the projection of the ground displacement vector along the LOS, say δR_d, as follows:

$$\varphi_d = \frac{4\pi}{\lambda}\delta R_d \tag{19.12}$$

Equation 19.12 simply states that, in the differential interferogram, a fringe cycle (of 2π radians) is associated with a LOS displacement of half the wavelength. An example of such an interferogram is presented in Figure 19.7.

This technique was first demonstrated for an agricultural region (Gabriel et al. 1989). The study of co-eruptive deflation of Mt. Etna in Italy (Massonnet et al. 1995) represented the first application of the technique to a volcano. Shortly after, at the same volcano, the first detection of a reversal from deflation to inflation and the first tracking of pre-eruptive inflation were also proved (Lanari et al. 1998).

Because of its capability to provide a comprehensive view of the ongoing ground displacements, the DInSAR technique has become an excellent complement to ground-based conventional geodetic measurements, such as GPS and levelling, which are carried out at a limited number of locations (Palano et al. 2007). Nowadays, the DInSAR technique represents a powerful tool for the detection and mapping of surface displacements related to a specific deformation episode and/or for capturing its spatial behaviour over a large area. Examples of the use of DInSAR at volcanoes include a comparative study of deformation at several volcanoes in the Galapagos archipelago (Amelung et al. 2000); the surface movements of emplaced lava flows (Stevens et al. 2001) and the modelling of the magmatic source of the 2001 flank eruption (Lundgren and Rosen 2003), both at Etna; the identification of dyke intrusion as the triggering cause of a slow fault-slip event on Kilauea volcano (Brooks et al. 2008); and the investigation of the dynamics of a shallow magma chamber at the Erta Ale (Pagli et al. 2012).

The DInSAR technique also presents some drawbacks. Apart from the possible decorrelation noise that may prevent measurements in some areas (vegetation or snow-covered areas), differential interferograms may be affected by several artefacts that may impair or mask the deformation signal. The main error sources are

1. *Atmospheric artefacts.* Changes in the refractive index between the two satellite passes, due to the presence of atmosphere, can cause a change in the length of the electromagnetic wave travel path that induces an error on the range measurements (Goldstein 1995). Since DInSAR interferograms are very sensitive to the range variation, even a very short propagation path change may cause significant disturbances in the interferogram, which can be particularly difficult to detect because they usually mimic a deformation signal, especially in the case of tropospheric artefacts.

2. *Residual topography.* If scene topography is not precise enough and/or the perpendicular baseline is too large, a residual topographic artefact will be present in the interferograms. Clearly, whenever possible, these disturbances can be reduced by choosing short-baseline interferograms.

3. *Orbital ramps.* Imperfect knowledge of the orbital trajectories of the satellites, that is, of the baseline parameter, may cause a wrong compensation of the topographic contribution. This basically results in a tilt of the interferometric phase image (orbital ramps) mainly in the range direction (Pepe et al. 2011).

These disturbances may significantly reduce the applicability of DInSAR techniques in monitoring active volcanoes, especially when ground deformations are relatively slow (less than 1 cm year^{-1}) or when volcanoes are particularly exposed to rapidly variable atmospheric conditions (i.e. volcanoes with high topography and/or located on islands or close to the sea). As an example, it was recognized that the first results on the inflation and deflation of Mt. Etna were affected by significant atmospheric artefacts, essentially related to the high topography of the volcano that is capable of influencing local weather (Beauducel et al. 2000). Moreover, anomalies up to two to three fringes in interferograms of Aleutian volcanoes, which are obviously subject to adverse weather, have been also reported (Lu et al. 2000). For a typical satellite C-band system, these artefacts can account for up to 6–9 cm, which may be unacceptable in several applications.

19.5.3 DEFORMATION TIME-SERIES TECHNIQUES

Mitigation of the DInSAR technique errors can be achieved by jointly exploiting a set of acquisitions instead of using single image pairs. Moreover, for active volcanoes, it is of great interest not only to study a single deformation episode, but also to analyze the temporal evolution of the detected displacements. Techniques that fulfil both these needs have been more recently developed. They aim at computing deformation time series from sequences of temporally separated SAR images and typically require a few tens of images to provide reliable results.

Two main categories of such techniques exist: those working on localized targets, referred to as Persistent scatterer (PS) methods (Ferretti et al. 2000; Strozzi et al. 2003; Hooper et al. 2004), and those also using extended targets, referred to as small-baseline (SB) methods (Berardino et al. 2002; Mora et al. 2003; Schmidt and Bürgmann 2003; Lanari et al. 2004; Prati et al. 2010; Fornaro et al. 2015); solutions that incorporate both approaches have been also proposed (Hooper 2008).

Regardless of the specific approach or its particular implementation, time-series techniques are based on a proper combination of interferograms whose result is the ground displacement history of each pixel of the investigated area with respect to its position at the time of the first acquisition. A mean deformation velocity map of the area is also achievable, which is particularly useful under linear deformation regimes. An example of such a result is presented in Figure 19.9, where the mean deformation velocity map of Naples Bay (Italy) is shown: green areas are (on average) stable, while red ones are dominated by subsidence. This area includes three active volcanoes: Campi Flegrei caldera, Mt. Vesuvius and Ischia Island. The velocity map, obtained by processing 155 images acquired by the ERS and ENVISAT satellites from 1992 to 2010, highlights that the most deformed area falls within the Campi Flegrei caldera, which is interested by the so called Bradyseism, a well known volcanic phenomenon characterized by alternate periods of uplift and subsidence. Figure 19.9 also shows the displacement time series of a point in such an area, near the harbour of Pozzuoli city. It is clear that the technique is able to highlight a subsidence phenomenon at a quite constant rate (about 3 cm year^{-1}) between 1992 and 2000, interrupted by a fast uplift episode that resolved several months later. Afterwards, the deformation signal shows alternation of slow uplift and subsidence periods, which demonstrates that the long-lasting subsidence trend was definitively interrupted in 2000 and the volcano entered a new phase. Undoubtedly, for a densely populated area experiencing repeated unrest, where more than 1 million people are exposed to volcanic hazard, this measurement

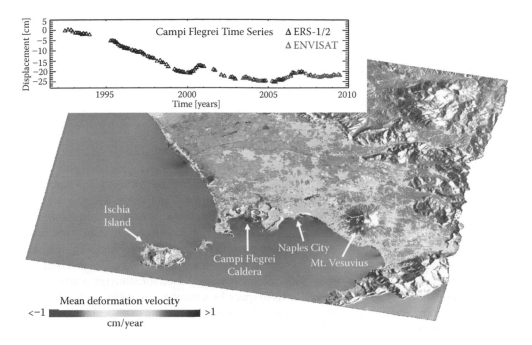

FIGURE 19.9 Measuring deformations of active volcanoes by DInSAR methodology: Results achieved on Campi Flegrei area by means of nearly 20-year SAR observations.

represents extremely valuable information. Moreover, the use of such spatially and temporally dense information makes possible a deeper geophysical analysis of the observed phenomena. For instance, at Campi Flegrei the large spatial coverage was exploited to better constrain, through analytical modelling, the deformation source of the 2004–2006 uplift episode (Trasatti et al. 2008), as well as that of the 2012–2013 period (D'Auria et al. 2015), to investigate the impact of mechanical heterogeneities of the lithosphere on analytic and numerical modelling (Manconi et al. 2010), and to infer possible connections between Campi Flegrei and Vesuvius deformation fields (Walter et al. 2014).

The capability to generate deformation time series strongly increases the usability of DInSAR techniques, especially for non-specialists in remote sensing. In addition, it also noticeably increases the measurement precision with respect to the single interferogram approach. Indeed, atmospheric artefacts can be detected and filtered out because of their different variability in space and time compared with that of the deformation signal (Ferretti et al. 2000; Hanssen 2001). Moreover, topography artefacts can be mitigated by using the space–time data redundancy (Ferretti et al. 2000; Hooper and Zebker 2007). Finally, possible orbital ramps can be effectively compensated for by using a limited number of GPS stations (Gourmelen et al. 2010).

All these characteristics increase the final precision of the measurements by up to about 0.5–1 mm year^{-1} for deformation velocity measurements, and by up to 5–10 mm for time-series displacements. These figures have been confirmed by extensive comparison between SAR and classical geodetic data (Casu et al. 2006; Lanari et al. 2006; Raucoules et al. 2009).

The application of time-series techniques to active volcanoes are countless. They include measurement of inflation associated with a dyke-like magma body at an intermediate depth in the southwest rift zone of Mauna Loa volcano, Hawaii, from 2002 to 2005 (Amelung et al. 2007); characterization of the active source at the Long Valley caldera (Tizzani et al. 2007); and modelling of the magmatic evolution at Santorini (Parks et al. 2012) and El Hierro (González et al. 2013) volcanoes. At Etna, DInSAR time-series techniques have been used to investigate the long-term characterization (almost two decades) of a complete inflation–deflation cycle and its relation to the eruptions that have occurred (Neri et al. 2009), the relation between flank instability and magmatic activity (Solaro et al. 2010), the controversial interpretation of the source of the eastern flank movement (Palano et al. 2009; Ruch et al. 2010; Currenti et al. 2012; Ruch et al. 2013) and the effects of gravity and magma-induced volcano spreading (Lundgren et al. 2004). Furthermore, still at Etna, the synergic use of DInSAR and gravity change measurements allowed the detection of deep magma uprising months to years before the onset of a new eruption (Del Negro et al. 2013).

19.6 CONCLUSION

Volcanoes differ from other natural hazards because they are located in fixed areas. This is an advantage for the accuracy of satellite remote sensing applications with regards to active volcanoes. Satellites are able to monitor volcanoes on a global scale, including the most isolated and inaccessible areas, at relatively low costs and with timeliness and continuity. Before the use of remote sensing in active volcanism, scientists had to gather on-site data and process them, and often these activities were impossible to achieve in short times or at all. Moreover, the collection of data through remote sensing techniques is undoubtedly safer than field work in the case of impelling eruptive episodes.

Although past and present EO missions were not specifically designed and conceived for volcanological applications, satellite observations are increasingly used for studying active volcanoes. The huge amount of data collected every day from space by a number of satellite instruments allows observing and interpreting most of the events and phenomena occurring in a volcano (thermal activity, gas and ash emissions and ground deformations). These phenomena can be observed and interpreted in a matter of minutes to hours and on the global scale. This valuable amount of data, thanks to the continuous research and insights in this field, becomes information and then knowledge, improving our understanding of physical processes and helping us in predicting volcano behaviour.

Until now, satellite remote sensing has helped in many aspects of volcanology, improving our knowledge of subterranean volcanic processes which are beneath eruptive events, and making possible a systematic monitoring and surveillance of volcanically active areas. Pre-eruptive signs, unresting phases and co- and post-eruptive phenomena were largely investigated by means of EO satellites, which actually offer scientific and user communities a wide range of observations and value-added products. With the new generations of EO space systems, with improved spatial, temporal and spectral resolutions, the use of satellite data for volcanological applications will continue to grow, further improving our capability to monitor and understand volcanoes.

REFERENCES

Amelung, F., Jonsson, S., Zebker, H., and Segall, P. (2000). Widespread uplift and "trapdoor" faulting on Galapagos volcanoes observed with radar interferometry. *Nature*, 407, 993–996.
Amelung, F., Yun, S., Walter, T. R., Segall, P., and Kim, S. (2007). Stress control of deep rift intrusion at Mauna Loa Volcano, Hawaii. *Science*, 316, 1026–1030. doi: 10.1126/science.1140035.
Arino, O., and Melinotte, J. M. (1995). Fire index atlas. *Earth Observation Quarterly*, 50, 11–16.
Bamler, R., and Hartl, P. (1998). Synthetic aperture radar interferometry. *Inverse Problem*, 14, R1–R54.
Baxter, P. (2005). Human impacts of volcanoes. In Martí, J., and Ernst, G. G. J. (eds.), *Volcanoes and the Environment*. Cambridge: Cambridge University Press, 273–303.
Beauducel, F., Briole, P., and Froger, J. L. (2000). Volcano-wide fringes in ERS synthetic aperture radar interferograms of Etna (1992–1998): Deformation or tropospheric effect? *Journal of Geophysical Research*, 105(7), 16391–16402.
Berardino, P., Fornaro, G., Lanari, R., and Sansosti, E. (2002). A new algorithm for surface deformation monitoring based on small baseline differential SAR interferograms. *IEEE Transaction Geosciences and Remote Sensing*, 40, 2375–2383.
Blackett, M., and Wooster, M. J. (2011). Evaluation of SWIR-based methods for quantifying active volcano radiant emissions using NASA EOS-ASTE *data. Geomatics, Natural Hazards and Risk*, 2, 51–79.
Bovensmann, H., Burrows, J. P., Buchwitz, M., Frerick, J., Noël, S., Rozanov, V. V., Chance, K. V., and Goede, A. P. G. (1999). SCIAMACHY: Mission objectives and measurement modes. *Journal of the Atmospheric Sciences*, 56, 127–150.

Brooks, B. A., Foster, J., Sandwell, D., Wolfe, C. J., Okubo, P., Poland, M., and Myer, D. (2008). Magmatically triggered slow slip at Kilauea Volcano, Hawaii. *Science*, 321, 29.

Burgi, P. Y., Caillet, M., and Haefeli, S. (2002). Field temperature measurements at Erta 'Ale lava lake, Ethiopia. *Bulletin of Volcanology*, 64, 472–485.

Burgmann, R., Rosen, P. A., and Fielding, E. J. (2000). Synthetic aperture radar interferometry to measure earth's surface topography and its deformation. *Annual Review of Earth and Planetary Sciences*, 28, 169–209.

Burrows, J., Weber, M., Buchwitz, M., Rozanov, V., Ladstätter-Weißenmayer, A., Richter, A., de Beek, R. et al. (1999). The Global Ozone Monitoring Experiment (GOME): Mission concept and first scientific results. *Journal of the Atmospheric Sciences*, 56, 151–175.

Caltabiano, T., Romano, R., and Budetta, G. (1994). SO2 flux measurements at Mount Etna, Italy. *Journal of Geophysical Research*, 99, 12809–12819.

Carn, S. A., Krueger, A. J., Krotkov, N. A., Arellano, S., and Yang, K. (2008). Daily monitoring of Ecuadorian volcanic degassing from space. *Journal of Volcanology and Geothermal Research*, 176(1), 141–150.

Carn, S. A., Krueger, A. J., Krotkov, N. A., Yang, K., and Evans, K. (2009). Tracking volcanic sulfur dioxide clouds for aviation hazard mitigation. *Natural Hazards*, 51(2), 325–343.

Carn, S. A., Strow, L. L., de Souza-Machado, S., Edmonds, Y., and Hannon, S. (2005). Quantifying tropospheric volcanic emissions with AIRS: The 2002 eruption of Mt. Etna (Italy). *Geophysical Research Letters*, 32, L02301. doi: 10.1029/2004GL021034.

Carranza, E. J. M., and Castro, O. T. (2006). Predicting lahar-inundation zones: Case study in West Mount Pinatubo, Philippines. *Natural Hazards*, 37, 331–372. doi: 10.1007/s11069-005-6141-y.

Casadevall, T. J. (ed.). (1994). *Volcanic Ash and Aviation Safety: Proceedings of the First International Symposium on Volcanic Ash and Aviation Safety*. U.S. Geological Survey Bulletin 2047. Reston, VA: U.S. Geological Survey.

Casadevall, T. J. (1996). The 1989–1990 eruption of Redoubt volcano, Alaska: Impacts on aircraft operations. *Journal of Volcanology and Geothermal Research*, 62, 301–316.

Casu, F., Manzo, M., Lanari, R. (2006). A quantitative assessment of the SBAS algorithm performance for surface deformation retrieval from DInSAR data. *Remote Sensing of Environment*, 102, 195–210. doi: 10.1016/j.rse.2006.01.023.

Chahine, M. T., Pagano, T. S., Aumann, H. H., Atlas, R., Barnet, C., Chen, L., Divakarla, M. et al. (2006). The Atmospheric Infrared Sounder (AIRS): Improving weather forecasting and providing new insights into climate. *Bulletin of the American Meteorological Society*, 87, 891–894. doi: 10.1175/BAMS-87-7-891.

Clarisse, L., Coheur, P. F., Prata, A. J., Hurtmans, D., Razavi, A., Phulpin, T., Hadji-Lazaro, J., and Clerbaux, C. (2008). Tracking and quantifying volcanic SO2 with IASI, the September 2007 eruption at Jebel at Tair. *Atmospheric Chemistry Physics*, 8, 7723–7734.

Clarisse, L., Hurtmans, D., Clerbaux, C., Hadji-Lazaro, J., Ngadi, Y., and Coheur, P. F. (2012). Retrieval of sulphur dioxide from the Infrared Atmospheric Sounding Interferometer (IASI). *Atmospheric Measurement Techniques*, 5, 581–594. doi: 10.5194/amt-5-581-2012.

Constantine, E. K., Bluth, G. J. S., and Rose, W. I. (2000). TOMS and AVHRR observations of drifting volcanic clouds from the August 1991 eruptions of Cerro Hudson. In Mouginis-Mark, P. J., Crisp, J. A., and Fink, J. H. (eds.), *Remote Sensing of Active Volcanism*. Geophysical Monograph 116. Washington, DC: American Geophysical Union, 45–64.

Coppola, D., Laiolo, M., and Cigolini, C. (2016). Fifteen years of thermal activity at Vanuatu's volcanoes (2000–2015) revealed by MIROVA. *Journal of Volcanology and Geothermal Research*, 322, 6–19.

Coppola, D., Piscopo, D., Staudacher, T., and Cigolini, C. (2009). Lava discharge rate and effusive pattern at Piton de la Fournaise from MODIS data. *Journal of Volcanology and Geothermal Research*, 184, 174–192.

Costa, A., and Macedonio, G. (2005). Computational modeling of lava flows: A review. *Geological Society of America Special Papers*, 396, 209–218. doi: 10.1130/0-8137-2396-5.209.

Crisp, J., and Baloga, S. (1990). A method for estimating eruption rates of planetary lava flows. *Icarus*, 85, 512–515.

Currenti, G., Solaro, G., Napoli, R., Pepe, A., Bonaccorso, A., Del Negro, C., and Sansosti, E. (2012). Modeling of ALOS and COSMO-SkyMed satellite data at Mt. Etna: Implications on relation between seismic activation of the Pernicana fault system and volcanic unrest. *Remote Sensing of Environment*, 125, 64–72. doi: 10.1016/j.rse.2012.07.008.

D'Auria, L., Pepe, S., Castaldo, R., Giudicepietro, F., Macedonio, G., Ricciolino, P., Tizzani, P. et al. (2015). Magma injection beneath the urban area of Naples: A new mechanism for the 2012–2013 volcanic unrest at Campi Flegrei caldera. *Scientific Reports*, 5(13100). doi: 10.1038/srep13100.

Davies, M. A., and Rose, W. (1998). Evaluating GOES imagery for volcanic cloud observations at the Soufriere Hills volcano, Montserrat. *Eos Transactions of the American Geophysical Union*, 79, 505–507.

Dean, K., Servilla, M., Roach, A., Foster, B., and Engle, K. (1998). Satellite monitoring of remote volcanoes improves study efforts in Alaska. *Eos Transactions of the American Geophysical Union*, 79, 413.

Dehn, J., Dean, K., and Engle, K. (2000). Thermal monitoring of North Pacific volcanoes from space. *Geology*, 28, 755–758.

Dehn, J., Dean, K. G., Engle, K., and Izbekov, P. (2002). Thermal precursors in satellite images of the 1999 eruption of Shishaldin volcano. *Bulletin of Volcanology*, 64(8), 525–534.

Delmelle, P. (2003). Environmental impacts of tropospheric volcanic gas plumes. *Geological Society Special Publication*, 213, 381–400.

Del Negro, C., Currenti, G., Solaro, G., Greco, F., Pepe, A., Napoli, R., Pepe, S., Casu, F., Sansosti, E. (2013). Capturing the fingerprint of Etna volcano activity in gravity and satellite radar data. *Scientific Reports*, 3(3089). doi: 10.1038/srep03089.

Di Bello, G., Filizzola, C., Lacava, T., Marchese, F., Pergola, N., Pietrapertosa, C., Piscitelli, S., Scaffidi, I., and Tramutoli, V. (2004). Robust satellite techniques for volcanic and seismic hazards monitoring. *Annals of Geophysics*, 47(1), 49–64.

Dietterich, H. R., Poland, M. P., Schmidt, D. A., Cashman, K. V., Sherrod, D. R., and Espinosa, A. T. (2012). Tracking lava flow emplacement on the east rift zone of Kīlauea, Hawai'i, with synthetic aperture radar coherence. *Geochemistry Geophysics Geosystems*, 13, Q05001. doi: 10.1029/2011GC004016.

Dozier, J. (1981). A method for satellite identification of surface temperature fields of subpixel resolution. *Remote Sensing of Environment*, 11, 221–229.

Dragoni, M., and Tallarico, A. (2009). Assumption in the evaluation of lava effusion rate from heat radiation. *Geophysical Research Letters*, 36, L08302. doi: 10.1029/2009GL037411.

Duffell, H. J., Oppenheimer, C., Pyle, D. M., Galle, B., McGonigle, A. J. S., and Burton, M. R. (2003). Changes in gas composition prior to a minor explosive eruption at Masaya volcano, Nicaragua. *Journal of Volcanology and Geothermal Research*, 126(3–4), 327–339.

Durant, A. J., Villarosa, G., Rose W. I., Delmelle, P., Prata A. J., and Viramonte J. G. (2011). Long-range volcanic ash transport and fallout during the 2008 eruption of Chaiten volcano, Chile. *Physics and Chemistry of the Earth*, 45–46, 50–64. doi: 10.1016/j.pce.2011.09.004.

Ebmeier, S. K., Biggs, J., Mather, T. A., Elliott, J. R., Wadge, G., and Amelung F. (2012). Measuring large topographic change with InSAR: Lava thicknesses, extrusion rate and subsidence rate at Santiaguito volcano, Guatemala. *Earth and Planetary Science Letters*, 335–336, 216–225. doi: 10.1016/j.epsl.2012.04.027.

Elachi, C., and van Zyl, J. (2006). *Introduction to the Physics and Techniques of Remote Sensing*. 2nd ed. Hoboken, NJ: Wiley-Interscience.

Ellrod, G., Connell, B., and Hillger, D. (2003). Improved detection of airborne volcanic ash using multispectral infrared satellite data. *Journal of Geophysical Research*, 108(D12), 4356.

ESA-EUMETSAT. (2010). Workshop on the 14 April to 23 May 2010 eruption at the Eyjafjöll volcano, South Iceland, "Monitoring volcanic ash from space." Paris: European Space Agency/European Space Research Institute.

Ferretti, A., Prati, C., and Rocca, F. (2000). Nonlinear subsidence rate estimation using permanent scatterers in differential SAR interferometry. *IEEE Transactions on Geoscience and Remote Sensing*, 38, 202–2212.

Flynn, L. P., Harris, A. J., and Wright, R. (2001). Improved identification of volcanic features using Landsat 7 ETM+. *Remote Sensing of Environment*, 78(1–2), 180–193.

Flynn, L. P., Harris, A. J. L., Rothery, D. A., and Oppenheimer, C. (2000). High-spatial-resolution thermal remote sensing of active volcanic features using Landsat and hyperspectral data. In Mouginis-Mark, P. J., Crisp, J. A., and Fink, J. H. (eds.), *Remote Sensing of Active Volcanism*. Washington, DC: American Geophysical Union, 161–177.

Fornaro, G., Verde, S., Reale, D., and Pauciullo, A. (2015). CAESAR: An approach based on covariance matrix decomposition to improve multibaseline–multitemporal interferometric SAR processing. *IEEE Transactions on Geoscience and Remote Sensing*, 53(4), 2050–2065. doi: 10.1109/TGRS.2014.2352853.

Franceschetti, G., and Lanari, R. (1999). *Synthetic Aperture Radar Processing*. Boca Raton, FL: CRC Press.

Francis, P., and Rothery, D. (2000). Remote sensing of active volcanoes. *Annual Review of Earth and Planetary Sciences*, 28, 81–106. doi: 10.1146/annurev.earth.28.1.81.

Francis, P. N., Cooke, M. C., and Saunders, R. W. (2012). Retrieval of physical properties of volcanic ash using Meteosat: A case study from the 2010 Eyjafjallajökull eruption. *Journal of Geophysical Research*, 117, D00U09. doi: 10.1029/2011JD016788.

Gabriel, A. K., Goldstein, R. M., and Zebker, H. A. (1989). Mapping small elevation changes over large areas: Differential interferometry. *Journal of Geophysical Research*, 94, 9183–9191.

Ganci, G., Harris, A. J. L., Del Negro, C., Guehenneux, Y., Cappello, A., Labazuy, P., Calvari, S., and Gouhier, M. (2012). A year of lava fountaining at Etna: Volumes from SEVIRI. *Geophysical Research Letters*, 39, L06305.

Gangale, G., Prata, A. J., and Clarisse, L. (2010). The infrared spectral signature of volcanic ash determined from high-spectral resolution satellite measurements. *Remote Sensing of Environment*, 114, 414–425.

Gawarecki, S. J., Lyon, R. J. P., and Nordberg, W. (1965). Infrared spectral returns and imagery of the earth from space and their application to geological problems. *Science and Technology Series, American Astronomical Society*, 4, 13–33.

Giglio, L., Descloitres, J., Justice, C. O., and Kaufman, Y. J. (2003). An enhanced contextual fire detection algorithm for MODIS. *Remote Sensing of Environment*, 87, 273–282.

Giglio, L., Kendall, J. D., and Justice, C. O. (1999). Evaluation of global fire detection using simulated AVHRR infrared data. *International Journal of Remote Sensing*, 20, 1947–1985.

Girina, O. A. (2012). On precursor of Kamchatkan volcanoes eruptions based on data from satellite monitoring. *Journal of Volcanology and Seismology*, 6(3), 142–149.

Glaze, L. S., Francis, P. W., Self, S., and Rothery, D. A. (1989). The 16 September 1986 eruption of Lascar volcano, north Chile: Satellite investigations. *Bulletin of Volcanology*, 51, 149–160.

Glaze, L. S., Wilson, L., and Mouginis-Mark, P. J. (1999). Volcanic eruption plume top topography and heights as determined from photoclinometric analysis satellite data. *Journal of Geophysical Research*, 104(B2), 2989–3001.

Global Volcanism Program. (2017). http://www.volcano.si.edu/index.cfm (accessed June 2017).

Goldstein, R. (1995). Atmospheric limitations to repeat-track radar interferometry. *Geophysical Research Letters*, 22(18), 2517–2520. doi: 10.1029/95GL02475.

González, P. J., Samsonov, S. V., Pepe, S., Tiampo, K. F., Tizzani, P., Casu, F., Fernández, J., Camacho, A. G., and Sansosti, E. (2013). Magma storage and migration associated with the 2011–2012 El Hierro eruption: Implications for crustal magmatic systems at oceanic island volcanoes. *Journal of Geophysical Research*, 118, 4361–4377. doi: 10.1002/jgrb.50289.

Gouhier, M., Harris, A., Calvari, S., Labazuy, P., Guéhenneux, Y., Donnadieu, F., and Valade, S. (2011). Lava discharge during Etna's January 2011 fire fountain tracked using MSGSEVIRI. *Bulletin of Volcanology*, 74, 787–793.

Gourmelen, N., Amelung, F., and Lanari, R. (2010). InSAR-GPS integration: Inter-seismic strain accumulation across the Hunter Mountain fault in the eastern California shear zone. *Journal of Geophysical Research*, 115, B09408. doi: 10.1029/2009JB007064.

Graham, L. C. (1974). Synthetic interferometric radar for topographic mapping. *Proceedings of the IEEE*, 62, 763–768.

Guéhenneux, Y., Gouhier, M., and Labazuy, P. (2015). Improved space borne detection of volcanic ash for real-time monitoring using 3-band method. *Journal of Volcanology and Geothermal Research*, 293, 25–45.

Hanssen, R. A. (2001). *Radar Interferometry: Data Interpretation and Error Analysis*. Dordrecht, the Netherlands: Kluwer Academic.

Hanstrum, B. N., and Watson, A. S. (1983). A case study of two eruptions of Mount Galunggung and an investigation of volcanic eruption cloud characteristics using remote sensing techniques. *Australian Meteorological Magazine*, 31, 131–177.

Harris, A. J. L., and Baloga, S. M. (2009). Lava discharge rates from satellite-measured heat flux. *Geophysical Research Letters*, 36, L19302. doi: 10.1029/2009GL039717.

Harris, A. J. L., Blake, S., Rothery, D. A., and Stevens, N. F. (1997). A chronology of the 1991 to 1993 Etna eruption using AVHRR data: Implications for real time thermal volcano monitoring. *Journal of Geophysical Research*, 102, 7985–8003.

Harris, A. J. L., Dehn, J., and Calvari, S. (2007). Lava effusion rate definition and measurement: A review. *Bulletin of Volcanology*, 70, 1–22.

Harris, A. J. L., Flynn, L. P., Keszthelyi, L., Mouginis-Mark, P. J., Rowland, S. K., and Resing, J. A. (1998). Calculation of lava effusion rates from Landsat TM data. *Bulletin of Volcanology*, 60, 52–71.

Harris, A. J. L., Flynn, L. P., Rothery, D. A., Oppenheimer, C., and Sherman, S. B. (1999). Mass flux measurements at active lava lakes: Implications for magma recycling. *Journal of Geophysical Research*, 104, 7117–7136.

Harris, A. J. L., Murray, J. B., Aries, S. E., Davies, M. A., Flynn, L. P., Wooster, M. J., Wright, R., and Rothery, D. A. (2000). Effusion rate trends at Etna and Krafla and their implication for eruptive mechanisms. *Journal of Volcanology and Geothermal Research*, 102, 237–270.

Harris, A. J. L., Pilger, E., and Flynn, L. P. (2002). Web-based hot spot monitoring using GOES: What it is and how it works. *Advances in Environmental Monitoring and Modeling*, 1(3), 5–36.

Harris, A. J. L., Swabey, S. E. J., and Higgins, J. (1995). Automated thresholding of active lavas using AVHRR data. *International Journal of Remote Sensing*, 16(18), 3681–3686.

Higgins, J., and Harris, A. J. L. (1997). VAST: A program to locate and analyse volcanic thermal anomalies automatically from remotely sensed data. *Computers & Geosciences*, 23(6), 627–645.

Hillger, D. W., and Clark, J. D. (2002). Principal component analysis of MODIS for volcanic ash: Part I. Most important bands and implications for future GOES imagers. *Journal of Applied Meteorology*, 41, 985–1001.

Holasek, R. E., and Rose, W. I. (1991). Anatomy of 1986 Augustine volcano eruptions as recorded by multispectral image processing of digital AVHRR weather satellite data. *Bulletin Volcanologique*, 53, 420–435.

Holasek, R. E., and Self, S. (1995). GOES weather satellite observations and measurements of the May 18, 1980, Mount St. Helens eruption. *Journal of Geophysical Research*, 100(B5), 8469–8487.

Holasek, R. E., Self, S., and Woods, A. W. (1996). Satellite observations and interpretation of the 1991 Mount Pinatubo eruption plumes. *Journal of Geophysical Research*, 100, 8469–8487.

Hooper, A. (2008). A multi-temporal InSAR method incorporating both persistent scatterer and small baseline approaches. *Geophysical Research Letters*, 35, L16302. doi: 10.1029/2008GL034654.

Hooper, A., and Zebker, H. (2007). Phase unwrapping in three dimension with application to InSAR time series. *Journal of the Optical Society of America*, 24, 2737–2747.

Hooper, A., Zebker, H., Segall, P., and Kampes, B. (2004). A new method for measuring deformation on volcanoes and other natural terrains using InSAR persistent scatterers. *Geophysical Research Letters*, 31, L23611.

IATA (International Air Transport Association). (2010). http://www.iata.org/about/Documents/IATA AnnualReport2010.pdf.

Kaneko, T., Takasaki, K., Yasuda, A., and Aoki, Y. (2006). Thermal surveillance of the Asama 2004–2005 activity using MODIS night-time infrared images [in Japanese]. *Bulletin of the Volcanological Society of Japan*, 51, 273–282.

Kaneko, T., Yasuda, A., Ishimaru, T., Takagi M., Wooster, M. J., and Kagiyama, T. (2002). Satellite hot spot monitoring of Japanese volcanoes: A prototype AVHRR based system. *Advances in Environmental Monitoring and Modelling*, 1(1), 125–133.

Kaufman, Y. J., Justice, C. O., Flynn, L. P., Kendall, J. D., Prins, E. M., Giglio, L., Ward, D. E., Menzel, W. P., and Setzer, A. W. (1998). Potential global fire monitoring from EOS-MODIS. *Journal of Geophysical Research*, 103, 32215–32238.

Kearney, C. S., and Watson, I. M. (2009). Correcting satellite-based infrared sulfur dioxide retrievals for the presence of silicate ash. *Journal of Geophysical Research—Atmosphere*, 114, D22208.

Kennedy, P. J., Belward, A. S., and Gregoire, J. M. (1994). An improved approach to fire monitoring in West Africa using AVHRR data. *International Journal of Remote Sensing*, 15, 2235–2255.

Kervyn, M., Ernst, G. G. J., Harris, A., Mbede, E., Belton, F., and Jacobs, P. (2008). Thermal remote sensing of the low-intensity carbonatite volcanism of Oldoinyo Lengai, Tanzania. *International Journal of Remote Sensing*, 6467–6499.

Klüser, L., Erbertseder, T., and Meyer-Arnek, J. (2012). Observation of volcanic ash from Puyehue-Cordón Caulle with IASI. *Atmospheric Measurement Techniques Discussion*, 5, 4249–4283. doi: 10.5194/amtd -5-4249-2012.

Koeppen, W. C., Pilger, E., and Wright, R. (2011). Time series analysis of infrared satellite data for detecting thermal anomalies: A hybrid approach. *Bulletin of Volcanology*, 73, 577–593. doi: 10.1007/s00445 -010-0427-y.

Krotkov, N. A., Carn, S. A., Krueger, A. J., Bhartia, P. K., and Yang, K. (2006). Band residual difference algorithm for retrieval of SO2 from the Aura Ozone Monitoring Instrument (OMI). *IEEE Transactions on Geoscience and Remote Sensing*, 44(5), 1259–1266. doi: 10.1109/TGRS.2005.861932.

Krotkov, N. A., Krueger, A. J., and Bhartia, P. K. (1997). Ultraviolet optical model of volcanic clouds for remote sensing of ash and sulfur dioxide. *Journal of Geophysical Research*, 102, 21891–21904.

Krueger, A. J. (1983). Sighting of El Chichon sulfur dioxide clouds with the nimbus 7 Total Ozone Mapping Spectrometer. *Science*, 220, 1377–1379.

Krueger, A. J., Schaefer, S. J., Krotkov, N., Bluth, G. J. S., and Baker, S. (2000). Ultraviolet remote sensing of volcanic emissions. In Mouginis-Marks, P. J., Crisp, J. A., and Fink, J. H. (eds.), *Remote Sensing of Active Volcanism*. Geophysical Monograph Series 116. Washington, DC: American Geophysical Union, 2543.

Kubanek, J., Westerhaus, M., Schenk, A., Aisyah, N., Brotopuspito, K. S., and Heck, B. (2015). Volumetric change quantification of the 2010 Merapi eruption using TanDEM-X InSAR. *Remote Sensing of Environment*, 164, 16–25. doi: 10.1016/j.rse.2015.02.027.

Lanari, R., Casu, F., Manzo, M., and Lundgren, P. (2006). Application of the SBAS-DInSAR technique to fault creep: A case study of the Hayward fault, California. *Remote Sensing of Environment*, 109(1), 20–28. doi: 10.1016/j.rse.2006.12.003.

Lanari, R., Lundgren, P., and Sansosti, E. (1998). Dynamic deformation of Etna volcano observed by satellite radar interferometry. *Geophysical Research Letters*, 25(10), 1541–1543.

Lanari, R., Mora, O., Manunta, M., Mallorqui, J. J., Berardino, P., and Sansosti, E. (2004). A small-baseline approach for investigating deformations on full-resolution differential SAR interferograms. *IEEE Transactions on Geosciences and Remote Sensing*, 42, 1377–1386.

Langaas, S. (1992). Temporal and spatial distribution of Savannah fires in Senegal and the Gambia, West Africa, 1989–90, derived from multi-temporal AVHRR night images. *International Journal of Wildland Fire*, 2, 21–36.

Li, F., and Goldstein, R. M. (1990). Studies of multibaseline spaceborne interferometric synthetic aperture radars. *IEEE Transactions on Geosciences and Remote Sensing*, 28, 88–97.

Li, Z., Kaufman, Y. J., Ichoku, C., Fraser, R., Trishchenko, A., Giglio, L., Jin, J., and Yu, X. (2001). A review of AVHRR-based active fire detection algorithms: Principles, limitations, and recommendations. In Ahern, F., Goldammer, J. G., and Justice, C. (eds.), *Global and Regional Vegetation Fire Monitoring from Space: Planning and Coordinated International Effort*. The Hague: SPB Academic Publishing, 199–225.

Livesey, N. J., Snyder, W. V., Read, W. G., and Wagner, P. A. (2006). Retrieval algorithms for the EOS Microwave Limb Sounder (MLS) instrument. *Journal of Geoscience and Remote Sensing*, 44(5), 1144–1155.

Lombardo, V., and Buongiorno, M. F. (2006). Lava flow thermal analysis using three infrared bands of remote-sensing imagery: A study case from Mount Etna 2001 eruption. *Remote Sensing of Environment*, 101, 141–149.

Lovallo, M., Marchese, F., Pergola, N., and Telesca, L. (2007). Fisher information analysis of volcano-related advanced, very-high-resolution radiometer (AVHRR) thermal products time series. *Physica A: Statistical Mechanics and Its Applications*, 384(2), 529–534.

Lovallo, M., Marchese, F., Pergola, N., and Telesca, L. (2009). Fisher information measure of temporal fluctuations in satellite Advanced Very High Resolution Radiometer (AVHRR) thermal signals recorded in the volcanic area of Etna (Italy). *Communications in Nonlinear Science and Numerical Simulation*, 14(1), 174–181. doi: 10.1016/j.cnsns.2007.07.006.

Lu, Z., Mann, D., Freymueller, J., and Meyer, D. (2000). Synthetic aperture radar interferometry of Okmok volcano, Alaska 1: Radar observations. *Journal of Geophysical Research*, 105, 10791–10806.

Lundgren, P., Casu, F., Manzo, M., Pepe, A., Berardino, P., Sansosti, E., and Lanari, R. (2004). Gravity and magma induced spreading of Mount Etna volcano revealed by satellite radar interferometry. *Geophysical Research Letters*, 31, L04602. doi: 10.1029/003GL018736.

Lundgren, P., and Rosen, P. A. (2003). Source model for the 2001 flank eruption of Mt. Etna volcano. *Geophysical Research Letters*, 30(7), 1388. doi: 10.1029/2002GL016774.

Lynch, J. S., and Stephens, G. (1996). Mount Pinatubo: A satellite perspective of the June 1991 eruption. In Newhall, C. G., and Punongbayan, R. S. (eds.), *Fire and Mud: Eruptions and Lahars of Mount Pinatubo*. Seattle: University of Washington Press, 637–645.

Malingreau, J., and Kaswanda, P. (1986). Monitoring volcanic eruptions in Indonesia using weather satellite data: The Colo eruption of July 28, 1983. *Journal of Volcanology and Geothermal Research*, 27(1–2), 179–194.

Manconi, A., Walter, T. R., Manzo, M., Zeni, G., Tizzani, P., Sansosti, E., and Lanari, R. (2010). On the effects of 3-D mechanical heterogeneities at Campi Flegrei caldera, southern Italy. *Journal of Geophysical Research*, 115, B08405. doi: 10.1029/2009JB007099.

Marchese, F., Ciampa, M., Filizzola, C., Lacava, T., Mazzeo, G., Pergola, N., and Tramutoli, V. (2010). On the exportability of robust satellite techniques (RST) for active volcano monitoring. *Remote Sensing*, 2, 1575–1588.

Marchese, F., Falconieri, A., Pergola, N., and Tramutoli, V. (2014). A retrospective analysis of Shinmoedake (Japan) eruption of 26–27 January 2011 by means of Japanese geostationary satellite data. *Journal of Volcanology and Geothermal Research*, 269, 1–13.

Marchese, F., Filizzola, C., Genzano, N., Mazzeo, G., Pergola, N., and Tramutoli, V. (2011). Assessment and improvement of a robust satellite technique (RST) for thermal monitoring of volcanoes. *Remote Sensing of Environment*, 115(6), 1556–1563.

Marchese, F., Lacava, T., Pergola, N., Hattori, K., Miraglia, E., and Tramutoli, V. (2012). Inferring phases of thermal unrest at Mt. Asama (Japan) from infrared satellite observations. *Journal of Volcanology and Geothermal Research*, 237–238, 10–18.

Marchese, F., Pergola, N., and Telesca L. (2006). Investigating the temporal fluctuations in satellite Advanced Very High Resolution Radiometer thermal signals measured in the volcanic area of Etna (Italy). *Fluctuation and Noise Letters*, 6, 305–316.

Massonnet, D., Briole, P., and Arnaud, A. (1995). Deflation of Mount Etna monitored by spaceborne radar interferometry. *Nature*, 375, 567–570.

Massonnet, D., and Feigl, K. L. (1998). Radar interferometry and its application to changes in the earth's surface. *Reviews of Geophysics*, 36, 4, 441–500.

Mayberry, G. C., Rose, W. I., and Bluth, G. J. S. (2003). Dynamics of the volcanic and meteorological clouds produced by the December 26, 1997 eruption of Soufrière Hills volcano, Montserrat, W.I. In Druitt, T., and Kokelaar, P. (eds.), *The Eruption of Soufrière Hills Volcano, Montserrat, 1995–99*. London: Geological Society of London, 539–555.

McGee, K. A., and Sutton, A. J. (1994). Eruptive activity at Mt. St. Helens, Washington, U.S.A. 1984–1988: A gas geochemistry perspective. *Bulletin of Volcanology*, 56, 435–446.

McPeters, R. D. (1993). The atmospheric SO2 budget for Pinatubo derived from NOAA-11 SBUV/2 spectral data. *Geophysical Research Letters*, 20(18), 1971–1974.

Menzel, W. P., Baum, B. A., Strabala, K. I., and Frey, R. A. (2002). Cloud top properties and cloud phase. Algorithm Theoretical Basis Document ATBD-MOD-04. Madison: CIMSS, University of Wisconsin–Madison, 61. http://modis-atmos.gsfc.nasa.gov/_docs/atbd_mod04.pdf.

Miller, T. P., and Casadevall, T. J. (2000). Volcanic ash hazards to aviation. In Sigurdsson, H. (ed.), *Encyclopedia of Volcanoes*. San Diego: Academic Press, 915–930.

MIROVA (Middle InfraRed Observation of Volcanic Activity). (2017). Near real time volcanic hotspot detection system. http://www.mirovaweb.it/?country_id=2036.

MODVOLC. (2017). Near real time thermal monitoring of global hot-spots. http://hotspot.higp.hawaii.edu/ (accessed June 2017).

Mora, O., Mallorqui, J., and Broquetas, A. (2003). Linear and nonlinear terrain deformation maps from a reduced set of interferometric SAR images. *IEEE Transactions on Geosciences and Remote Sensing*, 41(10), 2243–2253.

Mosher, F. R. (2000). Four channel volcanic ash detection algorithm. In *10th Conference on Satellite Meteorology and Oceanography*, Long Beach, CA, 9–14 January, 457–460.

Neri, M., Casu, F., Acocella, V., Solaro, G., Pepe, S., Berardino, P., Sansosti, E., Caltabiano, T., Lundgren, P., and Lanari, R. (2009). Deformation and eruptions at Mt. Etna (Italy): A lesson from 15 years of observations. *Geophysical Research Letters*, 36, L02309. doi: 10.1029/2008GL036151.

Oppenheimer, C. (1998). Volcanological applications of meteorological satellites. *International Journal of Remote Sensing*, 19(15), 2829–2864.

Oppenheimer, C., Francis, P. W., Rothery, D. A., Carlton, R. W. T., and Glaze, L. (1993). Infrared image analysis of volcanic thermal features: Làscar Volcano, Chile, 1984–1992. *Journal of Geophysical Research*, 98, 4269–4286.

Pagli, C., Wright, T. J., Ebinger, C. J., Yun, S.-H., Cann, J. R., Barnie, T., and Ayele, A. (2012). Shallow axial magma chamber at the slow-spreading Erta Ale Ridge. *Nature Geoscience*, 5, 284–288. doi: 10.1038/ngeo1414.

Palano, M., Gresta, S., and Puglisi, G. (2009). Time-dependent deformation of the eastern flank of Mt. Etna: After-slip or viscoelastic relaxation? *Tectonophysics*, 473(3–4), 300–311. doi: 10.1016/j.tecto.2009.02.047.

Palano, M., Puglisi, G., and Gresta, S. (2007). Ground deformation at Mt. Etna: A joint interpretation of GPS and InSAR data from 1993 to 2000. *Bollettino di Geofisica Teorica ed Applicata*, 48, 81–98.

Parks, M. M., Biggs, J., England, P., Mather, T. A., Nomikou, P., Palamartchouk, K., Papanikolaou, X. et al. (2012). Evolution of Santorini volcano dominated by episodic and rapid fluxes of melt from depth. *Nature Geoscience*, 5, 749–754. doi: 10.1038/ngeo1562.

Pavolonis, M. J., Wayne, F. F., Heidinger, A. K., and Gallina, G. M. (2006). A daytime complement to the reverse absorption technique for improved automated detection of volcanic ash. *Journal of Atmospheric and Oceanic Technology*, 23, 1422–1444.

Pepe, A., Berardino, P., Bonano, M., Euillades, L. D., Lanari, R., and Sansosti, E. (2011). SBAS-based satellite orbit correction for the generation of DInSAR time-series: Application to RADARSAT-1 data. *IEEE Transactions on Geoscience and Remote Sensing*, 49(12), 5150–5165. doi: 10.1109/TGRS.2011.2155069.

Pergola, N., D'Angelo, G., Lisi, M., Marchese, F., Mazzeo, G., and Tramutoli, V. (2009). Time domain analysis of robust satellite techniques (RST) for near real-time monitoring of active volcanoes and thermal precursor identification. *Physics and Chemistry of the Earth*, 34, 380–385.

Pergola, N., Marchese, F., Lacava, T., Filizzola, C., Coviello, I., Paciello, R., and Tramutoli, V. (2015). A review of RST-VOLC, an original algorithm for automatic detection and near real-time monitoring of volcanic hot spots from space. *Geological Society Special Publication*, 426(1), 55. http://doi.org/10.1144/SP426.1.

Pergola, N., Marchese, F., and Tramutoli, V. (2004a). Automated detection of thermal features of active volcanoes by means of infrared AVHRR records. *Remote Sensing of Environment*, 93, 311–327.

Pergola, N., Marchese, F., Tramutoli, V., Filizzola, C., and Ciampa, M. (2008). Advanced satellite technique for volcanic activity monitoring and early warning. *Annals of Geophysics*, 51, 287–301.

Pergola, N., Tramutoli, V., Marchese, F., Scaffidi, I., and Lacava, T. (2004b). Improving volcanic ash cloud detection by a robust satellite technique. *Remote Sensing of Environment*, 90, 1–22.

Philibosian, B., and Simons, M. (2011). A survey of volcanic deformation on Java using ALOS PALSAR interferometric time series. *Geochemistry, Geophysics, Geosystems*, 12, Q11004. doi: 10.1029/2011GC003775.

Picchiani, M., Chini, M., Corradini, S., Merucci, L., Sellitto, P., Del Frate, F., and Stramondo, S. (2011). Volcanic ash detection and retrievals using MODIS data by means of neural networks. *Atmospheric Measurement Techniques*, 4, 2619–2631.

Pieri, D., and Abrams, M. (2005). ASTER observations of thermal precursors to the April 2003 eruption of Chikurachki volcano, Kurile Islands, Russia. *Remote Sensing of Environment*, 99, 84–94.

Pieri, D. C., and Baloga, S. M. (1986). Eruption rate, area, and length relationships for some Hawaiian lava flows. *Journal of Volcanology and Geothermal Research*, 30, 29–45.

Piscini, A., and Lombardo, V. (2014). Volcanic hot spot detection from optical multispectral remote sensing data using artificial neural networks. *Geophysical Journal International*, 196(3), 1525–1535.

Prata, A. J. (1989a). Observations of volcanic ash clouds in the 10–12 μm window using AVHRR/2 data. *International Journal of Remote Sensing*, 10, 751–761.

Prata, A. J. (1989b). Radiative transfer calculations for volcanic ash clouds. *Geophysical Research Letters*, 16(11), 1293–1296.

Prata, A. J., Bluth, G. J. S., Rose, W. I., Schneider, D. J., and Tupper, A. (2001). Comments on "Failures in detecting volcanic ash from a satellite-based technique." *Remote Sensing of Environment*, 78, 341–346.

Prata, A. J., and Grant, I. F. (2001). Retrieval of microphysical and morphological properties of volcanic ash plumes from satellite data: Application to Mt. Ruapehu, New Zealand. *Quarterly Journal of the Royal Meteorological Society*, 127, 2153–2179.

Prata, A. J., Rose, W. I., Self, S., and O'Brien, D. M. (2003). Global, long-term sulphur dioxide measurements from the TOVS data: A new tool for studying explosive volcanism and climate. *Geophysical Monograph*, 139, 75–92.

Prata, A. J., and Turner, P. J. (1997). Cloud top height determination from the ATSR. *Remote Sensing of Environment*, 59, 1–13.

Prata, A. J., and Bernardo, C. (2007). Retrieval of volcanic SO2 column abundance from Atmospheric Infrared Sounder data. *Journal of Geophysical Research Atmosphere*, 112(D20204), 17. doi: 10.1029/2006JD007955.

Prati, C., Ferretti, A., and Perissin, D. (2010). Recent advances on surface ground deformation measurement by means of repeated space-borne SAR observations. *Journal of Geodynamics*, 49, 161–170.

Pritchard, M. E., and Simons, M. (2002). A satellite geodetic survey of large-scale deformation of volcanic centres in the central Andes. *Nature*, 418, 167–171. doi: 10.1038/nature00872.

Raucoules, D., Bourgine, B., de Michele, M., Le Cozannet, G., Closset, L., Bremmer, C., Veldkamp, H. et al. (2009). Validation and intercomparison of persistent scatterers interferometry: PSIC4 project results. *Journal of Applied Geophysics*, 68, 335–347.

Read, W. G., Shippony, Z., and Van Snyder, W. (2003). EOS MLS forward model algorithm theoretical basis document. Technical Report JPL D-18130. Pasadena, CA: Jet Propulsion Laboratories.

Realmuto, V. J., Abrams, M. J., Buongiorno, M. F., and Pieri, D. C. (1994). The use of multispectral thermal infrared image data to estimate the sulfur dioxide flux from volcanoes: A case study from Mount Etna, Sicily, July 29, 1986. *Journal of Geophysical Research*, 99(B1), 481–488.

REALVOLC. (2017). Near realtime monitoring of active volcanoes in East Asia using satellite data. http//vrsserv.eri.u-tokyo.ac.jp/REALVOLC/index.html

Richards, M. S., Ackerman, S. A., Pavolonis, M. J., Feltz, W. F., and Tupper, A. (2006). Volcanic ash cloud heights using the MODIS CO2-slicing algorithm. Madison: CIMSS, University of Wisconsin–Madison, Department of Atmospheric and Oceanic Sciences.

Robock, A. (2000). Volcanic eruptions and climate. *Reviews of Geophysics*, 38(2), 191–219.

Rose, S., and Ramsey, M. (2009). The 2005 eruption of Kliuchevskoi volcano: Chronology and processes derived from ASTER spaceborne and field-based data. *Journal of Volcanology and Geothermal Research*, 184(3–4), 367–380.

Rose, W. I., Delene, D. J., Schneider, D. J., Bluth, G. J. S., Krueger, A. J., Sprod, I., McKee, C., Davies, H. L., and Ernst, G. G. J. (1995). Ice in the 1994 Rabaul eruption cloud: Implications for volcano hazard and atmospheric effects. *Nature*, 375, 477–479.

Rosen, P. A., Hensley, S., Joughin, I. R., Li, F. K., Madsen, S. N., Rodriguez, E., and Goldstein, R. (2000). Synthetic aperture radar interferometry. *Proceedings of the IEEE*, 88, 333–376.

Rothery, D. A., Francis, P. W., and Wood, C. A. (1988). Volcano monitoring using short wavelength infrared data from satellites. *Journal of Volcanology and Geothermal Research*, 93, 7993–8008.

Ruch, J., Acocella, V., Storti, F., Neri, M., Pepe, S., Solaro, G., and Sansosti, E. (2010). Detachment depth revealed by rollover deformation: An integrated approach at Mount Etna. *Geophysical Research Letters*, 37, L16304. doi: 10.1029/2010GL044131.

Ruch, J., Pepe, S., Casu, F., Acocella, V., Neri, M., Solaro, G., and Sansosti, E. (2012). How do rift zones relate to volcano flank instability? Evidence from collapsing rifts at Etna. *Geophysical Research Letters*, 39(L20311). doi: 10.1029/2012GL053683.

Ruch, J., Pepe, S., Casu, F., Solaro, G., Pepe, A., Acocella, V., Neri, M., and Sansosti, E. (2013). Seismo-tectonic behavior of the Pernicana Fault System (Mt Etna): A gauge for volcano flank instability? *Journal of Geophysical Research*, 118, 4398–4409. doi: 10.1002/jgrb.50281.

Sansosti, E. (2004). A simple and exact solution for the interferometric and stereo SAR geolocation problem. *IEEE Transactions on Geoscience and Remote Sensing*, 42, 1625–1634. doi: 10.1109/TGRS.2004.831442.

Sansosti, E., Berardino, P., Bonano, M., Calò, F., Castaldo, R., Casu, F., Manunta, M. et al. (2014). How second generation SAR systems are impacting the analysis of ground deformation. *International Journal of Applied Earth Observation and Geoinformation*, 28, 1–11. doi: 10.1016/j.jag.2013.10.007.

Sansosti, E., Casu, F., Manzo, M., and Lanari, R. (2010). Space-borne radar interferometry techniques for the generation of deformation time series: An advanced tool for Earth's surface displacement analysis. *Geophysical Research Letters*, 37, L20305. doi: 10.1029/2010GL044379.

Sansosti, E., Lanari, R., Fornaro, G., Franceschetti, G., Tesauro, M., Puglisi, G., and Coltelli, M. (1999). Digital elevation model generation using ascending and descending ERS-1/ERS-2 tandem data. *International Journal of Remote Sensing*, 20(8), 1527–1547. doi: 10.1080/014311699212597.

Sawada, Y. (1987). Study on analysis of volcanic eruptions based on eruption cloud image data obtained by the Geostationary Meteorological Satellite (GMS). Tokyo: Meteorology Research Institute.

Sawada, Y. (2002). Analysis of eruption cloud with Geostationary Meteorological Satellite imagery (Himawari). *Journal of Geography (Japan)*, 111, 374–394.

Schmidt, D. A., and Bürgmann, R. (2003). Time-dependent land uplift and subsidence in the Santa Clara Valley, California, from a large interferometric synthetic aperture radar data set. *Journal of Geophysical Research*, 108(B9), 2416. doi: 10.1029/2002JB002267.

Schneider, D. J., Rose, W. I., Coke, L. R., Bluth, G. J. S., Sprod, I., and Krueger, A. J. (1999). Early evolution of a stratospheric volcanic eruption cloud as observed with TOMS and AVHRR. *Journal of Geophysical Research*, 104(D4), 4037–4050.

Schneider, D. J., Rose, W. I., and Kelley, L. (1995). Tracking of 1992 eruption clouds from Crater Peak vent of Mount Spurr Volcano, Alaska, using AVHRR. In Keith, T. E. C. (ed.), *The 1992 Eruption of Crater Peak Vent, Mount Spurr Volcano Alaska*. U.S. Geological Survey Bulletin 2139. Reston, VA: U.S. Geological Survey, 27–35.

Simkin, T., Siebert, L., and Blong, R. (2001). Volcano fatalities – Lessons from the historical record. *Science*, 291(5502), 255. doi: 10.1126/science.291.5502.255.

Simpson, J. J., Hufford, G., Pieri, D., and Berg, J. (2000). Failures in detecting volcanic ash from a satellite-based technique. *Remote Sensing of Environment*, 72, 191–217.

Simpson, J. J., Hufford, G., Pieri, D., and Berg, J. (2001). Response to "Comments of 'Failures in detecting volcanic ash from a satellite based technique.'" *Remote Sensing of Environment*, 78, 347–357.

Solaro, G., Acocella, V., Pepe, S., Ruch, J., Neri, M., and Sansosti, E. (2010). Anatomy of an unstable volcano from InSAR: Multiple processes affecting flank instability at Mt. Etna, 1994–2008. *Journal of Geophysical Research*, 115, B10405. doi: 10.1029/2009JB000820.

Sparks, R. S. J., Bursik, M. I., Carey, S. N., Gilbert, J. E., Glaze, L., Sigurdsson, H., and Woods, A. W. (1987). *Volcanic Plumes*. Chichester, UK: Wiley.

Steffke, A. M., Fee, D., Garces, M., and Harris, A. (2010). Eruption chronologies, plume heights and eruption styles at Tungurahua volcano: Integrating remote sensing techniques and infrasound. *Journal of Volcanology and Geothermal Research*, 193, 143–160.

Steffke, A. M., and Harris, A. J. (2011). A review of algorithms for detecting volcanic hot spots in satellite infrared data. *Bulletin of Volcanology*, 73(9), 1109–1137.

Stevens, N. F., and Wadge, G. (2004). Towards operational repeat-pass SAR interferometry at active volcanoes. *Natural Hazards*, 33, 47–76.

Stevens, N. F., Wadge, G., Williams, C. A., Morely, J. G., Muller, J.-P., Murray, J. B., and Upton, U. (2001). Surface movements of emplaced lava flows measured by synthetic aperture radar interferometry. *Journal of Geophysical Research*, 106, 11293–11313.

Stix, J., and Gaonac'h, H. (2000). Gas, plume, and thermal monitoring. In Sigurdsson, H., Houghton, B. F., McNutt, S. R., Rymer, H., and Stix, J. (eds.), *Encyclopedia of Volcanoes*. San Diego: Academic Press, 1141–1163.

Stohl, A., Haimberger, L., Scheele, M. P., and Wernli, H. (2001). An intercomparison of results from three trajectory models. *Meteorological Applications*, 8, 127–135.

Strozzi, T., Wegmuller, U., Werner, C. L., Wiesmann, A., and Spreckels, V. (2003). JERS SAR interferometry for land subsidence monitoring. *IEEE Transactions on Geoscience and Remote Sensing*, 41(7), 1702–1708. doi: 10.1109/TGRS.2003.813273.

Taylor, I., Mackie, S., and Watson, M. (2015). Investigating the use of the Saharan dust index as a tool for the detection of volcanic ash in SEVIRI imagery. *Journal of Volcanology and Geothermal Research*, 304, 126–141.

Thomas, H. E., and Prata, A. J. (2011). Sulphur dioxide as a volcanic ash proxy during the April–May 2010 eruption of Eyjafjallajökull Volcano, Iceland. *Atmospheric Chemistry and Physics*, 11, 6871–6880. doi: 10.5194/acp-11-6871-2011.

Thomas, H. E., and Watson, I. M. (2010). Observations of volcanic emissions from space: Current and future perspectives. *Natural Hazards*, 54, 323–354.

Tilling, R. I, Topinka, L., and Swanson, D. A. (1990). Eruptions of Mount St. Helens: Past, present and future. U.S. Geological Survey Special Interest Publication. Reston, VA: U.S. Geological Survey.

Tizzani, P., Berardino, P., Casu, F., Euillades, P., Manzo, M., Ricciardi, G. P., Zeni, G., and Lanari R. (2007). Surface deformation of Long Valley caldera and Mono Basin, California, investigated with the SBAS-InSAR approach. *Remote Sensing of Environment*, 108, 277–289. doi: 10.1016/j.rse.2006.11.015.

Trasatti, E., Casu, F., Giunchi, C., Pepe, S., Solaro, G., Tagliaventi, S., Berardino, P. et al. (2008). The 2004–2006 uplift episode at Campi Flegrei caldera (Italy): Constraints from SBAS-DInSAR ENVISAT data and Bayesian source inference. *Geophysical Research Letters*, 35, L07308. doi: 10.1029/2007GL033091.

Trunk, L., and Bernard, A. (2008). Investigating crater lake warming using ASTER thermal imagery: Case studies at Ruapehu, Poás, Kawah Ijen, and Copahué Volcanoes. *Journal of Volcanology and Geothermal Research*, 178(2), 259–270.

Tuck, B. H., and Huskey, L. (1994). Economic disruptions by Redoubt volcano: Assessment methodology and anecdotal empirical evidence. In Casadevall, T. J. (ed.), *Volcanic Ash and Aviation Safety: Proceedings of the First International Symposium on Volcanic Ash and Aviation Safety*. U.S. Geological Survey Bulletin 2047. Reston, VA: U.S. Geological Survey, 137–140.

Tupper, A., Carn, S., Davey, J., Kamada, Y., Potts, R., Prata, F., and Tokuno, M. (2004). An evaluation of volcanic cloud detection techniques during recent significant eruptions in the western 'Ring of Fire'. *Remote Sensing of Environment*, 91, 27–46.

Tupper, A., Textor, C., Herzog, M., Graf, H. F., and Richards M. S. (2012). Tall clouds from small eruptions: The sensitivity of eruption height and fine ash content to tropospheric instability. *Natural Hazards*, 2, 375–401. doi: 10.1007/s11069-009-9433-9.

UNEP (United Nations Environment Programme). (2004). IGOS geohazards theme report. https://homepage.univie.ac.at/thomas.glade/Publications/MarshEtAl2004.pdf.

Van Manen, S. M., Blake, S., and Dehn, J. (2012). Satellite thermal infrared data of Shiveluch, Kliuchevskoi and Karymsky, 1993–2008: Effusion, explosions and the potential to forecast ash plumes. *Bulletin of Volcanology*, 74, 1313–1335.

Van Manen, S. M., Dehn, J., and Blake, S. (2010). Satellite thermal observations of the Bezymianny lava dome 1993–2008: Precursory activity, large explosion, and dome growth. *Journal of Geophysical Research*, 115, B8. doi: 10.1029/2009JB006966.

Vogel, L., Sihler, H., Lampel, J., Wagner, T., and Platt, U. (2012). Retrieval interval mapping, a tool to optimize the spectral retrieval range in differential optical absorption spectroscopy. *Atmospheric Measurement Techniques*, 5, 4195–4247. doi: 10.5194/amtd-5-4195-2012.

Walter, T. R., Shirzaei, M., Manconi, A., Solaro, G., Pepe, A., Manzo, M., and Sansosti, E. (2014). Possible coupling of Campi Flegrei and Vesuvius as revealed by InSAR time series, correlation analysis and time dependent modelling. *Journal of Volcanology and Geothermal Research*, 280, 104–110. doi: 10.1016/j.jvolgeores.2014.05.006.

Watson, I. M., Realmuto, V. J., Rose, W. I., Prata, A. J., Bluth, G. J. S., Gu, Y., Bader C. E., and Yu, T. (2004). Thermal infrared remote sensing of volcanic emissions using the moderate resolution imaging spectro-radiometer. *Journal of Volcanology and Geothermal Research*, 135, 75–89.

Webley, P. W., and Mastin, L. G. (2009). Improved prediction and tracking of volcanic ash clouds. *Journal of Volcanology and Geothermal Research*, 186, 1–2, 1–9.

Wen, S., and Rose, W. I. (1994). Retrieval of sizes and total masses of particles in volcanic clouds using AVHRR bands 4 and 5. *Journal of Geophysical Research*, 99(D3), 5421–5431.

Williams, R. S., and Friedman, J. D. (1970). Satellite observation of effusive volcanism. *Journal of the British Interplanetary Society*, 23, 441–450.

Woods, A. W., and Self, S. (1992). Thermal disequilibrium at the top of volcanic eruption columns. *Nature*, 355, 628–630.

Wooster, M. J. (2001). Long-term infrared surveillance of Lascar volcano: Contrasting activity cycles and cooling pyroclastics. *Geophysical Research Letters*, 28(5), 847–850.

Wooster, M. J., and Rothery, D. A. (1997a). Thermal monitoring of Lascar volcano, Chile, using infrared data from the along-track scanning radiometer: A 1992–1995 time series. *Bulletin of Volcanology*, 58, 566–579.

Wooster, M. J., and Rothery, D. A. (1997b). Time series analysis of effusive volcanic activity using the ERS along track scanning radiometer: The 1995 eruption of Fernandina volcano, Galapagos Island. *Remote Sensing of Environment*, 69, 109–117.

Wooster, M. J., Zhukov, B., and Oertel, D. (2003). Fire radiative energy for quantitative study of biomass burning: Derivation from the BIRD experimental satellite and comparison to MODIS fire products. *Remote Sensing of Environment*, 86, 83–107.

Wright, R. (2016). MODVOLC: 14 years of autonomous observations of effusive volcanism from space. *Geological Society Special Publication*, 426(1), 23–53.

Wright, R., Carn, S. A., and Flynn, L. P. (2005). A satellite chronology of the May–June 2003 eruption of Anatahan volcano. *Journal of Volcanology and Geothermal Research*, 146, 102–116.

Wright, R., Flynn, L., Garbeil, H., Harris, A., and Pilger, E. (2002). Automated volcanic eruption detection using MODIS. *Remote Sensing of Environment*, 82, 135–155.

Yu, T., Rose, W. I., and Prata, A. J. (2002). Atmospheric correction for satellite-based volcanic ash mapping and retrievals using "split window" IR data from GOES and AVHRR. *Journal of Geophysical Research*, 107(D16), 10.1029.

Zebker, H. A., and Goldstein, R. M. (1986). Topographic mapping from interferometric SAR observations. *Journal of Geophysical Research*, 91, 4993–4999.

Zebker, H. A., and Villasenor, J. (1992). Decorrelation in interferometric radar echoes. *IEEE Transactions on Geoscience and Remote Sensing*, 30, 950–959.

20 Application of Thermal Remote Sensing to the Observation of Natural Hazards

Matthew Blackett

CONTENTS

20.1 INTRODUCTION

Natural hazards are events which have the potential to adversely affect large numbers of people, whether the event is geological (earthquakes or volcanoes), meteorological (storms) or hydrological (floods). Fortunately, many such events can be observed remotely from space, and one such method utilizes thermal sensors which are found onboard many orbiting satellite platforms and which can be used for observing such events based on temperature differences. This chapter discusses the basic principles and applications of thermal remote sensing.

Natural hazards often constitute events that are manifested over a wide area. For example, a lava flow may extend many kilometres from the source volcano, an earthquake may be felt hundreds of kilometres from the epicentre and a hurricane may display a diameter of more than 1000 km. The large, synoptic scale of natural hazards presents the ideal opportunity for their observation using remotely sensed observations from orbiting satellites which have a large field of view of the Earth surface–atmosphere system. When utilized with thermal sensors, remotely sensed data display additional utility in being able to detect the relative temperatures of different objects – something which is of use in relation to many natural hazard phenomena. Here the fundamental concepts of thermal remote sensing are explained, and following this, examples are presented detailing how such data are used more widely in the observation of natural hazard events.

20.2   FUNDAMENTAL CONCEPTS

All objects with a temperature above absolute zero (0 K or –273°C) emit electromagnetic radiation as a function of associated molecular vibrations. The quantity of radiation, or radiance, is measured in watts (W) that are emitted from a surface (m²) within a particular solid angle (steradian, Sr) and in a particular direction towards a sensor (Figure 20.1). As the temperature of an object increases, so too do molecular vibrations and associated electromagnetic emissions, according to the Stefan–Boltzmann law (Stefan 1879; Boltzman, 1884):

$$E = \sigma T^4 \tag{20.1}$$

where E = emitted radiant energy (W m⁻²), T = temperature of blackbody (K) and σ = Stefan–Boltzmann constant (5.6697×10^{-8} W m⁻² K⁻⁴).

The radiation emitted travels in the form of electromagnetic waves, and as an object's temperature increases, the main wavelength of these emissions decreases, according to Wien's displacement law (Wien 1896):

$$\lambda_{max} = \frac{b}{T} \tag{20.2}$$

where λ_{max} = peak wavelength (μm), T = temperature of a blackbody (K) and b = Wien's displacement constant (2.8978×10^{-3} m K).

Given the temperature of most terrestrial objects, around 20°C (293 K), the chief wavelength of emission will be at 9.99 μm, which forms part of the thermal infrared (TIR) portion of the electromagnetic spectrum (Figure 20.2). A hotter object, such as a frying pan at 100°C (373 K), will emit chiefly at 7.77 μm, again in the TIR portion of the spectrum, whereas fresh lavas, which may reach temperatures of 1000°C (1273 K), emit chiefly at 1.97 μm in what is known as the short-wave

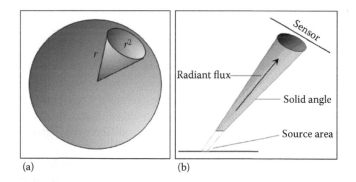

(a)   (b)

FIGURE 20.1   (a) Solid angle (Steradian) whereby r = radius and the solid angle constitutes that which, when projected onto a sphere from its centre, has an area of r^2. (b) Radiant flux from an emitting surface, through a solid angle, to the detecting sensor.

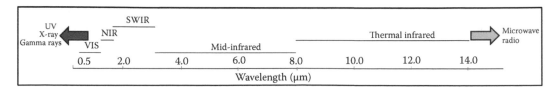

FIGURE 20.2   Electromagnetic spectrum from 0.0 μm to the TIR region, with key regions highlighted. UV, ultraviolet; SWIR, short-wave infrared; NIR, near infrared; VIS, visible.

infrared portion of the spectrum. With the increase of an object's temperature, quantities of energy increase at shorter wavelengths, which ultimately become visible to the human eye, with hot objects often appearing red. Strictly speaking, thermal remote sensing would only include that conducted with TIR bands, but here what is considered is the remote sensing of objects based on their radiant *emissions* (that are a function of an object's temperature), as opposed to visual remote sensing, which detects only the radiation that is *reflected* from a particular object.

The spectral radiance (the radiance emitted at a particular wavelength, L_λ, in units of Wm^{-2} sr^{-1} m^{-1}) emitted from an object, assuming it is a blackbody (i.e. that it absorbs and emits radiation perfectly), can be determined using the Planck function (below), which integrates the area under the Planck curve derived (Figure 20.3):

$$L_\lambda(T) = \frac{C_1}{\lambda^5 \left(\exp\left(\frac{C_2}{\lambda T}\right) - 1 \right)} \qquad (20.3)$$

where T = temperature (K), λ = wavelength (m) and C_1 and C_2 = constants of 1.19×10^{-16} Wm^{-2} sr^{-1} and 1.44×10^{-2} m K, respectively (Wooster 2002). Evidently, the hotter an object, the greater the quantity of energy it will emit and the shorter its peak wavelength of emission.

TIR sensors are sensitive to electromagnetic radiation in the thermal parts of the spectrum, and hence to the emissions from surfaces which are ambient or hot. The additional consideration required when it comes to utilizing satellite thermal imagery, however, is that the sensors are in orbit, and hence are far from the emitting object. Consequently, once a signal (R_λ) has arrived at the sensor, it has travelled through much of Earth's atmosphere, which both attenuates (*atmospheric transmittance*) and augments (*atmospheric radiance*) it. Additionally, the emissivity of the radiating surface and the reflectance of any solar radiation must be considered. These influences can be expressed using the following equation:

$$R_\lambda = \left\{ \varepsilon_\lambda L_\lambda(T) + [1 - \varepsilon_\lambda] \frac{R_{d\lambda}}{\pi} \right\} \tau_\lambda + R_{atm\lambda} \qquad (20.4)$$

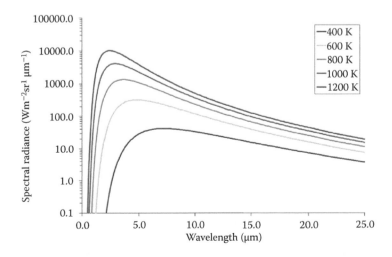

FIGURE 20.3 Planck curve drawn in relation to surfaces at a range of temperatures (400–1200 K). Note that the hotter a surface, the shorter the chief wavelength of emission.

TABLE 20.1

Details of the Satellite Sensors Discussed in This Chapter

Polar Orbiting Satellite Sensors

Sensor	Satellite Platform(s)	Range of Bands	Spatial Resolution	Revisit Frequency	Altitude
ASTER	Earth Observing System Terra	Visual–near infrared (VNIR)	15 m	16 days	705 km
		Short-wave infrared (SWIR)	30 m		
		Thermal infrared (TIR)	90 m		
ATSR	European Remote Sensing	SWIR-TIR	1030 m	3 days	780 km
AVHRR	National Oceanic and Atmospheric Administration (various)	VNIR-TIR	1.1 km	2× daily	830–870 km
ETM+	Landsat-7	VNIR-TIR	15–60 m	16 days	705 km
Hyperion	Earth Observing-1	VNIR-SWIR	30 m	16 days	705 km
MODIS	Earth Observing System Terra and Aqua	VNIR	250 m	4× daily	705 km
		VNIR-SWIR	500 m		
		VNIR-TIR	1000 m		
TM	Landsat-5	VNIR-TIR	30–120 m	16 days	705 km
TIRS	Landsat-8	TIR	100 m	16 days	705 km

Geostationary Satellite Sensors

Sensor	Satellite Platform(s)	Range of Bands	Spatial Resolution	Image Frequency	Altitude
Imager	GOES	VNIR	1 km	15 minutes	35,800 km
		Mid-infrared-TIR	4 km		
SEVIRI	Meteosat	VNIR	1 km	15 minutes	35,800 km
			4 km		

where the absence of sunlight (i.e. nighttime data) is assumed and ε_λ = spectral emissivity of the surface, $[1 - \varepsilon_\lambda]$ relates to the proportion of downwelling radiance reflected from the surface, $R_{d,\lambda}$ = downwelling atmospheric radiance (Wm^{-2} sr^{-1} m^{-1}), τ_λ = atmospheric transmittance (unitless) and $R_{atm,\lambda}$ = upwelling atmospheric path radiance (Wm^{-2} sr^{-1} m^{-1}) (Buongiorno et al. 2002). Using this equation allows us to determine the true conditions at the emitting surface (in terms of its temperature and/or radiant emissions), based on the 'top-of-atmosphere' remotely sensed observations. The absence of cloud is assumed since cloud will absorb the thermal signals emitted from the surface and prevent their detection at the sensor. Satellite sensors which have the capability to detect thermal electromagnetic emissions are given in Table 20.1 and are discussed in detail in this chapter.

20.3   SATELLITE IMAGERY

Satellite imagery, just like imagery from a digital camera, is made up of many pixels which form the smallest portion that can be resolved within an image. The size of these pixels (the spatial resolution), and the area of the Earth's surface they therefore view, varies from sensor to sensor (Table 20.1). In general, the shorter the wavelength being detected, the smaller the pixels can be. Satellites imaging Earth's surface can generally be grouped into one of two categories: polar orbiting satellites (or sun synchronous), which orbit the Earth pole to pole and pass over a particular point on Earth's surface at the same local solar time every day and view all of Earth's surface, and geostationary satellites, which orbit high and directly above the equator but do not move relative to the ground and, as such, are always directly over the same surface.

In terms of thermal remote sensing, some of the largest pixels relate to geostationary satellites due to their high altitudes, including the National Oceanic and Atmospheric Administration (NOAA) Geostationary Operational Environmental Satellite (GOES) (~4 km) and the Spinning Enhanced Visible and Infrared Imager (SEVIRI) instrument of the European Space Agency Meteosat satellite (~3 km) (Deneke et al. 2008; Koeppen et al. 2011). The spatial resolution of polar orbiting satellites is better, with the TIR pixels of the National Aeronautics and Space Administration (NASA) Moderate Resolution Imaging Spectroradiometer (MODIS) being 1 km × 1 km and those of the Advanced Spaceborne Thermal Emission and Reflection Radiometer (ASTER) being 90 × 90 m. Clearly, a surface is likely to change considerably over 90^2 m^2, and even more so over 1^2 or 4^2 km^2. Indeed, the signal detected for one particular pixel is actually an average of the whole area viewed: a pixel integrated value. For example, if viewing an active lava flow, it might be that the pixel area in question consists of an active, fresh lava flow, cooling lava and areas completely unaffected by volcanic activity, all contributing to the one pixel value. This being the case, only the hottest and most widespread lava flows are likely to be detected within the largest pixels, while smaller pixels will generally be more sensitive to less widespread or cooler phenomena.

However, it is not just spatial resolution which must be considered when assessing the usefulness of satellite imagery. Considerations of spectral and temporal resolution are also important, and these must each be traded off against each other. ASTER, for example, has a high spatial resolution (e.g. 90 m in the TIR) compared with MODIS (1 km), but this is at the expense of temporal resolution with its smaller field of view (60 km), resulting in a smaller proportion of Earth's surface being imaged during any orbit. MODIS does have a wider field of view (2330 km), and hence a greater temporal resolution, but due to data transfer requirements, the viewing pixels are larger (i.e. the spatial resolution is adversely affected). Similar considerations apply for spectral resolution, with the presence of many bands generating large volumes of data that can only be provided at the expense of the spatial or temporal resolution. Examples of the same volcanic phenomena as viewed by sensors with varying spatial resolution and spectral bands are shown in Figure 20.4.

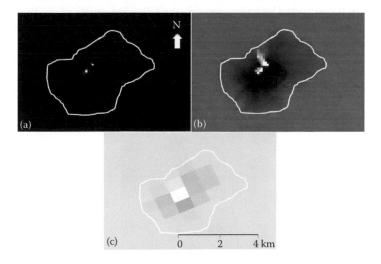

FIGURE 20.4   TIR imagery of Stromboli on 23 July 2003. (a) ASTER band 6 (2.185–2.225 μm) nighttime imagery at 30 m resolution. (b) ASTER band 12 (8.925–9.275 μm) nighttime imagery at 90 m resolution. (c) MODIS band 20 (3.660–3.840 μm) imagery at 1000 m resolution.

20.4 NATURAL HAZARD APPLICATIONS

In the case of some natural hazards, for example, volcanoes, the potential utility of thermal remote sensing will be obvious, with volcanism being associated with the transfer of heat from the Earth's interior to the surface (Oppenheimer 1998). The potential utility for other hazards, however, is more subtle. With regard to earthquakes, some scientists have suggested that they display precursory phenomena, which might include heating of Earth's surface (Ouzounov and Freund 2004). Thermal remote sensing is widely used to monitor clouds and has also shown utility in the observation of other natural hazard phenomena, including heat waves, landslides and floods.

20.4.1 VOLCANOES

The very nature of volcanism means that thermal remote sensing is optimal for monitoring it. The temperature of volcanic surfaces may be anywhere from ambient, for a cooled lava crust, to more than 1000°C for fresh basaltic lava flows (Tallarico and Dragoni 1999), highlighting the utility of all parts of the infrared spectrum and not just that in the thermal portion. An example of ASTER TIR imagery of a lava flow is shown in Figure 20.5.

Using remotely sensed thermal data, many aspects of volcanic activities have been explored. For example, Glaze et al. (1989) were some of the first scientists to use TIR remotely sensed data (from the Landsat Thematic Mapper) to quantify variations in the energy emission from volcanoes. Later, Oppenheimer et al. (1993) used TIR imagery from the NASA Landsat-5 Thematic Mapper to follow the growth of a lava dome at Mount Lascar, Chile, identifying its presence 15 months before locals could observe it. For a detailed history of volcanic applications, see Blackett (2017).

One of the most common ways of quantifying such observations has been to convert detected radiance values to temperature, thereby indicating the average temperature of the surface of each pixel observed. This can be done by taking the radiance value corrected for emissivity and atmospheric effects and inverting the Planck function (Equation 20.3). From an estimation of the surface temperature, the actual quantity of energy, or heat flux, that the surface emits can be derived, this time using the Stefan–Boltzman law (Equation 20.1). It has been shown that variations in volcanic activity are reflected in variations in power emission, and as such, we are able to derive relationships between remotely sensed thermal observations and actual activity on the ground. Using the detected power, for example, the lava effusion rate (E_r [m^3 s^{-1}]) can be calculated by parameterizing the following equation:

$$E_r = \frac{Q_{lava} + Q_{conv}}{\rho_{lava} C_p \Delta T + \phi C_L} \tag{20.5}$$

where Q_{lava} = total radiant heat flux (MW), Q_{conv} = total convective heat flux (MW), ρ_{lava} = lava density (kg m^{-3}), C_p = specific heat capacity (J kg^{-1} K^{-1}), ΔT = average temperature decrease through

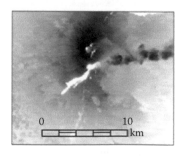

FIGURE 20.5   Nighttime imagery of Mt. Etna during a 2002 eruption viewed with ASTER thermal band 14 (11.3 μm). The lava flow is intensely hot and, as such, emits huge quantities of energy which are manifested in the thermal image as the very bright region.

active flow (K), ø = average mass fraction of crystals grown during cooling and C_L = latent heat of crystallization (J kg^{-1}) (Harris et al. 1998).

Many other examples have related actual styles of activity to remotely sensed radiant and/or temperature observations. Wooster and Rothery (1997), for example, isolated a cyclical pattern in the thermal emissions of Lascar volcano, Chile, as detected using the Along Track Scanning Radiometer (ATSR), which was explained in terms of physical processes (Matthews et al. 1997). Flynn et al. (2001) have discussed in detail how the improved spatial resolution of the Landsat-7 Enhanced Thematic Mapper thermal bands provided for an enhanced means of identifying volcanic features. Wright et al. (2011) have shown that Hyperion data can be used to discern different eruption styles, and even different lava types, based on derived temperature distributions. Lava fountains were shown to display the highest temperatures, but as the lava cooled reaching the ground, cooler surfaces were also observed, creating a bimodal lava surface temperature; pāhoehoe lava flows, in contrast, were shown to display a unimodal temperature distribution, with a tail of trailing-off temperatures caused by the rapid formation of a continuous skin with just some core exposures. Urai and Ishizuka (2011) demonstrated the use of ASTER data in estimating the volcanic explosivity index of Russian volcanic eruptions while Blackett (2013) showed how this sensor could be used to derive a chronology of the behaviour of a remote and inaccessible volcano. From a hazard management perspective too, the practical use of thermal remote sensing for the monitoring of remote volcanoes was confirmed by Vaughan et al. (2008), in which exaggerated reports of volcanic activity at Oldoinyo Lengai, Tanzania, were quashed by examining thermal ASTER and MODIS imagery. Additionally, hazard maps have been created of volcanic regions by combining infrared-derived lava effusion rates and lava flow models (Wright et al. 2008). Indeed, so useful have satellite thermal observations proven to be in monitoring volcanic activity that a number of automated or semi-automated systems now exist to monitor global volcanic activity using such data. The MODVOLC system (http://modis.higp.hawaii.edu/), for example, analyzes MODIS imagery on a daily basis and isolates pixels which appear to display thermal anomalies in the region of both known volcanoes and areas unrelated to volcanic activity but which display thermally anomalous signals. MODVOLC outputs are available within 12 hours following each overpass of the MODIS sensors, and are posted online to view for various time periods (Wright et al. 2004, 2005) (Figure 20.6). Wright et al. (2015) were able to isolate periodic behaviour in some volcanoes using data from the MODVOLC system.

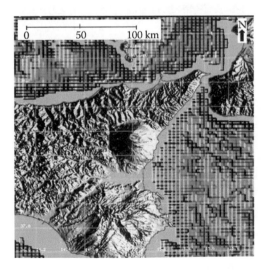

FIGURE 20.6   Output from the MODVOLC website for Mt. Etna, for the period 23 March–22 April 2013. The red-orange areas represent intense heat, and hence infrared emissions. (From Hawaii Institute of Geophysics and Planetology, MODVOLC, http://modis.higp.hawaii.edu/.)

One consideration which is vital in terms of the remote sensing of volcanic activity is the regularity of observations. A satellite image provides only an instantaneous 'snapshot' of the surface, while in reality, a volcanic surface may change significantly on the order of minutes (Glaze et al. 1989; Pieri and Abrams 2004). This means that for true volcanic monitoring, a high temporal resolution is optimal, rendering geostationary satellites such as GOES, with image acquisition every 15 minutes, and the Advanced Very High Resolution Radiometer (AVHRR), with multiple image acquisitions daily as of greater utility than sensors such as ASTER, for which image acquisition is every 16 days. In recent years, systems have been experimented with that consist of the triggering of higher-spatial-resolution sensors by lower-spatial but higher-temporal-resolution sensors, such as the AVHRR triggering of ASTER (Carter et al. 2008). More sophisticated systems have also been proposed, including the ESA Sentinel Convoy. This would consist of a proposed VISible-to-thermal IR micro-SATellite (VISIR-SAT) being flown with the Sentinel series of satellites, thus providing higher-spatial-resolution observations of thermal anomalies initially detected by the more widely viewing, but coarser, Sentinel satellites (Ruecker et al. 2015).

20.4.2 WILDFIRES

Largely in the same ways as volcanoes, by virtue of being hot, wildfires can be monitored using thermal imagery, and the associated information can be used to inform relevant authorities of hazard management tasks or for quantifying any environmental impacts. Wooster et al. (2003), for example, determined that if the power emission of a wildfire can be calculated from remotely detected thermal emissions (e.g. using the Stefan–Boltzman equation, Equation 20.1), then it can be used to estimate the quantity of biomass combustion. Using thermal satellite data, scientists have also been able to isolate seasonal and interannual variations in fires and their intensities (Duncan et al. 2003).

One of the key sensors used for fire monitoring is MODIS. Even before the launch of this sensor, now termed the 'workhorse' for global fire monitoring, its potential for monitoring global fires had been promoted (Kaufman et al. 1998). Today, data from the MODIS thermal bands are manipulated automatically, using an algorithm developed by Giglio et al. (2003), to produce the MODIS fire products which map the locations of fires globally, on a daily, and composite daily, basis, and which are available for free download from NASA (Figure 20.7). The MODVOLC system provides

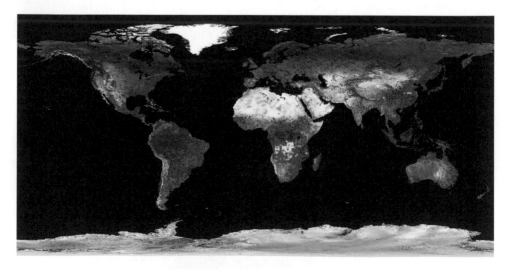

FIGURE 20.7 Global 10-day composite fire map, derived from MODIS imagery for 31 May–9 June 2013 (NASA 2017a). (From Land, Atmosphere Near real-time Capability for EOS [LANCE] system operated by the NASA/GSFC/Earth Observing System Data and Information System [EOSDIS], with funding provided by NASA/HQ, https://earthdata.nasa.gov/earth-observation-data/near-real-time/firms.)

also provides utility in identifying fires, although as it uses a different algorithm than the 'official' MODIS products, it often detects different events. However, despite its widespread use, MODIS is not the only sensor with the appropriate thermal detection characteristics for monitoring wildfires. The Landsat range of sensors are also equipped with thermal bands which provide utility in the monitoring of fires, and even the early sensors of the Landsat range were shown to be so sensitive as to be able to detect subsurface coal fires (Cracknell and Mansor 1992).

20.4.3 EARTHQUAKES

Earthquakes are known for the damage they inflict on the built environment. Such damage can be monitored and assessed using various remote sensing techniques involving the use of optical imagery to 'see' the damage, or radar data to detect any associated changes in the built or physical environment. However, the use of thermal remote sensing for monitoring earthquakes is something which has only relatively recently been explored. Some scientists have claimed that prior to earthquakes, the land heats up (by up to 10°C) (Saraf and Choudhury 1994), and have attributed such phenomena to various factors, including the generation of electric currents in the rock as it becomes compressed (Ouzounov and Freund 2004). Others have presented evidence that fault lines themselves are anomalously warm (Cervone et al. 2005). The precise mechanism of heat generation, and indeed the very presence of such precursory thermal anomalies, is controversial (Blackett et al. 2011), but the theory is clear: if the surface does heat prior to an earthquake, the sensitive thermal sensors on satellites may observe this, and a number of workers have made such claims (Choudhury et al. 2006; Pergola et al. 2010). The sort of thermal remote sensing data used to monitor such subtle temperature changes include that of the MODIS land surface temperature product (Figure 20.8, discussed later). Until the mechanisms that might produce such observations are proven, however, the true utility of thermal remote sensing for earthquake studies will remain uncertain.

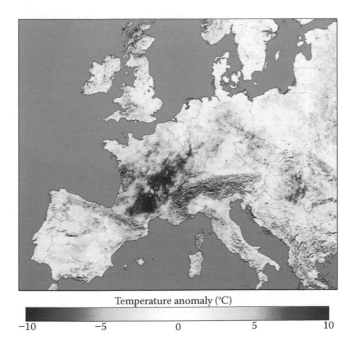

Temperature anomaly (°C)

−10   −5   0   5   10

FIGURE 20.8   Heat anomalies experienced in Europe in 2003, as derived from MODIS land surface temperature data (NASA 2017b). (From NASA's Earth Observatory, https://earthobservatory.nasa.gov/IOTD/view.php?id=3714.)

20.4.4   Landslides

The proven utility of thermal remote sensing for monitoring volcanic activity, and potential for earthquakes, is clearly also be useful for landslide monitoring where landslides are induced by these phenomena. However, in terms of the thermal remote detection and/or monitoring of landslides and slope stability, work is limited, focusing on the detection of the temperature of the land surface, which is influenced by surface moisture: a key indicator of slope stability. The presence of moisture in the ground influences the surface energy balance, increasing the amount of evaporation that occurs and, in turn, reducing the surface temperature (Small and Kurc 2001). The detection of subtle temperature changes using thermal remote sensing might indicate changing moisture levels. For example, Mondini et al. (2009) and Reichenbach et al. (2012) used ASTER TIR data to show the temperature of landslide-prone areas in Umbria, Italy, to be cooler than that of stable areas, which is consistent with such regions also being wetter. Similarly, Kosugi et al. (2012) and Kinoshita et al. (2012) confirmed the utility of thermal remote sensing for detecting the moist areas of a slope, via a temperature approximation, and related these areas to those of the highest susceptibility to slope failure.

20.4.5   Heat Waves

Given its intrinsic capacity to measure temperatures, thermal remote sensing has the real capacity to detect and monitor temperatures of the Earth's surface and, as such, temperature extremes. Such monitoring is all the more pertinent given contemporary concerns over climate change and seemingly increasing numbers of heat waves over the past decade (Barriopedro et al. 2011). The advantageous characteristic of remotely sensed thermal imagery for heat wave hazards is the synoptic view that is provided, meaning that entire continents can be monitored instantaneously. One product commonly used for this purpose is the MODIS land surface temperature product, which is derived by converting the thermal signals emitted from the surface and detected at the satellite to values of temperature (Wan and Li 1998; NASA 2017b). An example of such data taken from the 2003 heat wave experienced in Europe is shown in Figure 20.8, and indeed, using such data, the effects of heat waves even on the natural environment have been examined (Albright et al. 2011).

20.4.6   Flooding

Water has the particular characteristic of a high specific heat capacity. This means that bodies of water require large quantities of energy to raise their temperature as compared with rock (and the land surface). The specific heat capacity of water is 4.2 kJ/kg °C compared with that of rock, of around 1.0 kJ/kg °C (Hasnain 1998), meaning that it takes more than four times the energy to raise a kilogram of water by 1°C as compared with rock. This also works in reverse: water loses its heat more slowly than rock and, as such, will remain warmer for longer. The implications of this are that during the daytime, the land surface will heat more rapidly and will appear warmer than surrounding bodies of water; however, at night the land will lose heat more rapidly and hence will appear cooler (as evident in Figure 20.9). Based on such observations, thermal remote sensing can be used to identify regions of surface water and flood, particularly in locales where there is a significant temperature contrast between floodwaters and the surrounding land (Gumley and King 1995). Indeed, even early studies demonstrated the utility of thermal AVHRR imagery for discerning floodwaters from the surrounding land, such as in relation to the 1978 Red River floods (Berg et al. 1981), and the utility was found to be enhanced at night, when the temperature contrast was most extreme (Barton and Bathols 1989). More recently, ASTER TIR imagery has been used to examine the extent of tidal flooding off the eastern seaboard of the United States (Allen 2012). However, given the submetre resolution of many optical remote sensors, such as the WorldView-2, the utility of thermal remote sensing for flood monitoring is perhaps less compelling.

20.4.7 STORMS

It is fortunate from the aspect of remote sensing that storms and other intense atmospheric phenomena are represented by the presence of clouds. Due to their altitude, clouds are particularly cold and, as such, emit smaller quantities of radiation than warmer bodies, such as the Earth's surface (Figure 20.9), and in fact, the higher the altitude of a cloud, the colder it will be, with storm clouds usually extending far into the stratosphere and exhibiting temperatures sometimes as low as −80°C (Ceccato and Dinku 2010). Such differences in temperature are evident in thermal imagery, allowing for the interpretation of the atmospheric conditions at a particular time. Sensors such as GOES (for the Americas) and Meteosat (for Europe and Africa) have been endowed with TIR bands in large part for their potential to observe atmospheric clouds, and despite a coarse spatial resolution, their subhourly temporal resolution means that they monitor the changes in such phenomena (Velden et al. 1998; Kossin 2002). This is of particular use in the case of potentially destructive atmospheric natural hazards such as hurricanes.

TIR GOES imagery of Hurricane Lili (of 2002) is shown in Figure 20.10, with higher (more intensely white) clouds towards the centre of the storm, lower clouds at the periphery and clear

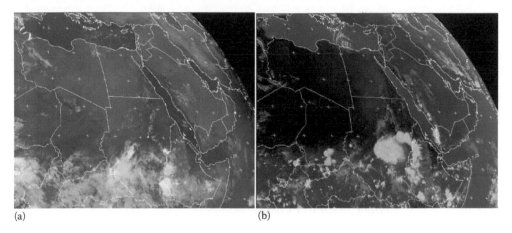

(a)   (b)

FIGURE 20.9   TIR Meteosat image of northwest Africa and the Middle East from 27 September 2012 at (a) 2 a.m. and (b) 2 p.m. (Universal Time Coordinated [UTC]). In these images, colder objects appear brighter (hence white clouds) and warmer objects appear darker. At night (a), the ocean is darker than the land surface and hence warmer; the situation reverses in the daytime (b).

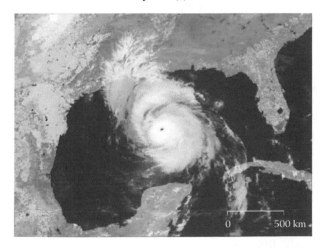

FIGURE 20.10   GOES TIR (11 μm) image of Hurricane Lili acquired on 2 October 2002 (NASA 2017c). (From NASA's Earth Observatory, https://earthobservatory.nasa.gov/NaturalHazards/view.php?id=10266.)

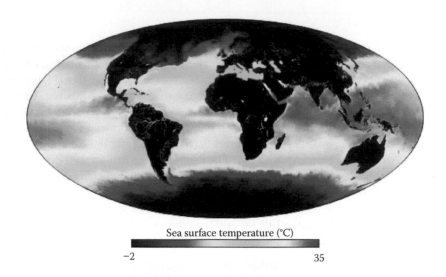

Sea surface temperature (°C)

−2   35

FIGURE 20.11   Average global sea surface temperature for May 2001, derived from the MODIS sea surface temperature product (NASA 2017d). (From NASA's Earth Observatory, http://earthobservatory.nasa.gov /IOTD/view.php?id=1869.)

skies at the eye all evident. It was initially thought that thermal imagery might even be useful for estimating rainfall volumes and intensity (Vicente et al. 1998), although more recently it has been suggested that this is likely to be inaccurate (Ceccato and Dinku 2010). An added advantage of the capacity for surface temperatures to be derived from thermal imagery is that it is known that for hurricanes to form, a sea surface temperature of 26.5°C is necessary (Goldenberg et al. 2001). Using thermal imagery, regions of the ocean meeting this requirement can be mapped and monitored for the development of tropical storms (Figure 20.11). The global sea surface temperature in the month of May 2001, as derived from MODIS TIR data, is shown in Figure 20.11.

20.5   PROSPECTS FOR THE FUTURE

Great strides have been made in recent decades in terms of the sensors available to obtain thermal observations, and in the quality of the data they produce. Given this enhanced accuracy, observations have moved on from being largely qualitative to being more quantitative, often indicating what is occurring on the ground without the requirement for potentially dangerous and/or expensive fieldwork. Prospects for the future are equally promising, as long as the scientific communities involved continue to lobby for the provision of thermal channels with adequate spatial and temporal resolution (Blackett 2013). Fortunately, such lobbies were not ignored in the design and production of the Landsat-8 sensor, launched in 2013 and possessing TIR bands (although with a reduced spatial resolution) (Reuter et al. 2010). In fact, only months after its launch, the thermal infrared sensor (TIRS) of Landsat-8 returned imagery depicting thermal activity at Paluweh volcano, Indonesia (Figure 20.12) (Blackett 2014). The trend today, however, is towards smaller, less costly satellites, launched privately, with an example being the Space Ultra-Compact Hyperspectral Imager (SUCHI) which was built by the University of Hawaii and launched in 2015 (Crites et al. 2013). A further and indeed necessary trend is for more regular monitoring, with increases in temporal resolution being necessary from a hazard management perspective. This might be achieved by the launch of more advanced satellites (perhaps at the expense of spatial resolution), but more realistically will be achieved with the enhanced integration of sensors with differing temporal and spatial resolutions, each being able to capture different aspects of natural hazards that are evident over different spatial or temporal scales (Koeppen et al. 2011). Additionally, integration between smaller, less costly

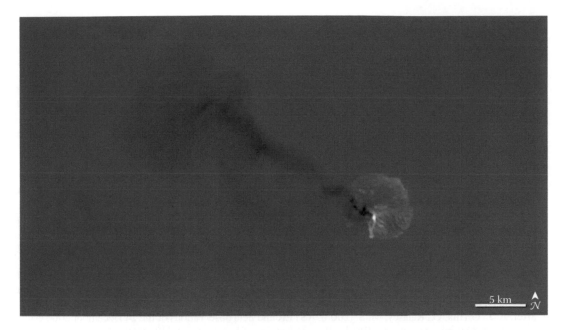

FIGURE 20.12   Landsat-8 TIR sensor imagery of Paluweh volcano, Indonesia, on 29 April 2013 (NASA 2017e). (Photo by Robert Simmon. From NASA's Earth Observatory, using data from USGS and NASA, http:// www.nasa.gov/mission_pages/landsat/news/indonesia-volcano.html.)

satellites will provide a temporal resolution much greater than would be possible using one such sensor alone, with the Disaster Monitoring Constellation, consisting of five international satellites, providing a successful example of such a system currently in operation (Gillespie et al. 2007).

20.6   CONCLUSION

The observation of terrestrial phenomena using thermal remote sensing is largely dependent on the temperature manifestation of observed features, and evidently, based on the short summary presented here, its utility for observing and monitoring various natural hazards is clear. Indeed, the use of such observations has become crucial in many aspects, particularly for the monitoring of remote, inaccessible regions, but also for the monitoring of more fundamental environmental phenomena, including the weather and its associated hazards. There is much potential for further developments in the discipline of thermal remote sensing as new sensors are developed with improved or enhanced capabilities, and as scientists discover more about natural hazard events and how they are manifested on a local and global scale.

REFERENCES

Albright, T. P., Pidgeona, A. M., Rittenhouse, C. D., Clayton, M. K., Flather, C. H., Culbert, P. D., Radeloff, V. S. (2011). Heat waves measured with MODIS land surface temperature data predict changes in avian community structure. *Remote Sensing of Environment*, 115, 245–254.

Allen, T. R. (2012). Estimating coastal lagoon tidal flooding and repletion with multidate ASTER thermal imagery. *Remote Sensing*, 4, 3110–3126.

Barriopedro, D., Fischer, E. M., Luterbacher, J., Trigo, R. M., García-Herrera, R. (2011). The hot summer of 2010: Redrawing the temperature record map of Europe. *Science*, 332, 220–224.

Barton, I. J., Bathols, J. M. (1989). Monitoring floods with AVHRR. *Remote Sensing of Environment*, 30 (1), 89–94.

Berg, C. P., Weisnet, D. R., Matson, M. (1981). Assessing the Red River of the north 1978 flooding from NOAA satellite data. In Deutsch, M., Weisnet, D. R., Rango, A., eds., *Satellite Hydrology, Proceedings of the Fifth Annual William T. Pecora Symposium on Remote Sensing, Sioux Falls, South Dakota, June 10–15, 1979*. Technical Publication Series TPS 81-1. Minneapolis: American Water Resources Association.

Blackett, M. (2013). Review of the utility of infrared remote sensing for detecting and monitoring volcanic activity with the case study of shortwave infrared data for Lascar volcano from 2001–2005. *Geological Society, London, Special Publications*, 380(1), 107–135. doi: 10.1144/SP380.10.

Blackett, M. (2014). Early analysis of Landsat-8 thermal infrared sensor imagery of volcanic activity. *Remote Sensing*, 6(3), 2282–2295. doi: 10.3390/rs6032282.

Blackett, M. (2017). An overview of infrared remote sensing of volcanic activity. *Journal of Imaging*, 3(2), 13. doi: 10.3390/jimaging3020013.

Blackett, M., Wooster, M. J. (2011). Evaluation of SWIR-based methods for quantifying active volcano radiant emissions using NASA EOS-ASTER data. *Geomatics, Natural Hazards and Risk*, 2, 51–79.

Blackett, M., Wooster, M. J., Malamud, B. (2011). Exploring claims of land surface temperature precursors to the 2001 Gujarat earthquake. *Geophysical Research Letters*, 38, L15303. doi: 10.1029/2011GL048282.

Boltzmann, L. (1884). Ableitung des Stefan'schen Gesetzes, betreffend die Abhängigkeit der Wärmestrahlung von der Temperatur aus der electromagnetischen Lichttheorie. *Annal Physik Chemie*, 22, 291–294.

Buongiorno, M., Realmuto, V., Doumaz, F. (2002). Recovery of spectral emissivity from Thermal Infrared Multispectral Scanner imagery acquired over a mountainous terrain. A case study from Mount Etna Sicily. *Remote Sensing of Environment*, 79, 123–133.

Carter, A. J., Girina, O., Ramsey, M. S. (2008). ASTER and field observations of the 24 December 2006 eruption of Bezymianny volcano, Russia. *Remote Sensing of Environment*, 112, 2569–2577.

Ceccato, P. N., Dinku, T. (2010). Introduction to remote sensing for monitoring rainfall, temperature, vegetation and water bodies. IRI Technical Report. Palisades, NY: International Research Institute for Climate and Society.

Cervone, G., Singh, R. P., Kafatos, M., Yu, C. (2005). Wavelet maxima curves of surface latent heat flux anomalies associated with Indian earthquakes. *Natural Hazards and Earth System Sciences*, 5, 87–99.

Choudhury, S., Dasgupta, S., Saraf, A., Panda, S. (2006). Remote sensing observations of pre-earthquake thermal anomalies in Iran. *International Journal of Remote Sensing*, 27, 4381–4396.

Cracknell, A. P., Mansor, S. B. (1992). Detection of sub-surface coal fires using Landsat Thematic Mapper data. In *XVIIth ISPRS Congress Technical Commission VII: Interpretation of Photographic and Remote Sensing Data, August 2–14, 1992*. Washington, D.C., USA, 750–753.

Crites, S. T., Lucey, P. G., Wright, R., Chan, J., Garbeil, H., Horton, K., Imai, A., Wood, M. and Yoneshige, L. (2013). SUCHI: The Space Ultra-Compact Hyperspectral Imager for small satellites. *Society of Photo-Optical Instrumentation Engineers Proceedings*, Sensors and Systems for Space Applications VI, 873902, Baltimore, MD, USA.

Deneke, H. M., Feijt, A. J., Roebeling, R. A. (2008). Estimating surface solar irradiance from METEOSAT SEVIRI-derived cloud properties. *Remote Sensing of Environment*, 112, 3131–3141.

Duncan, B. N., Martin, R. V., Staudt, A. C. (2003). Interannual and seasonal variability of biomass burning emissions constrained by satellite observations. *Journal of Geophysical Research*, 108, 4100.

Flynn, L. P., Harris, A. J. L., Wright, R. (2001). Improved identification of volcanic features using Landsat 7 ETM+. *Remote Sensing of Environment*, 78, 180–193.

Giglio, L., Descloitres, J., Justice, C. O., Kaufman, Y. J. (2003). An enhanced contextual fire detection algorithm for MODIS. *Remote Sensing of Environment*, 87, 273–282.

Gillespie, T. W., Chu, J. Frankenberg, E. (2007). Assessment and prediction of natural hazards from satellite imagery. *Progress in Physical Geography*, 31, 459–470.

Glaze, L., Francis, P. W., Rothery, D. A. (1989). Measuring thermal budgets of active volcanoes by satellite remote sensing. *Nature*, 338, 144–146.

Goldenberg, S. B., Landsea, C. W., Mestas-Nuñez, A. M. (2001). The recent increase in Atlantic hurricane activity: Causes and implications. *Science*, 293, 474–479.

Gumley, L., King, M. D. (1995). Remote sensing of flooding in the US upper Midwest during the summer of 1993. *Bulletin of the American Meteorological Society*, 76, 933–943.

Harris, A. J. L., Flynn, L. P., Keszthelyi, L. (1998). Calculation of lava effusion rates from Landsat TM data. *Bulletin of Volcanology*, 60, 52–71.

Hasnain, S. M. (1998). Review on sustainable thermal energy storage technologies, Part I: Heat storage materials and techniques. *Energy Conversion and Management*, 39(11), 1127–1138.

Kaufman, Y. J., Justice, C. O., Flynn, L. P. (1998). Potential global fire monitoring from EOS-MODIS. *Journal of Geophysical Research*, 103, D24.

Kinoshita, A., Okamoto, A., Kawano, T. (2012). Study on the analysis of soil moisture distribution characteristics on slopes using a thermal infrared sensor. *Journal of the Japan Society of Erosion Control Engineering*, 64, 6.

Koeppen, W. C., Pilger, E., Wright, R. (2011). Time series analysis of infrared satellite data for detecting thermal anomalies: A hybrid approach. *Bulletin of Volcanology*, 73, 577–593.

Kossin, J. P. (2002). Daily hurricane variability inferred from GOES infrared imagery. *Monthly Weather Review*, 130, 2260–2270.

Kosugi, K., Yamakawa, Y., Masaoka, N. (2012). Application of thermal infrared remote sensing for detecting slopes with high landslide vulnerability. *Journal of the Japan Society of Erosion Control Engineering*, 64, 32–37.

Matthews, S., Gardeweg, M., Sparks, R. (1997). The 1984 to 1996 cyclic activity of Lascar volcano, northern Chile: Cycles of dome growth, dome subsidence, degassing and explosive eruptions. *Bulletin of Volcanology*, 59, 72–82.

MODVOLC. http://modis.higp.hawaii.edu/ (accessed 12 July 2017).

Mondini, A., Carlà, R., Reichenbach, P. (2009). Use of a remote sensing approach to detect landslide thermal behaviour. *Geophysical Research Abstracts*, 11, EGU2009-7790.

NASA (National Aeronautics and Space Administration). (2017a). Fire Information for Resource Management System (FIRMS) MODIS hotspot/active fire detections. Data set. https://earthdata.nasa.gov/earth-observation-data/near-real-time/firms (accessed 28 November 2017).

NASA (National Aeronautics and Space Administration) (2017b). Earth Observatory. https://earthobservatory.nasa.gov/IOTD/view.php?id=3714 (accessed 28 November 2017).

NASA (National Aeronautics and Space Administration) Visible Earth. (2017c). Earth Observatory. https://earthobservatory.nasa.gov/NaturalHazards/view.php?id=10266 (accessed 28 November 2017).

NASA (National Aeronautics and Space Administration) Earth Observatory. (2017d). http://earthobservatory.nasa.gov/IOTD/view.php?id=1869 (accessed 28 November 2017).

NASA (National Aeronautics and Space Administration). (2017e). Mission pages. http://www.nasa.gov/mission_pages/landsat/news/indonesia-volcano.html (accessed 28 November 2017).

Oppenheimer, C. (1998). Volcanological applications of meteorological satellites. *International Journal of Remote Sensing*, 19, 2829–2864.

Oppenheimer, C., Francis, P. W., Rothery, D. A., Carlton, R. W. T., Glaze, L. S. (1993). Infrared image analysis of volcanic thermal features: Láscar Volcano, Chile, 1984–1992. *Journal of Geophysical Research: Solid Earth*, 98, 4269–4286.

Ouzounov, D., Freund, F. (2004). Mid-infrared emission prior to strong earthquakes analyzed by remote sensing data. *Advances in Space Research*, 33, 268–273.

Pergola, N., Aliano, C., Coviello, I. (2010). Using RST approach and EOS-MODIS radiances for monitoring seismically active regions: A study on the 6 April 2009 Abruzzo earthquake. *Natural Hazards and Earth System Sciences*, 10, 239–249.

Pieri, D., Abrams, M. (2004). ASTER watches the world's volcanoes: A new paradigm for volcanological observations from orbit. *Journal of Volcanology and Geothermal Research*, 135, 13–28.

Reichenbach, P., Mondini, A. C., Rossi, M. (2012). Analysis of the thermal behavior of landslides using remote sensing data. Presented at AOGS-EGU Joint Meeting, Singapore, 13–17 August.

Reuter, D., Richardson, C., Irons, J., Allen, R., Anderson, M., Budinoff, J., Casto, G. et al. (2010). The thermal infrared sensor on the Landsat data continuity mission. In *IEEE International Geoscience and Remote and Sensing Symposium (IGARSS), July 25–30, 2010*. Honolulu, USA, 754–757.

Ruecker, G., Menz, G., Heinemann, S., Hartmann, M., Oertel, D. (2015). VISIR-SAT – A prospective microsatellite based multi-spectral thermal mission for land applications. In *Proceedings of the ISPRS International Archives of the Photogrammetry, Remote Sensing and Spatial Information Sciences*, Berlin, 11–15 May, vol. XL-7/W3, 1283–1289.

Saraf, A. K., Choudhury, S. (2004). Satellite detects pre-earthquake thermal anomalies associated with past major earthquakes. Presented at Map Asia Conference, Beijing, August.

Small, E. E., Kurc, S. (2001). The influence of soil moisture on the surface energy balance in semiarid environments. Technical Completion Report. Sorocco: New Mexico Water Resources Research Institute in cooperation with the Department of Earth and Environmental Science, New Mexico Tech, June.

Stefan, J. (1879). Über die Beziehung zwischen der Wärmestrahlung und der Temperatur. *Sitzungsberichte der mathematisch-naturwissenschaftlichen Classe der kaiserlichen Akademie der Wissenschaften*, 79, 391–428.

Tallarico, A., Dragoni, M. (1999). Viscous Newtonian laminar flow in a rectangular channel: Application to Etna lava flows. *Bulletin of Volcanology*, 61, 40–47.

Urai, M., Ishizuka, Y. (2011). Advantages and challenges of space-borne remote sensing for Volcanic Explosivity Index (VEI): The 2009 eruption of Sarychev Peak on Matua Island, Kuril Islands, Russia. *Journal of Volcanology and Geothermal Research*, 208, 163–168.

Vaughan, R., Kervyn, M., Realmuto, V. (2008). Satellite measurements of recent volcanic activity at Oldoinyo Lengai, Tanzania. *Journal of Volcanology and Geothermal Research*, 173, 196–206.

Velden, C., Olander, C., Zehr, R. M. (1998). Development of an objective scheme to estimate tropical cyclone intensity from digital geostationary satellite infrared imagery. *Weather Forecasting*, 13, 172–186.

Vicente, G. A., Scofield, R. A., Menzel, W. P. (1998). The operational GOES infrared rainfall estimation technique. *Bulletin of the American Meteorological Society*, 79, 1883–1898.

Wan, Z., Li, Z.-L. (1998). A physics-based algorithm for retrieving land-surface emissivity and temperature from EOS/MODIS data. *IEEE Transactions on Geoscience and Remote Sensing*, 35, 980–996.

Wien, W. (1896). Uber die Energieverteilung in Emissionspektrum eines schwarzen Korpers. *Annalen der Physik*, 58, 662–669.

Wooster, M. J. (2002). Small-scale experimental testing of fire radiative energy for quantifying mass combusted in natural vegetation fires. *Geophysical Research Letters*, 2921. doi: 10.1029/2002GL015487.

Wooster, M. J., Rothery, D. A. (1997). Thermal monitoring of Lascar volcano, Chile, using infrared data from the along-track scanning radiometer: A 1992–1995 time series. *Bulletin of Volcanology*, 58, 566–579.

Wooster, M., Zhukov, B., Oertel, D. (2003). Fire radiative energy release for quantitative study of biomass burning: Derivation from the BIRD experimental satellite and comparison to MODIS fire products. *Remote Sensing of Environment*, 86, 83–107.

Wright, R., Blackett, M., Hill-Butler, C. (2015). Some observations regarding the thermal flux from Earth's erupting volcanoes for the period of 2000 to 2014. *Geophysical Research Letters*, 42(2), 282–289. doi: 10.1002/2014GL061997.

Wright, R., Carn, S. A., Flynn, L. P. (2005). A satellite chronology of the May–June 2003 eruption of Anatahan volcano. *Journal of Volcanology and Geothermal Research*, 146, 102–116.

Wright, R., Flynn, L. P., Garbeil, H. (2004). MODVOLC: Near-real-time thermal monitoring of global volcanism. *Journal of Volcanology and Geothermal Research*, 135, 29–49.

Wright, R., Garbeil, H., Harris, A. (2008). Using infrared satellite data to drive a thermo-rheological/stochastic lava flow emplacement model: A method for near-real-time volcanic hazard assessment. *Geophysical Research Letters*, 35, L19307. doi: 10.1029/2008GL035228.

Wright, R., Glaze, L., Baloga, S. M. (2011). Constraints on determining the eruption style and composition of terrestrial lavas from space. *Geology*, 39, 1127–1130.

Index

T - #0165 - 111024 - C526 - 254/178/24 - PB - 9780367571917 - Gloss Lamination